C++
服务器开发精髓

张远龙 著

电子工业出版社
Publishing House of Electronics Industry
北京·BEIJING

内 容 简 介

本书从操作系统原理角度讲解进行 C++服务器开发所需掌握的技术栈。全书总计 9 章，第 1～2 章讲解 C++ 11/14/17 新标准中的常用特性、新增类库，以及 C++开发者必须熟练掌握的开发调试工具链；第 3～6 章详细讲解 C++服务器开发中的多线程编程技术、网络编程重难点知识、网络故障调试与排查常用工具，以及通信协议的设计思路、技巧；第 7～8 章详细讲解一个带网络通信组件的高性能服务的基本设计思路和注意事项；第 9 章进一步补充服务相关的常用模块设计思路和方法。本书秉承的思想是，通过掌握技术原理，可以轻松制造"轮子"，灵活设计出优雅、鲁棒的服务，并快速学习新技术。

无论是对于 C/C++开发者、计算机专业的学生，还是对于想了解操作系统原理的读者，本书都极具参考价值。

未经许可，不得以任何方式复制或抄袭本书之部分或全部内容。
版权所有，侵权必究。

图书在版编目（CIP）数据

C++服务器开发精髓 / 张远龙著. —北京：电子工业出版社，2021.7
ISBN 978-7-121-41263-9

Ⅰ. ①C… Ⅱ. ①张… Ⅲ. ①C++语言－程序设计 Ⅳ. ①TP312.8

中国版本图书馆 CIP 数据核字(2021)第 098889 号

责任编辑：张国霞
印　　刷：三河市良远印务有限公司
装　　订：三河市良远印务有限公司
出版发行：电子工业出版社
　　　　　北京市海淀区万寿路 173 信箱　邮编 100036
开　　本：787×1092　1/16　印张：47　字数：1060 千字
版　　次：2021 年 7 月第 1 版
印　　次：2021 年 7 月第 2 次印刷
印　　数：5001～10000 册　定价：168.00 元

凡所购买电子工业出版社图书有缺损问题，请向购买书店调换。若书店售缺，请与本社发行部联系，联系及邮购电话：(010) 88254888，88258888。
质量投诉请发邮件至 zlts@phei.com.cn，盗版侵权举报请发邮件至 dbqq@phei.com.cn。
本书咨询联系方式：010-51260888-819，faq@phei.com.cn。

前 言

为什么写作本书

笔者自学生时代便开始接触 C++，工作以后先后负责过 C++客户端和服务端的开发工作。时至今日，C++仍然是笔者最喜欢的编程语言。在笔者看来，C++一旦学成，奇妙无穷，还可以快速学习其他编程语言和技术。

本书讲解了笔者近十年来使用 C++的一些经验和技巧，着重讲解基于 C++的操作系统原理和服务器开发技术，希望读者通过学习本书，可以了解如何学习 C++，以及如何成为一名合格的 C++开发者。

C/C++的当前应用领域

需要注意的是，本书不细分 C 语言与 C++的区别。在通常情况下，我们可以将 C++看作 C 语言的一个超集。C++虽然从功能层面来看，离 C 语言越来越"远"，但从语法层面来看，其大多数语法与 C 语言基本一致。对于 C++面向对象的特性，如果仔细探究的话，我们会发现 C++类方法的具体语法还是 C 语言的过程式语法，虽然这种现状正在不断改变。

C 语言目前主要用于操作系统类偏底层的应用开发，比如 Windows、Linux 这样的大型商业操作系统，以及嵌入式操作系统、嵌入式设备。有些开源软件也会选择 C 语言进行开发，主要是考虑程序执行效率和生成的可执行文件的体积（C 代码生成的可执行文件体积相对较小），当然，其中不乏一些历史技术选型的原因，比如 Redis、libevent、Nginx 等。

在将高级语言翻译成机器二进制码时，C++编译器生成了大量的额外机器码，而这种机器码相对于 C 语言来说不是必需的。例如，对于一个 C++类的实例方法，编译器在生成这个方法的机器码时，会将函数的第 1 个参数设置为对象的 this 指针地址，以此来实现对象与函数的绑定。正因如此，许多开发者都会优化和调整编译器生成的汇编代码。

C++当前的常见应用领域有：①我们目前见到的各种桌面应用软件，尤其是Windows桌面软件，例如QQ、安全类杀毒类软件、浏览器等；②一些基础软件和高级语言的运行时环境，例如大型数据库软件、Java虚拟机、C#的CLR、Python编译器和运行时环境等；③业务型应用软件的后台，例如大型网络游戏的服务端和一些企业内部的应用系统等。

C++与操作系统

虽然Java、Python等的SDK或运行时环境最终也会调用操作系统API，但其自带的SDK或者运行时环境都提供了常见的操作系统功能。而C++的运行时环境一般是操作系统自身，因此C++是离操作系统更近的一种编程语言，执行效率更高。

但是，C++的整套语法不具备"功能完备性"，在大多数情况下，单纯地使用其本身提供的功能无法创建出任何有意义的程序，还必须借助操作系统API来实现。例如，C++本身不直接提供网络通信功能的SDK，必须借助操作系统提供的套接字API才能实现网络通信；而对于Java来说，JDK自带的java.net、java.io等包则提供了完整的网络通信功能。所以，熟悉操作系统相关原理和API是用好C++的前提，这也是C++难学、对新手不友好的主要原因之一。

不过，随着C++标准和版本的不断迭代，这种现状正在改变：在C++标准库中引入了越来越多的功能，避免直接调用操作系统API。

不管怎样，应用直接使用操作系统API，程序执行效率高，控制力度大，开发能力仅仅限制于操作系统本身，这是C++的优势之一。比如对于Java，假设操作系统提供了某个功能，但Java虚拟机不提供该功能，则开发人员也无法使用该功能。

编程大师Charles Petzold曾说过，操作系统是一个非常复杂的系统，在API之上加一层编程语言并不能消除其复杂性，最多将复杂性隐藏起来而已，而懂得系统API能让我们更快地挣脱困境。

如何看待C++ 11/14/17/20 标准

C++既支持面向对象设计（OOP），也支持以模板语法为代表的泛型编程（GP）。从最初业界和开发者翘首以盼的C++ 11标准开始，历经C++ 14、C++ 17，到今天的C++ 20，版本差别越来越大，原来需要使用的第三库的功能也被陆续添加到C++标准库中。C++标准不断发展，遵循C++最新标准的编译器层出不穷，C++变化越来越大、越来越快。

对于C++ 11、C++ 14、C++ 17乃至C++ 20的学习，笔者建议以实用为主，不必太纠结新标准中的一些高级特性和复杂模板，更应该学习其中实用的语法和工具库。

如何学好 C++ 和后端开发

首先，我们应该打好基础。我们要熟练使用 C++，还要结合具体的操作系统学习 C++，熟悉某操作系统的 API 函数，以及与系统 API 关联的各类技术，比如各种进程与线程函数、多线程资源同步函数、文件操作函数、系统时间函数、内存分配与管理函数、网络编程、PE 或 ELF 文件的编译、链接原理等。

如果已打好基础，就可以找一些高质量的开源项目去实战。最好找一些没有复杂业务的开源项目，或者是自己熟悉其业务的开源项目（如 IM 系统）。如果不熟悉其业务，那么不但要学习其业务（软件功能），还要学习其源码，最终两者难以兼顾。

因此，在学习这些项目之前，应该先确定自己的学习目的。如果学习目的是学习和借鉴这款软件的架构设计，那么建议先进行整体把握，不要一开始就迷失在细枝末节中，这叫作"粗读"。如果学习目的是学习开源软件在一些细节上的处理方法，那么可以有针对性地阅读自己感兴趣的模块，深入每一行代码。当然，学习适合自己当前阶段的项目源码才是最好的。

学习的过程一般是接触、熟悉、模仿、创造。不管对什么开源项目，在没有任何思路或者解决方案时，我们都应该先接触、熟悉、不断模仿，做到至少心中有一套对某场景的解决方案，再来谈创新、批判及改造。

笔者在学习陌生的开源项目时，喜欢先将程序用调试器正常"跑"起来；然后中断，统计当前的线程数，通过 main 函数从主线程追踪其他工作线程是如何创建的；接着分析和研究各线程的用途和线程之间的交互，这样可以做到整体性把握；最后找感兴趣的细节去学习。

总之，C++ 是一门讲究深度的编程语言，其"深度"不体现在掌握多少 C++ 语法，而在于是否熟悉所写的 C++ 代码背后的系统原理，这是需要长期积累的，当然，一旦学成，就可以快速学习其他编程语言和框架。

本书概要

本书总计 9 章，主要基于 C++，详细讲解服务器开发中基础且重要的技术栈，以期读者掌握"造轮子"的方法。

第 1 章讲解 C++ 新标准中新增的常用语言特性和类库。

第 2 章讲解 C++ 开发者应该掌握的各类开发工具和工作环境，详细、深入地讲解 Linux gdb 调试方面的内容。毫不夸张地说，掌握了 gdb 调试，就等于拿到了学习各种 C++ 开源项目的"钥匙"。

第 3 章详细讲解多线程的原理，涵盖 Windows 和 Linux 的各类线程同步原语，以及基于线程同步技术、生产者/消费者模型衍生的队列系统。

第 4 章进行操作系统层面的网络编程重难点解析，讲解 Linux 上的常用网络通信模型，通过大量详尽的代码实例和测试，深入浅出地探究和验证网络通信编程的重难点技术。

第 5 章讲解排查和定位网络通信问题的常用开发工具。

第 6 章详细讲解网络通信协议的设计思想，并从"造轮子"的角度讲解常用网络通信协议的格式、使用方法和注意事项，讲解设计网络通信协议时需要考虑的各类问题，最后对几种常用的通信协议逐一剖析并给出具体的实现逻辑。

第 7 章详细讲解如何设计一个高性能的带网络通信组件的服务，并结合一些经典案例进行分析，还详细讲解经典服务框架的设计思路和各个模块的具体实现方法。

第 8 章以 redis-server 源码为例，论证第 7 章讲解的服务设计原理。

第 9 章是对第 7 章内容的补充，详细讲解一个服务的常用模块设计思路。

相关资源

本书提供源码下载、读者交流群等服务，详情请参见本书封底的读者服务信息。

若想获取关于高性能服务器开发的更多知识，可以关注笔者的两个微信公众号："高性能服务器开发"和"程序员小方"。

致谢

感谢笔者的妻子承担家务及照顾笔者的生活，让笔者可以集中精力写作本书。

感谢各位同事帮助笔者成长与提高。

感谢王旭东等同学为本书的校对和勘误做出贡献，感谢"高性能服务器开发"群内小伙伴们的支持。

感谢电子工业出版社工作严谨、高效的张国霞编辑，她在成书过程中对笔者的指导、协助和鞭策，是本书得以完成的重要助力。

目　录

第1章　C++必知必会 1

1.1　C++ RAII 惯用法 1
1.1.1　版本 1：最初的写法 1
1.1.2　版本 2：使用 goto 语句 3
1.1.3　版本 3：使用 do...while(0)循环 5
1.1.4　版本 4：使用 RAII 惯用法 7
1.1.5　小结 12

1.2　pimpl 惯用法 12

1.3　C++ 11/14/17 新增的实用特性 17

1.4　统一的类成员初始化语法与 std::initializer_list<T> 19

1.5　C++ 17 注解标签（attributes） 24
1.5.1　C++ 98/03 的 enumeration 和 C++ 11 的 enumerator 25
1.5.2　C++ 17 的注解标签 25

1.6　final、override 关键字和=default、=delete 语法 28
1.6.1　final 关键字 28
1.6.2　override 关键字 29
1.6.3　=default 语法 31
1.6.4　=delete 语法 32

1.7　auto 关键字的用法 34

1.8　Range-based 循环语法 35
1.8.1　自定义对象如何支持 Range-based 循环语法 37
1.8.2　for-each 循环的实现原理 38

1.9　C++ 17 结构化绑定 39

1.10　stl 容器新增的实用方法 43
1.10.1　原位构造与容器的 emplace 系列函数 43
1.10.2　std::map 的 try_emplace 方法与 insert_or_assign 方法 44

1.11 stl 中的智能指针类详解 52
1.11.1 C++ 98/03 的尝试——std::auto_ptr 52
1.11.2 std::unique_ptr 55
1.11.3 std::shared_ptr 59
1.11.4 std::enable_shared_from_this 61
1.11.5 std::weak_ptr 63
1.11.6 智能指针对象的大小 67
1.11.7 使用智能指针时的注意事项 68

第 2 章 C++ 后端开发必备的工具和调试知识 71
2.1 SSH 工具与 FTP 工具 71
2.1.1 Xshell 71
2.1.2 FTP 75
2.2 makefile 与 CMake 76
2.3 使用 Visual Studio 管理和阅读开源项目代码 83
2.4 gdb 调试 87
2.4.1 被调试的程序需要带调试信息 87
2.4.2 启动 gdb 调试的方法 89
2.5 gdb 常用命令详解——利用 gdb 调试 Redis 94
2.5.1 gdb 常用调试命令概览和说明 94
2.5.2 用 gdb 调试 Redis 前的准备工作 96
2.5.3 run 命令 97
2.5.4 continue 命令 98
2.5.5 break 命令 98
2.5.6 tbreak 命令 101
2.5.7 backtrace 与 frame 命令 101
2.5.8 info break、enable、disable、delete 命令 102
2.5.9 list 命令 104
2.5.10 print 与 ptype 命令 107
2.5.11 info 与 thread 命令 109
2.5.12 next、step、until、finish、return、jump 命令 112
2.5.13 disassemble 命令 122
2.5.14 set args 与 show args 命令 122
2.5.15 watch 命令 123
2.5.16 display 命令 124
2.5.17 dir 命令 125

2.6	使用 gdb 调试多线程程序	126
	2.6.1 调试多线程程序的方法	126
	2.6.2 在调试时控制线程切换	128
2.7	使用 gdb 调试多进程程序——以调试 Nginx 为例	137
2.8	gdb 实用调试技巧	143
	2.8.1 将 print 输出的字符串或字符数组完整显示	144
	2.8.2 让被 gdb 调试的程序接收信号	144
	2.8.3 函数明明存在，添加断点时却无效	145
	2.8.4 调试中的断点	146
	2.8.5 自定义 gdb 调试命令	147
2.9	gdb tui——gdb 图形化界面	148
	2.9.1 开启 gdb TUI 模式	149
	2.9.2 gdb TUI 模式下的 4 个窗口	149
	2.9.3 解决 tui 窗口不自动更新内容的问题	150
	2.9.4 窗口焦点切换	150
2.10	gdb 的升级版——cgdb	151
2.11	使用 VisualGDB 调试	154
	2.11.1 使用 VisualGDB 调试已经运行的程序	155
	2.11.2 使用 VisualGDB 从头调试程序	156

第 3 章 多线程编程与资源同步 159

3.1	线程的基本概念及常见问题	159
	3.1.1 主线程退出，支线程也将退出吗	159
	3.1.2 某个线程崩溃，会导致进程退出吗	160
3.2	线程的基本操作	160
	3.2.1 创建线程	160
	3.2.2 获取线程 ID	166
	3.2.3 等待线程结束	173
3.3	惯用法：将 C++类对象实例指针作为线程函数的参数	178
3.4	整型变量的原子操作	184
	3.4.1 为什么给整型变量赋值不是原子操作	185
	3.4.2 Windows 平台上对整型变量的原子操作	186
	3.4.3 C++ 11 对整型变量原子操作的支持	187
3.5	Linux 线程同步对象	190
	3.5.1 Linux 互斥体	190
	3.5.2 Linux 信号量	198

- 3.5.3 Linux 条件变量 ... 202
- 3.5.4 Linux 读写锁 ... 208
- 3.6 Windows 线程同步对象 ... 217
  - 3.6.1 WaitForSingleObject 与 WaitForMultipleObjects 函数 ... 217
  - 3.6.2 Windows 临界区对象 ... 219
  - 3.6.3 Windows Event 对象 ... 224
  - 3.6.4 Windows Mutex 对象 ... 229
  - 3.6.5 Windows Semaphore 对象 ... 231
  - 3.6.6 Windows 读写锁 ... 235
  - 3.6.7 Windows 条件变量 ... 238
  - 3.6.8 在多进程之间共享线程同步对象 ... 243
- 3.7 C++ 11/14/17 线程同步对象 ... 244
  - 3.7.1 std::mutex 系列 ... 244
  - 3.7.2 std::shared_mutex ... 248
  - 3.7.3 std::condition_variable ... 253
- 3.8 如何确保创建的线程一定能运行 ... 256
- 3.9 多线程使用锁经验总结 ... 258
  - 3.9.1 减少锁的使用次数 ... 258
  - 3.9.2 明确锁的范围 ... 259
  - 3.9.3 减少锁的使用粒度 ... 259
  - 3.9.4 避免死锁的一些建议 ... 260
  - 3.9.5 避免活锁的一些建议 ... 262
- 3.10 线程局部存储 ... 262
  - 3.10.1 Windows 的线程局部存储 ... 262
  - 3.10.2 Linux 的线程局部存储 ... 264
  - 3.10.3 C++ 11 的 thread_local 关键字 ... 267
- 3.11 C 库的非线程安全函数 ... 268
- 3.12 线程池与队列系统的设计 ... 270
  - 3.12.1 线程池的设计原理 ... 270
  - 3.12.2 环形队列 ... 275
  - 3.12.3 消息中间件 ... 275
- 3.13 纤程（Fiber）与协程（Routine） ... 277
  - 3.13.1 纤程 ... 277
  - 3.13.2 协程 ... 280

第 4 章 网络编程重难点解析 282

- 4.1 学习网络编程时应该掌握的 socket 函数 282
  - 4.1.1 在 Linux 上查看 socket 函数的帮助信息 283
  - 4.1.2 在 Windows 上查看 socket 函数的帮助信息 285
- 4.2 TCP 网络通信的基本流程 286
- 4.3 设计跨平台网络通信库时的一些 socket 函数用法 290
  - 4.3.1 socket 数据类型 290
  - 4.3.2 在 Windows 上调用 socket 函数 290
  - 4.3.3 关闭 socket 函数 291
  - 4.3.4 获取 socket 函数的错误码 291
  - 4.3.5 套接字函数的返回值 293
  - 4.3.6 select 函数第 1 个参数的问题 293
  - 4.3.7 错误码 WSAEWOULDBLOCK 和 EWOULDBLOCK 294
- 4.4 bind 函数重难点分析 294
  - 4.4.1 对 bind 函数如何选择绑定地址 294
  - 4.4.2 bind 函数的端口号问题 295
- 4.5 select 函数的用法和原理 302
  - 4.5.1 Linux 上的 select 函数 302
  - 4.5.2 Windows 上的 select 函数 317
- 4.6 socket 的阻塞模式和非阻塞模式 318
  - 4.6.1 如何将 socket 设置为非阻塞模式 318
  - 4.6.2 send 和 recv 函数在阻塞和非阻塞模式下的表现 320
  - 4.6.3 非阻塞模式下 send 和 recv 函数的返回值总结 331
  - 4.6.4 阻塞与非阻塞 socket 的各自适用场景 333
- 4.7 发送 0 字节数据的效果 333
- 4.8 connect 函数在阻塞和非阻塞模式下的行为 339
- 4.9 连接时顺便接收第 1 组数据 343
- 4.10 如何获取当前 socket 对应的接收缓冲区中的可读数据量 346
  - 4.10.1 分析 346
  - 4.10.2 注意事项 350
- 4.11 Linux EINTR 错误码 351
- 4.12 Linux SIGPIPE 信号 352
- 4.13 Linux poll 函数的用法 353
- 4.14 Linux epoll 模型 361
  - 4.14.1 基本用法 361

4.14.2　epoll_wait 与 poll 函数的区别 363
　　4.14.3　LT 模式和 ET 模式 363
　　4.14.4　EPOLLONESHOT 选项 380
4.15　高效的 readv 和 writev 函数 386
4.16　主机字节序和网络字节序 387
　　4.16.1　主机字节序 387
　　4.16.2　网络字节序 388
　　4.16.3　操作系统提供的字节转换函数汇总 389
4.17　域名解析 API 介绍 390

第 5 章　网络通信故障排查常用命令 397
5.1　ifconfig 命令 397
5.2　ping 命令 401
5.3　telnet 命令 402
5.4　netstat 命令 407
5.5　lsof 命令 409
5.6　nc 命令 412
5.7　curl 命令 415
5.8　tcpdump 命令 416

第 6 章　网络通信协议设计 422
6.1　理解 TCP 422
6.2　如何解决粘包问题 423
6.3　解包与处理 425
6.4　从 struct 到 TLV 430
　　6.4.1　协议的演化 430
　　6.4.2　协议的分类 434
　　6.4.3　协议设计工具 434
6.5　整型数值的压缩 435
6.6　设计通信协议时的注意事项 437
　　6.6.1　字节对齐 437
　　6.6.2　显式地指定整型字段的长度 438
　　6.6.3　涉及浮点数时要考虑精度问题 438
　　6.6.4　大小端问题 438
　　6.6.5　协议与自动升级功能 438

6.7	包分片		439
6.8	XML 与 JSON 格式的协议		444
6.9	一个自定义协议示例		445
6.10	理解 HTTP		460
	6.10.1	HTTP 格式介绍	460
	6.10.2	GET 与 POST 方法	461
	6.10.3	HTTP chunk 编码	465
	6.10.4	HTTP 客户端的编码实现	466
	6.10.5	HTTP 服务端的实现	466
	6.10.6	HTTP 与长连接	471
	6.10.7	libcurl	471
	6.10.8	Restful 接口与 Java Spring MVC	477
6.11	SMTP、POP3 与邮件客户端		478
	6.11.1	邮件协议简介	478
	6.11.2	SMTP	479
	6.11.3	POP3	494
	6.11.4	邮件客户端	499
6.12	WebSocket 协议		499
	6.12.1	WebSocket 协议的握手过程	500
	6.12.2	WebSocket 协议的格式	503
	6.12.3	WebSocket 协议的压缩格式	506
	6.12.4	WebSocket 协议装包与解包示例	508
	6.12.5	解析握手协议	512

第 7 章 单个服务的基本结构 ... 515

7.1	网络通信组件的效率问题		515
	7.1.1	高效网络通信框架的设计原则	515
	7.1.2	连接的被动关闭与主动关闭	519
	7.1.3	长连接和短连接	519
7.2	原始的服务器结构		520
7.3	一个连接对应一个线程模型		522
7.4	Reactor 模式		523
7.5	one thread one loop 思想		524
	7.5.1	one thread one loop 程序的基本结构	524
	7.5.2	线程的分工	525
	7.5.3	唤醒机制的实现	527

· XIII ·

7.5.4　handle_other_things 方法的实现逻辑 532
　　　7.5.5　带定时器的程序结构 533
　　　7.5.6　one thread one loop 的效率保障 534
　7.6　收发数据的正确做法 534
　　　7.6.1　如何收取数据 534
　　　7.6.2　如何发送数据 535
　　　7.6.3　不要多个线程同时利用一个 socket 收（发）数据 538
　7.7　发送、接收缓冲区的设计要点 538
　　　7.7.1　为什么需要发送缓冲区和接收缓冲区 539
　　　7.7.2　如何设计发送缓冲区和接收缓冲区 539
　　　7.7.3　服务端发送数据时对端一直不接收的问题 543
　7.8　网络库的分层设计 544
　　　7.8.1　网络库设计中的各个层 544
　　　7.8.2　将 Session 进一步分层 550
　　　7.8.3　连接信息与 EventLoop/Thread 的对应关系 551
　7.9　后端服务中的定时器设计 551
　　　7.9.1　最简单的定时器 551
　　　7.9.2　定时器设计的基本思路 552
　　　7.9.3　定时器逻辑的性能优化 561
　　　7.9.4　对时间的缓存 564
　7.10　处理业务数据时是否一定要单独开线程 565
　7.11　非侵入式结构与侵入式结构 570
　　　7.11.1　非侵入式结构 570
　　　7.11.2　侵入式结构 571
　7.12　带有网络通信模块的服务器的经典结构 578
　　　7.12.1　为何要将 listenfd 设置成非阻塞模式 578
　　　7.12.2　基于 one thread one loop 结构的经典服务器结构 584
　　　7.12.3　服务器的性能瓶颈 586

第 8 章　Redis 网络通信模块源码分析 587

　8.1　调试 Redis 环境与准备 587
　　　8.1.1　Redis 源码编译与启动 587
　　　8.1.2　通信示例与术语约定 589
　8.2　探究 redis-server 端的网络通信模块 589
　　　8.2.1　监听 fd 的初始化工作 589
　　　8.2.2　接受客户端连接 592

		8.2.3 epollfd 的创建 600
		8.2.4 监听 fd 与客户端 fd 是如何挂载到 epollfd 上的 601
		8.2.5 readQueryFromClient 函数 611
		8.2.6 如何处理可写事件 613
		8.2.7 Redis 6.0 多线程网络 I/O 620
		8.2.8 Redis 对客户端的管理 635
		8.2.9 客户端断开流程 646
		8.2.10 Redis 中收发缓冲区的设计 653
		8.2.11 定时器逻辑 659
		8.2.12 钩子函数 662
		8.2.13 redis-server 端网络通信模块小结 662
	8.3	探究 redis-cli 端的网络通信模型 663
	8.4	Redis 的通信协议格式 673
		8.4.1 请求命令格式 673
		8.4.2 应答命令格式 674
		8.4.3 多命令和流水线 677
		8.4.4 特殊的 redis-cli 与内联命令 677
		8.4.5 Redis 对协议数据的解析逻辑 678
第 9 章	服务器开发中的常用模块设计 681
	9.1	断线自动重连的应用场景和逻辑设计 681
	9.2	保活机制与心跳包 683
		9.2.1 TCP keepalive 选项 683
		9.2.2 应用层的心跳包机制设计 684
		9.2.3 有代理的心跳包机制设计 689
		9.2.4 带业务数据的心跳包 690
		9.2.5 心跳包与流量 690
		9.2.6 心跳包与调试 691
		9.2.7 心跳包与日志 691
	9.3	日志模块的设计 692
		9.3.1 为什么需要日志 692
		9.3.2 日志系统的技术实现 692
		9.3.3 在 C/C++中输出网络数据包日志 716
		9.3.4 调试时的日志 719
		9.3.5 统计程序性能日志 719
		9.3.6 根据类型将日志写入不同的文件中 725

- 9.3.7 集中式日志服务与分布式日志服务 ... 725
- 9.3.8 从业务层面看在一条日志中应该包含什么内容 ... 727
- 9.3.9 在日志中不要出现敏感信息 ... 729
- 9.3.10 开发过程中的日志递进缩减策略 ... 730
- 9.4 错误码系统的设计 ... 730
  - 9.4.1 错误码的作用 ... 730
  - 9.4.2 错误码系统设计实践 ... 731
- 9.5 监控端口 ... 733

第 1 章
C++必知必会

本章讲解 C++中的一些常用技巧和 C++ 11/14/17 标准中的一些常用语言特性，但不包括新标准中多线程部分的内容，这部分内容将在第 3 章讲解。

1.1　C++ RAII 惯用法

什么是 RAII？容笔者先卖个关子，给大家讲个故事。

1.1.1　版本 1：最初的写法

在笔者刚学习服务器开发的时候，公司给笔者安排了一个练习：在 Windows 系统上写一个 C++程序，用该程序实现一个简单的服务，在客户端连接上来时，给客户端发一条"HelloWorld"消息后关闭连接，不用保证客户端一定能收到。

如果熟悉基础网络编程知识，那么你会觉得这很容易，因为这个程序描述的就是 TCP 网络通信的基本流程，其程序实现流程如下。

（1）创建 socket。

（2）绑定 IP 地址和端口号。

（3）在该 IP 地址和端口号上启动监听，循环等待客户端连接的到来，在客户端连接成功后，向其发送一条"HelloWorld"消息，然后断开连接。

　　在 Windows 上使用网络通信 API 之前，需要使用 WSAStartup 函数初始化 socket 库；在程序结束时需要使用 WSACleanup 函数清理 socket 库。

笔者很快就将程序写出来了：

```
#include <winsock2.h>
#include <stdio.h>

//链接 Windows 的 socket 库
#pragma comment(lib, "ws2_32.lib")
```

```cpp
int main(int argc, char* argv[])
{
    //初始化socket库
    WORD wVersionRequested = MAKEWORD(2, 2);
    WSADATA wsaData;
    int err = WSAStartup(wVersionRequested, &wsaData);
    if (err != 0)
        return 1;

    if (LOBYTE(wsaData.wVersion) != 2 || HIBYTE(wsaData.wVersion) != 2)
    {
        WSACleanup();
        return 1;
    }

    //创建用于监听的socket
    SOCKET sockSrv = socket(AF_INET, SOCK_STREAM, 0);
    if (sockSrv == -1)
    {
        WSACleanup();
        return 1;
    }

    SOCKADDR_IN addrSrv;
    addrSrv.sin_addr.S_un.S_addr = htonl(INADDR_ANY);
    addrSrv.sin_family = AF_INET;
    addrSrv.sin_port = htons(6000);
    //绑定socket,监听6000端口
    if (bind(sockSrv, (SOCKADDR*)&addrSrv, sizeof(SOCKADDR)) == -1)
    {
        closesocket(sockSrv);
        WSACleanup();
        return 1;
    }

    //启动监听,准备接受客户端的连接请求
    if (listen(sockSrv, 15) == -1)
    {
        closesocket(sockSrv);
        WSACleanup();
        return 1;
    }

    SOCKADDR_IN addrClient;
    int len = sizeof(SOCKADDR);
    char msg[] = "HelloWorld";
    while (true)
    {
        //等待客户端请求的到来,如果有客户端连接,则接受连接
        SOCKET sockClient = accept(sockSrv, (SOCKADDR*)&addrClient, &len);
        if (sockClient == -1)
            break;

        //向客户端发送"HelloWorld"消息
```

```
        send(sockClient, msg, strlen(msg), 0);
        closesocket(sockClient);
}// end while-loop

closesocket(sockSrv);
WSACleanup();

return 0;
}
```

以上代码虽然满足了公司的要求，但是有些地方不太令人满意，因为代码中充斥着用于避免出错的重复资源清理逻辑：closesocket(sockSrv)和WSACleanup()。

这样的场景在实际开发过程中经常存在，例如下面这段伪代码：

```
char* p = new char[1024];
if (操作1不成功)
{
    delete[] p;
    p = NULL;
    return;
}

if (操作2不成功)
{
    delete[] p;
    p = NULL;
    return;
}

if (操作3不成功)
{
    delete[] p;
    p = NULL;
    return;
}
delete[] p;
p = NULL;
```

我们可以将上述场景理解为"先分配资源，再进行相关操作，在任意中间步骤出错时都对相应的资源进行回收，如果中间步骤没有出错，则在资源使用完毕后对其进行回收"。上述伪代码片段中释放资源的重要性不言而喻：因为分配了堆内存，所以不释放会造成内存泄露。但是这样编写代码太容易出错了！我们必须时刻保持警惕，在任意出错的步骤中都要记得加上回收资源的代码。这样的编码方式不仅容易出错，还会导致大量的代码重复。那有没有办法解决这类问题呢？有，使用goto语句。

1.1.2 版本2：使用goto语句

还是以前面网络通信的代码为例，如果使用goto语句，则该代码可以简化如下：

```cpp
#include <winsock2.h>
#include <stdio.h>

//链接Windows的socket库
#pragma comment(lib, "ws2_32.lib")

int main(int argc, char* argv[])
{
    //注意:由于goto语句不能跳过变量定义,所以在这里提前定义下文需要用到的变量
    SOCKET sockSrv;
    SOCKADDR_IN addrSrv;
    SOCKADDR_IN addrClient;
    int len = sizeof(SOCKADDR);
    char msg[] = "HelloWorld";

    //初始化socket库
    WORD wVersionRequested = MAKEWORD(2, 2);
    WSADATA wsaData;
    int err = WSAStartup(wVersionRequested, &wsaData);
    if (err != 0)
        return 1;

    if (LOBYTE(wsaData.wVersion) != 2 || HIBYTE(wsaData.wVersion) != 2)
    {
        goto cleanup2;
    }

    //创建用于监听的socket
    sockSrv = socket(AF_INET, SOCK_STREAM, 0);
    if (sockSrv == -1)
    {
        goto cleanup2;
    }

    addrSrv.sin_addr.S_un.S_addr = htonl(INADDR_ANY);
    addrSrv.sin_family = AF_INET;
    addrSrv.sin_port = htons(6000);
    //绑定socket,监听6000端口
    if (bind(sockSrv, (SOCKADDR*)&addrSrv, sizeof(SOCKADDR)) == -1)
    {
        goto cleanup1;
    }

    //启动监听,准备接受客户端的连接请求
    if (listen(sockSrv, 15) == -1)
    {
        goto cleanup1;
    }

    while (true)
    {
        //等待客户端请求的到来,如果有客户端连接,则接受连接
```

```
        SOCKET sockClient = accept(sockSrv, (SOCKADDR*)&addrClient, &len);
        if (sockClient == -1)
            break;

        //向客户端发送"HelloWorld"消息
        send(sockClient, msg, strlen(msg), 0);
        closesocket(sockClient);
    }// end while-loop
cleanup1:
    closesocket(sockSrv);
cleanup2:
    WSACleanup();

    return 0;
}
```

使用 goto 语句后，一旦某个中间步骤出错，则跳转到统一的清理点进行资源清理操作。

但是，我们总被告知要慎用 goto 语句，因为它会让程序的结构变得混乱和难以维护。姑且不论这是否正确，如果不用 goto 语句，那么有没有更好的实现方式呢？有，使用 do...while(0) 循环。

1.1.3　版本 3：使用 do...while(0) 循环

以上代码使用 do...while(0) 循环改进后如下：

```
#include <winsock2.h>
#include <stdio.h>

//链接 Windows 的 socket 库
#pragma comment(lib, "ws2_32.lib")

int main(int argc, char* argv[])
{
    //初始化 socket 库
    WORD wVersionRequested = MAKEWORD(2, 2);
    WSADATA wsaData;
    int err = WSAStartup(wVersionRequested, &wsaData);
    if (err != 0)
        return 1;

    SOCKET sockSrv = -1;
    do
    {
        if (LOBYTE(wsaData.wVersion) != 2 || HIBYTE(wsaData.wVersion) != 2)
            break;

        //创建用于监听的 socket
        sockSrv = socket(AF_INET, SOCK_STREAM, 0);
        if (sockSrv == -1)
```

```
            break;

        SOCKADDR_IN addrSrv;
        addrSrv.sin_addr.S_un.S_addr = htonl(INADDR_ANY);
        addrSrv.sin_family = AF_INET;
        addrSrv.sin_port = htons(6000);
        //绑定 socket，监听 6000 端口
        if (bind(sockSrv, (SOCKADDR*)&addrSrv, sizeof(SOCKADDR)) == -1)
            break;

        //启动监听，准备接受客户端的连接请求
        if (listen(sockSrv, 15) == -1)
            break;

        SOCKADDR_IN addrClient;
        int len = sizeof(SOCKADDR);
        char msg[] = "HelloWorld";
        while (true)
        {
            //等待客户端的请求到来，如果有客户端连接，则接受连接
            SOCKET sockClient = accept(sockSrv, (SOCKADDR*)&addrClient, &len);
            if (sockClient == -1)
                break;

            //向客户端发送"HelloWorld"消息
            send(sockClient, msg, strlen(msg), 0);
            closesocket(sockClient);
        }// end inner-while-loop
    } while (0); //end outer-while-loop

    if (sockSrv != -1)
        closesocket(sockSrv);

    WSACleanup();

    return 0;
}
```

以上代码利用 do...while(0)循环中的 break 特性巧妙地将资源回收操作集中到一个地方，使用 for 循环也能达到同样的效果。我们同样可以使用 do...while(0)改造上面堆内存分配与释放的示例，伪代码如下：

```
char* p = NULL;
do
{
    p = new char[1024];
    if (操作 1 不成功)
        break;

    if (操作 2 不成功)
        break;

    if (操作 3 不成功)
        break;
} while (0);
```

```
delete[] p;
p = NULL;
```

这是 do...while(0) 的一个妙用。但是，在 C++ 中有更好的写法来代替 do...while(0)，即 RAII 惯用法。

1.1.4　版本 4：使用 RAII 惯用法

RAII（Resource Acquisition Is Initialization，资源获取就是初始化）指资源在我们拿到时就已经初始化，一旦不再需要该资源，就可以自动释放该资源。

对于 C++ 来说，资源在构造函数中初始化（可以在构造函数中调用单独的初始化函数），在析构函数中释放或清理。常见的情形就是在函数调用中创建 C++ 对象时分配资源，在 C++ 对象出了作用域时将其自动清理和释放（不管这个对象是如何出作用域的，不管是否因为某个中间步骤不满足条件而导致提前返回，也不管是否正常走完全部流程后返回）。

还是以上面网络通信的例子来说，初始化程序时需要分配两种资源：Windows 的 socket 网络库和一个用于监听的 socket。首先，初始化好 Windows socket 网络库；然后创建一个用于监听的 socket。在程序结束时，我们需要清理这两种资源。

使用 RAII 惯用法改进后的代码如下：

```
#include <winsock2.h>
#include <stdio.h>

//链接 Windows 的 socket 网络库
#pragma comment(lib, "ws2_32.lib")
class ServerSocket
{
public:
    ServerSocket()
    {
        m_bInit = false;
        m_ListenSocket = -1;
    }

    ~ServerSocket()
    {
        if (m_ListenSocket != -1)
            ::closesocket(m_ListenSocket);

        if (m_bInit)
            ::WSACleanup();
    }

    bool DoInit()
    {
        //初始化 socket 库
        WORD wVersionRequested = MAKEWORD(2, 2);
```

```cpp
    WSADATA wsaData;
    int err = ::WSAStartup(wVersionRequested, &wsaData);
    if (err != 0)
        return false;

    if (LOBYTE(wsaData.wVersion) != 2 || HIBYTE(wsaData.wVersion) != 2)
        return false;

    m_bInit = true;

    //创建用于监听的socket
    m_ListenSocket = ::socket(AF_INET, SOCK_STREAM, 0);
    if (m_ListenSocket == -1)
        return false;

    return true;
}

bool DoBind(const char* ip, short port = 6000)
{
    SOCKADDR_IN addrSrv;
    addrSrv.sin_addr.S_un.S_addr = inet_addr(ip);
    addrSrv.sin_family = AF_INET;
    addrSrv.sin_port = htons(port);
    if (::bind(m_ListenSocket, (SOCKADDR*)&addrSrv, sizeof(SOCKADDR)) == -1)
        return false;

    return true;
}

bool DoListen(int backlog = 15)
{
    if (::listen(m_ListenSocket, backlog) == -1)
        return false;

    return true;
}

bool DoAccept()
{
    SOCKADDR_IN addrClient;
    int len = sizeof(SOCKADDR);
    char msg[] = "HelloWorld";
    while (true)
    {
        //等待客户端的请求到来，如果有客户端连接，则接受连接
        SOCKET sockClient = ::accept(m_ListenSocket, (SOCKADDR*)&addrClient, &len);
        if (sockClient == -1)
            break;

        //向客户端发送"HelloWorld"消息
        ::send(sockClient, msg, strlen(msg), 0);
        ::closesocket(sockClient);
    }// end inner-while-loop
```

```cpp
        return false;
    }
private:
    bool       m_bInit;
    SOCKET     m_ListenSocket;
};

int main(int argc, char* argv[])
{
    ServerSocket serverSocket;
    if (!serverSocket.DoInit())
        return false;

    if (!serverSocket.DoBind("0.0.0.0", 6000))
        return false;

    if (!serverSocket.DoListen(15))
        return false;

    if (!serverSocket.DoAccept())
        return false;

    return 0;
}
```

以上代码并没有在构造函数中分配资源,而是单独使用一个 DoInit 方法初始化资源,并在析构函数中回收相应的资源。这样在 main 函数中就不用担心任何中间步骤失败而忘记释放资源了,因为一旦 main 函数调用结束,serverSocket 对象就会自动调用其析构函数回收相应的资源。这就是 RAII 惯用法的原理!

> 严格来说,以上代码中 ServerSocket 的成员变量 m_bInit 应该被设计成类静态成员,调用 WSAStartup 和 WSACleanup 的函数应该被设计成类的静态方法,因为它们只需在程序初始化和退出时各调用一次就可以了。

希望读者能理解 RAII 惯用法,因为它在 C++ 中太常用了。我们也可以使用 RAII 惯用法再次改写上文中分配堆内存的伪代码示例:

```cpp
//定义 HeapObjectWrapper RAII 对象
class HeapObjectWrapper
{
public:
    HeapObjectWrapper(int size)
    {
        m_p = new char[size];
    }

    ~HeapObjectWrapper()
    {
        delete[] m_p;
        m_p = NULL;
    }
```

```
private:
    char*  m_p;
};

//使用 HeapObjectWrapper RAII 对象
HeapObjectWrapper heapObj(1024);
if (操作 1 不成功)
    return;

if (操作 2 不成功)
    return;

if (操作 3 不成功)
    return;
```

其中，heapObj 对象一旦出了其作用域，该程序就会自动调用其析构函数释放堆内存。当然，RAII 惯用法中对资源分配和释放的定义可以延伸出各种外延和内涵，例如对多线程锁的获取和释放。我们在实际开发中也常常遇到以下情形：

```
void SomeFunction()
{
    得到某把锁；
    if (条件 1)
    {
        if (条件 2)
        {
            操作 1
            释放锁；
            return;
        }
        else (条件 3)
        {
            操作 2
            释放锁；
            return;
        }
    }

    if (条件 3)
    {
        操作 3
        释放锁；
        return;
    }

    操作 4
    释放锁；
}
```

这是一段很常见的逻辑：为了避免死锁，我们必须在每个可能退出的分支上都释放锁。随着逻辑写得越来越复杂，我们忘记在某个退出的分支上释放锁的可能性也越来越大。而 RAII 惯用法正好解决了这个问题：我们可以将锁包裹成一个对象，在构造函数中获取锁，在析构函数中释放锁。伪代码如下：

```cpp
class SomeLockGuard
{
public:
    SomeLockGuard()
    {
        //加锁
        m_lock.lock();
    }

    ~SomeLockGuard()
    {
        //解锁
        m_lock.unlock();
    }

private:
    SomeLock  m_lock;
};

void SomeFunction()
{
    SomeLockGuard lockWrapper;
    if (条件1)
    {
        if (条件2)
        {
            操作1
            return;
        }
        else (条件3)
        {
            操作2
            return;
        }
    }

    if (条件3)
    {
        操作3
        return;
    }

    操作4
}
```

使用 RAII 惯用法之后，我们就再也不必在每个函数出口处都加上释放锁的代码了，因为在函数调用结束后会自动释放锁。

对于以上代码，有经验的读者可能一眼就看出来了：这不就是 C++ 11 中 std::lock_guard 和 boost 库中 boost::mutex::scoped_lock 的实现原理吗？确实是，本书后续章节会详细介绍操作系统和 C++ 11 提供的各类锁的用法。

1.1.5 小结

资源泄露和死锁等问题具有非常强的隐蔽性，如果在生产环境中出现这些问题，则难以复现、排查和定位问题。理解并熟练使用 RAII 惯用法不仅能让我们的代码更加简洁和模块化，也能让我们在开发阶段避免一部分资源泄漏和死锁问题。

1.2 pimpl 惯用法

这里有一个名为 CSocketClient 的网络通信类，定义如下：

```
/**
 * 网络通信的基础类，SocketClient.h
 * zhangyl 2017.07.11
 */
class CSocketClient
{
public:
    CSocketClient();
    ~CSocketClient();

public:
    void SetProxyWnd(HWND hProxyWnd);

    bool    Init(CNetProxy* pNetProxy);
    bool    Uninit();

    int Register(const char* pszUser, const char* pszPassword);
    void GuestLogin();

    BOOL    IsClosed();
    BOOL    Connect(int timeout = 3);
    void    AddData(int cmd, const std::string& strBuffer);
    void    AddData(int cmd, const char* pszBuff, int nBuffLen);
    void    Close();

    BOOL    ConnectServer(int timeout = 3);
    BOOL    SendLoginMsg();
    BOOL    RecvLoginMsg(int& nRet);
    BOOL    Login(int& nRet);

private:
    void LoadConfig();
    static UINT CALLBACK SendDataThreadProc(LPVOID lpParam);
    static UINT CALLBACK RecvDataThreadProc(LPVOID lpParam);
    bool Send();
    bool Recv();
    bool CheckReceivedData();
    void SendHeartbeatPackage();

private:
    //这里暴露了太多与实现细节相关的成员对象
```

```
    SOCKET                  m_hSocket;
    short                   m_nPort;
    char                    m_szServer[64];
    long                    m_nLastDataTime;
    long                    m_nHeartbeatInterval;
    CRITICAL_SECTION        m_csLastDataTime;
    HANDLE                  m_hSendDataThread;
    HANDLE                  m_hRecvDataThread;
};
```

CSocketClient 类的 public 方法提供了对外接口供第三方使用，每个函数的具体实现都在 SocketClient.cpp 中，对第三方不可见。对于在 Windows 系统上提供给第三方使用的库，库作者一般需要提供.h、.lib 和.dll 文件给库使用者，对于 Linux 系统则需要提供.h、.a 或.so 文件。

不管在哪种操作系统上，提供像 SocketClient.h 这样的头文件给第三方使用时，库作者大多会隐隐不安——因为 SocketClient.h 文件中 CSocketClient 类的大量成员变量和私有函数都暴露了这个类的太多实现细节，很容易让使用者看出其实现原理。这样的头文件对于一些涉及核心技术实现的库和 SDK，是非常敏感的。

那有没有办法既能保持对外接口不变，又能尽量不暴露一些关键的成员变量和私有函数的实现方法呢？有，我们可以将代码稍微修改一下：

```
/**
 * 网络通信的基础类，SocketClient.h
 * zhangyl 2017.07.11
 */
//Impl 前置声明
class Impl;

class CSocketClient
{
public:
    CSocketClient();
    ~CSocketClient();

public:
    void SetProxyWnd(HWND hProxyWnd);

    bool    Init(CNetProxy* pNetProxy);
    bool    Uninit();

    int Register(const char* pszUser, const char* pszPassword);
    void GuestLogin();

    BOOL    IsClosed();
    BOOL    Connect(int timeout = 3);
    void    AddData(int cmd, const std::string& strBuffer);
    void    AddData(int cmd, const char* pszBuff, int nBuffLen);
    void    Close();

    BOOL    ConnectServer(int timeout = 3);
    BOOL    SendLoginMsg();
```

```
    BOOL        RecvLoginMsg(int& nRet);
    BOOL        Login(int& nRet);

private:
    Impl*    m_pImpl;
};
```

在以上代码中，所有的关键成员变量都已经不存在了，取而代之的是一个类型为 Impl 的指针成员变量 m_pImpl。

> 具体采用什么名称，读者完全可以根据自己的实际情况来定，不一定非要使用 "Impl" 和 "m_pImpl" 这样的名称。

Impl 类现在对使用者完全透明，为了在 CSocketClient 类中引用 Impl 类，我们在 SocketClient.h 文件中使用了一个前置声明（以上加粗代码行），然后就可以将原来属于 CSocketClient 类的成员变量转移到 Impl 类中了：

```
/**
 * 网络通信的基础类, SocketClient.cpp
 * zhangyl 2017.07.11
 */
class Impl
{
public:
    Impl()
    {
        //TODO：可以在这里对成员变量做一些初始化工作
    }

    ~Impl()
    {
        //TODO：可以在这里做一些清理工作
    }

public:
    SOCKET                  m_hSocket;
    short                   m_nPort;
    char                    m_szServer[64];
    long                    m_nLastDataTime;
    long                    m_nHeartbeatInterval;
    CRITICAL_SECTION        m_csLastDataTime;
    HANDLE                  m_hSendDataThread;
    HANDLE                  m_hRecvDataThread;
};
```

我们接着在 CSocketClient 构造函数中创建这个 m_pImpl 对象，在 CSocketClient 析构函数中释放这个对象：

```
//SocketClient.cpp
CSocketClient::CSocketClient()
{
    m_pImpl = new Impl();
}
```

```
CSocketClient::~CSocketClient()
{
    delete m_pImpl;
}
```

这样，在 CSocketClient 类内部，对于我们原来直接引用的成员变量，现在可以使用 m_pImpl->变量名来引用了。

> 这里仅以演示隐藏 CSocketClient 的成员变量为例，隐藏类的私有方法与隐藏成员变量的做法相同，即将原来属于 CSocketClient 的方法变成 Impl 的方法。

需要强调的是，在实际开发中，由于 Impl 类是 CSocketClient 的辅助类，没有独立存在的必要，所以一般会将 Impl 类定义成 CSocketClient 的内部类。即采用如下形式：

```
/**
 * 网络通信的基础类，SocketClient.h
 * zhangyl 2017.07.11
 */
class CSocketClient
{
public:
    CSocketClient();
    ~CSocketClient();

    //省略重复的代码

private:
    class    Impl;
    Impl*    m_pImpl;
};
```

然后在 ClientSocket.cpp 中定义 Impl 类的实现：

```
/**
 * 网络通信的基础类，SocketClient.cpp
 * zhangyl 2017.07.11
 */
class CSocketClient::Impl
{
public:
    void LoadConfig()
    {
        //省略具体的实现
    }

    //省略其他方法

public:
    SOCKET                  m_hSocket;
    short                   m_nPort;
    char                    m_szServer[64];
    long                    m_nLastDataTime;
    long                    m_nHeartbeatInterval;
    CRITICAL_SECTION        m_csLastDataTime;
    HANDLE                  m_hSendDataThread;
```

```
    HANDLE                      m_hRecvDataThread;
}

CSocketClient::CSocketClient()
{
    m_pImpl = new Impl();
}

CSocketClient::~CSocketClient()
{
    delete m_pImpl;
}
```

现在 CSocketClient 这个类除了保留对外的接口，其内部实现用到的变量和方法基本对使用者不可见了。C++中对类的这种封装方法被称为 pimpl 惯用法，即 Pointer to Implementation（也有人认为是 Private Implementation）。

> 在实际开发中，Impl 类的声明和定义既可以使用 class 关键字，也可以使用 struct 关键字。在 C++中，struct 类型可以用于定义成员方法，但 struct 所有的成员变量和方法默认都是 public 的。

现在总结该方法的优点，如下所述。

◎ 核心数据成员被隐藏，不必暴露在头文件中，对使用者透明，提高了安全性。
◎ 降低了编译依赖，提高了编译速度。原来头文件中的一些私有成员变量可能是非指针、非引用类型的自定义类型，需要在当前类的头文件中包含这些类型的头文件。在使用了 pimpl 惯用法以后，这些私有成员变量就被移动到当前类的 cpp 文件中，因此头文件不再需要包含这些成员变量的类型头文件，当前头文件变得"干净"，其他文件在引用这个头文件时，依赖的类型变少，加快了编译速度。
◎ 接口与实现分离。使用了 pimpl 惯用法之后，即使 CSocketClient 或者 Impl 类的实现细节发生了变化，对使用者都透明，对外的 CSocketClient 类声明却仍然可以保持不变。例如，我们可以增、删、改 Impl 的成员变量和成员方法，而保持 SocketClient.h 文件的内容不变；如果不使用 pimpl 惯用法，则我们做不到不改变 SocketClient.h 文件而增、删、改 CSocketClient 类的成员。

C++ 11 标准引入了智能指针对象，我们可以使用 std::unique_ptr 对象来管理上述用于隐藏具体实现的 m_pImpl 指针。可以将 SocketClient.h 文件修改如下：

```
//for std::unique_ptr
#include <memory>

class CSocketClient
{
public:
    CSocketClient();
    ~CSocketClient();

    //省略重复的代码
```

```
private:
    struct                          Impl;
    std::unique_ptr<Impl>           m_pImpl;
};
```

在 SocketClient.cpp 中修改 CSocketClient 对象的构造函数和析构函数，如果编译器仅支持 C++ 11 标准，则可以这么修改：

```
CSocketClient::CSocketClient()
{
    //C++ 11 标准并未提供 std::make_unique 方法，该方法是 C++ 14 提供的
    m_pImpl.reset(new Impl());
}
```

如果编译器支持 C++ 14 及以上标准，则可以这么修改：

```
CSocketClient::CSocketClient() : m_pImpl(std::make_unique<Impl>())
{
}
```

由于已经使用了智能指针来管理 m_pImpl 指向的堆内存，所以在析构函数中不再需要显式地释放堆内存：

```
CSocketClient::~CSocketClient()
{
    //不再需要显式删除了
    //delete m_pImpl;
}
```

pimp 惯用法是 C/C++项目开发中一种非常实用的代码编写策略，建议读者掌握它。

1.3 C++ 11/14/17 新增的实用特性

C++ 11 标准是 C++发展史上具有里程碑意义的一个版本，主要改进了 C++ 98/03 标准存在的两大问题。

◎ 废弃了 C++ 98/03 标准中一些不实用的语法和库（如 std::auto_ptr），改进或者增强了 C++ 98/03 标准中一些语法元素的用法（如 auto 关键字、统一的类的初始化列表语法），新增了其他编程语言早已支持的关键字和语法（如 final 关键字、=default 语法、=delete 语法）。

◎ 开始在语法和自带标准库层面增加对操作系统功能的支持（如线程库、时间库）。之前很多功能的实现，C++本身是无法支持的，必须依赖和使用原生操作系统的 API 函数，导致开发者需要编写大量的平台相关的代码。

随着 C++ 11 标准的发展，后续又出现了 C++ 14、C++ 17 及 C++ 20 标准，但它们都是对 C++ 11 做小范围的修改和扩展，其主要内容还是继续完善一些特性和进一步提高一些标准库的性能。由于 C++ 11 新增了大量方便开发的功能与特性，所以支持 C++ 11 标准的编译器（如 VC++ 12、g++ 4.8）一经发布，就立即被广大开发者和使用 C++的企业广泛采用。C++ 11 新增的特性确实大大提高了开发效率，使用起来非常方便。

对于 MSVC 编译器，支持 C++ 11 新标准的最低版本是 VC++ 12（随 Visual Studio 2013 一起发布，Visual Studio 简称 VS）。VS 2015 支持部分 C++ 14 特性，VS 2017 支持 C++ 14 和部分 C++ 17 特性，VS 2019 支持 C++ 17 大多数语言特性。

对于 gcc/g++编译器，支持 C++ 11 新标准的最低版本是 gcc/g++ 4.8，支持 C++ 14 的版本是 gcc/g++ 4.9，gcc/g++ 7.3 支持 C++ 17 的大多数语言特性。

对于某个支持 C++ 11/14/17 标准的 Visual Studio 版本，我们一般不需要做任何特殊设置即可使用新语言标准支持的语法特性和库功能，当然，也可以指定具体的语言规范版本。以 Visual Studio 2019 为例，新建一个 C++项目之后，选中该项目，在弹出的右键菜单中选择 Properties 菜单项，打开该项目的属性设置对话框，如下图所示。

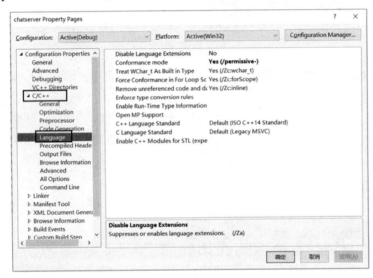

然后选择菜单"C/C++"→"Language"→"C++ Language Standard"，在弹出的下拉菜单中选择需要的 C++标准即可，如下图所示。

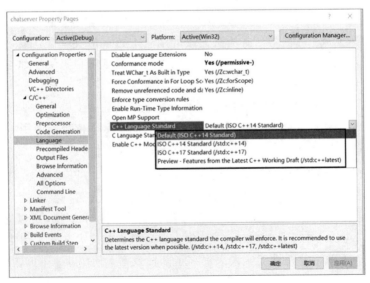

这里需要说明：Default 选项在 VS 2019 中对应使用 C++ 14 规范标准；对 Preview-Features from the latest C++ Working Draft(/std:c++ latest)选项使用当前最新的 C++标准（这里是 C++ 20），在微软完成 C++ 20 标准中所有的实现时，这里会多出一个选项，即 ISO C++ 20 Standard（/std:c++20）。

对于 gcc/g++编译器（最低版本 4.8），如果要使用 C++ 11/14/17 语言规范，则需要在编译时设置相应的选项值。例如将 test.cpp 文件编译成名为 test 的可执行文件，并需要 C++ 11 语言规范的支持，则使用：

```
g++ -g -o test test.cpp -std=c++11
```

在使用 Makefile 文件编译项目时可以这样指定：

```
make CXXFLAGS="-g -O0 -std=c++11"
```

在使用 cmake 编译项目时可以在 CMakeLists.txt 文件中添加如下行：

```
set(CMAKE_CXX_FLAGS "${CMAKE_CXX_FLAGS} -std=c++11 -g -Wall -O0 -Wno-unused-variable")
```

对于-std 选项的值，可以将其按需设置成 c++14 或 c++17。

对于 gcc/g++ 7.3 及以上版本，由于编译器本身就支持包括 C++ 17 在内的所有语言特性，因此如果在代码中用到某个特性，则由于编译器本身就支持，可以不必使用-std 选项指定具体的语言标准。

C++ 11/14/17 新增的特性非常多，本章接下来会介绍实际开发中高频使用的新语法和新特性。

1.4 统一的类成员初始化语法与 std::initializer_list<T>

假设类 A 有一个类型为 int 数组的成员变量，在 C++ 98/03 标准中，如果我们要在构造函数中对这个成员变量进行初始化，则需要这样写：

```cpp
//C++ 98/03 类成员变量是数组时的初始化语法
class A
{
   public:
      A()
      {
          arr[0] = 2;
          arr[1] = 0;
          arr[2] = 1;
          arr[3] = 9;
      }
   public:
      int arr[4];
};
```

对于字符数组，我们可能就要在构造函数中使用 strcpy、memcpy 这一类函数了；再者，如果数组元素足够多，初始值又没什么规律，则这种赋值代码会有很多行。但是，如果 arr 是一个局部变量，则我们在定义 arr 时可以使用如下语法对其进行初始化：

```
int arr[4] = {2, 0, 1, 9};
```

既然在 C++ 98/03 标准中局部变量数组支持这种语法，那么为什么在类成员变量语法中不支持这种语法呢？这是旧语法不合理的一个地方。因此在 C++ 11 语法中，对类成员变量也可以使用这种语法进行初始化：

```
//在 C++ 11 中可以使用大括号语法初始化数组类型的成员变量
class A
{
    public:
        A() : arr{2, 0, 1, 9}
        {
        }

    public:
        int arr[4];
};
```

可以看出，新语法比旧语法更加简洁。

在 Java 这类语言中定义一个类时，可以为其成员变量设置一个初始值，语法如下：

```
class A
{
    public int a = 1;
    public String string = "helloworld";
};
```

但在 C++ 98/03 标准中，对类的成员必须使用 static const 修饰，而且类型必须是整型（包括 bool、char、int、long 等），这样才能使用这种初始化语法：

```
//C++ 98/03 在类定义处初始化成员变量
class A
{
    public:
        //T 的类型必须是整型，且必须使用 static const 修饰
        static const T t = 某个整型值;
};
```

在 C++ 11 标准中就没有这种限制了，我们可以使用花括号（即{}）对任意类型的变量进行初始化，而且不用是 static 类型：

```
//C++ 11 在类定义处初始化成员变量
class A
{
    public:
        bool        ma{true};
        int         mb{2019};
        std::string mc{"helloworld"};
};
```

当然，在实际开发中，建议还是将这些成员变量的初始化统一写到构造函数的初始化列表中，方便阅读和维护代码。

综上所述，在 C++ 11 标准中，无论是局部变量还是类变量，使用花括号（{}）初始化的语法被统一起来，写法也变得简洁。

那么这种语法是如何实现的呢？如何在自定义类中也支持这种花括号呢？这就需要用到 C++ 11 新引入的对象 std::initializer_list<T> 了。它是一个模板对象，接收一个自定义参数类型 T，T 既可以是基础数据类型（如编译器内置的 bool、char、int 等），也可以是自定义的复杂数据类型。为了使用 std::initializer_list<T>，需要包含头文件，下面是一个例子：

```cpp
#include <iostream>
#include <initializer_list>
#include <vector>

class A
{
public:
    A(std::initializer_list<int> integers)
    {
        m_vecIntegers.insert(m_vecIntegers.end(), integers.begin(), integers.end());
    }

    ~A()
    {
    }

    void append(std::initializer_list<int> integers)
    {
        m_vecIntegers.insert(m_vecIntegers.end(), integers.begin(), integers.end());
    }

    void print()
    {
        size_t size = m_vecIntegers.size();
        for (size_t i = 0; i < size; ++i)
        {
            std::cout << m_vecIntegers[i] << std::endl;
        }
    }
private:
    std::vector<int> m_vecIntegers;
};

int main()
{
    A a{ 1, 2, 3 };
    a.print();

    std::cout << "After appending..." << std::endl;

    a.append({ 4, 5, 6 });
```

```
        a.print();

        return 0;
}
```

在以上代码中自定义了一个类 A，为了让 A 的构造函数和 append 方法同时支持花括号语法，给这两个方法同时设置了一个参数 integers，参数的类型均为 std::initializer_list<int>，程序执行结果如下：

```
[root@myaliyun testcpp]# ./test_initializer_list
1
2
3
After appending...
1
2
3
4
5
6
```

再来看一个例子，网上某 C++ JSON 库支持采用如下语法创建一个 JSON 对象：

```
// a way to express an _array_ of key/value pairs
// [["currency", "USD"], ["value", 42.99]]
json array_not_object = json::array({ {"currency", "USD"}, {"value", 42.99} });
```

那么，这个 json::array 方法是如何实现的呢？这利用 std::initializer_list<T>也很容易实现，首先在花括号中有两个元素{"currency", "USD"}和{"value", 42.99}，且这两个元素的值不一样，前者是两个字符串类型，后者是一个字符串类型和一个浮点类型。因此，我们可以创建两个构造函数分别支持这两种类型的构造函数，构造的对象类型为 jsonNode；然后创建一个 JSON 对象，实现其 array 方法，该方法接收一个参数，参数类型为 std::initializer_list<jsonNode>。完整的代码如下：

```
#include <iostream>
#include <string>
#include <initializer_list>
#include <vector>

//简单地模拟 JSON 支持的几种数据类型
enum class jsonType
{
    jsonTypeNull,
    jsonTypeInt,
    jsonTypeLong,
    jsonTypeDouble,
    jsonTypeBool,
    jsonTypeString,
    jsonTypeArray,
    jsonTypeObject
};

struct jsonNode
```

```cpp
{
    jsonNode(const char* key, const char* value) :
        m_type(jsonType::jsonTypeString),
        m_key(key),
        m_value(value)
    {
        std::cout << "jsonNode contructor1 called." << std::endl;
    }

    jsonNode(const char* key, double value) :
        m_type(jsonType::jsonTypeDouble),
        m_key(key),
        m_value(std::to_string(value))
    {
        std::cout << "jsonNode contructor2 called." << std::endl;
    }

    //省略其他类型的构造函数

    jsonType        m_type;
    std::string     m_key;
    //始终使用 string 类型保存值，避免浮点类型因为精度问题而显示不同的结果
    std::string     m_value;
};
class json
{
public:
    static json& array(std::initializer_list<jsonNode> nodes)
    {
        m_json.m_nodes.clear();
        m_json.m_nodes.insert(m_json.m_nodes.end(), nodes.begin(), nodes.end());

        std::cout << "json::array() called." << std::endl;

        return m_json;
    }

    json()
    {
    }

    ~json()
    {
    }

    std::string toString()
    {
        size_t size = m_nodes.size();
        for (size_t i = 0; i < size; ++i)
        {
            switch (m_nodes[i].m_type)
            {
            //根据类型组装成一个 JSON 字符串，代码省略
            case jsonType::jsonTypeDouble:
```

```
                break;
            }
        }
    }
private:
    std::vector<jsonNode> m_nodes;
    static json            m_json;
};

json json::m_json;
int main()
{
    json array_not_object = json::array({ {"currency", "USD"}, {"value", 42.99} });

    return 0;
}
```

程序执行结果如下：

```
[root@myaliyun testcpp]# ./construct_complex_objects
jsonNode contructor1 called.
jsonNode contructor2 called.
json::array() called.
```

希望读者通过上面两个例子理解 std::initializer_list<T>的使用场景。std::initializer_list<T>除了提供了构造函数，还提供了三个成员函数，这和 stl 其他容器的同名方法用法一样：

```
//返回列表中的元素个数
size_type size() const;
//返回第 1 个元素的指针
const T* begin() const;
//返回最后一个元素的下一个位置，代表结束
const T* end() const;
```

1.5 C++ 17 注解标签（attributes）

在 C++ 98/03 时代，不同的编译器使用不同的注解为代码增加了一些额外的说明，读者可能在各种 C/C++代码中见过#pragma、__declspec、__attribute 等注解。然而，不同的编译器对于同一功能可能使用不同的注解，导致我们需要为不同的编译器编写不同的注解代码。从 C++ 11 开始，新的语言标准统一制定了一些常用的注解标签，本节介绍一些比较常用的注解标签。使用注解标签的语法如下：

```
[[attribute]] types/functions/enums/etc
```

这些标签可用于修饰任意类型、函数或者 enumeration，在 C++ 17 之前不能用于修饰命名空间（namespace）和 enumerator，在 C++ 17 标准中，这个限制也被取消了。

1.5.1　C++ 98/03 的 enumeration 和 C++ 11 的 enumerator

读者可能对 enumeration 和 enumerator 这两个词感到困惑，前者指的是从 C 时代就存在的不限定作用域的枚举。例如，下面的枚举类型就是一个 enumeration：

```
//一个 enumeration 例子
enum Color
{
   black,
   white,
   red
};

//无法编译通过
bool white = true;
```

这种枚举类型之所以被称为不限定作用域的枚举，是因为一旦定义了这样一种枚举，在其所在的作用域内就不能再定义与之同名的变量了。例如，如果定义了上述 Color 枚举，此时再定义一个名为 white 的变量，就无法编译通过了。而 enumerator 指的是从 C++ 11 开始引入的以如下形式定义的枚举变量：

```
//一个 enumerator 的例子
enum class Color
{
   black,
   white,
   red
};

//可以编译通过
bool white = true;
```

此时，由于枚举值 white 对外部不可见（必须通过 Color::white 引用），所以可以定义一个同名的 white 变量。这种枚举变量被称为限定作用域的枚举。

1.5.2　C++ 17 的注解标签

在分清楚 enumeration 和 enumerator 之后，让我们回到正题上来。

C++ 11 引入的常用注解标签有[[noreturn]]，这个注解的含义是告诉编译器某个函数没有返回值，例如：

```
[[noreturn]] void terminate();
```

这个标签一般在设计一些系统函数时使用，例如 std::abort()和 std::exit()。

C++ 14 引入了[[deprecated]]标签来表示一个函数或者类型等已被弃用，在使用这些被弃用的函数或者类型并编译时，编译器会给出相应的警告，有的编译器直接生成编译错误：

```
[[deprecated]] void funcX();
```

这个标签在实际开发中非常有用，尤其在设计一些库代码时，如果库作者希望某个函数或者类型不想再被用户使用，则可以使用该标注标记。当然，我们也可以使用如下语法

给出编译时的具体警告或者出错信息:

```
[[deprecated("use funY instead")]] void funcX();
```

有如下代码:

```
#include <iostream>

[[deprecated("use funcY instead")]] void funcX()
{
    //实现省略
}

int main()
{
    funcX();
    return 0;
}
```

若在 main 函数中调用被标记为 deprecated 的函数 funcX，则在 gcc/g++ 7.3 中编译时会得到如下警告信息:

```
[root@myaliyun testmybook]# g++ -g -o test_attributes test_attributes.cpp
test_attributes.cpp: In function 'int main()':
test_attributes.cpp:10:11: warning: 'void funcX()' is deprecated: use funcY instead
[-Wdeprecated-declarations]
     funcX();
           ^
test_attributes.cpp:3:42: note: declared here
 [[deprecated("use funcY instead")]] void funcX()
                                          ^~~~~
```

> Java 开发者对这个标注应该再熟悉不过了。在 Java 中使用@Deprecated 标注可以达到同样的效果，这大概是 C++标准委员"拖欠"广大 C++开发者太久的一个特性吧。

C++ 17 提供了三个实用注解: [[fallthrough]]、[[nodiscard]]和[[maybe_unused]]，这里逐一介绍它们的用法。

[[fallthrough]]用于 switch-case 语句中，在某个 case 分支执行完毕后如果没有 break 语句，则编译器可能会给出一条警告。但有时这可能是开发者有意为之的。为了让编译器明确知道开发者的意图，可以在需要某个 case 分支被"贯穿"的地方（上一个 case 没有 break 语句）显式设置[[fallthrough]]标记。代码示例如下:

```
switch (type)
{
case 1:
    func1();
    //这个位置缺少 break 语句，且没有 fallthrough 标注，
    //可能是一个逻辑错误，在编译时编译器可能会给出警告，以提醒修改
case 2:
    func2();
    //这里也缺少 break 语句，但是使用了 fallthrough 标注，
    //说明是开发者有意为之的，编译器不会给出任何警告
```

```
[[fallthrough]];
case 3:
    func3();
}
```

注意：在 gcc/g++ 中，[[fallthrough]]后面的分号不是必需的，在 Visual Studio 中必须加上分号，否则无法编译通过。

熟悉 Golang 的读者，可能对 fallthrough 这一语法特性非常熟悉，Golang 中在 switch-case 后加上 fallthrough，是一个常用的告诉编译器意图的语法规则。代码示例如下。

```
//以下是Golang语法
s := "abcd"
switch s[3] {
    case 'a':
        fmt.Println("The integer was <= 4")
        fallthrough
    case 'b':
        fmt.Println("The integer was <= 5")
        fallthrough
    case 'c':
        fmt.Println("The integer was <= 6")
    default:
        fmt.Println("default case")
}
```

[[nodiscard]]一般用于修饰函数，告诉函数调用者必须关注该函数的返回值（即不能丢弃该函数的返回值）。如果函数调用者未将该函数的返回值赋值给一个变量，则编译器会给出一个警告。例如，假设有一个网络连接函数 connect，我们通过返回值明确说明了连接是否建立成功，则为了防止调用者在使用时直接将该值丢弃，我们可以将该函数使用[[nodiscard]]标记：

```
[[nodiscard]] int connect(const char* address, short port)
{
    //实现省略
}

int main()
{
    connect("127.0.0.1", 8888);

    return 0;
}
```

在 C++ 20 中，对于诸如 operator new()、std::allocate()等库函数均使用了[[nodiscard]]进行标记，以强调必须使用这些函数的返回值。

在通常情况下，编译器会对程序代码中未使用的函数或变量给出警告，另一些编译器干脆不允许通过编译。在 C++ 17 之前，程序员为了消除这些未使用的变量带来的编译警告或者错误，要么修改编译器的警告选项设置，要么定义一个类似于 UNREFERENCED_PARAMETER 的宏来显式调用这些未使用的变量一次，以消除编译警告或错误：

```
#define UNREFERENCED_PARAMETER(x) x
int APIENTRY wWinMain(HINSTANCE hInstance,
                      HINSTANCE hPrevInstance,
                      LPWSTR    lpCmdLine,
                      int       nCmdShow)
{
    UNREFERENCED_PARAMETER(hPrevInstance);
    UNREFERENCED_PARAMETER(lpCmdLine);

    //无关代码省略
}
```

以上代码节选自一个标准 Win32 程序的结构，其中的函数参数 hPrevInstance 和 lpCmdLine 一般不会被用到，编译器会给出警告。为了消除这类警告，这里定义了一个宏 UNREFERENCED_PARAMETER 并进行调用，造成这两个参数被使用的假象。

有了[[maybe_unused]]注解之后，我们就再也不需要这类宏来"欺骗"编译器了。以上代码使用该注解后可以修改如下：

```
int APIENTRY wWinMain(HINSTANCE hInstance,
                      [[maybe_unused]] HINSTANCE hPrevInstance,
                      [[maybe_unused]] LPWSTR    lpCmdLine,
                      int       nCmdShow)
{
    //无关代码省略
}
```

读者可以通过 C++官网了解 C++新标准中更多注解的用法。

1.6　final、override 关键字和=default、=delete 语法

final、override、=default、=delete 是 C++ 11 新增的一组非常具有标记意义的关键字和语法，这里逐一介绍它们。

1.6.1　final 关键字

final 关键字修饰一个类，这个类将不允许被继承，这在其他语言（如 Java）中早就实现了。在 C++ 11 中，final 关键字要写在类名的后面，这在其他语言中是写在 class 关键字前面的。示例如下：

```
class A final
{

};

class B : A
```

```
{

};
```

由于类 A 被声明成 final，B 继承 A，所以编译器会报如下错误提示类 A 不能被继承：

`error C3246: 'B' : cannot inherit from 'A' as it has been declared as 'final'`

1.6.2 override 关键字

C++语法规定，在父类中加了 virtual 关键字的方法可以被子类重写，子类重写该方法时可以加或不加 virtual 关键字，例如下面这样：

```
class A
{
protected:
    virtual void func(int a, int b)
    {
    }
};

class B : A
{
protected:
    virtual void func(int a, int b)
    {
    }
};

class C : B
{
protected:
    void func(int a, int b)
    {
    }
};
```

这种宽松的规定可能会带来以下两个问题。

◎ 当我们阅读代码时，无论子类重写的方法是否添加了 virtual 关键字，我们都无法直观地确定该方法是否是重写的父类方法。

◎ 如果我们在子类中不小心写错了需要重写的方法的函数签名（可能是参数类型、个数或返回值类型），这个方法就会变成一个独立的方法，这可能会违背我们重写父类某个方法的初衷，而编译器在编译时并不会检查到这个错误。

为了解决以上两个问题，C++ 11 引进了 override 关键字，其实 override 关键字并不是新语法，在 Java 等其他编程语言中早就支持。类方法被 override 关键字修饰，表明该方法重写了父类的同名方法，加了该关键字后，编译器会在编译阶段做相应的检查，如果其父类不存在相同签名格式的类方法，编译器就会给出相应的错误提示。

情形一，父类不存在，子类标记了 override 的方法：

```
class A
{

};

class B : A
{
protected:
    void func(int k, int d) override
    {
    }
};
```

由于在父类 A 中没有 func 方法，所以编译器会提示错误：

```
error C3668: 'B::func' : method with override specifier 'override' did not override any base class methods
```

情形二，父类存在，子类标记了 override 的方法，但函数签名不一致：

```
class A
{
protected:
    virtual int func(int k, int d)
    {
        return 0;
    }
};

class B : A
{
protected:
    virtual void func(int k, int d) override
    {
    }
};
```

这时编译器会报同样的错误。

正确的代码如下：

```
class A
{
protected:
    virtual void func(int k, int d)
    {
    }
};

class B : A
{
protected:
    virtual void func(int k, int d) override
    {
    }
};
```

1.6.3 =default 语法

如果一个 C++类没有显式给出构造函数、析构函数、拷贝构造函数、operator= 这几类函数的实现，则在需要它们时，编译器会自动生成；或者，在给出这些函数的声明时，如果没有给出其实现，则编译器在链接时会报错。如果使用=default 标记这类函数，则编译器会给出默认的实现。来看一个例子：

```
class A
{

};

int main()
{
    A a;
    return 0;
}
```

这样的代码是可以编译通过的，因为编译器默认生成 A 的一个无参构造函数，假设我们现在向 A 提供一个有参构造函数：

```
class A
{
public:
    A(int i)
    {
    }

};

int main()
{
    A a;
    return 0;
}
```

这时，编译器就不会自动生成默认的无参构造函数了，这段代码会编译出错，提示 A 没有合适的无参构造函数：

```
error C2512: 'A' : no appropriate default constructor available
```

我们这时可以手动为 A 加上无参构造函数，也可以使用=default 语法强行让编译器自己生成：

```
class A
{
public:
    A() = default;

    A(int i)
    {
    }

};
```

```
int main()
{
    A a;
    return 0;
}
```

　　=default 最大的作用可能是在开发中简化了构造函数中没有实际初始化代码的写法，尤其是声明和实现分别属于.h 和.cpp 文件。例如，对于类 A，其头文件为 a.h，其实现文件为 a.cpp，则正常情况下我们需要在 a.cpp 文件中写其构造函数和析构函数的实现（可能没有实际的构造和析构逻辑）：

```
//a.h
class A
{
public:
    A();
    ~A();
};
//a.cpp
#include "a.h"

A::A()
{
}

A::~A()
{
}
```

　　可以发现，即使在 A 的构造函数和析构函数中什么逻辑也没有，我们还是不得不在 a.cpp 中写上构造函数和析构函数的实现。有了=default 关键字，我们就可以在 a.h 中直接写成：

```
//a.h
class A
{
public:
    A() = default;
    ~A() = default;
};
//a.cpp
#include "a.h"
//在 cpp 文件中就不用再写 A 的构造函数和析构函数的实现了
```

1.6.4　=delete 语法

　　既然有强制让编译器生成构造函数、析构函数、拷贝构造函数、operator=的语法，那么也应该有禁止编译器生成这些函数的语法，没错，就是=delete。

　　在 C++ 98/03 规范中，如果我们想让一个类不能被拷贝（即不能调用其拷贝构造函数），则可以将其拷贝构造函数和 operator=函数定义成 private 的：

```cpp
class A
{
public:
    A() = default;
    ~A() = default;

private:
    A(const A& a)
    {
    }

    A& operator =(const A& a)
    {
    }
};

int main()
{
    A a1;
    A a2(a1);
    A a3;
    a3 = a1;

    return 0;
}
```

通过以上代码利用 a1 构造 a2 时，编译器会提示错误：

```
error C2248: 'A::A' : cannot access private member declared in class 'A'
error C2248: 'A::operator =' : cannot access private member declared in class 'A'
```

我们利用这种方式间接实现了一个类不能被拷贝的功能，这也是继承自 boost::noncopyable 的类不能被拷贝的实现原理。现在有了=delete 语法，我们直接使用该语法禁止编译器生成这两个函数即可：

```cpp
class A
{
public:
    A() = default;
    ~A() = default;

public:
    A(const A& a) = delete;

    A& operator =(const A& a) = delete;
};

int main()
{
    A a1;
    //A a2(a1);
    A a3;
    //a3 = a1;

    return 0;
}
```

一般在一些工具类中，我们不需要用到构造函数、析构函数、拷贝构造函数、operator=这 4 个函数，为了防止编译器自己生成，同时为了减小生成的可执行文件的体积，建议使用=delete 语法禁止编译器为这 4 个函数生成默认的实现代码，例如：

```cpp
class EncodeUtil
{
public:
    static std::wstring AnsiiToUnicode(const std::string& strAnsii);
    static std::string UnicodeToAnsii(const std::wstring& strUnicode);
    static std::string AnsiiToUtf8(const std::string& strAnsii);
    static std::string Utf8ToAnsii(const std::string& strUtf8);
    static std::string UnicodeToUtf8(const std::wstring& strUnicode);
    static std::wstring Utf8ToUnicode(const std::string& strUtf8);

private:
    EncodeUtil() = delete;
    ~EncodeUtil() = delete;

    EncodeUtil(const EncodeUtil& rhs) = delete;
    EncodeUtil& operator=(const EncodeUtil& rhs) = delete;
};
```

1.7 auto 关键字的用法

auto 关键字在 C++ 98/03 标准中与 static 关键字用途相反，用于修饰所有局部变量，即这个变量具有"自动"的生命周期，但是这个规定没有任何实际用处。因而在 C++ 11 新标准中修改了其用法，让编译器自己推导一些变量的数据类型，例如：

```cpp
int a = 1;
auto b = a;
```

这里变量 b 的类型被声明为 auto，编译器根据变量 a 的类型推导出变量 b 的类型也是 int。但是这样的写法在实际开发中实用性不高，所以 auto 一般用于让编译器自动推导一些复杂的模板数据类型，以简化语法，例如：

```cpp
std::map<std::string, std::string> seasons;
seasons["spring"] = "123";
seasons["summer"] = "456";
seasons["autumn"] = "789";
seasons["winter"] = "101112";

for (std::map<std::string, std::string>::iterator iter = seasons.begin(); iter != seasons.end(); ++iter)
{
    std::cout << iter->second << std::endl;
}
```

在上面的代码中，迭代器变量 iter 的类型是 std::map<std::string, std::string>::iterator，类型名太长，写起来很麻烦，在 C++ 11 语法中可以使用 auto 关键字达到同样的效果，这样会方便很多：

```
std::map<std::string, std::string> seasons;
seasons["spring"] = "123";
seasons["summer"] = "456";
seasons["autumn"] = "789";
seasons["winter"] = "101112";

for (auto iter = seasons.begin(); iter != seasons.end(); ++iter)
{
    std::cout << iter->second << std::endl;
}
```

1.8　Range-based 循环语法

大多数语言都支持 for-each 语法遍历一个数组或集合中的元素。在 C++ 98/03 规范中，对于一个数组 int arr[10]，如果我们想要遍历这个数组，则只能使用递增的计数去引用数组中的每个元素：

```
int arr[10] = {0};
for (int i = 0; i < 10; ++i)
{
    std::cout << arr[i] << std::endl;
}
```

在 C++ 11 规范中有了 for-each 语法，可以这么写：

```
int arr[10] = {0};
for (int i : arr)
{
    std::cout << i << std::endl;
}
```

对于 1.7 节中遍历 std::map 的内容，我们也可以使用这种语法：

```
std::map<std::string, std::string> seasons;
seasons["spring"] = "123";
seasons["summer"] = "456";
seasons["autumn"] = "789";
seasons["winter"] = "101112";

for (auto iter : seasons)
{
    std::cout << iter.second << std::endl;
}
```

for-each 语法虽然很强大，但是有两个需要注意的地方。

◎ for-each 中的迭代器类型与数组或集合中元素的类型完全一致，而原来使用老式语法迭代 stl 容器（如 std::map）时，迭代器 iter 的类型是 stl 容器中元素的指针类型。因此，在上面例子的老式语法中，iter 是一个指针类型（std::pair<std::string, std::string>*），使用 iter->second 去引用键值；而在 for-each 语法中，iter 与容器中元素的数据类型（std::pair<std::string, std::string>）相同，因此使用 iter.second 可直接引用键值。

◎ 在 for-each 语法中，对于复杂的数据类型，迭代器是原始数据的拷贝，而不是原始数据的引用。什么意思呢？来看一个例子：

```
std::vector<std::string> v;
v.push_back("zhangsan");
v.push_back("lisi");
v.push_back("maowu");
v.push_back("maliu");
for (auto iter : v)
{
    iter = "hello";
}
```

我们遍历容器 v，试图将 v 中元素的值都修改成 "hello"，在实际执行时却达不到我们想要的效果。这就是上文说的 for-each 中的迭代器是元素的拷贝，所以这里只是将每次的拷贝都修改成 "hello"，原始数据并不会被修改。我们可以将迭代器修改成原始数据的引用：

```
std::vector<std::string> v;
v.push_back("zhangsan");
v.push_back("lisi");
v.push_back("maowu");
v.push_back("maliu");
for (auto& iter : v)
{
    iter = "hello";
}
```

这样就达到修改原始数据的目的了。这是使用 for-each 比较容易出错的地方。对于容器中的复杂数据类型，我们应该尽量使用这种引用原始数据的方式，减少不必要的拷贝构造函数调用开销：

```
class A
{
public:
    A()
    {
    }
    ~A() = default;

    A(const A& rhs)
    {
    }

public:
    int m;
};

int main()
{
    A a1;
    A a2;
    std::vector<A> v;
    v.push_back(a1);
```

```
        v.push_back(a2);
        for (auto iter : v)
        {
            //在每一轮循环中,iter 都是 v 中元素的拷贝(通过调用 A 的拷贝构造函数生成)
            //实际使用 for-each 循环时应该尽量使用 v 中元素的引用,减少不必要的拷贝构造函数调用开销
            iter.m = 9;
        }

        return 0;
}
```

1.8.1 自定义对象如何支持 Range-based 循环语法

介绍了这么多,如何让自定义对象支持 Range-based 循环语法呢?为了支持这种语法,这个对象至少需要实现如下两个方法:

```
//需要返回第 1 个迭代子的位置
Iterator begin();
//需要返回最后一个迭代子的下一个位置
Iterator end();
```

上面的 Iterator 是自定义数据类型的迭代子类型,这里的 Iterator 类型必须支持如下三种操作(原因在下文中会解释)。

◎ operator++(自增)操作,可以在自增之后返回下一个迭代子的位置。
◎ operator!=(判不等操作)操作。
◎ operator*(解引用,dereference)操作。

下面是一个自定义对象支持 for-each 循环的例子:

```
#include <iostream>
#include <string>

template<typename T, size_t N>
class A
{
public:
    A()
    {
        for (size_t i = 0; i < N; ++i)
        {
            m_elements[i] = i;
        }
    }
    ~A()
    {
    }

    T* begin()
    {
        return m_elements + 0;
    }

    T* end()
```

```
        {
            return m_elements + N;
        }

private:
    T       m_elements[N];
};

int main()
{
    A<int, 10> a;
    for (auto iter : a)
    {
        std::cout << iter << std::endl;
    }

    return 0;
}
```

注意：在以上代码中，迭代子 Iterator 是 T*，是指针类型，本身就支持 operator++和 operator!=操作，所以这里并没有提供这两个方法的实现。那么为什么迭代子要支持 operator++和 operator!=操作呢？我们来看一下编译器是如何实现这种 for-each 循环的。

1.8.2　for-each 循环的实现原理

上述 for-each 循环可被抽象成如下公式：

```
for (for-range-declaration : for-range-initializer)
    statement;
```

C++ 14 标准是这样解释上面的公式的：

```
auto && __range = for-range-initializer;
for ( auto __begin = begin-expr, __end = end-expr; __begin != __end; ++__begin )
{
    for-range-declaration = *__begin;
    statement;
}
```

在这个循环中，begin-expr 返回的迭代子 __begin 需要支持自增操作，且每次循环时都会与 end-expr 返回的迭代子 __end 做判不等比较，在循环内部通过调用迭代子的解引用（*）操作取得实际的元素。这就是上文说的迭代子对象需要支持 operator++、operator != 和 operator*的原因了。

但是在上面的公式中，一个逗号表达式中的 "auto __begin = begin-expr, __end = end-expr;" 只使用了一个类型符号 auto，导致起始迭代子 __begin 和结束迭代子 __end 是同一类型，这样不太灵活。在某些设计中，可能希望结束迭代子是另一种类型。

因此在 C++ 17 标准中要求编译器解释 for-each 循环为如下形式：

```
1 auto && __range = for-range-initializer;
2 auto __begin = begin-expr;
3 auto __end = end-expr;
4 for ( ; __begin != __end; ++__begin ) {
```

```
5      for-range-declaration = *__begin;
6      statement;
7  }
```

看到了吧，代码第 2 行和第 3 行将获取起始迭代子__begin 和结束迭代子__end 分开写，这样这两个迭代子就可以是不同的类型了。虽然类型可以不一样，但这两种类型之间仍然支持 operator!=操作。C++ 17 对 C++ 14 的这种改变，对旧的代码不会产生任何影响，但可以让之后的开发更加灵活。

1.9 C++ 17 结构化绑定

对于 std::map 容器，很多读者应该都很熟悉。map 容器提供了一个 insert 方法，用于向 map 中插入元素，但是很少有人记得 insert 方法的返回值是什么类型。让我们看一下 C++ 98/03 提供的 insert 方法的签名：

```
std::pair<iterator,bool> insert( const value_type& value );
```

这里我们仅关心其返回值，这个返回值是 std::pair<T1,T2>类型。由于 map 中元素的 key 不允许重复，所以如果 insert 方法调用成功，则 T1 是被成功插入 map 中的元素的迭代器，T2 的类型为 bool，此时其值为 true（表示插入成功）；如果 insert 由于 key 重复，T1 是造成插入失败、已经存在于 map 中的元素的迭代器，则此时 T2 的值为 false（表示插入失败）。

在 C++ 98/03 标准中可以使用 std::pair<T1,T2>的 first 和 second 属性分别引用 T1 和 T2 的值。如下面的代码所示：

```
#include <iostream>
#include <string>
#include <map>

int main()
{
    std::map<std::string, int> cities;
    cities["beijing"]   = 0;
    cities["shanghai"]  = 1;
    cities["shenzhen"]  = 2;
    cities["guangzhou"] = 3;

    //for (const auto& [key, value] : m)
    //{
    //    std::cout << key << ": " << value << std::endl;
    //}

    //这一行在 C++ 11 之前的写法实在太麻烦了！
    //std::pair<std::map<std::string, int>::iterator, int> insertResult =
    cities.insert(std::pair<std::string, int>("shanghai", 2));
    //在 C++ 11 中写成:
    auto insertResult = cities.insert(std::pair<std::string, int>("shanghai", 2));

    std::cout << "Is insertion successful ? " << (insertResult.second ? "true" :
```

```
"false")
            << ", element key: " << insertResult.first->first << ", value: " <<
insertResult.first->second << std::endl;

    return 0;
}
```

以上代码太繁复了，我们可以使用 auto 关键字让编译器自动推导类型。

std::pair 一般只能表示两个元素，在 C++ 11 标准中引入了 std::tuple 类型，有了这个类型，我们就可以放任意数量的元素了，原来需要被定义成结构体的 POD 对象，我们可以直接使用 std::tuple 表示。例如，对于下面表示用户信息的结构体：

```
struct UserInfo
{
    std::string    username;
    std::string    password;
    int            gender;
    int            age;
    std::string    address;
};

int main()
{
    UserInfo userInfo = { "Tom", "123456", 0, 25, "Pudong Street" };
    std::string username = userInfo.username;
    std::string password = userInfo.password;
    int gender = userInfo.gender;
    int age = userInfo.age;
    std::string address = userInfo.address;

    return 0;
}
```

我们不再需要定义 struct UserInfo 这样的对象，可以直接使用 std::tuple 表示：

```
int main()
{
    std::tuple<std::string, std::string, int, int, std::string> userInfo("Tom",
 "123456", 0, 25, "Pudong Street");

    std::string username = std::get<0>(userInfo);
    std::string password = std::get<1>(userInfo);
    int gender = std::get<2>(userInfo);
    int age = std::get<3>(userInfo);
    std::string address = std::get<4>(userInfo);

    return 0;
}
```

从 std::tuple 中获取对应位置的元素时，可以使用 std::get<N>，其中 N 是元素的序号（从 0 开始）。

与定义结构体相比，无论是通过 std::pair 的 first、second 还是 std::tuple 的 std::get<N> 方法来获取元素子属性，这些代码都是难以维护的，其根本原因是 first 和 second 这样的命名不能做到见名知意。

C++ 17 引入的结构化绑定（Structured Binding）将我们从这类代码中"解放"出来。结构化绑定使用的语法如下：

```
auto [a, b, c, ...] = expression;
auto [a, b, c, ...] { expression };
auto [a, b, c, ...] ( expression );
```

右边的 expression 可以是一个函数调用、花括号表达式或者支持结构化绑定的某个类型的变量。例如：

```
//形式1
auto [iterator, inserted] = someMap.insert(...);
//形式2
double myArray[3] = { 1.0, 2.0, 3.0 };
auto [a, b, c] = myArray;
//形式3
struct Point
{
    double x;
    double y;
};
Point myPoint(10.0, 20.0);
auto [myX, myY] = myPoint;
```

这样，我们可以给用于绑定到目标的变量名（语法中的 a、b、c）起一个有意义的名称。

需要注意的是，绑定名称 a、b、c 是绑定目标的一份拷贝，当绑定类型不是基础数据类型时，如果你的本意不是想要得到绑定目标的副本，则为了避免拷贝带来的不必要开销，建议使用引用；如果不需要修改绑定目标，则建议使用 const 引用。示例如下：

```
double myArray[3] = { 1.0, 2.0, 3.0 };
auto& [a, b, c] = myArray;
//形式3
struct Point
{
    double x;
    double y;
};
Point myPoint(10.0, 20.0);
const auto& [myX, myY] = myPoint;
```

结构化绑定（Structured Binding）是 C++ 17 引入的一个非常好用的语法特性。有了这种语法，在遍历像 map 这样的容器时，我们可以使用更简洁和清晰的代码去遍历这些容器：

```
std::map<std::string, int> cities;
cities["beijing"] = 0;
cities["shanghai"] = 1;
cities["shenzhen"] = 2;
cities["guangzhou"] = 3;

for (const auto& [cityName, cityNumber] : cities)
```

```
{
    std::cout << cityName << ": " << cityNumber << std::endl;
}
```

在以上代码中，cityName 和 cityNumber 可以更好地反映这个 map 容器的元素内容。再来看一个例子，在某 WebSocket 网络库中有如下代码：

```
std::pair<int, bool> uncork(const char *src = nullptr, int length = 0, bool optionally = false) {
    LoopData *loopData = getLoopData();

    if (loopData->corkedSocket == this) {
        loopData->corkedSocket = nullptr;

        if (loopData->corkOffset) {
            //使用结构化绑定语法：
            //failed 表示是否写入发送成功，written 表示实际发送的字节数量
            auto [written, failed] = write(loopData->corkBuffer, loota>corkOffset, false, length);
            loopData->corkOffset = 0;

            if (failed) {
                return {0, true};
            }
        }

        return write(src, length, optionally, 0);
    } else {
        return {0, false};
    }
}
```

在以上加粗代码行中，write 函数的返回类型是 std::pair<int, bool>，被绑定到 written、failed 这两个变量中。前者在写入成功的情况下表示实际写入的字节数，后者表示是否写入成功：

```
std::pair<int, bool> write(const char *src, int length, bool optionally = false, int nextLength = 0) {
    //具体实现省略
}
```

结构化绑定的限制

用于结构化绑定的变量不能使用 constexpr 修饰或声明为 static，例如：

```
//正常编译
auto [first, second] = std::pair<int, int>(1, 2);
//无法编译通过
//constexpr auto [first, second] = std::pair<int, int>(1, 2);
//无法编译通过
//static auto [first, second] = std::pair<int, int>(1, 2);
```

有些编译器也不支持在 Lamda 表达式捕获列表中使用结构化绑定的语法。

1.10 stl 容器新增的实用方法

下面讲解 stl 容器新增的实用方法。

1.10.1 原位构造与容器的 emplace 系列函数

在介绍 emplace 和 emplace_back 方法之前,我们先看一段代码:

```cpp
#include <iostream>
#include <list>

class Test
{
public:
    Test(int a, int b, int c)
    {
        ma = a;
        mb = b;
        mc = c;
        std::cout << "Test constructed." << std::endl;
    }

    ~Test()
    {
        std::cout << "Test destructed." << std::endl;
    }

    Test(const Test& rhs)
    {
        if (this == &rhs)
            return;

        this->ma = rhs.ma;
        this->mb = rhs.mb;
        this->mc = rhs.mc;

        std::cout << "Test copy-constructed." << std::endl;
    }
private:
    int ma;
    int mb;
    int mc;
};

int main()
{
    std::list<Test> collections;
    for (int i = 0; i < 10; ++i)
    {
        Test t(1 * i, 2 * i, 3 * i);
        collections.push_back(t);
```

```
    }
    return 0;
}
```

以上代码在一个循环里产生一个对象，然后将这个对象放入集合中，这样的代码在实际开发中太常见了。但是这样的代码存在严重的效率问题：循环中的 t 对象在每次循环时，都分别调用了一次构造函数、拷贝构造函数和析构函数，如下所示。

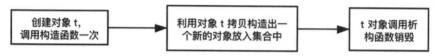

以上总共循环 10 次，调用 30 次。但实际上，我们的初衷是创建一个对象 t，将其直接放入集合中，而不是将 t 作为一个中间临时产生的对象，这样的话，总共需要调用 t 的构造函数 10 次就可以了。C++ 11 提供了一个在这种情形下替代 push_back 的方法——emplace_back。通过使用 emplace_back，可以将 main 函数中的代码改写如下：

```
std::list<Test> collections;
for (int i = 0; i < 10; ++i)
{
    collections.emplace_back(1 * i, 2 * i, 3 * i);
}
```

经过以上改写，在实际执行时只需调用 Test 类的构造函数 10 次，大大提高了执行效率。

同理，在这种情形下，对于像 std::list、std::vector 这样的容器，其 push、push_front 方法在 C++ 11 中也有对应的改进方法，即 emplace/emplace_front 方法。在 C++ Reference 上将这里的 emplace 操作称为 "原位构造元素（EmplaceConstructible）" 是非常贴切的。

原 方 法	C++ 11 改进方法	方法含义
push/insert	emplace	在容器的指定位置原位构造元素
push_front	emplace_front	在容器首部原位构造元素
push_back	emplace_back	在容器尾部原位构造元素

除了使用 emplace 系列的函数原位构造元素，我们也可以为 Test 类添加移动构造函数（Move Constructor），复用产生的临时对象 t 以提高效率。

1.10.2 std::map 的 try_emplace 方法与 insert_or_assign 方法

因为 std::map 中元素的 key 是唯一的，所以在实际开发中经常会有这样一类需求：向某个 map 中插入元素时需要先检测 map 中指定的 key 是否存在，不存在时做插入操作，存在时直接取来使用；或者在指定的 key 不存在时做插入操作，存在时做更新操作。

以 PC 版的 QQ 为例，好友列表中的每个好友都对应一个 userid，当我们双击某个 QQ 好友头像时，如果与该好友的聊天对话框（这里使用 ChatDialog 表示）已经存在，则直接将其激活并显示，如果不存在，则将其创建并激活、显示。假设我们使用 std::map 来管理这些聊天对话框，则在 C++ 17 之前的版本中，必须编写额外的逻辑去判断元素是否存在。可以将上述逻辑编写如下：

```cpp
class ChatDialog
{
//其他实现省略
public:
    void activate()
    {
        //实现省略
    }
};

//用于管理所有聊天对话框的map, key 是好友 id, ChatDialog 是聊天对话框指针
std::map<int64_t, ChatDialog*> m_ChatDialogs;

//双击好友头像后
void onDoubleClickFriendItem(int64_t userid)
{
    auto targetChatDialog = m_ChatDialogs.find(userid);
    //好友对话框不存在，则创建它并激活、显示
    if (targetChatDialog == m_ChatDialogs.end())
    {
        ChatDialog* pChatDialog = new ChatDialog();
        m_ChatDialogs.insert(std::pair<int64_t, ChatDialog*>(userid, pChatDialog));
        pChatDialog->activate();
    }
    //好友对话框存在，直接激活并显示
    else
    {
        targetChatDialog->second->activate();
    }
}
```

在 C++ 17 中，map 提供了一个 try_emplace 方法，该方法会检测指定的 key 是否存在，如果存在，则什么也不做。函数签名如下：

```cpp
template <class... Args>
pair<iterator, bool> try_emplace(const key_type& k, Args&&... args);

template <class... Args>
pair<iterator, bool> try_emplace(key_type&& k, Args&&... args);

template <class... Args>
iterator try_emplace(const_iterator hint, const key_type& k, Args&&... args);

template <class... Args>
iterator try_emplace(const_iterator hint, key_type&& k, Args&&... args);
```

在以上函数签名中，参数 k 表示需要插入的 key；args 参数是一个不定参数，表示构造 value 对象需要传给构造函数的参数；通过 hint 参数可以指定插入的位置。

在前两种签名形式中，try_emplace 的返回值是一个 std::pair<T1, T2>类型，其中 T2 是一个 bool 类型，表示元素是否成功插入 map 中；T1 是一个 map 的迭代器，如果插入

成功，则返回指向插入位置的元素的迭代器，如果插入失败，则返回 map 中已存在的相同 key 元素的迭代器。我们用 try_emplace 改写上面的代码（这里不关心插入的位置，因此使用前两个签名）：

```cpp
#include <iostream>
#include <map>

class ChatDialog
{
//其他实现省略
public:
    void activate()
    {
        //实现省略
    }
};

//用于管理所有聊天对话框的 map, key 是好友 id, ChatDialog 是聊天对话框指针
std::map<int64_t, ChatDialog*> m_ChatDialogs;

//普通版本
void onDoubleClickFriendItem(int64_t userid)
{
    auto targetChatDialog = m_ChatDialogs.find(userid);
    //好友对话框不存在，创建它并激活
    if (targetChatDialog == m_ChatDialogs.end())
    {
        ChatDialog* pChatDialog = new ChatDialog();
        m_ChatDialogs.insert(std::pair<int64_t, ChatDialog*>(userid, pChatDialog));
        pChatDialog->activate();
    }
    //好友对话框存在，直接激活
    else
    {
        targetChatDialog->second->activate();
    }
}

//C++ 17 版本 1
void onDoubleClickFriendItem2(int64_t userid)
{
    //结构化绑定和 try_emplace 都是 C++ 17 语法
    auto [iter, inserted] = m_ChatDialogs.try_emplace(userid);
    if (inserted)
        iter->second = new ChatDialog();

    iter->second->activate();
}

int main()
{
    //测试用例
    //906106643 是 userid
    onDoubleClickFriendItem2(906106643L);
```

```
    //906106644 是 userid
    onDoubleClickFriendItem2(906106644L);
    //906106643 是 userid
    onDoubleClickFriendItem2(906106643L);

    return 0;
}
```

使用 try_emplace 改写后的代码简洁了许多。但是在以上代码中需要注意：由于 std::map<int64_t, ChatDialog*> m_ChatDialogs 的 value 是指针类型（ChatDialog*），而 try_emplace 的第 2 个参数支持的是构造一个 ChatDialog 对象，而不是指针类型，因此在某个 userid 不存在时，成功插入 map 后会导致相应的 value 为空指针。因此，我们利用 inserted 的值按需新建一个 ChatDialog。当然，在新的 C++规范（C++ 11 及后续版本）提供了灵活而强大的智能指针以后，我们不应该再有任何理由去使用裸指针了，因此可以对以上代码使用 std::unique_ptr 智能指针类型来重构：

```
/**
 * std::map::try_emplace用法演示
 * zhangyl 2019.10.06
 */

#include <iostream>
#include <map>
#include <memory>

class ChatDialog
{
//其他实现省略
public:
    ChatDialog()
    {
        std::cout << "ChatDialog constructor" << std::endl;
    }

    ~ChatDialog()
    {
        std::cout << "ChatDialog destructor" << std::endl;
    }

    void activate()
    {
        //实现省略
    }
};

//用于管理所有聊天对话框的map，key是好友id，value是聊天对话框智能指针对象
std::map<int64_t, std::unique_ptr<ChatDialog>> m_ChatDialogs;

//C++ 17 版本2
void onDoubleClickFriendItem3(int64_t userid)
{
    //结构化绑定和try_emplace都是C++ 17语法
    auto spChatDialog = std::make_unique<ChatDialog>();
```

```
    auto [iter, inserted] = m_ChatDialogs.try_emplace(userid,
std::move(spChatDialog));
    iter->second->activate();
}

int main()
{
    //测试用例
    //906106643 是 userid
    onDoubleClickFriendItem3(906106643L);
    //906106644 是 userid
    onDoubleClickFriendItem3(906106644L);
    //906106643 是 userid
    onDoubleClickFriendItem3(906106643L);

    return 0;
}
```

以上代码将 map 的类型从 std::map<int64_t, ChatDialog*>改为 std::map<int64_t, std::unique_ptr<ChatDialog>>，让程序自动管理聊天对话框对象。程序在 gcc/g++ 7.3 中编译并运行输出如下：

```
[root@mydev test]# g++ -g -o test_map_try_emplace_with_smartpointer
test_map_try_emplace_with_smartpointer.cpp -std=c++17
[root@mydev test]# ./test_map_try_emplace_with_smartpointer
ChatDialog constructor
ChatDialog constructor
ChatDialog constructor
ChatDialog destructor
ChatDialog destructor
ChatDialog destructor
```

在以上代码中，构造函数和析构函数均被调用了 3 次，实际上，按最原始的逻辑（上文中普通版本）来讲，ChatDialog 应该只被构造和析构 2 次，多出来的一次是因为在 try_emplace 时，无论某个 userid 是否存在于 map 中，均创建一个 ChatDialog 对象（这是额外的用不上的对象）。由于这个对象并没有被用上，所以在出了 onDoubleClickFriendItem3 函数的作用域后，智能指针对象 spChatDialog 被析构，进而导致这个额外的、用不上的 ChatDialog 对象被析构。这相当于做了一次无用功。为此，我们可以继续优化代码如下：

```
#include <iostream>
#include <map>
#include <memory>

class ChatDialog
{
//其他实现省略
public:
    ChatDialog()
    {
        std::cout << "ChatDialog constructor" << std::endl;
    }
```

```cpp
    ~ChatDialog()
    {
        std::cout << "ChatDialog destructor" << std::endl;
    }

    void activate()
    {
        //实现省略
    }
};

//用于管理所有聊天对话框的 map, key 是好友 id, value 是聊天对话框智能指针对象
std::map<int64_t, std::unique_ptr<ChatDialog>> m_ChatDialogs;

//C++ 17 版本 3
void onDoubleClickFriendItem3(int64_t userid)
{
    //结构化绑定和 try_emplace 都是 C++ 17 语法
    auto [iter, inserted] = m_ChatDialogs.try_emplace(userid, nullptr);
    if (inserted)
    {
        //这样就按需创建了
        auto spChatDialog = std::make_unique<ChatDialog>();
        iter->second = std::move(spChatDialog);
    }

    iter->second->activate();
}

int main()
{
    //测试用例
    //906106643 是 userid
    onDoubleClickFriendItem3(906106643L);
    //906106644 是 userid
    onDoubleClickFriendItem3(906106644L);
    //906106643 是 userid
    onDoubleClickFriendItem3(906106643L);

    return 0;
}
```

以上代码按照之前裸指针版本的思路，按需创建了一个智能指针对象，避免了一次 ChatDialog 对象无用的构造和析构。再次编译程序，执行结果如下：

```
[root@mydev test]# g++ -g -o test_map_try_emplace_with_smartpointer2 test_map_try_emplace_with_smartpointer2.cpp -std=c++17
[root@mydev test]# ./test_map_try_emplace_with_smartpointer2
ChatDialog constructor
ChatDialog constructor
ChatDialog destructor
ChatDialog destructor
```

> 在 auto [iter, inserted] = m_ChatDialogs.try_emplace(userid, nullptr);语句中，m_ChatDialogs.try_emplace(userid, nullptr)函数返回两个值，第 2 个值 inserted 是一个布尔变量，表示操作是否成功，如果成功，则在第 1 个返回值 iter 中含有函数调用成功后的数据。这种函数存在多个返回值且其中一个值表示函数是否调用成功，我们称这种模式为 ok-idiom 模式，Golang 开发者应该很熟悉这种 ok-idiom 模式。

为了方便验证 try_emplace 函数支持原位构造（上文已经介绍），我们将 map 的 value 类型改成 ChatDialog 类型。在实际开发中，对于非 POD 类型的复杂数据类型，在 stl 容器中应该存储其指针或者智能指针类型，而不是对象本身。修改后的代码如下：

```cpp
#include <iostream>
#include <map>

class ChatDialog
{
//其他实现省略
public:
    ChatDialog(int64_t userid) : m_userid(userid)
    {
        std::cout << "ChatDialog constructor" << std::endl;
    }

    ~ChatDialog()
    {
        std::cout << "ChatDialog destructor" << std::endl;
    }

    void activate()
    {
        //实现省略
    }

private:
    int64_t     m_userid;
};

//用于管理所有聊天对话框的 map，key 是好友 id，value 是聊天对话框对象
std::map<int64_t, ChatDialog>   m_ChatDialogs;

//C++ 17 版本 4
void onDoubleClickFriendItem3(int64_t userid)
{
    //第 2 个 userid 是传给 ChatDialog 构造函数的参数
    auto [iter, inserted] = m_ChatDialogs.try_emplace(userid, userid);
    iter->second.activate();
}

int main()
{
    //测试用例
```

```
    //906106643 是 userid
    onDoubleClickFriendItem3(906106643L);
    //906106644 是 userid
    onDoubleClickFriendItem3(906106644L);
    //906106643 是 userid
    onDoubleClickFriendItem3(906106643L);

    return 0;
}
```

在以上代码中，我们为 ChatDialog 类的构造函数增加了一个 userid 参数，因此当调用 try_emplace 方法时，需要传递一个参数，这样 try_emplace 就会根据 map 中是否已存在同样的 userid 按需构造 ChatDialog 对象了。程序的执行结果和上一个代码示例应该是一样的：

```
[root@mydev test]# g++ -g -o test_map_try_emplace_with_directobject
test_map_try_emplace_with_directobject.cpp -std=c++17
[root@mydev test]# ./test_map_try_emplace_with_directobject
ChatDialog constructor
ChatDialog constructor
ChatDialog destructor
ChatDialog destructor
```

对于智能指针对象 std::unique_ptr，在后面的小节中将详细介绍。

上文介绍了 map 中指定的 key 不存在则插入相应的 value，存在则直接使用该 key 对应的 value 的情形。这里再来介绍 map 中指定的 key 不存在则插入相应的 value，存在则更新其 value 的情形。C++ 17 为 map 容器新增了一个 insert_or_assign 方法，让我们不再像 C++ 17 标准之前一样额外编写先判断是否存在，不存在则插入，存在则更新的代码了，这次我们可以一步到位。insert_or_assign 的函数签名如下：

```
template <class M>
pair<iterator, bool> insert_or_assign(const key_type& k, M&& obj);

template <class M>
pair<iterator, bool> insert_or_assign(key_type&& k, M&& obj);

template <class M>
iterator insert_or_assign(const_iterator hint, const key_type& k, M&& obj);

template <class M>
iterator insert_or_assign(const_iterator hint, key_type&& k, M&& obj);
```

其各个函数参数的含义与 try_emplace 一样，这里不再赘述。

再来看一个例子：

```
int main()
{
    std::map<std::string, int> mapUsersAge{ { "Alex", 45 }, { "John", 25 } };
    mapUsersAge.insert_or_assign("Tom", 26);
    mapUsersAge.insert_or_assign("Alex", 27);

    for (const auto& [userName, userAge] : mapUsersAge)
```

```
        {
            std::cout << "userName: " << userName << ", userAge: " << userAge << std::endl;
        }
}
```

在以上代码中尝试插入名为 Tom 的用户，由于该人名在 map 中不存在，因此插入成功；当插入人名为 Alex 的用户时，由于在 map 中已经存在该人名，因此只对其年龄进行更新，将 Alex 的年龄从 45 更新为 27。程序执行结果如下：

```
[root@mydev test]# g++ -g -o test_map_insert_or_assign
test_map_insert_or_assign.cpp -std=c++17
[root@mydev test]# ./test_map_insert_or_assign
userName: Alex, userAge: 27
userName: John, userAge: 25
userName: Tom, userAge: 26
```

本节介绍了 C++ 11/17 为 stl 容器新增的几个实用方法，合理利用这些新增的方法会让我们的程序变得更简洁、高效。其实，新的 C++标准一直在不断改进和优化现有的 stl 容器，如果经常需要与这些容器打交道，则建议留意 C++新标准中这些容器的新动态。

1.11 stl 中的智能指针类详解

C/C++最为人诟病的是内存泄露问题，后来的大多数语言都内置了内存分配与释放功能，有的甚至对语言的使用者屏蔽了内存指针这一概念。这里对此不置褒贬，手动分配与释放内存有利有弊，自动分配与释放内存亦如此，这是两种不同的设计哲学。有人认为，内存如此重要，怎能放心将其交给用户去管理呢？另外一些人则认为，内存如此重要，怎能放心将其交给系统去管理呢？在 C/C++中，内存泄露的问题一直困扰着广大开发者，因此各类库和工具也一直在努力尝试各种方法去检测和避免内存泄露，例如 boost，因此智能指针技术应运而生。

1.11.1 C++ 98/03 的尝试——std::auto_ptr

现在讨论 std::auto_ptr 不免让人怀疑是不是有点过时了，确实如此，C++ 11 标准废弃了 std::auto_ptr（在 C++ 17 标准中被移除），取而代之的是 std::unique_ptr。这里之所以介绍 std::auto_ptr 的用法及它在设计上的不足之处，是想让读者了解 C++中智能指针的发展历程。我们在了解一项技术过去的样子和发展轨迹后，就能更好地掌握它。

std::auto_ptr 的基本用法如下：

```
#include <memory>

int main()
{
    //初始化方式1
    std::auto_ptr<int> sp1(new int(8));
    //初始化方式2
```

```
    std::auto_ptr<int> sp2;
    sp2.reset(new int(8));

    return 0;
}
```

智能指针对象 sp1 和 sp2 均持有一个在堆上分配的 int 对象,值都是 8,这两块堆内存都在 sp1 和 sp2 释放时得到释放。这是 std::auto_ptr 的基本用法。

> sp 是 smart pointer(智能指针)的简写。

std::auto_ptr 容易让人误用的地方是其不常用的复制语义,即当复制一个 std::auto_ptr 对象时(拷贝复制或 operator=复制),原 std::auto_ptr 对象所持有的堆内存对象也会被转移给复制出来的新 std::auto_ptr 对象。示例代码如下:

```cpp
#include <iostream>
#include <memory>

int main()
{
    //测试拷贝复制
    std::auto_ptr<int> sp1(new int(8));
    std::auto_ptr<int> sp2(sp1);
    if (sp1.get() != NULL)
    {
        std::cout << "sp1 is not empty." << std::endl;
    }
    else
    {
        std::cout << "sp1 is empty." << std::endl;
    }

    if (sp2.get() != NULL)
    {
        std::cout << "sp2 is not empty." << std::endl;
    }
    else
    {
        std::cout << "sp2 is empty." << std::endl;
    }

    //测试赋值复制
    std::auto_ptr<int> sp3(new int(8));
    std::auto_ptr<int> sp4;
    sp4 = sp3;
    if (sp3.get() != NULL)
    {
        std::cout << "sp3 is not empty." << std::endl;
    }
    else
    {
        std::cout << "sp3 is empty." << std::endl;
    }
```

```
    if (sp4.get() != NULL)
    {
        std::cout << "sp4 is not empty." << std::endl;
    }
    else
    {
        std::cout << "sp4 is empty." << std::endl;
    }

    return 0;
}
```

在以上代码中分别利用了拷贝构造（sp1=>sp2）和赋值构造（sp3=>sp4）来创建新的 std::auto_ptr 对象，因此 sp1 持有的堆对象被转移给 sp2，sp3 持有的堆对象被转移给 sp4。示意图如下。

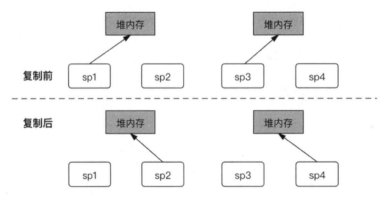

程序执行结果如下：

```
[root@iZ238vnojlyZ testx]# g++ -g -o test_auto_ptr test_auto_ptr.cpp
[root@iZ238vnojlyZ testx]# ./test_auto_ptr
sp1 is empty.
sp2 is not empty.
sp3 is empty.
sp4 is not empty.
```

因为 std::auto_ptr 是不常用的复制语义，所以我们应该避免在 stl 容器中使用 std::auto_ptr，例如不应该写出如下代码：

`std::vector<std::auto_ptr<int>> myvectors;`

当用算法对容器进行操作时（如最常见的容器元素遍历），很难避免不对容器中的元素进行赋值传递，这样便会使容器中的多个元素被置为空指针，这不是我们希望看到的，可能会造成一些意想不到的错误。

作为 std::auto_ptr 的替代者，std::unique_ptr 吸取了这个教训，下文会详细介绍。

正因为 std::auto_ptr 的设计存在缺陷，所以 C++ 11 标准充分借鉴和吸收了 boost 库中智能指针的设计思想，引入了三种新类型的智能指针，即 std::unique_ptr、std::shared_ptr 和 std::weak_ptr。

C++ 11 没有全部照搬 boost 智能指针类型，而是选择了其中三个最实用的类型。boost 还有 scoped_ptr，在 C++ 11 中可以通过 std::unique_ptr 达到与 boost::scoped_ptr 一样的效果。

所有智能指针类（包括 std::unique_ptr）均被定义于头文件<memory>中。

在 C++ 11 及后续语言规范中，std::auto_ptr 已被废弃，在我们的代码中不应再使用它。

1.11.2　std::unique_ptr

std::unique_ptr 对其持有的堆内存具有唯一拥有权，也就是说该智能指针对资源（即其管理的堆内存）的引用计数永远是 1，std::unique_ptr 对象在销毁时会释放其持有的堆内存。可以采用以下方式初始化一个 std::unique_ptr 对象：

```cpp
//初始化方式 1
std::unique_ptr<int> sp1(new int(123));

//初始化方式 2
std::unique_ptr<int> sp2;
sp2.reset(new int(123));

//初始化方式 3
std::unique_ptr<int> sp3 = std::make_unique<int>(123);
```

我们应该尽量采用初始化方式 3 去创建一个 std::unique_ptr，而不是采用初始化方式 1 和 2，因为初始化方式 3 更安全。

让很多人对 C++ 11 规范吐槽的地方之一是，C++ 11 新增了 std::make_shared 方法创建一个 std::shared_ptr 对象，却没有提供相应的 std::make_unique 方法创建一个 std::unique_ptr 对象，该方法直到 C++ 14 时才被添加进来。当然，在 C++ 11 中很容易实现一个这样的方法：

```cpp
template<typename T, typename... Ts>
std::unique_ptr<T> make_unique(Ts&& ...params)
{
    return std::unique_ptr<T>(new T(std::forward<Ts>(params)...));
}
```

鉴于 std::auto_ptr 的前车之鉴，std::unique_ptr 禁止复制语义，为了达到这个效果，std::unique_ptr 类的拷贝构造函数和赋值运算符（operator =）均被标记为=delete。

```cpp
template <class T>
class unique_ptr
{
    //其他代码省略

    //拷贝构造函数和赋值运算符被标记为 delete
    unique_ptr(const unique_ptr&) = delete;
    unique_ptr& operator=(const unique_ptr&) = delete;
};
```

因此，以下代码是无法通过编译的：

```cpp
std::unique_ptr<int> sp1(std::make_unique<int>(123));;

//以下代码无法通过编译
//std::unique_ptr<int> sp2(sp1);
std::unique_ptr<int> sp3;
//以下代码无法通过编译
//sp3 = sp1;
```

不过禁止复制语义也存在特例,例如可以通过一个函数返回一个 std::unique_ptr:

```cpp
#include <memory>

std::unique_ptr<int> func(int val)
{
    std::unique_ptr<int> up(new int(val));
    return up;
}

int main()
{
    std::unique_ptr<int> sp1 = func(123);

    return 0;
}
```

以上代码从 func 函数中得到一个 std::unique_ptr 对象,然后返回给 sp1。

既然 std::unique_ptr 不能被复制,那么如何将一个 std::unique_ptr 对象持有的堆内存转移给另外一个呢?答案是使用移动构造,示例代码如下:

```cpp
#include <memory>

int main()
{
    std::unique_ptr<int> sp1(std::make_unique<int>(123));

    std::unique_ptr<int> sp2(std::move(sp1));

    std::unique_ptr<int> sp3;
    sp3 = std::move(sp2);

    return 0;
}
```

以上代码利用了 std::move 将 sp1 持有的堆内存(值为 123)转移给 sp2,再将 sp2 转移给 sp3。最后,sp1 和 sp2 不再持有堆内存的引用,变成一个空的智能指针对象。并不是所有对象的 std::move 操作都有意义,只有实现了移动构造函数(Move Constructor)或移动赋值运算符(operator =)的类才行,而 std::unique_ptr 正好实现了二者。以下是该实现的伪代码:

```cpp
template<typename T, typename Deletor>
class unique_ptr
{
    //其他函数省略
public:
```

```cpp
    unique_ptr(unique_ptr&& rhs)
    {
        this->m_pT = rhs.m_pT;
        //源对象释放
        rhs.m_pT = nullptr;
    }

    unique_ptr& operator=(unique_ptr&& rhs)
    {
        this->m_pT = rhs.m_pT;
        //源对象释放
        rhs.m_pT = nullptr;
        return *this;
    }

private:
    T*    m_pT;
};
```

这就是 std::unique_ptr 具有移动语义的原因。

std::unique_ptr 不仅可以持有一个堆对象,也可以持有一组堆对象,示例如下:

```cpp
#include <iostream>
#include <memory>

int main()
{
    //创建 10 个 int 类型的堆对象
    //形式 1
    std::unique_ptr<int[]> sp1(new int[10]);

    //形式 2
    std::unique_ptr<int[]> sp2;
    sp2.reset(new int[10]);
    //形式 3
    std::unique_ptr<int[]> sp3(std::make_unique<int[]>(10));

    for (int i = 0; i < 10; ++i)
    {
        sp1[i] = i;
        sp2[i] = i;
        sp3[i] = i;
    }

    for (int i = 0; i < 10; ++i)
    {
        std::cout << sp1[i] << ", " << sp2[i] << ", " << sp3[i] << std::endl;
    }

    return 0;
}
```

程序执行结果如下:

```
[root@myaliyun testmybook]# g++ -g -o test_unique_ptr_with_array
test_unique_ptr_with_array.cpp -std=c++17
```

```
[root@myaliyun testmybook]# ./test_unique_ptr_with_array
0, 0, 0
1, 1, 1
2, 2, 2
3, 3, 3
4, 4, 4
5, 5, 5
6, 6, 6
7, 7, 7
8, 8, 8
9, 9, 9
```

std::shared_ptr 和 std::weak_ptr 也可以持有一组堆对象，用法与 std::unique_ptr 相同，下文不再赘述。

在默认情况下，智能指针对象在析构时只会释放其持有的堆内存（调用 delete 或者 delete[]），但是假设这块堆内存代表的对象还对应一种需要回收的资源（如操作系统的套接字句柄、文件句柄等），我们则可以通过给智能指针自定义资源回收函数来实现资源回收。假设现在有一个 Socket 类，对应操作系统的套接字句柄，在回收时需要关闭该对象，我们则可以这样自定义智能指针对象的资源释放函数，以 std::unique_ptr 为例：

```cpp
#include <iostream>
#include <memory>

class Socket
{
public:
    Socket()
    {
    }
    ~Socket()
    {
    }

    //关闭资源句柄
    void close()
    {

    }
};

int main()
{
    auto deletor = [](Socket* pSocket) {
        //关闭句柄
        pSocket->close();
        //TODO:你甚至可以在这里打印一行日志
        delete pSocket;
    };

    std::unique_ptr<Socket, void(*)(Socket * pSocket)> spSocket(new Socket(), deletor);
```

```
    return 0;
}
```

自定义 std::unique_ptr 的资源释放函数的语法规则如下：

```
std::unique_ptr<T, DeletorFuncPtr>
```

其中，T 是我们要释放的对象类型，DeletorFuncPtr 是一个自定义函数指针。以上加粗代码行表示 DeletorFuncPtr 的写法有点复杂，我们可以使用 decltype(deletor) 让编译器自己推导 deletor 的类型，因此可以将加粗代码行修改如下：

```
std::unique_ptr<Socket, decltype(deletor)> spSocket(new Socket(), deletor);
```

1.11.3　std::shared_ptr

std::unique_ptr 对其持有的资源具有独占性，而 std::shared_ptr 持有的资源可以在多个 std::shared_ptr 之间共享，每多一个 std::shared_ptr 对资源的引用，资源引用计数就会增加 1，在每一个指向该资源的 std::shared_ptr 对象析构时，资源引用计数都会减少 1，最后一个 std::shared_ptr 对象析构时，若发现资源计数为 0，则将释放其持有的资源。多个线程之间递增和减少资源的引用计数都是安全的（注意：这不意味着多个线程同时操作 std::shared_ptr 管理的资源是安全的）。std::shared_ptr 提供了一个 use_count 方法来获取当前管理的资源的引用计数。除了上面描述的内容，std::shared_ptr 的用法和 std::unique_ptr 基本相同。

下面是一个初始化 std::shared_ptr 的示例：

```
//初始化方式 1
std::shared_ptr<int> sp1(new int(123));

//初始化方式 2
std::shared_ptr<int> sp2;
sp2.reset(new int(123));

//初始化方式 3
std::shared_ptr<int> sp3;
sp3 = std::make_shared<int>(123);
```

和 std::unique_ptr 一样，我们应该优先使用 std::make_shared 初始化一个 std::shared_ptr 对象。

再来看另外一段代码：

```
#include <iostream>
#include <memory>

class A
{
public:
    A()
    {
        std::cout << "A constructor" << std::endl;
    }
```

```cpp
    ~A()
    {
        std::cout << "A destructor" << std::endl;
    }
};

int main()
{
    {
        //初始化方式1
        //在 sp1 构造时触发对象 A 的构造,因此 A 的构造函数被执行
        //此时只有一个 sp1 对象引用新建的 A 对象(下面统称为资源对象 A)
        //因此打印出来的引用计数值为 1
        std::shared_ptr<A> sp1(new A());

        std::cout << "use count: " << sp1.use_count() << std::endl;

        //初始化方式2
        //利用 sp1 拷贝一份 sp2,导致打印出来的引用计数为 2
        std::shared_ptr<A> sp2(sp1);
        std::cout << "use count: " << sp1.use_count() << std::endl;

        //调用 sp2 的 reset 方法,sp2 释放对资源对象 A 的引用
        //因此打印出来的引用计数值再次变为 1
        sp2.reset();
        std::cout << "use count: " << sp1.use_count() << std::endl;

        {
            //利用 sp1 再次创建 sp3,因此打印出来的引用计数变为 2
            std::shared_ptr<A> sp3 = sp1;
            std::cout << "use count: " << sp1.use_count() << std::endl;
        }

        //程序执行到这里,sp3 因出了其作用域被析构
        //资源 A 的引用计数递减 1,因此接下来打印的引用计数为 1
        std::cout << "use count: " << sp1.use_count() << std::endl;
    }

    //在这里,sp1 出了其作用域被析构,在其析构时递减资源 A 的引用计数至 0,并析构资源 A 对象
    //因此类 A 的析构函数被调用

    return 0;
}
```

整个程序的执行结果如下:

```
[root@myaliyun testmybook]# ./test_shared_ptr_use_count
A constructor
use count: 1
use count: 2
use count: 1
use count: 2
use count: 1
A destructor
```

1.11.4　std::enable_shared_from_this

在实际开发中有时需要在类中返回包裹当前对象（this）的一个 std::shared_ptr 对象给外部使用，C++新标准也为我们考虑到了这一点，有如此需求的类只要继承自 std::enable_shared_from_this<T>模板对象即可。用法如下：

```cpp
#include <iostream>
#include <memory>
class A : public std::enable_shared_from_this<A>
{
public:
    A()
    {
        std::cout << "A constructor" << std::endl;
    }

    ~A()
    {
        std::cout << "A destructor" << std::endl;
    }

    std::shared_ptr<A> getSelf()
    {
        return shared_from_this();
    }
};

int main()
{
    std::shared_ptr<A> sp1(new A());

    std::shared_ptr<A> sp2 = sp1->getSelf();

    std::cout << "use count: " << sp1.use_count() << std::endl;

    return 0;
}
```

在以上代码中，类 A 继承自 std::enable_shared_from_this<A>并提供了一个 getSelf 方法返回自身的 std::shared_ptr 对象，在 getSelf 方法中调用了 shared_from_this 方法。

std::enable_shared_from_this 使用起来比较方便，但也存在很多注意事项。

注意事项一：不应该共享栈对象的 this 指针给智能指针对象。

假设我们将上面代码中 main 函数的第 1 行生成 A 对象的方式改成一个栈变量：

```cpp
//其他相同代码省略

int main()
{
    A a;

    std::shared_ptr<A> sp2 = a.getSelf();
```

```
        std::cout << "use count: " << sp2.use_count() << std::endl;

        return 0;
}
```

则运行修改后的代码，会发现程序在"std::shared_ptr<A> sp2 = a.getSelf();"处崩溃。这是因为，智能指针管理的是堆对象，栈对象会在函数调用结束后自行销毁，因此不能通过 shared_from_this()将该对象交由智能指针对象管理。切记：智能指针最初设计的目的就是管理堆对象。

注意事项二：**std::enable_shared_from_this** 的循环引用问题。

再来看另外一段代码：

```
// test_std_enable_shared_from_this.cpp
#include <iostream>
#include <memory>

class A : public std::enable_shared_from_this<A>
{
public:
    A()
    {
        m_i = 9;

        std::cout << "A constructor" << std::endl;
    }

    ~A()
    {
        m_i = 0;

        std::cout << "A destructor" << std::endl;
    }

    void func()
    {
        //比较好的做法是在构造函数中调用 shared_from_this()为 m_SelfPtr 赋值
        //但很遗憾不能这么做，如果将其写在构造函数里面，则程序会直接崩溃
        m_SelfPtr = shared_from_this();
    }

public:
    int                   m_i;
    std::shared_ptr<A>    m_SelfPtr;

};

int main()
{
    {
        std::shared_ptr<A> spa(new A());
        spa->func();
    }//spa 在这里出了作用域
```

```
    return 0;
}
```

乍一看上面的代码好像没有问题，让我们实际运行一下看看输出结果：

```
[root@myaliyun testmybook]# g++ -g -o test_std_enable_shared_from_this_problem
test_std_enable_shared_from_this_problem.cpp
[root@myaliyun testmybook]# ./test_std_enable_shared_from_this_problem
A constructor
```

我们会发现，在程序的整个生命周期内，只有 A 类构造函数的调用输出，没有 A 类析构函数的调用输出，这意味着新建的 A 对象产生内存泄漏了！

我们来分析一下新建的 A 对象为什么得不到释放。程序在执行完加粗代码行后，spa 出了其作用域准备析构，在析构时其发现仍然有另一个 std::shared_ptr 对象即 A::m_SelfPtr 引用了 A，因此 spa 只会将对 A 的引用计数递减为 1，然后销毁自身。现在很矛盾：必须销毁 A 才能销毁其成员变量 m_SelfPtr，而销毁 A 前必须先销毁 m_SelfPtr。这就是 std::enable_shared_from_this 的循环引用问题。我们在实际开发中应该避免做出这样的逻辑设计，在这种情形下即使使用了智能指针也会造成内存泄漏。也就是说，一个资源的生命周期可以交给一个智能指针对象来管理，但是该智能指针的生命周期不可以再交给该资源来管理。

1.11.5　std::weak_ptr

std::weak_ptr 是一个不控制资源生命周期的智能指针，是对对象的一种弱引用，只提供了对其管理的资源的一个访问手段，引入它的目的是协助 std::shared_ptr 工作。

std::weak_ptr 可以从一个 std::shared_ptr 或另一个 std::weak_ptr 对象构造，std::shared_ptr 可以直接赋值给 std::weak_ptr，也可以通过 std::weak_ptr 的 lock 函数来获得 std::shared_ptr。它的构造和析构不会引起引用计数的增加或减少。std::weak_ptr 可用来解决 std::shared_ptr 相互引用时的死锁问题，即两个 std::shared_ptr 相互引用，这两个智能指针的资源引用计数就永远不可能减少为 0，资源永远不会被释放。

示例代码如下：

```cpp
#include <iostream>
#include <memory>

int main()
{
    //创建一个 std::shared_ptr 对象
    std::shared_ptr<int> sp1(new int(123));
    std::cout << "use count: " << sp1.use_count() << std::endl;

    //通过构造函数得到一个 std::weak_ptr 对象
    std::weak_ptr<int> sp2(sp1);
    std::cout << "use count: " << sp1.use_count() << std::endl;
```

```cpp
    //通过赋值运算符得到一个 std::weak_ptr 对象
    std::weak_ptr<int> sp3 = sp1;
    std::cout << "use count: " << sp1.use_count() << std::endl;

    //通过一个 std::weak_ptr 对象得到另一个 std::weak_ptr 对象
    std::weak_ptr<int> sp4 = sp2;
    std::cout << "use count: " << sp1.use_count() << std::endl;

    return 0;
}
```

程序执行结果如下：

```
[root@myaliyun testmybook]# g++ -g -o test_weak_ptr test_weak_ptr.cpp
[root@myaliyun testmybook]# ./test_weak_ptr
use count: 1
use count: 1
use count: 1
use count: 1
```

无论通过何种方式创建 std::weak_ptr，都不会增加资源的引用计数，因此每次输出的 sp1 的引用计数值都是 1。

既然 std::weak_ptr 不管理所引用资源的生命周期，其引用的资源就可能在某个时刻失效。我们需要使用 std::weak_ptr 引用该资源时，如何得知该资源是否有效呢？std::weak_ptr 提供了一个 expired 方法来做这项检测，该方法返回 true，说明其引用的资源已失效；返回 false，说明该资源仍然有效，这时可以使用 std::weak_ptr 的 lock 方法得到一个 std::shared_ptr 对象后继续操作资源。以下代码演示了该用法：

```cpp
//m_tmpConn 是一个 std::weak_ptr<TcpConnection>对象
//m_tmpConn 引用的 TcpConnection 已被销毁，直接返回
if (m_tmpConn.expired())
    return;

std::shared_ptr<TcpConnection> conn = m_tmpConn.lock();
if (conn)
{
    //省略对 conn 的后续操作代码
}
```

有读者可能对以上代码产生疑问，既然使用了 std::weak_ptr 的 expired 方法判断了对象是否有效，那么为什么不直接使用 std::weak_ptr 对象对引用资源进行操作呢？这实际上是行不通的，std::weak_ptr 类没有重写 operator-> 和 operator*方法，因此不能像 std::shared_ptr 或 std::unique_ptr 一样直接操作对象，std::weak_ptr 类也没有重写 operator bool()操作，因此不能通过 std::weak_ptr 对象直接判断其引用的资源是否存在：

```cpp
#include <memory>

class A
{
public:
    void doSomething()
```

```cpp
        {
        }
};

int main()
{
    std::shared_ptr<A> sp1(new A());

    std::weak_ptr<A> sp2(sp1);

    //正确的代码
    if (sp1)
    {
        //正确的代码
        sp1->doSomething();
        (*sp1).doSomething();
    }

    //正确的代码
    if (!sp1)
    {

    }

    //错误的代码,无法编译通过
    //if (sp2)
    //{
    //    //错误的代码,无法编译通过
    //    sp2->doSomething();
    //    (*sp2).doSomething();
    //}

    //错误的代码,无法编译通过
    //if (!sp2)
    //{

    //}

    return 0;
}
```

在多线程场景下,即使刚刚使用了 std::weak_ptr 的 expired 方法判断其引用的资源是否有效,但若不加一些多线程保护策略,却接着调用 std::weak_ptr 的 lock 方法尝试得到一个 std::shared_ptr 对象去操作资源,则仍然可能存在安全隐患。其原因是,在调用 lock 方法期间,引用的资源可能恰好被销毁,这可能会造成比较棘手的问题。示例代码如下:

```cpp
//线程1的部分逻辑
void TcpSession::SendData(const std::string& msg)
{
    if (m_tmpConn.expired())
        return;

    //下面一行存在安全隐患,虽然已经使用 expired 方法确认 m_tmpConn 引用的资源有效
    //但是可能在 expired 方法调用之后、lock 方法调用之前,m_tmpConn 引用的资源刚好失效
```

```cpp
    std::shared_ptr<TcpConnection> conn = m_tmpConn.lock();
    if (conn)
    {
        conn->send(msg);
    }
}

//线程 2 的部分逻辑
void TcpSession::OnClose()
{
    //让 m_tmpConn 引用的资源失效
    m_tmpConn.reset();
}
```

在以上代码中，m_tmpConn 是 TcpSession 的成员变量，其类型是 std::weak_ptr<TcpConnection>。由于两个线程同时操作 m_tmpConn，所以即使线程 1 在希望使用 m_tmpConn 引用的资源时，先用 expired 方法判断对应的 TcpConnection 是否有效，再用 lock 方法尝试得到一个 std::shared_ptr 对象，但如果在 lock 方法调用前，线程 2 释放了 m_tmpConn 引用的资源，那么线程 1 接下来的逻辑仍然存在安全隐患。在多线程场景下，这是一种常见的错误，需要避免。

因此，std::weak_ptr 的正确使用场景是引用的资源如果可用就使用，不可用就不使用，不参与资源的生命周期管理。例如在网络分层结构中，Session 对象（会话对象）利用 Connection 对象（连接对象）提供的服务进行工作，但是 Session 对象不管理 Connection 对象的生命周期，Session 管理 Connection 的生命周期是不合理的，因为网络底层出错时会导致 Connection 对象被销毁，此时 Session 对象如果强行持有 Connection 对象，则与事实矛盾。

std::weak_ptr 应用场景中的经典例子是订阅者模式或者观察者模式。这里以订阅者为例来说明，消息发布器只有在某个订阅者存在的情况下才会向其发布消息，不能管理订阅者的生命周期。

```cpp
class Subscriber
{
    //具体实现省略
};

class SubscribeManager
{
public:
    void publish()
    {
        for (const auto& iter : m_subscribers)
        {
            if (!iter.expired())
            {
                //TODO: 向订阅者发送消息
            }
        }
```

```
private:
    std::vector<std::weak_ptr<Subscriber>>  m_subscribers;
};
```

1.11.6 智能指针对象的大小

一个 std::unique_ptr 对象的大小与裸指针的大小相同（即 sizeof(std::unique_ptr<T>) == sizeof(void*)），而 std::shared_ptr 的大小是 std::unique_ptr 的两倍。以下是分别在 Visual Studio 2019 和 gcc/g++ 4.8 上（将二者都编译成 x64 程序）测试的结果。

测试代码如下：

```
#include <iostream>
#include <memory>
#include <string>

int main()
{
    std::shared_ptr<int> sp0;
    std::shared_ptr<std::string> sp1;
    sp1.reset(new std::string());
    std::unique_ptr<int> sp2;
    std::weak_ptr<int> sp3;

    std::cout << "sp0 size: " << sizeof(sp0) << std::endl;
    std::cout << "sp1 size: " << sizeof(sp1) << std::endl;
    std::cout << "sp2 size: " << sizeof(sp2) << std::endl;
    std::cout << "sp3 size: " << sizeof(sp3) << std::endl;

    return 0;
}
```

Visual Studio 2019 的运行结果如下图所示。

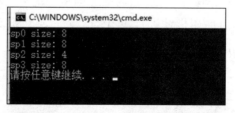

gcc/g++的运行结果如下图所示。

在 32 位机器上，std::unique_ptr 占 4 字节，std::shared_ptr 和 std::weak_ptr 均占 8 字节；在 64 位机器上，std_unique_ptr 占 8 字节，std::shared_ptr 和 std::weak_ptr 均占 16 字节。也就是说，std_unique_ptr 的大小总是和原始指针的大小一样，std::shared_ptr 和 std::weak_ptr 的大小是原始指针的大小的两倍。

1.11.7 使用智能指针时的注意事项

C++新标准提倡的理念之一是不再手动调用 delete 或者 free 函数去释放堆内存，而是把它们交给新标准提供的各种智能指针对象。C++新标准中的各种智能指针是如此实用与强大，在现在的 C++项目开发中，我们应该尽量使用它们。智能指针虽然好用，但稍不注意，也可能存在许多难以发现的 bug，下面总结了一些实用经验。

1. 一旦使用了智能指针管理一个对象，就不该再使用原始裸指针去操作它

看一段代码：

```
#include <memory>

class Subscriber
{
    //具体实现省略
};

int main()
{
    Subscriber* pSubscriber = new Subscriber();

    std::unique_ptr<Subscriber> spSubscriber(pSubscriber);

    delete pSubscriber;

    return 0;
}
```

这段代码创建了一个堆对象 pSubscriber，然后利用智能指针 spSubscriber 去管理它，私下却利用原始指针 pSubscriber 销毁了该对象，这让智能指针对象 spSubscriber "情何以堪"！

注意，一旦智能指针对象接管了我们的资源，对资源的所有操作就都应该通过智能指针对象进行，不建议再通过原始指针进行。当然，除了 std::weak_ptr，std::unique_ptr 和 std::shared_ptr 都提供了用于获取原始指针的 get 函数。

```
int main()
{
    Subscriber* pSubscriber = new Subscriber();

    std::unique_ptr<Subscriber> spSubscriber(pSubscriber);

    //pTheSameSubscriber 和 pSubscriber 指向同一个对象
    Subscriber* pTheSameSubscriber= spSubscriber.get();

    return 0;
}
```

2. 知道在哪些场合使用哪种类型的智能指针

在通常情况下，如果我们的资源不需要在其他地方共享，就应该优先使用 std::unique_ptr，反之使用 std::shared_ptr。当然，这是在该智能指针需要管理资源的生命周期的情况下进行的；如果不需要管理对象的生命周期，则请使用 std::weak_ptr。

3. 认真考虑，避免操作某个引用资源已经释放的智能指针

通过前面的例子，我们很容易知道一个智能指针持有的资源是否有效，但还是建议在不同的场景下谨慎一些，因为在某些场景下很容易误判。例如下面的代码：

```cpp
#include <iostream>
#include <memory>

class T
{
public:
    void doSomething()
    {
        std::cout << "T do something..." << m_i << std::endl;
    }

private:
    int     m_i;
};

int main()
{
    std::shared_ptr<T> sp1(new T());
    const auto& sp2 = sp1;

    sp1.reset();

    //由于 sp2 不再持有对象的引用，所以程序在这里的行为是不确定的
    sp2->doSomething();

    return 0;
}
```

在以上代码中，sp2 是 sp1 的引用，sp1 被置空后，sp2 也一同为空。这时调用 sp2->doSomething()，sp2->（即 operator->）在内部会调用 get 方法获取原始指针对象，这时得到一个空指针（地址为 0），继续调用 doSomething()会导致程序崩溃。

有些读者可能觉得以上代码片段存在的问题是显而易见的，让我们把这个例子放到实际项目中再看一下：

```cpp
//连接断开
void MonitorServer::OnClose(const std::shared_ptr<TcpConnection>& conn)
{
    std::lock_guard<std::mutex> guard(m_sessionMutex);
    for (auto iter = m_sessions.begin(); iter != m_sessions.end(); ++iter)
    {
        //通过比对 connection 对象找到对应的 session
        if ((*iter)->GetConnectionPtr() == conn)
        {
            m_sessions.erase(iter);
            //注意，程序在如下行崩溃
            LOGI("monitor client disconnected: %s",
conn->peerAddress().toIpPort().c_str());
```

```
            break;
        }
    }
}
```

以上代码来自实际的商业项目，其崩溃的原因是调用了 conn->peerAddress 方法。为什么这个方法的调用可能会引发程序崩溃呢？现在还可以显而易见地看出问题吗？

其崩溃的原因是传入的 conn 对象和上一个例子中的 sp2 一样都是另一个 std::shared_ptr 的引用，当连接断开时，对应的 TcpConnection 对象可能早已被销毁，而 conn 引用会变成空指针（严格来说，是不再持有一个 TcpConnection 对象的引用），此时调用 TcpConnection 的 peerAddress 方法就会产生和上一个示例一样的错误。

4. 作为类成员变量，应该优先使用前置声明（forward declarations）

我们知道，为了减少编译依赖、加快编译速度和减少生成的二进制文件的大小，C/C++ 项目一般在 *.h 文件中对指针类型尽量使用前置声明，而不是直接包含对应类的头文件。例如：

```
//Test.h
//这里使用 A 的前置声明，而不是直接包含 A.h 文件
class A;

class Test
{
public:
    Test();
    ~Test();

private:
    A* m_pA;
};
```

同样的道理，在头文件中使用智能指针对象作为类成员变量时，也应该优先使用前置声明去引用智能指针对象的包裹类，而不是直接包含包裹类的头文件。

```
//Test.h
#include <memory>

//智能指针包裹类 A，这里优先使用 A 的前置声明，而不是直接包含 A.h 文件
class A;

class Test
{
public:
    Test();
    ~Test();

private:
    std::unique_ptr<A> m_spA;
};
```

Modern C/C++ 已经成为 C/C++ 的开发趋势，建议读者善用和熟练使用本节介绍的后三种智能指针对象。

第 2 章
C++后端开发必备的工具和调试知识

本章讲解 C++后端开发必备的工具和调试知识。

2.1 SSH 工具与 FTP 工具

在 C/C++开发中，我们大多数使用的开发机器是 Windows 和 Mac 机器，在程序开发完成后拿到专门的 Linux 机器上去编译和调试。公司内部通常不大可能给每个开发人员都分配一台 Linux 机器，而是准备一台配置较高的开发机器，给每个开发人员都分配一个 Linux 账号；或者干脆让员工自己安装虚拟机，并在虚拟机上安装 Linux 系统使用。

无论采用以上哪种方式，我们通常都不会直接登录 Linux 机器进行操作，而是通过一些支持 SSH 协议的工具远程连接到目标 Linux 机器上。目前常用的两大 SSH 工具分别是 Xshell 和 SecureCRT，这两种工具的用法基本相似，这里以 Xshell 为例来说明。

2.1.1 Xshell

虽然 Xshell 是一款商业软件，但也提供了免费的个人使用版本（Free License for Home and School Users），在其官网可以找到下载链接。

进入个人免费版本，填写自己的姓名和邮箱后单击 DOWNLOAD 按钮，相应的邮箱就会收到一封含有下载 Xshell 免费版本链接的邮件。

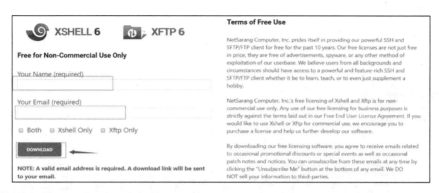

下载并安装 Xshell 后，我们就可以使用 Xshell 连接到 Linux 机器上了。

单击主界面的菜单"文件"→"新建…"，或者单击工具栏的"新建"按钮，会弹出"新建会话属性"对话框。

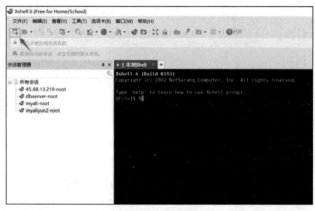

在相应的编辑框中输入我们的 Linux 机器名称、主机地址、端口号（默认是 22），然后单击"连接"按钮即可。

在接下来弹出的对话框中输入我们的 Linux 机器的用户名和密码后单击"确定"按钮，即可连接上 Linux 机器，出现如下图所示的界面即表示连接成功。

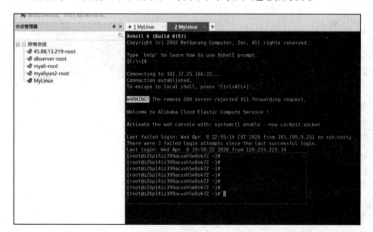

在 Xshell 的左侧有一个会话管理界面，记录了我们连接过的所有会话，我们双击其中某一项即可使用对应的配置连接对应的 Linux 机器。

我们有时需要在目标 Linux 机器与本地机器之间相互发送文件，对于 CentOS 系统来说，可以使用 sz 和 rz 命令，如果在读者的机器上没有这两个命令，则可以使用如下命令安装：

```
yum install lrzsz
```

安装完毕之后，如果需要向目标 Linux 机器发送一个文件，则可以在连接 Linux 机器后在 Shell 终端执行：

```
rz -y
```

此时就会弹出"选择文件"对话框，我们选择需要上传的文件，便可以将本机上的该文件上传到远端的 Linux 机器上。

反过来，如果我们想将远端的 Linux 机器上的文件发送到本机上，则执行：

```
sz 文件全饰路径
```

会弹出一个对话框让我们选择文件在本机上的保存位置。

2.1.2　FTP

上面介绍的 rz 和 sz 命令虽然方便，但只适用于传输单个文件，不适用批量传输文件和文件夹，为此我们可以使用支持 SFTP 的工具远程连接到目标 Linux 机器上来进行文件和文件夹传输。

这样的工具有很多，例如 WinSCP、Xshell 配套的 XFTP。笔者喜欢的一款工具是开源的 FTP 软件 FileZilla。FileZilla 界面如下图所示。

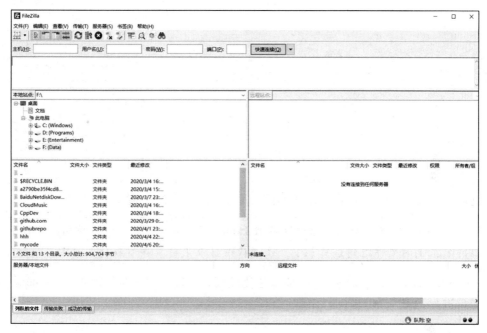

单击菜单"文件"→"站点管理器…"或者工具栏的"打开站点管理器"按钮打开站点管理器界面。

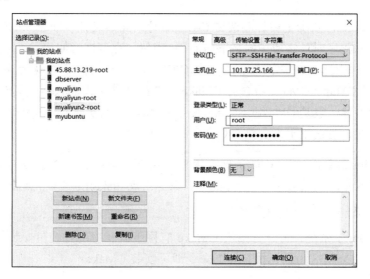

将协议改成 SFTP，输入主机名、用户名和密码信息之后单击"连接"按钮即可连接上远程 Linux。

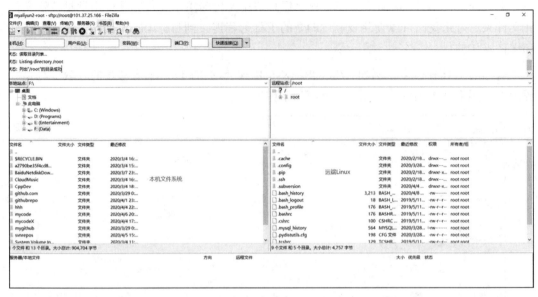

此时我们就可以自由地将文件在本机和远端 Linux 系统中拖曳了。

需要注意的是，如果登录的账号对 Linux 下的某个目录无权限访问，则不可以将本机上的文件拖到该目录下，或者从该目录下将文件下载到本机上。

FileZilla 的源码是开源的，源码质量也比较高，有兴趣的读者可以从 FileZilla 官网下载其源码进行学习。

2.2 makefile 与 CMake

我们在 Windows 机器上编译和调试 C/C++程序时可以使用 Visual Studio，在 Linux 机器上编译 C/C++程序时最终使用的是 gcc/g++，当然，在调试时使用 gdb。我们一般使用 makefile 文件组织大型 C/C++或者含有多个 C/C++文件的项目，有人认为 makefile 不太方便，于是发明了 CMake。CMake 将含有 CMake 指令的文件生成 makefile 文件，含有 CMake 指令的文件的名称一般是 CMakeLists.txt，使用 CMake 是在实际开发中组织和管理 C/C++ 项目的常用方式，也适用于大多数 C/C++开源项目。

对 Linux 下的一些 C/C++开源项目，执行 configure 命令后一般会生成 CMakeLists.txt 文件，接着执行 CMake 命令会生成 makefile 文件，然后执行 make 命令利用 gcc/g++对项目进行编译。

对于 Windows 系统，可以直接从 CMake 官网下载相应的安装包安装 CMake；对于 Linux 系统，以 CentOS 为例，执行如下命令即可安装 CMake：

```
yum install cmake
```

安装好 CMake 之后需要编写 CMakeList.txt。CMakeLists.txt 的样例如下。

样例一：

```
cmake_minimum_required(VERSION 2.6)

project (FLAMGINGO_SERVER)

set(CMAKE_CXX_FLAGS "${CMAKE_CXX_FLAGS} -std=c++0x -g -Wall -O0
-Wno-unused-variable -pthread")

link_directories(
    ${PROJECT_SOURCE_DIR}/lib
    /usr/lib64/mysql/
)

set(net_srcs
base/AsyncLog.cpp
base/ConfigFileReader.cpp
base/Platform.cpp

net/Acceptor.cpp
net/ByteBuffer.cpp
net/Channel.cpp
)

set(mysqlapi_srcs
mysqlapi/DatabaseMysql.cpp
mysqlapi/Field.cpp
mysqlapi/QueryResult.cpp
)

set(chatserver_srcs
chatserversrc/main.cpp
chatserversrc/ChatServer.cpp
chatserversrc/ChatSession.cpp
)

set(fileserver_srcs
fileserversrc/main.cpp
fileserversrc/FileServer.cpp
fileserversrc/FileSession.cpp
)

set(imgserver_srcs
imgserversrc/main.cpp
fileserversrc/FileServer.cpp
fileserversrc/FileSession.cpp
)

add_executable(chatserver ${net_srcs} ${chatserver_srcs} ${mysqlapi_srcs})
#只包含库目录是没用的，还必须使用 TARGET_LINK_LIBRARIES 链接该库
TARGET_LINK_LIBRARIES(chatserver mysqlclient)

add_executable(fileserver ${net_srcs} ${fileserver_srcs})
```

```
TARGET_LINK_LIBRARIES(fileserver)

add_executable(imgserver ${net_srcs} ${imgserver_srcs})
TARGET_LINK_LIBRARIES(imgserver)
```

样例二：

```
PROJECT(TRADE)

AUX_SOURCE_DIRECTORY(./ SRC_LIST)
SET(EXECUTABLE_OUTPUT_PATH ../bin)

ADD_DEFINITIONS(-g -O0 -W -Wall -D_REENTRANT -D_FILE_OFFSET_BITS=64 -DAC_HAS_INFO
-DAC_HAS_WARNING -DAC_HAS_ERROR -DAC_HAS_CRITICAL -DTIXML_USE_STL
-DHAVE_CXX_STDHEADERS -Wno-deprecated ${CMAKE_CXX_FLAGS})

INCLUDE_DIRECTORIES(
./
/usr/local/include/commonlib/json2
/usr/local/include/commonlib/mysql
)

LINK_DIRECTORIES(
./
/usr/local/lib/commonlib
)

ADD_EXECUTABLE(trade ${SRC_LIST})

TARGET_LINK_LIBRARIES(trade pthread netutil json2 mysqlclient dbapi )
```

下面对以上代码中的一些内容进行解释。

（1）首行的 cmake_minimum_required 指令指定支持该 CMakeLists.txt 文件的 CMake 最低版本号。

（2）project 指令指定该项目的名称，注意，项目名称不是最终生成的二进制程序名，在一个项目下面可以生成多个二进制程序名。

（3）set 定义和设置各种变量，set 括号后的第 1 个名称是定义的变量名称，其后是变量的值，例如上述文件定义了 CMAKE_CXX_FLAGS、net_srcs、mysqlapi_src、chatserver_srcs、fileserver_srcs、imgserver_srcs 共 6 个变量，之后可以使用${变量名}引用这些变量。这些变量可以是内置的变量，例如 CMAKE_CXX_FLAGS 指定 g++编译选项，EXECUTABLE_OUTPUT_PATH 指定输出的二进制文件路径，也可以是自定义变量如 chatserver_srcs、fileserver_srcs 等。

（4）CMake 使用 aux_source_directory 指令指定源码目录，使用 include_directories 指令指定 include 目录，使用 link_directories 指定 lib 目录。

（5）CMake 使用指令指定生成的动态库或静态库的名称，格式如下：

```
add_library(libname [SHARED|STATIC|MODULE] [EXCLUDE_FROM_ALL] source1 source2 ...
sourceN)
```

例如：

```
add_library(hello hello1.cpp hello2.cpp)
```

我们不需要写全 libhello.so 或 libhello.a，只需填写 hello 即可，CMake 会自动生成 libhello.X。类型有如下三种。

◎ SHARED：动态库（扩展名为.so）。
◎ STATIC：静态库（扩展名为.a）。
◎ MODULE：在使用 dyld 的系统中有效，若不支持 dyld，则被当作 SHARED 对待。

EXCLUDE_FROM_ALL 参数的意思是这个库不会被默认构建，除非有其他组件依赖或者手工构建。如下命令会生成一个 libkafkawrapper.so 文件，且 libkafkawrapper.so 文件的生成依赖 librdkafka.so、librdkafka++.so、libcrypto.so、libssl.so 这 4 个库：

```
add_library(kafkawrapper SHARED ${kafka_wrapper_srcs})
TARGET_LINK_LIBRARIES(kafkawrapper rdkafka rdkafka++ crypto ssl)
```

（6）TARGET_LINK_LIBRARIES 指定生成的二进制文件依赖的其他库，上文已有介绍。

编写完 CMakeLists.txt 文件后，进入 CMakeLists.txt 文件所在的目录，依次执行如下命令即可生成最终的二进制文件：

```
# 利用 cmake 生成 makefile
cmake .
# 执行 make 命令，利用 gcc/g++编译生成最终的二进制文件
make
```

CMakeLists.txt 也支持递归执行，先执行父目录的 CMakeLists.txt 文件，再执行子目录的 CMakeLists.txt 文件。

若想了解 CMake 的更多信息，则可以参考 CMake 官网。

接下来看看如何利用 CMake 生成 Visual Studio 工程文件。

对于习惯了 Visual Studio 强大的管理项目、编码和调试功能的读者来说，在 Linux 下使用 gcc/g++编译和使用 gdb 调试是痛苦的事情。对于大多数 C/C++开源项目，如果我们不在意 Windows 和 Linux 在一些底层 API 接口上的使用差别，想熟悉该项目的执行脉络和原理，则在 Windows 上使用 Visual Studio 调试该项目也未尝不可。凡是可以使用 CMake 编译的 Linux 程序(即提供了 CMakeLists.txt 文件)，我们都可以利用 CMake 生成 Windows 上的 Visual Studio 工程文件。

这里以著名的开源网络库 libuv 为例。从 libuv 官网下载最新的 libuv 的源码，得到 libuv-1.x.zip 文件并解压。笔者在自己的机器上将代码解压缩至 E:\libuv-1.x，在解压缩后的目录中确实存在一个 CMakeLists.txt 文件，如下图所示。

启动 Windows 机器上的 CMake 图形化工具（cmake-gui），按下图所示进行设置。

设置完成之后，单击界面上的 Configure 按钮，会提示 vsprojects 目录不存在，提示是否创建，单击 Yes 按钮进行创建。

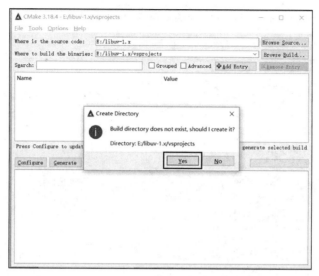

如果在机器上安装了多个版本的 Visual Studio，则接下来会弹出对话框让我们选择要生成的工程文件对应的 Visual Studio 版本号，可以根据自己的实际情况按需选择，这里选择 Visual Studio 16 2019。

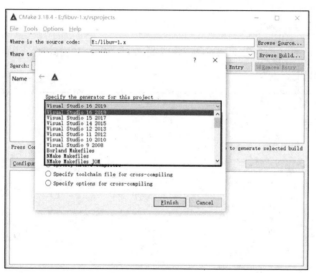

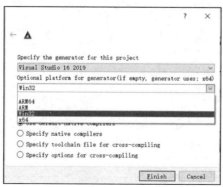

单击 Finish 按钮后开始启动 CMake 的检测和配置工作。等待一会儿，在 CMake 底部的输出框中提示 "Configuring Done"，表示配置工作已经完成。

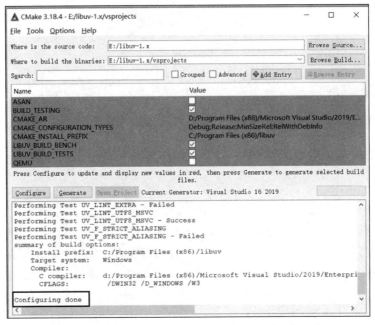

接下来单击 Generate 按钮即可生成所选版本的 Visual Studio 工程文件，生成的文件位于 vsprojects 目录下。

我们可以在界面上单击 Open Project 按钮直接打开工程文件,也可以找到对应目录下的 libuv.sln 文件并打开。打开后的界面如下图所示。

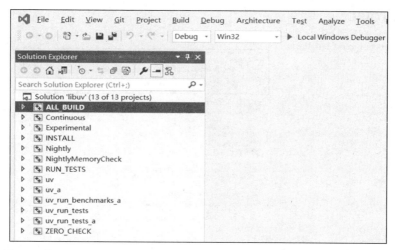

接下来就可以使用 Visual Studio 对 libuv 进行编译和调试了。

再次深入回顾上述过程:单击 Configure 按钮之后,和在 Linux 机器上执行 cmake 命令一样,CMake 也在检测所在的系统环境是否匹配在 CMakeLists.txt 中定义的各种环境,其本质上是生成了一份可以在 Windows 机器上编译和运行的代码(也就是说该源码支持在 Windows 机器上运行)。因此,对于很多虽然提供了 CMakeLists.txt 文件但并不支持在 Windows 机器上运行的 Linux 工程,虽然利用上述方法也能最终生成 Visual Studio 工程文件,但这些文件并不能直接在 Windows 机器上无错地编译和调试。

> 由于不同的 CMake 版本支持的 CMakeLists.txt 中的语法可能略有差别,有些 CMakeLists.txt 文件在使用 configure 方法时可能会产生一些错误,所以需要做些修改才能通过。

2.3 使用 Visual Studio 管理和阅读开源项目代码

Visual Studio 提供了强大的 C/C++ 项目开发和管理能力,本节介绍如何使用 Visual Studio 管理 C/C++ 开源项目,这里以 Redis 项目为例。

启动 Visual Studio,新建一个空的 Win32 控制台程序,在工程建好后关闭该工程,因为接下来需要移动这些文件。

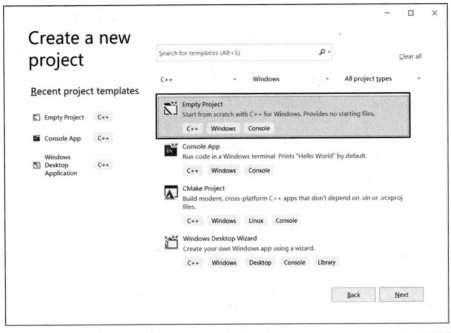

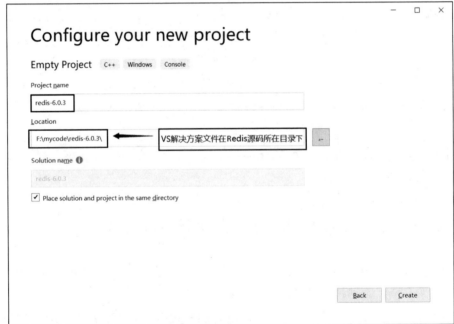

这样会在Redis源码目录下根据我们设置的名称生成一个文件夹（这里是redis-6.0.3），将该文件夹中的所有文件都复制到Redis源码根目录下，然后删掉生成的这个文件夹。

第 2 章 C++后端开发必备的工具和调试知识

再次用 Visual Studio 打开 redis-6.0.3.sln 文件，然后在解决方案资源管理器视图中单击显示 ALL Files 按钮并保持该按钮为选中状态（如果找不到解决方案资源管理器视图，则可以在"视图"菜单中打开，组合键为 Ctrl+Alt+L）。

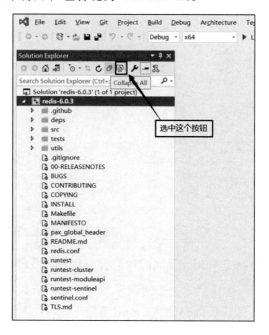

然后选中所有需要添加到解决方案中的文件，单击鼠标右键选择"Include In Project"菜单即可，如果文件较多，则 Visual Studio 可能需要一会儿才能完成。为了减少等待时间，也可以一批一批地添加。

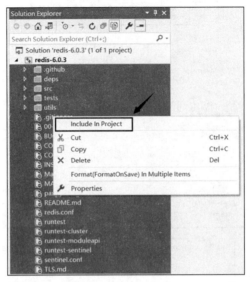

接着选择"文件"菜单的"全部保存"菜单项保存即可（组合键为 Ctrl+Shift+S），最终效果如下图所示。

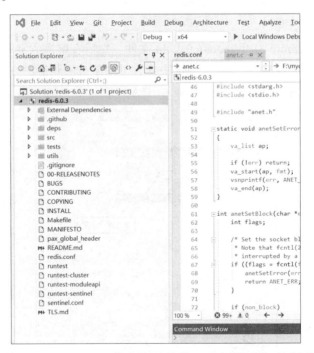

这样我们就能利用 Visual Studio 强大的功能管理和阅读我们的源码了。

对于 Linux 系统下的一些项目，虽然阅读和编辑代码时可以使用 Visual Studio，但是编译和调试这些项目时还是建议将其放到 Linux 系统上。例如笔者虽然使用 Visual Studio

管理和阅读 Redis 源码，但编译和调试 Redis 还是使用 Linux 系统（使用 gcc 编译并使用 gdb 调试）。

> 注意：在 C/C++ 开源项目中一般使用各种宏去条件编译一些代码，在实际生成的二进制文件中不一定包含这些代码，所以在 Visual Studio 中看到的某段代码的行号与在 gdb 中调试的实际代码行号不一定相同，在给某一行代码设置断点时，请以 gdb 中通过 list 命令看到的代码行号为准。

2.4 gdb 调试

调试是开发过程中不可或缺的工作，在 Linux 编程中通常使用 gdb 来调试 C/C++ 程序。本节以调试 Redis 代码为例，讲解 gdb 调试的常见知识和经验总结。

2.4.1 被调试的程序需要带调试信息

在调试某个程序时，为了能清晰地看到调试的每一行代码、调用的堆栈、变量名、函数名等信息，我们需要在调试程序中加上调试符号信息，即在使用 gcc 编译程序时需要加上 -g 选项。举个例子，通过以下命令将生成一个带调试信息的 hello_server 程序（hello_server.c 为任意 cpp 文件）：

```
gcc -g -o hello_server hello_server.c
```

那么如何判断 hello_server 是否带有调试信息呢？我们使用 gdb 来调试这个程序，gdb 会显示正确读取该程序的调试信息：

```
[root@localhost testclient]# gdb hello_server
GNU gdb (GDB) Red Hat Enterprise Linux 7.6.1-100.el7_4.1
Copyright (C) 2013 Free Software Foundation, Inc.
License GPLv3+: GNU GPL version 3 or later <http://gnu.org/licenses/gpl.html>
This is free software: you are free to change and redistribute it.
There is NO WARRANTY, to the extent permitted by law.  Type "show copying"
and "show warranty" for details.
This GDB was configured as "x86_64-redhat-linux-gnu".
For bug reporting instructions, please see:
<http://www.gnu.org/software/gdb/bugs/>...
Reading symbols from /root/testclient/hello_server...done.
(gdb)
```

在 gdb 加载成功以后，会显示一行 "Reading symbols from/root/testclient/hello_server...done." 信息，表示读取符号文件完毕，说明该程序带有调试信息。我们不加-g 选项再试试：

```
[root@localhost testclient]# gcc -o hello_server2 hello_server.c
[root@localhost testclient]# gdb hello_server2
GNU gdb (GDB) Red Hat Enterprise Linux 7.6.1-100.el7_4.1
Copyright (C) 2013 Free Software Foundation, Inc.
License GPLv3+: GNU GPL version 3 or later <http://gnu.org/licenses/gpl.html>
This is free software: you are free to change and redistribute it.
```

```
There is NO WARRANTY, to the extent permitted by law. Type "show copying"
and "show warranty" for details.
This GDB was configured as "x86_64-redhat-linux-gnu".
For bug reporting instructions, please see:
<http://www.gnu.org/software/gdb/bugs/>...
Reading symbols from /root/testclient/hello_server2...(no debugging symbols
found)...done.
(gdb)
```

细心的读者应该看出差别了，这次不加-g 选项，用 gdb 调试生成的 hello_server2 程序，读取调试符号信息时提示：

```
Reading symbols from /root/testclient/hello_server2...(no debugging symbols
found)...done.
```

当然，这里顺便提一下，除了不加-g 选项，也可以使用 Linux 的 strip 命令移除某个程序中的调试信息。这里对 hello_server 使用 strip 命令试试：

```
[root@localhost testclient]# strip hello_server
##使用 strip 命令前
-rwxr-xr-x. 1 root root 12416 Sep  8 09:45 hello_server
##使用 strip 命令后
-rwxr-xr-x. 1 root root 6312 Sep  8 09:55 hello_server
```

可以发现，我们对 hello_server 使用 strip 命令之后，这个程序明显变小了（由 12416 字节减小为 6312 字节）。我们通常会在程序测试没问题后，将其发布到生产环境或者正式环境中，生成不带调试符号信息的程序，以减小程序体积或提高程序执行效率。

再用 gdb 验证这个程序的调试信息是否被移除：

```
[root@localhost testclient]# gdb hello_server
GNU gdb (GDB) Red Hat Enterprise Linux 7.6.1-100.el7_4.1
Copyright (C) 2013 Free Software Foundation, Inc.
License GPLv3+: GNU GPL version 3 or later <http://gnu.org/licenses/gpl.html>
This is free software: you are free to change and redistribute it.
There is NO WARRANTY, to the extent permitted by law. Type "show copying"
and "show warranty" for details.
This GDB was configured as "x86_64-redhat-linux-gnu".
For bug reporting instructions, please see:
<http://www.gnu.org/software/gdb/bugs/>...
Reading symbols from /root/testclient/hello_server...(no debugging symbols
found)...done.
(gdb)
```

这里需要补充两个说明。

（1）这里举的例子虽然以 gcc 为例，但-g 选项实际上同样适用于使用 makefile、cmake 等工具编译生成的 Linux 程序。

（2）在实际生成调试程序时，我们一般不仅要加上-g 选项，也被建议关闭编译器程序的优化选项。编译器程序的优化选项一般有 5 个级别，即 O0 ~ O4（O0 是字母 O 加上数字 0），其中，O0 表示不优化（关闭优化），从 O1 到 O4，优化级别越来越高，O4 级别最高。关闭优化的目的是在调试时符号文件显示的调试变量等能与源代码完全对应。举个例子，假设有以下代码：

```
int func()
{
    int a = 1;
    int b = a + 1;
    int c = a + b;
    return a + b + c;
}

int main()
{
    int a = func();
    printf("%d\n", a);
}
```

在以上代码中，因为在 main 函数中调用了 func 函数，由于调用 func 函数得到的返回值可以在编译期间直接算出来，所以如果开启了优化选项，则可能在实际调试时这个函数中的局部变量 a、b、c 被编译器优化，取而代之的是直接的值，甚至连 func 函数也可能被优化。如果出现这种情况，则我们在调试时看到的代码和实际的代码可能有所差异，这会给排查和定位问题带来困难。当然，上面说的优化选项是否一定会出现，不同版本的编译器可能有不同的表现。总之，在生成调试文件时建议关闭编译器优化选项（使用 O0 选项）。

2.4.2 启动 gdb 调试的方法

使用 gdb 调试一个程序一般有三种方法：gdb filename、gdb attach pid、gdb filename corename，接下来逐一进行介绍。

1. 方法一 直接调试目标程序

使用如下命令：

```
gdb filename
```

其中，filename 是我们需要启动的调试程序文件名，这种方式会直接使用 gdb 启动一个程序进行调试，也就是说这个程序还没有启动。前面使用 gdb 调试 hello_server 时就使用了这种方式。

2. 方法二 attach 到进程

在某些情况下，一个程序已经启动，我们想调试这个程序，但又不想重启这个程序。假设有这样一个场景，我们的聊天测试服务器程序运行了一段时间，却再也无法接受新的客户端连接，那么我们肯定不能为此重启程序，如果重启，则当前程序的各种状态信息就丢失了。这时我们只需使用 gdb attach 程序的进程 ID 将 gdb 调试器 attach 到我们的聊天测试服务器程序上。假设我们的聊天程序叫作 chatserver，则使用 ps 命令获取该进程的 PID，然后 gdb attach 上去，就可以调试了：

```
[zhangyl@localhost flamingoserver]$ ps -ef | grep chatserver
zhangyl   42921     1 17 11:18 ?        00:00:04 ./chatserver -d
zhangyl   42936 42898  0 11:18 pts/0    00:00:00 grep --color=auto chatserver
```

我们得到 chatserver 的 PID 为 42921，然后使用 gdb attach 42921 把 gdb attach 到 chatserver 进程上：

```
[zhangyl@localhost flamingoserver]$ gdb attach 42921
attach: No such file or directory.
Attaching to process 42921
Reading symbols from /home/zhangyl/flamingoserver/chatserver...done.
Reading symbols from /usr/lib64/mysql/libmysqlclient.so.18...Reading symbols from /usr/lib64/mysql/libmysqlclient.so.18...(no debugging symbols found)...done.
Reading symbols from /lib64/libpthread.so.0...(no debugging symbols found)...done.
[New LWP 42931]
[New LWP 42930]
[New LWP 42929]
[New LWP 42928]
[New LWP 42927]
[New LWP 42926]
[New LWP 42925]
[New LWP 42924]
[New LWP 42922]
[Thread debugging using libthread_db enabled]
Using host libthread_db library "/lib64/libthread_db.so.1".
Loaded symbols for /lib64/libpthread.so.0
Reading symbols from /lib64/libc.so.6...(no debugging symbols found)...done.
```

为了节约篇幅，在以上代码中删掉了一些无关的信息。当出现提示"Attaching to process 42921"时，说明我们已经成功地将 gdb attach 到目标进程中了。需要注意的是，由于我们的程序使用了一些系统库（如 libc.so），而这是发行版本的 Linux 系统，是没有调试符号的，所以 gdb 会提示找不到这些库的调试符号。我们的目的是调试 chatserver，并不关注对系统 API 调用的内部实现，所以我们可以不关注这些提示。只要在 chatserver 文件中有调试信息即可。

在使用 gdb attach 附加上目标进程后，调试器会暂停下来，此时我们可以使用 continue 命令让程序继续运行，或者加上相应的断点再继续运行程序（不熟悉这里提到的 continue 命令也没有关系，下面会详细介绍这些命令的用法）。

若想调试完程序后结束此次调试，且不对当前进程 chatserver 有任何影响，也就是说想让这个程序继续运行，则可以在 gdb 的命令行界面输入 detach 命令让程序与 gdb 调试器分离，这样 chatserver 也可以继续运行：

```
(gdb) detach
Detaching from program: /home/zhangyl/flamingoserver/chatserver, process 42921
```

然后退出 gdb 就可以了：

```
(gdb) quit
[zhangyl@localhost flamingoserver]$
```

3. 方法三 调试 core 文件——定位进程崩溃问题

有时，我们的服务器程序在运行一段时间后会突然崩溃。这当然不是我们所希望看到的，我们需要解决这个问题。只要程序在崩溃时有 core 文件产生，我们就可以使用这个 core 文件定位崩溃的原因。当然，Linux 系统默认是不会开启程序崩溃时产生 core 文件这一功能的，我们可以使用 ulimit -c 来查看系统是否开启了这一功能（顺便提一句，通过 ulimit -a 命令不仅可以查看 core 文件的生成功能是否开启，还可以查看其他功能，例如系统允许的最大文件描述符的数量等）：

```
[zhangyl@localhost flamingoserver]$ ulimit -a
core file size          (blocks, -c) 0
data seg size           (kbytes, -d) unlimited
scheduling priority             (-e) 0
file size               (blocks, -f) unlimited
pending signals                 (-i) 15045
max locked memory       (kbytes, -l) 64
max memory size         (kbytes, -m) unlimited
open files                      (-n) 1024
pipe size            (512 bytes, -p) 8
POSIX message queues     (bytes, -q) 819200
real-time priority              (-r) 0
stack size              (kbytes, -s) 8192
cpu time               (seconds, -t) unlimited
max user processes              (-u) 4096
virtual memory          (kbytes, -v) unlimited
file locks                      (-x) unlimited
```

如上所示，core file size 所在的行默认是 0，表示关闭生成 core 文件的选项，如果我们需要修改某个选项的值，则可以通过命令形式 "ulimit 选项名 设置值" 来修改。例如，可以将 core 文件生成的大小改成具体的某个值（最大允许的字节数）或不限制大小（unlimited），这里直接改成不限制大小，则执行命令 ulimit -c unlimited 即可：

```
[zhangyl@localhost flamingoserver]$ ulimit -c unlimited
[zhangyl@localhost flamingoserver]$ ulimit -a
core file size          (blocks, -c) unlimited
data seg size           (kbytes, -d) unlimited
scheduling priority             (-e) 0
file size               (blocks, -f) unlimited
pending signals                 (-i) 15045
max locked memory       (kbytes, -l) 64
max memory size         (kbytes, -m) unlimited
open files                      (-n) 1024
pipe size            (512 bytes, -p) 8
POSIX message queues     (bytes, -q) 819200
real-time priority              (-r) 0
stack size              (kbytes, -s) 8192
cpu time               (seconds, -t) unlimited
max user processes              (-u) 4096
virtual memory          (kbytes, -v) unlimited
file locks                      (-x) unlimited
```

注意，这个命令很容易出错，第 1 个 ulimit 是 Linux 命令，-c 选项后面的 unlimited 是选项的值，表示不限制大小，当然，我们也可以将其改成具体的数值大小。

还有一个问题就是，这样进行修改，关闭这个 Linux 会话后，这个设置项的值会被还原成 0，而我们的服务器程序一般以后台程序（守护进程）长周期地运行，也就是说当前会话虽然被关闭，但服务器程序仍然在后台运行。这样，这个程序在某个时刻崩溃后，是无法产生 core 文件的，不利于排查问题。所以我们希望这个选项永久生效。设置永久生效的方式有以下两种。

（1）在/etc/security/limits.conf 中增加一行：

```
#<domain>      <type>      <item>          <value>
*              soft        core            unlimited
```

这里设置的是不限制 core 文件的大小，也可以将其设置成具体的数值，例如 1024 表示生成的 core 文件最大为 1024KB。

（2）把 ulimit -c unlimited 行加到/etc/profile 文件中，放到这个文件最后一行即可，修改成功后执行 source /etc/profile 可以让配置立即生效。当然，这只作用于 root 用户，如果想仅仅作用于某一用户，则可以把 ulimit -c unlimited 行加到该用户对应的~/.bashrc 或 ~/.bash_profile 文件中。

生成的 core 文件的默认命名方式是 core.pid，其位置是崩溃程序所在的目录。举个例子，某个程序在运行时其进程 ID 是 16663，则其崩溃时产生的 core 文件的名称是 core.16663。比如我们服务器上的 msg_server 崩溃，在当前目录下产生了如下 core 文件：

```
-rw------- 1 root root 10092544 Sep  9 15:14 core.21985
```

我们就可以通过这个 core.21985 文件排查程序崩溃的原因了。调试 core 文件的命令如下：

```
gdb filename corename
```

其中，filename 是程序名，这里是 msg_server；corename 是 core.21985。

我们输入 gdb msg_server core.21985 启动调试：

```
[root@myaliyun msg_server]# gdb msg_server core.21985
Reading symbols from /root/teamtalkserver/src/msg_server/msg_server...done.
[New LWP 21985]
[Thread debugging using libthread_db enabled]
Using host libthread_db library "/lib64/libthread_db.so.1".
Core was generated by `./msg_server -d'.
Program terminated with signal 11, Segmentation fault.
#0  0x00000000004ceb1f in std::less<CMsgConn*>::operator() (this=0x2283878,
    __x=@0x7ffca83563a0: 0x2284430, __y=@0x51: <error reading variable>)
    at /usr/include/c++/4.8.2/bits/stl_function.h:235
235             { return __x < __y; }
```

可以看到程序在 stl_function.h 的第 235 行崩溃，然后通过 bt 命令查看程序崩溃时的调用堆栈，进一步分析，就能找到崩溃的原因：

```
(gdb) bt
#0  0x00000000004ceb1f in std::less<CMsgConn*>::operator() (this=0x2283878,
    __x=@0x7ffca83563a0: 0x2284430, __y=@0x51: <error reading variable>)
    at /usr/include/c++/4.8.2/bits/stl_function.h:235
#1  0x00000000004cdd70 in std::_Rb_tree<CMsgConn*, CMsgConn*,
std::_Identity<CMsgConn*>, std::less<CMsgConn*>,
std::allocator<CMsgConn*> >::_M_get_insert_unique_pos
    (this=0x2283878, __k=@0x7ffca83563a0: 0x2284430) at
/usr/include/c++/4.8.2/bits/stl_tree.h:1324
#2  0x00000000004cd18a in std::_Rb_tree<CMsgConn*, CMsgConn*,
std::_Identity<CMsgConn*>, std::less<CMsgConn*>,
```

```
    std::allocator<CMsgConn*> >::_M_insert_unique<CMsgConn* const&> (this=0x2283878,
    __v=@0x7ffca83563a0: 0x2284430) at /usr/include/c++/4.8.2/bits/stl_tree.h:1377
#3  0x00000000004cc8bd in std::set<CMsgConn*, std::less<CMsgConn*>,
    std::allocator<CMsgConn*> >::insert (this=0x2283878, __x=@0x7ffca83563a0: 0x2284430)
    at /usr/include/c++/4.8.2/bits/stl_set.h:463
#4  0x00000000004cb011 in CImUser::AddUnValidateMsgConn (this=0x2283820,
    pMsgConn=0x2284430) at /root/teamtalkserver/src/msg_server/ImUser.h:42
#5  0x00000000004c64ae in CDBServConn::_HandleValidateResponse (this=0x227f6a0,
    pPdu=0x22860d0) at /root/teamtalkserver/src/msg_server/DBServConn.cpp:319
#6  0x00000000004c5e3d in CDBServConn::HandlePdu (this=0x227f6a0, pPdu=0x22860d0)
    at /root/teamtalkserver/src/msg_server/DBServConn.cpp:203
#7  0x00000000005022b3 in CImConn::OnRead (this=0x227f6a0) at
    /root/teamtalkserver/src/base/imconn.cpp:148
#8  0x0000000000501db3 in imconn_callback (callback_data=0x7f4b20
    <g_db_server_conn_map>, msg=3 '\003', handle=8, pParam=0x0)
    at /root/teamtalkserver/src/base/imconn.cpp:47
#9  0x0000000000504025 in CBaseSocket::OnRead (this=0x227f820) at
    /root/teamtalkserver/src/base/BaseSocket.cpp:178
#10 0x0000000000502f8a in CEventDispatch::StartDispatch (this=0x2279990,
    wait_timeout=100) at /root/teamtalkserver/src/base/EventDispatch.cpp:386
#11 0x00000000004fddbe in netlib_eventloop (wait_timeout=100) at
    /root/teamtalkserver/src/base/netlib.cpp:160
#12 0x00000000004d18c2 in main (argc=2, argv=0x7ffca8359978) at
    /root/teamtalkserver/src/msg_server/msg_server.cpp:213
(gdb)
```

堆栈#0~#3 是系统库函数的调用序列，是经过反复测试的，一般不存在问题；堆栈#4~#12 是我们自己的业务逻辑调用序列，我们可以排查这部分代码进而定位问题。

细心的读者会发现这样一个问题：一个程序在运行时，其 PID 是可以获取的，但是在程序崩溃后产生了 core 文件，尤其是多个程序同时崩溃，我们无法通过 core 文件名中的 PID 来判断对应哪个服务。有以下两种方法解决这个问题。

（1）在程序启动时记录 PID：

```
void writePID()
{
    uint32_t curPid = (uint32_t) getpid();
    FILE* f = fopen("xxserver.pid", "w");
    if (f != NULL)
    {
        char szPid[32];
        snprintf(szPid, sizeof(szPid), "%d", curPid);
        fwrite(szPid, strlen(szPid), 1, f);
        fclose(f);
    }
}
```

我们在程序启动时调用上述 writePID 函数，将程序当时的 PID 记录到 xxserver.pid 文件中，这样在程序崩溃时，我们就可以从这个文件中得到进程当时运行的 PID 了，并且可以与默认的 core 文件名后面的 PID 匹配。

（2）自定义 core 文件的名称和目录。/proc/sys/kernel/core_uses_pid 可以控制在产生

的 core 文件的文件名中是否添加 PID 作为扩展，如果添加，则文件的内容为 1，否则为 0；/proc/sys/kernel/core_pattern 可以设置格式化的 core 文件保存位置或文件名。修改方式如下：

```
echo "/corefile/core-%e-%p-%t" > /proc/sys/kernel/core_pattern
```

对各个参数的说明如下表所示。

参数名称	参数含义（英文）	参数含义（中文）
%p	insert pid into filename	添加 pid 到 core 文件名中
%u	insert current uid into filename	添加当前 uid 到 core 文件名中
%g	insert current gid into filename	添加当前 gid 到 core 文件名中
%s	insert signal that caused the coredump into the filename	添加导致产生 core 的信号到 core 文件名中
%t	insert UNIX time that the coredump occurred into filename	添加 core 文件的生成时间（UNIX）到 core 文件名中
%h	insert hostname where the coredump happened into filename	添加主机名到 core 文件名中
%e	insert coredumping executable name into filename	添加程序名到 core 文件名中

假设我们现在的程序是 test，我们设置该程序崩溃时的 core 文件名如下：

```
echo "/root/testcore/core-%e-%p-%t" > /proc/sys/kernel/core_pattern
```

那么最终会在/root/testcore/目录下生成的 test 的 core 文件名格式如下：

```
-rw-------. 1 root root 409600 Jan 14 13:54 core-test-13154-1547445291
```

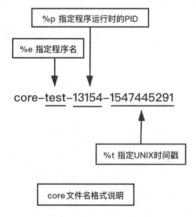

core 文件名格式说明

需要注意的是，用户必须对指定的 core 文件目录有写权限，否则会因为权限不足无法生成 core 文件。

2.5　gdb 常用命令详解——利用 gdb 调试 Redis

为了结合实践，这里以使用 gdb 调试 Redis 源码为例介绍每个命令，当然，本节只讲解 gdb 中一些常用命令的基础用法，下节会讲解 gdb 中的一些高级用法。

2.5.1　gdb 常用调试命令概览和说明

下面给出了一个常用命令列表，后面会结合具体的例子详细介绍每个命令的用法。

命令名称	命令缩写	命令说明
run	r	运行一个程序
continue	c	让暂停的程序继续运行
break	b	添加断点
tbreak	tb	添加临时断点
backtrace	bt	查看当前线程的调用堆栈
frame	f	切换到当前调用线程的指定堆栈
info	info	查看断点、线程等信息
enable	enable	启用某个断点
disable	disable	禁用某个断点
delete	del	删除断点
list	l	显示源码
print	p	打印或修改变量、寄存器的值
ptype	ptype	查看变量的类型
thread	thread	切换到指定的线程
next	n	运行到下一行
step	s	如果有调用函数，则进入调用的函数内部，相当于 step into
until	u	运行到指定的行停下来
finish	fi	结束当前调用函数，到上一层函数调用处
return	return	结束当前调用函数并返回指定的值，到上一层函数调用处
jump	j	将当前程序的执行流跳转到指定的行或地址
disassemble	dis	查看汇编代码
set args		设置程序启动命令行参数
show args		查看设置的命令行参数
watch	watch	监视某个变量或内存地址的值是否发生了变化
display	display	监视的变量或者内存地址，在程序中断后自动输出监控的变量或内存地址
dir	dir	重定向源码文件的位置

上表只列举了一些常用命令，未列举一些不常用的命令（如 file 命令）。不建议读者刻意记忆这些命令，因为命令较多，建议读者找几个程序实际练习一下，这样就容易记住了。表中"命令缩写"那一栏，是笔者平时对命令的简写输入，读者可以采用，也可以不采用。一个命令可以简写成什么样子，gdb 没有强行规定，但读者在简写 gdb 命令时需要遵循如下两个规则。

（1）一个命令在简写时，不能让 gdb 出现多个选择，若出现多个选择，gdb 就不知道对应哪个命令了。举个例子，输入 th 命令，而 th 命令对应的命令有 thread 和 thbreak（上表没有列出），这样 gdb 就不知道要使用哪个命令了。需要更具体地输入，gdb 才能识别：

```
(gdb) th
Ambiguous command "th": thbreak, thread.
```

（2）gdb 的某些命令虽然对应多个选择，但是有些命令的简写是确定的，比如 r 是 run 命令的简写，虽然有人输入 r 时可能是想使用 return 命令。

总之，如果记不清楚某个命令的简写，就可以直接使用该命令的全写，每个命令都是很常见的英文单词，通俗易懂且不难记忆。

2.5.2 用 gdb 调试 Redis 前的准备工作

本节逐一介绍上面每个命令的使用方法，会介绍一些很有用的调试细节和使用技巧。如果还不熟悉 gdb 调试，则建议认真阅读本节。

为了结合实践，这里仍以调试 Redis 源码为例来介绍每个命令。Redis 的最新源码下载地址可以在 Redis 官网获得，本书用到的 Redis 版本是 6.0.3。使用 wget 命令将 Redis 源码文件下载下来：

```
[root@myaliyun ~]# wget http://download.redis.io/releases/redis-6.0.3.tar.gz
--2020-05-27 11:55:15--  http://download.redis.io/releases/redis-6.0.3.tar.gz
Resolving download.redis.io (download.redis.io)... 109.74.203.151
Connecting to download.redis.io (download.redis.io)|109.74.203.151|:80...
connected.
HTTP request sent, awaiting response... 200 OK
Length: 2210882 (2.1M) [application/x-gzip]
Saving to: 'redis-6.0.3.tar.gz'

39%
[==========================================>
                            ] 869,940    29.0KB/s  eta 39s
```

下载完成后解压缩：

```
[root@myaliyun ~]# tar zxvf redis-6.0.3.tar.gz
```

进入生成的 redis-6.0.3 目录并使用 makefile 进行编译，为了方便调试，我们需要生成调试符号并且关闭编译器优化选项，操作如下：

```
[root@myaliyun ~]# cd redis-6.0.3
[root@myaliyun redis-6.0.3]# make CFLAGS="-g -O0" -j 4
```

-g 选项表示生成调试符号，-o0 选项表示关闭优化，-j 4 选项表示同时开启 4 个进程进行编译，加快编译速度。Redis 是纯 C 项目，使用的编译器是 gcc，所以这里设置编译器的选项时使用的是 CFLAGS 选项；对于 C++项目，使用的编译器一般是 g++，相应的编译器选项是 CXXFLAGS，请注意区别。

如果在编译过程中出现如下错误：

```
[root@myaliyun redis-6.0.3]# make CFLAGS="-g -O0"
cd src && make all
make[1]: Entering directory `/root/redis-6.0.3/src'
    CC adlist.o
In file included from adlist.c:34:0:
zmalloc.h:50:31: fatal error: jemalloc/jemalloc.h: No such file or directory
 #include <jemalloc/jemalloc.h>
                               ^
compilation terminated.
make[1]: *** [adlist.o] Error 1
make[1]: Leaving directory `/root/redis-6.0.3/src'
```

```
make: *** [all] Error 2
```

则可以改用以下命令来编译,这是由于系统没有安装 jemalloc 库,可以修改编译参数,让 Redis 使用系统默认的 malloc 而不是 jemalloc:

```
make MALLOC=libc CFLAGS="-g -O0" -j 4
```

编译成功后,进入 src 目录,使用 gdb 启动 redis-server 程序:

```
[root@myaliyun src]# gdb redis-server
Reading symbols from redis-server...
```

2.5.3 run 命令

在默认情况下,gdb+filename 只是 attach 到一个调试文件,并没有启动这个程序,我们需要输入 run 命令启动这个程序(run 命令被简写成 r):

```
(gdb) r
Starting program: /root/redis-6.0.3/src/redis-server
[Thread debugging using libthread_db enabled]
Using host libthread_db library "/usr/lib64/libthread_db.so.1".
31306:C 27 May 2020 12:47:56.624 # oO0OoO0OoO0Oo Redis is starting oO0OoO0OoO0Oo
31306:C 27 May 2020 12:47:56.624 # Redis version=6.0.3, bits=64, commit=00000000, modified=0, pid=31306, just started
31306:C 27 May 2020 12:47:56.624 # Warning: no config file specified, using the default config. In order to specify a config file use /root/redis-6.0.3/src/redis-server /path/to/redis.conf
```

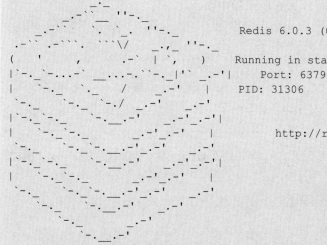

```
31306:M 27 May 2020 12:47:56.626 # WARNING: The TCP backlog setting of 511 cannot be enforced because /proc/sys/net/core/somaxconn is set to the lower value of 128.
31306:M 27 May 2020 12:47:56.626 # Server initialized
31306:M 27 May 2020 12:47:56.626 # WARNING overcommit_memory is set to 0! Background save may fail under low memory condition. To fix this issue add 'vm.overcommit_memory = 1' to /etc/sysctl.conf and then reboot or run the command 'sysctl vm.overcommit_memory=1' for this to take effect.
31306:M 27 May 2020 12:47:56.626 # WARNING you have Transparent Huge Pages (THP) support enabled in your kernel. This will create latency and memory usage issues with Redis. To fix this issue run the command 'echo never >
```

```
/sys/kernel/mm/transparent_hugepage/enabled' as root, and add it to your
/etc/rc.local in order to retain the setting after a reboot. Redis must be restarted
after THP is disabled.
[New Thread 0x7ffff0bb9700 (LWP 31310)]
[New Thread 0x7ffff03b8700 (LWP 31311)]
[New Thread 0x7fffefbb7700 (LWP 31312)]
31306:M 27 May 2020 12:47:56.626 * Ready to accept connections
```

这就是 redis-server 的启动界面。如果程序已经启动，则再次输入 run 命令就会重启程序。我们在 gdb 界面按 Ctrl+C 组合键（界面中的^C）让程序中断，再次输入 r 命令，gdb 会提示我们是否重启程序，输入 y 确认重启：

```
^C
Thread 1 "redis-server" received signal SIGINT, Interrupt.
0x00007ffff71e2603 in epoll_wait () from /usr/lib64/libc.so.6
(gdb) r
The program being debugged has been started already.
Start it from the beginning? (y or n) y
Starting program: /root/redis-6.0.3/src/redis-server
...省略重复的输出...
```

2.5.4　continue 命令

在程序触发断点或者使用 Ctrl+C 组合键中断后，如果我们想让程序继续运行，则只需输入 continue 命令即可（简写成 c）。当然，如果通过 continue 命令让程序在继续运行的过程中触发设置的程序断点，则程序会在断点处中断：

```
^C
Thread 1 "redis-server" received signal SIGINT, Interrupt.
0x00007ffff71e2603 in epoll_wait () from /usr/lib64/libc.so.6
(gdb) c
Continuing.
```

2.5.5　break 命令

break 命令即我们添加断点的命令，可以将其简写成 b。我们可以使用以下方式添加断点：

```
## 在 functionname 函数的入口处添加一个断点
break functionname
## 在当前文件行号为 LineNo 的地方添加一个断点
break LineNo
## 在 filename 文件行号为 LineNo 的地方添加一个断点
break filename:LineNo
```

在以上程序中提到的三种方式是添加断点的常用方式。举个例子，对于一般的 Linux 程序来说，main 函数是程序入口函数，redis-server 也不例外，如果我们知道了函数的名称，就可以直接利用函数的名称添加一个断点。这里以在 main 函数处设置断点为例，执行如下命令：

```
(gdb) b main
Breakpoint 1 at 0x436abd: file server.c, line 5001.
```

添加好后，使用 run 命令重启程序，就可以触发这个断点了，gdb 会停在断点处：

```
(gdb) r
The program being debugged has been started already.
Start it from the beginning? (y or n) y
Starting program: /root/redis-6.0.3/src/redis-server
[Thread debugging using libthread_db enabled]
Using host libthread_db library "/usr/lib64/libthread_db.so.1".

Breakpoint 1, main (argc=1, argv=0x7fffffffe308) at server.c:5001
5001        spt_init(argc, argv);
(gdb)
```

redis-server 默认的端口号是 6379。我们知道，无论上层如何封装，这个端口号最终肯定是通过操作系统的 socket API bind 函数绑定上去的，我们通过文件搜索，可以找到调用这个函数的文件，它位于 anet.c 文件第 455 行，如下图所示。

```
C anet.c            ×
src > C anet.c > ⊕ anetListen(char *, int, sockaddr *, socklen_t, int)
454     static int anetListen(char *err, int s, struct sockaddr *sa, socklen_t len, int backlog) {
455         if (bind(s,sa,len) == -1) {
456             anetSetError(err, "bind: %s", strerror(errno));
457             close(s);
458             return ANET_ERR;
459         }
460
461         if (listen(s, backlog) == -1) {
462             anetSetError(err, "listen: %s", strerror(errno));
463             close(s);
464             return ANET_ERR;
465         }
466         return ANET_OK;
467     }
468
```

使用 break 命令在这个地方添加一个断点：

```
(gdb) b anet.c:455
Breakpoint 2 at 0x42aab2: file anet.c, line 455.
```

由于程序绑定的端口号是在 redis-server 启动时初始化的，所以为了能触发这个断点，我们再次使用 run 命令重启这个程序，程序第 1 次会触发 main 函数处的断点，输入 continue 命令继续运行，接着触发 anet.c 文件第 455 行的断点：

```
(gdb) r
The program being debugged has been started already.
Start it from the beginning? (y or n) y
Starting program: /root/redis-6.0.3/src/redis-server
[Thread debugging using libthread_db enabled]
Using host libthread_db library "/usr/lib64/libthread_db.so.1".

Breakpoint 1, main (argc=1, argv=0x7fffffffe308) at server.c:5001
5001        spt_init(argc, argv);
(gdb) c
Continuing.
32219:C 27 May 2020 13:06:17.326 # oO0OoO0OoO0Oo Redis is starting oO0OoO0OoO0Oo
32219:C 27 May 2020 13:06:17.326 # Redis version=6.0.3, bits=64, commit=00000000,
```

```
modified=0, pid=32219, just started
32219:C 27 May 2020 13:06:17.326 # Warning: no config file specified, using the default
config. In order to specify a config file use /root/redis-6.0.3/src/redis-server
/path/to/redis.conf

Breakpoint 2, anetListen (err=0x568c48 <server+680> "", s=6, sa=0x6173f0, len=28,
backlog=511) at anet.c:455
455         if (bind(s,sa,len) == -1) {
(gdb)
```

现在断点停在 anet.c 文件第 455 行，我们可以直接使用"break + 行号"添加断点，例如可以在第 458、464、466 行分别添加一个断点，看看这个函数执行完毕后走哪个 return 语句退出。通过"b 行号"形式添加三个断点，操作如下：

```
454     static int anetListen(char *err, int s, struct sockaddr *sa, socklen_t len,
int backlog) {
455         if (bind(s,sa,len) == -1) {
456             anetSetError(err, "bind: %s", strerror(errno));
457             close(s);
458             return ANET_ERR;
459         }
(gdb) l
460
461         if (listen(s, backlog) == -1) {
462             anetSetError(err, "listen: %s", strerror(errno));
463             close(s);
464             return ANET_ERR;
465         }
466         return ANET_OK;
467     }
468
469     static int anetV6Only(char *err, int s) {
(gdb) b 458
Breakpoint 3 at 0x42aafc: file anet.c, line 458.
(gdb) b 464
Breakpoint 4 at 0x42ab48: file anet.c, line 464.
(gdb) b 466
Breakpoint 5 at 0x42ab4f: file anet.c, line 466.
(gdb)
```

添加三个断点后，通过 continue 命令继续运行程序，程序运行到第 466 行中断，说明该函数执行了第 466 行的 return 语句：

```
(gdb) c
Continuing.

Breakpoint 5, anetListen (err=0x568c48 <server+680> "", s=6, sa=0x6173f0, len=28,
backlog=511) at anet.c:466
466         return ANET_OK;
(gdb)
```

至此，先调用 bind 函数再调用 listen 函数，会发现 redis-server 已绑定端口并成功开启了监听。我们可以再打开一个 Shell 窗口验证一下，结果证实 6379 端口已经处于监听状态：

```
[root@myaliyun redis-6.0.3]# lsof -i -Pn | grep redis
redis-ser 32219    root      6u  IPv6 36564164       0t0  TCP *:6379 (LISTEN)
```

2.5.6 tbreak 命令

break 命令用于添加一个永久断点；tbreak 命令用于添加一个临时断点，其第 1 个字母"t"表示"temporarily（临时的）"，也就是说，通过这个命令添加的断点是临时的。临时断点，指的是一旦该断点触发了一次，就会被自动删除。添加断点的方法与上面介绍的 break 命令一样，这里不再赘述：

```
(gdb) tbreak main
Temporary breakpoint 1 at 0x436abd: file server.c, line 5001.
(gdb) r
Starting program: /root/redis-6.0.3/src/redis-server
[Thread debugging using libthread_db enabled]
Using host libthread_db library "/usr/lib64/libthread_db.so.1".

Temporary breakpoint 1, main (argc=1, argv=0x7fffffffe308) at server.c:5001
5001        spt_init(argc, argv);
(gdb) r
The program being debugged has been started already.
Start it from the beginning? (y or n) y
Starting program: /root/redis-6.0.3/src/redis-server
//省略 redis-server 启动成功的部分输出
24068:M 27 May 2020 21:19:31.141 # Server initialized
```

我们在以上代码中使用 tbreak 命令在 main 函数处添加了一个临时断点，第 1 次启动程序并触发断点后，再次重新运行程序，不再触发断点，因为这个临时断点已被删除，此时 redis-server 直接启动成功。

2.5.7 backtrace 与 frame 命令

backtrace 可简写成 bt，用于查看当前所在线程的调用堆栈。现在我们的 redis-server 中断在 anet.c 文件第 466 行，可以通过 backtrace 命令查看当前的调用堆栈。

```
(gdb) bt
#0  anetListen (err=0x568c48 <server+680> "", s=6, sa=0x617410, len=28, backlog=511) at anet.c:466
#1  0x000000000042adcc in _anetTcpServer (err=0x568c48 <server+680> "", port=6379, bindaddr=0x0,
    af=10, backlog=511) at anet.c:501
#2  0x000000000042aea7 in anetTcp6Server (err=0x568c48 <server+680> "", port=6379, bindaddr=0x0,
    backlog=511) at anet.c:524
#3  0x0000000000430c74 in listenToPort (port=6379, fds=0x568b14 <server+372>,
    count=0x568b54 <server+436>) at server.c:2648
#4  0x00000000004313a9 in initServer () at server.c:2792
#5  0x00000000004371e4 in main (argc=1, argv=0x7fffffffe318) at server.c:5128
(gdb)
```

这里一共有 6 层堆栈，堆栈编号为#0～#5，顶层是 main 函数，底层是断点所在的 anetListen 函数。如果我们想切换到其他堆栈处，则可以使用 frame 命令，frame 命令的使用方法如下：

```
frame 堆栈编号（编号不用加#）
```

frame 命令可被简写成 f。这里依次切换至堆栈#1、#2、#3、#4、#5，然后切换回#0，操作如下图所示。

```
(gdb) bt
#0  anetListen (err=0x568c48 <server+680> "", s=6, sa=0x617410, len=28, backlog=511) at anet.c:466
#1  0x000000000042adcc in _anetTcpServer (err=0x568c48 <server+680> "", port=6379, bindaddr=0x0,
    af=10, backlog=511) at anet.c:501
#2  0x00000000042aea7 in anetTcp6Server (err=0x568c48 <server+680> "", port=6379, bindaddr=0x0,
    backlog=511) at anet.c:524
#3  0x0000000000430c74 in listenToPort (port=6379, fds=0x568b14 <server+372>,
    count=0x568b54 <server+436>) at server.c:2648
#4  0x00000000004313a9 in initServer () at server.c:2792
#5  0x00000000004371e4 in main (argc=1, argv=0x7fffffffe318) at server.c:5128
(gdb) f 1
#1  0x000000000042adcc in _anetTcpServer (err=0x568c48 <server+680> "", port=6379, bindaddr=0x0,
    af=10, backlog=511) at anet.c:501
501                 if (anetListen(err,s,p->ai_addr,p->ai_addrlen,backlog) == ANET_ERR) s = ANET_ERR;
(gdb) f 2
#2  0x00000000042aea7 in anetTcp6Server (err=0x568c48 <server+680> "", port=6379, bindaddr=0x0,
    backlog=511) at anet.c:524
524             return _anetTcpServer(err, port, bindaddr, AF_INET6, backlog);
(gdb) f 3
#3  0x0000000000430c74 in listenToPort (port=6379, fds=0x568b14 <server+372>,
    count=0x568b54 <server+436>) at server.c:2648
2648                fds[*count] = anetTcp6Server(server.neterr,port,NULL,
(gdb) f 4
#4  0x00000000004313a9 in initServer () at server.c:2792
2792            listenToPort(server.port,server.ipfd,&server.ipfd_count) == C_ERR)
(gdb) f 5
#5  0x00000000004371e4 in main (argc=1, argv=0x7fffffffe318) at server.c:5128
5128            initServer();
(gdb) f 0
#0  anetListen (err=0x568c48 <server+680> "", s=6, sa=0x617410, len=28, backlog=511) at anet.c:466
466             return ANET_OK;
(gdb)
```

通过对上面的各个堆栈进行查看，我们可以得出这里的调用层级关系，即 main 函数在第 5128 行调用了 initServer 函数，initServer 函数在第 2792 行调用了 listenToPort 函数，listenToPort 函数在第 2648 行调用了 anetTcp6Server 函数，anetTcp6Server 函数在第 524 行调用了 _anetTcpServer 函数，_anetTcpServer 函数在第 501 行调用了 anetListen 函数，当前断点正好位于 anetListen 函数中。

2.5.8　info break、enable、disable、delete 命令

在程序中加了很多断点以后，若想查看加了哪些断点，则可以使用 info break 命令（简写成 info b），如下图所示。

```
(gdb) info b
Num     Type           Disp Enb Address            What
1       breakpoint     keep y   0x0000000000436b77 in main at server.c:5001
2       breakpoint     keep y   0x000000000042ab5c in anetListen at anet.c:455
3       breakpoint     keep y   0x000000000042aba6 in anetListen at anet.c:458
4       breakpoint     keep y   0x000000000042abf2 in anetListen at anet.c:464
5       breakpoint     keep y   0x000000000042abf9 in anetListen at anet.c:466
(gdb)
```

通过上图，我们可以得到如下信息：目前一共添加了 5 个断点，断点 1、2、5 已经触发一次，其他断点未触发；每个断点的位置（所在的文件和行号）、内存地址、断点启用和禁用状态信息也一目了然地展示出来。

如果我们想禁用某个断点，则使用 disable 断点编号就可以禁用这个断点了，被禁用的断点不会再被触发；被禁用的断点可以使用 enable 断点编号并重新开启：

```
(gdb) disable 1
(gdb) info b
Num     Type           Disp Enb Address            What
1       breakpoint     keep n   0x0000000000436b77 in main at server.c:5001
2       breakpoint     keep y   0x000000000042ab5c in anetListen at anet.c:455
```

```
         breakpoint already hit 1 time
3        breakpoint     keep y  0x000000000042aba6 in anetListen at anet.c:458
4        breakpoint     keep y  0x000000000042abf2 in anetListen at anet.c:464
5        breakpoint     keep y  0x000000000042abf9 in anetListen at anet.c:466
```

使用 disable 1 后，第 1 个断点 Enb 一栏的值由 y 变成 n，断点 1 不会再触发，即程序不会在 main 函数处中断，程序一直到断点 2 处才会停下来：

```
(gdb) r
The program being debugged has been started already.
Start it from the beginning? (y or n) y
Starting program: /root/redis-6.0.3/src/redis-server
[Thread debugging using libthread_db enabled]
Using host libthread_db library "/usr/lib64/libthread_db.so.1".
6572:C 25 Feb 2021 11:44:16.727 # oO0OoO0OoO0Oo Redis is starting oO0OoO0OoO0Oo
6572:C 25 Feb 2021 11:44:16.727 # Redis version=6.0.3, bits=64, commit=00000000,
modified=0, pid=6572, just started
6572:C 25 Feb 2021 11:44:16.727 # Warning: no config file specified, using the default
config. In order to specify a config file use /root/redis-6.0.3/src/redis-server
/path/to/redis.conf

Breakpoint 2, anetListen (err=0x568c48 <server+680> "", s=6, sa=0x6173f0, len=28,
backlog=511)
    at anet.c:455
455         if (bind(s,sa,len) == -1) {
(gdb)
```

如果 disable 和 enable 命令不加断点编号，则分别表示禁用和启用所有断点：

```
(gdb) disable
(gdb) info b
Num      Type           Disp Enb Address            What
1        breakpoint     keep n  0x0000000000436b77 in main at server.c:5001
2        breakpoint     keep n  0x000000000042ab5c in anetListen at anet.c:455
         breakpoint already hit 1 time
3        breakpoint     keep n  0x000000000042aba6 in anetListen at anet.c:458
4        breakpoint     keep n  0x000000000042abf2 in anetListen at anet.c:464
5        breakpoint     keep n  0x000000000042abf9 in anetListen at anet.c:466
(gdb) enable
(gdb) info b
Num      Type           Disp Enb Address            What
1        breakpoint     keep y  0x0000000000436b77 in main at server.c:5001
2        breakpoint     keep y  0x000000000042ab5c in anetListen at anet.c:455
         breakpoint already hit 1 time
3        breakpoint     keep y  0x000000000042aba6 in anetListen at anet.c:458
4        breakpoint     keep y  0x000000000042abf2 in anetListen at anet.c:464
5        breakpoint     keep y  0x000000000042abf9 in anetListen at anet.c:466
(gdb)
```

使用 delete 编号可以删除某个断点，例如 delete 2 3 表示要删除断点 2 和断点 3：

```
(gdb) info b
Num      Type           Disp Enb Address            What
1        breakpoint     keep y  0x0000000000436b77 in main at server.c:5001
2        breakpoint     keep y  0x000000000042ab5c in anetListen at anet.c:455
         breakpoint already hit 1 time
3        breakpoint     keep y  0x000000000042aba6 in anetListen at anet.c:458
```

```
4       breakpoint     keep y   0x000000000042abf2 in anetListen at anet.c:464
5       breakpoint     keep y   0x000000000042abf9 in anetListen at anet.c:466
(gdb) delete 2 3
1       breakpoint     keep y   0x0000000000436b77 in main at server.c:5001
4       breakpoint     keep y   0x000000000042abf2 in anetListen at anet.c:464
5       breakpoint     keep y   0x000000000042abf9 in anetListen at anet.c:466
```

同样，如果输入 delete 时不加命令号，则表示删除所有断点。

2.5.9 list 命令

list 命令和后面介绍的 print 命令是 gdb 调试中使用频率最高的命令。list 命令用于查看当前断点附近的代码，可以简写成 l。我们使用 frame 命令切换到上文的堆栈#4 处，然后输入 list 命令查看下效果。

```
(gdb) bt
#0  anetListen (err=0x568c48 <server+680> "", s=6, sa=0x617400, len=28, backlog=511) at anet.c:466
#1  0x000000000042adcc in _anetTcpServer (err=0x568c48 <server+680> "", port=6379, bindaddr=0x0,
    af=10, backlog=511) at anet.c:501
#2  0x000000000042aea7 in anetTcp6Server (err=0x568c48 <server+680> "", port=6379, bindaddr=0x0,
    backlog=511) at anet.c:524
#3  0x0000000000430c74 in listenToPort (port=6379, fds=0x568b14 <server+372>,
    count=0x568b54 <server+436>) at server.c:2648
#4  0x00000000004313a9 in initServer () at server.c:2792
#5  0x00000000004371e4 in main (argc=1, argv=0x7fffffffe308) at server.c:5128
(gdb) f 4
#4  0x00000000004313a9 in initServer () at server.c:2792
2792            listenToPort(server.port,server.ipfd,&server.ipfd_count) == C_ERR)
(gdb) l
2787        }
2788        server.db = zmalloc(sizeof(redisDb)*server.dbnum);
2789
2790        /* Open the TCP listening socket for the user commands. */
2791        if (server.port != 0 &&
2792            listenToPort(server.port,server.ipfd,&server.ipfd_count) == C_ERR)
2793            exit(1);
2794        if (server.tls_port != 0 &&
2795            listenToPort(server.tls_port,server.tlsfd,&server.tlsfd_count) == C_ERR)
2796            exit(1);
(gdb)
```

可以发现断点"停在"第 2792 行，输入 list 命令以后，会显示第 2792 行前后的 10 行代码（第 2787~2796 行）。

再次输入 list 命令试一下：

```
(gdb) l
2797
2798        /* Open the listening Unix domain socket. */
2799        if (server.unixsocket != NULL) {
2800            unlink(server.unixsocket); /* don't care if this fails */
2801            server.sofd = anetUnixServer(server.neterr,server.unixsocket,
2802                server.unixsocketperm, server.tcp_backlog);
2803            if (server.sofd == ANET_ERR) {
2804                serverLog(LL_WARNING, "Opening Unix socket: %s", server.neterr);
2805                exit(1);
2806            }
(gdb) l
2807            anetNonBlock(NULL,server.sofd);
2808        }
2809
2810        /* Abort if there are no listening sockets at all. */
```

```
2811            if (server.ipfd_count == 0 && server.tlsfd_count == 0 && server.sofd <
0) {
2812                serverLog(LL_WARNING, "Configured to not listen anywhere, exiting.");
2813                exit(1);
2814            }
2815
2816        /* Create the Redis databases, and initialize other internal state. */
(gdb)
```

代码继续往后显示 10 行（第 2797～2806 行），也就是说，第 1 次输入 list 命令时会显示断点前后的 10 行代码，继续输入 list 命令时每次都会接着向后显示 10 行代码，一直到文件结束。

list +命令（即 list 加号）可以从当前代码位置向下显示 10 行代码（向文件末尾方向），这和连续输入多条 list 命令的效果是一样的。list -命令（即 list 减号）可以从当前代码位置向上显示 10 行代码（往文件开始方向）。操作效果如下：

```
(gdb) l
2797
2798        /* Open the listening Unix domain socket. */
2799            if (server.unixsocket != NULL) {
2800                unlink(server.unixsocket); /* don't care if this fails */
2801                server.sofd = anetUnixServer(server.neterr,server.unixsocket,
2802                    server.unixsocketperm, server.tcp_backlog);
2803                if (server.sofd == ANET_ERR) {
2804                    serverLog(LL_WARNING, "Opening Unix socket: %s", server.neterr);
2805                    exit(1);
2806                }
(gdb) list +
2807                anetNonBlock(NULL,server.sofd);
2808            }
2809
2810        /* Abort if there are no listening sockets at all. */
2811            if (server.ipfd_count == 0 && server.tlsfd_count == 0 && server.sofd <
0) {
2812                serverLog(LL_WARNING, "Configured to not listen anywhere, exiting.");
2813                exit(1);
2814            }
2815
2816        /* Create the Redis databases, and initialize other internal state. */
(gdb) list -
2797
2798        /* Open the listening Unix domain socket. */
2799            if (server.unixsocket != NULL) {
2800                unlink(server.unixsocket); /* don't care if this fails */
2801                server.sofd = anetUnixServer(server.neterr,server.unixsocket,
2802                    server.unixsocketperm, server.tcp_backlog);
2803                if (server.sofd == ANET_ERR) {
2804                    serverLog(LL_WARNING, "Opening Unix socket: %s", server.neterr);
2805                    exit(1);
2806                }
(gdb) list -
2787            }
2788            server.db = zmalloc(sizeof(redisDb)*server.dbnum);
```

```
2789
2790            /* Open the TCP listening socket for the user commands. */
2791            if (server.port != 0 &&
2792                listenToPort(server.port,server.ipfd,&server.ipfd_count) == C_ERR)
2793                exit(1);
2794            if (server.tls_port != 0 &&
2795                listenToPort(server.tls_port,server.tlsfd,&server.tlsfd_count) == C_ERR)
2796                exit(1);
(gdb)
```

list 默认显示的行数量可以通过修改 gdb 的相关配置来实现，由于我们一般不会修改这个配置值，因此这里就不介绍了。

list 不仅可以显示当前断点处的代码，也可以显示其他文件某一行的代码，读者可以在 gdb 中输入 help list 命令来查看更多的用法：

```
(gdb) help list
List specified function or line.
With no argument, lists ten more lines after or around previous listing.
"list -" lists the ten lines before a previous ten-line listing.
One argument specifies a line, and ten lines are listed around that line.
Two arguments with comma between specify starting and ending lines to list.
Lines can be specified in these ways:
  LINENUM, to list around that line in current file,
  FILE:LINENUM, to list around that line in that file,
  FUNCTION, to list around beginning of that function,
  FILE:FUNCTION, to distinguish among like-named static functions.
  *ADDRESS, to list around the line containing that address.
With two args if one is empty it stands for ten lines away from the other arg.
```

在上面的帮助信息中介绍了可以使用 list FILE:LINENUM 显示某个文件某一行处的代码。我们使用 gdb 的目的是调试，所以我们更关心的是断点附近的代码，而不是通过 gdb 阅读代码。对于阅读代码，gdb 并不是一个好工具。比如，笔者用 gdb 调试 Redis，用 VSCode 或者 Visual Studio 阅读代码，界面如下图所示。

在 VSCode 中阅读 Redis 源码

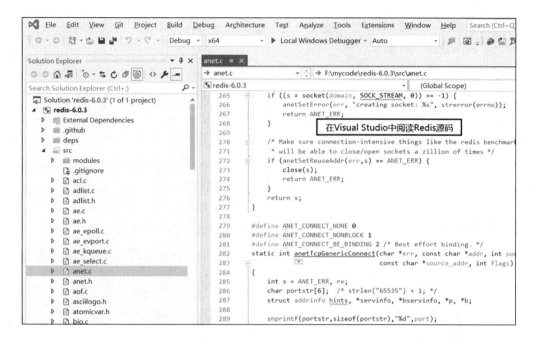

在Visual Studio中阅读Redis源码

2.5.10　print 与 ptype 命令

print 命令可以被简写成 p。通过 print 命令可以在调试过程中方便地查看变量的值，也可以修改当前内存中的变量值。我们切换到堆栈#4，打印以下三个变量：

```
(gdb) bt
#0  anetListen (err=0x568c48 <server+680> "", s=6, sa=0x6173e0, len=28, backlog=511)
at anet.c:455
#1  0x000000000042ad26 in _anetTcpServer (err=0x568c48 <server+680> "", port=6379,
bindaddr=0x0, af=10, backlog=511) at anet.c:501
#2  0x000000000042adf8 in anetTcp6Server (err=0x568c48 <server+680> "", port=6379,
bindaddr=0x0, backlog=511) at anet.c:524
#3  0x0000000000430bba in listenToPort (port=6379, fds=0x568b14 <server+372>,
count=0x568b54 <server+436>) at server.c:2648
#4  0x00000000004312ef in initServer () at server.c:2792
#5  0x000000000043712a in main (argc=1, argv=0x7fffffffe308) at server.c:5128
(gdb) f 4
#4  0x00000000004312ef in initServer () at server.c:2792
2792            listenToPort(server.port,server.ipfd,&server.ipfd_count) == C_ERR)
(gdb) p server.port
$1 = 6379
(gdb) p server.ipfd
$2 = {0 <repeats 16 times>}
(gdb) p server.ipfd_count
$3 = 0
(gdb)
```

这里使用 print 命令分别打印出 server.port、server.ipfd、server.ipfd_count 的值。其中，server.ipfd 显示 {0 <repeats 16 times>}，这是 gdb 显示字符串和字符数组的特有方式，当一个字符串变量、字符数组或者连续的内存值重复若干次时，gdb 就会以这种模式显示，以节约显示空间。

通过 print 命令不仅可以输出变量的值，也可以输出特定表达式的计算结果，甚至可以输出一些函数的执行结果。

举个例子，我们可以输入 p &server.port 来输出 server.port 的地址。对于 C++对象，我们可以通过 p this 显示当前对象的地址，也可以通过 p *this 列出当前对象的各个成员变量的值，如果有三个变量可以相加（假设变量名分别为 a、b、c），则可以使用 p a+b+c 来打印这三个变量的值。

假设 func 是一个可以执行的函数，则通过 p func()命令就可以输出该变量的执行结果。以一种常见的情形为例，在某个时刻，某个系统函数执行失败了，通过系统变量 errno 得到一个错误码，我们可以使用 p strerror(errno)将这个错误码对应的文字信息打印出来，这样就不用费劲地在 man 手册上查找这个错误码对应的错误含义了。

通过 print 命令不仅可以输出表达式的结果，还可以修改变量的值，我们尝试将上文中的端口号从 6379 改成 6400 试试：

```
(gdb) p server.port=6400
$4 = 6400
(gdb) p server.port
$5 = 6400
(gdb)
```

当然，一个变量的值在修改后能否起作用，要看这个变量的具体位置和作用。举个例子，对于表达式 int a = b / c，如果将 c 修改成 0，程序就会产生除零异常。又如，对于如下代码：

```
int j = 100;
for (int i = 0; i < j; ++i) {
    printf("i = %d\n", i);
}
```

如果在循环过程中通过 print 命令将 j 的大小由 100 改成 1000，那么这个循环将输出 i 的值 1000 次。

print 输出变量值时可以指定输出格式，命令使用格式如下：

```
print /format variable
```

format 常见的取值如下。

◎ o octal，八进制显示。

◎ x hex，十六进制显示。

◎ d decimal，十进制显示。

◎ u unsigned decimal，无符号十进制显示。

◎ t binary，二进制显示。

◎ f float，浮点值显示。

◎ a address，内存地址格式显示（与十六进制相似）。

◎ i instruction，指令格式显示。

◎ s string，字符串格式显示。

◎ z hex, zero padded on the left，十六进制左侧补 0 显示。

对于完整的格式和用法，可以在 gdb 中输入 help x 查看，演示如下：

```
(gdb) p /x server.port
$6 = 0x1900
(gdb) p /s server.port
$7 = 6400
(gdb) p /o server.port
$8 = 014400
(gdb) p /i server.port
Format letter "i" is meaningless in "print" command.
(gdb) p /t server.port
$9 = 1100100000000
(gdb) p /f server.port
$10 = 8.96831017e-42
(gdb) p /t server.port
$11 = 1100100000000
(gdb) p /a server.port
$12 = 0x1900
(gdb)
```

总结起来，通过 print 命令，我们不仅可以查看程序运行过程中各个变量的状态值，也可以通过临时修改变量的值来控制程序的行为。

gdb 还有另一个命令 ptype，顾名思义，其含义是 "print type"，就是输出一个变量的类型。例如试着输出 redis 堆栈#4 处的变量 server 和变量 server.port 的类型：

```
(gdb) ptype server
type = struct redisServer {
    pid_t pid;
    char *configfile;
    char *executable;
    char **exec_argv;
    int dynamic_hz;
    int config_hz;
    ...省略部分字段...
(gdb) ptype server.port
type = int
```

可以看到，对于一个复合数据类型的变量，ptype 不仅列出了这个变量的类型（这里是一个名为 redisServer 的结构体），而且详细列出了每个成员变量的字段名，有了这个功能，我们在调试时就不必去代码文件中翻看某个变量的类型定义了。

2.5.11 info 与 thread 命令

info 命令是一个复合指令，可以用来查看当前进程所有线程的运行情况。这里还是以 redis-server 进程为例进行演示，使用 delete 命令删掉所有断点，然后使用 run 命令重启 redis-server，等程序正常启动后，我们按 Ctrl+C 组合键（代码中的^C）将程序中断，然后使用 info threads 查看进程当前的所有线程信息和这些线程分别中断在何处。

```
9022:M 25 Feb 2021 12:30:13.585 * Loading RDB produced by version 6.0.3
9022:M 25 Feb 2021 12:30:13.585 * RDB age 15 seconds
9022:M 25 Feb 2021 12:30:13.585 * RDB memory usage when created 0.84 Mb
9022:M 25 Feb 2021 12:30:13.585 * DB loaded from disk: 0.000 seconds
9022:M 25 Feb 2021 12:30:13.585 * Ready to accept connections
^C
Thread 1 "redis-server" received signal SIGINT, Interrupt.
0x00007ffff71e2603 in epoll_wait () from /usr/lib64/libc.so.6
(gdb) info threads
  Id   Target Id                          Frame
* 1    Thread 0x7ffff7feb740 (LWP 9022) "redis-server"  0x00007ffff71e2603 in epoll_wait ()
  from /usr/lib64/libc.so.6
  2    Thread 0x7ffff0bb9700 (LWP 9026) "bio_close_file" 0x00007ffff74bc965 in pthread_cond_wait@@GLIBC_2.3.2 () from /usr/li
b64/libpthread.so.0
  3    Thread 0x7ffff03b8700 (LWP 9027) "bio_aof_fsync" 0x00007ffff74bc965 in pthread_cond_wait@@GLIBC_2.3.2 () from /usr/li
b64/libpthread.so.0
  4    Thread 0x7fffefbb7700 (LWP 9028) "bio_lazy_free" 0x00007ffff74bc965 in pthread_cond_wait@@GLIBC_2.3.2 () from /usr/li
b64/libpthread.so.0
(gdb)
```

通过 info threads 的输出，我们知道 redis-server 正常启动后一共产生了 4 个线程，其中有 1 个主线程和 3 个工作线程，线程编号（Id 那一列）分别是 1、2、3、4。3 个工作线程（2、3、4）分别阻塞在 Linux API pthread_cond_wait 处，而主线程 1 阻塞在 epoll_wait 处。注意，第 1 栏的名称虽然叫作 Id，但第 1 栏的数值并不是线程的 Id；第 3 栏有个括号，内容为 LWP 5029，这个 5029 才是当前线程真正的线程 Id。那么 LWP 是什么意思呢？在 Linux 系统早期的内核里其实不存在真正的线程实现，当时所有的线程都是用进程实现的，这些模拟线程的进程被称为 Light Weight Process（轻量级进程），之后版本的 Linux 系统内核有了真正的线程实现，但该名称仍被保留了下来。

读者可能有个疑问：怎么知道线程 1 是主线程，线程 2、3、4 是工作线程呢？是不是因为线程 1 前面有个星号（*）？错，线程编号前面的星号表示 gdb 当前作用于哪个线程，不是说标了星号就是主线程。当前有 4 个线程，也就有 4 个调用堆栈，如果此时输入 backtrace 命令查看调用堆栈，则由于 gdb 当前作用于线程 1，所以通过 backtrace 命令显示的是线程 1 的调用堆栈：

```
(gdb) bt
#0  0x00007ffff71e2603 in epoll_wait () from /usr/lib64/libc.so.6
#1  0x0000000000428a9e in aeApiPoll (eventLoop=0x5e5770, tvp=0x7fffffffe140) at
ae_epoll.c:112
#2  0x00000000004297e2 in aeProcessEvents (eventLoop=0x5e5770, flags=27) at
ae.c:447
#3  0x0000000000429ab6 in aeMain (eventLoop=0x5e5770) at ae.c:539
#4  0x00000000004372bb in main (argc=1, argv=0x7fffffffe308) at server.c:5175
(gdb)
```

看到了吧！堆栈#4 的 main 函数也证明了线程编号为 1 的线程是主线程。

那么如何切换到其他线程呢？我们可以通过 thread 线程编号命令切换到指定的线程。例如我们想切换到线程 2，则只需输入 thread 2 命令即可，接着输入 bt 命令就能查看这个线程的调用堆栈了：

```
(gdb) thread 2
[Switching to thread 2 (Thread 0x7ffff0bb9700 (LWP 5030))]
#0  0x00007ffff74bc965 in pthread_cond_wait@@GLIBC_2.3.2 () from
/usr/lib64/libpthread.so.0
(gdb) bt
#0  0x00007ffff74bc965 in pthread_cond_wait@@GLIBC_2.3.2 () from
/usr/lib64/libpthread.so.0
#1  0x00000000004991c0 in bioProcessBackgroundJobs (arg=0x0) at bio.c:190
```

```
#2  0x00007ffff74b8dd5 in start_thread () from /usr/lib64/libpthread.so.0
#3  0x00007ffff71e202d in clone () from /usr/lib64/libc.so.6
(gdb)
```

所以利用 info threads 命令可以调试多线程程序。当然，使用 gdb 调试多线程程序存在一个很麻烦的问题，在之后的章节中会有介绍。

将 gdb 切换到哪个线程，哪个线程前面就被加上星号标记，例如我们把 gdb 当前作用的线程切换到线程 2 上之后，线程 2 前面就被加上了星号。

```
(gdb) thread 2
[Switching to thread 2 (Thread 0x7ffff0bb9700 (LWP 9026))]
#0  0x00007ffff74bc965 in pthread_cond_wait@@GLIBC_2.3.2 () from /usr/lib64/libpthread.so.0
(gdb) info threads
  Id   Target Id                                          Frame
  1    Thread 0x7ffff7feb740 (LWP 9022) "redis-server"    0x00007ffff71e2603 in epoll_wait () from /usr/lib64/
* 2    Thread 0x7ffff0bb9700 (LWP 9026) "bio_close_file"  0x00007ffff74bc965 in pthread_cond_wait@@GLIBC_2.3.2
    from /usr/lib64/libpthread.so.0
  3    Thread 0x7ffff03b8700 (LWP 9027) "bio_aof_fsync"   0x00007ffff74bc965 in pthread_cond_wait@@GLIBC_2.3.2
    from /usr/lib64/libpthread.so.0
  4    Thread 0x7fffefbb7700 (LWP 9028) "bio_lazy_free"   0x00007ffff74bc965 in pthread_cond_wait@@GLIBC_2.3.2
    from /usr/lib64/libpthread.so.0
(gdb)
```

（切换到线程 2）

info 命令还可以用来查看当前函数的参数值，组合命令是 info args。我们找个函数来试一下这个命令：

```
(gdb) info threads
  Id   Target Id                                          Frame
  1    Thread 0x7ffff7feb740 (LWP 5029) "redis-server"    0x00007ffff71e2603 in epoll_wait () from /usr/lib64/libc.so.6
* 2    Thread 0x7ffff0bb9700 (LWP 5030) "bio_close_file"  0x00007ffff74bc965 in pthread_cond_wait@@GLIBC_2.3.2 () from /usr/lib64/libpthread.so.0
  3    Thread 0x7ffff03b8700 (LWP 5031) "bio_aof_fsync"   0x00007ffff74bc965 in pthread_cond_wait@@GLIBC_2.3.2 () from /usr/lib64/libpthread.so.0
  4    Thread 0x7fffefbb7700 (LWP 5032) "bio_lazy_free"   0x00007ffff74bc965 in pthread_cond_wait@@GLIBC_2.3.2 () from /usr/lib64/libpthread.so.0
(gdb) thread 1
[Switching to thread 1 (Thread 0x7ffff7feb740 (LWP 5029))]
#0  0x00007ffff71e2603 in epoll_wait () from /usr/lib64/libc.so.6
(gdb) bt
#0  0x00007ffff71e2603 in epoll_wait () from /usr/lib64/libc.so.6
#1  0x0000000000428a9e in aeApiPoll (eventLoop=0x5e5770, tvp=0x7fffffffe140) at ae_epoll.c:112
#2  0x00000000004297e2 in aeProcessEvents (eventLoop=0x5e5770, flags=27) at ae.c:447
#3  0x0000000000429ab6 in aeMain (eventLoop=0x5e5770) at ae.c:539
#4  0x00000000004372bb in main (argc=1, argv=0x7fffffffe308) at server.c:5175
(gdb) f 2
#2  0x00000000004297e2 in aeProcessEvents (eventLoop=0x5e5770, flags=27) at ae.c:447
447             numevents = aeApiPoll(eventLoop, tvp);
(gdb) info args
eventLoop = 0x5e5770
flags = 27
(gdb)
```

以上代码片段先切回到主线程 1，然后切回到堆栈#2。堆栈#2 调用处的函数是 aeProcessEvents，这个函数一共有两个参数，分别是 eventLoop 和 tvp，使用 info args 命令可以输出这两个参数的值。eventLoop 是一个指针类型的参数，对于指针类型的参数，gdb 默认会输出该变量的指针值。如果想输出该指针指向的对象的值，则可以在变量名前面加上星号，即解引用操作符（*），这里使用 p *eventLoop 即可：

```
(gdb) p *eventLoop
$16 = {maxfd = 7, setsize = 10128, timeEventNextId = 1, lastTime = 1590562419, events
= 0x7ffff7f9b010, fired = 0x5e57d0, timeEventHead = 0x617440, stop = 0, apidata =
0x5f9460, beforesleep = 0x42f866 <beforeSleep>,
 aftersleep = 0x42fa28 <afterSleep>, flags = 0}
(gdb)
```

如果还要查看其成员值，则继续使用变量名->字段名即可（如 p eventLoop->maxfd），我们在前面介绍 print 命令时已经介绍过这种用法，这里不再赘述。

info 命令的功能远非上面介绍的三种，读者可以在 gdb 中输入 help info 查看更多的 info 组合命令的用法。

2.5.12　next、step、until、finish、return、jump 命令

之所以把这几个命令放在一起，是因为它们是用 gdb 调试程序时最常用的几个控制流命令。next 命令可被简写为 n，用于让 gdb 跳到下一行代码。这里跳到下一行代码不是说一定要跳到离代码最近的下一行，而是根据程序逻辑跳转到相应的位置。举个例子：

```
int a = 0;
if (a == 9)
{
   print("a is equal to 9.\n");
}

int b = 10;
print("b = %d.\n", b);
```

如果 gdb 中断在以上代码第 2 行，此时输入 next 命令，则 gdb 将跳到第 7 行，因为这里 if 条件不满足。

在 gdb 命令行界面直接按下回车键，默认是将最近一条命令重新执行一遍。所以，当我们使用 next 命令单步调试时，不必反复输入 n 命令，在输入一次 n 命令之后想再次输入 next 命令时直接按回车键就可以了。

```
Breakpoint 7, main (argc=1, argv=0x7fffffffe308) at server.c:5001
5001        spt_init(argc, argv);
(gdb) n
5003        setlocale(LC_COLLATE,"");
(gdb) n
5004        tzset(); /* Populates 'timezone' global. */
(gdb) n
5005        zmalloc_set_oom_handler(redisOutOfMemoryHandler);
(gdb) n
5006        srand(time(NULL)^getpid());
(gdb) n
5007        gettimeofday(&tv,NULL);
```

```
(gdb) n
5008        crc64_init();
(gdb) n
5011        getRandomBytes(hashseed,sizeof(hashseed));
(gdb) n
5012        dictSetHashFunctionSeed(hashseed);
```

上面的执行过程等价于输入第 1 个 n 后直接回车:

```
Breakpoint 7, main (argc=1, argv=0x7fffffffe308) at server.c:5001
5001        spt_init(argc, argv);
(gdb) n
5003        setlocale(LC_COLLATE,"");
(gdb)
5004        tzset(); /* Populates 'timezone' global. */
(gdb)
5005        zmalloc_set_oom_handler(redisOutOfMemoryHandler);
(gdb)
5006        srand(time(NULL)^getpid());
(gdb)
5007        gettimeofday(&tv,NULL);
(gdb)
5008        crc64_init();
(gdb)
5011        getRandomBytes(hashseed,sizeof(hashseed));
(gdb)
5012        dictSetHashFunctionSeed(hashseed);
```

next 命令的调试术语叫"单步步过（step over）"，即遇到函数调用时不进入函数体内部，而是直接跳过。下面说的 step 命令就是"单步步入（step into）"，顾名思义，就是遇到函数调用时进入函数内部。step 可被简写为 s。举个例子，在 redis-server 的 main 函数中有个 spt_init(argc, argv)函数调用，我们停在这一行时，输入 s 将进入这个函数内部：

```
//为了说明问题本身，这里除去了不相关的干扰，代码有删减
int main(int argc, char **argv) {
    struct timeval tv;
    int j;

    spt_init(argc, argv);

    setlocale(LC_COLLATE,"");
    tzset(); /* Populates 'timezone' global. */
    zmalloc_set_oom_handler(redisOutOfMemoryHandler);
    srand(time(NULL)^getpid());
    gettimeofday(&tv,NULL);
    //省略部分无关代码
}
```

这里演示一下，先用 b main 在 main 函数处加一个断点，然后使用 r 命令重新跑程序，会触发刚才加在 main 函数处的断点，然后使用 n 命令让程序走到 spt_init(argc, argv)函数调用处，再输入 s 命令就可以进入该函数中了：

```
(gdb) b main
Breakpoint 8 at 0x436abd: file server.c, line 5001.
(gdb) r
```

```
The program being debugged has been started already.
Start it from the beginning? (y or n) y
Starting program: /root/redis-6.0.3/src/redis-server
[Thread debugging using libthread_db enabled]
Using host libthread_db library "/usr/lib64/libthread_db.so.1".

Breakpoint 8, main (argc=1, argv=0x7fffffffe308) at server.c:5001
5001        spt_init(argc, argv);
(gdb) s
spt_init (argc=1, argv=0x7fffffffe308) at setproctitle.c:153
153         char **envp = environ;
(gdb) l
148         return 0;
149     } /* spt_copyargs() */
150
151
152     void spt_init(int argc, char *argv[]) {
153         char **envp = environ;
154         char *base, *end, *nul, *tmp;
155         int i, error;
156
157         if (!(base = argv[0]))
(gdb)
```

说到 step 命令，我们还有一个需要注意的地方，就是当函数的参数也是函数调用时，使用 step 命令会依次进入各个函数中，顺序是怎样的呢？举个例子，看看下面这段代码：

```
01  int func1(int a, int b)
02  {
03      int c = a + b;
04      c += 2;
05      return c;
06  }
07
08  int func2(int p, int q)
09  {
10      int t = q * p;
11      return t * t;
12  }
13
14  int func3(int m, int n)
15  {
16      return m + n;
17  }
18
19  int main()
20  {
21      int c;
22      c = func3(func1(1, 2), func2(8, 9));
23      printf("c=%d.\n", c);
24      return 0;
25  }
```

以上代码的程序入口函数是 main 函数。在第 22 行中，func3 使用 func1 和 func2 的返回值作为自己的参数。我们在第 22 行输入 step 命令，会先进入哪个函数中呢？这就需要补充一个知识点了——函数调用方式。

常用的函数调用方式有__cdecl 和__stdcall，C++的非静态成员函数的调用方式是__thiscall，函数参数的传递在本质上是函数参数的入栈。对于这三种函数调用方式，参数的入栈顺序都是从右向左的。在这段代码中，由于没有显式标明函数的调用方式，所以采用的函数调用方式是__cdecl，这是 C/C++中全局函数和类静态方法的默认调用方式。

因此，当我们在第 22 行代码处输入 step 时，先进入的是 func2；当从 func2 返回时再次输入 step 命令，会接着进入 func1；当从 func1 返回时，两个参数已经计算出来了，这时会最终进入 func3。读者只有理解这一点，在遇到这样的代码时，才能根据需要进入自己想要的函数中调试。

在实际调试时，我们在某个函数中调试一会儿后，不希望再一步步地执行到函数返回处，而是希望直接执行完当前函数并回到上一层调用处，这时可以使用 finish 命令。与 finish 命令类似的还有 return 命令。return 命令用于结束执行当前函数，同时指定该函数的返回值。这里需要注意二者的区别：finish 命令用于执行完整的函数体，然后正常返回到上层调用中；return 命令用于立即从函数的当前位置结束并返回到上层调用中，也就是说，如果使用了 return 命令，则在当前函数还有剩余的代码未执行完毕时，也不会再执行了。我们用一个例子来验证一下：

```
01  #include <stdio.h>
02
03  int func()
04  {
05      int a = 9;
06      printf("a=%d.\n", a);
07
08      int b = 8;
09      printf("b=%d.\n", b);
10      return a + b;
11  }
12
13  int main()
14  {
15      int c = func();
16      printf("c=%d.\n", c);
17
18      return 0;
19  }
```

在 main 函数处加一个断点，然后运行程序；在第 15 行使用 step 命令进入 func 函数中；接着使用 next 命令单步运行到代码第 8 行，直接输入 return 命令，这样 func 函数剩余的代码就不会接着执行了，所以 printf("b=%d.\n", b);行没有输出。同时，由于我们没有在 return 命令中指定这个函数的返回值，所以最终在 main 函数中得到的变量 c 的值是一个脏数据。这就验证了上面的结论：return 命令立即从函数当前位置结束并返回到上一层调用中。验证过程如下：

```
(gdb) b main
Breakpoint 1 at 0x40057d: file test.c, line 15.
```

```
(gdb) r
Starting program: /root/testreturn/test

Breakpoint 1, main () at test.c:15
15          int c = func();
Missing separate debuginfos, use: debuginfo-install glibc-2.17-196.el7_4.2.x86_64
(gdb) s
func () at test.c:5
5           int a = 9;
(gdb) n
6           printf("a=%d.\n", a);
(gdb) n
a=9.
8           int b = 8;
(gdb) return
Make func return now? (y or n) y
#0  0x0000000000400587 in main () at test.c:15
15          int c = func();
(gdb) n
16          printf("c=%d.\n", c);
(gdb) n
c=-134250496.
18          return 0;
(gdb)
```

我们再次用 return 命令指定一个值试一下，这样我们得到的变量 c 的值应该就是我们指定的值了。验证过程如下：

```
(gdb) r
The program being debugged has been started already.
Start it from the beginning? (y or n) y
Starting program: /root/testreturn/test

Breakpoint 1, main () at test.c:15
15          int c = func();
(gdb) s
func () at test.c:5
5           int a = 9;
(gdb) n
6           printf("a=%d.\n", a);
(gdb) n
a=9.
8           int b = 8;
(gdb) return 9999
Make func return now? (y or n) y
#0  0x0000000000400587 in main () at test.c:15
15          int c = func();
(gdb) n
16          printf("c=%d.\n", c);
(gdb) n
c=9999.
18          return 0;
(gdb) p c
$1 = 9999
(gdb)
```

仔细观察以上代码，我们用 return 命令修改了函数的返回值，当使用 print 命令打印 c 的值时，c 的值也确实被修改成了 9999。

再次对比使用 finish 命令结束函数执行的结果：

```
(gdb) r
The program being debugged has been started already.
Start it from the beginning? (y or n) y
Starting program: /root/testreturn/test

Breakpoint 1, main () at test.c:15
15          int c = func();
(gdb) s
func () at test.c:5
5           int a = 9;
(gdb) n
6           printf("a=%d.\n", a);
(gdb) n
a=9.
8           int b = 8;
(gdb) finish
Run till exit from #0  func () at test.c:8
b=8.
0x0000000000400587 in main () at test.c:15
15          int c = func();
Value returned is $3 = 17
(gdb) n
16          printf("c=%d.\n", c);
(gdb) n
c=17.
18          return 0;
(gdb)
```

结果和我们预期的一样，finish 正常结束了我们的函数，剩余的代码也会被正常执行。因此 c 的值是 17。

实际调试时，还有一个 until 命令，可被简写为 u，我们使用这个命令让程序运行到指定的行停下来。还是以 redis-server 的代码为例：

```
Breakpoint 1, initServer () at server.c:2742
2742        signal(SIGHUP, SIG_IGN);
(gdb) l
2737    }
2738
2739    void initServer(void) {
2740        int j;
2741
2742        signal(SIGHUP, SIG_IGN);
2743        signal(SIGPIPE, SIG_IGN);
2744        setupSignalHandlers();
2745
2746        if (server.syslog_enabled) {
(gdb) l
2747            openlog(server.syslog_ident, LOG_PID | LOG_NDELAY | LOG_NOWAIT,
2748                server.syslog_facility);
```

```
2749        }
2750
2751        /* Initialization after setting defaults from the config system. */
2752        server.aof_state = server.aof_enabled ? AOF_ON : AOF_OFF;
2753        server.hz = server.config_hz;
2754        server.pid = getpid();
2755        server.current_client = NULL;
2756        server.fixed_time_expire = 0;
(gdb) l
2757        server.clients = listCreate();
2758        server.clients_index = raxNew();
2759        server.clients_to_close = listCreate();
2760        server.slaves = listCreate();
2761        server.monitors = listCreate();
2762        server.clients_pending_write = listCreate();
2763        server.clients_pending_read = listCreate();
2764        server.clients_timeout_table = raxNew();
2765        server.slaveseldb = -1; /* Force to emit the first SELECT command. */
2766        server.unblocked_clients = listCreate();
(gdb) l
2767        server.ready_keys = listCreate();
2768        server.clients_waiting_acks = listCreate();
2769        server.get_ack_from_slaves = 0;
2770        server.clients_paused = 0;
2771        server.events_processed_while_blocked = 0;
2772        server.system_memory_size = zmalloc_get_memory_size();
2773
2774        if (server.tls_port && tlsConfigure(&server.tls_ctx_config) == C_ERR) {
2775            serverLog(LL_WARNING, "Failed to configure TLS. Check logs for more info.");
2776            exit(1);
(gdb)
```

以上是 redis-server 中 initServer 函数的部分代码，位于文件 server.c 中。当 gdb 停在第 2740 行时（注意：这里的行号以在 gdb 调试器中显示的为准，不是源码文件中的行号，由于存在条件编译，部分代码可能不会被编译到可执行文件中，所以实际的调试符号文件中的行号与源码文件中的行号可能不会完全一致），我们可以通过输入 u 2774 命令让 gdb 直接跳到第 2774 行，这样就能快速执行完第 2740～2774 行中间的代码（不包括第 2774 行）。当然，我们可以先在第 2774 行加一个断点，然后使用 continue 命令运行到这一行来达到同样的效果，但是使用 until 命令显然更方便：

```
(gdb) b 2740
Note: breakpoint 1 also set at pc 0x4310d4.
Breakpoint 2 at 0x4310d4: file server.c, line 2742.
(gdb) r
The program being debugged has been started already.
Start it from the beginning? (y or n) y
Starting program: /root/redis-6.0.3/src/redis-server
...省略部分输出...
Breakpoint 1, initServer () at server.c:2742
2742        signal(SIGHUP, SIG_IGN);
(gdb) u 2774
initServer () at server.c:2774
```

```
2774            if (server.tls_port && tlsConfigure(&server.tls_ctx_config) == C_ERR) {
(gdb)
```

jump 命令的基本用法如下：

```
jump <location>
```

location 可以是程序的行号或者函数的地址，jump 会让程序执行流跳转到指定的位置执行，其行为也是不可控的，例如跳过了某个对象的初始化代码，直接执行操作该对象的代码，可能会导致程序崩溃或其他意外操作。jump 命令可被简写为 j，但是不可被简写为 jmp，使用该命令时有一个注意事项：如果 jump 跳转到的位置没有设置断点，那么 gdb 执行完跳转操作后会继续向下执行。举个例子：

```
01  int somefunc()
02  {
03      //代码 A
04      //代码 B
05      //代码 C
06      //代码 D
07      //代码 E
08      //代码 F
09  }
```

假设我们的断点的初始位置在第 3 行（代码 A），那么这时我们使用 jump 6，程序会跳过代码 B 和 C 的执行，执行完代码 D（跳转点），程序并不会停在第 6 行，而是继续执行后续代码，因此如果我们想查看执行跳转处代码的结果，就需要在第 6、7 或 8 行代码处设置断点。

通过 jump 命令除了可以跳过一些代码的执行，还可以执行一些我们想执行的代码，而这些代码在正常逻辑下可能并不会被执行。当然，根据实际的程序逻辑，可能会产生一些非预期结果，这需要我们自行斟酌使用。举个例子，假设现在有如下代码：

```
01  #include <stdio.h>
02  int main()
03  {
04      int a = 0;
05      if (a != 0)
06      {
07          printf("if condition\n");
08      }
09      else
10      {
11          printf("else condition\n");
12      }
13
14      return 0;
15  }
```

我们在第 4、14 行设置一个断点，在触发第 4 行的断点后，在正常情况下程序执行流会走 else 分支，我们可以使用 jump 7 强行让程序执行 if 分支，接着 gdb 会因触发第 14 行的断点停下来，此时我们接着执行 jump 11，程序会将 else 分支中的代码重新执行一遍。整个操作过程如下：

```
[root@localhost testcore]# gdb test
Reading symbols from /root/testcore/test...done.
(gdb) b main
Breakpoint 1 at 0x400545: file main.cpp, line 4.
(gdb) b 14
Breakpoint 2 at 0x400568: file main.cpp, line 14.
(gdb) r
Starting program: /root/testcore/test

Breakpoint 1, main () at main.cpp:4
4       int a = 0;
Missing separate debuginfos, use: debuginfo-install glibc-2.17-260.el7.x86_64
libgcc-4.8.5-36.el7.x86_64 libstdc++-4.8.5-36.el7.x86_64
(gdb) jump 7
Continuing at 0x400552.
if condition

Breakpoint 2, main () at main.cpp:14
14      return 0;
(gdb) jump 11
Continuing at 0x40055e.
else condition

Breakpoint 2, main () at main.cpp:14
14      return 0;
(gdb) c
Continuing.
[Inferior 1 (process 13349) exited normally]
(gdb)
```

　　redis-server 在入口函数 main 处调用了 initServer，我们使用 b initServer、b 2753、b 2755 分别在这个函数入口处、第 2753 行、第 2755 行增加 3 个断点，然后使用 run 命令重新运行程序。触发第 1 个断点后，输入 c 命令继续运行，然后触发第 2753 行的断点，接着输入 jump 2755。以下是操作过程：

```
Breakpoint 3, initServer () at server.c:2742
2742        signal(SIGHUP, SIG_IGN);
(gdb) l
2737    }
2738
2739    void initServer(void) {
2740        int j;
2741
2742        signal(SIGHUP, SIG_IGN);
2743        signal(SIGPIPE, SIG_IGN);
2744        setupSignalHandlers();
2745
2746        if (server.syslog_enabled) {
(gdb) l
2747            openlog(server.syslog_ident, LOG_PID | LOG_NDELAY | LOG_NOWAIT,
2748                server.syslog_facility);
2749        }
2750
2751        /* Initialization after setting defaults from the config system. */
```

```
2752            server.aof_state = server.aof_enabled ? AOF_ON : AOF_OFF;
2753            server.hz = server.config_hz;
2754            server.pid = getpid();
2755            server.current_client = NULL;
2756            server.fixed_time_expire = 0;
(gdb) b 2753
Breakpoint 4 at 0x43112f: file server.c, line 2753.
(gdb) b 2755
Breakpoint 5 at 0x431146: file server.c, line 2755.
(gdb) r
The program being debugged has been started already.
Start it from the beginning? (y or n) y
Starting program: /root/redis-6.0.3/src/redis-server
...省略部分输出...
Breakpoint 3, initServer () at server.c:2742
2742            signal(SIGHUP, SIG_IGN);
(gdb) c
Continuing.

Breakpoint 4, initServer () at server.c:2753
2753            server.hz = server.config_hz;
(gdb) jump 2755
Continuing at 0x431146.

Breakpoint 5, initServer () at server.c:2755
2755            server.current_client = NULL;
(gdb)
```

程序跳过了第 2754 行的代码。第 2754 行的代码用于获取当前的进程 id：

```
2754 server.pid = getpid();
```

由于这一行被跳过，所以 server.pid 的值应该是一个无效的值，我们可以使用 print 命令将这个值打印出来看一下：

```
(gdb) p server.pid
$1 = 0
```

结果是 0，这个 0 值是 Redis 初始化时设置的。

---

gdb 的 jump 命令的作用，与使用 Visual Studio 调试时通过鼠标将程序当前的执行点从一个位置拖到另一个位置的作用一样。

在Visual Studio中可以用鼠标将当前执行点拖动至指定位置

2.5.13 disassemble 命令

在某些场景下，我们可能要通过查看某段代码的汇编指令去排查问题，或者在调试一些不含调试信息的 release 版程序时，只能通过反汇编代码定位问题。在此类场景下，disassemble 命令就派上用场了。disassemble 会输出当前函数的汇编指令，例如在 Redis 的 initServer 函数中执行该命令，会输出 initServer 函数的汇编指令，操作如下：

```
Breakpoint 1, initServer () at server.c:2742
2742            signal(SIGHUP, SIG_IGN);
(gdb) disassemble
Dump of assembler code for function initServer:
   0x00000000004310cb <+0>:     push   %rbp
   0x00000000004310cc <+1>:     mov    %rsp,%rbp
   0x00000000004310cf <+4>:     push   %rbx
   ...代码太多，此处省略...
   0x0000000000431a93 <+2504>:  add    $0x18,%rsp
--Type <RET> for more, q to quit, c to continue without paging--
   0x0000000000431a97 <+2508>:  pop    %rbx
   0x0000000000431a98 <+2509>:  pop    %rbp
   0x0000000000431a99 <+2510>:  retq
End of assembler dump.
(gdb)
```

gdb 的反汇编格式默认为 AT&T 格式，可以通过 show disassembly-flavor 查看当前的反汇编格式。如果习惯阅读 intel 汇编格式，则可以使用 set disassembly-flavor intel 命令来设置。操作如下：

```
(gdb) set disassembly-flavor intel
(gdb) disassemble
Dump of assembler code for function initServer:
   0x00000000004310cb <+0>:     push   rbp
   0x00000000004310cc <+1>:     mov    rbp,rsp
   0x00000000004310cf <+4>:     push   rbx
   0x00000000004310d0 <+5>:     sub    rsp,0x18
   ...代码太多，此处省略...
   0x0000000000431a93 <+2504>:  add    rsp,0x18
--Type <RET> for more, q to quit, c to continue without paging--
   0x0000000000431a97 <+2508>:  pop    rbx
   0x0000000000431a98 <+2509>:  pop    rbp
   0x0000000000431a99 <+2510>:  ret
End of assembler dump.
(gdb)
```

disassemble 命令在程序崩溃后产生 core 文件，且在无对应的调试符号时非常有用，此时可以通过分析汇编代码排查一些问题。

2.5.14 set args 与 show args 命令

很多程序都需要我们传递命令行参数。在 gdb 调试中使用 gdb filename args 这种形式给被调试的程序传递命令行参数是行不通的。正确的做法是在用 gdb attach 程序后，使用 run 命令之前，使用 set args 命令行参数来指定被调试程序的命令行参数。还是以

redis-server 为例，redis 在启动时可以指定一个命令行参数，即它的配置文件，位于 redis-server 文件的上一层目录下。所以我们可以在 gdb 中这样传递这个参数：set args ../redis.conf，可以通过 show args 查看命令行参数是否设置成功：

```
(gdb) set args ../redis.conf
(gdb) show args
Argument list to give program being debugged when it is started is "../redis.conf".
(gdb)
```

如果在单个命令行参数之间有空格，则可以使用引号将参数包裹起来：

```
(gdb) set args "999 xx" "hu jj"
(gdb) show args
Argument list to give program being debugged when it is started is ""999 xx" "hu jj"".
(gdb)
```

如果想清除已经设置好的命令行参数，则使用 set args 不加任何参数即可：

```
(gdb) set args
(gdb) show args
Argument list to give program being debugged when it is started is "".
(gdb)
```

2.5.15　watch 命令

watch 是一个强大的命令，可以用来监视一个变量或者一段内存，当这个变量或者该内存处的值发生变化时，gdb 就会中断。监视某个变量或者某个内存地址会产生一个观察点（watch point）。

比如有一个面试题：有一个变量的值被意外修改，单步调试或者挨个检查使用该变量的代码，工作量非常大，那么如何快速定位该变量被修改的位置呢？其实，面试官想要的答案是"通过数据断点"。watch 命令可能通过添加硬件断点来达到监视数据变化的目的。watch 命令的使用方式是 watch 变量名或内存地址，一个观察点一般有以下几种格式。

（1）整形变量：

```
int i;
watch i
```

（2）指针类型：

```
char *p;
watch p 与 watch *p
```

注意：watch p 与 watch *p 是有区别的，前者是查看*(&p)，是 p 变量本身；后者是 p 所指的内存的内容，一般是我们所需的，我们在大多数情况下要看某内存地址上的数据如何变化。

（3）监视一个数组或内存区间：

```
char buf[128];
watch buf
```

这里是对 buf 的 128 个数据进行了监视。

需要注意的是：当设置的观察点是一个局部变量时，局部变量失效后，观察点也会失效。例如在观察点失效时，gdb 可能会提示如下信息：

```
Watchpoint 2 deleted because the program has left the block in which its expression is valid.
```

2.5.16　display 命令

display 命令用于监视变量或者内存的值，每次 gdb 中断，都会自动输出这些被监视变量或内存的值。例如，某个程序有一些全局变量，在每次触发断点后 gdb 中断下来时，我们都希望自动输出这些全局变量的最新值，这时就可以使用 display 命令了。display 命令的使用格式是 display 变量名/内存地址/寄存器名：

```
0x00007ffff71e2603 in epoll_wait () from /usr/lib64/libc.so.6
(gdb) display $ebx
1: $ebx = 24068
(gdb) display /x $ebx
2: /x $ebx = 0x5e04
(gdb) display $eax
3: $eax = -4
(gdb) b main
Breakpoint 2 at 0x436abd: file server.c, line 5001.
(gdb) r
The program being debugged has been started already.
Start it from the beginning? (y or n) y
Starting program: /root/redis-6.0.3/src/redis-server
...省略部分输出...
Breakpoint 2, main (argc=1, argv=0x7fffffffe308) at server.c:5001
5001        spt_init(argc, argv);
1: $ebx = 0
2: /x $ebx = 0x0
3: $eax = 4418219
(gdb)
```

在以上代码中，我们使用 display 命令分别监视寄存器 ebp 和寄存器 eax，要求 ebp 寄存器分别使用十进制和十六进制两种形式输出其值，这样每次 gdb 中断下来时，都会自动将这些寄存器的值输出。我们可以使用 info display 查看当前已经监视了哪些值，使用 delete display 清除全部被监视的变量，使用 delete display 编号移除对指定变量的监视。操作演示如下：

```
(gdb) delete display
Delete all auto-display expressions? (y or n) n
(gdb) delete display 3
(gdb) info display
Auto-display expressions now in effect:
Num Enb Expression
2:   y  $ebp
1:   y  $eax
```

2.5.17 dir 命令

读者可能会遇到这样的场景：使用 gdb 调试时，生成可执行文件的机器和实际执行该可执行程序的机器不是同一台机器，例如大多数企业产生目标服务程序的机器是编译机器，即发版机，然后把发版机产生的可执行程序拿到生产机器上执行。这时如果可执行程序崩溃，我们用 gdb 调试 core 文件时，gdb 就会提示 "No such file or directory"，如下所示：

```
Program received signal SIGSEGV, Segmentation fault.
0x00000000004d5662 in CAsyncLog::crash () at
/home/flamingoserver/base/AsyncLog.cpp:475
475        /home/flamingoserver/base/AsyncLog.cpp: No such file or directory.
```

或者由于一些原因，编译时的源码文件被挪动了位置，使用 gdb 调试时也会出现上述情况。

gcc/g++ 编译出来的可执行程序并不包含完整的源码，-g 只是加了一个可执行程序与源码之间的位置映射关系，我们可以通过 dir 命令重新定位这种关系。

dir 命令的使用格式如下：

```
# 加一个源文件路径到当前路径的前面，指定多个路径，可以使用 ":"
dir SourcePath1:SourcePath2:SourcePath3
```

SourcePath1、SourcePath2、SourcePath3 指的就是需要设置的源码目录，gdb 会依次到这些目录下搜索相应的源文件。

以上面的错误提示为例，原来的 AsyncLog.cpp 文件位于 /home/flamingoserver/base/ 目录下，由于这个目录被挪动，所以 gdb 提示找不到该文件。现在假设该文件被移动到 /home/zhangyl/flamingoserver/base/ 目录下，我们只需在 gdb 调试中执行 dir /home/zhangyl/flamingoserver/base/，即可重定向可执行程序与源码的位置关系：

```
(gdb) dir /home/zhangyl/flamingoserver/base/
Source directories searched: /home/zhangyl/flamingoserver/base:$cdir:$cwd
(gdb) r
The program being debugged has been started already.
Start it from the beginning? (y or n) y
Starting program: /home/zhangyl/chatserver
...省略部分输出...
Program received signal SIGSEGV, Segmentation fault.
0x00000000004d5662 in CAsyncLog::crash () at
/home/flamingoserver/base/AsyncLog.cpp:475
475        *p = 0;
(gdb) list
470    }
471
472    void CAsyncLog::crash()
473    {
474        char* p = nullptr;
475        *p = 0;
476    }
477
478    void CAsyncLog::writeThreadProc()
479    {
```

```
(gdb)
```

使用 dir 命令重新定位源文件的位置之后,gdb 就不会再提示这样的错误了,我们此时也可以使用 gdb 的其他命令(如 list 命令)查看源码。

如果要查看当前设置了哪些源码搜索路径,则可以使用 show dir 命令:

```
(gdb) show dir
Source directories searched: /home/zhangyl/flamingoserver/base:$cdir:$cwd
(gdb)
```

dir 命令不加参数时,表示清空当前已设置的源码搜索路径:

```
(gdb) dir
Reinitialize source path to empty? (y or n) y
Source directories searched: $cdir:$cwd
(gdb) show dir
Source directories searched: $cdir:$cwd
(gdb)
```

2.6 使用 gdb 调试多线程程序

前面实际上已经介绍了调试多线程程序的方法,本节进行总结。

当然,调试多线程程序的前提是我们熟悉多线程的基础知识,包括线程的创建和退出、线程之间的各种同步原语等。

2.6.1 调试多线程程序的方法

使用 gdb 将程序跑起来,然后按 Ctrl+C 组合键将程序中断,使用 info threads 命令查看当前进程有多少线程:

```
31995:M 25 Aug 2020 14:35:43.964 * Ready to accept connections
^C
Thread 1 "redis-server" received signal SIGINT, Interrupt.
0x00007ffff71e2603 in epoll_wait () from /usr/lib64/libc.so.6
(gdb) info threads
  Id   Target Id                                         Frame
* 1    Thread 0x7ffff7feb740 (LWP 31995) "redis-server"  0x00007ffff71e2603 in
epoll_wait () from /usr/lib64/libc.so.6
  2    Thread 0x7ffff0bb9700 (LWP 31999) "bio_close_file" 0x00007ffff74bc965 in
pthread_cond_wait@@GLIBC_2.3.2 () from /usr/lib64/libpthread.so.0
  3    Thread 0x7ffff03b8700 (LWP 32000) "bio_aof_fsync" 0x00007ffff74bc965 in
pthread_cond_wait@@GLIBC_2.3.2 () from /usr/lib64/libpthread.so.0
  4    Thread 0x7fffefbb7700 (LWP 32001) "bio_lazy_free" 0x00007ffff74bc965 in
pthread_cond_wait@@GLIBC_2.3.2 () from /usr/lib64/libpthread.so.0
(gdb)
```

还是以 redis-server 为例,使用 gdb 将程序运行起来后,我们按 Ctrl+C 组合键将程序中断,此时可以使用 info threads 命令查看 redis-server 有多少线程,每个线程正在执行哪里的代码。

使用 thread 线程编号可以切换到对应的线程，然后使用 bt 命令查看对应的线程从顶层到底层的函数调用，以及上层调用下层对应源码的位置。当然，也可以使用 frame 栈函数编号（栈函数编号即下图中的#0 ~ #4，使用 frame 命令时不需要加"#"）切换到当前函数调用堆栈的任何一层函数调用中，然后分析该函数的执行逻辑，使用 print 等命令输出各种变量和表达式的值，或者进行单步调试：

```
(gdb) thread 1
[Switching to thread 1 (Thread 0x7ffff7feb740 (LWP 31995))]
#0  0x00007ffff71e2603 in epoll_wait () from /usr/lib64/libc.so.6
(gdb) bt
#0  0x00007ffff71e2603 in epoll_wait () from /usr/lib64/libc.so.6
#1  0x0000000000428b0e in aeApiPoll (eventLoop=0x5e5760, tvp=0x7fffffffe140) at ae_epoll.c:112
#2  0x0000000000429864 in aeProcessEvents (eventLoop=0x5e5760, flags=27) at ae.c:447
#3  0x0000000000429b44 in aeMain (eventLoop=0x5e5760) at ae.c:539
#4  0x0000000000437375 in main (argc=1, argv=0x7fffffffe308) at server.c:5175
(gdb)
```

如上所示，我们切换到了 redis-server 的 1 号线程，然后输入 bt 命令查看该线程的调用堆栈，发现顶层是 main 函数，说明这是主线程程序，同时得到从 main 开始往下各个函数调用对应的源码位置，我们可以通过这些源码的位置学习和研究调用处的逻辑。对每个线程都进行这样的分析之后，我们基本上就可以搞清楚整个程序运行中的执行逻辑了。

接着我们分别通过得到的各个线程的线程函数名去源码中搜索，找到创建这些线程的函数（下文为了叙述方便，以 f 代称这个函数），再接着通过搜索 f 或者给 f 加断点重启程序，看函数 f 是如何被调用的，这些操作一般在程序初始化阶段进行。

redis-server 1 号线程是在 main 函数中创建的，我们再看下 2 号线程的创建，使用 thread 2 切换到 2 号线程，然后使用 bt 命令查看 2 号线程的调用堆栈，得到 2 号线程的线程函数为 bioProcessBackgroundJobs。注意，顶层的 clone 和 start_thread 是系统函数，我们找的线程函数应该是项目中的自定义线程函数。

```
(gdb) thread 2
[Switching to thread 2 (Thread 0x7ffff0bb9700 (LWP 9026))]
#0  0x00007ffff74bc965 in pthread_cond_wait@@GLIBC_2.3.2 () from /usr/lib64/libpthread.so.0
(gdb) info threads
  Id   Target Id                                          Frame
  1    Thread 0x7ffff7feb740 (LWP 9022) "redis-server"    0x00007ffff71e2603 in epoll_wait () from /usr/lib64/libc
* 2    Thread 0x7ffff0bb9700 (LWP 9026) "bio_close_file"  0x00007ffff74bc965 in pthread_cond_wait@@GLIBC_2.3.2 () 
    from /usr/lib64/libpthread.so.0
  3    Thread 0x7ffff03b8700 (LWP 9027) "bio_aof_fsync"   0x00007ffff74bc965 in pthread_cond_wait@@GLIBC_2.3.2 ()
    from /usr/lib64/libpthread.so.0
  4    Thread 0x7fffefbb7700 (LWP 9028) "bio_lazy_free"   0x00007ffff74bc965 in pthread_cond_wait@@GLIBC_2.3.2 ()
    from /usr/lib64/libpthread.so.0
(gdb) bt
#0  0x00007ffff74bc965 in pthread_cond_wait@@GLIBC_2.3.2 () from /usr/lib64/libpthread.so.0
#1  0x000000000049927a in bioProcessBackgroundJobs (arg=0x0) at bio.c:190
#2  0x00007ffff74b8dd5 in start_thread () from /usr/lib64/libpthread.so.0
#3  0x00007ffff71e202d in clone () from /usr/lib64/libc.so.6
(gdb)
```

通过在项目中搜索 bioProcessBackgroundJobs 函数，我们发现 bioProcessBackgroundJobs 函数在 bioInit 中被调用，而且确实是在 bioInit 函数中创建了线程 2，因此我们看到了 pthread_create(&thread,&attr, bioProcessBackgroundJobs,arg) != 0)这样的调用：

```c
//bio.c 96 行
void bioInit(void) {
    //省略部分代码

    for (j = 0; j < BIO_NUM_OPS; j++) {
        void *arg = (void*)(unsigned long) j;
        //这里创建了 bioProcessBackgroundJobs 线程
        if (pthread_create(&thread,&attr,bioProcessBackgroundJobs,arg) != 0) {
            serverLog(LL_WARNING,"Fatal: Can't initialize Background Jobs.");
            exit(1);
        }
        bio_threads[j] = thread;
    }
}
```

此时，我们可以继续在项目中查找 bioInit 函数，看看它在哪里被调用，或者直接给 bioInit 函数加上断点，然后重启 redis-server，等到断点被触发再使用 bt 命令查看此时的调用堆栈，就知道 bioInit 函数在何处被调用了：

```
(gdb) b bioInit
Breakpoint 1 at 0x498e5e: file bio.c, line 103.
(gdb) r
The program being debugged has been started already.
Start it from the beginning? (y or n) y
Starting program: /root/redis-6.0.3/src/redis-server
[Thread debugging using libthread_db enabled]
...省略部分无关输出...
Breakpoint 1, bioInit () at bio.c:103
103         for (j = 0; j < BIO_NUM_OPS; j++) {
(gdb) bt
#0  bioInit () at bio.c:103
#1  0x0000000000431b5d in InitServerLast () at server.c:2953
#2  0x000000000043724f in main (argc=1, argv=0x7fffffffe318) at server.c:5142
(gdb)
```

至此，我们发现 2 号线程在 main 函数中调用了 InitServerLast 函数，后者又调用了 bioInit 函数，然后在 bioInit 函数中创建了新的线程 bioProcessBackgroundJobs，我们只要分析这个执行流，就能搞清楚其逻辑了。

同样，redis-server 还有 3 号线程和 4 号线程，我们也可以按分析 2 号线程的方式分析 3 号线程和 4 号线程。

以上是笔者阅读不熟悉的 C/C++ 项目时的常用方法，当然，对于一些特殊的项目源码，我们还需要了解该项目的业务内容，因为只有结合业务，才能看懂各个线程调用栈及初始化各个线程函数中的业务逻辑。

2.6.2 在调试时控制线程切换

在调试多线程程序时，我们可能希望执行流一直在某个线程中执行，而不是切换到其他线程，有办法做到这样吗？

为了说明清楚这个问题，我们假设现在调试的程序有 5 个线程，除了主线程，其他 4

个工作线程的线程函数都如下所示：

```
void* worker_thread_proc(void* arg)
{
    while (true)
    {
        //代码第 1 行
        //代码第 2 行
        //代码第 3 行
        //代码第 4 行
        //代码第 5 行
        //代码第 6 行
        //代码第 7 行
        //代码第 8 行
        //代码第 9 行
        //代码第 10 行
        //代码第 11 行
        //代码第 12 行
        //代码第 13 行
        //代码第 14 行
        //代码第 15 行
    }
}
```

为了方便表述，我们把 4 个工作线程分别叫作线程 A、线程 B、线程 C、线程 D。

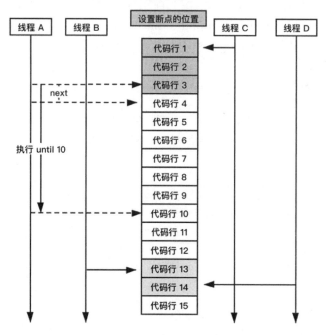

如上图所示，假设某个时刻，线程 A 停在第 3 行代码处，线程 B、C、D 停在代码第 1~15 行的任一位置，此时线程 A 是 gdb 当前的调试线程，此时我们输入 next 命令，期望调试器跳转到代码第 4 行；或者输入 util 10 命令，期望调试器跳转到代码第 10 行。但是在实际情况下，如果在代码第 1、2、13 或 14 行设置了断点，则 gdb 再次停下来时，可能会停在代码第 1、2、13、14 行。

这是多线程程序的特点：当我们从代码第 4 行让程序继续运行时，线程 A 虽然会继续往下执行，下一次应该在代码第 14 行停下来，但是线程 B、C、D 也在同步运行，如果此时系统的线程调度将 CPU 时间片切换到线程 B、C 或者 D，那么 gdb 最终停下来时，可能是线程 B、C、D 触发了代码第 1、2、13、14 行的断点，此时调试的线程会变为 B、C 或者 D，打印相关的变量值时可能就不是我们期望的线程 A 函数中的相关变量值了。

还存在一种情况，单步调试线程 A 时，我们不希望线程 A 函数中的值被其他线程改变。针对调试多线程程序存在的这些情况，gdb 提供了一个将程序执行流锁定在当前调试线程中的命令选项——scheduler-locking，这个选项有三个值，分别是 on、step 和 off，使用方法如下：

```
set scheduler-locking on/step/off
```

set scheduler-locking on 可以用来锁定当前线程，只观察这个线程的运行情况，锁定这个线程时，其他线程处于暂停状态，也就是说，在当前线程执行 next、step、until、finish、return 命令时，其他线程是不会运行的。

需要注意的是，在使用 set scheduler-locking on/step 选项时要确认当前线程是否是我们期望锁定的线程，如果不是，则可以使用 thread +线程编号切换到我们需要的线程，再调用 set scheduler-locking on/step 锁定。

set scheduler-locking step 也用来锁定当前线程，当且仅当使用 next 或 step 命令做单步调试时会锁定当前线程，如果使用 until、finish、return 等线程内的调试命令（它们不是单步控制命令），则其他线程还是有机会运行的。与 on 选项的值相比，step 选项的值为单步调试提供了更加精细化的控制，因为在某些场景下，我们希望单步调试时其他线程不要对所属的当前线程的变量值造成影响。

set scheduler-locking off 用于释放锁定当前线程。

下面以一个小示例说明这三个选项的用法。编写如下代码：

```
01 #include <stdio.h>
02 #include <pthread.h>
03 #include <unistd.h>
04
05 long g = 0;
06
07 void* worker_thread_1(void* p)
08 {
09     while (true)
10     {
11         g = 100;
12         printf("worker_thread_1\n");
13         usleep(300000);
14     }
15
16     return NULL;
17 }
18
19 void* worker_thread_2(void* p)
```

```
20  {
21      while (true)
22      {
23          g = -100;
24          printf("worker_thread_2\n");
25          usleep(500000);
26      }
27
28      return NULL;
29  }
30
31  int main()
32  {
33      pthread_t thread_id_1;
34      pthread_create(&thread_id_1, NULL, worker_thread_1, NULL);
35      pthread_t thread_id_2;
36      pthread_create(&thread_id_2, NULL, worker_thread_2, NULL);
37
38      while (true)
39      {
40          g = -1;
41          printf("g=%d\n", g);
42          g = -2;
43          printf("g=%d\n", g);
44          g = -3;
45          printf("g=%d\n", g);
46          g = -4;
47          printf("g=%d\n", g);
48
49          usleep(1000000);
50      }
51
52      return 0;
53  }
```

以上代码在主线程（main 函数所在的线程）中创建了两个工作线程，主线程接下来的逻辑是在一个循环里面依次将全局变量 g 修改成-1、-2、-3、-4，然后休眠 1 秒；工作线程 worker_thread_1、worker_thread_2 分别在自己的循环里将全局变量 g 修改成 100 和 -100。

我们编译程序后将程序使用 gdb 跑起来，三个线程同时运行，交错输出：

```
[root@myaliyun xx]# g++ -g -o main main.cpp -lpthread
[root@myaliyun xx]# gdb main
...省略部分无关输出...
Reading symbols from main...
(gdb) r
Starting program: /root/xx/main
[Thread debugging using libthread_db enabled]
...省略部分无关输出...
[New Thread 0x7ffff6f56700 (LWP 402)]
worker_thread_1
[New Thread 0x7ffff6755700 (LWP 403)]
g=-1
```

```
g=-2
g=-3
g=-4
worker_thread_2
worker_thread_1
worker_thread_2
worker_thread_1
worker_thread_1
g=-1
g=-2
g=-3
g=-4
worker_thread_2
worker_thread_1
worker_thread_1
worker_thread_2
worker_thread_1
g=-1
g=-2
g=-3
g=-4
worker_thread_2
worker_thread_1
worker_thread_1
worker_thread_2
```

我们按 Ctrl+C 组合键将程序中断,如果当前线程不在主线程中,则可以先使用 info threads 和 thread id 切换到主线程:

```
^C
Thread 1 "main" received signal SIGINT, Interrupt.
0x00007ffff701bfad in nanosleep () from /usr/lib64/libc.so.6
(gdb) info threads
  Id   Target Id                                       Frame
* 1    Thread 0x7ffff7feb740 (LWP 1191) "main" 0x00007ffff701bfad in nanosleep ()
from /usr/lib64/libc.so.6
  2    Thread 0x7ffff6f56700 (LWP 1195) "main" 0x00007ffff701bfad in nanosleep () from
/usr/lib64/libc.so.6
  3    Thread 0x7ffff6755700 (LWP 1196) "main" 0x00007ffff701bfad in nanosleep () from
/usr/lib64/libc.so.6
(gdb) thread 1
[Switching to thread 1 (Thread 0x7ffff7feb740 (LWP 1191))]
#0  0x00007ffff701bfad in nanosleep () from /usr/lib64/libc.so.6
(gdb)
```

然后在代码第 11 行和第 41 行各加一个断点。反复执行 until 48 命令,发现工作线程 1 和 2 还是有机会被执行的:

```
(gdb) b main.cpp:41
Breakpoint 1 at 0x401205: file main.cpp, line 41.
(gdb) b main.cpp:11
Breakpoint 2 at 0x40116e: file main.cpp, line 11.
(gdb) until 48
0x00007ffff704c884 in usleep () from /usr/lib64/libc.so.6
(gdb)
```

```
worker_thread_2
[Switching to Thread 0x7ffff6f56700 (LWP 1195)]

Thread 2 "main" hit Breakpoint 2, worker_thread_1 (p=0x0) at main.cpp:11
11                    g = 100;
(gdb)
worker_thread_2
[Switching to Thread 0x7ffff7feb740 (LWP 1191)]

Thread 1 "main" hit Breakpoint 1, main () at main.cpp:41
41                    printf("g=%d\n", g);
(gdb)
worker_thread_1
worker_thread_2
g=-1
g=-2
g=-3
g=-4
main () at main.cpp:49
49                    usleep(1000000);
(gdb)
worker_thread_2
[Switching to Thread 0x7ffff6f56700 (LWP 1195)]

Thread 2 "main" hit Breakpoint 2, worker_thread_1 (p=0x0) at main.cpp:11
11                    g = 100;
(gdb)
```

现在再次将线程切换到主线程（如果在 gdb 中断后，当前线程不是主线程），执行 set scheduler-locking on 命令，然后继续反复执行 until 48 命令：

```
(gdb) set scheduler-locking on
(gdb) until 48

Thread 1 "main" hit Breakpoint 1, main () at main.cpp:41
41                    printf("g=%d\n", g);
(gdb) until 48
g=-1
g=-2
g=-3
g=-4
main () at main.cpp:49
49                    usleep(1000000);
(gdb) until 48

Thread 1 "main" hit Breakpoint 1, main () at main.cpp:41
41                    printf("g=%d\n", g);
(gdb)
g=-1
g=-2
g=-3
g=-4
main () at main.cpp:49
49                    usleep(1000000);
(gdb) until 48
```

```
Thread 1 "main" hit Breakpoint 1, main () at main.cpp:41
41                  printf("g=%d\n", g);
(gdb)
g=-1
g=-2
g=-3
g=-4
main () at main.cpp:49
49                  usleep(1000000);
(gdb) until 48

Thread 1 "main" hit Breakpoint 1, main () at main.cpp:41
41                  printf("g=%d\n", g);
(gdb)
```

再次使用 until 命令时，gdb 锁定了主线程，其他两个工作线程再也不会被执行了，因此两个工作线程无任何输出。

我们再使用 set scheduler-locking step 模式锁定主线程，然后反复执行 until 48 命令：

```
(gdb) set scheduler-locking step
(gdb) until 48
worker_thread_2
worker_thread_1
g=-100
g=-2
g=-3
g=-4
main () at main.cpp:49
49                  usleep(1000000);
(gdb) until 48
worker_thread_2
[Switching to Thread 0x7ffff6f56700 (LWP 1195)]

Thread 2 "main" hit Breakpoint 2, worker_thread_1 (p=0x0) at main.cpp:11
11                  g = 100;
(gdb) until 48
worker_thread_2
worker_thread_1

Thread 2 "main" hit Breakpoint 2, worker_thread_1 (p=0x0) at main.cpp:11
11                  g = 100;
(gdb) until 48
worker_thread_2
[Switching to Thread 0x7ffff7feb740 (LWP 1191)]

Thread 1 "main" hit Breakpoint 1, main () at main.cpp:41
41                  printf("g=%d\n", g);
(gdb) until 48
worker_thread_1
worker_thread_2
g=-100
g=-2
g=-3
```

```
g=-4
main () at main.cpp:49
49                  usleep(1000000);
(gdb) until 48
worker_thread_2
[Switching to Thread 0x7ffff6f56700 (LWP 1195)]

Thread 2 "main" hit Breakpoint 2, worker_thread_1 (p=0x0) at main.cpp:11
11                  g = 100;
(gdb) until 48
worker_thread_2
worker_thread_1

Thread 2 "main" hit Breakpoint 2, worker_thread_1 (p=0x0) at main.cpp:11
11                  g = 100;
(gdb)
```

可以看到，使用 step 模式锁定主线程后，使用 until 命令时，另外两个工作线程仍然有执行的机会。我们再次切换到主线程，使用 next 命令单步调试下试试：

```
(gdb) info threads
  Id   Target Id                                         Frame
  1    Thread 0x7ffff7feb740 (LWP 1191) "main" 0x00007ffff701bfad in nanosleep () from /usr/lib64/libc.so.6
* 2    Thread 0x7ffff6f56700 (LWP 1195) "main" worker_thread_1 (p=0x0) at main.cpp:11
  3    Thread 0x7ffff6755700 (LWP 1196) "main" 0x00007ffff701bfad in nanosleep () from /usr/lib64/libc.so.6
(gdb) thread 1
[Switching to thread 1 (Thread 0x7ffff7feb740 (LWP 1191))]
#0  0x00007ffff701bfad in nanosleep () from /usr/lib64/libc.so.6
(gdb) set scheduler-locking step
(gdb) next
Single stepping until exit from function nanosleep,
which has no line number information.
0x00007ffff704c884 in usleep () from /usr/lib64/libc.so.6
(gdb) next
Single stepping until exit from function usleep,
which has no line number information.
main () at main.cpp:40
40                  g = -1;
(gdb) next

Thread 1 "main" hit Breakpoint 1, main () at main.cpp:41
41                  printf("g=%d\n", g);
(gdb) next
g=-1
42                  g = -2;
(gdb) next
43                  printf("g=%d\n", g);
(gdb) next
g=-2
44                  g = -3;
(gdb) next
45                  printf("g=%d\n", g);
(gdb) next
```

```
g=-3
46                  g = -4;
(gdb) next
47                  printf("g=%d\n", g);
(gdb) next
g=-4
49                  usleep(1000000);
(gdb) next
40                  g = -1;
(gdb) next

Thread 1 "main" hit Breakpoint 1, main () at main.cpp:41
41                  printf("g=%d\n", g);
(gdb) next
g=-1
42                  g = -2;
(gdb) next
43                  printf("g=%d\n", g);
(gdb) next
g=-2
44                  g = -3;
(gdb) next
45                  printf("g=%d\n", g);
(gdb) next
g=-3
46                  g = -4;
(gdb) next
47                  printf("g=%d\n", g);
(gdb) next
g=-4
49                  usleep(1000000);
(gdb) next
40                  g = -1;
(gdb) next

Thread 1 "main" hit Breakpoint 1, main () at main.cpp:41
41                  printf("g=%d\n", g);
(gdb)
```

此时发现设置了以 step 模式锁定主线程,工作线程不会在单步调试主线程时被执行,即使在工作线程中设置了断点。

最后,我们使用 set scheduler-locking off 取消对主线程的锁定,然后继续使用 next 命令单步调试:

```
(gdb) set scheduler-locking off
(gdb) next
worker_thread_2
worker_thread_1
g=-100
42                  g = -2;
(gdb) next
worker_thread_2
[Switching to Thread 0x7ffff6f56700 (LWP 1195)]
```

```
Thread 2 "main" hit Breakpoint 2, worker_thread_1 (p=0x0) at main.cpp:11
11                  g = 100;
(gdb) next
g=100
g=-3
g=-4
worker_thread_2
12                  printf("worker_thread_1\n");
(gdb) next
worker_thread_1
13                  usleep(300000);
(gdb) next
worker_thread_2
[Switching to Thread 0x7ffff7feb740 (LWP 1191)]

Thread 1 "main" hit Breakpoint 1, main () at main.cpp:41
41                  printf("g=%d\n", g);
(gdb) next
[Switching to Thread 0x7ffff6f56700 (LWP 1195)]

Thread 2 "main" hit Breakpoint 2, worker_thread_1 (p=0x0) at main.cpp:11
11                  g = 100;
(gdb) next
g=-1
g=-2
g=-3
g=-4
worker_thread_2
12                  printf("worker_thread_1\n");
(gdb)
```

取消锁定之后，单步调试时三个线程都有机会被执行，线程 1 的断点也被正常触发。

2.7 使用 gdb 调试多进程程序——以调试 Nginx 为例

这里说的多进程程序指的是一个进程使用 Linux 系统调用 fork 函数产生的子进程，没有相互关联的进程调试指的是 gdb 调试单个进程，前面已经详细讲解过了。

在实际应用中，有一类应用会通过 Linux 函数 fork 出新的子进程。以 Nginx 为例，Nginx 对客户端的连接采用了多进程模型，在接受客户端的连接后，会创建一个新的进程来处理该连接上的信息来往，新产生的进程与原进程互为父子关系。那么如何用 gdb 调试这样的父子进程呢？一般有两种方法，下面详细讲解。

1. 方法一

先在一个 Shell 窗口中用 gdb 调试父进程，等子进程被 fork 出来后，再新开一个 Shell 窗口使用 gdb attach 命令将 gdb attach 到子进程上。

这里以调试 Nginx 服务为例。从 Nginx 官网下载最新的 Nginx 源码（本书采用的版本是 1.18.0），然后编译和安装：

下载Nginx源码

```
[root@iZbp14iz399acush5e8ok7Z zhangyl]# wget
http://nginx.org/download/nginx-1.18.0.tar.gz
--2020-07-05 17:22:10--  http://nginx.org/download/nginx-1.18.0.tar.gz
Resolving nginx.org (nginx.org)... 95.211.80.227, 62.210.92.35,
2001:1af8:4060:a004:21::e3
Connecting to nginx.org (nginx.org)|95.211.80.227|:80... connected.
HTTP request sent, awaiting response... 200 OK
Length: 1039530 (1015K) [application/octet-stream]
Saving to: 'nginx-1.18.0.tar.gz'

nginx-1.18.0.tar.gz
100%[=======================================================================
========================>]   1015K   666KB/s    in 1.5s

2020-07-05 17:22:13 (666 KB/s) - 'nginx-1.18.0.tar.gz' saved [1039530/1039530]

## 解压缩Nginx
[root@iZbp14iz399acush5e8ok7Z zhangyl]# tar zxvf nginx-1.18.0.tar.gz

## 编译Nginx
[root@iZbp14iz399acush5e8ok7Z zhangyl]# cd nginx-1.18.0
[root@iZbp14iz399acush5e8ok7Z nginx-1.18.0]# ./configure
--prefix=/usr/local/nginx
[root@iZbp14iz399acush5e8ok7Z nginx-1.18.0]make CFLAGS="-g -O0"

##这样Nginx就被安装到/usr/local/nginx/目录下了
[root@iZbp14iz399acush5e8ok7Z nginx-1.18.0]make install
```

> 注意：使用make命令编译时，我们为了让生成的Nginx带有调试符号信息同时关闭编译器优化，设置了"-g -O0"选项。

启动Nginx：

```
[root@iZbp14iz399acush5e8ok7Z sbin]# cd /usr/local/nginx/sbin
[root@iZbp14iz399acush5e8ok7Z sbin]# ./nginx -c /usr/local/nginx/conf/nginx.conf
[root@iZbp14iz399acush5e8ok7Z sbin]# lsof -i -Pn | grep nginx
nginx     5246              root    9u  IPv4 22252908      0t0  TCP *:80 (LISTEN)
nginx     5247              nobody  9u  IPv4 22252908      0t0  TCP *:80 (LISTEN)
```

如上所示，Nginx默认开启两个进程，在笔者的机器上以root用户运行的Nginx进程是父进程，进程号是5246，以nobody用户运行的进程是子进程，进程号是5247。我们在当前窗口中使用gdb attach 5246命令将gdb attach到Nginx主进程上：

```
[root@iZbp14iz399acush5e8ok7Z sbin]# gdb attach 5246
...省略部分输出...
0x00007fd42a103c5d in sigsuspend () from /lib64/libc.so.6
Missing separate debuginfos, use: yum debuginfo-install
glibc-2.28-72.el8_1.1.x86_64 libxcrypt-4.1.1-4.el8.x86_64 pcre-8.42-4.el8.x86_64
sssd-client-2.2.0-19.el8.x86_64 zlib-1.2.11-10.el8.x86_64
(gdb)
```

此时就可以调试Nginx父进程了，例如使用bt命令查看当前调用堆栈：

```
(gdb) bt
#0  0x00007fd42a103c5d in sigsuspend () from /lib64/libc.so.6
```

```
#1  0x000000000044ae32 in ngx_master_process_cycle (cycle=0x1703720) at
src/os/unix/ngx_process_cycle.c:164
#2  0x000000000040bc05 in main (argc=3, argv=0x7ffe49109d68) at
src/core/nginx.c:382
(gdb) f 1
#1  0x000000000044ae32 in ngx_master_process_cycle (cycle=0x1703720) at
src/os/unix/ngx_process_cycle.c:164
164             sigsuspend(&set);
(gdb) l
159             }
160         }
161
162         ngx_log_debug0(NGX_LOG_DEBUG_EVENT, cycle->log, 0, "sigsuspend");
163
164         sigsuspend(&set);
165
166         ngx_time_update();
167
168         ngx_log_debug1(NGX_LOG_DEBUG_EVENT, cycle->log, 0,
(gdb)
```

使用 f 1 命令切换到当前调用堆栈#1，可以发现 Nginx 父进程的主线程挂起在 src/core/nginx.c:382 处。

此时可以使用 c 命令让程序继续运行，也可以添加断点或者进行其他调试操作。

再开一个 Shell 窗口，使用 gdb attach 5247 命令将 gdb attach 到 nginx 子进程上：

```
[root@iZbp14iz399acush5e8ok7Z sbin]# gdb attach 5247
...省略部分输出...
0x00007fd42a1c842b in epoll_wait () from /lib64/libc.so.6
Missing separate debuginfos, use: yum debuginfo-install
glibc-2.28-72.el8_1.1.x86_64 libblkid-2.32.1-17.el8.x86_64
libcap-2.26-1.el8.x86_64 libgcc-8.3.1-4.5.el8.x86_64
libmount-2.32.1-17.el8.x86_64 libselinux-2.9-2.1.el8.x86_64
libuuid-2.32.1-17.el8.x86_64 libxcrypt-4.1.1-4.el8.x86_64 pcre-8.42-4.el8.x86_64
pcre2-10.32-1.el8.x86_64 sssd-client-2.2.0-19.el8.x86_64
systemd-libs-239-18.el8_1.2.x86_64 zlib-1.2.11-10.el8.x86_64
(gdb)
```

使用 bt 命令查看子进程的主线程的当前调用堆栈：

```
(gdb) bt
#0  0x00007fd42a1c842b in epoll_wait () from /lib64/libc.so.6
#1  0x000000000044e546 in ngx_epoll_process_events (cycle=0x1703720,
timer=18446744073709551615, flags=1) at src/event/modules/ngx_epoll_module.c:800
#2  0x000000000043f317 in ngx_process_events_and_timers (cycle=0x1703720) at
src/event/ngx_event.c:247
#3  0x000000000044c38f in ngx_worker_process_cycle (cycle=0x1703720, data=0x0) at
src/os/unix/ngx_process_cycle.c:750
#4  0x000000000044926f in ngx_spawn_process (cycle=0x1703720, proc=0x44c2e1
<ngx_worker_process_cycle>, data=0x0, name=0x4cfd70 "worker process", respawn=-3)
    at src/os/unix/ngx_process.c:199
#5  0x000000000044b5a4 in ngx_start_worker_processes (cycle=0x1703720, n=1, type=-3)
at src/os/unix/ngx_process_cycle.c:359
#6  0x000000000044acf4 in ngx_master_process_cycle (cycle=0x1703720) at
src/os/unix/ngx_process_cycle.c:131
```

```
#7  0x000000000040bc05 in main (argc=3, argv=0x7ffe49109d68) at
src/core/nginx.c:382
(gdb) f 1
#1  0x000000000044e546 in ngx_epoll_process_events (cycle=0x1703720,
timer=18446744073709551615, flags=1) at src/event/modules/ngx_epoll_module.c:800
800             events = epoll_wait(ep, event_list, (int) nevents, timer);
(gdb)
```

可以发现，子进程挂起在 src/event/modules/ngx_epoll_module.c:800 的 epoll_wait 函数处。我们在 epoll_wait 函数返回后（src/event/modules/ngx_epoll_ module.c:804）加一个断点，然后使用 c 命令让 Nginx 子进程继续运行：

```
800             events = epoll_wait(ep, event_list, (int) nevents, timer);
(gdb) list
795         /* NGX_TIMER_INFINITE == INFTIM */
796
797         ngx_log_debug1(NGX_LOG_DEBUG_EVENT, cycle->log, 0,
798                        "epoll timer: %M", timer);
799
800         events = epoll_wait(ep, event_list, (int) nevents, timer);
801
802         err = (events == -1) ? ngx_errno : 0;
803
804         if (flags & NGX_UPDATE_TIME || ngx_event_timer_alarm) {
(gdb) b 804
Breakpoint 1 at 0x44e560: file src/event/modules/ngx_epoll_module.c, line 804.
(gdb) c
Continuing.
```

接着在浏览器中访问 Nginx 网站，这里的 IP 地址是笔者的云主机地址，读者在实际调试时将其改成自己的 Nginx 服务器所在的地址即可，如果是本机，那么地址就是 127.0.0.1，由于默认端口是 80，所以不用指定端口号：

```
http://你的 IP 地址:80
等价于
http://你的 IP 地址
```

此时回到 nginx 子进程的调试界面，发现断点被触发：

```
Breakpoint 1, ngx_epoll_process_events (cycle=0x1703720,
timer=18446744073709551615, flags=1) at src/event/modules/ngx_epoll_module.c:804
804         if (flags & NGX_UPDATE_TIME || ngx_event_timer_alarm) {
(gdb)
```

使用 bt 命令可以获得此时的调用堆栈：

```
(gdb) bt
#0  ngx_epoll_process_events (cycle=0x1703720, timer=18446744073709551615, flags=1)
at src/event/modules/ngx_epoll_module.c:804
#1  0x000000000043f317 in ngx_process_events_and_timers (cycle=0x1703720) at
src/event/ngx_event.c:247
#2  0x000000000044c38f in ngx_worker_process_cycle (cycle=0x1703720, data=0x0) at
src/os/unix/ngx_process_cycle.c:750
#3  0x000000000044926f in ngx_spawn_process (cycle=0x1703720, proc=0x44c2e1
<ngx_worker_process_cycle>, data=0x0, name=0x4cfd70 "worker process", respawn=-3)
    at src/os/unix/ngx_process.c:199
#4  0x000000000044b5a4 in ngx_start_worker_processes (cycle=0x1703720, n=1, type=-3)
```

```
                                             at src/os/unix/ngx_process_cycle.c:359
#5  0x000000000044acf4 in ngx_master_process_cycle (cycle=0x1703720) at
src/os/unix/ngx_process_cycle.c:131
#6  0x000000000040bc05 in main (argc=3, argv=0x7ffe49109d68) at
src/core/nginx.c:382
(gdb)
```

使用 info threads 命令可以查看子进程的所有线程信息，发现 Nginx 子进程只有一个主线程：

```
(gdb) info threads
  Id   Target Id                                        Frame
* 1    Thread 0x7fd42b17c740 (LWP 5247) "nginx" ngx_epoll_process_events
(cycle=0x1703720, timer=18446744073709551615, flags=1) at
src/event/modules/ngx_epoll_module.c:804
(gdb)
```

Nginx 父进程不处理客户端的请求，处理客户端请求的逻辑在子进程中，当单个子进程的客户端请求数达到一定数量时，父进程会重新 fork 一个新的子进程来处理新的客户端请求，也就是说子进程数量可以有多个，我们可以开多个 Shell 窗口，使用 gdb attach 到各个子进程上调试。

总之，我们可以使用这种方法添加各种断点调试 Nginx 的功能，慢慢地就能熟悉 Nginx 的各个内部逻辑了。

然而，该方法存在一个缺点，即程序已经启动了，我们只能使用 gdb 观察程序在这之后的行为，如果想调试程序从启动到运行的执行流程，则可能不太适用。有些读者可能会说：用 gdb attach 到进程后，加好断点，然后使用 run 命令重启进程，这样就可以调试程序从启动到运行的执行流程了。问题是这种方法并不通用，因为对于多进程服务模型，有些父子进程有一定的依赖关系，是不方便在运行过程中重启的。这时方法二就比较合适了。

2. 方法二

gdb 调试器提供了一个 follow-fork 选项，通过 set follow-fork mode 设置一个进程 fork 出新的子进程时，gdb 是继续调试父进程（取值是 parent）还是继续调试子进程（取值是 child），默认继续调试父进程（取值是 parent）：

```
# fork 之后 gdb attach 到子进程
set follow-fork child
# fork 之后 gdb attach 到父进程，这是默认值
set follow-fork parent
```

可以使用 show follow-fork mode 查看当前值：

```
(gdb) show follow-fork mode
Debugger response to a program call of fork or vfork is "child".
```

还是以调试 nginx 为例，先进入 nginx 可执行文件所在的目录，将方法一中的 Nginx 服务停下来：

```
[root@iZbp14iz399acush5e8ok7Z sbin]# cd /usr/local/nginx/sbin/
[root@iZbp14iz399acush5e8ok7Z sbin]# ./nginx -s stop
```

在 Nginx 源码中存在这样的逻辑，这个逻辑会在程序 main 函数处被调用：

```c
//src/os/unix/ngx_daemon.c:13 行
ngx_int_t
ngx_daemon(ngx_log_t *log)
{
    int  fd;

    switch (fork()) {
    case -1:
        ngx_log_error(NGX_LOG_EMERG, log, ngx_errno, "fork() failed");
        return NGX_ERROR;

    //fork 出来的子进程走这个 case
    case 0:
        break;

    //父进程中的 fork 返回值是子进程的 PID，大于 0，因此走这个 case，主进程会退出
    default:
        exit(0);
    }

    ...省略部分代码...
}
```

如以上代码中的注释所示，为了不让主进程退出，我们在 Nginx 的配置文件中增加一行，这样 Nginx 就不会调用 ngx_daemon 函数了：

```
daemon off;
```

接下来执行 gdb nginx，通过设置参数将配置文件 nginx.conf 传给待调试的 Nginx 进程：

```
Quit anyway? (y or n) y
[root@iZbp14iz399acush5e8ok7Z sbin]# gdb nginx
...省略部分输出...
Reading symbols from nginx...done.
(gdb) set args -c /usr/local/nginx/conf/nginx.conf
(gdb)
```

接着输入 run 命令尝试运行 Nginx：

```
(gdb) run
Starting program: /usr/local/nginx/sbin/nginx -c /usr/local/nginx/conf/nginx.conf
[Thread debugging using libthread_db enabled]
...省略部分输出...
[Detaching after fork from child process 7509]
```

如前文所述，gdb 遇到 fork 指令时默认会 attach 到父进程，因此在以上输出中有一行提示"Detaching after fork from child process 7509"，我们按 Ctrl+C 组合键将程序中断，然后输入 bt 命令查看当前调用堆栈，输出的堆栈信息和我们在方法一中看到的父进程的调用堆栈一样，说明 gdb 在程序 fork 之后确实 attach 到父进程了：

```
^C
Program received signal SIGINT, Interrupt.
0x00007ffff6f73c5d in sigsuspend () from /lib64/libc.so.6
(gdb) bt
#0  0x00007ffff6f73c5d in sigsuspend () from /lib64/libc.so.6
#1  0x000000000044ae32 in ngx_master_process_cycle (cycle=0x71f720) at src/os/unix/ngx_process_cycle.c:164
```

```
#2  0x000000000040bc05 in main (argc=3, argv=0x7fffffffe4e8) at
src/core/nginx.c:382
(gdb)
```

如果想让 gdb 在 fork 之后 attach 到子进程,则可以在程序运行之前设置 set follow-fork child,然后使用 run 命令重新运行程序:

```
(gdb) set follow-fork child
(gdb) run
The program being debugged has been started already.
Start it from the beginning? (y or n) y
Starting program: /usr/local/nginx/sbin/nginx -c /usr/local/nginx/conf/nginx.conf
[Thread debugging using libthread_db enabled]
Using host libthread_db library "/lib64/libthread_db.so.1".
[Attaching after Thread 0x7ffff7fe7740 (LWP 7664) fork to child process 7667]
[New inferior 2 (process 7667)]
[Detaching after fork from parent process 7664]
[Inferior 1 (process 7664) detached]
[Thread debugging using libthread_db enabled]
Using host libthread_db library "/lib64/libthread_db.so.1".
^C
Thread 2.1 "nginx" received signal SIGINT, Interrupt.
[Switching to Thread 0x7ffff7fe7740 (LWP 7667)]
0x00007ffff703842b in epoll_wait () from /lib64/libc.so.6
(gdb) bt
#0  0x00007ffff703842b in epoll_wait () from /lib64/libc.so.6
#1  0x000000000044e546 in ngx_epoll_process_events (cycle=0x71f720,
timer=18446744073709551615, flags=1) at src/event/modules/ngx_epoll_module.c:800
#2  0x000000000043f317 in ngx_process_events_and_timers (cycle=0x71f720) at
src/event/ngx_event.c:247
#3  0x000000000044c38f in ngx_worker_process_cycle (cycle=0x71f720, data=0x0) at
src/os/unix/ngx_process_cycle.c:750
#4  0x000000000044926f in ngx_spawn_process (cycle=0x71f720, proc=0x44c2e1
<ngx_worker_process_cycle>, data=0x0, name=0x4cfd70 "worker process", respawn=-3)
    at src/os/unix/ngx_process.c:199
#5  0x000000000044b5a4 in ngx_start_worker_processes (cycle=0x71f720, n=1, type=-3)
at src/os/unix/ngx_process_cycle.c:359
#6  0x000000000044acf4 in ngx_master_process_cycle (cycle=0x71f720) at
src/os/unix/ngx_process_cycle.c:131
#7  0x000000000040bc05 in main (argc=3, argv=0x7fffffffe4e8) at
src/core/nginx.c:382
(gdb)
```

接着按 Ctrl+C 组合键将程序中断,然后使用 bt 命令查看当前线程的调用堆栈,结果显示它确实是方法一中子进程的主线程所在的调用堆栈,这说明 gdb 确实 attach 到子进程了。

我们可以利用方法二调试程序 fork 之前和之后的任何逻辑,这是一种较为通用的多进程调试方法,建议掌握。

2.8 gdb 实用调试技巧

这里讲解 gdb 的一些实用调试技巧。

2.8.1 将 print 输出的字符串或字符数组完整显示

当我们使用 print 命令打印一个字符串或者字符数组时，如果该字符串太长，则使用 print 命令默认显示不完整，这时可以通过在 gdb 中输入 set print element 0 进行设置，这样再次使用 print 命令就能完整地显示该变量的所有字符串了：

```
void ChatSession::OnGetFriendListResponse(const std::shared_ptr<TcpConnection>& conn)
{
    std::string friendlist;
    MakeUpFriendListInfo(friendlist, conn);
    std::ostringstream os;
    os << "{\"code\": 0, \"msg\": \"ok\", \"userinfo\":" << friendlist << "}";
    Send(msg_type_getofriendlist, m_seq, os.str());
}
```

通过以上代码，在第 1 次使用 print 命令输出 friendlist 变量值时，只能显示部分字符串。使用 set print element 0 设置后就能完整显示了：

```
(gdb) n
563         os << "{\"code\": 0, \"msg\": \"ok\", \"userinfo\":" << friendlist << "}";
(gdb) p friendlist
$1 = "[{\"members\":[{\"address\":\"\",\"birthday\":19900101,\"clienttype\":0,\"customface\":\"\",\"facetype\":2,\"gender\":0,\"mail\":\"\",\"markname\":\"\",\"nickname\":\"bj_man\",\"phonenumber\":\"\",\"signature\":\"\",\"status\":0,\"userid\":4,"...
(gdb) set print element 0
(gdb) p friendlist
$2 = "[{\"members\":[{\"address\":\"\",\"birthday\":19900101,\"clienttype\":0,\"customface\":\"\",\"facetype\":2,\"gender\":0,\"mail\":\"\",\"markname\":\"\",\"nickname\":\"bj_man\",\"phonenumber\":\"\",\"signature\":\"\",\"status\":0,\"userid\":4,\"username\":\"13811411052\"},{\"address\":\"\",\"birthday\":19900101,\"clienttype\":0,\"customface\":\"\",\"facetype\":0,\"gender\":0,\"mail\":\"\",\"markname\":\"\",\"nickname\":\"Half\",\"phonenumber\":\"\",\"signature\":\"\",\"status\":0,\"userid\":5,\"username\":\"15618326596\"},{\"address\":\"\",\"birthday\":19900101,\"clienttype\":0,\"customface\":\"\",\"facetype\":34,\"gender\":0,\"mail\":\"\",\"markname\":\"\",\"nickname\":\"云淡风轻\",\"phonenumber\":\"\",\"signature\":\"\",\"status\":0,\"userid\":7,\"username\":\"china001\"},...代码太多，此处省略...
```

2.8.2 让被 gdb 调试的程序接收信号

先看看下面这段程序：

```
void prog_exit(int signo)
{
    std::cout << "program recv signal [" << signo << "] to exit." << std::endl;
}

int main(int argc, char* argv[])
{
    //设置信号处理
```

```cpp
    signal(SIGCHLD, SIG_DFL);
    signal(SIGPIPE, SIG_IGN);
    signal(SIGINT, prog_exit);
    signal(SIGTERM, prog_exit);

    int ch;
    bool bdaemon = false;
    while ((ch = getopt(argc, argv, "d")) != -1)
    {
        switch (ch)
        {
        case 'd':
            bdaemon = true;
            break;
        }
    }

    if (bdaemon)
        daemon_run();

    //省略无关代码
}
```

在上面这段程序中，我们让程序在接收到 Ctrl+C 信号（对应的信号值是 SIGINT）时简单打印一行信息。我们用 gdb 调试这个程序时，由于 Ctrl+C 信号默认会被 gdb 接收（让调试器中断），所以导致我们无法模拟程序接收这一信号。有以下两种方法解决这个问题。

（1）在 gdb 中使用 signal 函数手动向我们的程序发送信号，这里就是 signal SIGINT。

（2）改变 gdb 信号处理的设置，通过 handle SIGINT nostop print 告诉 gdb 在接收到 SIGINT 时不要停止，并把该信号传递给目标调试程序：

```
(gdb) handle SIGINT nostop print pass
SIGINT is used by the debugger.
Are you sure you want to change it? (y or n) y

Signal        Stop    Print   Pass to program Description
SIGINT        No      Yes     Yes             Interrupt
(gdb)
```

2.8.3 函数明明存在，添加断点时却无效

有时，一个函数明明存在，在我们的程序中也存在调试符号，但使用 break functionName 添加断点时，gdb 却提示如下：

```
Make breakpoint pending on future shared library load? y/n
```

这时即使输入 y，添加的断点可能也不会正确触发。我们需要改变添加断点的策略，使用该函数所在的代码文件和行号添加断点，这样就能添加同样效果的断点了。

2.8.4 调试中的断点

在实际调试中，我们一般会用到 3 种断点：普通断点、条件断点和数据断点。

普通断点就是我们添加的断点除去条件断点和硬件断点的断点。

数据断点是被监视的内存值或者变量值发生变化时触发的断点，前面介绍 watch 命令时添加的部分断点就是数据断点。

下面重点介绍条件断点。条件断点就是满足某个条件才会触发的断点。这里举个直观的例子：

```c
void do_something_func(int i)
{
    i ++;
    i = 100 * i;
}

int main()
{
    for(int i = 0; i < 10000; ++i)
    {
        do_something_func(i);
    }

    return 0;
}
```

在以上代码中，假如我们希望在变量 i 等于 5000 时，进入 do_something_func 函数中追踪这个函数的执行细节，则可以修改代码，增加一个 i=5000 的 if 条件，然后重新编译链接调试。这样显然比较麻烦，尤其是对于一些大型项目，每次重新编译链接都需要花一定的时间，而且调试完了还得把程序修改回来。有了条件断点，我们就不需要这么麻烦了，直接添加一个条件断点即可。添加条件断点的命令是 break [lineNo] if [condition]，其中 lineNo 是程序触发断点后需要停的位置，condition 是断点触发的条件。这里可以将其写成 break 11 if i==5000，11 就是调用 do_something_fun 函数所在的行号。当然这里的行号必须是合理的行号，如果行号非法或者行号位置不合理，则也不会触发这个断点：

```
(gdb) break 11 if i==5000
Breakpoint 2 at 0x400514: file test1.c, line 10.
(gdb) r
The program being debugged has been started already.
Start it from the beginning? (y or n) y
Starting program: /root/testgdb/test1

Breakpoint 1, main () at test1.c:9
9           for(int i = 0; i < 10000; ++i)
(gdb) c
Continuing.

Breakpoint 2, main () at test1.c:11
11              do_something_func(i);
(gdb) p i
$1 = 5000
```

把 i 打印出来，gdb 确实在 i=5000 时停了下来。

添加条件断点还有一种方法，就是先添加一个普通断点，然后使用"condition 断点编号 断点触发条件"这样的格式来添加。我们通过这种方法添加上述断点：

```
(gdb) b 11
Breakpoint 1 at 0x400514: file test1.c, line 11.
(gdb) info b
Num     Type           Disp Enb Address            What
1       breakpoint     keep y   0x0000000000400514 in main at test1.c:11
(gdb) condition 1 i==5000
(gdb) r
Starting program: /root/testgdb/test1
y

Breakpoint 1, main () at test1.c:11
11              do_something_func(i);
Missing separate debuginfos, use: debuginfo-install glibc-2.17-196.el7_4.2.x86_64
(gdb) p i
$1 = 5000
(gdb)
```

同样，如果断点编号不存在，也无法添加成功，gdb 就会提示断点不存在：

```
(gdb) condition 2 i==5000
No breakpoint number 2.
```

2.8.5 自定义 gdb 调试命令

在某些场景下，我们需要根据自己的程序自定义一些可以在调试时输出程序特定信息的命令。这在 gdb 中很容易做到，只要在 Linux 用户根目录下，root 用户就对应/root 目录，非 root 用户对应/home/用户名目录，在上述目录下自定义一个.gdbinit 文件即可。注意，在 Linux 系统中这是一个隐藏文件，可以使用 ls -a 命令查看；如果文件不存在，则新建一个即可，然后在这个文件中写上自定义的 gdb 命令。

这里以 Apache Web Server 源码为例（可从 Apache 官网下载该源码），在其源码根目录下有个.gdbinit 文件，在这个文件中存储了 Apache Web Server 自定义的 gdb 命令：

```
# gdb macros which may be useful for folks using gdb to debug
# apache.  Delete it if it bothers you.

define dump_table
    set $t = (apr_table_entry_t *)((apr_array_header_t *)$arg0)->elts
    set $n = ((apr_array_header_t *)$arg0)->nelts
    set $i = 0
    while $i < $n
    if $t[$i].val == (void *)0L
       printf "[%u] '%s'=>NULL\n", $i, $t[$i].key
    else
       printf "[%u] '%s'='%s' [%p]\n", $i, $t[$i].key, $t[$i].val, $t[$i].val
    end
    set $i = $i + 1
    end
end
```

```
# 省略部分代码

# Set sane defaults for common signals:
handle SIGPIPE noprint pass nostop
handle SIGUSR1 print pass nostop
```

当然，在这个文件的底部已配置不让 gdb 调试器处理 SIGPIPE 和 SIGUSR1 这两个信号，而是将这两个信号直接传递给被调试的程序本身，即如果在使用 gdb 调试 Apache Web Server 时产生了 SIGPIPE 或 SIGUSR1 信号，则 gdb 本身不处理这两个信号，而是将这两个信号传递给 Apache Web Server 程序。

2.9　gdb tui——gdb 图形化界面

gdb 调试令很多开发者头疼的问题之一，是很多 Linux 用户或者刚从其他图形化 IDE 转过来的开发者，都习惯了有强大的源码显示窗口的调试器，可能对 gdb 用 list 命令显示源码的方式非常不习惯，这主要是因为 gdb 在调试时不能很好地展示源码。gdb 在调试时可以通过 list 命令显示源码，但是通过 list 命令显示的代码不会高亮显示当前正在执行的代码行，这时可以使用 gdb 自带的 gdbtui 来解决问题。

如下所示为使用 gdbtui 调试 redis-server 的截图（图中的 Shell 工具为 SecureCRT），这样看代码比使用 list 命令方便了很多。

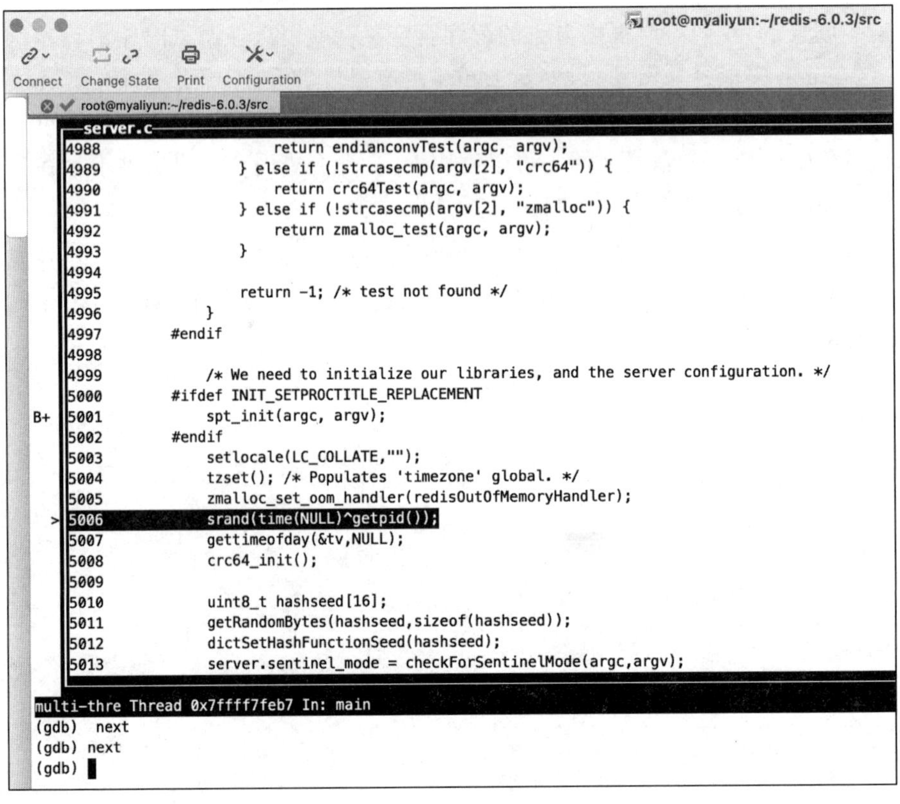

2.9.1　开启 gdb TUI 模式

开启 gdb TUI 模式有以下两种方法。

（1）使用 gdbtui 或者 gdb-tui 开启一个调试：

```
gdbtui -q 需要调试的程序名
```

（2）在 gdb 调试过程中程序被 gdb 中断时，按键盘上的 Ctrl+X+A 组合键调出 gdbtui。

2.9.2　gdb TUI 模式下的 4 个窗口

在默认情况下，在 gdbtui 模式下会显示 command 窗口和 source 窗口，如上图所示。在 TUI 模式下还有其他窗口，其中 4 个常用的窗口如下。

（1）command 窗口：命令窗口，可以键入调试命令，窗口类型为 cmd。

（2）source 窗口：源代码窗口，显示当前行、断点等信息，窗口类型为 src。

（3）assembly 窗口：汇编代码窗口，窗口类型为 asm。

（4）register 窗口：寄存器窗口，窗口类型为 reg。

可以通过在 cmd 窗口中输入 layout+窗口类型的命令选择自己需要的窗口，例如在 cmd 窗口中输入 layout asm 可以切换到 assembly 窗口。

layout 命令还可以用来修改窗口布局，该命令支持的窗口类型参数如下：

```
Usage: layout prev | next | <layout_name>
Layout names are:
```

```
    src   : Displays source and command windows.
    asm   : Displays disassembly and command windows.
    split : Displays source, disassembly and command windows.
    regs  : Displays register window. If existing layout
            is source/command or assembly/command, the
            register window is displayed. If the
            source/assembly/command (split) is displayed,
            the register window is displayed with
            the window that has current logical focus.
```

另外，可以通过 winheight 命令修改各个窗口的大小：

```
(gdb) help winheight
Set the height of a specified window.
Usage: winheight <win_name> [+ | -] <#lines>
Window names are:
src  : the source window
cmd  : the command window
asm  : the disassembly window
regs : the register display

##将代码窗口的高度扩大 5 行代码
winheight src + 5
##将代码窗口的高度减小 4 行代码
winheight src - 4
```

2.9.3　解决 tui 窗口不自动更新内容的问题

在当前 gdb tui 窗口扩大或者缩小以后，gdbtui 窗口中的内容不会自己刷新以适应新的窗口尺寸，我们可以通过 space 键强行让 gdbtui 窗口刷新。

2.9.4　窗口焦点切换

在默认设置下，方向键和 PageUp/PageDown 都是用来控制 gdbtui 的 src 窗口的，所以通过上下键显示前一条命令和后一条命令的功能不存在了，不过可以通过 Ctrl+N 或 Ctrl+P 组合键来获取这个功能。

注意：通过方向键调整 gdbtui 的 src 窗口后，可以通过 update 命令重新把焦点定位到当前执行的代码上。

我们可以通过 focus 命令调整焦点位置，默认在 src 窗口中通过 focus next 命令将焦点移到 cmd 窗口，这时就可以像以前一样，通过方向键来切换到上一条命令和下一条命令。同理，可以使用 focus prev 切回到 src 窗口。如果焦点不在 src 窗口，就不能通过方向键来浏览源码了：

```
(gdb) help focus
help focus
Set focus to named window or next/prev window.
Usage: focus {<win> | next | prev}
Valid Window names are:
src  : the source window
asm  : the disassembly window
```

```
regs : the register display
cmd  : the command window
```

2.10 gdb 的升级版——cgdb

在使用 gdb 单步调试时，代码每执行一行才接着显示一行，很多用惯了图形界面 IDE 调试的朋友可能会觉得非常不方便，而 gdbtui 可能看起来不错，但是存在经常花屏的问题，让很多人不胜其烦。那么在 Linux 下有没有既能在调试时动态显示当前调试处的文件代码又不花屏的工具呢？有，它就是 cgdb。

cgdb 在本质上是对 gdb 做了一层"包装"，在 gdb 中可以使用的所有命令，在 cgdb 中也可以使用。

从 cgdb 官网下载最新版的 cgdb，执行以下命令将 cgdb 压缩包下载到本地：

```
wget https://cgdb.me/files/cgdb-0.7.0.tar.gz
```

然后进行解压缩、编译、安装：

```
tar -xvfz cgdb-0.7.0.tar.gz
cd cgdb-0.7.0
./configure
make
make install
```

cgdb 在编译过程中会依赖一些第三方库，如果这些库在系统中不存在，就会报错，对其进行安装就可以了。

安装 cgdb 成功以后，在命令行中输入命令 cgdb 启动 cgdb，启动后的界面如下图所示。

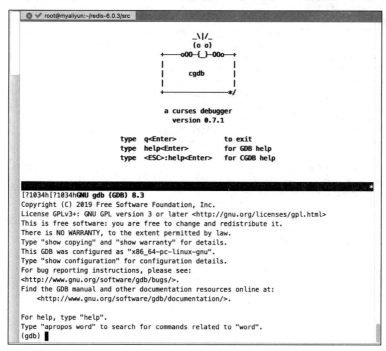

界面分为上下两部分：上半部分为代码窗口，显示处于调试过程中的代码；下半部分显示 gdb 原来的命令窗口。默认的窗口焦点在命令窗口中，如果想将窗口焦点切换到代码窗口，则按键盘上的 Esc 键，再次按键盘上的字母 i 键即可。注意："窗口焦点"的概念很重要，它决定着我们当前可以操作的是代码窗口还是命令窗口（其实和 gdbtui 一样）。

以 Redis 自带的客户端程序 redis-cli 为例，输入以下命令启动调试：

```
cgdb redis-cli
```

启动后的界面如下图所示。

```
                       exit(0);
                   }
               }
           }

           static sds askPassword() {
               linenoiseMaskModeEnable();
               sds auth = linenoise("Please input password: ");
               linenoiseMaskModeDisable();
               return auth;
           }

           /*------------------------------------------------
            * Program main()
            *------------------------------------------------*/

           int main(int argc, char **argv) {
               int firstarg;

               config.hostip = sdsnew("127.0.0.1");
               config.hostport = 6379;
/root/redis-6.0.3/src/redis-cli.c
[?1034h[?1034hGNU gdb (GDB) 8.3
Copyright (C) 2019 Free Software Foundation, Inc.
License GPLv3+: GNU GPL version 3 or later <http://gnu.org/licenses/gpl.html>
This is free software: you are free to change and redistribute it.
There is NO WARRANTY, to the extent permitted by law.
Type "show copying" and "show warranty" for details.
This GDB was configured as "x86_64-pc-linux-gnu".
Type "show configuration" for configuration details.
For bug reporting instructions, please see:
<http://www.gnu.org/software/gdb/bugs/>.
Find the GDB manual and other documentation resources online at:
    <http://www.gnu.org/software/gdb/documentation/>.

For help, type "help".
Type "apropos word" to search for commands related to "word"...
Reading symbols from redis-cli...
(gdb)
```

然后加两个断点，如下图所示。

```
 root@myaliyun:~/redis-6.0.3/src
7914        linenoiseMaskModeDisable();
7915        return auth;
7916    }
7917
7918    /*------------------------------------------------------------------------------
7919     * Program main()
7920     *------------------------------------------------------------------------------ */
7921
7922    int main(int argc, char **argv) {
7923        int firstarg;
7924
7925 |──> config.hostip = sdsnew("127.0.0.1");
7926        config.hostport = 6379;
7927        config.hostsocket = NULL;
7928        config.repeat = 1;
7929        config.interval = 0;
7930        config.dbnum = 0;
7931        config.interactive = 0;
7932        config.shutdown = 0;
7933        config.monitor_mode = 0;
7934        config.pubsub_mode = 0;
7935        config.latency_mode = 0;
/root/redis-6.0.3/src/redis-cli.c
Type "show copying" and "show warranty" for details.
This GDB was configured as "x86_64-pc-linux-gnu".
Type "show configuration" for configuration details.
For bug reporting instructions, please see:
<http://www.gnu.org/software/gdb/bugs/>.
Find the GDB manual and other documentation resources online at:
    <http://www.gnu.org/software/gdb/documentation/>.

For help, type "help".
Type "apropos word" to search for commands related to "word"...
Reading symbols from redis-cli...
(gdb) b main      ← 添加第 1 个断点
(gdb) r
Starting program: /root/redis-6.0.3/src/redis-cli
[Thread debugging using libthread_db enabled]
Using host libthread_db library "/usr/lib64/libthread_db.so.1".

Breakpoint 1, main (argc=1, argv=0x7fffffffe318) at redis-cli.c:7925
(gdb) b 7934      ← 添加第 2 个断点
(gdb)
```

如上图所示，我们在程序的 main（第 7922 行）和第 7934 行分别加了一个断点，添加断点以后，断点未触发前，代码视图中断点处的行号将以深色显示，另有一个箭头指向当前执行的行。我们调试时，可以看到代码视图中相应的代码也发生了变化，并且箭头始终指向当前执行的行数，如下图所示。

cgdb 虽然已经比原始的 gdb 和 gdbtui 模式在代码显示方面改进许多，但在 cgdb 中调用 gdb 的 print 命令无法显示字符串类型的中文字符，要么显示乱码，要么不显示，这给程序调试带来了很大的困扰。

```
   851        }
   852        return REDIS_ERR;
   853  }
   854
   855  /* Connect to the server. It is possible to pass certain flags to the function:
   856   *      CC_FORCE: The connection is performed even if there is already
   857   *                a connected socket.
   858   *      CC_QUIET: Don't print errors if connection fails. */
   859  static int cliConnect(int flags) {
   860 ─▶   if (context == NULL || flags & CC_FORCE) {
   861          if (context != NULL) {
   862              redisFree(context);
   863          }
   864
   865          if (config.hostsocket == NULL) {
   866              context = redisConnect(config.hostip,config.hostport);
   867          } else {
   868              context = redisConnectUnix(config.hostsocket);
   869          }
/root/redis-6.0.3/src/redis-cli.c
(gdb) r
The program being debugged has been started already.
Start it from the beginning? (y or n) y
The program being debugged has been started already.
Starting program: /root/redis-6.0.3/src/redis-cli
[Thread debugging using libthread_db enabled]
Using host libthread_db library "/usr/lib64/libthread_db.so.1".

Breakpoint 1, main (argc=1, argv=0x7fffffffe318) at redis-cli.c:7925
(gdb) c
Continuing.

Breakpoint 2, main (argc=1, argv=0x7fffffffe318) at redis-cli.c:7934
(gdb) c
Continuing.

Breakpoint 3, cliConnect (flags=0) at redis-cli.c:860
(gdb)
```

总体来说，cgdb 仍然能满足我们大多数场景下的调试需求，与 gdb 相比，有了 cgdb，我们在 Linux 系统中调试程序就方便多了。

2.11 使用 VisualGDB 调试

VisualGDB 是一款 Visual Studio 插件，安装以后可以在 Windows 系统上使用 Visual Studio 调试远程 Linux 程序，这样做的好处就是可以利用 Visual Studio 强大的代码阅读和调试功能。可能有读者会说，从 Visual Studio 2015 开始，Visual Studio 不是已经自带调试 Linux 程序的功能了么？很遗憾，Visual Studio 2015 或者 2017 自带的调试 Linux 程序的功能很鸡肋，调试一些简单的 Linux 小程序还可以，调试复杂的或者有多个源文件的 Linux 程序就难了。VisualGDB 是一款功能强大的商业软件，在本质上是利用 SSH 协议连接到远程 Linux 机器，然后利用 Visual Studio 产生相应的 gdb 命令，通过远程机器上的 gdbserver 传递给远程 gdb 调试器。其代码阅读功能建立在 samba 文件服务器之上。

使用这个工具远程调试 Linux 程序的方法有两种，下面一一进行讲解。

2.11.1 使用 VisualGDB 调试已经运行的程序

如果一个 Linux 程序已经运行，则可以使用 VisualGDB 的远程 attach 功能。为了方便演示，我们将 Linux 机器上的 redis-server 运行起来：

```
[root@localhost src]# ./redis-server
```

安装好 VisualGDB 插件以后，我们在 Visual Studio 的 Tools 菜单中选择 Linux Source Cache Manager 菜单项，将弹出如下对话框。

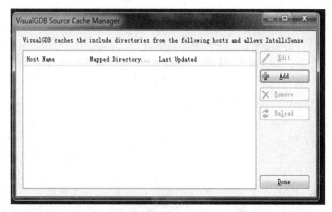

单击 Add 按钮，配置需要调试的 Linux 程序所在的 Linux 机器地址、用户名和密码。

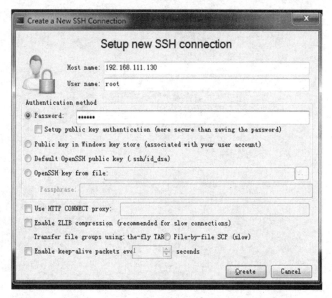

然后在 Debug 菜单中选择 Attach to Process...菜单项，将弹出 Attach To Process 对话框，对 Transport 类型选 "VisualGDB"，对 Qualifier 选择我们刚才配置的 Linux 主机信息。如果连接没问题，则在下面的进程列表中会弹出远程主机的进程列表，选择刚才启动的 redis-server，然后单击 Attach 按钮。

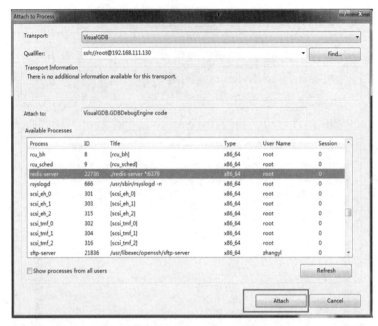

这样就可以在 Visual Studio 中调试这个 Linux 进程了。

2.11.2 使用 VisualGDB 从头调试程序

更多的时候，我们需要从一个程序启动处（main 函数处）调试程序，例如学习 Redis 源码时使用 VisualGDB 也是很方便的。在 Visual Studio 的 DEBUG 菜单中选择 Quick Debug With GDB 菜单项，在弹出的对话框中配置 Linux 程序所在的地址和目录。

第 2 章　C++后端开发必备的工具和调试知识

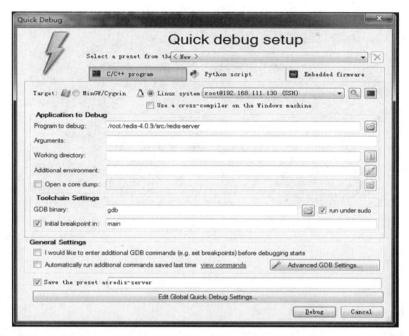

再单击图中 Debug 按钮，就可以启动调试了。

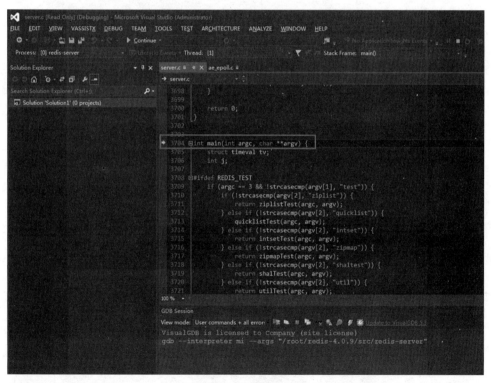

我们的程序会自动停在 main 函数处，这样就能利用强大的 Visual Studio 对 redis-server 进行调试了。当然，也可以在 VisualGDB 提供的 GDB Session 窗口中直接输入 gdb 的原始命令进行调试。

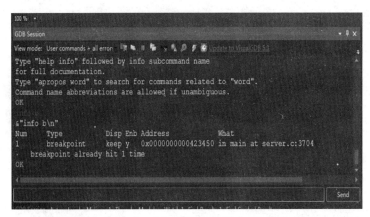

在 VisualGDB 中也存在一些缺点，调试 Linux 程序时可能会存在卡顿、延迟等现象。

第 3 章

多线程编程与资源同步

本章将结合操作系统的原理介绍多线程的方方面面，涉及 Windows 和 Linux 两个平台的线程技术。

3.1 线程的基本概念及常见问题

提到线程，就不得不提到进程。一个进程代表计算机中实际运行的一个程序。在现代操作系统的保护模式下，每个进程都拥有自己独立的进程地址空间和上下文堆栈。但就一个程序本身执行的操作来说，进程其实什么也不做（不执行任何进程代码），只提供一个大环境容器，进程中的实际执行体是线程（thread）。因此在一个进程中至少得有一个线程，我们把这个线程称为"主线程"。

通俗地说，线程是进程中实际执行代码的最小单元，由操作系统安排调度(何时启动、运行、暂停及消亡)。

这里重点强调我们在实际开发中使用多线程时需要弄明白的问题。

3.1.1 主线程退出，支线程也将退出吗

在 Windows 系统中，当一个进程存在多个线程时，如果主线程执行结束，那么这时支线程（也可以叫工作线程）即使还没有执行完相关代码，也会退出。也就是说，主线程一旦退出，整个进程也就结束了。之所以强调这一点，是因为很多初学者经常在工作线程中写了很多逻辑代码，但没有注意到主线程已经提前退出，导致这些工作线程的代码来不及执行。解决这一问题的方案有很多，核心是让主线程不退出，或者主线程至少在工作线程完成工作之前不要退出。常见的解决方案有让主线程启动一个循环或者主线程等待工作线程退出后再退出。

在 Linux 系统中，如果主线程退出，则工作线程一般不会受到影响，还会继续运行，但此时这个进程就会变成僵尸进程。这是一种不好的做法，在实际开发中应该避免产生僵尸进程。

使用 ps -ef 命令查看系统进程列表时，带有<defunct>字样的进程即僵尸进程：

```
[root@localhost ~]# ps -ef
UID        PID    PPID   C STIME TTY       TIME CMD
root         2       0   0 Jan18 ?     00:00:01 [kthreadd]
root         3       2   0 Jan18 ?     00:00:25 [ksoftirqd/0]
root         5       2   0 Jan18 ?     00:00:00 [kworker/0:0H]
root     60928      -1   0 14:48 pts/1  00:00:00 [linuxtid] <defunct>
```

Linux 版本众多，在某些 Linux 版本的实现中，主线程退出也会导致支线程退出，这就和在 Windows 上的行为一样了。我们在实际开发中应以自己的机器测试结果为准。

3.1.2 某个线程崩溃，会导致进程退出吗

这是一道常见的面试题，还有一种问法是：进程中的某个线程崩溃，是否会对其他线程造成影响？

一般来说，每个线程都是独立执行的单位，都有自己的上下文堆栈，一个线程崩溃不会对其他线程造成影响。但是在通常情况下，一个线程崩溃也会导致整个进程退出。例如在 Linux 操作系统中可能会产生一个 Segment Fault 错误，这个错误会产生一个信号，操作系统对这个信号的默认处理就是结束进程，这样整个进程都被销毁，在这个进程中存在的其他线程自然也就不存在了。

3.2 线程的基本操作

本节介绍对线程的一些基本操作。

3.2.1 创建线程

在使用线程之前，我们首先要学会创建一个新的线程。不管是哪个库还是哪种高级语言（如 Java），线程的创建最终还是通过调用操作系统的 API 进行的。这里先介绍操作系统的接口，分 Linux 和 Windows 两种常用的操作系统来介绍。当然，这里并不是照本宣科地把 Linux man 手册或者 msdn 上的函数签名搬过来，而是介绍实际开发中常用的参数和需要注意的重难点。

1. Linux 线程的创建

在 Linux 平台上使用 pthread_create 这个 API 来创建线程，其函数签名如下：

```
int pthread_create(pthread_t *thread,
           const pthread_attr_t *attr,
           void *(*start_routine) (void *),
           void *arg);
```

其中，参数 thread 是一个输出参数，如果线程创建成功，则通过这个参数可以得到创建成功的线程 ID（本章很快会介绍线程 ID 的知识）；参数 attr 指定了该线程的属性，一般被设置为 NULL，表示使用默认的属性；参数 start_routine 指定了线程函数，这个函数的调用方式必须是 __cdecl，这是 C Declaration 的缩写，__cdecl 是在 C/C++ 中定义全局函数时默认的调用方式。在 Windows 操作系统上使用 CreateThread 创建线程函数时要求线程函数必须使用 __stdcall 调用方式，但是定义的函数默认是 __cdecl 调用方式，所以必须显式声明 Windows 的线程函数为 __stdcall 调用方式（下文很快会介绍到）。也就是说，定义出来的全局函数可以作为 Linux pthread_create 的线程函数，但不能作为 Windows 的 CreateThread 的线程函数，这是因为如下函数调用方式是等价的：

```
//代码片段1：不显式指定函数调用方式，其调用方式为默认的__cdecl
void* start_routine (void* args)
{
}

//代码片段2：显式指定函数调用方式为默认的__cdecl，等价于代码片段1
void* __cdecl start_routine (void* args)
{
}
```

参数 arg 用于在创建线程时将某个参数传入线程函数中，由于它是 void* 类型，所以可以方便我们最大化地传入任意信息给线程函数；对于代码的返回值，如果成功创建线程，则返回 0；如果创建失败，则返回响应的错误码，常见的错误码有 EAGAIN、EINVAL、EPERM。

下面是一个使用 pthread_create 创建线程的简单示例：

```
#include <pthread.h>
#include <unistd.h>
#include <stdio.h>

void* threadfunc(void* arg)
{
    while (1)
    {
        //睡眠1秒
        sleep(1);

        printf("I am a New Thread!\n");
    }

    return NULL;
}

int main()
{
    pthread_t threadid;
    pthread_create(&threadid, NULL, threadfunc, NULL);

    //权宜之计，让主线程不要退出
    while (1)
```

```
        {
        }
        return 0;
}
```

2. Windows 线程的创建

在 Windows 上创建线程要用到 CreateThread，其函数签名如下：

```
HANDLE CreateThread(
  LPSECURITY_ATTRIBUTES     lpThreadAttributes,
  SIZE_T                    dwStackSize,
  LPTHREAD_START_ROUTINE    lpStartAddress,
  LPVOID                    lpParameter,
  DWORD                     dwCreationFlags,
  LPDWORD                   lpThreadId
);
```

其中，参数 lpThreadAttributes 表示线程的安全属性，一般被设置为 NULL；参数 dwStackSize 指线程的栈空间大小，单位为字节数，一般被指定为 0，表示使用默认的大小；参数 lpStartAddress 为线程函数，其类型是 LPTHREAD_START_ROUTINE（函数指针类型），其定义如下：

```
typedef DWORD (__stdcall *LPTHREAD_START_ROUTINE)(LPVOID lpThreadParameter);
```

上文提到过，在 Windows 上使用 CreateThread 创建的线程函数，其调用方式必须是 __stdcall，因此将如下函数设置成 CreateThread 的线程函数是不行的：

```
DWORD threadfunc(LPVOID lpThreadParameter);
```

上文对其原因进行了解释：如果不指定函数的调用方式，则默认使用 __cdecl 调用方式，而这里的线程函数要求是 __stdcall 调用方式，因此必须在函数名前面显式指定函数调用方式为 __stdcall：

```
DWORD __stdcall threadfunc(LPVOID lpThreadParameter);
```

在 Windows 上，WINAPI 和 CALLBACK 这两个宏的值都是 __stdcall，因此在很多项目中看到的线程函数签名大多有如下两种写法：

```
//写法 1
DWORD WINAPI threadfunc(LPVOID lpThreadParameter);
//写法 2
DWORD CALLBACK threadfunc(LPVOID lpThreadParameter);
```

参数 lpParameter 是传给线程函数的参数，和 Linux 下 pthread_create 函数的 arg 一样，实际上都是 void* 类型（LPVOID 类型实际上是用 typedef 包装后的 void* 类型）：

```
typedef void* LPVOID;
```

参数 dwCreationFlags 是一个 32 位的无符号整型（DWORD），一般被设置为 0，表示创建好线程后立即启动线程的运行；对于一些特殊情况，比如我们不希望创建线程后立即开始执行，则可以将这个值设置为 4（对应 Windows 定义的宏 CREATE_SUSPENDED），在需要时再使用 ResumeThread 这个 API 运行线程。

参数 lpThreadId 表示线程创建成功时返回的线程 ID，也表示一个 32 位无符号整数（DWORD）的指针（LPDWORD）。

在 Windows 上使用句柄（HANDLE 类型）来管理线程对象，句柄在本质上是内核句柄表中的索引值。如果成功创建线程，则返回该线程的句柄，否则返回 NULL。

下面的代码片段演示了如何在 Windows 上创建一个线程：

```
#include <Windows.h>
#include <stdio.h>
DWORD WINAPI ThreadProc(LPVOID lpParameters)
{
   while (true)
   {
      //睡眠1秒，Windows 上的 Sleep 函数的时间参数单位为毫秒
      Sleep(1000);

      printf("I am New Thread!\n");
   }

   return 0;
}
int main()
{
   DWORD dwThreadID;
   HANDLE hThread = CreateThread(NULL, 0, ThreadProc, NULL, 0, &dwThreadID);
   if (hThread == NULL)
   {
      printf("Failed to CreateThread.\n");
      return -1;
   }

   //权宜之计，让主线程不要退出
   while (true)
   {
   }

   return 0;
}
```

3. Windows CRT 提供的线程创建函数

CRT（C Runtime，C 运行时）通俗地说就是 C 函数库。在 Windows 上，微软实现的 C 库也提供了一套用于创建线程的函数（当然，这个函数底层还是调用相应的操作系统的线程创建 API）。在实际项目开发中推荐使用这个函数来创建线程，而不是使用 CreateThread 函数。

Windows C 库创建线程时常用的函数是 _beginthreadex，其声明位于 process.h 头文件中，签名如下：

```
uintptr_t _beginthreadex(
    void *security,
    unsigned stack_size,
    unsigned ( __stdcall *start_address )( void * ),
    void *arglist,
    unsigned initflag,
    unsigned *thrdaddr
);
```

_beginthreadex 函数签名和 Windows 的 CreateThread API 函数签名基本一致，这里不再赘述。

以下是一个使用_beginthreadex 创建线程的例子：

```
#include <process.h>
#include <stdio.h>

unsigned int __stdcall threadfun(void* args)
{
    while (true)
    {
        printf("I am New Thread!\n");
    }
}

int main(int argc, char* argv[])
{
    unsigned int threadid;
    _beginthreadex(0, 0, threadfun, 0, 0, &threadid);

    //权宜之计，让主线程不要退出
    while (true)
    {
    }

    return 0;
}
```

4. C++ 11 提供的 std::thread 类

无论是在 Linux 上还是在 Windows 上创建线程的 API，都有一个非常不方便的地方，就是线程函数签名必须使用规定的格式（对参数的个数和类型、返回值类型都有要求）。C++ 11 新标准引入了一个新的类 std::thread（需要包含头文件<thread>），使用这个类可以将任意签名形式的函数作为线程函数。以下代码分别创建了两个线程，线程函数签名不一样：

```
#include <stdio.h>
#include <thread>

void threadproc1()
{
    while (true)
    {
        printf("I am New Thread 1!\n");
```

```
    }
}

void threadproc2(int a, int b)
{
    while (true)
    {
        printf("I am New Thread 2!\n");
    }
}

int main()
{
    //创建线程t1
    std::thread t1(threadproc1);
    //创建线程t2
    std::thread t2(threadproc2, 1, 2);

    //权宜之计，让主线程不要退出
    while (true)
    {
    }

    return 0;
}
```

当然，std::thread 在使用上容易出错，即 std::thread 对象在线程函数运行期间必须是有效的。什么意思呢？我们来看一个例子：

```
#include <stdio.h>
#include <thread>

void threadproc()
{
    while (true)
    {
        printf("I am New Thread!\n");
    }
}

void func()
{
    std::thread t(threadproc);
}

int main()
{
    func();

    //权宜之计，让主线程不要退出
    while (true)
    {
    }
```

```
    return 0;
}
```

以上代码在 main 函数中调用了 func 函数，在 func 函数中创建了一个线程，乍一看好像没什么问题，但在实际运行时会崩溃。崩溃的原因是，在 func 函数调用结束后，func 中的局部变量 t（线程对象）被销毁，而此时线程函数仍在运行。所以在使用 std::thread 类时，必须保证在线程函数运行期间其线程对象有效。std::thread 对象提供了一个 detach 方法，通过这个方法可以让线程对象与线程函数脱离关系，这样即使线程对象被销毁，也不影响线程函数的运行。我们只需在 func 函数中调用 detach 方法即可，代码如下：

```
//其他代码保持不变，这里仅贴出修改后的 func 函数
void func()
{
    std::thread t(threadproc);
    t.detach();
}
```

然而在实际开发中并不推荐这么做，原因是我们可能需要使用线程对象去控制和管理线程的运行和生命周期。所以，我们的代码应该尽量保证线程对象在线程运行期间有效，而不是单纯地调用 detach 方法使线程对象与线程函数的运行分离。

3.2.2 获取线程 ID

在 1 个线程创建成功以后，我们可以拿到一个线程 ID。线程 ID 在整个操作系统范围内是唯一的。我们可以使用线程 ID 来标识和区分线程，例如在日志文件中输出日志的同时将输出日志的线程 ID 一起输出，这样可以方便我们判断和排查问题。上面也介绍了创建线程时可以通过 pthread_create 函数的第 1 个参数 thread（Linux 平台）和 CreateThread 函数的最后一个参数 lpThreadId（Windows 平台）得到线程的 ID。大多数时候，我们需要在当前调用线程中获取当前线程的 ID，在 Linux 平台上可以调用 pthread_self 函数获取，在 Windows 平台上可以调用 GetCurrentThreadID 函数获取，其函数签名分别如下：

```
pthread_t pthread_self(void);

DWORD GetCurrentThreadId();
```

这两个函数都比较简单，这里不再介绍。pthread_t 和 DWORD 类型在本质上都是 32 位无符号整型。

在 Windows 7 中可以在任务管理器中查看某个进程的线程数量。在下图中框住的是每个进程的线程数量，例如对于 vmware-tray.exe 进程一共有三个线程。如果在打开任务管理器时没有看到线程数这一列，则可以单击任务管理器的"查看"→"选择列"菜单，在弹出的对话框中勾选线程数即可显示。

1. pstack 命令

在 Linux 系统中可以通过 pstack 命令查看一个进程的线程数量和每个线程的调用堆栈情况：

```
pstack pid
```

这时将 pid 设置为要查看的进程 ID 即可。以笔者机器上 Nginx 的 worker 进程为例，首先使用 ps 命令查看 Nginx 进程 ID，然后使用 pstack 即可查看该进程每个线程的调用堆栈（这里演示的 Nginx 只有 1 个线程，如果有多个线程，则会显示每个线程的调用堆栈）：

```
[root@iZ238vnojlyZ ~]# ps -ef | grep nginx
root      2150     1  0 May22 ?        00:00:00 nginx: master process /usr/sbin/nginx
-c /etc/nginx/nginx.conf
nginx     2151  2150  0 May22 ?        00:00:07 nginx: worker process
root     16621 16541  0 18:53 pts/0    00:00:00 grep --color=auto nginx
[root@iZ238vnojlyZ ~]# pstack 2151
#0  0x00007f70a61ca2a3 in __epoll_wait_nocancel () from /lib64/libc.so.6
#1  0x0000000000437313 in ngx_epoll_process_events ()
#2  0x0000000000042efc3 in ngx_process_events_and_timers ()
#3  0x0000000000435681 in ngx_worker_process_cycle ()
#4  0x0000000000434104 in ngx_spawn_process ()
#5  0x0000000000435854 in ngx_start_worker_processes ()
#6  0x000000000043631b in ngx_master_process_cycle ()
#7  0x0000000000412229 in main ()
```

使用 pstack 命令查看的程序必须有调试符号，用户必须有相应的查看权限。

2. 利用 pstack 命令排查 Linux 进程 CPU 使用率过高的问题

在实际开发中，我们经常需要排查和定位一个进程 CPU 占用率过高的问题，这时可以结合使用 Linux top 和 pstack 命令来排查。来看一个具体的例子。如下图所示，我们使用 top 命令后发现机器上进程 ID 为 4427 的 qmarket 进程的 CPU 使用率达到 22.8%。

```
top - 14:35:26 up 5:37, 11 users, load average: 0.41, 0.54, 0.65
Tasks: 260 total,   1 running, 259 sleeping,   0 stopped,   0 zombie
%Cpu(s):  8.6 us,  5.4 sy,  0.0 ni, 85.8 id,  0.1 wa,  0.0 hi,  0.2 si,  0.0 st
KiB Mem : 65672324 total, 42604132 free, 18366676 used,  4701516 buff/cache
KiB Swap: 12582908 total, 12582908 free,        0 used. 46646976 avail Mem

  PID USER      PR  NI    VIRT    RES    SHR S  %CPU %MEM     TIME+ COMMAND
 4427 root      20   0  584792  16492   3232 S  22.8  0.0   2:49.84 qmarket
27509 root      20   0 1417184  12672   3352 S  10.6  0.0  12:06.72 marketserver
 7989 mysql     20   0 2668868 441644   9388 S   7.3  0.7  15:13.32 mysqld
10472 root      20   0 5824956   1.6g  14724 S   5.3  2.5  19:58.47 java
  439 root      20   0   52384   1952   1372 S   5.0  0.0  16:19.33 plymouthd
12505 root      20   0 6244776   1.7g  14672 S   5.0  2.8  17:46.60 java
11079 root      20   0 6095056   1.7g  14744 S   4.3  2.8  15:50.41 java
10210 root      20   0 8359228   3.3g  14608 S   2.6  5.3   7:10.00 java
10627 root      20   0 5775332   1.6g  14740 S   2.0  2.5   7:00.54 java
12114 root      20   0 5952936   1.7g  14604 S   2.0  2.7   9:22.55 java
12640 root      20   0 5957720   1.7g  14852 S   1.7  2.7   7:52.70 java
 1096 root      20   0 5724136 683120  13636 S   1.0  1.0   0:51.79 java
 5561 root      20   0  162176   2468   1588 R   0.7  0.0   0:00.07 top
 6888 consul    20   0  289064  54848  31872 S   0.7  0.1   2:56.53 consul
 8240 prometh+  20   0  290744  48932  16980 S   0.7  0.1   0:41.29 prometheus
32327 root      20   0  171112   5632   4264 S   0.7  0.0   0:24.81 sshd
12102 redis     20   0  142956   7384   1588 S   0.3  0.0   0:33.92 redis-server
    1 root      20   0  194360   7568   4212 S   0.0  0.0   0:10.10 systemd
    2 root      20   0       0      0      0 S   0.0  0.0   0:00.05 kthreadd
    3 root      20   0       0      0      0 S   0.0  0.0   0:05.91 ksoftirqd/0
    5 root       0 -20       0      0      0 S   0.0  0.0   0:00.00 kworker/0:0H
    7 root      rt   0       0      0      0 S   0.0  0.0   0:00.02 migration/0
    8 root      20   0       0      0      0 S   0.0  0.0   0:00.00 rcu_bh
    9 root      20   0       0      0      0 S   0.0  0.0   0:30.27 rcu_sched
   10 root       0 -20       0      0      0 S   0.0  0.0   0:00.00 lru-add-drain
   11 root      rt   0       0      0      0 S   0.0  0.0   0:00.10 watchdog/0
   12 root      rt   0       0      0      0 S   0.0  0.0   0:00.09 watchdog/1
   13 root      rt   0       0      0      0 S   0.0  0.0   0:00.01 migration/1
   14 root      20   0       0      0      0 S   0.0  0.0   0:00.21 ksoftirqd/1
   16 root       0 -20       0      0      0 S   0.0  0.0   0:00.00 kworker/1:0H
   17 root      rt   0       0      0      0 S   0.0  0.0   0:00.08 watchdog/2
   18 root      rt   0       0      0      0 S   0.0  0.0   0:00.02 migration/2
   19 root      20   0       0      0      0 S   0.0  0.0   0:00.21 ksoftirqd/2
   21 root       0 -20       0      0      0 S   0.0  0.0   0:00.00 kworker/2:0H
   22 root      rt   0       0      0      0 S   0.0  0.0   0:00.08 watchdog/3
   23 root      rt   0       0      0      0 S   0.0  0.0   0:00.03 migration/3
   24 root      20   0       0      0      0 S   0.0  0.0   0:00.56 ksoftirqd/3
   26 root       0 -20       0      0      0 S   0.0  0.0   0:00.00 kworker/3:0H
   28 root      20   0       0      0      0 S   0.0  0.0   0:00.00 kdevtmpfs
   29 root       0 -20       0      0      0 S   0.0  0.0   0:00.00 netns
```

我们使用 top -H 命令再次输出系统的进程列表，top 命令的 -H 选项的作用是显示每个进程各个线程的运行状态（线程模式）。执行结果如下图所示。

```
top - 14:23:22 up 5:25, 11 users, load average: 0.47, 0.84, 0.77
Threads: 2028 total,   3 running, 2025 sleeping,   0 stopped,   0 zombie
%Cpu(s):  8.6 us,  6.2 sy,  0.0 ni, 84.9 id,  0.1 wa,  0.0 hi,  0.2 si,  0.0 st
KiB Mem : 65672324 total, 42696728 free, 18333176 used,  4642420 buff/cache
KiB Swap: 12582908 total, 12582908 free,        0 used. 46681488 avail Mem

  PID USER      PR  NI    VIRT    RES    SHR S  %CPU %MEM     TIME+ COMMAND
13690 mysql     20   0 2668868 441396   9388 S   6.5  0.7   0:26.87 mysqld
  439 root      20   0   52384   1952   1372 S   4.9  0.0  15:43.84 plymouthd
 4433 root      20   0  440804  10044   3228 S   4.2  0.0   0:02.24 qmarket
13035 root      20   0 5824956   1.6g  14716 S   3.6  2.5  11:39.28 java
 4434 root      20   0  440804  10044   3228 S   3.3  0.0   0:01.33 qmarket
 4429 root      20   0  440804  10044   3228 S   2.9  0.0   0:01.14 qmarket
 4430 root      20   0  440804  10044   3228 S   2.9  0.0   0:01.33 qmarket
 4431 root      20   0  440804  10044   3228 S   2.9  0.0   0:01.33 qmarket
 4432 root      20   0  440804  10044   3228 S   2.9  0.0   0:01.35 qmarket
 4445 root      20   0  440804  10044   3228 S   2.9  0.0   0:01.12 qmarket
27509 root      20   0 1417184  12760   3352 S   2.9  0.0   3:47.66 marketserver
11180 root      20   0 6095056   1.7g  14744 R   2.6  2.8  10:36.47 java
27532 root      20   0 1417184  12760   3352 S   2.6  0.0   3:47.29 marketserver
 4439 root      20   0  164132   4356   1596 R   2.0  0.0   0:00.81 top
27530 root      20   0 1417184  12760   3352 S   2.0  0.0   1:22.08 marketserver
10574 root      20   0 5824956   1.6g  14716 S   1.6  2.5   4:26.59 java
 4218 mysql     20   0 2668868 441396   9388 S   1.0  0.7   0:01.16 mysqld
10687 root      20   0 5775332   1.6g  14740 S   1.0  2.5   4:19.63 java
 8777 nfsnobo+  20   0  114572  17692   5152 S   0.7  0.0   0:04.88 node_exporter
 9299 nfsnobo+  20   0  114572  17692   5152 S   0.7  0.0   0:07.53 node_exporter
10570 root      20   0 5824956   1.6g  14716 S   0.7  2.5   0:01.09 java
11085 root      20   0 6095056   1.7g  14744 S   0.7  2.8   0:27.57 java
12594 root      20   0 6242728   1.7g  14672 S   0.7  2.7   1:33.18 java
12596 root      20   0 6242728   1.7g  14672 S   0.7  2.7   1:33.37 java
16408 root      20   0 6242728   1.7g  14672 S   0.7  2.7   0:56.55 java
27529 root      20   0 1417184  12760   3352 S   0.7  0.0   0:30.82 marketserver
29758 root      20   0 7905760 651544  13632 S   0.7  1.0   0:06.58 java
32327 root      20   0  171112   5632   4264 S   0.7  0.0   0:20.31 sshd
    9 root      20   0       0      0      0 S   0.3  0.0   0:29.14 rcu_sched
 1145 root      20   0 5724136 678220  13636 S   0.3  1.0   0:01.52 java
 1167 root      20   0 5724136 678220  13636 S   0.3  1.0   0:00.79 java
 7418 consul    20   0  209064  54608  31872 S   0.3  0.1   0:10.71 consul
 8013 mysql     20   0 2668868 441396   9388 S   0.3  0.7   0:05.11 mysqld
 8148 mysql     20   0 2668868 441396   9388 S   0.3  0.7   0:10.60 mysqld
 8776 nfsnobo+  20   0  114572  17692   5152 S   0.3  0.0   0:08.62 node_exporter
13146 nfsnobo+  20   0  114572  17692   5152 S   0.3  0.0   0:07.49 node_exporter
10224 root      20   0 8348956   3.3g  14608 S   0.3  5.3   0:16.61 java
10239 root      20   0 8348956   3.3g  14608 S   0.3  5.3   0:01.42 java
10249 root      20   0 8348956   3.3g  14608 S   0.3  5.3   0:12.86 java
10292 root      20   0 8348956   3.3g  14608 S   0.3  5.3   0:05.78 java
```

如上图所示，top 命令第 1 栏的输出虽然还是 PID，但显示的实际上是每个线程的线程 ID。我们可以发现 qmarket 线程 ID 为 4429、4430、4431、4432、4433、4434、4445 的线程其 CPU 使用率较高。那么这几个线程到底做了什么导致 CPU 使用率高呢？我们使用 pstack 4427 命令来看一下这几个线程（4427 是 qmarket 的进程 ID）：

```
[root@js-dev2 ~]# pstack 4427
Thread 9 (Thread 0x7f315cb39700 (LWP 4428)):
#0  0x00007f315db3d965 in pthread_cond_wait@@GLIBC_2.3.2 () from
/lib64/libpthread.so.0
#1  0x00007f315d8dc82c in
std::condition_variable::wait(std::unique_lock<std::mutex>&) () from
/lib64/libstdc++.so.6
#2  0x0000000000467a89 in CAsyncLog::WriteThreadProc ()
at ../../sourcebase/utility/AsyncLog.cpp:300
#3  0x0000000000469a0f in std::_Bind_simple<void
(*())()>::_M_invoke<>(std::_Index_tuple<>) (this=0xddeb60) at
/usr/include/c++/4.8.2/functional:1732
#4  0x0000000000469969 in std::_Bind_simple<void (*())()>::operator()()
(this=0xddeb60) at /usr/include/c++/4.8.2/functional:1720
#5  0x0000000000469902 in std::thread::_Impl<std::_Bind_simple<void
(*())() > >::_M_run() (this=0xddeb48) at /usr/include/c++/4.8.2/thread:115
#6  0x00007f315d8e0070 in ?? () from /lib64/libstdc++.so.6
#7  0x00007f315db39dd5 in start_thread () from /lib64/libpthread.so.0
#8  0x00007f315d043ead in clone () from /lib64/libc.so.6
Thread 8
//省略线程 8 的堆栈输出
Thread 7
//省略线程 7 的堆栈输出
Thread 6
//省略线程 6 的堆栈输出
Thread 5
//省略线程 5 的堆栈输出
Thread 4
//省略线程 4 的堆栈输出
Thread 3
//省略线程 3 的堆栈输出
Thread 2 (Thread 0x7f3154bf8700 (LWP 4445)):
#0  0x00007f315d00ae2d in nanosleep () from /lib64/libc.so.6
#1  0x00007f315d03b704 in usleep () from /lib64/libc.so.6
#2  0x000000000043ed67 in CThread::SleepMs (this=0x7f3150001b00,
nMilliseconds=1000) at ../../sourcebase/event/Thread.cpp:106
#3  0x0000000000441f82 in CEventDispatcher::Run (this=0x7f3150001b00)
at ../../sourcebase/event/EventDispatcher.cpp:63
#4  0x000000000043eb33 in CThread::_ThreadEntry (pParam=0x7f3150001b00)
at ../../sourcebase/event/Thread.cpp:26
#5  0x00007f315db39dd5 in start_thread () from /lib64/libpthread.so.0
#6  0x00007f315d043ead in clone () from /lib64/libc.so.6
Thread 1 (Thread 0x7f315f2ca3c0 (LWP 4427)):
#0  0x00007f315db3af47 in pthread_join () from /lib64/libpthread.so.0
#1  0x000000000043edc7 in CThread::Join (this=0x7ffc5eed32e0)
at ../../sourcebase/event/Thread.cpp:130
#2  0x000000000040cc61 in main (argc=1, argv=0x7ffc5eed3668)
at ../../sourceapp/qmarket/qmarket.cpp:309
```

在 pstack 输出的各个线程中，只要逐一对照我们的程序源码，梳理该线程中是否有大多数时间处于空转状态的逻辑，然后修改和优化这些逻辑，就可以解决 CPU 使用率过高的问题。在一般情况下，对不工作的线程应尽量使用锁对象让其挂起，而不是空转，这样可以提高系统资源利用率。

3. Linux 系统线程 ID 的本质

在 Linux 系统中有三种方法可以获取 1 个线程的 ID。

方法一，调用 pthread_create 函数时，在函数调用成功后，通过第 1 个参数可以得到线程 ID：

```
#include <pthread.h>

pthread_t tid;
pthread_create(&tid, NULL, thread_proc, NULL);
```

方法二，在需要获取 ID 的线程中调用 pthread_self 函数获取：

```
#include <pthread.h>

pthread_t tid = pthread_self();
```

方法三，通过系统调用获取线程 ID：

```
#include <sys/syscall.h>
#include <unistd.h>

int tid = syscall(SYS_gettid);
```

方法一和方法二获取线程 ID 的结果是一样的，都是 pthread_t 类型，输出的是一块内存空间地址，示意图如下。

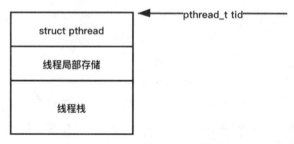

由于不同的进程可能有同样地址的内存块，因此通过方法一和方法二获取的线程 ID 可能不是全系统唯一的，一般是一个很大的数字（内存地址）。而通过方法三获取的线程 ID 是系统范围内全局唯一的，一般是一个不太大的整数，也就是 LWP（Light Weight Process，轻量级进程，早期 Linux 系统的线程是通过进程实现的，这种线程被称为轻量级进程）的 ID。

来看一段具体的代码：

```
#include <sys/syscall.h>
#include <unistd.h>
#include <stdio.h>
```

```c
#include <pthread.h>

void* thread_proc(void* arg)
{
    pthread_t* tid1 = (pthread_t*)arg;
    int tid2 = syscall(SYS_gettid);
    pthread_t tid3 = pthread_self();

    while(true)
    {
        printf("tid1: %ld, tid2: %ld, tid3: %ld\n", *tid1, tid2, tid3);
        sleep(1);
    }

}

int main()
{
    pthread_t tid;
    pthread_create(&tid, NULL, thread_proc, &tid);

    pthread_join(tid, NULL);

    return 0;
}
```

以上代码在新开的线程中使用上面介绍的三种方式获取线程 ID 并打印，输出结果如下：

```
tid1: 140185007511296, tid2: 60837, tid3: 140185007511296
```

tid2 即 LWP 的 ID，而 tid1 和 tid3 是一个内存地址，转换成 16 进制是 0x7F7F5D935700。

这与我们使用 pstack 命令看到的线程 ID 是一样的：

```
[root@localhost ~]# ps -efL | grep linuxtid
root      60712  60363  60712  0    2 13:25 pts/1    00:00:00 ./linuxtid
root      60712  60363  60713  0    2 13:25 pts/1    00:00:00 ./linuxtid
root      60720  60364  60720  0    1 13:25 pts/3    00:00:00 grep --color=auto linuxtid
[root@localhost ~]# pstack 60712
Thread 2 (Thread 0x7fd897a50700 (LWP 60713)):
#0  0x00007fd897b15e2d in nanosleep () from /lib64/libc.so.6
#1  0x00007fd897b15cc4 in sleep () from /lib64/libc.so.6
#2  0x0000000000400746 in thread_proc (arg=0x7fff390921c8) at linuxtid.cpp:15
#3  0x00007fd898644dd5 in start_thread () from /lib64/libpthread.so.0
#4  0x00007fd897b4eead in clone () from /lib64/libc.so.6
Thread 1 (Thread 0x7fd898a6e740 (LWP 60712)):
#0  0x00007fd898645f47 in pthread_join () from /lib64/libpthread.so.0
#1  0x000000000040077e in main () at linuxtid.cpp:25
[root@localhost ~]# ps -ef | grep linuxtid
root      60838  60363  0 14:27 pts/1    00:00:00 ./linuxtid
root      60846  60364  0 14:28 pts/3    00:00:00 grep --color=auto linuxtid
[root@localhost ~]# pstack 60838
Thread 2 (Thread 0x7f7f5d935700 (LWP 60839)):
#0  0x00007f7f5d9fae2d in nanosleep () from /lib64/libc.so.6
#1  0x00007f7f5d9facc4 in sleep () from /lib64/libc.so.6
```

```
#2  0x0000000000400746 in thread_proc (arg=0x7fff0523ae68) at linuxtid.cpp:15
#3  0x00007f7f5e529dd5 in start_thread () from /lib64/libpthread.so.0
#4  0x00007f7f5da33ead in clone () from /lib64/libc.so.6
Thread 1 (Thread 0x7f7f5e953740 (LWP 60838)):
#0  0x00007f7f5e52af47 in pthread_join () from /lib64/libpthread.so.0
#1  0x000000000040077e in main () at linuxtid.cpp:25
```

4. C++ 11 获取当前线程 ID 的方法

C++ 11 的线程库可以使用 std::this_thread 类的 get_id 获取当前线程 ID，这是一个类静态方法。

当然，也可以使用 std::thread 的 get_id 获取指定线程的 ID，这是一个类实例方法。

但 get_id 方法返回的是一个 std::thread::id 的包装类型，该类型不可以被直接强转成整型，C++ 11 线程库也没有为该对象提供任何转换成整型的接口。所以，我们一般使用 std::cout 这样的输出流来输出，或者先转换为 std::ostringstream 对象，再转换为字符串类型，然后把字符串类型转换为我们需要的整型，这算是 C++ 11 线程库获取线程 ID 一个不太方便的地方：

```cpp
//test_cpp11_thread_id.cpp
#include <thread>
#include <iostream>
#include <sstream>

void worker_thread_func()
{
    while (true)
    {
    }
}

int main()
{
    std::thread t(worker_thread_func);
    //获取线程 t 的 ID
    std::thread::id worker_thread_id = t.get_id();
    std::cout << "worker thread id: " << worker_thread_id << std::endl;

    //获取主线程的线程 ID
    std::thread::id main_thread_id = std::this_thread::get_id();
    //先将 std::thread::id 转换为 std::ostringstream 对象
    std::ostringstream oss;
    oss << main_thread_id;
    //再将 std::ostringstream 对象转换为 std::string
    std::string str = oss.str();
    std::cout << "main thread id: " << str << std::endl;
    //最后将 std::string 转换为整型值
    unsigned long long threadid = std::stoull(str);
    std::cout << "main thread id: " << threadid << std::endl;

    while (true)
    {
```

```
        //权宜之计，让主线程不要退出
    }
    return 0;
}
```

在 Linux x64 系统上编译并运行程序，输出结果如下：

```
[root@myaliyun codes]# g++ -g -o test_cpp11_thread_id test_cpp11_thread_id.cpp
-lpthread
[root@myaliyun codes]# ./test_cpp11_thread_id
worker thread id: 139875808245504
main thread id: 139875825641280
main thread id: 139875825641280
```

编译成 Windows x86 程序，运行结果如下图所示。

3.2.3 等待线程结束

在实际项目开发中，我们常常会有这样一种需求，即一个线程需要等待另一个线程执行完任务并退出后再继续执行。这在 Linux 和 Windows 中都提供了相应的 API，下面分别介绍一下。

1. 在 Linux 下等待线程结束

Linux 线程库提供了 pthread_join 函数，用来等待某线程的退出并接收它的返回值。这种操作被称为汇接（join）。pthread_join 函数签名如下：

```
int pthread_join(pthread_t thread, void** retval);
```

参数 thread 是需要等待的线程 ID；参数 retval 是输出参数，用于接收等待退出的线程的退出码（Exit Code），可以在调用 pthread_exit 退出线程时指定线程退出码，也可以在线程函数中通过 return 语句返回线程退出码。pthread_exit 函数签名如下：

```
#include <pthread.h>

void pthread_exit(void* value_ptr);
```

参数 value_ptr 的值可以通过 pthread_join 函数的第 2 个参数得到，如果不需要使用这个参数，则可以将其设置为 NULL。

pthread_join 函数在等待目标线程退出期间会挂起当前线程（调用 pthread_join 的线程），被挂起的线程处于等待状态，不会消耗任何 CPU 时间片。直到目标线程退出后，调用 pthread_join 的线程才会被唤醒，继续执行接下来的逻辑。这里通过一个实例演示这个函数的使用方法，实例功能为：在程序启动时开启一个工作线程，工作线程将当前系统时间写入文件后退出,主线程等待工作线程退出后,从文件中读取时间并将其显示在屏幕上。

相应的代码如下:

```c
#include <stdio.h>
#include <string.h>
#include <pthread.h>

#define TIME_FILENAME "time.txt"

void* fileThreadFunc(void* arg)
{
    time_t now = time(NULL);
    struct tm* t = localtime(&now);
    char timeStr[32] = {0};
    snprintf(timeStr, 32, "%04d/%02d/%02d %02d:%02d:%02d",
            t->tm_year+1900,
            t->tm_mon+1,
            t->tm_mday,
            t->tm_hour,
            t->tm_min,
            t->tm_sec);
    //文件不存在则创建;存在则覆盖
    FILE* fp = fopen(TIME_FILENAME, "w");
    if (fp == NULL)
    {
        printf("Failed to create time.txt.\n");
        return NULL;
    }

    size_t sizeToWrite = strlen(timeStr) + 1;
    size_t ret = fwrite(timeStr, 1, sizeToWrite, fp);
    if (ret != sizeToWrite)
    {
        printf("Write file error.\n");
    }

    fclose(fp);
    return NULL;
}

int main()
{
    pthread_t fileThreadID;
    int ret = pthread_create(&fileThreadID, NULL, fileThreadFunc, NULL);
    if (ret != 0)
    {
        printf("Failed to create fileThread.\n");
        return -1;
    }

    int* retval;
    pthread_join(fileThreadID, (void**)&retval);

    //使用 r 选项,要求文件必须存在
```

```c
    FILE* fp = fopen(TIME_FILENAME, "r");
    if (fp == NULL)
    {
        printf("open file error.\n");
        return -2;
    }

    char buf[32] = {0};
    int sizeRead = fread(buf, 1, 32, fp);
    if (sizeRead == 0)
    {
        printf("read file error.\n");
        fclose(fp);
        return -3;
    }

    printf("Current Time is: %s.\n", buf);

    fclose(fp);

    return 0;
}
```

程序执行结果如下：

```
[root@localhost threadtest]# ./test
Current Time is: 2018/09/24 21:06:01.
```

2. Windows 等待线程结束

在 Windows 下有两个非常重要的函数 API：WaitForSingleObject 和 WaitForMultipleObjects，前者用于等待 1 个线程结束，后者可以同时等待多个线程结束。这两个函数不仅可以用于等待线程退出，还可以用于等待其他线程同步对象。与 Linux 的 pthread_join 函数不同，Windows 的 WaitForSingleObject 函数提供了对可选择的等待时间的精细控制。

这里仅演示等待线程退出。将上面的 Linux 示例代码改写成 Windows 版本：

```c
#include <stdio.h>
#include <string.h>
#include <time.h>
#include <Windows.h>

#define TIME_FILENAME "time.txt"

DWORD WINAPI FileThreadFunc(LPVOID lpParameters)
{
    time_t now = time(NULL);
    struct tm* t = localtime(&now);
    char timeStr[32] = { 0 };
    sprintf_s(timeStr, 32, "%04d/%02d/%02d %02d:%02d:%02d",
        t->tm_year + 1900,
        t->tm_mon + 1,
        t->tm_mday,
        t->tm_hour,
        t->tm_min,
```

```cpp
            t->tm_sec);
    //文件不存在则创建；存在则覆盖
    FILE* fp = fopen(TIME_FILENAME, "w");
    if (fp == NULL)
    {
        printf("Failed to create time.txt.\n");
        return 1;
    }

    size_t sizeToWrite = strlen(timeStr) + 1;
    size_t ret = fwrite(timeStr, 1, sizeToWrite, fp);
    if (ret != sizeToWrite)
    {
        printf("Write file error.\n");
    }

    fclose(fp);

    return 2;
}

int main()
{
    DWORD dwFileThreadID;
    HANDLE hFileThread = CreateThread(NULL, 0, FileThreadFunc, NULL, 0,
                                &dwFileThreadID);
    if (hFileThread == NULL)
    {
        printf("Failed to create fileThread.\n");
        return -1;
    }

    //无限等待，直到文件线程退出，否则程序将一直挂起
    WaitForSingleObject(hFileThread, INFINITE);

    //使用 r 选项，要求文件必须存在
    FILE* fp = fopen(TIME_FILENAME, "r");
    if (fp == NULL)
    {
        printf("open file error.\n");
        return -2;
    }

    char buf[32] = { 0 };
    int sizeRead = fread(buf, 1, 32, fp);
    if (sizeRead == 0)
    {
        printf("read file error.\n");
        fclose(fp);
        return -3;
    }

    printf("Current Time is: %s.\n", buf);

    fclose(fp);
```

```
    return 0;
}
```

程序执行结果如下图所示。

```
C:\Windows\system32\cmd.exe
Current Time is: 2018/09/24 21:34:22.
请按任意键继续. . .
```

3. C++ 11 提供的等待线程结果的函数

可以想到，既然 C++ 11 的 std::thread 统一了 Linux 和 Windows 的线程创建函数，那么它应该也提供了等待线程退出的接口。确实如此，std::thread 的 join 方法就是用来等待线程退出的方法。当然，使用这个函数时，必须保证该线程处于运行状态，也就是说等待的线程必须是可以 join 的，如果需要等待的线程已经退出，则此时调用 join 方法，程序就会崩溃。因此，C++ 11 的线程库同时提供了一个 joinable 方法来判断某个线程是否可以 join。

将上面的代码改写成 C++ 11 版本：

```
#include <stdio.h>
#include <string.h>
#include <time.h>
#include <thread>

#define TIME_FILENAME "time.txt"

void FileThreadFunc()
{
    time_t now = time(NULL);
    struct tm* t = localtime(&now);
    char timeStr[32] = { 0 };
    sprintf_s(timeStr, 32, "%04d/%02d/%02d %02d:%02d:%02d",
            t->tm_year + 1900,
            t->tm_mon + 1,
            t->tm_mday,
            t->tm_hour,
            t->tm_min,
            t->tm_sec);
    //文件不存在则创建；存在则覆盖
    FILE* fp = fopen(TIME_FILENAME, "w");
    if (fp == NULL)
    {
        printf("Failed to create time.txt.\n");
        return;
    }

    size_t sizeToWrite = strlen(timeStr) + 1;
    size_t ret = fwrite(timeStr, 1, sizeToWrite, fp);
    if (ret != sizeToWrite)
    {
        printf("Write file error.\n");
```

```cpp
    }
    fclose(fp);
}

int main()
{
    std::thread t(FileThreadFunc);
    if (t.joinable())
        t.join();

    //使用 r 选项，要求文件必须存在
    FILE* fp = fopen(TIME_FILENAME, "r");
    if (fp == NULL)
    {
        printf("open file error.\n");
        return -2;
    }

    char buf[32] = { 0 };
    int sizeRead = fread(buf, 1, 32, fp);
    if (sizeRead == 0)
    {
        printf("read file error.\n");
        fclose(fp);
        return -3;
    }

    printf("Current Time is: %s.\n", buf);

    fclose(fp);

    return 0;
}
```

3.3　惯用法：将 C++ 类对象实例指针作为线程函数的参数

前面的章节介绍了，除 C++ 11 的线程库提供的 std::thread 类对线程函数签名没有特殊要求外，无论是 Linux 还是 Windows 的线程函数签名，都必须是指定的格式，即参数和返回值必须是规定的形式。如果使用 C++ 面向对象的方式对线程函数进行封装，线程函数就不能是类的实例方法了，即必须是类的静态方法。那么为什么不能是类的实例方法呢？以 Linux 的线程函数签名为例：

```cpp
void* threadFunc(void* arg);
```

假设将线程的基本功能都封装到一个 Thread 类中，部分代码如下：

```cpp
class Thread
{
public:
    Thread();
    ~Thread();
```

```
    void start();
    void stop();

    void* threadFunc(void* arg);
};
```

由于 threadFunc 是一个类实例方法,无论是类的实例方法还是静态方法,C++编译器在编译时都会将这些函数"翻译"成全局函数,即去掉类的域限制。对于实例方法,为了保证类方法的正常功能,C++编译器在翻译时会将类的实例对象地址(也就是 this 指针)作为第 1 个参数传递给该方法,也就是说,翻译后的 threadFunc 的签名变成了如下形式(伪代码):

```
void* threadFunc(Thread* this, void* arg);
```

这样的话就不符合线程函数的签名要求了。因此,如果使用类方法作为线程函数,则只能是类的静态方法,不能是类的实例方法。

当然,如果使用 C++ 11 的 std::thread 类,就没有这个限制,即使类成员函数是类的实例方法也可以,但必须显式地将线程函数所属的类对象实例指针(在类的内部就是 this 指针)作为构造函数参数传递给 std::thread,这时还需要传递类的 this 指针,这在本质上是一样的,代码实例如下:

```cpp
#include <thread>
#include <memory>
#include <stdio.h>

class Thread
{
public:
    Thread()
    {
    }

    ~Thread()
    {
    }

    void Start()
    {
        m_stopped = false;
        //threadFunc 是类的非静态方法,
        //所以作为线程函数,第 1 个参数必须传递类实例的地址,即 this 指针
        m_spThread.reset(new std::thread(&Thread::threadFunc, this, 8888, 9999));
    }

    void Stop()
    {
        m_stopped = true;
        if (m_spThread)
        {
            if (m_spThread->joinable())
                m_spThread->join();
```

```cpp
        }
    }

private:
    void threadFunc(int arg1, int arg2)
    {
        while (!m_stopped)
        {
            printf("Thread function use instance method.\n");
        }
    }

private:
    std::shared_ptr<std::thread>  m_spThread;
    bool                          m_stopped;
};

int main()
{
    Thread mythread;
    mythread.Start();

    //权宜之计,让主线程不要提前退出
    while (true)
    {
    }

    return 0;
}
```

在以上代码中使用了 C++ 11 新增的智能指针 std::shared_ptr 类包裹了新建的 std::thread 对象,这样我们就不需要自己手动删除这个 std::thread 对象了。

综上所述,如果不使用 C++ 11 的语法,那么线程函数只能使用类的静态方法,且函数签名必须符合线程函数的签名要求。如果是类的静态方法,就无法访问类的实例方法了。为了解决这个问题,我们在实际开发中往往会在创建线程时将当前对象的地址(this 指针)传递给线程函数,然后在线程函数中将该指针转换为原来的类实例,再通过这个实例就可以访问类的所有方法了。

Thread.h 文件的代码如下:

```cpp
/**
 * Thread.h
 */
#ifdef WIN32
typedef HANDLE THREAD_HANDLE ;
#else
typedef pthread_t THREAD_HANDLE ;
#endif

/**
 * 定义一个线程对象
 */
class CThread
```

```cpp
{
public:
    CThread();
    virtual ~CThread();

    /**创建一个线程
     * @return true:创建成功; false:创建失败
     */
    virtual bool Create();

    /**获得本线程对象存储的线程句柄
     * @return 本线程对象存储的线程句柄
     */
    THREAD_HANDLE GetHandle();

    /**线程睡眠 nSeconds 秒
     * @param nSeconds, 睡眠秒数
     */
    void OSSleep(int nSeconds);
    void SleepMs(int nMilliseconds);
    bool Join();
    bool IsCurrentThread();
    void ExitThread();

private:
#ifdef WIN32
    static DWORD WINAPI _ThreadEntry(LPVOID pParam);
#else
    static void* _ThreadEntry(void* pParam);
#endif

    /**虚函数，子类可做一些实例化工作
     * @return true:创建成功 false:创建失败
     */
    virtual bool InitInstance();

    /**虚函数，子类清除实例
     */
    virtual void ExitInstance();

    /**存放线程函数的实际业务逻辑，纯虚函数，子类必须重写该函数
     */
    virtual void Run() = 0;

private:
    //线程句柄
    THREAD_HANDLE   m_hThread;
    DWORD           m_IDThread;
};
```

Thread.cpp 文件的代码如下：

```cpp
/**
 * Thread.cpp
 */
```

```cpp
#include "Thread.h"

#ifdef WIN32
DWORD WINAPI CThread::_ThreadEntry(LPVOID pParam)
#else
void* CThread::_ThreadEntry(void* pParam)
#endif
{
    CThread *pThread = (CThread *)pParam;
    if(pThread->InitInstance())
    {
        pThread->Run();
    }

    pThread->ExitInstance();

    return NULL;
}

CThread::CThread()
{
    m_hThread = (THREAD_HANDLE)0;
    m_IDThread = 0;
}

CThread::~CThread()
{
}

bool CThread::Create()
{
    if (m_hThread != (THREAD_HANDLE)0)
        return true;

    bool ret = true;
#ifdef WIN32
    m_hThread = ::CreateThread(NULL,0,_ThreadEntry,this,0,&m_IDThread);
    if(m_hThread==NULL)
    {
        ret = false;
    }
#else
    ret = (::pthread_create(&m_hThread,NULL,&_ThreadEntry , this) == 0);
#endif
    return ret;
}

bool CThread::InitInstance()
{
    return true;
}

void CThread::ExitInstance()
{
}
```

```cpp
void CThread::OSSleep(int seconds)
{
#ifdef WIN32
    ::Sleep(seconds*1000);
#else
    ::sleep(seconds);
#endif
}

void CThread::SleepMs(int nMilliseconds)
{
#ifdef WIN32
    ::Sleep(nMilliseconds);
#else
    ::usleep(nMilliseconds);
#endif
}

bool CThread::IsCurrentThread()
{
#ifdef WIN32
    return ::GetCurrentThreadId() == m_IDThread;
#else
    return ::pthread_self() == m_hThread;
#endif
}

bool CThread::Join()
{
    THREAD_HANDLE hThread = GetHandle();
    if(hThread == (THREAD_HANDLE)0)
    {
        return true;
    }
#ifdef WIN32
    return (::WaitForSingleObject(hThread,INFINITE) != 0);
#else
    return (::pthread_join(hThread, NULL) == 0);
#endif
}

void CThread::ExitThread()
{
#ifdef WIN32
    ::ExitThread(0);
#else
#endif
}
```

以上代码中的 CThread 类封装了一个线程的常用操作，使用 WIN32 宏区分 Windows 和 Linux 两个操作系统的线程操作代码。其中，InitInstance 和 ExitInstance 方法为虚函数，在继承了 CThread 的子类中可以改写这两个方法，根据实际需要在线程函数的 Run 方法前后做一些初始化和反初始化工作，而 Run 方法是纯虚函数，必须改写，我们应该在改写的 Run 方法中加入自己实际需要的业务逻辑。

在线程函数中创建线程（调用 CreateThread 或 pthread_create 方法）时，将当前对象的 this 指针作为线程函数的唯一参数传入，这样在线程函数中就可以通过线程函数的参数得到对象的指针了，通过这个指针可以自由访问类的实例方法。这一技巧被广泛用于各类 C++开源项目和实际的 C++商业项目中。

那么，类的实例方法就一定不能作为线程函数了吗？不一定，在 C++ 11 语法中可以使用 std::bind 工具给线程函数绑定一个 this 指针，这样就能使用类实例方法作为线程函数了，示例代码如下：

```
#include <thread>
#include <memory>

class CIUSocket
{
    //省略构造、析构等函数
public:
    void Init();
    void Uninit();

    void Join();

private:
    void    SendThreadProc();
    void    RecvThreadProc();

private:
    std::unique_ptr<std::thread>    m_spSendThread;
    std::unique_ptr<std::thread>    m_spRecvThread;
};
```

重点看看 CIUSocket::Init 函数：

```
void CIUSocket::Init()
{
    //SendThreadProc 和 RecvThreadProc 都是类的实例方法
    m_spSendThread.reset(new std::thread(std::bind(&CIUSocket::SendThreadProc, this)));
    m_spRecvThread.reset(new std::thread(std::bind(&CIUSocket::RecvThreadProc, this)));
}
```

3.4 整型变量的原子操作

线程同步技术，指的是多个线程同时操作某个资源（从程序的术语来说，这里的资源可能是一个简单的整型变量，也可能是一个复杂的 C++对象）。多线程同时操作资源指的是多线程同时对资源进行读写。我们需要采取一些特殊的措施去保护这些资源，以免引起一些资源访问冲突（如死锁）或者得到意料外的结果。

当然，最简单的资源类型应该是整型变量。

3.4.1 为什么给整型变量赋值不是原子操作

常见的整型变量操作有如下三种。

（1）给整型变量赋一个确定的值，如：

```
int a = 1;
```

这条指令操作一般是原子的，因为对应一条计算机指令，所以 CPU 将立即数 1 搬运到变量 a 的内存地址中即可，汇编指令如下：

```
mov dword ptr [a], 1
```

这确实是最不常见的情形。由于现代编译器一般存在优化策略，所以如果变量 a 的值在编译期间就可以计算出来（例如在这里的例子中 a 的值就是 1），那么 a 这个变量本身在正式版本的软件中（release 版）就很有可能被编译器优化，凡是使用 a 的地方，直接使用常量 1 来代替。所以在实际的执行指令中，这样的指令存在的可能性较低。

（2）变量自身增加或者减去一个值，如：

```
a++;
```

从 C/C++ 语法的级别来看，这是一条语句，应该是原子的；但从编译得到的汇编指令来看，其实不是原子的，其一般对应三条指令，首先将变量 a 对应的内存值搬运到某个寄存器（如 eax）中，然后将该寄存器中的值自增 1，再将该寄存器中的值搬运回 a 代表的内存中：

```
mov eax, dword ptr [a]
inc eax
mov dword ptr [a], eax
```

现在假设 a 的值是 0，有两个线程，每个线程对变量 a 的值都递增 1，预想一下，其结果应该是 2，可实际运行结果可能是 1！是不是很奇怪？分析如下：

```
int a = 0;

//线程 1
void thread_func1()
{
    a++;
}

//线程 2
void thread_func2()
{
    a++;
}
```

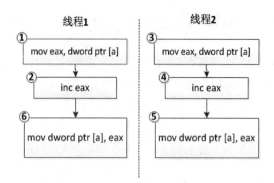

我们预想的结果是线程 1 和线程 2 的三条指令各自执行，最终 a 的值变为 2，但是由于操作系统线程调度的不确定性，线程 1 执行完指令①和②后，eax 寄存器中的值变为 1，此时操作系统切换到线程 2 执行，执行指令③④⑤，此时 eax 的值变为 1；接着操作系统切回线程 1 继续执行，执行指令⑥，得到 a 的最终结果 1。

（3）把一个变量的值赋给另一个变量，或者把一个表达式的值赋给另一个变量，如：
int a = b;

从 C/C++ 语法的级别来看，这条语句应该是原子的；但是从编译得到的汇编指令来看，由于现代计算机 CPU 架构体系的限制，数据不能直接从内存某处搬运到内存另外一处，必须借助寄存器中转，因此这条语句一般对应两条计算机指令，即将变量 b 的值搬运到某个寄存器（如 eax）中，再从该寄存器搬运到变量 a 的内存地址中：

```
mov eax, dword ptr [b]
mov dword ptr [a], eax
```

既然是两条指令，那么多个线程在执行这两条指令时，某个线程可能会在第 1 条指令执行完毕后被剥夺 CPU 时间片，切换到另一个线程而出现不确定的情况。这和上一种情况类似，就不再详细分析了。

很多人强调某些特殊的整型数值类型（如 bool 类型）的操作是原子的，这是由于某些 CPU 生产商有意识地从硬件平台保证这类操作的原子性，但并不是对每种类型的 CPU 架构都会保证。在这一事实成为标准之前，建议在多个线程中同时操作整型变量时还是使用原子操作函数或相应的线程同步技术对这些整型值进行保护。

3.4.2　Windows 平台上对整型变量的原子操作

对整型变量的原子操作是很常用且实用的操作，因此 Windows 也提供了 API 级别的支持，使用这些 API 可以直接对整型变量进行原子操作，而不用借助专门的锁对象。在 Windows 平台上，它们是 Interlocked 系列函数。这里给出 Interlocked 常用的 API 的一个列表。在该表中仅列出了与 32 位（bit）整型相关的 API 函数。Windows 还提供了对 8 位、16 位及 64 位的整型变量进行原子操作的 API，在实际使用时可以自行参考 MSDN。

函 数 名	函数说明
InterlockedIncrement	将 32 位整型变量自增 1
InterlockedDecrement	将 32 位整型变量自减 1
InterlockedExchangeAdd	将 32 位整型值增加 n（n 可以是负值）
InterlockedXor	将 32 位整型值与 n 进行异或操作
InterlockedCompareExchange	将 32 位整型值与 n_1 进行比较，如果相等，则将其替换成 n_2

以上表中的 InterlockedIncrement 为例来说明这类函数的用法。InterlockedIncrement 函数签名如下：

```
LONG InterlockedIncrement(LONG volatile *Addend);
```

这个函数的作用是将变量 Addend 自增 1，并返回自增后的值。

> 注意：这里的 LONG 类型即 long 类型，在 32 位系统中，LONG 占 4 字节。

写个例子验证一下：

```
#include <Windows.h>

int main()
{
    LONG nPreValue = 99;
    LONG nPostValue = InterlockedIncrement(&nPreValue);

    printf("nPreValue=%d, nPostValue=%d\n", nPreValue, nPostValue);

    return 0;
}
```

程序执行结果如下图所示。

3.4.3　C++ 11 对整型变量原子操作的支持

在 C++ 98/03 标准中，如果想对整型变量进行原子操作，则要么利用操作系统提供的相关原子操作 API，要么利用对应操作系统提供的锁对象对变量进行保护。无论是上述哪种方式，编写的代码都无法实现跨平台操作，例如前面介绍的 Interlocked 系列的 API 代码仅能运行于 Windows 系统中，无法被移植到 Linux 系统中。C++ 11 新标准发布以后改变了这种困境，新标准提供了对整型变量原子操作的相关库，即 std::atomic，这是一个模板类型：

```
template<class T>
struct atomic;
```

我们可以传入具体的整型类型（如 bool、char、short、int、uint 等）对模板进行实例化。实际上 stl 库也提供了这些实例化的模板类型。

类型别名	定义
std::atomic_bool	std::atomic<bool>
std::atomic_char	std::atomic<char>
std::atomic_schar	std::atomic<signed char>
std::atomic_uchar	std::atomic<unsigned char>
std::atomic_short	std::atomic<short>
std::atomic_ushort	std::atomic<unsigned short>
std::atomic_int	std::atomic<int>
std::atomic_uint	std::atomic<unsigned int>
std::atomic_long	std::atomic<long>
std::atomic_ulong	std::atomic<unsigned long>
std::atomic_llong	std::atomic<long long>
std::atomic_ullong	std::atomic<unsigned long long>
std::atomic_char16_t	std::atomic<char16_t>
std::atomic_char32_t	std::atomic<char32_t>
std::atomic_wchar_t	std::atomic<wchar_t>
std::atomic_int8_t	std::atomic<std::int8_t>
std::atomic_uint8_t	std::atomic<std::uint8_t>
std::atomic_int16_t	std::atomic< std::int16_t>
std::atomic_uint16_t	std::atomic<std::uint16_t>
std::atomic_int32_t	std::atomic<std::int32_t>
std::atomic_uint32_t	std::atomic<std::uint32_t>
std::atomic_int64_t	std::atomic<std::int64_t>
std::atomic_uint64_t	std::atomic<std::uint64_t>

上表中仅列出了 C++ 11 支持的常用的整型原子变量，可以参考 cppreference 官方网站获取完整的列表。

有了 C++语言本身对原子变量的支持以后，我们就可以方便地写出跨平台的代码了，来看一段代码：

```
//代码片段 3-4
#include <atomic>
#include <stdio.h>

int main()
{
    std::atomic<int> value;
    value = 99;
    printf("%d\n", (int)value);

    //自增 1，原子操作
    value++;
    printf("%d\n", (int)value);
```

```
    return 0;
}
```

以上代码可以同时在 Windows 和 Linux 平台上运行，但可能有读者根据个人习惯将以上代码写成如下形式：

```
#include <atomic>
#include <stdio.h>

int main()
{
    std::atomic<int> value = 99;
    printf("%d\n", (int)value);

    //自增1，原子操作
    value++;
    printf("%d\n", (int)value);

    return 0;
}
```

以上代码仅仅做了一些简单的改动，在 Windows 平台上运行良好，但在 Linux 平台上无法编译通过（这里指的是在支持 C++ 11 语法的 g++编译中编译），提示的错误如下：

```
error: use of deleted function 'std::atomic<int>::atomic(const std::atomic<int>&)'
```

产生这个错误的原因如下：

```
std::atomic<int> value = 99;
```

行代码调用的是 std::atomic 的拷贝构造函数。对于 int 类型，其形式一般如下：

```
std::atomic<int>::atomic(const std::atomic<int>& rhs);
```

而根据 C++ 11 语言规范，std::atomic 的拷贝构造函数默认使用=delete 标记禁止编译器自动生成，g++在这条规则上遵循 C++ 11 语言规范：

```
atomic& operator=(const atomic&) = delete;
```

所以在 Linux 平台上，以上代码因为无法使用 std::atomic 的这一拷贝构造函数，所以编译器会产生编译错误，而 Windows 的 VC++编译器在这条规则上并没有遵循 C++ 11 语言规范。

对于代码：

```
std::atomic<int> value;
value = 99;
```

由于 g++和 VC++同时实现了 C++新标准规范中的 operator=操作：

```
T operator=(T desired)
```

所以，以上代码可以同时在 g++和 VC++中编译通过。

鉴于 g++和 VC++在实现 std::atomic 时遵循 C++新标准规范的程度不一样，所以如果想利用 C++ 11 提供的 std::atomic 库编写跨平台的代码，则建议在使用 std::atomic 的相关方法时先参考官方提供的接口说明文档，而不是想当然地认为某个方法在一个平台上可以运行，那么在另外一个平台也能有相同的表现。切记这一点。

在代码片段 3-4 中之所以可以对 value 变量进行自增（++）操作，是因为 std::atomic 类重载了 operator++运算符，除此以外，std::atomic 还提供了大量有用的方法。

方 法 名	方法说明
operator=	存储值于原子对象中
store	原子地以非原子对象替换原子对象的值
load	原子地获得原子对象的值
exchange	原子地替换原子对象的值并获得它先前持有的值
compare_exchange_weak compare_exchange_strong	原子地比较原子对象与非原子参数的值，若相等则进行交换，若不相等则进行加载
fetch_add	原子地相加参数并返回相加之前的值
fetch_sub	原子地相减参数并返回相减之前的值
fetch_and	原子地与参数进行按位与操作，并返回操作前的值
fetch_or	原子地与参数进行按位或操作，并返回操作前的值
fetch_xor	原子地与参数进行按位异或操作，并返回操作前的值
operator++ operator++(int) operator-- operator--(int)	令值原子地增加或减少 1
operator += operator -= operator &= operator |= operator ^=	原子地与右边的参数进行加、减、按位与、按位或、按位异或操作并将操作结果存储于原子对象中

3.5 Linux 线程同步对象

本节详细介绍 Linux 线程同步对象的原理和使用场景。

3.5.1 Linux 互斥体

Linux 互斥体与 Windows 的临界区对象用法很相似，一般也是通过限制多个线程同时执行某段代码来保护资源的。和接下来要介绍的信号量、条件变量一样，Linux 互斥体都在 NPTL（Native POSIX Thread Library）中实现。在 NPTL 中，我们使用数据结构 pthread_mutex_t 表示一个互斥体对象（定义于 pthread.h 头文件中）。我们可以使用以下两种方式初始化互斥体对象。

（1）使用 PTHREAD_MUTEX_INITIALIZER 直接给互斥体变量赋值，示例代码如下：

```
#include <pthread.h>
pthread_mutex_t mymutex = PTHREAD_MUTEX_INITIALIZER;
```

（2）若互斥体是动态分配的或者需要设置互斥体的相关属性，则需要使用 pthread_mutex_init 函数初始化互斥体，其函数签名如下：

```
int pthread_mutex_init(pthread_mutex_t* restrict mutex,
                const pthread_mutexattr_t* restrict attr);
```

参数 mutex 即我们需要初始化的 mutex 对象的指针,参数 attr 是需要设置的互斥体属性。在通常情况下,我们使用默认属性将这个参数设置为 NULL,后面会详细介绍每一种属性的用法。如果函数执行成功,则会返回 0;如果执行失败,则会返回 1 个错误码。pthread_mutex_init 代码示例如下:

```
#include <pthread.h>

pthread_mutex_t mymutex;
pthread_mutex_init(&mymutex, NULL);
```

当我们不再需要一个互斥体对象时,可以使用 pthread_mutex_destroy 函数销毁它,pthread_mutex_destroy 函数签名如下:

```
int pthread_mutex_destroy(pthread_mutex_t* mutex);
```

参数 mutex 即我们需要销毁的互斥体对象,如果函数执行成功,则会返回 0;如果执行失败,则会返回一个错误码表明出错的原因。这里需要注意两点。

(1)无须销毁使用 PTHREAD_MUTEX_INITIALIZER 初始化的互斥体。

(2)不要销毁一个已经加锁或正在被条件变量使用的互斥体对象,当互斥体处于已加锁状态或者正在和条件变量配合使用时,调用 pthread_mutex_destroy 函数会返回 EBUSY 错误。

以下代码演示了尝试销毁一个被锁定的互斥体对象的过程:

```
//test_destroy_locked_mutex.cpp
#include <pthread.h>
#include <stdio.h>
#include <errno.h>

int main()
{
    pthread_mutex_t mymutex;
    pthread_mutex_init(&mymutex, NULL);
    int ret = pthread_mutex_lock(&mymutex);

    //尝试销毁被锁定的mutex对象
    ret = pthread_mutex_destroy(&mymutex);
    if (ret != 0)
    {
        if (ret == EBUSY)
            printf("EBUSY\n");
        printf("Failed to destroy mutex.\n");
    }

    ret = pthread_mutex_unlock(&mymutex);
    //尝试销毁已经解锁的mutex对象
    ret = pthread_mutex_destroy(&mymutex);
    if (ret == 0)
        printf("Succeeded to destroy mutex.\n");
```

```
    return 0;
}
```

编译以上代码并执行,得到我们期望的结果:

```
[root@myaliyun codes]# g++ -g -o test_destroy_locked_mutex
test_destroy_locked_mutex.cpp -lpthread
[root@myaliyun codes]# ./test_destroy_locked_mutex
EBUSY
Failed to destroy mutex.
Succeed to destroy mutex.
```

在实际开发中,如果我们遵循正确的互斥体使用规范,例如创建互斥体对象后再对其加锁,加锁后才对其进行解锁操作,解锁后才进行销毁操作,那么编码时一般不用考虑 pthread_mutex_init/pthread_mutex_destroy/pthread_mutex_lock/pthread_mutex_unlock 等函数的返回值。

对于互斥体的加锁和解锁操作,我们一般使用以下三个函数:

```
int pthread_mutex_lock(pthread_mutex_t* mutex);
int pthread_mutex_trylock(pthread_mutex_t* mutex);
int pthread_mutex_unlock(pthread_mutex_t* mutex);
```

参数 mutex 是我们需要加锁或解锁的互斥体对象,上述函数如果执行成功则返回 0,如果执行失败则返回一个错误码以表示具体的出错原因。具体的错误码随着互斥体对象属性类型的不同而不同。

在设置互斥体对象的属性时需要创建一个 pthread_mutexattr_t 类型的对象,和互斥体对象一样,需要使用 pthread_mutexattr_init 函数初始化它,当不需要这个属性对象时,记得使用 **pthread_mutexattr_destroy** 销毁它,其函数签名分别如下:

```
int pthread_mutexattr_init(pthread_mutexattr_t* attr);
int pthread_mutexattr_destroy(pthread_mutexattr_t* attr);
```

使用 **pthread_mutexattr_settype/pthread_mutexattr_gettype** 设置或获取想要的属性类型:

```
int pthread_mutexattr_settype(pthread_mutexattr_t* attr, int type);
int pthread_mutexattr_gettype(const pthread_mutexattr_t* restrict attr, int* restrict type);
```

属性类型一般有如下取值。

1. PTHREAD_MUTEX_NORMAL(普通锁)

这是互斥体对象的默认属性(即将上文介绍的 pthread_mutex_init 函数的第 2 个参数设置为 NULL)。在一个线程对一个普通锁加锁以后,其他线程会阻塞在 pthread_mutex_lock 调用处,直到对互斥体加锁的线程释放了锁。下面用一段实例代码验证一下:

```
#include <pthread.h>
#include <stdio.h>
#include <errno.h>
#include <unistd.h>

pthread_mutex_t mymutex;
```

```c
int              resourceNo = 0;
void* worker_thread(void* param)
{
    pthread_t threadID = pthread_self();

    printf("thread start, ThreadID: %d\n", threadID);

    while (true)
    {
        pthread_mutex_lock(&mymutex);

        printf("Mutex lock, resourceNo: %d, ThreadID: %d\n", resourceNo, threadID);
        resourceNo++;

        printf("Mutex unlock, resourceNo: %d, ThreadID: %d\n", resourceNo, threadID);

        pthread_mutex_unlock(&mymutex);

        //休眠1秒
        sleep(1);
    }

    return NULL;
}
int main()
{
    pthread_mutexattr_t mutex_attr;
    pthread_mutexattr_init(&mutex_attr);
    pthread_mutexattr_settype(&mutex_attr, PTHREAD_MUTEX_NORMAL);
    pthread_mutex_init(&mymutex, &mutex_attr);

    //创建了5个工作线程
    pthread_t threadID[5];

    for (int i = 0; i < 5; ++i)
        pthread_create(&threadID[i], NULL, worker_thread, NULL);

    for (int i = 0; i < 5; ++i)
        pthread_join(threadID[i], NULL);

    pthread_mutex_destroy(&mymutex);
    pthread_mutexattr_destroy(&mutex_attr);

    return 0;
}
```

以上代码共创建了 5 个工作线程，由于使用了互斥体保护资源 resourceNo，所以 pthread_mutex_lock 与 pthread_mutex_unlock 之间的输出每次都是连续的，一个线程必须完成了这个工作，其他线程才有机会获得执行这段代码的机会，在一个线程拿到锁后，其他线程会阻塞在 pthread_mutex_lock 处。

程序执行结果如下：

```
[root@localhost testmultithread]# ./test
thread start, ThreadID: 520349440
Mutex lock, resourceNo: 0, ThreadID: 520349440
Mutex unlock, resourceNo: 1, ThreadID: 520349440
thread start, ThreadID: 545527552
Mutex lock, resourceNo: 1, ThreadID: 545527552
Mutex unlock, resourceNo: 2, ThreadID: 545527552
thread start, ThreadID: 511956736
Mutex lock, resourceNo: 2, ThreadID: 511956736
Mutex unlock, resourceNo: 3, ThreadID: 511956736
thread start, ThreadID: 537134848
Mutex lock, resourceNo: 3, ThreadID: 537134848
Mutex unlock, resourceNo: 4, ThreadID: 537134848
thread start, ThreadID: 528742144
Mutex lock, resourceNo: 4, ThreadID: 528742144
Mutex unlock, resourceNo: 5, ThreadID: 528742144
Mutex lock, resourceNo: 5, ThreadID: 545527552
Mutex unlock, resourceNo: 6, ThreadID: 545527552
Mutex lock, resourceNo: 6, ThreadID: 537134848
Mutex unlock, resourceNo: 7, ThreadID: 537134848
Mutex lock, resourceNo: 7, ThreadID: 528742144
Mutex unlock, resourceNo: 8, ThreadID: 528742144
Mutex lock, resourceNo: 8, ThreadID: 520349440
Mutex unlock, resourceNo: 9, ThreadID: 520349440
Mutex lock, resourceNo: 9, ThreadID: 511956736
Mutex unlock, resourceNo: 10, ThreadID: 511956736
Mutex lock, resourceNo: 10, ThreadID: 545527552
Mutex unlock, resourceNo: 11, ThreadID: 545527552
Mutex lock, resourceNo: 11, ThreadID: 537134848
Mutex unlock, resourceNo: 12, ThreadID: 537134848
Mutex lock, resourceNo: 12, ThreadID: 520349440
Mutex unlock, resourceNo: 13, ThreadID: 520349440
Mutex lock, resourceNo: 13, ThreadID: 528742144
Mutex unlock, resourceNo: 14, ThreadID: 528742144
Mutex lock, resourceNo: 14, ThreadID: 511956736
...省略更多的输出结果...
```

一个线程如果对一个已经加锁的普通锁再次使用了 pthread_mutex_lock 加锁，那么程序会阻塞在第 2 次调用 pthread_mutex_lock 代码处。测试代码如下：

```
#include <pthread.h>
#include <stdio.h>
#include <errno.h>
#include <unistd.h>

int main()
{
    pthread_mutex_t mymutex;
    pthread_mutexattr_t mutex_attr;
    pthread_mutexattr_init(&mutex_attr);
    pthread_mutexattr_settype(&mutex_attr, PTHREAD_MUTEX_NORMAL);
    pthread_mutex_init(&mymutex, &mutex_attr);

    int ret = pthread_mutex_lock(&mymutex);
    printf("ret = %d\n", ret);
```

```
    ret = pthread_mutex_lock(&mymutex);
    printf("ret = %d\n", ret);

    pthread_mutex_destroy(&mymutex);
    pthread_mutexattr_destroy(&mutex_attr);

    return 0;
}
```

编译并使用 gdb 将程序运行起来，程序只输出了一行，我们按 Ctrl+C 组合键（下面的^C 字符）将 gdb 中断，然后使用 bt 命令，发现程序确实阻塞在第 2 个 pthread_mutex_lock 函数调用处：

```
[root@localhost testmultithread]# g++ -g -o test test.cpp -lpthread
[root@localhost testmultithread]# gdb test
Reading symbols from /root/testmultithread/test...done.
(gdb) r
Starting program: /root/testmultithread/test
[Thread debugging using libthread_db enabled]
Using host libthread_db library "/lib64/libthread_db.so.1".
ret = 0
^C
Program received signal SIGINT, Interrupt.
0x00007ffff7bcd4ed in __lll_lock_wait () from /lib64/libpthread.so.0
Missing separate debuginfos, use: debuginfo-install glibc-2.17-260.el7.x86_64
libgcc-4.8.5-36.el7.x86_64 libstdc++-4.8.5-36.el7.x86_64
(gdb) bt
#0  0x00007ffff7bcd4ed in __lll_lock_wait () from /lib64/libpthread.so.0
#1  0x00007ffff7bc8dcb in _L_lock_883 () from /lib64/libpthread.so.0
#2  0x00007ffff7bc8c98 in pthread_mutex_lock () from /lib64/libpthread.so.0
#3  0x00000000004007f4 in main () at ConsoleApplication10.cpp:17
(gdb)
```

在这种场景下，pthread_mutex_trylock 函数如果拿不到锁，则也不会阻塞，而是会立即返回 EBUSY 错误码。

2. PTHREAD_MUTEX_ERRORCHECK（检错锁）

如果一个线程使用了 pthread_mutex_lock 对已经加锁的互斥体对象再次加锁，则 pthread_mutex_lock 会返回 EDEADLK。我们使用下面的代码片段验证同一个线程对同一个互斥体对象多次加锁的效果：

```
#include <pthread.h>
#include <stdio.h>
#include <errno.h>
#include <unistd.h>

int main()
{
    pthread_mutex_t mymutex;
    pthread_mutexattr_t mutex_attr;
    pthread_mutexattr_init(&mutex_attr);
    pthread_mutexattr_settype(&mutex_attr, PTHREAD_MUTEX_ERRORCHECK);
```

```cpp
    pthread_mutex_init(&mymutex, &mutex_attr);

    int ret = pthread_mutex_lock(&mymutex);
    printf("ret = %d\n", ret);

    ret = pthread_mutex_lock(&mymutex);
    printf("ret = %d\n", ret);
    if (ret == EDEADLK)
        printf("EDEADLK\n");

    pthread_mutex_destroy(&mymutex);
    pthread_mutexattr_destroy(&mutex_attr);

    return 0;
}
```

编译并运行程序，程序输出结果确实如上所述：

```
[root@localhost testmultithread]# g++ -g -o test11 test.cpp -lpthread
[root@localhost testmultithread]# ./test11
ret = 0
ret = 35
EDEADLK
```

再来验证一个线程对某个互斥体加锁，其他线程再次对该互斥体加锁的效果：

```cpp
#include <pthread.h>
#include <stdio.h>
#include <errno.h>
#include <unistd.h>

pthread_mutex_t mymutex;

void* worker_thread(void* param)
{
    pthread_t threadID = pthread_self();

    printf("thread start, ThreadID: %d\n", threadID);

    while (true)
    {
        int ret = pthread_mutex_lock(&mymutex);
        if (ret == EDEADLK)
            printf("EDEADLK, ThreadID: %d\n", threadID);
        else
            printf("ret = %d, ThreadID: %d\n", ret, threadID);

        //休眠1秒
        sleep(1);
    }

    return NULL;
}

int main()
```

```
{
    pthread_mutexattr_t mutex_attr;
    pthread_mutexattr_init(&mutex_attr);
    pthread_mutexattr_settype(&mutex_attr, PTHREAD_MUTEX_ERRORCHECK);
    pthread_mutex_init(&mymutex, &mutex_attr);

    int ret = pthread_mutex_lock(&mymutex);
    printf("ret = %d\n", ret);

    //创建 5 个工作线程
    pthread_t threadID[5];
    for (int i = 0; i < 5; ++i)
        pthread_create(&threadID[i], NULL, worker_thread, NULL);

    for (int i = 0; i < 5; ++i)
        pthread_join(threadID[i], NULL);

    pthread_mutex_destroy(&mymutex);
    pthread_mutexattr_destroy(&mutex_attr);

    return 0;
}
```

编译程序，然后使用 gdb 运行起来，发现程序并没有任何输出，按 Ctrl+C 组合键中断，输入 info thread 命令，发现工作线程均阻塞在 pthread_mutex_lock 函数调用处，操作及输出结果如下：

```
[root@localhost testmultithread]# g++ -g -o test8 ConsoleApplication8.cpp -lpthread
[root@localhost testmultithread]# ./test8
ret = 0
thread start, ThreadID: -1821989120
thread start, ThreadID: -1830381824
thread start, ThreadID: -1838774528
thread start, ThreadID: -1847167232
thread start, ThreadID: -1813596416
^C
[root@localhost testmultithread]# gdb test8
Reading symbols from /root/testmultithread/test8...done.
(gdb) r
Starting program: /root/testmultithread/test8
[Thread debugging using libthread_db enabled]
Using host libthread_db library "/lib64/libthread_db.so.1".
ret = 0
[New Thread 0x7ffff6fd2700 (LWP 3276)]
thread start, ThreadID: -151181568
[New Thread 0x7ffff67d1700 (LWP 3277)]
thread start, ThreadID: -159574272
[New Thread 0x7ffff5fd0700 (LWP 3278)]
thread start, ThreadID: -167966976
[New Thread 0x7ffff57cf700 (LWP 3279)]
thread start, ThreadID: -176359680
[New Thread 0x7ffff4fce700 (LWP 3280)]
thread start, ThreadID: -184752384
^C
Program received signal SIGINT, Interrupt.
```

```
0x00007ffff7bc7f47 in pthread_join () from /lib64/libpthread.so.0
Missing separate debuginfos, use: debuginfo-install glibc-2.17-260.el7.x86_64
libgcc-4.8.5-36.el7.x86_64 libstdc++-4.8.5-36.el7.x86_64
(gdb) bt
#0  0x00007ffff7bc7f47 in pthread_join () from /lib64/libpthread.so.0
#1  0x00000000004009e9 in main () at ConsoleApplication8.cpp:50
(gdb) inf threads
  Id   Target Id         Frame
  6    Thread 0x7ffff4fce700 (LWP 3280) "test8" 0x00007ffff7bcd4ed in __lll_lock_wait
() from /lib64/libpthread.so.0
  5    Thread 0x7ffff57cf700 (LWP 3279) "test8" 0x00007ffff7bcd4ed in __lll_lock_wait
() from /lib64/libpthread.so.0
  4    Thread 0x7ffff5fd0700 (LWP 3278) "test8" 0x00007ffff7bcd4ed in __lll_lock_wait
() from /lib64/libpthread.so.0
  3    Thread 0x7ffff67d1700 (LWP 3277) "test8" 0x00007ffff7bcd4ed in __lll_lock_wait
() from /lib64/libpthread.so.0
  2    Thread 0x7ffff6fd2700 (LWP 3276) "test8" 0x00007ffff7bcd4ed in __lll_lock_wait
() from /lib64/libpthread.so.0
* 1    Thread 0x7ffff7fee740 (LWP 3272) "test8" 0x00007ffff7bc7f47 in pthread_join
() from /lib64/libpthread.so.0
(gdb)
```

通过上面的实验，如果互斥体的属性是 PTHREAD_MUTEX_ERRORCHECK，则当前线程重复调用 pthread_mutex_lock 会直接返回 EDEADLOCK，其他线程如果对这个互斥体再次调用 pthread_mutex_lock，则会阻塞在该函数的调用处。

3. PTHREAD_MUTEX_RECURSIVE（可重入锁）

该属性允许同一个线程对其持有的互斥体重复加锁，每成功调用 pthread_mutex_lock 一次，该互斥体对象的锁引用计数就会增加 1，相反，每成功调用 pthread_mutex_unlock 一次，锁引用计数就会减少 1。当锁引用计数值为 0 时，允许其他线程获得该锁，否则当其他线程调用 pthread_mutex_lock 尝试获取锁时，会阻塞在 pthread_mutex_lock 调用处。

下面总结 Linux 上互斥体对象的使用要点。

（1）上面演示了同一个线程对一个互斥体对象反复加锁的过程，在实际开发中这种场景非常少见。

（2）与 Windows 的临界区对象一样，在一些有很多出口的逻辑中，为了避免因忘记调用 pthread_mutex_unlock 出现死锁或者在逻辑出口处有大量解锁的重复代码出现，建议使用 RAII 技术将互斥体对象封装起来，具体操作方法在第 1 章介绍 RAII 技术时已经介绍过了。

3.5.2 Linux 信号量

信号量代表一定的资源数量，可以根据当前资源的数量按需唤醒指定数量的资源消费者线程，资源消费者线程一旦获取信号量，就会让资源减少指定的数量，如果资源数量减少为 0，则消费者线程将全部处于挂起状态；当有新的资源到来时，消费者线程将继续被唤醒。另外，信号量有"资源有多份，可以同时被多个线程访问"的意思。

Linux 信号量常用的一组 API 函数如下：

```
#include <semaphore.h>
int sem_init(sem_t* sem, int pshared, unsigned int value);
int sem_destroy(sem_t* sem);
int sem_post(sem_t* sem);
int sem_wait(sem_t* sem);
int sem_trywait(sem_t* sem);
int sem_timedwait(sem_t* sem, const struct timespec* abs_timeout);
```

sem_init 函数用于初始化一个信号量，第 1 个参数 sem 传入需要初始化的信号量对象的地址；第 2 个参数 pshared 表示该信号量是否可以被共享，取值为 0 表示该信号量只能在同一个进程的多个线程之间共享，取值为非 0 表示可以在多个进程之间共享；第 3 个参数 value 用于设置信号量初始状态下的资源数量。sem_init 函数调用成功则返回 0，失败则返回-1，在实际编码中只要写法得当，则一般不用关心该函数的返回值。

sem_destroy 函数用于销毁一个信号量。sem_post 函数用于将信号量的资源计数递增 1，并解锁该信号量对象，这样因使用 sem_wait 函数被阻塞的其他线程就会被唤醒。

如果当前信号量的资源计数为 0，则 sem_wait 函数会阻塞调用线程，直到信号量对象的资源计数大于 0 时被唤醒，唤醒后将资源计数递减 1，然后立即返回；sem_trywait 函数是 sem_wait 函数的非阻塞版本，如果当前信号量对象的资源计数等于 0，则 sem_trywait 函数会立即返回，不会阻塞调用线程，返回值是-1，错误码 errno 被设置成 EAGAIN；sem_timedwait 函数是带有等待时间的版本，等待时间在第 2 个参数 abs_timeout 中被设置，这是个结构体类型，其定义如下：

```
struct timespec
{
    time_t tv_sec;  /* 秒 */
    long   tv_nsec; /* 纳秒，取值范围是[0～999999999] */
};
```

sem_timedwait 函数在 abs_timeout 参数设置的时间内等待信号量对象的资源计数大于 0，否则超时返回，返回值为-1，错误码 errno 是 ETIMEDOUT。在使用 sem_timedwait 函数时，abs_timeout 参数不能被设置为 NULL，否则程序会在调用 sem_timedwait 函数处崩溃。abs_timeout 参数指的是绝对时间，也就是说，如果打算让函数等待 5 秒，那么应该先得到当前系统的时间，然后加上 5 秒计算出最终的时间作为参数 abs_timeout 的值。

使用以上几个函数时还有几个需要注意的地方。

◎ sem_wait、sem_trywait、sem_timedwait 函数将资源计数递减 1 时会同时锁定信号量对象，因此当资源计数为 1 时，如果有多个线程调用 sem_wait 等函数等待该信号量，则只会有一个线程被唤醒。sem_wait 函数在返回时会释放对该信号量的锁。

◎ sem_wait、sem_trywait、sem_timedwait 函数在调用成功后返回值均为 0，调用失败时返回-1，可以通过错误码 errno 获得失败的原因。

◎ sem_wait、sem_trywait、sem_timedwait 函数可以被 Linux 信号中断，被信号中断后，函数立即返回，返回值为-1，错误码 errno 为 EINTR。

◎ 虽然上述函数名没有以 pthread_为前缀,但是实际使用这个系列的函数时仍然需要链接 pthread 库。

看一个信号量的具体使用示例:

```cpp
#include <pthread.h>
#include <errno.h>
#include <unistd.h>
#include <list>
#include <semaphore.h>
#include <iostream>

class Task
{
public:
    Task(int taskID)
    {
        this->taskID = taskID;
    }

    void doTask()
    {
        std::cout << "handle a task, taskID: " << taskID << ", threadID: " << pthread_self() << std::endl;
    }

private:
    int taskID;
};

pthread_mutex_t     mymutex;
std::list<Task*>    tasks;
sem_t               mysemaphore;

void* consumer_thread(void* param)
{
    Task* pTask = NULL;
    while (true)
    {
        if (sem_wait(&mysemaphore) != 0)
            continue;

        if (tasks.empty())
            continue;

        pthread_mutex_lock(&mymutex);
        pTask = tasks.front();
        tasks.pop_front();
        pthread_mutex_unlock(&mymutex);

        pTask->doTask();
        delete pTask;
    }

    return NULL;
```

```cpp
}

void* producer_thread(void* param)
{
    int taskID = 0;
    Task* pTask = NULL;

    while (true)
    {
        pTask = new Task(taskID);

        pthread_mutex_lock(&mymutex);
        tasks.push_back(pTask);
        std::cout << "produce a task, taskID: " << taskID << ", threadID: " << pthread_self() << std::endl;

        pthread_mutex_unlock(&mymutex);

        //释放信号量，通知消费者线程
        sem_post(&mysemaphore);

        taskID ++;

        //休眠 1 秒
        sleep(1);
    }

    return NULL;
}

int main()
{
    pthread_mutex_init(&mymutex, NULL);
    //初始信号量资源计数为 0
    sem_init(&mysemaphore, 0, 0);

    //创建 5 个消费者线程
    pthread_t consumerThreadID[5];
    for (int i = 0; i < 5; ++i)
        pthread_create(&consumerThreadID[i], NULL, consumer_thread, NULL);

    //创建 1 个生产者线程
    pthread_t producerThreadID;
    pthread_create(&producerThreadID, NULL, producer_thread, NULL);

    pthread_join(producerThreadID, NULL);

    for (int i = 0; i < 5; ++i)
        pthread_join(consumerThreadID[i], NULL);

    sem_destroy(&mysemaphore);
    pthread_mutex_destroy(&mymutex);

    return 0;
}
```

在以上代码中创建了 1 个生产者线程和 5 个消费者线程，初始信号量计数为 0 代表一开始没有可执行任务，所以 5 个消费线程均被阻塞在 sem_wait 调用处。接着生产者每隔 1 秒产生 1 个任务，然后通过调用 sem_post 将信号量的资源计数增加 1，此时其中 1 个线程会被唤醒。我们从任务队列中取出任务并执行，由于任务对象是新建的，所以我们需要将其删掉以免内存泄露。

在调用 sem_wait 和 sem_post 时会对信号量对象进行加锁和解锁，为什么这里还需要使用一个互斥体呢？其实，这个互斥体是用来保护队列 tasks 的，因为多个线程会同时读写它。

编译、生成可执行文件 semaphore 并运行，输出结果如下：

```
[root@localhost testsemaphore]# g++ -g -o semaphore semaphore.cpp -lpthread
[root@localhost testsemaphore]# ./semaphore
produce a task, taskID: 0, threadID: 140055260595968
handle a task, taskID: 0, threadID: 140055277381376
produce a task, taskID: 1, threadID: 140055260595968
handle a task, taskID: 1, threadID: 140055277381376
produce a task, taskID: 2, threadID: 140055260595968
handle a task, taskID: 2, threadID: 140055268988672
produce a task, taskID: 3, threadID: 140055260595968
handle a task, taskID: 3, threadID: 140055294166784
produce a task, taskID: 4, threadID: 140055260595968
handle a task, taskID: 4, threadID: 140055302559488
produce a task, taskID: 5, threadID: 140055260595968
handle a task, taskID: 5, threadID: 140055285774080
produce a task, taskID: 6, threadID: 140055260595968
handle a task, taskID: 6, threadID: 140055277381376
produce a task, taskID: 7, threadID: 140055260595968
handle a task, taskID: 7, threadID: 140055268988672
produce a task, taskID: 8, threadID: 140055260595968
handle a task, taskID: 8, threadID: 140055294166784
produce a task, taskID: 9, threadID: 140055260595968
handle a task, taskID: 9, threadID: 140055302559488
...省略更多的输出结果...
```

3.5.3　Linux 条件变量

有人说 Linux 条件变量（Condition Variable）是最不容易用错的一种线程同步对象，确实是这样，但必须建立在对条件变量熟练应用的基础之上。我们先来讨论一下为什么存在条件变量这样一种机制。

1. 为什么需要使用条件变量

在实际应用中，我们常常会有类似如下的需求：

```
//以下是伪代码，m 的类型是 pthread_mutex_t，已经初始化过了
int WaitForTrue()
{
    do {
        pthread_mutex_lock(&m);
```

```
        //验证 condition 是否为 true

        //解锁,让其他线程有机会改变 condition
        pthread_mutex_unlock(&m);

        //睡眠 n 秒
        sleep(n);
    } while (condition is false);

    return 1;
}
```

可以将以上逻辑表示成如下所示的流程图。

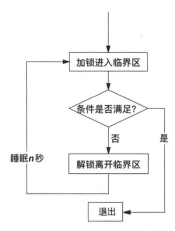

这段逻辑用于反复判断一个多线程的共享条件是否满足,一直到该条件满足为止。由于该条件被多个线程操作,因此在每次判断之前都需要进行加锁操作,判断完毕后需要进行解锁操作。但是,该逻辑存在严重的效率问题,假设解锁离开临界区后,其他线程修改了条件导致条件满足,则此时程序仍然需要睡眠 n 秒才能得到反馈。因此我们需要这样一种机制:某个线程 A 在条件不满足的情况下主动让出互斥体,让其他线程操作,线程 A 在此处等待条件满足;一旦条件满足,线程 A 就可以被立刻唤醒。线程 A 之所以可以安心等待,依赖的是其他线程的协作,它确认会有一个线程在发现条件满足以后,向它发送信号并且让出互斥体。如果其他线程不配合(不发送信号、不让出互斥体),就无法实现上述效果。

这就是为什么需要条件等待,但条件等待不是条件变量的全部功能。

2. 条件变量为什么要与互斥体对象结合使用

为什么条件变量一定要与一个互斥体对象结合使用呢?我们假设条件变量不与互斥体对象结合,来看看是什么效果。以下是伪代码:

```
1  //m 的类型是 pthread_mutex_t,并且已经初始化过了,cv 是条件变量
2  pthread_mutex_lock(&m)
3  while(condition_is_false)
4  {
5      pthread_mutex_unlock(&m);
```

```
6       //解锁之后，等待之前，可能条件已经满足，信号已经发出，但是该信号可能被错过
7       cond_wait(&cv);
8       pthread_mutex_lock(&m);
9   }
```

在以上代码中，假设线程 A 在执行完第 5 行代码后 CPU 时间片被剥夺，此时另一个线程 B 获得该互斥体对象 m，然后发送条件信号，等线程 A 重新获得时间片后，由于该信号已经被错过，可能会导致线程 A 在代码第 7 行无限阻塞下去。

造成这个问题的根源是释放互斥体对象与条件变量等待唤醒不是原子操作，即解锁和等待这两个步骤必须在同一个原子操作中，才能确保 cond_wait 在唤醒之前不会有其他线程获得这个互斥体对象。

3. 条件变量的使用

对条件变量的初始化和销毁，可以使用如下 API 函数：

```
int pthread_cond_init(pthread_cond_t* cond, const pthread_condattr_t* attr);
int pthread_cond_destroy(pthread_cond_t* cond);
```

在 Linux 系统中，pthread_cond_t 表示条件变量的数据类型，所以也可以使用如下方式初始化一个条件变量：

```
pthread_cond_t cond = PTHREAD_COND_INITIALIZER;
```

可以使用如下 API 函数等待条件变量被唤醒：

```
int pthread_cond_wait(pthread_cond_t* restrict cond, pthread_mutex_t* restrict mutex);
int pthread_cond_timedwait(pthread_cond_t* restrict cond, pthread_mutex_t* restrict mutex, const struct timespec* restrict abstime);
```

如果条件变量等待的条件没有被满足，则调用 pthread_cond_wait 的线程会一直等待下去。pthread_cond_timedwait 是 pthread_cond_wait 的非阻塞版本，它会在指定的时间内等待条件满足，超过参数 abstime 设置的时间后，pthread_cond_timedwait 函数会立即返回。

注意：abstime 参数同样是一个绝对时间，其设置方法与 sem_timedwait 函数的超时参数 abs_timeout 的设置方法一样。

因调用 pthread_cond_wait 而等待的线程可以被以下 API 函数唤醒：

```
int pthread_cond_signal(pthread_cond_t* cond);
int pthread_cond_broadcast(pthread_cond_t* cond);
```

pthread_cond_signal 一次唤醒一个线程，如果有多个线程调用 pthread_cond_wait 等待，则具体哪个线程被唤醒是不确定的（可以认为是随机的）；pthread_cond_broadcast 可以同时唤醒所有调用 pthread_cond_wait 等待的线程。前者相当于发送一次条件通知，后者相当于广播一次条件通知。在函数调用成功时，pthread_cond_signal 和 pthread_cond_broadcast 均返回 0，反之均返回具体的错误码值。

这里将前面介绍信号量的示例代码用条件变量改写一下：

```
//代码片段 3-5-3
#include <pthread.h>
#include <errno.h>
```

```cpp
#include <unistd.h>
#include <list>
#include <semaphore.h>
#include <iostream>

class Task
{
public:
    Task(int taskID)
    {
        this->taskID = taskID;
    }

    void doTask()
    {
        std::cout << "handle a task, taskID: " << taskID << ", threadID: " <<
pthread_self() << std::endl;
    }

private:
    int taskID;
};

pthread_mutex_t mymutex;
std::list<Task*> tasks;
pthread_cond_t   mycv;

void* consumer_thread(void* param)
{
    Task* pTask = NULL;
    while (true)
    {
        pthread_mutex_lock(&mymutex);
        while (tasks.empty())
        {
            //如果获得了互斥锁，但是条件不合适，则pthread_cond_wait会释放锁，不往下执行
            //发生变化后，如果条件合适，则pthread_cond_wait将直接获得锁
            pthread_cond_wait(&mycv, &mymutex);
        }

        pTask = tasks.front();
        tasks.pop_front();

        pthread_mutex_unlock(&mymutex);

        if (pTask == NULL)
            continue;

        pTask->doTask();
        delete pTask;
        pTask = NULL;
    }

    return NULL;
}
```

```cpp
void* producer_thread(void* param)
{
    int taskID = 0;
    Task* pTask = NULL;

    while (true)
    {
        pTask = new Task(taskID);

        pthread_mutex_lock(&mymutex);
        tasks.push_back(pTask);
        std::cout << "produce a task, taskID: " << taskID << ", threadID: " <<
pthread_self() << std::endl;

        pthread_mutex_unlock(&mymutex);

        //释放信号量，通知消费者线程
        pthread_cond_signal(&mycv);

        taskID ++;

        //休眠1秒
        sleep(1);
    }

    return NULL;
}

int main()
{
    pthread_mutex_init(&mymutex, NULL);
    pthread_cond_init(&mycv, NULL);

    //创建5个消费者线程
    pthread_t consumerThreadID[5];
    for (int i = 0; i < 5; ++i)
        pthread_create(&consumerThreadID[i], NULL, consumer_thread, NULL);

    //创建1个生产者线程
    pthread_t producerThreadID;
    pthread_create(&producerThreadID, NULL, producer_thread, NULL);

    pthread_join(producerThreadID, NULL);

    for (int i = 0; i < 5; ++i)
        pthread_join(consumerThreadID[i], NULL);

    pthread_cond_destroy(&mycv);
    pthread_mutex_destroy(&mymutex);

    return 0;
}
```

编译并执行上述程序，输出结果如下：

```
[root@localhost testsemaphore]# g++ -g -o cv cv.cpp -lpthread
[root@localhost testsemaphore]# ./cv
produce a task, taskID: 0, threadID: 140571200554752
handle a task, taskID: 0, threadID: 140571242518272
produce a task, taskID: 1, threadID: 140571200554752
handle a task, taskID: 1, threadID: 140571225732864
produce a task, taskID: 2, threadID: 140571200554752
handle a task, taskID: 2, threadID: 140571208947456
produce a task, taskID: 3, threadID: 140571200554752
handle a task, taskID: 3, threadID: 140571242518272
produce a task, taskID: 4, threadID: 140571200554752
handle a task, taskID: 4, threadID: 140571234125568
produce a task, taskID: 5, threadID: 140571200554752
handle a task, taskID: 5, threadID: 140571217340160
produce a task, taskID: 6, threadID: 140571200554752
handle a task, taskID: 6, threadID: 140571225732864
produce a task, taskID: 7, threadID: 140571200554752
handle a task, taskID: 7, threadID: 140571208947456
produce a task, taskID: 8, threadID: 140571200554752
handle a task, taskID: 8, threadID: 140571242518272
...省略更多的输出结果...
```

条件变量最关键的一个地方，就是需要弄清楚 pthread_cond_wait 在条件满足与不满足时的两种行为，这是难点，也是重点。

（1）pthread_cond_wait 函数在阻塞时，会释放其绑定的互斥体并阻塞线程。因此在调用该函数前应该对互斥体有个加锁操作（对应代码片段 3-5-3 第 1 个加粗行）。

（2）收到条件信号时，pthread_cond_wait 会返回并对其绑定的互斥体进行加锁，因此在其下面一定有个对互斥体进行解锁的操作（对应代码片段 3-5-3 第 2 个加粗行）。

4. 条件变量的虚假唤醒

在将互斥体和条件变量配合使用的示例代码 3-5-3 中有个很有意思的地方，就是使用了 while 语句，条件变量醒来之后再次判断条件是否满足：

```
while (tasks.empty())
{
    pthread_cond_wait(&mycv, &mymutex);
}
```

为什么不写成如下所示的代码呢？

```
if (tasks.empty())
{
    pthread_cond_wait(&mycv, &mymutex);
}
```

答案是不得不如此。因为某次操作系统唤醒 pthread_cond_wait 时 tasks.empty()可能仍为 true，即操作系统可能在某些情况下唤醒条件变量，也就是说存在没有其他线程向条件变量发送信号，但等待此条件变量的线程有可能醒来的情形。我们将条件变量的这种行为称为虚假唤醒（spurious wakeup）。因此将条件（判断 tasks.empty()为 true）放在一个 while 循环中意味着光唤醒条件变量不行，还必须满足条件，程序才能继续执行正常的逻辑。

这看起来像个 Bug，但它在 Linux 系统中是实实在在存在的。为什么会存在虚假唤醒呢？一个原因是 pthread_cond_wait 是 futex 系统调用，属于阻塞型的系统调用，当系统调用被信号中断时，会返回-1，并且把 errno 错误码置为 EINTR。很多这种系统调用在被信号中断后，都会自行重启（即再次调用一次这个函数），代码如下：

```
pid_t r_wait(int *stat_loc)
{
    int retval;
    //wait 函数因为被信号中断导致调用失败，会返回-1，错误码是 EINTR
    //注意：这里的 while 循环体是一条空语句
    while(((retval = wait(stat_loc)) == -1 && (errno == EINTR));

    return retval;
}
```

但是 pthread_cond_wait 函数的用途有点不一样，假设 pthread_cond_wait 函数被信号中断，则在 pthread_cond_wait 函数返回之后，到重新调用之前，pthread_cond_signal 或 pthread_cond_broadcast 函数可能已被调用。一旦错失该信号，则可能由于条件信号不再产生，再次调用 pthread_cond_wait 函数将导致程序无限等待。为了避免这种情况发生，我们宁可虚假唤醒，也不能再次调用 pthread_cond_wait 函数，以免陷入无穷等待中。

除了上面的信号因素，还存在一些情况：在条件满足时发送信号，但等到调用 pthread_cond_wait 函数的线程得到 CPU 时间片时，条件又再次不满足了。

好在无论是哪种情况，条件变量醒来之后再次测试条件是否满足就可以解决虚假唤醒问题，这就是使用 while 循环而不是 if 语句来判断条件的原因。

5. 条件变量信号丢失的问题

前面介绍了，如果一个条件变量信号在产生时（调用 pthread_cond_signal 或 pthread_cond_broadcast），没有相关线程调用 pthread_cond_wait 捕获该信号，该信号就会永久丢失，再次调用 pthread_cond_wait 会导致永久阻塞。我们在设计条件变量信号只会产生一次的逻辑中尤其需要注意这种情况。举个例子，假设现在某个程序中有一批等待条件变量的线程，和一个只产生一次条件变量信号的线程。为了让等待条件变量的线程能够正常运行而不阻塞，编写这段逻辑时一定要确保等待的线程在产生条件变量信号的线程发送条件信号之前调用 pthread_cond_wait。

3.5.4　Linux 读写锁

本节详细讲解 Linux 读写锁的应用场景、应用方法、属性及应用示例。

1. 读写锁的应用场景

在实际应用中，对共享变量的访问大多有个特点：在大多数情况下，线程只是读取共享变量的值，只在极少数情况下才会真正修改共享变量的值。对于这种情况，读请求之间无须同步，它们之间的并发访问是安全的。然而写请求必须锁住读请求和其他写请求。

这在实际应用中是存在的,例如读取一个全局对象的状态属性,这个状态属性的值一般不会变化,偶尔才会被修改。如果使用互斥体完全阻止读请求并发,则会造成性能的损失。

2. 读写锁的应用方法

读写锁在 Linux 系统中使用 pthread_rwlock_t 类型表示,对读写锁的初始化和销毁使用如下系统 API 函数实现:

```
#include <pthread.h>

int pthread_rwlock_init(pthread_rwlock_t* rwlock, const pthread_rwlockattr_t* attr);
int pthread_rwlock_destroy(pthread_rwlock_t* rwlock);
```

rwlock 参数是需要初始化和销毁的读写锁对象的地址,attr 参数用于设置读写锁的属性,一般将其设置为 NULL,表示使用默认的属性。若函数调用成功则返回 0,若调用失败则返回非 0 值,我们可以通过检测错误码 errno 获取错误成因。

当然,如果不需要动态创建或者设置非默认属性的读写锁对象,则也可以使用如下语法初始化一个读写锁对象:

```
pthread_rwlock_t myrwlock = PTHREAD_RWLOCK_INITIALIZER;
```

下面是 3 个请求读锁的系统 API 接口:

```
int pthread_rwlock_rdlock(pthread_rwlock_t* rwlock);
int pthread_rwlock_tryrdlock(pthread_rwlock_t* rwlock);
int pthread_rwlock_timedrdlock(pthread_rwlock_t* rwlock, const struct timespec* abstime);
```

下面是 3 个请求写锁的系统 API 接口:

```
int pthread_rwlock_wrlock(pthread_rwlock_t* rwlock);
int pthread_rwlock_trywrlock(pthread_rwlock_t* rwlock);
int pthread_rwlock_timedwrlock(pthread_rwlock_t* rwlock, const struct timespec* abstime);
```

读锁用于共享模式:如果当前读写锁已经被某线程以读模式占有,则其他线程调用 pthread_rwlock_rdlock(请求读锁)时会立刻获得读锁;如果当前读写锁已经被某线程以读模式占有,则其他线程调用 pthread_rwlock_wrlock(请求写锁)时会陷入阻塞;

写锁用于独占模式:如果当前读写锁被某线程以写模式占有,则无论是调用 pthread_rwlock_rdlock 还是调用 pthread_rwlock_wrlock,都会陷入阻塞。即在写模式下不允许任何读锁请求通过,也不允许任何写锁请求通过,读锁请求和写锁请求都要陷入阻塞中,直到线程释放写锁。

可以将上述读写锁逻辑总结成如下表格。

锁的当前状态、其他线程请求锁类型	请求读锁	请求写锁
无锁	通过	通过
已经获得读锁	通过	阻止
已经获得写锁	阻止	阻止

释放读锁或写锁使用的都是同一接口：

```
int pthread_rwlock_unlock (pthread_rwlock_t* rwlock);
```

无论是请求读锁还是写锁，都提供了 trylock 的功能（pthread_rwlock_tryrdlock 和 pthread_rwlock_trywrlock），调用线程不会阻塞，而是立即返回。如果能成功获得读锁或者写锁，则函数返回 0，如果不能获得读锁或写锁，则函数返回非 0 值，此时错误码 errno 是 EBUSY。

当然，无论是请求读锁还是写锁，都提供了限时等待功能。如果不能获取读写锁，则会陷入阻塞，最多等待到参数 abstime 设置的时间，如果此时仍然无法获得锁，则返回，错误码 errno 是 ETIMEOUT。

3. 读写锁的属性

前面介绍 pthread_rwlock_init 函数时，提到其第 2 个参数可以设置读写锁的属性，读写锁的属性类型是 pthread_rwlockattr_t。glibc 引入了如下接口来查询和改变读写锁的类型：

```
#include <pthread.h>

int pthread_rwlockattr_setkind_np(pthread_rwlockattr_t* attr, int pref);
int pthread_rwlockattr_getkind_np(const pthread_rwlockattr_t* attr, int* pref);
```

pthread_rwlockattr_setkind_np 的第 2 个参数 pref 用于设置读写锁的类型，其取值如下：

```
enum
{
    //读者优先（即同时请求读锁和写锁时，请求读锁的线程优先获得锁）
    PTHREAD_RWLOCK_PREFER_READER_NP,
    //不要被名字迷惑，也是读者优先
    PTHREAD_RWLOCK_PREFER_WRITER_NP,
    //写者优先（即同时请求读锁和写锁时，请求写锁的线程优先获得锁）
    PTHREAD_RWLOCK_PREFER_WRITER_NONRECURSIVE_NP,
    PTHREAD_RWLOCK_DEFAULT_NP = PTHREAD_RWLOCK_PREFER_READER_NP
};
```

当然，为得到一个有效的 pthread_rwlockattr_t 对象，需要先调用 pthread_rwlockattr_init 函数初始化这样一个属性对象，在不需要时使用 pthread_rwlockattr_destroy 销毁它：

```
int pthread_rwlockattr_init(pthread_rwlockattr_t* attr);
int pthread_rwlockattr_destroy(pthread_rwlockattr_t* attr);
```

以下代码片段演示了如何初始化一个写者优先的读写锁：

```
pthread_rwlockattr_t attr;
pthread_rwlockattr_init(&attr);
pthread_rwlockattr_setkind_np(&attr,
PTHREAD_RWLOCK_PREFER_WRITER_NONRECURSIVE_NP);
pthread_rwlock_t rwlock;
pthread_rwlock_init(&rwlock, &attr);
```

4. 读写锁应用示例

读写锁应用示例如下：

```cpp
//代码片段3-5-4
#include <pthread.h>
#include <unistd.h>
#include <iostream>

int resourceID = 0;
pthread_rwlock_t myrwlock;

void* read_thread(void* param)
{
    while (true)
    {
        //请求读锁
        pthread_rwlock_rdlock(&myrwlock);

        std::cout << "read thread ID: " << pthread_self() << ", resourceID: " << resourceID << std::endl;

        //使用睡眠模拟读线程读的过程花了很长时间
        sleep(1);

        pthread_rwlock_unlock(&myrwlock);
    }

    return NULL;
}

void* write_thread(void* param)
{
    while (true)
    {
        //请求写锁
        pthread_rwlock_wrlock(&myrwlock);

        ++resourceID;
        std::cout << "write thread ID: " << pthread_self() << ", resourceID: " << resourceID << std::endl;
        //使用睡眠模拟读线程读的过程花了很长时间
        sleep(1);

        pthread_rwlock_unlock(&myrwlock);
    }

    return NULL;
}

int main()
{
    pthread_rwlock_init(&myrwlock, NULL);

    //创建5个请求读锁线程
    pthread_t readThreadID[5];
    for (int i = 0; i < 5; ++i)
        pthread_create(&readThreadID[i], NULL, read_thread, NULL);
```

```cpp
    //创建1个请求写锁线程
    pthread_t writeThreadID;
    pthread_create(&writeThreadID, NULL, write_thread, NULL);

    pthread_join(writeThreadID, NULL);

    for (int i = 0; i < 5; ++i)
        pthread_join(readThreadID[i], NULL);

    pthread_rwlock_destroy(&myrwlock);

    return 0;
}
```

在上述程序中创建了5个请求读锁的读线程和1个请求写锁的写线程，共享的资源是一个整形变量resourceID，我们编译并执行它以得到输出结果：

```
[root@localhost testmultithread]# g++ -g -o rwlock rwlock.cpp -lpthread
[root@localhost testmultithread]# ./rwlock
read thread ID: 140575861593856, resourceID: 0
read thread ID: 140575878379264, resourceID: 0
read thread ID: 140575853201152, resourceID: 0
read thread ID: 140575869986560, resourceID: 0
read thread ID: 140575886771968, resourceID: 0
read thread ID: read thread ID: read thread ID: read thread ID:
140575861593856140575886771968, resourceID: 0, resourceID:
0
140575878379264read thread ID: 140575869986560, resourceID: 0
, resourceID: 0
140575853201152, resourceID: 0
read thread ID: read thread ID: read thread ID:
140575861593856140575853201152140575886771968, resourceID: , resourceID: 0,
resourceID: 00

read thread ID: 140575869986560, resourceID: 0
...省略更多的输出结果...
```

以上输出结果验证了以下两个结论。

（1）由于读写锁对象myrwlock使用了默认属性，其行为是请求读锁的线程优先获得锁，请求写锁的线程write_thread很难获得锁，因此在其结果中基本没有请求写锁线程的输出结果。

（2）由于多个请求读锁的线程read_thread可以自由获得读锁，代码片段3-5-4中加粗行的输出不是原子的，所以多个读线程的输出可能交替，出现"错乱"。

我们将读写锁对象myrwlock的属性修改成请求写锁优先，再来试一试：

```cpp
//代码片段3-5-5
#include <pthread.h>
#include <unistd.h>
#include <iostream>

int resourceID = 0;
```

```cpp
pthread_rwlock_t myrwlock;

void* read_thread(void* param)
{
    while (true)
    {
        //请求读锁
        pthread_rwlock_rdlock(&myrwlock);

        std::cout << "read thread ID: " << pthread_self() << ", resourceID: " << resourceID << std::endl;

        //使用睡眠模拟读线程读的过程花了很长时间
        sleep(1);

        pthread_rwlock_unlock(&myrwlock);
    }

    return NULL;
}

void* write_thread(void* param)
{
    while (true)
    {
        //请求写锁
        pthread_rwlock_wrlock(&myrwlock);

        ++resourceID;
        std::cout << "write thread ID: " << pthread_self() << ", resourceID: " << resourceID << std::endl;

        //使用睡眠模拟读线程读的过程花了很长时间
        sleep(1);

        pthread_rwlock_unlock(&myrwlock);
    }

    return NULL;
}

int main()
{
    pthread_rwlockattr_t attr;
    pthread_rwlockattr_init(&attr);
    //设置成请求写锁优先
    pthread_rwlockattr_setkind_np(&attr,
PTHREAD_RWLOCK_PREFER_WRITER_NONRECURSIVE_NP);
    pthread_rwlock_init(&myrwlock, &attr);

    //创建5个请求读锁线程
    pthread_t readThreadID[5];
    for (int i = 0; i < 5; ++i)
        pthread_create(&readThreadID[i], NULL, read_thread, NULL);
```

```cpp
    //创建1个请求写锁线程
    pthread_t writeThreadID;
    pthread_create(&writeThreadID, NULL, write_thread, NULL);

    pthread_join(writeThreadID, NULL);

    for (int i = 0; i < 5; ++i)
        pthread_join(readThreadID[i], NULL);

    pthread_rwlock_destroy(&myrwlock);

    return 0;
}
```

编译程序并运行，输出结果如下：

```
[root@localhost testmultithread]# g++ -g -o rwlock2 rwlock2.cpp -lpthread
[root@localhost testmultithread]# ./rwlock2
read thread ID: 140122217539328, resourceID: 0
read thread ID: 140122242717440, resourceID: 0
read thread ID: 140122209146624, resourceID: 0
write thread ID: 140122200753920, resourceID: 1
read thread ID: 140122234324736, resourceID: 1
write thread ID: 140122200753920, resourceID: 2
write thread ID: 140122200753920, resourceID: 3
write thread ID: 140122200753920, resourceID: 4
write thread ID: 140122200753920, resourceID: 5
write thread ID: 140122200753920, resourceID: 6
write thread ID: 140122200753920, resourceID: 7
write thread ID: 140122200753920, resourceID: 8
write thread ID: 140122200753920, resourceID: 9
write thread ID: 140122200753920, resourceID: 10
write thread ID: 140122200753920, resourceID: 11
write thread ID: 140122200753920, resourceID: 12
write thread ID: 140122200753920, resourceID: 13
read thread ID: 140122217539328, resourceID: 13
write thread ID: 140122200753920, resourceID: 14
write thread ID: 140122200753920, resourceID: 15
write thread ID: 140122200753920, resourceID: 16
write thread ID: 140122200753920, resourceID: 17
write thread ID: 140122200753920, resourceID: 18
write thread ID: 140122200753920, resourceID: 19
write thread ID: 140122200753920, resourceID: 20
write thread ID: 140122200753920, resourceID: 21
write thread ID: 140122200753920, resourceID: 22
write thread ID: 140122200753920, resourceID: 23
...省略更多的输出结果...
```

由于将myrwlock设置成请求写锁优先，所以以上几乎都是write_thread的输出结果。

我们将代码片段3-5-5中write_thread线程函数的sleep语句（加粗代码行）挪到pthread_rwlock_unlock(&myrwlock);语句后面，增加请求写锁线程的睡眠时间：

```cpp
#include <pthread.h>
#include <unistd.h>
#include <iostream>
```

```
int resourceID = 0;
pthread_rwlock_t myrwlock;

void* read_thread(void* param)
{
    while (true)
    {
        //请求读锁
        pthread_rwlock_rdlock(&myrwlock);

        std::cout << "read thread ID: " << pthread_self() << ", resourceID: " << resourceID << std::endl;

        //使用睡眠模拟读线程读的过程花了很长时间
        sleep(1);

        pthread_rwlock_unlock(&myrwlock);
    }

    return NULL;
}

void* write_thread(void* param)
{
    while (true)
    {
        //请求写锁
        pthread_rwlock_wrlock(&myrwlock);

        ++resourceID;
        std::cout << "write thread ID: " << pthread_self() << ", resourceID: " << resourceID << std::endl;

        pthread_rwlock_unlock(&myrwlock);

        //放在这里，增加请求读锁线程获得锁的几率
        sleep(1);
    }

    return NULL;
}

int main()
{
    pthread_rwlockattr_t attr;
    pthread_rwlockattr_init(&attr);
    //设置成请求写锁优先
    pthread_rwlockattr_setkind_np(&attr,
PTHREAD_RWLOCK_PREFER_WRITER_NONRECURSIVE_NP);
    pthread_rwlock_init(&myrwlock, &attr);

    //创建 5 个请求读锁线程
    pthread_t readThreadID[5];
    for (int i = 0; i < 5; ++i)
```

```cpp
        pthread_create(&readThreadID[i], NULL, read_thread, NULL);

    //创建1个请求写锁线程
    pthread_t writeThreadID;
    pthread_create(&writeThreadID, NULL, write_thread, NULL);

    pthread_join(writeThreadID, NULL);

    for (int i = 0; i < 5; ++i)
        pthread_join(readThreadID[i], NULL);

    pthread_rwlock_destroy(&myrwlock);

    return 0;
}
```

再次编译程序并执行,得到输出结果:

```
[root@localhost testmultithread]# g++ -g -o rwlock3 rwlock3.cpp -lpthread
[root@localhost testmultithread]# ./rwlock3
read thread ID: 140315524790016, resourceID: 0
read thread ID: 140315549968128, resourceID: 0
read thread ID: 140315541575424, resourceID: 0
write thread ID: 140315508004608, resourceID: 1
read thread ID: 140315549968128, resourceID: 1
read thread ID: 140315541575424, resourceID: 1
read thread ID: 140315524790016, resourceID: 1
read thread ID: 140315516397312, resourceID: 1
read thread ID: 140315533182720, resourceID: 1
write thread ID: 140315508004608, resourceID: 2
read thread ID: 140315541575424, resourceID: 2
read thread ID: 140315524790016, resourceID: 2
read thread ID: 140315533182720, resourceID: 2
read thread ID: 140315516397312, resourceID: 2
read thread ID: 140315549968128, resourceID: 2
read thread ID: 140315516397312, resourceID: 2
write thread ID: 140315508004608, resourceID: 3
read thread ID: 140315549968128, resourceID: 3
read thread ID: 140315541575424, resourceID: 3
read thread ID: 140315533182720, resourceID: 3read thread ID: read thread ID:
140315524790016, resourceID: 3
140315516397312, resourceID: 3

read thread ID: read thread ID: read thread ID: 140315524790016140315549968128,
resourceID: , resourceID: 33
140315516397312, resourceID: 3
read thread ID: 140315541575424, resourceID: read thread ID: 140315533182720,
resourceID: 3
3

write thread ID: 140315508004608, resourceID: 4
read thread ID: 140315516397312, resourceID: 4
read thread ID: 140315541575424, resourceID: 4
read thread ID: 140315524790016, resourceID: 4
read thread ID: 140315549968128, resourceID: 4
read thread ID: 140315533182720, resourceID: 4
```

```
read thread ID: 140315524790016, resourceID: 4
read thread ID: 140315541575424, resourceID: 4
write thread ID: 140315508004608, resourceID: 5
read thread ID: 140315516397312, resourceID: 5
read thread ID: 140315541575424, resourceID: 5
read thread ID: 140315524790016, resourceID: 5
read thread ID: 140315533182720, resourceID: 5
read thread ID: 140315549968128, resourceID: 5
```

这次请求读锁的线程和请求写锁的线程的输出结果分布就比较均匀了。

以上例子比较简单，建议实际运行代码实验一下。

3.6 Windows 线程同步对象

本节详细介绍 Windows 线程同步对象的原理和使用场景。

3.6.1 WaitForSingleObject 与 WaitForMultipleObjects 函数

在介绍 Windows 线程同步对象之前，先来介绍两个与之相关的、非常重要的函数：WaitForSingleObject 和 WaitForMultipleObjects。WaitForSingleObject 函数签名如下：

```
DWORD WaitForSingleObject(HANDLE hHandle, DWORD dwMilliseconds);
```

该函数的作用是等待一个内核对象，在 Windows 系统上，一个内核对象通常使用其句柄来操作。参数 hHandle 是需要等待的内核对象；参数 dwMilliseconds 是等待这个内核对象的最大时间，时间单位是毫秒，其类型是 DWORD，这是一个 unsigned long 类型。如果我们需要无限等待下去，则可以将这个参数值设置为 INFINITE 宏。

在 Windows 上可以调用 WaitForSingleObject 等待的常见对象如下表所示。

可以被等待的对象	等待对象成功的含义	对象类型
线程	等待线程结束	HANDLE
进程	等待进程结束	HANDLE
Event（事件）	等待 Event 有信号	HANDLE
Mutex（互斥体）	等待持有 Mutex 的线程释放该 Mutex，等待成功，拥有该 Mutex	HANDLE
Semaphore（信号量）	等待该 Semaphore 对象有信号	HANDLE

下面详细介绍 Event、Mutex、Semaphore 这三种类型的线程同步对象。这里先接着介绍 WaitForSingleObject 函数的用法，该函数的返回值一般有以下类型。

（1）WAIT_FAILED，表示 WaitForSingleObject 函数调用失败，调用失败时，可以通过 GetLastError 函数得到具体的错误码。

（2）WAIT_OBJECT_0，表示 WaitForSingleObject 成功"等待"到设置的对象。

（3）WAIT_TIMEOUT，表示等待超时。

（4）WAIT_ABANDONED，表示等待的对象是 Mutex 类型，如果持有某个 Mutex 的线程已经结束运行，但该线程在结束运行前未调用 ReleaseMutex 函数释放对该 Mutex 的持有权，则此时我们调用 WaitForSingleObject 函数等待该 Mutex，就会返回 WAIT_ABANDONED 值，该值表明此 Mutex 处于废弃状态。处于废弃状态的 Mutex 的表现是未知的，因此不建议再使用该 Mutex。

WaitForSingleObject 只能"等待"单个对象，如果需要同时等待多个对象，则可以使用 WaitForMultipleObjects，除了对象的数量变多，其用法基本和 WaitForSingleObject 一样。WaitForMultipleObjects 函数签名如下：

```
DWORD WaitForMultipleObjects(
    DWORD           nCount,
    const HANDLE    *lpHandles,
    BOOL            bWaitAll,
    DWORD           dwMilliseconds
);
```

参数 lpHandles 是需要等待的对象数组指针。参数 nCount 指定了该数组的长度。参数 bWaitAll 表示是否等待数组 lpHandles 中的所有对象都有信号：取值为 TRUE 时，WaitForMultipleObjects 会等待所有对象有信号才会返回；取值为 FALSE 时，当其中一个对象有信号时立即返回，其返回值表示哪个对象有信号。

在参数 bWaitAll 被设置为 FALSE 的情况下，返回值除了上面介绍的 WAIT_FAILED 和 WAIT_TIMEOUT，还有另外两种情形（分别对应 WaitForSingleObject 返回值 WAIT_OBJECT_0 和 WAIT_ABANDONED）。

（1）WAIT_OBJECT_0 ~ (WAIT_OBJECT_0 + nCount – 1)。举个例子，假设现在等待三个对象 A1、A2、A3，它们在数组 lpHandles 中的下标依次是 0、1、2，WaitForMultipleObjects 的返回值是 Wait_OBJECT_0 + 1，则表示对象 A2 有信号，导致 WaitForMultipleObjects 调用成功后返回。

伪代码如下：

```
HANDLE waitHandles[3];
waitHandles[0] = hA1Handle;
waitHandles[1] = hA2Handle;
waitHandles[2] = hA3Handle;

DWORD dwResult = WaitForMultipleObjects(3, waitHandles, FALSE, 3000);
switch(dwResult)
{
    case WAIT_OBJECT_0 + 0:
        //A1 有信号
        break;

    case WAIT_OBJECT_0 + 1:
        //A2 有信号
        break;

    case WAIT_OBJECT_0 + 2:
        //A3 有信号
        break;
```

```
    default:
        //出错或超时
        break;
}
```

（2）WAIT_ABANDONED_0 ~ (WAIT_ABANDONED_0 + nCount – 1)。这种情形与上面的使用方法相同，通过 nCount–1 可以知道等待对象数组中的哪个对象始终没有被其他线程释放使用权。

3.6.2　Windows 临界区对象

在所有 Windows 资源同步对象中，CriticalSection（临界区对象，有时被翻译成"关键段"）都是最简单、易用的，能用于防止多线程同时执行其保护的那段代码（临界区代码），即临界区的代码在某一时刻只允许 1 个线程执行，示意图如下。

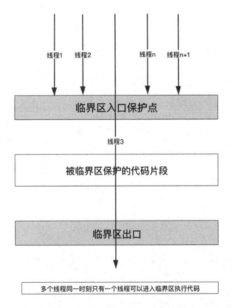

Windows 没有公开 CriticalSection 数据结构的定义，我们一般使用如下 5 个 API 函数操作临界区对象：

```
void InitializeCriticalSection(LPCRITICAL_SECTION lpCriticalSection);
void DeleteCriticalSection(LPCRITICAL_SECTION lpCriticalSection);

BOOL TryEnterCriticalSection(LPCRITICAL_SECTION lpCriticalSection);
void EnterCriticalSection(LPCRITICAL_SECTION lpCriticalSection);
void LeaveCriticalSection(LPCRITICAL_SECTION lpCriticalSection);
```

InitializeCriticalSection 和 DeleteCriticalSection 用于初始化和销毁 1 个 CRITICAL_SECTION 对象；位于 EnterCriticalSection 和 LeaveCriticalSection 之间的代码即临界区代码；调用 EnterCriticalSection 的线程会尝试进入临界区，如果进入不了，则会阻塞调用线程，直到成功进入或者超时。TryEnterCriticalSection 会尝试进入临界区，如果可以进入，

则函数返回 TRUE；如果无法进入，则立即返回，不会阻塞调用线程，函数返回 FALSE。LeaveCriticalSection 函数让调用的线程离开临界区，离开临界区以后，临界区的代码允许其他线程调用 EnterCriticalSection 进入。

EnterCriticalSection 的超时时间很长，可以在注册表的 HKEY_LOCAL_MACHINE\SYSTEM\CurrentControlSet\Control\Session Manager 位置修改参数 CriticalSectionTimeout 的值进行调整，当然，在实际开发中我们从来不会修改这个值，如果代码等待时间较长且最终超时，就需要检查逻辑设计是否合理。

来看一段实例代码：

```
01 #include <Windows.h>
02 #include <list>
03 #include <iostream>
04 #include <string>
05
06 CRITICAL_SECTION        g_cs;
07 int                     g_number = 0;
08
09 DWORD __stdcall WorkerThreadProc(LPVOID lpThreadParameter)
10 {
11     DWORD dwThreadID = GetCurrentThreadId();
12
13     while (true)
14     {
15         EnterCriticalSection(&g_cs);
16         std::cout << "EnterCriticalSection, ThreadID: " << dwThreadID << std::endl;
17         g_number++;
18         SYSTEMTIME st;
19         //获取当前系统时间
20         GetLocalTime(&st);
21         char szMsg[64] = { 0 };
22         sprintf(szMsg,
23             "[%04d-%02d-%02d %02d:%02d:%02d:%03d]NO.%d, ThreadID: %d.",
24             st.wYear, st.wMonth, st.wDay,
25             st.wHour, st.wMinute, st.wSecond, st.wMilliseconds,
26             g_number, dwThreadID);
27
28         std::cout << szMsg << std::endl;
29         std::cout << "LeaveCriticalSection, ThreadID: " << dwThreadID << std::endl;
30         LeaveCriticalSection(&g_cs);
31
32         //睡眠1秒
33         Sleep(1000);
34     }
35
36     return 0;
37 }
38
39 int main()
40 {
41     InitializeCriticalSection(&g_cs);
42
```

```
43    HANDLE hWorkerThread1 = CreateThread(NULL, 0, WorkerThreadProc, NULL, 0, NULL);
44    HANDLE hWorkerThread2 = CreateThread(NULL, 0, WorkerThreadProc, NULL, 0, NULL);
45
46    WaitForSingleObject(hWorkerThread1, INFINITE);
47    WaitForSingleObject(hWorkerThread2, INFINITE);
48
49    //关闭线程句柄
50    CloseHandle(hWorkerThread1);
51    CloseHandle(hWorkerThread2);
52
53    DeleteCriticalSection(&g_cs);
54
55    return 0;
56  }
```

以上代码的执行结果如下：

```
EnterCriticalSection, ThreadID: 1224
[2019-01-19 22:25:41:031]NO.1, ThreadID: 1224.
LeaveCriticalSection, ThreadID: 1224
EnterCriticalSection, ThreadID: 6588
[2019-01-19 22:25:41:031]NO.2, ThreadID: 6588.
LeaveCriticalSection, ThreadID: 6588
EnterCriticalSection, ThreadID: 6588
[2019-01-19 22:25:42:031]NO.3, ThreadID: 6588.
LeaveCriticalSection, ThreadID: 6588
EnterCriticalSection, ThreadID: 1224
[2019-01-19 22:25:42:031]NO.4, ThreadID: 1224.
LeaveCriticalSection, ThreadID: 1224
EnterCriticalSection, ThreadID: 1224
[2019-01-19 22:25:43:031]NO.5, ThreadID: 1224.
LeaveCriticalSection, ThreadID: 1224
EnterCriticalSection, ThreadID: 6588
[2019-01-19 22:25:43:031]NO.6, ThreadID: 6588.
LeaveCriticalSection, ThreadID: 6588
EnterCriticalSection, ThreadID: 1224
[2019-01-19 22:25:44:031]NO.7, ThreadID: 1224.
LeaveCriticalSection, ThreadID: 1224
EnterCriticalSection, ThreadID: 6588
[2019-01-19 22:25:44:031]NO.8, ThreadID: 6588.
LeaveCriticalSection, ThreadID: 6588
```

我们在以上代码中新建了两个工作线程，线程函数都是 WorkerThreadProc。线程函数在第 15 行调用 EnterCriticalSection 进入临界区；在第 30 行调用 LeaveCriticalSection 离开临界区；第 16～29 行的代码即临界区的代码，这段代码由于受到临界区对象 g_cs 的保护，每次只允许 1 个工作线程执行这段代码。虽然在临界区代码中有多个输出，但这些输出一定都是连续的，不会交叉输出。

在以上输出中有同一个线程连续两次进入临界区，这是有可能的。也就是说，当其中 1 个线程离开临界区时，即使此时有其他线程在这个临界区外面等待，由于线程调度的不确定性，此时正在等待的线程也不会有先进入临界区的优势，它和刚离开这个临界区的线程再次竞争进入临界区的机会均等。我们来看一张图，如下所示。

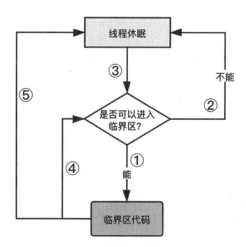

如上所示将线程函数的执行流程绘制成一个流程图,两个线程竞争进入临界区可能存在如下情形。

情形一:线程 A 被唤醒并获得 CPU 时间片进入临界区,执行流程①;执行临界区代码输出→线程 B 获得 CPU 时间片,执行流程②;失去 CPU 时间片,进入休眠→线程 A 执行完临界区代码,离开临界区后执行流程⑤;失去 CPU 时间片,进入休眠→线程 B 被唤醒,获得 CPU 时间片执行流程③、①,执行临界区代码输出。在这种情形下,线程 A 和线程 B 会轮流进入临界区执行代码。

情形二:线程 A 被唤醒并获得 CPU 时间片进入临界区,执行流程①;执行临界区代码输出→线程 B 获得 CPU 时间片,执行流程③;执行流程②,在临界区外面失去 CPU 时间片,进入休眠→线程 A 执行完临界区代码,离开临界区后执行流程④、①。在这种情形下,会出现某个线程连续两次甚至更多次进入临界区执行代码。

如果某个线程在尝试进入临界区时因无法阻塞而进入睡眠状态,则其他线程离开这个临界区后,之前因为这个临界区而阻塞的线程可能会被唤醒并再次竞争,也可能不被唤醒。但是存在这样一种特例,假设现在存在两个线程 A 和 B,线程 A 是离开临界区的线程,且不需要再次进入临界区,那么线程 B 在被唤醒时一定可以进入临界区。线程 B 从睡眠状态被唤醒,涉及一次线程的切换,有时这种开销是不必要的。我们可以让线程 B 执行一个简单的循环,等待一段时间后再进入临界区,而不是先睡眠再唤醒。前者与后者相比,执行这个循环的消耗更小。这就是自旋。在这种情形下,Windows 提供了另一个初始化临界区的函数 InitializeCriticalSectionAndSpinCount,这个函数比 InitializeCriticalSection 多了一次自旋:

```
BOOL InitializeCriticalSectionAndSpinCount(
    LPCRITICAL_SECTION  lpCriticalSection,
    DWORD               dwSpinCount
);
```

dwSpinCount 参数表示自旋的次数,利用自旋来避免线程因为等待而进入睡眠并再次被唤醒,消除线程上下面切换带来的消耗,提高效率。当然,在实际开发中,这种方式是靠不住的,线程调度是操作系统内核的策略,应用层上的应用不应该假设线程的调度策略按预想的来执行,但是理解线程与临界区之间的原理有助于我们编写更高效的代码。

需要说明的是,临界区对象通过保护一段代码不被多个线程同时执行,来保证多个线程读写一个对象是安全的。由于同一时刻只有一个线程可以进入临界区,因此这种对资源的操作是排他的,即对于同一个临界区对象,不会出现多个线程同时操作该资源,哪怕资源本身可以支持在同一时刻被多个线程操作(例如多个线程对资源进行读操作),这会带来资源操作效率低的问题。

我们一般将进入临界区的线程称为该临界区的拥有者(owner),即临界区持有者。

最后,为了避免死锁,EnterCriticalSection 和 LeaveCriticalSection 需要成对使用,尤其是在具有多个出口的函数中,记得在每个分支处都加上 LeaveCriticalSection。伪代码如下:

```
void someFunction()
{
    EnterCriticalSection(&someCriticalSection);
    if (条件A)
    {
        if (条件B)
        {
            LeaveCriticalSection(&someCriticalSection);
            //出口1
            return;
        }

        LeaveCriticalSection(&someCriticalSection);
        //出口2
        return;
    }

    if (条件C)
    {
        LeaveCriticalSection(&someCriticalSection);
        //出口3
        return;
    }

    if (条件C)
    {
        LeaveCriticalSection(&someCriticalSection);
        //出口4
        return;
    }
}
```

在以上代码中,为了能让临界区对象被正常释放,在函数的每个出口都加上了 LeaveCriticalSection 调用,如果函数的出口非常多,则这样的代码很难维护。所以一般建议使用 RAII 技术将临界区 API 封装成对象,该对象在其作用域内调用构造函数进入临界区,在出了其作用域后调用析构函数离开临界区。示例代码如下:

```cpp
class CCriticalSection
{
public:
    CCriticalSection(CRITICAL_SECTION& cs) : m_CS(cs)
    {
        EnterCriticalSection(&m_CS);
    }

    ~CCriticalSection()
    {
        LeaveCriticalSection(&m_CS);
    }

private:
    CRITICAL_SECTION& m_CS;
};
```

利用 CCriticalSection 类，我们可以对以上伪代码进行优化：

```cpp
void someFunction()
{
    CCriticalSection autoCS(someCriticalSection);
    if (条件 A)
    {
        if (条件 B)
        {
            //出口 1
            return;
        }

        //出口 2
        return;
    }

    if (条件 C)
    {
        //出口 3
        return;
    }

    if (条件 C)
    {
        //出口 4
        return;
    }
}
```

在以上代码中，变量 autoCS 会在出了函数作用域后调用其析构函数，在析构函数中调用 LeaveCriticalSection 自动离开临界区。

3.6.3　Windows Event 对象

本节讨论的 Event 对象不是 Windows UI 事件驱动机制中的事件，而是多线程同步中的 Event 对象，也是 Windows 的内核对象之一，在 Windows 多线程程序设计中使用频率较高。这里先讲解如何创建 Event 对象，然后逐步展开介绍。创建 Event 的 Windows API 函数签名如下：

```
HANDLE CreateEvent(
    LPSECURITY_ATTRIBUTES       lpEventAttributes,
    BOOL                        bManualReset,
    BOOL                        bInitialState,
    LPCTSTR                     lpName
);
```

对其中的参数和返回值说明如下。

（1）参数 lpEventAttributes 设置 Event 对象的安全属性，Windows 中的所有内核对象都可以设置这个属性，我们一般将其设置为 NULL，即使用默认的安全属性。

（2）参数 bManualReset 设置 Event 对象受信（变成有信号状态）时的行为，当设置其为 TRUE 时，表示需要手动调用 ResetEvent 函数将 Event 重置成无信号状态；当设置其为 FALSE 时，Event 事件对象在受信后会被自动重置为无信号状态。

（3）参数 bInitialState 设置 Event 事件对象的初始状态是否受信，TRUE 表示有信号，FALSE 表示无信号。

（4）参数 lpName 可以设置 Event 对象的名称，如果不需要设置名称，则可以将该参数设置为 NULL。一个 Event 对象根据是否设置了名称，分为具名对象（具有名称的对象）和匿名对象。Event 对象是可以通过名称在不同的进程之间共享的，通过这种方式共享很有用。

（5）返回值，如果成功创建了 Event 对象，则返回对象的句柄；如果创建失败，则返回 NULL。

对于一个无信号的 Event 对象，我们可以通过 SetEvent 将其变成受信状态，SetEvent 函数签名如下：

```
BOOL SetEvent(HANDLE hEvent);
```

我们将参数 hEvent 设置为我们需要设置信号的 Event 句柄即可。

同理，对于一个已经受信的 Event 对象，我们可以使用 ResetEvent 对象将其变成无信号状态。ResetEvent 函数签名如下，参数 hEvent 即我们需要重置的 Event 对象句柄：

```
BOOL ResetEvent(HANDLE hEvent);
```

来看一个具体的例子，假设现在有两个线程，其中一个是主线程，主线程等待工作线程执行某一项耗时的任务后，将任务结果显示出来。代码如下：

```
#include <Windows.h>
#include <string>
#include <iostream>

bool            g_bTaskCompleted = false;
std::string g_TaskResult;

DWORD __stdcall WorkerThreadProc(LPVOID lpThreadParameter)
```

```
{
    //使用Sleep函数模拟一个很耗时的操作
    //睡眠3秒
    Sleep(3000);
    g_TaskResult = "task completed";
    g_bTaskCompleted = true;

    return 0;
}

int main()
{
    HANDLE hWorkerThread = CreateThread(NULL, 0, WorkerThreadProc, NULL, 0, NULL);
    while (true)
    {
        if (g_bTaskCompleted)
        {
            std::cout << g_TaskResult << std::endl;
            break;
        }
        else
            std::cout << "Task is in progress..." << std::endl;
    }

    CloseHandle(hWorkerThread);
    return 0;
}
```

程序执行结果如下图所示。

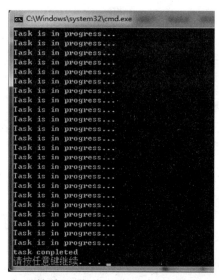

在以上代码中，主线程为了等待工作线程完成任务后获取结果，使用了一个循环去不断查询任务完成标识，这是一种低效的做法：等待的线程（主线程）做了很多无用功，对CPU时间片也是一种浪费。我们使用Event对象改写以上代码：

```
01 #include <Windows.h>
02 #include <string>
03 #include <iostream>
```

```cpp
04
05  bool          g_bTaskCompleted = false;
06  std::string   g_TaskResult;
07  HANDLE        g_hTaskEvent = NULL;
08
09  DWORD __stdcall WorkerThreadProc(LPVOID lpThreadParameter)
10  {
11      //使用Sleep函数模拟1个很耗时的操作
12      //睡眠3秒
13      Sleep(3000);
14      g_TaskResult = "task completed";
15      g_bTaskCompleted = true;
16
17      //设置事件信号
18      SetEvent(g_hTaskEvent);
19
20      return 0;
21  }
22
23  int main()
24  {
25      //创建1个匿名的手动重置初始无信号的事件对象
26      g_hTaskEvent = CreateEvent(NULL, TRUE, FALSE, NULL);
27      HANDLE hWorkerThread = CreateThread(NULL, 0, WorkerThreadProc, NULL, 0, NULL);
28
29      DWORD dwResult = WaitForSingleObject(g_hTaskEvent, INFINITE);
30      if (dwResult == WAIT_OBJECT_0)
31          std::cout << g_TaskResult << std::endl;
32
33      CloseHandle(hWorkerThread);
34      CloseHandle(g_hTaskEvent);
35      return 0;
36  }
```

在以上代码中，主线程在工作线程完成任务前会一直阻塞在代码第29行，没有任何消耗，在工作线程完成任务后调用SetEvent让事件对象受信，这样主线程会立即得到通知，从WaitForSingleObject处返回，此时任务已经完成，可以得到任务结果了。

在实际开发中，我们可以利用等待的时间做其他事情，在需要时检测事件对象是否有信号即可。另外，Event对象有以下两个显著特点。

（1）与临界区对象（以及Mutex对象）相比，Event对象没有让持有者线程变成其owner，所以Event对象可以同时唤醒多个等待的工作线程。

（2）手动重置的Event对象一旦变成受信状态，其信号就不会丢失。也就是说，当Event从无信号变成有信号时，即使某个线程当时没有调用WaitForSingleObject等待该Event对象受信，而是在这之后才调用WaitForSingleObject，则仍然能检测到事件的受信状态，即不会丢失信号，而后面要介绍的条件变量可能会丢失信号。

蘑菇街开源的即时通信工具Teamtalk的PC版在使用socket连接服务器时，使用Event对象设计了一个超时做法。传统的做法是将socket设置为非阻塞的，调用完connect函数之后，调用select函数检测socket是否可写，在select函数里面设置超时时间。Teamtalk

中的做法如下：

```cpp
//TcpClientModule_Impl.cpp 文件第145行
IM::Login::IMLoginRes* TcpClientModule_Impl::doLogin(CString &linkaddr, UInt16 port
    ,CString& uName,std::string& pass)
{
    //在imcore::IMLibCoreConnect 中通过 connect 函数连接服务器
    m_socketHandle = imcore::IMLibCoreConnect(util::cStringToString(linkaddr), port);
    imcore::IMLibCoreRegisterCallback(m_socketHandle, this);
    if(util::waitSingleObject(m_eventConnected, 5000))
    {
        IM::Login::IMLoginReq imLoginReq;
        string& name = util::cStringToString(uName);
        imLoginReq.set_user_name(name);
        imLoginReq.set_password(pass);
        imLoginReq.set_online_status(IM::BaseDefine::USER_STATUS_ONLINE);
        imLoginReq.set_client_type(IM::BaseDefine::CLIENT_TYPE_WINDOWS);
        imLoginReq.set_client_version("win_10086");

        if (TCPCLIENT_STATE_OK != m_tcpClientState)
            return 0;

        sendPacket(IM::BaseDefine::SID_LOGIN,
IM::BaseDefine::CID_LOGIN_REQ_USERLOGIN, ++g_seqNum
            , &imLoginReq);
        m_pImLoginResp->Clear();
        util::waitSingleObject(m_eventReceived, 10000);
    }

    return m_pImLoginResp;
}
```

util::waitSingleObject 是封装 API WaitForSingleObject 的函数：

```cpp
//utilCommonAPI.cpp 文件第197行
BOOL waitSingleObject(HANDLE handle, Int32 timeout)
{
    int t = 0;
    DWORD waitResult = WAIT_FAILED;
    do
    {
        int timeWaiter = 500;
        t += timeWaiter;
        waitResult = WaitForSingleObject(handle, timeWaiter);
    } while ((WAIT_TIMEOUT == waitResult) && (t < timeout));

    return (WAIT_OBJECT_0 == waitResult);
}
```

等待的 m_eventConnected 对象是 Event 类型：

```cpp
//定义
HANDLE m_eventConnected;
//在 TcpClientModule_Impl 构造函数中初始化
//m_eventConnected = CreateEvent(NULL, FALSE, FALSE, NULL);
```

这个 WaitForSingleObejct 何时会返回呢？如果网络线程中的 connect 函数可以正常连接服务器，则会让 m_eventConnected 受信，这样 WaitForSingleObejct 函数就会返回了，接着组装登录数据包发送数据：

```
void TcpClientModule_Impl::onConnectDone()
{
    m_tcpClientState = TCPCLIENT_STATE_OK;
    ::SetEvent(m_eventConnected);

    m_bDoReloginServerNow = FALSE;
    if (!m_pServerPingTimer)
    {
        _startServerPingTimer();
    }
}
```

总结一下，这里利用一个 Event 对象实现了一个同步登录的过程，网络连接最大超时事件被设置为 5000 毫秒（5 秒）：

```
util::waitSingleObject(m_eventConnected, 5000)
```

3.6.4　Windows Mutex 对象

Windows 中的 Mutex（Mmutual Exclusive，互斥体）对象在同一时刻最多只能属于 1 个线程，也可以也不属于任何线程。获得 Mutex 对象的线程成为该 Mutex 的拥有者（owner）。我们可以在创建 Mutex 对象时设置 Mutex 是否属于创建它的线程，其他线程如果希望获得该 Mutex，则可以调用 WaitForSingleObject 进行申请。创建 Mutex 的 API 是 CreateMutex，其函数签名如下：

```
HANDLE CreateMutex(
    LPSECURITY_ATTRIBUTES  lpMutexAttributes,
    BOOL                   bInitialOwner,
    LPCTSTR                lpName
);
```

对其参数和返回值说明如下。

（1）参数 lpMutexAttributes 的用法同 CreateEvent，一般将其设置为 NULL。

（2）参数 bInitialOwner 设置调用 CreateMutex 的线程是否立即拥有该 Mutex 对象，TRUE 指立即拥有，FALSE 指不立即拥有。不立即拥有时，其他线程通过调用 WaitForSingleObject 可以获得该 Mutex 对象。

（3）参数 lpName 指 Mutex 对象的名称。Mutex 对象和 Event 对象一样，也可以通过名称在多个线程之间共享，如果不需要名称，则可以将该参数设置为 NULL，根据是否具有名称，Mutex 对象可分为具名 Mutex 和匿名 Mutex。

（4）返回值，函数调用成功时返回 Mutex 的句柄，调用失败时返回 NULL。

当一个线程不再需要该 Mutex 时，可以使用 ReleaseMutex 函数释放 Mutex，让其他需要等待该 Mutex 的线程有机会获得该 Mutex。ReleaseMutex 函数签名如下：

```cpp
BOOL ReleaseMutex(HANDLE hMutex);
```

参数 hMutex 即需要释放所有权的 Mutex 对象句柄。

来看一段具体的实例代码：

```cpp
#include <Windows.h>
#include <string>
#include <iostream>

HANDLE      g_hMutex = NULL;
int         g_iResource = 0;

DWORD __stdcall WorkerThreadProc(LPVOID lpThreadParameter)
{
    DWORD dwThreadID = GetCurrentThreadId();
    while (true)
    {
        if (WaitForSingleObject(g_hMutex, 1000) == WAIT_OBJECT_0)
        {
            g_iResource++;
            std::cout << "Thread: " << dwThreadID << " becomes mutex owner, ResourceNo: " << g_iResource << std::endl;
            ReleaseMutex(g_hMutex);
        }
        Sleep(1000);
    }

    return 0;
}

int main()
{
    //创建1个匿名的Mutex对象并设置在默认情况下主线程不拥有该Mutex
    g_hMutex = CreateMutex(NULL, FALSE, NULL);

    HANDLE hWorkerThreads[5];
    for (int i = 0; i < 5; ++i)
    {
        hWorkerThreads[i] = CreateThread(NULL, 0, WorkerThreadProc, NULL, 0, NULL);
    }

    for (int i = 0; i < 5; ++i)
    {
        //等待工作线程退出
        WaitForSingleObject(hWorkerThreads[i], INFINITE);
        CloseHandle(hWorkerThreads[i]);
    }

    CloseHandle(g_hMutex);
    return 0;
}
```

在以上代码中，主线程创建了一个 Mutex，并且设置不拥有它，然后 5 个工作线程通过竞争获得这个 Mutex 的使用权，拿到这个 Mutex 后就可以操作共享资源 g_iResource 了。

程序的执行效果是 5 个工作线程随机获得该资源的使用权。

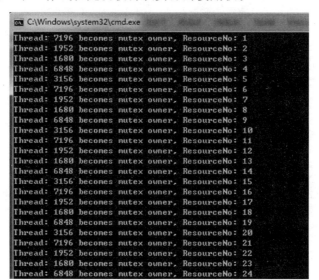

互斥体对象的排他性有点类似于公共汽车上的座位，如果一个座位已经被别人占用，其他人则需要等待，如果该座位没被占用，则人们先到先得。当不再需要占用座位时，需要把座位腾出来让其他人使用。假设某个线程在退出后仍然没有释放其持有的 Mutex 对象，则这时使用 WaitForSingleObject 等待该 Mutex 对象的线程，也会立即返回，返回值是 WAIT_ABANDONED，表示该 Mutex 处于废弃状态（abandoned），处于废弃状态的 Mutex 不能再被使用，其行为是未定义的。

3.6.5　Windows Semaphore 对象

Semaphore 也是 Windows 多线程同步常用的对象之一。与上面介绍的 Event、Mutex 不同，信号量存在一个资源计数的概念。Event 对象虽然可以同时唤醒多个线程，却不能精确地控制同时唤醒指定数量的线程，而 Semaphore 能。创建 Semaphore 对象的 API 函数签名如下：

```
HANDLE CreateSemaphore(
    LPSECURITY_ATTRIBUTES   lpSemaphoreAttributes,
    LONG                    lInitialCount,
    LONG                    lMaximumCount,
    LPCTSTR                 lpName
);
```

对其参数和返回值介绍如下。

（1）参数 lpSemaphoreAttributes 指定了 Semaphore 对象的安全属性，一般将其设置为 NULL，表示使用默认的安全属性。

（2）参数 lInitialCount 指定初始可用的资源数量，假设初始资源数量为 2，如果有 5 个线程正在调用 WaitForSingleObject 函数等待该信号量，则会有两个线程被唤醒，每调用一次 WaitForSingleObject 获得 Semaphore 对象，该对象的资源计数就会减少 1。

· 231 ·

（3）参数 lMaximumCount 是最大资源数量上限，如果使用 ReleaseSemaphore 不断增加资源计数，资源数量最大不能超过这个值，则这个值必须大于 0。

（4）参数 lpName 指定 Semaphore 对象的名称，Semaphore 对象也是可以通过名称跨进程共享的，如果不需要设置名称，则可以将该参数设置为 NULL，设置了名称的 Semaphore 对象被称为具名信号量，反之被称为匿名信号量。

（5）返回值：在函数调用成功时返回 Semaphore 对象的句柄，反之返回 NULL。

如果需要增加信号量的资源计数，则可以使用 ReleaseSemaphore 函数，其函数签名如下：

```
BOOL ReleaseSemaphore(
    HANDLE hSemaphore,
    LONG   lReleaseCount,
    LPLONG lpPreviousCount
);
```

对其参数和返回值介绍如下。

（1）参数 hSemaphore 是需要操作的信号量句柄。

（2）参数 lReleaseCount 是需要增加的资源数量。

（3）参数 lpPreviousCount 是一个 long 类型（在 32 位系统上为 4 字节）的指针，在函数执行成功后，返回上一次资源的数量，如果用不到该参数，则可以将其设置为 NULL。

Windows 信号量与 Linux 信号量的用法基本相同，来看一个具体的例子。

假设现在有一个即时通信程序，网络线程不断地从网络上收到聊天消息，其他 4 个消息处理线程需要对收到的聊天信息进行加工。我们需要根据当前消息的数量唤醒其中 4 个工作线程中的一个或多个，这正是信号量使用的典型案例，代码如下：

```
#include <Windows.h>
#include <string>
#include <iostream>
#include <list>
#include <time.h>

HANDLE                   g_hMsgSemaphore = NULL;
std::list<std::string>   g_listChatMsg;
//保护 g_listChatMsg 的临界区对象
CRITICAL_SECTION         g_csMsg;

DWORD __stdcall NetThreadProc(LPVOID lpThreadParameter)
{
    int nMsgIndex = 0;
    while (true)
    {
        EnterCriticalSection(&g_csMsg);
        //随机产生 1~4 条消息
        int count = rand() % 4 + 1;
        for (int i = 0; i < count; ++i)
        {
```

```cpp
            nMsgIndex++;
            SYSTEMTIME st;
            GetLocalTime(&st);
            char szChatMsg[64] = { 0 };
            sprintf_s(szChatMsg, 64, "[%04d-%02d-%02d %02d:%02d:%02d:%03d] A new msg, NO.%d.",
                st.wYear,
                st.wMonth,
                st.wDay,
                st.wHour,
                st.wMinute,
                st.wSecond,
                st.wMilliseconds,
                nMsgIndex);
            g_listChatMsg.emplace_back(szChatMsg);
        }
        LeaveCriticalSection(&g_csMsg);

        //增加 count 个资源数量
        ReleaseSemaphore(g_hMsgSemaphore, count, NULL);
    }// end while-loop

    return 0;
}

DWORD __stdcall ParseThreadProc(LPVOID lpThreadParameter)
{
    DWORD dwThreadID = GetCurrentThreadId();
    std::string current;
    while (true)
    {
        if (WaitForSingleObject(g_hMsgSemaphore, INFINITE) == WAIT_OBJECT_0)
        {
            EnterCriticalSection(&g_csMsg);
            if (!g_listChatMsg.empty())
            {
                current = g_listChatMsg.front();
                g_listChatMsg.pop_front();
                std::cout << "Thread: " << dwThreadID << " parse msg: " << current << std::endl;
            }
            LeaveCriticalSection(&g_csMsg);
        }
    }

    return 0;
}

int main()
{
    //初始化随机数种子
    srand(time(NULL));
    InitializeCriticalSection(&g_csMsg);

    //创建 1 个匿名的 Semaphore 对象，初始资源数量为 0
```

```
    g_hMsgSemaphore = CreateSemaphore(NULL, 0, INT_MAX, NULL);

    HANDLE hNetThread = CreateThread(NULL, 0, NetThreadProc, NULL, 0, NULL);

    HANDLE hWorkerThreads[4];
    for (int i = 0; i < 4; ++i)
        hWorkerThreads[i] = CreateThread(NULL, 0, ParseThreadProc, NULL, 0, NULL);

    for (int i = 0; i < 4; ++i)
    {
        //等待工作线程退出
        WaitForSingleObject(hWorkerThreads[i], INFINITE);
        CloseHandle(hWorkerThreads[i]);
    }

    WaitForSingleObject(hNetThread, INFINITE);
    CloseHandle(hNetThread);

    CloseHandle(g_hMsgSemaphore);

    DeleteCriticalSection(&g_csMsg);
    return 0;
}
```

在以上代码中，网络线程每次都随机产生 1～4 条聊天消息放入消息容器 g_listChatMsg 中，然后根据当前新产生的消息数量调用 ReleaseSemaphore 增加相应的资源计数，这样就有相应的处理线程被唤醒，从容器 g_listChatMsg 中取出消息进行处理。

> **注意**：由于会涉及多个线程操作消息容器 g_listChatMsg，所以这里使用了一个临界区对象 g_csMsg 对其进行保护。

程序执行效果如下：

```
//这里截取输出中间部分...输出太多，部分结果省略
Thread: 3704 parse msg: [2019-01-20 16:31:47:568] A new msg, NO.26.
Thread: 3704 parse msg: [2019-01-20 16:31:47:568] A new msg, NO.27.
Thread: 3704 parse msg: [2019-01-20 16:31:47:568] A new msg, NO.28.
Thread: 3704 parse msg: [2019-01-20 16:31:47:568] A new msg, NO.29.
Thread: 3704 parse msg: [2019-01-20 16:31:47:568] A new msg, NO.30.
Thread: 3704 parse msg: [2019-01-20 16:31:47:568] A new msg, NO.31.
Thread: 3704 parse msg: [2019-01-20 16:31:47:568] A new msg, NO.32.
Thread: 3704 parse msg: [2019-01-20 16:31:47:568] A new msg, NO.33.
Thread: 3704 parse msg: [2019-01-20 16:31:47:568] A new msg, NO.34.
Thread: 3704 parse msg: [2019-01-20 16:31:47:568] A new msg, NO.35.
Thread: 3704 parse msg: [2019-01-20 16:31:47:568] A new msg, NO.36.
Thread: 3704 parse msg: [2019-01-20 16:31:47:568] A new msg, NO.37.
Thread: 3704 parse msg: [2019-01-20 16:31:47:568] A new msg, NO.38.
Thread: 3704 parse msg: [2019-01-20 16:31:47:568] A new msg, NO.39.
Thread: 3704 parse msg: [2019-01-20 16:31:47:568] A new msg, NO.40.
Thread: 3704 parse msg: [2019-01-20 16:31:47:568] A new msg, NO.41.
Thread: 3704 parse msg: [2019-01-20 16:31:47:568] A new msg, NO.42.
Thread: 3704 parse msg: [2019-01-20 16:31:47:568] A new msg, NO.43.
Thread: 3704 parse msg: [2019-01-20 16:31:47:569] A new msg, NO.44.
Thread: 3704 parse msg: [2019-01-20 16:31:47:569] A new msg, NO.45.
```

```
Thread: 3704 parse msg: [2019-01-20 16:31:47:569] A new msg, NO.46.
Thread: 3704 parse msg: [2019-01-20 16:31:47:569] A new msg, NO.47.
Thread: 5512 parse msg: [2019-01-20 16:31:47:569] A new msg, NO.48.
Thread: 6676 parse msg: [2019-01-20 16:31:47:569] A new msg, NO.49.
Thread: 6676 parse msg: [2019-01-20 16:31:47:569] A new msg, NO.50.
```

3.6.6　Windows 读写锁

与 Linux 读写锁的原理相同，Windows 系统也有读写锁。Windows 系统上的读写锁叫作 Slim Reader/Writer (SRW) Locks，对应的数据类型叫作 SRWLOCK，微软没有公开这个数据结构的细节，只提供了一些 API 函数对其进行操作：

```
//初始化1个读写锁，PSRWLOCK 的定义是 SRWLOCK*
void InitializeSRWLock(PSRWLOCK SRWLock);

//以共享模式获得读写锁
void AcquireSRWLockShared(PSRWLOCK SRWLock);
//释放共享模式的读写锁
void ReleaseSRWLockShared(PSRWLOCK SRWLock);

//以排他模式获得读写锁
void AcquireSRWLockExclusive(PSRWLOCK SRWLock);
//释放排他模式的读写锁
void ReleaseSRWLockExclusive(PSRWLOCK SRWLock);
```

还可以使用如下语法初始化一个 SRWLOCK 对象：

```
SRWLOCK mySRWLock = SRWLOCK_INIT;
```

与临界区对象不同的是，Windows 不需要显式销毁一个读写锁对象，因此不存在 DeleteSRWLock 这样的函数用于销毁一个读写锁。

综合前面介绍的 Linux 读写锁，以下是对 Windows 和 Linux 平台的读写锁对象进行简单封装的一个工具类，用于模拟 C++ 17 的 std::shared_mutex。

SharedMutex.h：

```cpp
/**
 * SharedMutex.h, C++ 11 没有 std::shared_mutex, 自己模拟一个
 * zhangyl 2019.11.08
 */

#ifndef __SHARED_MUTEX_H__
#define __SHARED_MUTEX_H__

#ifdef WIN32
#include <Windows.h>
#else
#include
#include <pthread.h>
#endif

//模拟 std::shared_mutex
class SharedMutex final
{
```

```cpp
public:
    SharedMutex();
    ~SharedMutex();

    void acquireReadLock();
    void acquireWriteLock();
    void unlockReadLock();
    void unlockWriteLock();

private:
    SharedMutex(const SharedMutex& rhs) = delete;
    SharedMutex& operator =(const SharedMutex& rhs) = delete;

private:
#ifdef WIN32
    SRWLOCK             m_SRWLock;
#else
    pthread_rwlock_t    m_SRWLock;
#endif
};

//模拟 std::shared_lock
class SharedLockGuard final
{
public:
    SharedLockGuard(SharedMutex& sharedMutex);
    ~SharedLockGuard();

private:
    SharedLockGuard(const SharedLockGuard& rhs) = delete;
    SharedLockGuard operator=(const SharedLockGuard& rhs) = delete;

private:
    SharedMutex&        m_SharedMutex;
};

//模拟 std::unique_lock
class UniqueLockGuard final
{
public:
    UniqueLockGuard(SharedMutex& sharedMutex);
    ~UniqueLockGuard();

private:
    UniqueLockGuard(const UniqueLockGuard& rhs) = delete;
    UniqueLockGuard operator=(const UniqueLockGuard& rhs) = delete;

private:
    SharedMutex& m_SharedMutex;
};

#endif //!__SHARED_MUTEX_H__
```

SharedMutex.cpp：

```cpp
/**
 * SharedMutex.cpp
 * zhangyl 2019.11.08
 */

#include "SharedMutex.h"

SharedMutex::SharedMutex()
{
#ifdef WIN32
    ::InitializeSRWLock(&m_SRWLock);
#else
    ::pthread_rwlock_init(&m_SRWLock, nullptr);
#endif
}

SharedMutex::~SharedMutex()
{
#ifdef WIN32
    //Windows 上的读写锁不需要显式销毁
#else
    ::pthread_rwlock_destroy(&m_SRWLock);
#endif
}

void SharedMutex::acquireReadLock()
{
#ifdef WIN32
    ::AcquireSRWLockShared(&m_SRWLock);
#else
    ::pthread_rwlock_rdlock(&m_SRWLock);
#endif
}

void SharedMutex::acquireWriteLock()
{
#ifdef WIN32
    ::AcquireSRWLockExclusive(&m_SRWLock);
#else
    ::pthread_rwlock_wrlock(&m_SRWLock);
#endif
}

void SharedMutex::unlockReadLock()
{
#ifdef WIN32
    ::ReleaseSRWLockShared(&m_SRWLock);
#else
    ::pthread_rwlock_unlock(&m_SRWLock);
#endif
}

void SharedMutex::unlockWriteLock()
{
#ifdef WIN32
```

```cpp
        ::ReleaseSRWLockExclusive(&m_SRWLock);
#else
        ::pthread_rwlock_unlock(&m_SRWLock);
#endif
}

SharedLockGuard::SharedLockGuard(SharedMutex& sharedMutex) :
    m_SharedMutex(sharedMutex)
{
    m_SharedMutex.acquireReadLock();
}

SharedLockGuard::~SharedLockGuard()
{
    m_SharedMutex.unlockReadLock();
}

UniqueLockGuard::UniqueLockGuard(SharedMutex& sharedMutex) :
    m_SharedMutex(sharedMutex)
{
    m_SharedMutex.acquireWriteLock();
}

UniqueLockGuard::~UniqueLockGuard()
{
    m_SharedMutex.unlockWriteLock();
}
```

3.6.7　Windows 条件变量

和 Linux 的条件变量作用一样，Windows 系统在 Windows XP 和 Windows Server 2003 版本以后也引入了条件变量（言下之意，Windows XP 及以前的版本不支持条件变量）。在 Windows 中代表条件变量的结构体对象是 CONDITION_VARIABLE，微软并没有给出对这个结构的具体定义，只提供了一系列 API 函数去操作这个对象。

可以使用如下 API 函数初始化一个 Windows 条件变量：

```cpp
//PCONDITION_VARIABLE 类型是 CONDITION_VARIABLE*，
//即结构体 CONDITION_VARIABLE 的指针类型
void InitializeConditionVariable(PCONDITION_VARIABLE ConditionVariable);
```

我们也可以使用如下语法初始化一个条件变量对象：

```cpp
CONDITION_VARIABLE myConditionVariable = CONDITION_VARIABLE_INIT;
```

与临界区（CriticalSection）、读写锁（Slim Reader/Writer Locks）一样，Windows 的条件变量也是 user-mode 对象，不可以跨进程共享。

在 Windows 上使用条件变量时需要配合一个临界区或读写锁，使用临界区的方式和 Linux 的条件变量类似，等待资源变为可用的线程调用 SleepConditionVariableCS() 或 SleepConditionVariableSRW 函数进行等待，这两个函数签名如下：

```
BOOL SleepConditionVariableCS(
    PCONDITION_VARIABLE    ConditionVariable,
    PCRITICAL_SECTION      CriticalSection,
    DWORD                  dwMilliseconds
);

BOOL SleepConditionVariableSRW(
    PCONDITION_VARIABLE    ConditionVariable,
    PSRWLOCK               SRWLock,
    DWORD                  dwMilliseconds,
    ULONG                  Flags
);
```

这两个函数分别使用了一个临界区（CriticalSection 参数）和一个读写锁（SRWLock 参数）。参数 dwMilliseconds 可用于设置一个等待时间，如果需要无限等待，则可以将其设置为 INFINITE。这两个函数在 dwMilliseconds 超时时间内阻塞当前调用线程，函数调用成功则返回非 0 值，调用失败则返回 0 值。如果由于 dwMilliseconds 超时导致调用失败，则调用 GetLastError() 会得到错误码 ERROR_TIMEOUT。

和 Linux 条件变量的使用方式一样，在调用 SleepConditionVariableCS 或者 SleepConditionVariableSRW 函数之前，调用线程必须持有对应的临界区或读写锁对象；调用 SleepConditionVariableCS 或 SleepConditionVariableSRW 函数后，线程进入睡眠会释放持有的临界区或读写锁对象；SleepConditionVariableCS 或者 SleepConditionVariableSRW 函数在成功返回后，将再次持有对应的互斥体或读写锁对象，此时资源变为可用状态，操作完成资源后，如果希望其他线程可以继续操作资源，则需要释放其持有的临界区或读写锁对象。

以上流程的伪代码如下（这里以临界区为例）：

```
CRITICAL_SECTION       CritSection;
CONDITION_VARIABLE     ConditionVar;

void PerformOperationOnSharedResource()
{
    //进入临界区，持有临界区锁
    EnterCriticalSection(&CritSection);

    //等待共享资源变为可用
    while( TestSharedResourceAvailable() == FALSE )
    {
        //线程无限等待，直到资源可用
        SleepConditionVariableCS(&ConditionVar, &CritSection, INFINITE);
    }

    //资源在这里已变为可用状态，对其进行操作
    OperateSharedResource();

    //离开临界区对象，让其他线程可以对共享资源进行操作
    LeaveCriticalSection(&CritSection);
```

```
    //其他操作
}
```

等待资源变为可用的线程在调用 SleepConditionVariableCS 或者 SleepConditionVariableSRW 函数后,释放其持有的临界区或读写锁对象,这样其他线程就有机会对共享资源进行修改了。这些线程通过修改共享资源让资源变为可用,可以调用 WakeConditionVariable 或 WakeAllConditionVariable 唤醒调用 SleepConditionVariableCS 或者 SleepConditionVariableSRW 函数等待的线程,前者只唤醒一个等待的线程,后者唤醒所有等待的线程。这两个函数签名如下:

```
//唤醒单个线程
void WakeConditionVariable(PCONDITION_VARIABLE ConditionVariable);

//唤醒多个线程
void WakeAllConditionVariable(PCONDITION_VARIABLE ConditionVariable);
```

需要注意的是,和 Linux 的条件变量一样,Windows 的条件变量也存在虚假唤醒这一行为,所以在条件变量被唤醒时,不一定是其他线程调用 WakeConditionVariable 或 WakeAllConditionVariable 唤醒的,而是操作系统造成的虚假唤醒。所以在以上伪代码中,即使 SleepConditionVariableCS 函数返回,也需要再次判断资源是否可用,所以将其放在一个 while 循环里面判断资源是否可用。

```
//等待共享资源变为可用
while( TestSharedResourceAvailable() == FALSE )
{
    //线程无限等待,直到资源可用
    //SleepConditionVariableCS 返回后会再次对 while 条件进行判断,
    //如果是虚假唤醒,则 while 条件仍是 FALSE,线程再次调用 SleepConditionVariableCS 等待
    SleepConditionVariableCS(&ConditionVar, &CritSection, INFINITE);
}
```

我们将 3.5.3 节条件变量的例子改写成 Windows 版本:

```
/**
 * 演示 Windows 条件变量的使用方法
 * zhangyl 2019.11.11
 */

#include <Windows.h>
#include <iostream>
#include <list>

class Task
{
public:
    Task(int taskID)
    {
        this->taskID = taskID;
    }

    void doTask()
    {
        std::cout << "handle a task, taskID: " << taskID << ", threadID: " <<
```

```cpp
        GetCurrentThreadId() << std::endl;
    }

private:
    int taskID;
};

CRITICAL_SECTION    myCriticalSection;
CONDITION_VARIABLE  myConditionVar;
std::list<Task*>    tasks;

DWORD WINAPI consumerThread(LPVOID param)
{
    Task* pTask = NULL;
    while (true)
    {
        //进入临界区
        EnterCriticalSection(&myCriticalSection);
        while (tasks.empty())
        {
            //如果SleepConditionVariableCS挂起,则挂起前会离开临界区,不往下执行
            //发生变化后,条件合适时,SleepConditionVariableCS将直接进入临界区
            SleepConditionVariableCS(&myConditionVar, &myCriticalSection, INFINITE);
        }

        pTask = tasks.front();
        tasks.pop_front();

        //SleepConditionVariableCS被唤醒后进入临界区,
        //为了让其他线程有机会操作tasks,这里需要再次离开临界区
        LeaveCriticalSection(&myCriticalSection);

        if (pTask == NULL)
            continue;

        pTask->doTask();
        delete pTask;
        pTask = NULL;
    }

    return 0;
}

DWORD WINAPI producerThread(LPVOID param)
{
    int taskID = 0;
    Task* pTask = NULL;

    while (true)
    {
        pTask = new Task(taskID);

        //进入临界区
        EnterCriticalSection(&myCriticalSection);
        tasks.push_back(pTask);
```

```cpp
        std::cout << "produce a task, taskID: " << taskID << ", threadID: " <<
GetCurrentThreadId() << std::endl;

        //离开临界区
        LeaveCriticalSection(&myCriticalSection);

        WakeConditionVariable(&myConditionVar);

        taskID++;

        //休眠 1 秒
        Sleep(1000);
    }

    return 0;
}

int main()
{
    InitializeCriticalSection(&myCriticalSection);
    InitializeConditionVariable(&myConditionVar);

    //创建 5 个消费者线程
    HANDLE consumerThreadHandles[5];
    for (int i = 0; i < 5; ++i)
        consumerThreadHandles[i] = CreateThread(NULL, 0, consumerThread, NULL, 0, NULL);

    //创建 1 个生产者线程
    HANDLE producerThreadHandle = CreateThread(NULL, 0, producerThread, NULL, 0, NULL);

    //等待生产者线程退出
    WaitForSingleObject(producerThreadHandle, INFINITE);

    //等待消费者线程退出
    for (int i = 0; i < 5; ++i)
        WaitForSingleObject(consumerThreadHandles[i], INFINITE);

    DeleteCriticalSection(&myCriticalSection);

    return 0;
}
```

以上代码创建了 1 个生产者线程和 5 个消费者线程，生产者产生 1 个 task 后，将 task 放入队列中，然后随机唤醒一个消费者线程，因此每次处理任务的消费者线程 ID 是不一样的，程序执行结果如下图所示。

```
produce a task, taskID: 0, threadID: 31128
handle a task, taskID: 0, threadID: 31064
produce a task, taskID: 1, threadID: 31128
handle a task, taskID: 1, threadID: 31096
produce a task, taskID: 2, threadID: 31128
handle a task, taskID: 2, threadID: 31120
produce a task, taskID: 3, threadID: 31128
handle a task, taskID: 3, threadID: 31124
produce a task, taskID: 4, threadID: 31128
handle a task, taskID: 4, threadID: 31116
produce a task, taskID: 5, threadID: 31128
handle a task, taskID: 5, threadID: 31064
produce a task, taskID: 6, threadID: 31128
handle a task, taskID: 6, threadID: 31096
produce a task, taskID: 7, threadID: 31128
handle a task, taskID: 7, threadID: 31120
produce a task, taskID: 8, threadID: 31128
handle a task, taskID: 8, threadID: 31124
produce a task, taskID: 9, threadID: 31128
handle a task, taskID: 9, threadID: 31116
produce a task, taskID: 10, threadID: 31128
handle a task, taskID: 10, threadID: 31064
produce a task, taskID: 11, threadID: 31128
handle a task, taskID: 11, threadID: 31096
produce a task, taskID: 12, threadID: 31128
handle a task, taskID: 12, threadID: 31120
```

3.6.8　在多进程之间共享线程同步对象

前面介绍了 Windows Event、Mutex、Semaphore 对象，通过其创建函数 CreateX（可以将 X 换成 Event、Mutex 或 Semaphore）都可以给这些对象指定一个名称，有了名称之后，这些线程资源同步对象就可以通过名称在不同的进程之间共享了。

在 Windows 上，对某些程序，无论双击其启动图标几次都只会启动一个实例，我们把这类程序叫作单实例程序（Single Instance）。我们可以利用命名的线程资源同步对象实现这个效果，这里以互斥体为例，示例代码如下：

```
int APIENTRY _tWinMain(_In_ HINSTANCE hInstance,
            _In_opt_ HINSTANCE hPrevInstance,
            _In_ LPTSTR    lpCmdLine,
            _In_ int       nCmdShow)
{
    //省略无关代码

    if (CheckInstance())
    {
        HWND hwndPre = FindWindow(szWindowClass, NULL);
        if (::IsWindow(hwndPre))
        {
            if (::IsIconic(hwndPre))
                ::SendMessage(hwndPre, WM_SYSCOMMAND, SC_RESTORE | HTCAPTION, 0);

            ::SetWindowPos(hwndPre, HWND_TOP, 0, 0, 0, 0, SWP_NOMOVE | SWP_NOSIZE | SWP_SHOWWINDOW | SWP_NOACTIVATE);
            ::SetForegroundWindow(hwndPre);
            ::SetFocus(hwndPre);
            return 0;
        }
    }
```

```
    //省略无关代码
}
```

以上代码在 WinMain 函数开始处先检查是否存在已运行的程序实例，如果存在，则找到运行中的实例程序主窗口并激活。这就是我们看到最小化很多单例程序后双击该程序图标会重新激活最小化程序的实现原理。

下面重点讲解 CheckInstance 函数的实现，代码如下：

```
bool CheckInstance()
{
    HANDLE hSingleInstanceMutex = ::CreateMutex(NULL, FALSE,
_T("MySingleInstanceApp"));
    if (hSingleInstanceMutex != NULL)
    {
        if (::GetLastError() == ERROR_ALREADY_EXISTS)
            return true;
    }

    return false;
}
```

假设首次启动这个进程，这个进程就会调用 CreateMutex 函数创建一个名称为 MySingleInstanceApp 的互斥体对象。再次准备启动一份这个进程时，将再次调用 CreateMutex 函数，由于该名称的互斥体对象已经存在，所以将返回已存在的互斥体对象地址，此时通过 GetLastError()得到的错误码是 ERROR_ALREADY_EXISTS，表示该名称的互斥体对象已存在，此时激活已存在的前一个实例，退出当前进程即可。

3.7　C++ 11/14/17 线程同步对象

在 C/C++中直接使用操作系统提供的多线程资源同步 API 虽然限制最少，但使用起来毕竟不方便，同样的代码不能同时兼容 Windows 和 Linux 两个平台。C++ 11 标准新增了很多现代编程语言的标配，线程资源同步对象就是其中很重要的部分。本节将讨论 C++ 11 标准中新增的用于线程同步的 std::mutex 和 std::condition_variable 对象的用法，有了它们，我们就可以写出跨平台的多线程程序了。

3.7.1　std::mutex 系列

在 C++ 11/14/17 中提供了如下 mutex 系列类型。

互斥量	版本	作用
mutex	C++ 11	基本的互斥量
timed_mutex	C++ 11	有超时机制的互斥量
recursive_mutex	C++ 11	可重入的互斥量
recursive_timed_mutex	C++ 11	结合 timed_mutex 和 recursive_mutex 特点的互斥量
shared_timed_mutex	C++ 14	具有超时机制的可共享互斥量
shared_mutex	C++ 17	共享的互斥量

这个系列类型的对象均提供了加锁（lock）、尝试加锁（trylock）和解锁（unlock）的方法，这里以 std::mutex 类为例，说明 std::mutex 系列锁的用法，示例代码如下：

```cpp
#include <iostream>
#include <chrono>
#include <thread>
#include <mutex>

//g_num 使用 g_num_mutex 进行保护
int          g_num = 0;
std::mutex   g_num_mutex;

void slow_increment(int id)
{
    for (int i = 0; i < 3; ++i)
    {
        g_num_mutex.lock();
        ++g_num;
        std::cout << id << " => " << g_num << std::endl;
        g_num_mutex.unlock();

        //睡眠 1 秒
        std::this_thread::sleep_for(std::chrono::seconds(1));
    }
}

int main()
{
    std::thread t1(slow_increment, 0);
    std::thread t2(slow_increment, 1);
    t1.join();
    t2.join();

    return 0;
}
```

在以上代码中创建了两个线程 t1 和 t2，在线程函数的 for 循环中调用 std::mutex.lock 方法和 std::mutex.unlock 方法对全局变量 g_num 进行保护。编译程序并输出如下结果：

```
[root@localhost testmultithread]# g++ -g -o mutex c11mutex.cpp -std=c++0x -lpthread
[root@localhost testmultithread]# ./mutex
0 => 1
1 => 2
0 => 3
1 => 4
1 => 5
0 => 6
```

注意：如果在 Linux 下编译和运行程序，则在编译时需要链接 pthread 库，否则虽然能够正常编译，但在运行时程序会崩溃，崩溃的原因为 "terminate called after throwing an instance of 'std::system_error' what(): Enable multithreading to use std::thread: Operation not permitted"。

为了避免死锁，std::mutex.lock 方法和 std::mutex::unlock 方法需要成对使用，但是如上面所介绍的，如果在一个函数中有很多出口，而互斥体对象又是需要在整个函数作用域被保护的资源，那么我们在编码时会因为忘记在某个出口处调用 std::mutex.unlock 而造成死锁。前面讲到，推荐通过 RAII 技术封装这两个接口。其实 C++新标准也为我们提供了如下封装。

互斥量管理	版本	作用
lock_guard	C++ 11	基于作用域的互斥量管理
unique_lock	C++ 11	更加灵活的互斥量管理
shared_lock	C++ 14	共享互斥量的管理
scoped_lock	C++ 17	多互斥量避免死锁的管理

这里以 std::lock_guard 为例：

```
void func()
{
    std::lock_guard<std::mutex> guard(mymutex);
    //在这里放被保护的资源操作
}
```

mymutex 的类型是 std::mutex，guard 对象的构造函数会自动调用 mymutex.lock 方法对 mymutex 进行加锁，在 guard 对象出了其作用域时，guard 对象的析构函数会自动调用 mymutex.unlock 方法对 mymutex 进行解锁。

注意：mymutex 生命周期必须长于 func 函数的作用域，很多人在初学这个利用 RAII 技术封装的 std::lock_guard 对象时，可能会写出这样的代码：

```
//错误的写法，这样是无法在多线程调用该函数时保护指定的数据的
void func()
{
    std::mutex m;
    std::lock_guard<std::mutex> guard(m);
    //这里放被保护的资源操作
}
```

另外，如果某个线程已经对一个 std::mutex 对象调用了 lock 方法，则该线程再次调用 lock 方法对这个 std::mutex 进行加锁时，其行为是未定义的，这是一个错误的做法。"行为未定义"指在不同的平台上可能会有不同的行为。

```
01  #include <mutex>
02
03  int main()
04  {
05      std::mutex m;
06      m.lock();
07      m.lock();
08      m.unlock();
09
10      return 0;
11  }
```

实际测试时，以上代码重复调用 std::mutex.lock 方法，在 Windows 平台上会引发程

序崩溃，如下图所示。

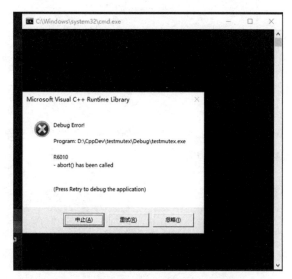

以上代码在 Linux 上运行时会阻塞在第 2 次调用 std::mutex.lock 方法处，验证结果如下：

```
[root@localhost testmultithread]# g++ -g -o mutexlock mutexlock.cpp -std=c++0x -lpthread
[root@localhost testmultithread]# gdb mutexlock
Reading symbols from /root/testmultithread/mutexlock...done.
(gdb) r
Starting program: /root/testmultithread/mutexlock
[Thread debugging using libthread_db enabled]
Using host libthread_db library "/lib64/libthread_db.so.1".
^C
Program received signal SIGINT, Interrupt.
0x00007ffff7bcd4ed in __lll_lock_wait () from /lib64/libpthread.so.0
Missing separate debuginfos, use: debuginfo-install glibc-2.17-260.el7.x86_64 libgcc-4.8.5-36.el7.x86_64 libstdc++-4.8.5-36.el7.x86_64
(gdb) bt
#0  0x00007ffff7bcd4ed in __lll_lock_wait () from /lib64/libpthread.so.0
#1  0x00007ffff7bc8dcb in _L_lock_883 () from /lib64/libpthread.so.0
#2  0x00007ffff7bc8c98 in pthread_mutex_lock () from /lib64/libpthread.so.0
#3  0x00000000004006f7 in __gthread_mutex_lock (__mutex=0x7fffffffe3e0)
    at /usr/include/c++/4.8.2/x86_64-redhat-linux/bits/gthr-default.h:748
#4  0x00000000004007a2 in std::mutex::lock (this=0x7fffffffe3e0) at /usr/include/c++/4.8.2/mutex:134
#5  0x0000000000400777 in main () at mutexlock.cpp:7
(gdb) f 5
#5  0x0000000000400777 in main () at mutexlock.cpp:7
7           m.lock();
(gdb) l
2
3       int main()
4       {
5           std::mutex m;
6           m.lock();
```

```
7       m.lock();
8       m.unlock();
9
10      return 0;
11  }
(gdb)
```

我们使用 gdb 运行程序，然后使用 bt 命令，看到程序确实阻塞在第 2 个 m.lock()处（代码第 7 行）。

总结一下：在实际开发中，我们应该尽量避免同一个线程对一个已经调用了 lock 方法的 std::mutex 对象再次调用 lock 方法。示例如下：

//错误的做法
线程 A 对 mutexM 加锁 => 线程 A 对 mutexM 加锁

//正确的做法
线程 A 对 mutexM 加锁 => 线程 A 对 mutexM 解锁 => 线程 A 对 mutexM 加锁

//正确的做法
线程 A 对 mutexM 加锁 => 线程 B 对 mutexM 加锁（会阻塞）

> 事实上，如果需要同一个线程多次对某个 mutex 进行加锁，则不能使用 std::mutex，应该使用 std::recursive_mutex，它也是在 C++ 11 标准中被引入的，同一个线程对某个 std::recursive_mutex 加锁多少次，就应该解锁多少次，这样其他线程才能持有该 std::recursive_mutex。

3.7.2　std::shared_mutex

C++ 11 标准让很多开发者诟病的原因之一是，它借鉴了 Boost 库的 boost::mutex、boost::shared_mutex 来引入 std::mutex 和 std::shared_mutex，但是在 C++ 11 中只引入了 std::mutex，直到 C++ 17 才有 std::shared_mutex，导致使用仅支持 C++ 11 标准的编译器（例如 Visual Studio 2013、gcc/g++ 4.8）进行开发非常不便。

> 在商业项目中一般不会轻易升级编译器，因为商业项目一般牵涉的代码范围较大，升级编译器后可能导致大量旧的文件需要修改。例如对于被广泛使用的 CentOS 7.0，其自带的 gcc 编译器是 4.8，在升级 gcc 的同时会导致系统自带的 glibc 库发生变化，使系统中大量的其他程序无法运行。因此在实际的商业项目中，升级旧的开发环境必须非常谨慎。

std::shared_mutex 的底层实现是操作系统提供的读写锁，也就是说，在有多个线程对共享资源读且少许线程对共享资源写的情况下，std::shared_mutex 比 std::mutex 效率更高。

std::shared_mutex 提供了 lock 方法和 unlock 方法分别用于获取写锁和解除写锁，提供了 lock_shared 方法和 unlock_shared 方法分别用于获取读锁和解除读锁。我们一般将写锁模式称为排他锁（Exclusive Locking），将读锁模式称为共享锁（Shared Locking）。

另外，在C++新标准中引入了与 std::shared_mutex 配合使用的两个对象——std::unique_lock 和 std::shared_lock，这两个对象在构造时自动对 std::shared_mutex 加锁，在析构时自动对 std::shared_mutex 解锁，前者用于加解 std::shared_mutex 的写锁，后者用于加解 std::shared_mutex 的读锁。

---

std::unique_lock 在 C++ 11 中引入，std::shared_lock 在 C++ 14 中引入。

下面是对共享资源存在多个读线程和一个写线程，分别使用 std::mutex 和 std::shared_mutex 做的一个性能测试，测试代码如下：

```cpp
/**
 * std::shared_mutex 与 std::mutex 的性能对比
 * zhangyl 2019.11.10
 */

//读线程数量
#define READER_THREAD_COUNT     8
//最大循环次数
#define LOOP_COUNT              5000000

#include <iostream>
#include <mutex>
#include <shared_mutex>
#include <thread>

class shared_mutex_counter
{
public:
    shared_mutex_counter() = default;
    ~shared_mutex_counter() = default;

    //使用 std::shared_mutex，同一时刻多个读线程可以同时访问 m_value 的值
    unsigned int get() const
    {
        //注意：这里使用 std::shared_lock
        std::shared_lock<std::shared_mutex> lock(m_mutex);
        return m_value;
    }

    //使用 std::shared_mutex，同一时刻仅有一个写线程可以修改 m_value 的值
    void increment()
    {
        //注意：这里使用 std::unique_lock
        std::unique_lock<std::shared_mutex> lock(m_mutex);
        m_value++;
    }

    //使用 std::shared_mutex，同一时刻仅有一个写线程可以重置 m_value 的值
    void reset()
    {
        //注意：这里使用 std::unique_lock
        std::unique_lock<std::shared_mutex> lock(m_mutex);
        m_value = 0;
```

```cpp
    }
private:
    mutable std::shared_mutex    m_mutex;
    unsigned int                 m_value = 0;  //m_value 是多个线程的共享资源
};

class mutex_counter
{
public:
    mutex_counter() = default;
    ~mutex_counter() = default;

    //使用 std::mutex，同一时刻仅有一个线程可以访问 m_value 的值
    unsigned int get() const
    {
        std::unique_lock<std::mutex> lk(m_mutex);
        return m_value;
    }

    //使用 std::mutex，同一时刻仅有一个线程可以修改 m_value 的值
    void increment()
    {
        std::unique_lock<std::mutex> lk(m_mutex);
        m_value++;
    }

private:
    mutable std::mutex    m_mutex;
    unsigned int          m_value = 0;  //m_value 是多个线程的共享资源
};

//测试 std::shared_mutex
void test_shared_mutex()
{
    shared_mutex_counter counter;
    int temp;

    //写线程函数
    auto writer = [&counter]() {
        for (int i = 0; i < LOOP_COUNT; i++)
        {
            counter.increment();
        }
    };

    //读线程函数
    auto reader = [&counter, &temp]() {
        for (int i = 0; i < LOOP_COUNT; i++)
        {
            temp = counter.get();
        }
    };

    //存放读线程对象指针的数组
```

```cpp
    std::thread** tarray = new std::thread * [READER_THREAD_COUNT];

    //记录起始时间
    clock_t start = clock();

    //创建 READER_THREAD_COUNT 个读线程
    for (int i = 0; i < READER_THREAD_COUNT; i++)
        tarray[i] = new std::thread(reader);

    //创建一个写线程
    std::thread tw(writer);

    for (int i = 0; i < READER_THREAD_COUNT; i++)
        tarray[i]->join();

    tw.join();

    //记录起始时间
    clock_t end = clock();
    printf("[test_shared_mutex]\n");
    printf("thread count: %d\n", READER_THREAD_COUNT);
    printf("result: %d cost: %dms temp: %d \n", counter.get(), end - start, temp);
}

//测试 std::mutex
void test_mutex()
{
    mutex_counter counter;

    int temp;

    //写线程函数
    auto writer = [&counter]() {
        for (int i = 0; i < LOOP_COUNT; i++)
        {
            counter.increment();
        }
    };

    //读线程函数
    auto reader = [&counter, &temp]() {
        for (int i = 0; i < LOOP_COUNT; i++)
        {
            temp = counter.get();
        }
    };

    //存放读线程对象指针的数组
    std::thread** tarray = new std::thread * [READER_THREAD_COUNT];

    //记录起始时间
    clock_t start = clock();

    //创建 READER_THREAD_COUNT 个读线程
    for (int i = 0; i < READER_THREAD_COUNT; i++)
```

```cpp
        tarray[i] = new std::thread(reader);

    //创建一个写线程
    std::thread tw(writer);

    for (int i = 0; i < READER_THREAD_COUNT; i++)
        tarray[i]->join();

    tw.join();

    //记录结束时间
    clock_t end = clock();
    printf("[test_mutex]\n");
    printf("thread count:%d\n", READER_THREAD_COUNT);
    printf("result:%d cost:%dms temp:%d \n", counter.get(), end - start, temp);
}

int main()
{
    //在每次测试时都只开启test_mutex或test_shared_mutex函数中一个
    test_mutex();
    //test_shared_mutex();
    return 0;
}
```

下图展示了对 Windows 上 Visual Studio 2019 的测试结果。

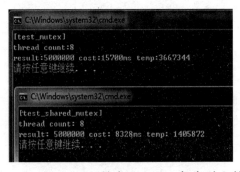

在 Linux 上，由于 std::shared_mutex 是在 C++ 17 中才引入的（gcc 7.0 及以上，这里使用的是 gcc 7.3），因此编译时需要加上编译参数--std=c++17，测试结果如下：

```
[root@myaliyun testmutexbenchmark]# g++ -g -o test_shared_mutex
TestSharedMutexBenchmark.cpp -std=c++17 -lpthread
[root@myaliyun testmutexbenchmark]# vi TestSharedMutexBenchmark.cpp
[root@myaliyun testmutexbenchmark]# g++ -g -o test_mutex
TestSharedMutexBenchmark.cpp -std=c++17 -lpthread
[root@myaliyun testmutexbenchmark]# ll
total 416
-rwxr-xr-x 1 root root 205688 Nov 10 22:35 test_mutex
-rwxr-xr-x 1 root root 205688 Nov 10 22:35 test_shared_mutex
-rw-r--r-- 1 root root   4112 Nov 10 22:35 TestSharedMutexBenchmark.cpp
[root@myaliyun testmutexbenchmark]# ./test_mutex
[test_mutex]
thread count:8
result:5000000 cost:2460000ms temp:4341759
```

```
[root@myaliyun testmutexbenchmark]# ./test_shared_mutex
[test_shared_mutex]
thread count: 8
result: 5000000 cost: 2620000ms temp: 735375
```

读者可以尝试修改 READER_THREAD_COUNT 的值来测试不同数量的读线程的输出结果。

> std::mutex 和 std::shared_mutex 分别对应 Java JDK 中的 ReentrantLock 和 ReentrantReadWriteLock。

建议在开发过程中认真甄别使用场景，合理使用 std::shared_mutex 替代部分 std::mutex，以提高程序执行效率。

3.7.3 std::condition_variable

C++ 11 提供了 std::condition_variable 类来代表条件变量，与 Linux 系统原生的条件变量一样，还提供了等待条件变量满足的 wait 系列方法（wait、wait_for、wait_until 方法），发送条件信号时使用 notify 方法（notify_one 和 notify_all 方法）。当然，使用 std::condition_variable 对象时需要绑定 1 个 std::unique_lock 或 std::lock_guard 对象。

> 与 Linux 或 Windows 自带的条件变量相比，对 C++ 11 的 std::condition_variable 不再需要显式地初始化和销毁。

我们将前面介绍 Linux 条件变量时的例子改写成 C++ 11 版本：

```
#include <thread>
#include <mutex>
#include <condition_variable>
#include <list>
#include <iostream>

class Task
{
public:
    Task(int taskID)
    {
        this->taskID = taskID;
    }

    void doTask()
    {
        std::cout << "handle a task, taskID: " << taskID << ", threadID: " << std::this_thread::get_id() << std::endl;
    }
private:
    int taskID;
};

std::mutex                  mymutex;
std::list<Task*>            tasks;
```

```cpp
std::condition_variable mycv;

void* consumer_thread()
{
    Task* pTask = NULL;
    while (true)
    {
        //使用括号减小 guard 锁的作用范围
        {
            std::unique_lock<std::mutex> guard(mymutex);
            while (tasks.empty())
            {
                //如果获得了互斥锁，但是条件不合适，
                //则 pthread_cond_wait 会释放锁，不向下执行
                //发生变化后，如果条件合适，则 pthread_cond_wait 将直接获得锁
                mycv.wait(guard);
            }

            pTask = tasks.front();
            tasks.pop_front();
        }

        if (pTask == NULL)
            continue;

        pTask->doTask();
        delete pTask;
        pTask = NULL;
    }

    return NULL;
}

void* producer_thread()
{
    int taskID = 0;
    Task* pTask = NULL;

    while (true)
    {
        pTask = new Task(taskID);

        //使用括号减小 guard 锁的作用范围
        {
            std::lock_guard<std::mutex> guard(mymutex);
            tasks.push_back(pTask);
            std::cout << "produce a task, taskID: " << taskID << ", threadID: " << std::this_thread::get_id() << std::endl;
        }

        //释放信号量，通知消费者线程
        mycv.notify_one();

        taskID ++;

        //休眠 1 秒
        std::this_thread::sleep_for(std::chrono::seconds(1));
```

```
    }
    return NULL;
}

int main()
{
    //创建5个消费者线程
    std::thread consumer1(consumer_thread);
    std::thread consumer2(consumer_thread);
    std::thread consumer3(consumer_thread);
    std::thread consumer4(consumer_thread);
    std::thread consumer5(consumer_thread);

    //创建1个生产者线程
    std::thread producer(producer_thread);

    producer.join();
    consumer1.join();
    consumer2.join();
    consumer3.join();
    consumer4.join();
    consumer5.join();

    return 0;
}
```

编译并执行程序,输出结果如下:

```
[root@localhost testmultithread]# g++ -g -o cpp11cv cpp11cv.cpp -std=c++0x -lpthread
[root@localhost testmultithread]# ./cpp11cv
produce a task, taskID: 0, threadID: 140427590100736
handle a task, taskID: 0, threadID: 140427623671552
produce a task, taskID: 1, threadID: 140427590100736
handle a task, taskID: 1, threadID: 140427632064256
produce a task, taskID: 2, threadID: 140427590100736
handle a task, taskID: 2, threadID: 140427615278848
produce a task, taskID: 3, threadID: 140427590100736
handle a task, taskID: 3, threadID: 140427606886144
produce a task, taskID: 4, threadID: 140427590100736
handle a task, taskID: 4, threadID: 140427598493440
produce a task, taskID: 5, threadID: 140427590100736
handle a task, taskID: 5, threadID: 140427623671552
produce a task, taskID: 6, threadID: 140427590100736
handle a task, taskID: 6, threadID: 140427632064256
produce a task, taskID: 7, threadID: 140427590100736
handle a task, taskID: 7, threadID: 140427615278848
produce a task, taskID: 8, threadID: 140427590100736
handle a task, taskID: 8, threadID: 140427606886144
produce a task, taskID: 9, threadID: 140427590100736
handle a task, taskID: 9, threadID: 140427598493440
...省略更多的输出结果...
```

3.8 如何确保创建的线程一定能运行

本章开头介绍了如何创建线程，但是如何确保创建的线程一定能运行？很多人会说，对于使用系统 API 创建的线程，只需判断创建的线程函数是否调用成功，但无法百分百保证线程函数一定能运行。

在一些严谨的项目中创建线程时，不仅会判断线程创建函数是否调用成功，还会在线程函数中利用线程同步对象通知创建者线程是否创建成功。来看一段代码：

```cpp
#include <thread>
#include <mutex>
#include <condition_variable>
#include <iostream>

std::mutex                mymutex;
std::condition_variable   mycv;
bool success = false;

void thread_func()
{
    {
        std::unique_lock<std::mutex> lock(mymutex);
        success = true;
        mycv.notify_all();
    }

    //实际的线程业务逻辑在下面为了模拟方便，简单地写个死循环
    while (true)
    {
    }
}

int main()
{
    std::thread t(thread_func);

    //使用花括号减小锁的粒度
    {
        std::unique_lock<std::mutex> lock(mymutex);
        while (!success)
        {
            mycv.wait(lock);
        }
    }

    std::cout << "start thread successfully." << std::endl;

    t.join();

    return 0;
}
```

以上代码在发出一个创建新线程的请求后，会立刻阻塞在一个条件变量上。工作线程

如果成功运行起来,则会发送条件变量信号告知主线程,这样主线程就知道新线程一定成功运行了。

基于以上思路,创建一组线程时可以逐个创建,成功运行一个新线程后再创建下一个线程,确保线程组中的每一个线程都可以运行。示例代码如下:

```cpp
#include <thread>
#include <mutex>
#include <condition_variable>
#include <iostream>
#include <vector>
#include <memory>

std::mutex                mymutex;
std::condition_variable   mycv;
bool success = false;

void thread_func(int no)
{
    {
        std::unique_lock<std::mutex> lock(mymutex);
        success = true;
        mycv.notify_all();
    }

    std::cout << "worker thread started, threadNO: " << no << std::endl;
    //实际的线程业务逻辑在下面,这里为了模拟方便,简单地写个死循环
    while (true)
    {
    }
}

int main()
{
    std::vector<std::shared_ptr<std::thread>> threads;

    for (int i = 0; i < 5; ++i)
    {
        success = false;
        std::shared_ptr<std::thread> spthread;
        spthread.reset(new std::thread(thread_func, i));

        //使用花括号减小锁的粒度
        {
            std::unique_lock<std::mutex> lock(mymutex);
            while (!success)
            {
                mycv.wait(lock);
            }
        }

        std::cout << "start thread successfully, index: " << i << std::endl;

        threads.push_back(spthread);
```

```
    }

    for (auto& iter : threads)
        iter->join();

    return 0;
}
```

编译上述程序并运行，运行结果如下：

```
[root@myaliyun codes]# g++ -g -o makesurethreadgroup makesurethreadgroup.cpp
-std=c++0x -lpthread
[root@myaliyun codes]# ./makesurethreadgroup
worker thread started, threadNO: 0
start thread successfully, index: 0
worker thread started, threadNO: 1
start thread successfully, index: 1
worker thread started, threadNO: 2
start thread successfully, index: 2
worker thread started, threadNO: 3
start thread successfully, index: 3
worker thread started, threadNO: 4
start thread successfully, index: 4
```

可以看到，新线程逐个运行起来了。当然，我们不一定要使用条件变量，也可以使用其他类型的线程同步对象，例如 Windows 平台的 Event 对象等。

注意，上文介绍的确保线程一定能运行的做法在新项目开发中可能很少被发现，这是因为多年前多线程技术开始流行的时候，由于软硬件限制，加之很多开发人员对多线程编程不熟悉，所以创建新线程时确保一个线程可以运行起来非常必要；而如今多线程编程已经司空见惯，加上操作系统和 CPU 普遍对多线程有良好的支持，所以再也不用写这样的"防御"代码了。现在只要正确使用线程创建函数，在实际编码时对线程函数的返回值一般也不必判断，基本可以认为新线程一定会创建成功，而且线程函数可以正常运行。

3.9 多线程使用锁经验总结

前面介绍了 Windows 和 Linux 操作系统提供的各种常用锁对象的使用原理和方法。多线程编程少不了与这些锁打交道，在使用锁时稍不注意就可能会造成死锁或者程序性能出现问题。这里总结了一些经验。

3.9.1 减少锁的使用次数

在实际开发中能不使用锁则尽量不使用锁，当然，这不是绝对的：如果使用锁也能满足性能要求，则使用锁也无妨。使用了锁的代码一般会存在如下性能损失：

（1）加锁和解锁操作，本身有一定的开销；

（2）临界区的代码不能并发执行；

（3）进入临界区的次数过于频繁，线程之间对临界区的争夺太过激烈，若线程竞争互斥体失败，就会陷入阻塞并让出 CPU，所以执行上下文切换的次数要远远多于不使用互斥体的次数。

替代锁的方式有很多，例如无锁队列。

3.9.2 明确锁的范围

来看一段代码：

```
if(my_hashtable.is_empty())
{
   pthread_mutex_lock(&my_mutex);
   htable_insert(my_hashtable, &my_elem);
   pthread_mutex_unlock(&my_mutex);
}
```

能看出这段代码的问题吗？以上代码虽然对 my_hashtable 的插入操作使用了锁进行保护，但是判断 my_hashtable 是否为空也需要使用锁进行保护，所以正确的写法应该如下：

```
pthread_mutex_lock(&my_mutex);
if(my_hashtable.is_empty())
{
   htable_insert(my_hashtable, &elem);
}
pthread_mutex_unlock(&my_mutex);
```

3.9.3 减少锁的使用粒度

减小锁的使用粒度，指的是尽量减小锁作用的临界区代码范围，临界区的代码范围越小，多个线程排队进入临界区的时间就会越短。

来看两个具体的示例。

示例一：

```
01 void TaskPool::addTask(Task* task)
02 {
03    std::lock_guard<std::mutex> guard(m_mutexList);
04    std::shared_ptr<Task> spTask;
05    spTask.reset(task);
06    m_taskList.push_back(spTask);
07
08    m_cv.notify_one();
09 }
```

在以上代码中，guard 锁保护 m_taskList。仔细分析这段代码会发现，代码第 4、5、8 行其实没必要作为临界区内的代码，建议将其挪到临界区外面，修改如下：

```
01 void TaskPool::addTask(Task* task)
02 {
03    std::shared_ptr<Task> spTask;
04    spTask.reset(task);
05
```

```
06      {
07          std::lock_guard<std::mutex> guard(m_mutexList);
08          m_taskList.push_back(spTask);
09      }
10
11      m_cv.notify_one();
12  }
```

修改之后，guard 锁的作用范围就是第 7、8 行了，仅对 m_taskList.push_back() 操作做保护，这样锁的粒度就变小了。

示例二：

```
void EventLoop::doPendingFunctors()
{
    std::unique_lock<std::mutex> lock(m_mutex);
    for (size_t i = 0; i < m_pendingFunctors.size(); ++i)
    {
        m_pendingFunctors[i]();
    }
}
```

在以上代码中，m_pendingFunctors 是被锁保护的对象，它的类型是 std::vector，这样的代码运行效率较低，必须等当前线程逐个处理完 m_pendingFunctors 中的元素后，其他线程才能操作 m_pendingFunctors。修改代码如下：

```
void EventLoop::doPendingFunctors()
{
    std::vector<Functor> localFunctors;

    {
        std::unique_lock<std::mutex> lock(m_mutex);
        localFunctors.swap(m_pendingFunctors);
    }

    for (size_t i = 0; i <localFunctors.size(); ++i)
    {
        localFunctors[i]();
    }
}
```

修改之后的代码使用了一个局部变量 localFunctors，然后把 m_pendingFunctors 中的内容倒换到 localFunctors 中，这样就可以释放锁并允许其他线程操作 m_pendingFunctors 了。现在只要继续操作本地对象 localFunctors 就可以了，提高了效率。

3.9.4 避免死锁的一些建议

（1）在一个函数中如果有一个加锁操作，那么一定要记得在函数退出时解锁，且在每个退出路径上都不要忘记解锁。例如：

```
void some_func()
{
    //加锁代码
```

```
    if (条件 1)
    {
        //其他代码
        //解锁代码
        return;
    }
    else
    {
        //其他代码
        //解锁代码
        return;
    }

    if (条件 2)
    {
        if (条件 3)
        {
            //其他代码
            //解锁代码
            return;
        }

       if (条件 4)
       {
            //其他代码
            //解锁代码
            return;
       }
    }

    while (条件 5)
    {
        if (条件 6)
        {
            //其他代码
            //解锁代码
            return;
        }
    }
}
```

在上述函数中的每个逻辑出口处都需要写上解锁代码。前面也说过，这种逻辑非常容易因为忘记在某个地方加上解锁代码而造成死锁，所以一般建议使用 RAII 技术将加锁和解锁代码封装起来，1.1.4 节已经详细地介绍了这个做法，这里不再重复介绍。

（2）在线程退出时一定要及时释放其持有的锁。在实际开发中会因为一些特殊需求创建了一些临时线程，这些线程在执行完相应的任务后会退出。对于这类线程，如果其持有了锁，则在线程退出时，一定要记得释放其持有的锁。

（3）多线程请求锁的方向要一致，避免死锁。假设现在有两个锁 A 和 B，线程 1 在请求了锁 A 后再请求锁 B，线程 2 在请求了锁 B 后再请求锁 A，这种线程请求锁的方向就不一致了：线程 1 的方向是从 A 到 B，线程 2 的方向是从 B 到 A，多个线程请求锁的方向不一致容易造成死锁。所以建议线程 1 和线程 2 请求锁的方向保持一致，要么都从 A

到 B，要么都从 B 到 A。

（4）当需要同一个线程重复请求一个锁时，就需要明白使用锁的行为是递增锁引用计数，还是阻塞或者直接获得锁。

3.9.5 避免活锁的一些建议

"活锁"指多个线程使用 trylock 系列的函数时，由于相互谦让，导致即使在某段时间内锁资源可用，也可能导致需要锁的线程拿不到锁。举个生活中的例子，马路上两个人迎面走来，两个人可能会同时向一个方向避让，其本意是给对方让路，但还是发生了碰撞。

我们在实际编码时，应该尽量避免让过多的线程使用 trylock 请求锁，以免出现活锁，这是对资源的一种浪费。

3.10 线程局部存储

对于 1 个存在多个线程的进程来说，有时需要每个线程都自己操作自己的这份数据。这有点类似于 C++ 类的实例属性，每个实例对象操作的都是自己的属性。我们把这样的数据称为线程局部存储（Thread Local Storage，TLS），将对应的存储区域称为线程局部存储区。

3.10.1 Windows 的线程局部存储

Windows 将线程局部存储区分成 TLS_MINIMUM_AVAILABLE 个块，每个块都通过 1 个索引值对外提供访问。

TLS_MINIMUM_AVAILABLE 默认是 64，在 winnt.h 文件中有如下定义：

```
#define TLS_MINIMUM_AVAILABLE 64
```

Windows TLS 结构示意图如下图所示。

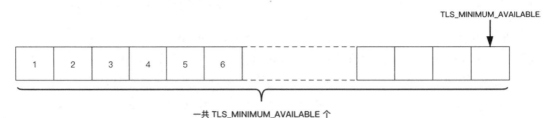

在 Windows 中使用 TlsAlloc 函数获得一个线程局部存储块的索引：

```
DWORD TlsAlloc();
```

如果这个函数调用失败，则返回值是 TLS_OUT_OF_INDEXES（0xFFFFFFFF）；如果这个函数调用成功，则会得到一个索引。接下来就可以利用如下两个 API 函数分别在这个索引指向的内存块中存储数据，或者在这个索引指向的内存块中取出数据了：

```
LPVOID TlsGetValue(DWORD dwTlsIndex);
BOOL TlsSetValue(DWORD dwTlsIndex, LPVOID lpTlsValue);
```

当不再需要索引指向的内存块时，可以使用如下函数来释放索引和内存块：

```
BOOL TlsFree(DWORD dwTlsIndex);
```

当然，在使用线程局部存储时除了可以使用 API 函数，还可以使用 Microsoft VC++ 编译器提供的如下方法定义一个线程局部变量：

```
__declspec(thread) int g_mydata = 1;
```

来看一个具体的例子：

```
#include <Windows.h>
#include <iostream>

__declspec(thread) int g_mydata = 1;

DWORD __stdcall WorkerThreadProc1(LPVOID lpThreadParameter)
{
    while (true)
    {
        ++g_mydata;

        Sleep(1000);
    }
    return 0;
}

DWORD __stdcall WorkerThreadProc2(LPVOID lpThreadParameter)
{
    while (true)
    {
        std::cout << "g_mydata = " << g_mydata << ", ThreadID = " << GetCurrentThreadId() << std::endl;

        Sleep(1000);
    }
    return 0;
}

int main()
{
    HANDLE hWorkerThreads[2];
    hWorkerThreads[0] = CreateThread(NULL, 0, WorkerThreadProc1, NULL, 0, NULL);
    hWorkerThreads[1] = CreateThread(NULL, 0, WorkerThreadProc2, NULL, 0, NULL);

    CloseHandle(hWorkerThreads[0]);
    CloseHandle(hWorkerThreads[1]);

    while (true)
    {
        Sleep(1000);
    }
```

```
        return 0;
}
```

在以上代码中，全局变量 g_mydata 是一个线程局部变量，因此在该进程中每一个线程都会拥有这样一个变量副本。由于是不同的副本，所以在 WorkerThreadProc1 中，这个变量不断递增，对 WorkerThreadProc2 的 g_mydata 不会造成任何影响，因此其值始终是 1。程序执行结果如下图所示。

```
D:\CppDev\testmutex\Debug\tls.exe
g_mydata = 1, ThreadID = 69492
g_mydata = 1, ThreadID = 69492
g_mydata = 1, ThreadID = 69492
g_mydata = 1, ThreadID = 69492
g_mydata = 1, ThreadID = 69492
g_mydata = 1, ThreadID = 69492
g_mydata = 1, ThreadID = 69492
g_mydata = 1, ThreadID = 69492
g_mydata = 1, ThreadID = 69492
g_mydata = 1, ThreadID = 69492
```

需要说明的是，在 Windows 中被声明成线程局部变量的对象，在编译器生成可执行文件时，会在最终的 PE 文件中专门生成一个叫作 tls 的节，用于存放这些线程局部变量。

3.10.2　Linux 的线程局部存储

Linux 上的 NTPL 提供了一套函数接口来实现线程局部存储功能：

```
int pthread_key_create(pthread_key_t* key, void (*destructor)(void*));
int pthread_key_delete(pthread_key_t key);

int pthread_setspecific(pthread_key_t key, const void* value);
void* pthread_getspecific(pthread_key_t key);
```

pthread_key_create 函数在调用成功时会返回 0，在调用失败时会返回非 0 值。函数若调用成功，则会为线程局部存储创建一个新的 key，用户通过这个 key 设置（调用 pthread_setspecific）和获取（pthread_getspecific）数据。因为进程中的所有线程都可以使用返回的键，所以这个 key 应该指向一个全局变量。

参数 destructor 是一个自定义函数指针，其签名如下：

```
void* destructor(void* value)
{
    //多是为了释放 value 指针指向的资源
}
```

线程终止时，如果 key 关联的值不是 NULL，那么 NTPL 会自动执行定义的 destructor 函数；如果无须解构，则可以将 destructor 设置为 NULL。

来看一个具体的例子：

```
#include <pthread.h>
#include <stdio.h>
#include <stdlib.h>
```

```c
//线程局部存储 key
pthread_key_t thread_log_key;

void write_to_thread_log(const char* message)
{
    if (message == NULL)
        return;

    FILE* logfile = (FILE*)pthread_getspecific(thread_log_key);
    fprintf(logfile, "%s\n", message);
    fflush(logfile);
}

void close_thread_log(void* logfile)
{
    char logfilename[128];
    sprintf(logfilename, "close logfile: thread%ld.log\n", (unsigned long)pthread_self());
    printf(logfilename);

    fclose((FILE *)logfile);
}

void* thread_function(void* args)
{
    char logfilename[128];
    sprintf(logfilename, "thread%ld.log", (unsigned long)pthread_self());

    FILE* logfile = fopen(logfilename, "w");
    if (logfile != NULL)
    {
        pthread_setspecific(thread_log_key, logfile);

        write_to_thread_log("Thread starting...");
    }

    return NULL;
}

int main()
{
    pthread_t threadIDs[5];
    pthread_key_create(&thread_log_key, close_thread_log);
    for(int i = 0; i < 5; ++i)
        pthread_create(&threadIDs[i], NULL, thread_function, NULL);

    for(int i = 0; i < 5; ++i)
        pthread_join(threadIDs[i], NULL);

    return 0;
}
```

上述程序一共创建了 5 个线程,每个线程都会自己生成一个日志文件,每个线程都将自己的日志写入自己的文件中。在线程执行结束时,会关闭打开的日志文件句柄。程序运行结果如下图所示。

```
[root@localhost testmultithread]# g++ -g -o linuxtls linuxtls.cpp -lpthread
[root@localhost testmultithread]# ./linuxtls
close logfile: thread139702125414144.log
close logfile: thread139702150592256.log
close logfile: thread139702133806848.log
close logfile: thread139702117021440.log
close logfile: thread139702142199552.log
```

可以看到，在生成的 5 个日志文件中都被写入了一行"Thread starting..."。

上面的程序首先调用 pthread_key_create 函数来申请 1 个槽位。在 NPTL 实现下，pthread_key_t 是无符号整型，pthread_key_create 调用成功时会将 1 个小于 1024 的值填入第 1 个入参指向的 pthread_key_t 类型的变量中。之所以小于 1024，是因为 NPTL 实现一共提供了 1024 个槽位。如下图所示，记录槽位分配情况的数据结构 pthread_keys 是进程唯一的，pthread_keys 结构的示意图如下所示。

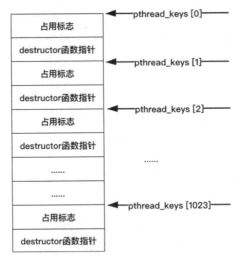

和 Windows 一样，Linux gcc 编译器也提供了一个关键字 __thread 用于定义线程局部变量。例如：

```
__thread int val = 0;
```

再来看一个示例：

```cpp
#include <pthread.h>
#include <iostream>
#include <unistd.h>

//线程局部存储 key
__thread int g_mydata = 99;

void* thread_function1(void* args)
{
    while (true)
    {
        g_mydata ++;
    }
```

```cpp
        return NULL;
}

void* thread_function2(void* args)
{
    while (true)
    {
        std::cout << "g_mydata = " << g_mydata << ", ThreadID: " << pthread_self() << std::endl;
        sleep(1);
    }

    return NULL;
}

int main()
{
    pthread_t threadIDs[2];
    pthread_create(&threadIDs[0], NULL, thread_function1, NULL);
    pthread_create(&threadIDs[1], NULL, thread_function2, NULL);

    for(int i = 0; i < 2; ++i)
        pthread_join(threadIDs[i], NULL);

    return 0;
}
```

由于 thread_function1 修改的是自己的 g_mydata，因此 thread_function2 输出 g_mydata 的值始终是 99。

```
[root@localhost testmultithread]# g++ -g -o linuxtls2 linuxtls2.cpp -lpthread
[root@localhost testmultithread]# ./linuxtls2
g_mydata = 99, ThreadID: 140243186276096
g_mydata = 99, ThreadID: 140243186276096
g_mydata = 99, ThreadID: 140243186276096
g_mydata = 99, ThreadID: 140243186276096
g_mydata = 99, ThreadID: 140243186276096
g_mydata = 99, ThreadID: 140243186276096
g_mydata = 99, ThreadID: 140243186276096
g_mydata = 99, ThreadID: 140243186276096
g_mydata = 99, ThreadID: 140243186276096
g_mydata = 99, ThreadID: 140243186276096
g_mydata = 99, ThreadID: 140243186276096
...省略更多的输出结果...
```

3.10.3　C++ 11 的 thread_local 关键字

C++ 11 标准提供了一个新的关键字 thread_local 来定义一个线程变量。使用方法如下：

```
thread_local int g_mydata = 1;
```

有了这个关键字，使用线程局部存储的代码就可以同时在 Windows 和 Linux 上运行了。示例如下：

```cpp
#include <thread>
#include <chrono>
#include <iostream>

thread_local int g_mydata = 1;

void thread_func1()
{
    while (true)
    {
        ++g_mydata;
    }
}

void thread_func2()
{
    while (true)
    {
        std::cout << "g_mydata = " << g_mydata << ", ThreadID = " << std::this_thread::get_id() << std::endl;
        std::this_thread::sleep_for(std::chrono::seconds(1));
    }
}

int main()
{
    std::thread t1(thread_func1);
    std::thread t2(thread_func2);

    t1.join();
    t2.join();

    return 0;
}
```

需要注意的是，在 Windows 上，虽然 thread_local 关键字在 C++ 11 标准中已引入，但是 Visual Studio 2013（支持 C++ 11 语法最低的一个 Visual Studio 版本）编译器并不支持这个关键字，建议在 Visual Studio 2015 及以上版本中测试以上代码。

最后，关于线程局部存储变量，需要再强调两点。

（1）对于线程变量，每个线程都会有该变量的一个拷贝，互不影响，该局部变量一直存在，直到线程退出。

（2）系统的线程局部存储区域的内存空间并不大，所以尽量不要用这个空间存储大的数据块，如果不得不使用大的数据块，则可以将大的数据块存储在堆内存中，再将该堆内存的地址指针存储在线程局部存储区域。

3.11　C 库的非线程安全函数

先来看一段代码：

```
#include <time.h>

int main()
{
    time_t tNow = time(NULL);
    time_t tEnd = tNow + 1800;
    //注意加粗代码行时的位置
    struct tm* ptm = localtime(&tNow);
    struct tm* ptmEnd = localtime(&tEnd);

    char szTmp[50] = { 0 };
    strftime(szTmp, 50, "%H:%M:%S", ptm);

    //struct tm* ptmEnd = localtime(&tEnd);
    char szEnd[50] = { 0 };
    strftime(szEnd, 50, "%H:%M:%S", ptmEnd);
    printf("%s\n", szTmp);
    printf("%s\n", szEnd);

    return 0;
}
```

程序执行结果如下：

```
20:53:48
20:53:48
```

是不是很奇怪？tNow 和 tEnd 明明相差 1800 秒，为什么输出的值一模一样？调整一加粗代码行在整段代码中的位置：

```
#include <time.h>

int main()
{
    time_t tNow = time(NULL);
    time_t tEnd = tNow + 1800;
    //注意加粗代码行时的位置
    struct tm* ptm = localtime(&tNow);

    char szTmp[50] = { 0 };
    strftime(szTmp, 50, "%H:%M:%S", ptm);

    struct tm* ptmEnd = localtime(&tEnd);
    char szEnd[50] = { 0 };
    strftime(szEnd, 50, "%H:%M:%S", ptmEnd);
    printf("%s\n", szTmp);
    printf("%s\n", szEnd);

    return 0;
}
```

这次的输出结果正确：

```
20:25:44
20:55:44
```

为什么会出现这种情况呢？来看看 localtime 函数签名：

```
struct tm* localtime(const time_t* timep);
```

这个函数返回值的类型是一个 tm 结构体指针，而外部并不需要释放这个指针指向的内存，因此这个函数内部一定使用了一个全局变量或函数内部的静态变量。这样再次调用这个函数时，前一次的调用结果就可能被后一次的调用结果覆盖了。简化该模型：

```
int* func(int k)
{
    static int result;
    result = k;
    return &result;
}
```

当多个线程甚至单个线程调用这个函数时，例如两个线程分别调用了以上函数：

```
//线程 1 调用
int* p1 = func(1);
//线程 2 调用
int* p2 = func(2);
```

那么 p1 和 p2 的结果会是什么呢？可能是 1，也可能是 2，甚至既不是 1 也不是 2。

像 localtime 这类 CRT 提供的具有上述行为的函数，我们称之为非线程安全函数，在实际开发中应避免在多线程程序中使用这类函数。这类函数如 strtok，甚至连操作系统提供的 socket 函数 gethostbyname 也不是线程安全的。

```
char* strtok(char* str, const char* delim);

struct hostent* gethostbyname(const char* name);
```

为什么会出现这类函数呢？因为最初编写很多 CRT 函数时还没有多线程技术，所以很多函数内部的实现都使用了函数内部的静态变量和全局变量。随着多线程技术的出现，很多函数都有了对应的多线程安全版本，例如 localtime_r 和 strtok_r。很多函数内部都改用了线程局部存储技术来替代原来使用静态变量或者全局变量的做法。

3.12 线程池与队列系统的设计

本节讲解线程池和队列系统设计方面的内容。

3.12.1 线程池的设计原理

线程池其实只是一组线程。在一般情况下，我们需要异步执行一些任务，这些任务的产生和执行是存在于程序的整个生命周期内的。与其让操作系统频繁地为我们创建和销毁线程，不如创建一组在程序生命周期内不会退出的线程。为了不浪费系统资源，我们的基本要求是当有任务需要执行时，这些线程可以自动拿到任务并执行，在没有任务时这些线程处于阻塞或者睡眠状态。

既然在程序生命周期内会产生很多任务，那么这些任务必须有一个存放的地方，这个地方就是队列，所以不要一提到队列就认为它是 1 个具体的 list，队列可以是 1 个全局变量或链表。而对于线程池中的线程如何从队列中取任务，也可以设计得非常灵活，例如从尾部放入任务并从头部取出，或者从头部放入并从尾部取出等。队列也可以根据实际应用设计得"丰富多彩"，例如，可以根据任务的优先级设计多个队列（分为高、中、低三个级别或者分为关键、普通两个级别）。

这在本质上就是生产者/消费者模式，产生任务的线程是生产者，线程池中的线程是消费者。当然，这不是绝对的，线程池中的线程在处理 1 个任务后可能产生 1 个新的关联任务，那么此时这个工作线程又是生产者角色。

既然会有多个线程同时操作这个队列，那么根据多线程程序的原则，我们一般需要对这个队列加锁，以免多线程竞争产生非预期的结果。当然，在技术上除了要解决线程池创建、向队列中投递任务、从队列中取任务并处理的问题，我们还需要做一些善后工作，例如线程池的清理，即如何退出线程池中的工作线程和清理任务队列，这也是线程池和任务队列的核心原理。具体的实现很容易，来看一个具体的例子：

TaskPool.h 文件的内容如下：

```cpp
/**
 * 任务池模型，TaskPool.h
 * zhangyl 2019.02.14
 */
#include <thread>
#include <mutex>
#include <condition_variable>
#include <list>
#include <vector>
#include <memory>
#include <iostream>

class Task
{
public:
    virtual void doIt()
    {
        std::cout << "handle a task..." << std::endl;
    }

    virtual ~Task()
    {
        //在析构函数中展示一个 task 销毁过程
        std::cout << "a task destructed..." << std::endl;
    }
};

class TaskPool final
{
public:
    TaskPool();
    ~TaskPool();
    TaskPool(const TaskPool& rhs) = delete;
    TaskPool& operator=(const TaskPool& rhs) = delete;
```

```cpp
public:
    void init(int threadNum = 5);
    void stop();

    void addTask(Task* task);
    void removeAllTasks();

private:
    void threadFunc();

private:
    std::list<std::shared_ptr<Task>>              m_taskList;
    std::mutex                                    m_mutexList;
    std::condition_variable                       m_cv;
    bool                                          m_bRunning;
    std::vector<std::shared_ptr<std::thread>>     m_threads;
};
```

TaskPool.cpp 文件的内容如下：

```cpp
/**
 * 任务池模型，TaskPool.cpp
 * zhangyl 2019.02.14
 */
#include "TaskPool.h"

TaskPool::TaskPool() : m_bRunning(false)
{
}

TaskPool::~TaskPool()
{
    removeAllTasks();
}

void TaskPool::init(int threadNum/* = 5*/)
{
    if (threadNum <= 0)
        threadNum = 5;

    m_bRunning = true;

    for (int i = 0; i < threadNum; ++i)
    {
        std::shared_ptr<std::thread> spThread;
        spThread.reset(new std::thread(std::bind(&TaskPool::threadFunc, this)));
        m_threads.push_back(spThread);
    }
}

void TaskPool::threadFunc()
{
    std::shared_ptr<Task> spTask;
    while (true)
    {
        {//减小 guard 锁的作用范围
```

```cpp
            std::unique_lock<std::mutex> guard(m_mutexList);
            while (m_taskList.empty())
            {
                if (!m_bRunning)
                    break;

                //如果获得了互斥锁,但是条件不满足,
                //则m_cv.wait()调用会释放锁,且挂起当前线程,因此不往下执行
                //发生变化后,条件满足时,m_cv.wait()将唤醒挂起的线程且获得锁
                m_cv.wait(guard);
            }

            if (!m_bRunning)
                break;

            spTask = m_taskList.front();
            m_taskList.pop_front();
        }

        if (spTask == NULL)
            continue;

        spTask->doIt();
        spTask.reset();
    }

    std::cout << "exit thread, threadID: " << std::this_thread::get_id() << std::endl;
}

void TaskPool::stop()
{
    m_bRunning = false;
    m_cv.notify_all();

    //等待所有线程退出
    for (auto& iter : m_threads)
    {
        if (iter->joinable())
            iter->join();
    }
}

void TaskPool::addTask(Task* task)
{
    std::shared_ptr<Task> spTask;
    spTask.reset(task);

    {
        std::lock_guard<std::mutex> guard(m_mutexList);
        m_taskList.push_back(spTask);
        std::cout << "add a Task." << std::endl;
    }

    m_cv.notify_one();
}
```

```cpp
void TaskPool::removeAllTasks()
{
    {
        std::lock_guard<std::mutex> guard(m_mutexList);
        for (auto& iter : m_taskList)
            iter.reset();

        m_taskList.clear();
    }
}
```

以上代码封装了一个简单的任务队列模型，我们可以这么使用这个 TaskPool 对象：

```cpp
#include "TaskPool.h"
#include <chrono>

int main()
{
    TaskPool threadPool;
    threadPool.init();

    Task* task = NULL;
    for (int i = 0; i < 10; ++i)
    {
        task = new Task();
        threadPool.addTask(task);
    }

    std::this_thread::sleep_for(std::chrono::seconds(5));

    threadPool.stop();

    return 0;
}
```

程序执行结果如下图所示。

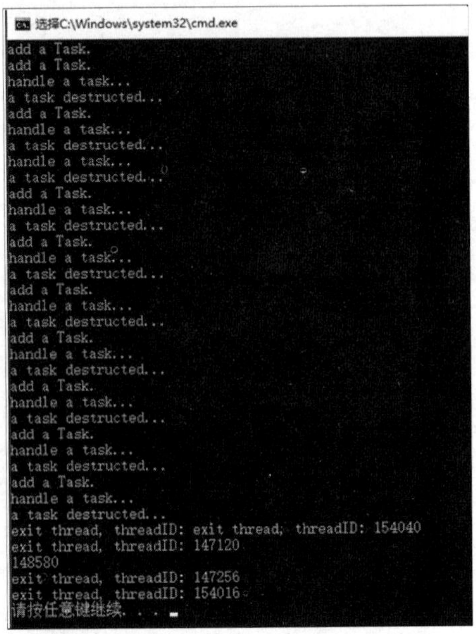

由于退出线程的输出提示不是原子的，而是多个线程并行执行的，因此上图中这部分的输出出现了错乱。

以上代码演示了一个基本的多线程队列模型，虽然简单，但是比较典型，可以应付实际生产中的一部分需求。我们可以基于这个基础模型进行扩展，其基本原理都是一样的。

3.12.2 环形队列

如果生产者和消费者（即产生任务者和处理任务者）的速度差不多，则可以将队列改成环形队列，以节省内存空间。另外，很多应用为了追求效率，都利用了一些技巧将队列无锁化，这里不再展开介绍。

3.12.3 消息中间件

基于生产者/消费者模型衍生的队列系统在实际开发中很常用，以至于在一组服务中可能每个进程都需要一个这样的队列系统。既然如此，出于复用和解耦的目的，业界出现许多独立的队列系统，这些队列系统或以一个独立的进程运行，或以支持分布式的一组服务运行。我们把这种独立的队列系统称为消息中间件。这些消息中间件在功能上做了丰富的扩展，例如消费的方式、主备切换、容灾容错、数据自动备份和过期数据自动清理等，比较典型的有 Kafka、ActiveMQ、RabbitMQ、RocketMQ 等。如下所示是 Kafka 官网提供的介绍 Kafka 作用的一张图。

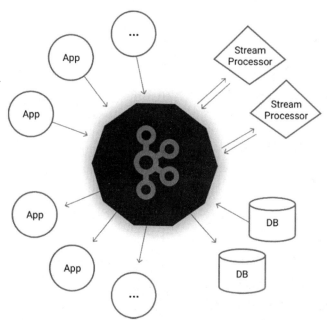

下图是某个金融交易系统的后台服务拓扑图，其大量使用了消息中间件 RabbitMQ。

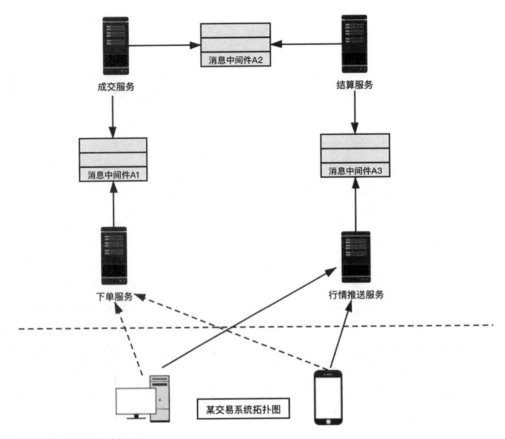

某交易系统拓扑图

其整个交易流程如下。

（1）前端通过 HTTP 请求向下单服务请求下单，下单服务在校验完数据后，向消息中间件 A1 投递一条下单请求。

（2）成交服务订阅消息中间件 A1 的消息，取出下单请求，结合自己的成交规则，如果可以成交，则向消息中间件 A2 投递一条成交后的消息。

（3）结算服务订阅消息中间件 A2，从其中拿到成交的消息后，对用户资金账户进行结算，结算完成后，用户的下单就正式完成了，然后产生一条行情消息投递给消息中间件 A3。

（4）行情推送服务器从消息中间件 A3 中拿到行情消息后，将其推送给所有已经连接的客户端。

在上述过程中，每个消息中间件（RabbitMQ）都有一个生产者和消费者，虚线箭头表示短连接，实线箭头表示长连接。当然，实际的金融交易系统要比这里的模型复杂得多，这里为了演示方便，做了大量简化。

有了这种专门的队列系统，生产者和消费者将最大化解耦。利用消息中间件提供的对外消息接口，生产者只需负责生产消息，不必关心谁是消费者；消费者不必关心生产者是谁、何时有数据；队列系统本身也不必关心自己有多少生产者和消费者。当然，这种消息中间件还有其他非常优秀的功能，例如对数据的备份、负载和容灾容错措施。建议读者适当了解一两种开源队列系统的用法，尤其是其设计思路。

3.13 纤程（Fiber）与协程（Routine）

本节讲解纤程与协程方面的内容。

3.13.1 纤程

纤程（Fiber）是 Windows 中的概念。当我们需要异步执行一些任务时，常见的一种做法就是开启一个工作线程，在工作线程中执行我们的任务。但是这样存在两个问题。

（1）由于线程的调度是由操作系统内核控制的，所以我们无法确定操作系统何时会运行或挂起该线程。

（2）对于一些轻量级的任务，创建一个新的线程去做，消耗较大。

那么有没有一种机制，既能新建线程执行任务，又没有新建线程消耗那么大呢？有，这就是纤程。

在 Windows 中，一个线程可以有多个纤程，用户可以根据需要在各个纤程之间自由切换。如果需要在某个线程中使用纤程，则必须先将该线程切换成纤程模式，这可以通过调用如下 API 函数实现：

```
LPVOID ConvertThreadToFiber(LPVOID lpParameter);
```

这个函数不仅将当前线程切换成纤程模式，也可以得到线程中的第 1 个纤程。我们可以通过这个函数的返回值来引用和操作纤程，这个纤程是线程中的主纤程。但是这个主纤程无法指定纤程函数，所以什么也做不了。我们可以通过参数 lpParameter 向主纤程传递数据，使用如下 API 函数获取当前纤程的数据：

```
PVOID GetFiberData();
```

当在不同的纤程之间切换时，也会涉及纤程上下面的切换，包括 CPU 寄存器数据的切换。在默认情况下，x86 系统的 CPU 浮点状态信息不属于 CPU 寄存器，不会为每个纤程都维护一份，因此如果在我们的纤程中执行浮点操作，则会导致数据被破坏。为了禁用这种行为，需要用到 ConvertThreadToFiberEx 函数，该函数签名如下：

```
LPVOID ConvertThreadToFiberEx(LPVOID lpParameter, DWORD dwFlags);
```

将第 2 个参数 dwFlags 设置为 FIBER_FLAG_FLOAT_SWITCH 即可。

将线程从纤程模式切回至默认的线程模式时，会用到 API 函数：

```
BOOL ConvertFiberToThread();
```

因为默认的主纤程什么都做不了，所以我们在需要时要创建新的纤程，这时会用到 API 函数：

```
LPVOID CreateFiber(SIZE_T                dwStackSize,
```

```
                       LPFIBER_START_ROUTINE       lpStartAddress,
                       LPVOID                      lpParameter);
```

和创建线程的函数类似,参数 dwStackSize 指定纤程栈的大小,如果使用默认的大小,则将该值设置为 0 即可。我们可以将 CreateFiber 函数的返回值作为操作纤程的句柄。

纤程函数签名如下:

```
VOID WINAPI FIBER_START_ROUTINE(LPVOID lpFiberParameter);
```

当不需要使用纤程时,记得调用 DeleteFiber 删除纤程对象:

```
void DeleteFiber(LPVOID lpFiber);
```

在不同的纤程之间切换时,会用到 API 函数:

```
void SwitchToFiber(LPVOID lpFiber);
```

参数 lpFiber 即前面所说的纤程句柄。

和线程存在线程局部存储一样,纤程也可以有自己的局部存储——纤程局部存储,获取和设置纤程局部存储数据时,会用到 API 函数:

```
DWORD WINAPI FlsAlloc(PFLS_CALLBACK_FUNCTION lpCallback);
BOOL WINAPI FlsFree(DWORD dwFlsIndex);

BOOL WINAPI FlsSetValue(DWORD dwFlsIndex, PVOID lpFlsData);
PVOID WINAPI FlsGetValue(DWORD dwFlsIndex);
```

这 4 个函数和在 3.10.1 节介绍的相应的 4 个函数的用法一样,这里不再赘述。

Windows 还提供了一个获取当前执行纤程的 API 函数,返回值也是纤程句柄:

```
PVOID GetCurrentFiber();
```

来看一个具体的例子:

```
#include <Windows.h>
#include <string>

char g_szTime[64] = { "time not set..." };
LPVOID mainWorkerFiber = NULL;
LPVOID pWorkerFiber = NULL;

void WINAPI workerFiberProc(LPVOID lpFiberParameter)
{
    while (true)
    {
        //假设这是一项很耗时的操作
        SYSTEMTIME st;
        GetLocalTime(&st);
        wsprintfA(g_szTime, "%04d-%02d-%02d %02d:%02d:%02d", st.wYear, st.wMonth, st.wDay, st.wHour, st.wMinute, st.wSecond);
        printf("%s\n", g_szTime);

        //切换回主纤程
        //SwitchToFiber(mainWorkerFiber);
    }
}
```

```
int main()
{
    mainWorkerFiber = ConvertThreadToFiber(NULL);

    int index = 0;
    while (index < 100)
    {
        ++index;

        pWorkerFiber = CreateFiber(0, workerFiberProc, NULL);
        if (pWorkerFiber == NULL)
            return -1;
        //切换至新的纤程
        **SwitchToFiber(pWorkerFiber);**

        **memset(g_szTime, 0, sizeof(g_szTime));**
        strncpy(g_szTime, "time not set...", strlen("time not set..."));

        printf("%s\n", g_szTime);

        Sleep(1000);
    }

    DeleteFiber(pWorkerFiber);

    //切换回线程模式
    ConvertFiberToThread();

    return 0;
}
```

以上代码只有一个主线程，主线程在第 1 个加粗行切换至新建的纤程 pWorkerFiber。由于在新建的纤程函数中是一个 while 无限循环，所以 main 函数中的第 2 个加粗行及后续的代码行永远不会执行。输出结果如下图所示。

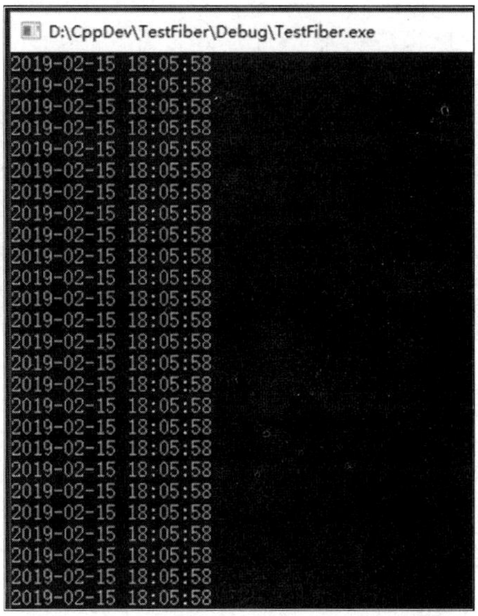

我们将纤程函数 workerFiberProc 加斜体行的注释放开，这样 main 函数的第 2 个加粗行就有机会执行了，输出结果如下图所示。

以上代码的跳跃步骤是先从 main 函数的第 1 个加粗行跳到 workerFiberProc 函数进行执行，在 workerFiberProc 函数中的加斜体行跳回 main 函数的第 2 个加粗行执行，接着周而复始地进行下一轮循环，直到 main 函数的 while 条件不再满足，退出程序。

> 纤程从本质上来说是协程（coroutine），Windows 的纤程技术让单个线程能按用户的意愿像线程一样自由切换，且没有线程切换那样的开销和不可控性。

Windows 最早引入纤程是为了方便地将 UNIX 单线程程序迁移到 Windows 上。有人提出，调试时，若程序的执行点在纤程函数内部，则调用堆栈对用户是割裂的，这对于习惯看连续性上下面堆栈的用户来说可能不太友好，如下图所示。

3.13.2 协程

线程是操作系统的内核对象，多线程编程时线程数过多，会导致上下面频繁切换，这对性能是一种额外的损耗。例如，在一些高并发的网络服务器编程中，使用一个线程服务一个 socket 连接是很不明智的做法，因此现在的主流做法是利用操作系统提供的基于事件模式的异步编程模型，用少量的线程来服务大量的网络连接和 I/O。但是采用异步和基于事件的编程模型,让程序代码变得复杂,非常容易出错,也增加了排查错误的难度。

协程,可被认为是在应用层模拟的线程。协程避免了线程上下面切换的部分额外损耗,同时具有并发运行的优点,降低了编写并发程序的复杂度。还是以上面的高并发网络服务器为例,可以为每个 socket 连接都使用一个协程来处理,在兼顾性能的情况下代码也清晰。

协程的概念早于线程提出,但它是非抢占式的调度,无法实现公平的任务调用,也无法直接利用多核 CPU 的优势,因此我们不能武断地说协程比线程更高级。目前主流的操作系统原生 API 并不支持协程技术,新兴的一些高级编程语言如 Golang 都是在语言的运行时环境中利用线程技术模拟了一套协程。

在这些语言中,协程的内部实现都基于线程,思路是维护一组数据结构和 n 个线程,真正的执行者还是线程,协程执行的代码被扔进一个待执行队列中,由这 n 个线程从队列中拉出来执行,这就解决了协程的执行问题。那么协程是怎么切换的呢?以 Golang 为例,Golang 对操作系统的各种 I/O 函数(如 Linux 的 epoll、select,Windows 的 IOCP 等)进行了封装,这些封装后的函数被提供给应用程序使用,其内部调用了操作系统的异步 I/O 函数,当这些异步函数返回 busy 或 blocking 时,Golang 会利用这个时机将现有的执行序列压栈,让线程拉取另一个协程的代码来执行。由于 Golang 从编译器和语言基础库等多个层面对协程做了实现,所以 Golang 的协程是目前各类存在协程概念的语言中实现最完整和成熟的,10 万个协程同时运行也毫无压力。其优势就是,程序员在编写 Golang 代码时,可以更多地关注业务逻辑的实现,尽量避免在这些基础构件上耗费太多精力。

协程技术之所以这么流行,是因为大多数业务系统都倾向于使用异步编程来提高系统的性能,这就强行地将线性的程序逻辑打乱,使程序逻辑变得非常复杂,对程序状态的管理也变得困难,例如 Node.js 的层层 Callback。而 Golang 作为名门之后开始进入广大开发者的视野,并且迅速在 Web 后端应用。例如以 Docker 及围绕 Docker 展开的整个容器生态圈为代表,其最大的卖点就是协程技术,至此协程技术开始真正流行。

腾讯开源了一套 C/C++版本的协程库 libco,有兴趣的读者可以研究一下其实现原理。

所以,协程技术从来不是什么新事物,只是人们为了从重复、复杂的底层技术中解脱出来,快速、专注地进行业务开发所带来的产物。万变不离其宗,只要我们掌握了多线程编程技术的核心原理,就可以快速地学会协程技术。

第 4 章
网络编程重难点解析

在了解网络通信的基本原理后，我们需要实际编写一些网络通信程序。目前主要的网络通信技术都是基于 TCP/IP 协议栈的，对应到应用层编码来说，就是使用操作系统提供的 socket API 来编写网络通信程序。然而遗憾的是，因为存在各种网络库和开发 IDE，很多开发者或者网络编程的初学者很容易忽视对这些基础的 socket API 的掌握。殊不知，在操作系统层面提供的 API 会在相当长的时间内保持接口不变，一旦学成，终生受用。理解和掌握这些基础的 socket API，不仅很容易理解常见的开源网络通信库的实现原理，而且可以按需定制适合自己业务的网络通信框架。最重要的是，理解和掌握这些基础的 socket API，能帮助我们在排查各种网络疑难杂症时游刃有余。

4.1 学习网络编程时应该掌握的 socket 函数

Windows 和 Linux 上常用的 socket API 函数并不多，除了特定操作系统提供的一些基于自身系统特性的 API，大多数 socket API 都源于 BSD Socket（Berkeley Sockets，伯克利套接字），因此这些 socket 函数在不同的平台上都有相似的签名和参数。

> 一个 socket 句柄，在 Windows 上叫作"socket"，在 Linux 上叫作"file descriptor"（即 fd），它们的含义是一样的，本书不刻意区分这两种叫法。
>
> 监听 socket（或叫监听 fd）即 listen socket（或 listenfd）。为了行文方便，本书将网络通信中服务端一方创建出来的、绑定 IP 地址和端口、调用 listen 函数启动监听功能的 socket，叫作"监听 socket"。

这里有一个简单的函数列表，其中给出了我们应该熟练掌握的 socket 函数。

函数名称	简单说明
socket	创造某种类型的套接字
bind	将一个 socket 绑定到一个 IP 与端口的二元组上
listen	将一个 socket 变为监听状态

续表

函数名称	简单说明
connect	试图建立一个 TCP 连接，一般用于客户端
accept	尝试接收一个连接，一般用于服务端
send	通过一个 socket 发送数据
recv	通过一个 socket 收取数据
select	判断一组 socket 上的读写和异常事件
gethostbyname	通过域名获取机器地址
close	关闭一个套接字，回收该 socket 对应的资源。在 Windows 中对应的是 closesocket
shutdown	关闭 socket 收发通道
setsockopt	设置一个套接字选项
getsockopt	获取一个套接字选项

4.1.1　在 Linux 上查看 socket 函数的帮助信息

在 Linux 上，我们可以通过 man 手册查看相应的函数签名和用法。举个例子，如果要查看 connect 函数的用法，则只需在 Linux Shell 终端输入 man connect 即可：

```
[root@localhost ~]# man connect
CONNECT(2)         Linux Programmer's Manual         CONNECT(2)

NAME
     connect - initiate a connection on a socket

SYNOPSIS
     #include <sys/types.h>          /* See NOTES */
     #include <sys/socket.h>

     int connect(int sockfd, const struct sockaddr *addr,
            socklen_t addrlen);

DESCRIPTION
The connect() system call connects the socket referred to by the file descriptor sockfd
to the address specified by addr.  The addrlen argument specifies the size of addr.
The format of the address in addr is determined by the address space of the socket
sockfd; see socket(2) for further details.

...省略部分内容...
RETURN VALUE
If the connection or binding succeeds, zero is returned.  On error, -1 is returned,
and errno is set appropriately.

ERRORS
The following are general socket errors only.  There may be other domain-specific
error codes.

...省略部分内容...
```

如上面的代码片段所示，man 手册对一个函数的说明一般包括如下几部分。

（1）函数声明及相关数据结构所在的头文件。在实际编码时如果需要使用这个函数，则必须包含该头文件。

（2）函数签名，即该函数的参数类型、个数和返回值。

（3）函数用法说明，可能包括一些注意事项。

（4）函数返回值说明。

（5）调用函数出错时可能得到的错误码值。

（6）一些相关函数在 man 手册中的位置索引（connect 没有该部分）。

需要注意的是，通过该方法不仅可以查看 socket 函数，也可以查看 Linux 下的其他通用函数（如 fread），甚至一个 Shell 命令（如 sleep）。以 sleep 为例，我们想查看程序中 sleep 函数的用法时，由于 Linux 内置了一个叫作 sleep 的 Shell 命令，所以若在 Shell 窗口直接输入 man sleep，则默认显示 sleep 命令而不是 sleep 函数的帮助信息。

我们可以通过 man man 命令查看 man 手册的组成部分：

```
[root@localhost ~]# man man
## 无关部分省略
The table below shows the section numbers of the manual followed by the types of
pages they contain.

1   Executable programs or shell commands
2   System calls (functions provided by the kernel)
3   Library calls (functions within program libraries)
4   Special files (usually found in /dev)
5   File formats and conventions eg /etc/passwd
6   Games
7   Miscellaneous (including macro packages and conventions), e.g. man(7), groff(7)
8   System administration commands (usually only for root)
9   Kernel routines [Non standard]

A manual page consists of several sections.
```

通过上面的输出，可以看到 man 手册由 9 部分组成，sleep 函数属于 3 Library calls（系统库函数），所以我们输入 man 3 sleep 就可以查看 sleep 函数（而不是 sleep 这个 Linux Shell 命令）的帮助信息了：

```
[root@localhost ~]# man 3 sleep
SLEEP(3)            Linux Programmer's Manual            SLEEP(3)

NAME
       sleep - sleep for the specified number of seconds

SYNOPSIS
       #include <unistd.h>

       unsigned int sleep(unsigned int seconds);

DESCRIPTION
       sleep() makes the calling thread sleep until seconds seconds have elapsed or
a signal arrives which is not ignored.
```

```
RETURN VALUE
    Zero if the requested time has elapsed, or the number of seconds left to sleep,
if the call was interrupted by a signal handler.
```

...省略更多信息...

4.1.2　在 Windows 上查看 socket 函数的帮助信息

Windows 也有类似 man 手册的帮助文档，早些年，Visual Studio 会自带一套离线的 MSDN 文档库，其优点是不需要计算机联网，缺点是占用磁盘空间较大、内容陈旧。在手机网络非常普及的今天，建议使用在线版本的 MSDN。

打开在线 MSDN 网站后，建议在页面底部将页面语言设置为 English，这样搜索出来的内容会更准确、丰富，如下图所示。

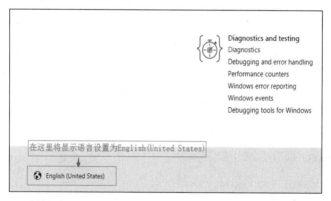

还是以 connect 函数为例，在以上页面的搜索框中输入 socket connect，然后按回车键，会得到一组搜索结果，选择我们需要的页面即可。与 man 手册相比，MSDN 关于 connect 函数的说明就比较详细了，大体分为以下几部分。

◎ Syntax：函数签名，即函数的参数类型、个数和返回值。
◎ Parameters：参数的用法详细说明。
◎ Return Value：函数的返回值说明，在返回值部分还有函数调用失败时的详细错误码说明。
◎ Remarks：函数的详细用法说明，某些函数还会给出示例代码。
◎ Requirements：使用该函数时对操作系统的版本要求、代码需要引入的头文件和库文件信息（如果有的话）。
◎ See Also：一些相关函数和知识点的链接信息。

需要注意的是，在 MSDN 上阅读相关 API 的帮助信息时，我们要辩证地对待其提供的信息，因为很多函数的实际工作原理和行为并不一定如 MSDN 介绍的那样。所以在某些 API 帮助信息下会有一些读者的评论信息，这些评论信息或对文档内容做一些补充、纠错，或给出一些代码示例。建议读者实际查阅时留意这部分信息，或许能得到一些很有用的帮助。

本书不会逐一介绍每个 socket 函数的基本用法，而是重点介绍关于这些函数的一些重要使用事项和原理，接下来一起学习吧！

4.2 TCP 网络通信的基本流程

不管是多么复杂的服务器或客户端程序，其网络通信的基本原理一定如下所述。

对于服务器，其通信流程一般如下所述。

（1）调用 socket 函数创建 socket（监听 socket）。

（2）调用 bind 函数将 socket 绑定到某个 IP 和端口的二元组上。

（3）调用 listen 函数开启监听。

（4）当有客户端请求连接上来时，调用 accept 函数接收连接，产生一个新的 socket（客户端 socket）。

（5）基于新产生的 socket 调用 send 或 recv 函数，开始与客户端进行数据交流。

（6）通信结束后，调用 close 函数关闭监听 socket。

对于客户端，其通信流程一般如下所述。

（1）调用 socket 函数创建客户端 socket。

（2）调用 connect 函数尝试连接服务器。

（3）连接成功后调用 send 或 recv 函数，开始与服务器进行数据交流。

（4）通信结束后，调用 close 函数关闭监听 socket。

上述流程可被绘制成下图所示。

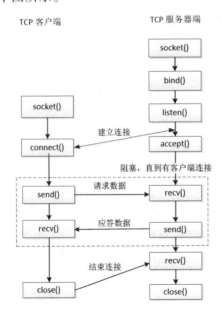

读者可能有疑问,为什么客户端调用 close() 会和服务端的 recv 函数有关?这涉及 recv 函数的返回值的意义,在下文中会有详细讲解。

服务端的实现代码如下:

```cpp
/**
 * TCP 服务器通信的基本流程
 * zhangyl 2018.12.13
 */
#include <sys/types.h>
#include <sys/socket.h>
#include <arpa/inet.h>
#include <unistd.h>
#include <iostream>
#include <string.h>
int main(int argc, char* argv[])
{
    //1.创建一个监听 socket
    int listenfd = socket(AF_INET, SOCK_STREAM, 0);
    if (listenfd == -1)
    {
        std::cout << "create listen socket error." << std::endl;
        return -1;
    }

    //2.初始化服务器地址
    struct sockaddr_in bindaddr;
    bindaddr.sin_family = AF_INET;
    bindaddr.sin_addr.s_addr = htonl(INADDR_ANY);
    bindaddr.sin_port = htons(3000);
    if (bind(listenfd, (struct sockaddr *)&bindaddr, sizeof(bindaddr)) == -1)
    {
        std::cout << "bind listen socket error." << std::endl;
        return -1;
    }

    //3.启动监听
    if (listen(listenfd, SOMAXCONN) == -1)
    {
        std::cout << "listen error." << std::endl;
        return -1;
    }

    while (true)
    {
        struct sockaddr_in clientaddr;
        socklen_t clientaddrlen = sizeof(clientaddr);
        //4.接受客户端连接
        int clientfd = accept(listenfd, (struct sockaddr *)&clientaddr, &clientaddrlen);
        if (clientfd != -1)
        {
            char recvBuf[32] = {0};
            //5.从客户端接收数据
```

```cpp
            int ret = recv(clientfd, recvBuf, 32, 0);
            if (ret > 0)
            {
                std::cout << "recv data from client, data: " << recvBuf << std::endl;
                //6.将收到的数据原封不动地发给客户端
                ret = send(clientfd, recvBuf, strlen(recvBuf), 0);
                if (ret != strlen(recvBuf))
                    std::cout << "send data error." << std::endl;
                else
                    std::cout << "send data to client successfully, data: " << recvBuf << std::endl;
            }
            else
            {
                std::cout << "recv data error." << std::endl;
            }

            close(clientfd);
        }
    }

    //7.关闭监听 socket
    close(listenfd);

    return 0;
}
```

客户端的实现代码如下：

```cpp
/**
 * TCP 客户端通信的基本流程
 * zhangyl 2018.12.13
 */
#include <sys/types.h>
#include <sys/socket.h>
#include <arpa/inet.h>
#include <unistd.h>
#include <iostream>
#include <string.h>

#define SERVER_ADDRESS  "127.0.0.1"
#define SERVER_PORT     3000
#define SEND_DATA       "helloworld"

int main(int argc, char* argv[])
{
    //1.创建一个 socket
    int clientfd = socket(AF_INET, SOCK_STREAM, 0);
    if (clientfd == -1)
    {
        std::cout << "create client socket error." << std::endl;
        return -1;
    }

    //2.连接服务器
```

```cpp
    struct sockaddr_in serveraddr;
    serveraddr.sin_family = AF_INET;
    serveraddr.sin_addr.s_addr = inet_addr(SERVER_ADDRESS);
    serveraddr.sin_port = htons(SERVER_PORT);
    if (connect(clientfd, (struct sockaddr *)&serveraddr, sizeof(serveraddr)) == -1)
    {
        std::cout << "connect socket error." << std::endl;
        return -1;
    }

    //3.向服务器发送数据
    int ret = send(clientfd, SEND_DATA, strlen(SEND_DATA), 0);
    if (ret != strlen(SEND_DATA))
    {
        std::cout << "send data error." << std::endl;
        return -1;
    }

    std::cout << "send data successfully, data: " << SEND_DATA << std::endl;

    //4.从服务器收取数据
    char recvBuf[32] = {0};
    ret = recv(clientfd, recvBuf, 32, 0);
    if (ret > 0)
    {
        std::cout << "recv data successfully, data: " << recvBuf << std::endl;
    }
    else
    {
        std::cout << "recv data error, data: " << recvBuf << std::endl;
    }

    //5. 关闭 socket
    close(clientfd);

    return 0;
}
```

通过以上代码，服务端在地址 0.0.0.0:3000 启动了一个监听，客户端连接服务器成功后，向服务器发送字符串 "helloworld"；服务器收到后，将收到的字符串原封不动地发给了客户端。

在 Linux Shell 界面输入以下命令编译服务端和客户端：

```
# 编译 server.cpp，生成可执行文件 server
[root@localhost testsocket]# g++ -g -o server server.cpp
# 编译 client.cpp，生成可执行文件 client
[root@localhost testsocket]# g++ -g -o client client.cpp
```

接下来看看执行效果，先启动服务器程序：

```
[root@localhost testsocket]# ./server
```

再启动客户端程序：

```
[root@localhost testsocket]# ./client
```

这时客户端输出：

```
send data successfully, data: helloworld
recv data successfully, data: helloworld
```

服务端输出：

```
recv data from client, data: helloworld
send data to client successfully, data: helloworld
```

以上就是 TCPsocket 网络通信的基本原理了。对于很多读者来说，以上两个代码片段可能很简单，但是深刻理解它们是进一步学习和开发复杂的网络通信程序的基础。看似简单的代码，隐藏了很多原理，接下来的章节将以这两段代码为蓝本，逐渐深入。

4.3 设计跨平台网络通信库时的一些 socket 函数用法

这里说的跨平台，指的是在 Windows 和 Linux 上使用相关 socket 函数，虽然二者的设计均参考了 Berkeley Sockets，但是在演化过程中仍然存在不少区别，本节还整理了一些常见的注意事项。

4.3.1 socket 数据类型

在 Windows 上，一个 socket 对象的类型是 SOCKET，它是一个句柄对象（本质上也是 int 类型）。在 Linux 上，一个 socket 对象的类型是 int。我们习惯将 socket 对象在 Windows 上称为 socket，在 Linux 上称为 fd。所以在很多网络库中使用了如下定义来包裹一个跨平台使用的 socket 类型：

```
#ifdef WIN32
typedef SOCKET    SOCKET_TYPE;
#else
typedef int       SOCKET_TYPE;
#endif
```

这样就可以同时在 Windows 和 Linux 上使用 SOCKET_TYPE 类型代表 socket 了。

无论是 Windows 还是 Linux，创建一个套接字的函数 socket 调用失败时均会返回-1。Windows 为这种情形定义了一个 INVALID_HANDLE_VALUE 宏：

```
#define INVALID_HANDLE_VALUE (-1)
```

在 Linux 上不存在 INVALID_HANDLE_VALUE 宏，我们可以通过上述语句定义它。

4.3.2 在 Windows 上调用 socket 函数

Linux 程序可以直接使用 socket 函数，但是对于 Windows 平台，必须先调用 WSAStartup 函数显式地将与 socket 函数相关的 dll 文件加载到进程地址空间中，在程序退出时需要调用 WSACleanup 函数卸载相关的 dll 文件。

这两个函数的用法示例如下：

```cpp
//在程序初始化时调用
bool InitSocket()
{
    //指定版本号
    WORD wVersionRequested = MAKEWORD(2, 2);
    WSADATA wsaData;
    int nErrorID = ::WSAStartup(wVersionRequested, &wsaData);
    if(nErrorID != 0)
        return false;

    if (LOBYTE(wsaData.wVersion) != 2 || HIBYTE(wsaData.wVersion) != 2)
    {
        UninitSocket();
        return false;
    }

    return true;
}
//在程序退出时调用
void UninitSocket()
{
    ::WSACleanup();
}
```

需要注意的是，WSAStartup 函数和 WSACleanup 函数是进程相关的，任何一个线程都可以调用。对于 WSAStartup 函数，某个线程调用一次之后，其他线程可以正常使用；反过来，如果某个线程不再使用相关的 socket 函数，则其调用 WSACleanup 函数时，会导致其他线程无法继续使用相关的 socket 函数。因此，我们在调用 WSACleanup 函数之前应该确保整个程序不再有使用 socket 函数的地方。鉴于此，一般在进程退出时才调用该函数。

4.3.3 关闭 socket 函数

出于历史原因，Windows 从来没有以良好兼容的方式实现 Berkeley 套接字 API。关闭套接字的接口时，在 Linux(Unix)中使用 close 函数，在 Windows 中使用 closesocket 函数，问题是 Windows 也定义了一个 close 函数，该函数还不能用于关闭 socket。如果调用该函数关闭 socket，则会导致程序崩溃。这是让很多开发者犯错的地方。可以这样进行包装：

```cpp
#ifndef WIN32
#define closesocket(s) close(s)
#endif
```

这样在 Linux 上就可以使用 closesocket 关闭 socket 了。

4.3.4 获取 socket 函数的错误码

若某个 socket 函数调用失败时，在 Windows 上则需要调用 WSAGetLastError()这个

API 获取错误码，在 Linux 上则直接使用 errno 变量获取错误码。所以我们可能会在一些网络库中看到如下代码：

```
#ifdef WIN32
#define GetSocketError() WSAGetLastError()
#else
#define GetSocketError() errno
#endif
```

这样就可以使用 GetSocketError 函数统一进行处理了。

看看 libevent 网络库在这一块的相关实现：

```
//在 Windows 上获取一个 socket 错误码的实现
#ifdef _WIN32
int evutil_socket_geterror(evutil_socket_t sock)
{
    int optval, optvallen=sizeof(optval);
    int err = WSAGetLastError();
    if (err == WSAEWOULDBLOCK && sock >= 0) {
        //不仅使用了WSAGetLastError，还通过 getsocketopt 选项来获取错误
        if (getsockopt(sock, SOL_SOCKET, SO_ERROR, (void*)&optval,
                    &optvallen))
            return err;
        if (optval)
            return optval;
    }
    return err;
}
#endif
//在 Linux 上设置错误码
#define EVUTIL_SET_SOCKET_ERROR(errcode)            \
        do { errno = (errcode); } while (0)
//在 Windows 上把错误码转换为错误描述信息的函数实现
const char *evutil_socket_error_to_string(int errcode)
{
    struct cached_sock_errs_entry *errs, *newerr, find;
    char *msg = NULL;

    EVLOCK_LOCK(windows_socket_errors_lock_, 0);

    find.code = errcode;
    errs = HT_FIND(cached_sock_errs_map, &windows_socket_errors, &find);
    if (errs) {
        msg = errs->msg;
        goto done;
    }

    if (0 != FormatMessageA(FORMAT_MESSAGE_FROM_SYSTEM |
                FORMAT_MESSAGE_IGNORE_INSERTS |
                FORMAT_MESSAGE_ALLOCATE_BUFFER,
                NULL, errcode, 0, (char *)&msg, 0, NULL))
        chomp (msg);
    else {
        size_t len = 50;
        msg = LocalAlloc(LMEM_FIXED, len);
```

```
    if (!msg) {
        msg = (char *)"LocalAlloc failed during Winsock error";
        goto done;
    }
    evutil_snprintf(msg, len, "winsock error 0x%08x", errcode);
}

newerr = (struct cached_sock_errs_entry *)
    mm_malloc(sizeof (struct cached_sock_errs_entry));

if (!newerr) {
    LocalFree(msg);
    msg = (char *)"malloc failed during Winsock error";
    goto done;
}

newerr->code = errcode;
newerr->msg = msg;
HT_INSERT(cached_sock_errs_map, &windows_socket_errors, newerr);

done:
    EVLOCK_UNLOCK(windows_socket_errors_lock_, 0);

    return msg;
}
//在 Linux 上就比较简单了，直接调用 strerror 函数传入错误码即可
#define evutil_socket_error_to_string(errcode)(strerror(errcode))
```

libevent 提供了这些宏访问和操作 socket 的错误代码，evutil_socket_geterror()则返回某特定套接字的错误号。在类 UNIX 系统中，EVUTIL_SET_SOCKET_ERROR()用于修改当前套接字的错误号，evutil_socket_error_to_string()用于返回代表某给定套接字错误号的字符串。

4.3.5 套接字函数的返回值

无论是 Windows 还是 Linux，大多数 socket 函数在调用失败后都会返回-1，在 Windows 上为这种情形专门定义了一个宏 SOCKET_ERROR：

```
#define SOCKET_ERROR (-1)
```

可以在 Linux 上也定义这样一个宏，这样就方便统一书写了：

```
#ifndef WIN32
#define SOCKET_ERROR (-1)
#endif
```

4.3.6 select 函数第 1 个参数的问题

select 函数的原型如下：

```
int select(int nfds,
       fd_set *readfds,
       fd_set *writefds,
       fd_set *exceptfds,
```

```
            struct timeval *timeout);
```

使用示例如下：

```
fd_set writeset;
FD_ZERO(&writeset);
FD_SET(m_hSocket, &writeset);
struct timeval tv;
tv.tv_sec = 3;
tv.tv_usec = 100;
select(m_hSocket + 1, NULL, &writeset, NULL, &tv);
```

无论是 Windows 还是 Linux，select 函数的参数 readfds、writefds 和 exceptfds 都是一个包含一组 socket 描述符数组的结构体。在 Linux 上，第 1 个参数 nfds 必须被设置为 readfds、writefds 或 exceptfds，所有 socket 描述符句柄中的最大值都加 1。但是在 Windows 上，select 函数不使用第 1 个参数 nfds，可以随意设置，这个参数用于保持与 Berkeley 套接字兼容。一般为了兼容，在 Windows 上也会将这个参数的值设置为这三个 fd_set 集合中的最大套接字值加 1。

4.3.7 错误码 WSAEWOULDBLOCK 和 EWOULDBLOCK

在某些套接字的函数操作不能立即完成时，在 Windows 上会返回错误码 WSAEWOULDBLOCK，该错误码在 Linux 上对应错误码 EWOULDBLOCK（在 Linux 还存在一个错误码 EAGAIN，与此同义）。

为了统一写法，我们可以在 Windows 上使用如下代码来兼容：

```
#ifdef WIN32
#define EWOULDBLOCK WSAEWOULDBLOCK
#endif
```

对于 I/O 复用技术，Windows 和 Linux 都支持：select 模型；Linux 特有的 poll、epoll 模型；Windows 特有的 WSAPoll 和完成端口模型（IOCP）。

此外，在 Windows 和 Linux 上有许多自己特有的 API 和网络通信模型，Windows 自我扩展的套接字函数一般以 WSA（Windows Socket API）开头，例如 WSASend、WSARecv 等。Windows 提供了方便使用的 WSAEventSelect 和 WSAAsyncSelect 等网络通信模型，Linux 则提供了 accept4、socketpair 等方便使用的 API。总之，二者的发展相互借鉴，无所谓孰优孰劣。

4.4　bind 函数重难点分析

本节对 bind 函数的重难点进行分析。

4.4.1　对 bind 函数如何选择绑定地址

4.2 节的服务器代码演示了 bind 函数的用法，让我们再看一下相关代码：

```
struct sockaddr_in bindaddr;
bindaddr.sin_family = AF_INET;
bindaddr.sin_addr.s_addr = htonl(INADDR_ANY);
bindaddr.sin_port = htons(3000);
if (bind(listenfd, (struct sockaddr *)&bindaddr, sizeof(bindaddr)) == -1)
{
    std::cout << "bind listen socket error." << std::endl;
    return -1;
}
```

其中 bind 的地址使用了一个宏，叫 INADDR_ANY，关于这个宏的解释为：如果应用程序不关心 bind 绑定的 IP，则可以使用 INADDR_ANY（如果是 IPv6，则对应 in6addr_any），底层的（协议栈）服务会自动选择一个合适的 IP 地址，这样在多个网卡机器上选择 IP 地址会变得简单。

也就是说，INADDR_ANY 相当于地址 0.0.0.0。再详细解释一下，假设我们在一台机器上开发一个服务器程序，则使用 bind 函数时，我们有多个 IP 地址可以选择。首先，这台机器的外网 IP 地址是 120.55.94.78，在当前局域网中的地址是 192.168.1.104。同时，这台机器有本地回环地址 127.0.0.1。

如果只想在本机上进行访问，那么对 bind 函数中的地址可以使用 127.0.0.1；如果服务只想被局域网中的内部机器访问，那么 bind 函数中的地址可以使用 192.168.1.104 这样的局域网地址；如果希望这个服务可以被公网访问，那么可以使用地址 0.0.0.0 或 INADDR_ANY。

4.4.2 bind 函数的端口号问题

网络通信程序的基本逻辑是客户端连接服务器，即从客户端的地址:端口连接到服务器的地址:端口上。以 4.2 节中的示例程序为例，服务端的端口号使用 3000，那客户端连接时的端口号是多少呢？在 TCP 通信双方中，一般服务端的端口号是固定的，而客户端的端口号是连接发起时由操作系统随机分配的（不会分配已被占用的端口）。端口号是一个 C short 类型的值，其范围是 0～65535，知道这点很重要，所以我们在编写压力测试程序时，由于端口数量的限制，在某台机器上网卡地址不变的情况下，压力测试程序在理论上最多只能发起 65 000 多个连接。在实际情况下，由于当时操作系统中的很多端口可能已被占用，所以实际可以使用的端口比这个更少，例如，一般规定端口号为 1024 以下的端口是保留端口，不建议用户程序使用。而对于 Windows 系统，MSDN 甚至明确地说明：Vista 及以后的 Windows，可用的动态端口范围是 49152～65535；而 Windows Server 及更早的系统，可用的动态端口范围是 1025～5000（我们可以通过修改注册表来改变这一设置）。

如果将 bind 函数中的端口号设置为 0，那么操作系统会随机为程序分配一个可用的监听端口。当然，服务器程序一般不会这么做，因为服务器程序是要对外服务的，必须让客户端知道确切的 IP 地址和端口号。

很多人觉得只有服务器程序可以调用 bind 函数绑定一个端口号，其实不然：在一些特殊的应用中，我们需要客户端程序以指定的端口号连接服务器，此时我们就可以在客户端程序中调用 bind 函数绑定一个具体的端口了。

我们用代码来验证以上内容。为了能看到连接状态，我们将客户端和服务器关闭 socket 的代码注释掉，这样连接会保持一段时间。

1. 客户端代码不绑定端口

修改后的服务器代码如下：

```cpp
/**
 * TCP 服务器通信的基本流程
 * zhangyl 2018.12.13
 */
#include <sys/types.h>
#include <sys/socket.h>
#include <arpa/inet.h>
#include <unistd.h>
#include <iostream>
#include <string.h>
#include <vector>

int main(int argc, char* argv[])
{
    //1.创建一个监听 socket
    int listenfd = socket(AF_INET, SOCK_STREAM, 0);
    if (listenfd == -1)
    {
        std::cout << "create listen socket error." << std::endl;
        return -1;
    }

    //2.初始化服务器地址
    struct sockaddr_in bindaddr;
    bindaddr.sin_family = AF_INET;
    bindaddr.sin_addr.s_addr = htonl(INADDR_ANY);
    bindaddr.sin_port = htons(3000);
    if (bind(listenfd, (struct sockaddr *)&bindaddr, sizeof(bindaddr)) == -1)
    {
        std::cout << "bind listen socket error." << std::endl;
        return -1;
    }

    //3.启动监听
    if (listen(listenfd, SOMAXCONN) == -1)
    {
        std::cout << "listen error." << std::endl;
        return -1;
    }

    //记录所有客户端连接的容器
    std::vector<int> clientfds;
    while (true)
    {
```

```cpp
    struct sockaddr_in clientaddr;
    socklen_t clientaddrlen = sizeof(clientaddr);
    //4.接受客户端连接
    int clientfd = accept(listenfd, (struct sockaddr *)&clientaddr, &clientaddrlen);
    if (clientfd != -1)
    {
        char recvBuf[32] = {0};
        //5.从客户端收取数据
        int ret = recv(clientfd, recvBuf, 32, 0);
        if (ret > 0)
        {
            std::cout << "recv data from client, data: " << recvBuf << std::endl;
            //6.将收到的数据原封不动地发送给客户端
            ret = send(clientfd, recvBuf, strlen(recvBuf), 0);
            if (ret != strlen(recvBuf))
                std::cout << "send data error." << std::endl;

            std::cout << "send data to client successfully, data: " << recvBuf << std::endl;
        }
        else
        {
            std::cout << "recv data error." << std::endl;
        }

        //close(clientfd);
        clientfds.push_back(clientfd);
    }
}

//7.关闭监听 socket
close(listenfd);

return 0;
}
```

修改后的客户端代码如下:

```cpp
/**
 * TCP 客户端通信的基本流程
 * zhangyl 2018.12.13
 */
#include <sys/types.h>
#include <sys/socket.h>
#include <arpa/inet.h>
#include <unistd.h>
#include <iostream>
#include <string.h>

#define SERVER_ADDRESS  "127.0.0.1"
#define SERVER_PORT     3000
#define SEND_DATA       "helloworld"

int main(int argc, char* argv[])
{
```

```cpp
//1.创建一个socket
int clientfd = socket(AF_INET, SOCK_STREAM, 0);
if (clientfd == -1)
{
    std::cout << "create client socket error." << std::endl;
    return -1;
}

//2.连接服务器
struct sockaddr_in serveraddr;
serveraddr.sin_family = AF_INET;
serveraddr.sin_addr.s_addr = inet_addr(SERVER_ADDRESS);
serveraddr.sin_port = htons(SERVER_PORT);
if (connect(clientfd, (struct sockaddr *)&serveraddr, sizeof(serveraddr)) == -1)
{
    std::cout << "connect socket error." << std::endl;
    return -1;
}

//3.向服务器发送数据
int ret = send(clientfd, SEND_DATA, strlen(SEND_DATA), 0);
if (ret != strlen(SEND_DATA))
{
    std::cout << "send data error." << std::endl;
    return -1;
}

std::cout << "send data successfully, data: " << SEND_DATA << std::endl;

//4.从客户端收取数据
char recvBuf[32] = {0};
ret = recv(clientfd, recvBuf, 32, 0);
if (ret > 0)
{
    std::cout << "recv data successfully, data: " << recvBuf << std::endl;
}
else
{
    std::cout << "recv data error, data: " << recvBuf << std::endl;
}

//5.关闭socket
//close(clientfd);
//这里仅仅是为了让客户端程序不退出
while (true)
{
    sleep(3);
}

return 0;
}
```

将程序编译好后（编译方法和上面一样），我们先启动 server，再启动 3 个客户端。然后通过 lsof 命令查看当前机器上的 TCP 连接信息。为了更清楚地显示结果，这里已经将不相关的连接信息去掉了，结果如下：

```
[root@localhost ~]# lsof -i -Pn
COMMAND    PID USER   FD   TYPE DEVICE SIZE/OFF NODE NAME
server    1445 root    3u  IPv4  21568      0t0  TCP *:3000 (LISTEN)
server    1445 root    4u  IPv4  21569      0t0  TCP 127.0.0.1:3000->127.0.0.1:40818
(ESTABLISHED)
server    1445 root    5u  IPv4  21570      0t0  TCP 127.0.0.1:3000->127.0.0.1:40820
(ESTABLISHED)
server    1445 root    6u  IPv4  21038      0t0  TCP 127.0.0.1:3000->127.0.0.1:40822
(ESTABLISHED)
client    1447 root    3u  IPv4  21037      0t0  TCP 127.0.0.1:40818->127.0.0.1:3000
(ESTABLISHED)
client    1448 root    3u  IPv4  21571      0t0  TCP 127.0.0.1:40820->127.0.0.1:3000
(ESTABLISHED)
client    1449 root    3u  IPv4  21572      0t0  TCP 127.0.0.1:40822->127.0.0.1:3000
(ESTABLISHED)
```

上面的结果显示，server 进程（进程 ID 是 1445）在 3000 端口开启了监听，有 3 个 client 进程（进程 ID 分别是 1447、1448、1449）分别通过端口号 40818、40820、40822 连到 server 进程上。作为客户端的一方，端口号是系统随机分配的。

2. 客户端绑定 0 号端口

将服务端的代码保持不变，修改客户端的代码如下：

```cpp
/**
 * TCP 客户端通信的基本流程
 * zhangyl 2018.12.13
 */
#include <sys/types.h>
#include <sys/socket.h>
#include <arpa/inet.h>
#include <unistd.h>
#include <iostream>
#include <string.h>

#define SERVER_ADDRESS "127.0.0.1"
#define SERVER_PORT    3000
#define SEND_DATA      "helloworld"

int main(int argc, char* argv[])
{
    //1.创建一个socket
    int clientfd = socket(AF_INET, SOCK_STREAM, 0);
    if (clientfd == -1)
    {
        std::cout << "create client socket error." << std::endl;
        return -1;
    }

    struct sockaddr_in bindaddr;
    bindaddr.sin_family = AF_INET;
    bindaddr.sin_addr.s_addr = htonl(INADDR_ANY);
    //将 socket 绑定到 0 号端口上
```

```cpp
    bindaddr.sin_port = htons(0);
    if (bind(clientfd, (struct sockaddr *)&bindaddr, sizeof(bindaddr)) == -1)
    {
        std::cout << "bind socket error." << std::endl;
        return -1;
    }

    //2.连接服务器
    struct sockaddr_in serveraddr;
    serveraddr.sin_family = AF_INET;
    serveraddr.sin_addr.s_addr = inet_addr(SERVER_ADDRESS);
    serveraddr.sin_port = htons(SERVER_PORT);
    if (connect(clientfd, (struct sockaddr *)&serveraddr, sizeof(serveraddr)) == -1)
    {
        std::cout << "connect socket error." << std::endl;
        return -1;
    }

    //3.向服务器发送数据
    int ret = send(clientfd, SEND_DATA, strlen(SEND_DATA), 0);
    if (ret != strlen(SEND_DATA))
    {
        std::cout << "send data error." << std::endl;
        return -1;
    }

    std::cout << "send data successfully, data: " << SEND_DATA << std::endl;

    //4.从客户端收取数据
    char recvBuf[32] = {0};
    ret = recv(clientfd, recvBuf, 32, 0);
    if (ret > 0)
    {
        std::cout << "recv data successfully, data: " << recvBuf << std::endl;
    }
    else
    {
        std::cout << "recv data error, data: " << recvBuf << std::endl;
    }

    //5.关闭socket
    //close(clientfd);
    //这里仅仅是为了让客户端程序不退出
    while (true)
    {
        sleep(3);
    }

    return 0;
}
```

再次编译客户端程序并启动 3 个 client 进程, 用 lsof 命令查看机器上的 TCP 连接情况, 结果如下:

```
[root@localhost ~]# lsof -i -Pn
COMMAND    PID USER    FD   TYPE DEVICE SIZE/OFF NODE NAME
server    1593 root    3u   IPv4  21807      0t0  TCP *:3000 (LISTEN)
server    1593 root    4u   IPv4  21808      0t0  TCP 127.0.0.1:3000->127.0.0.1:44220
(ESTABLISHED)
server    1593 root    5u   IPv4  19311      0t0  TCP 127.0.0.1:3000->127.0.0.1:38990
(ESTABLISHED)
server    1593 root    6u   IPv4  21234      0t0  TCP 127.0.0.1:3000->127.0.0.1:42365
(ESTABLISHED)
client    1595 root    3u   IPv4  22626      0t0  TCP 127.0.0.1:44220->127.0.0.1:3000
(ESTABLISHED)
client    1611 root    3u   IPv4  21835      0t0  TCP 127.0.0.1:38990->127.0.0.1:3000
(ESTABLISHED)
client    1627 root    3u   IPv4  21239      0t0  TCP 127.0.0.1:42365->127.0.0.1:3000
(ESTABLISHED)
```

通过上面的结果，我们发现 3 个 client 进程使用的端口号仍然是系统随机分配的，也就是说，绑定 0 号端口和不绑定的效果是一样的。

3. 客户端绑定一个固定端口

这里使用了 20000 端口，当然，读者可以根据自己的喜好选择，只要保证所选择的端口号当前没有被其他程序占用即可。服务器代码保持不变，将客户端绑定代码中的端口号从 0 改成 20000。这里为了节省篇幅，只贴出修改处的代码：

```
struct sockaddr_in bindaddr;
bindaddr.sin_family = AF_INET;
bindaddr.sin_addr.s_addr = htonl(INADDR_ANY);
//将 socket 绑定到 20000 端口上
bindaddr.sin_port = htons(20000);
if (bind(clientfd, (struct sockaddr *)&bindaddr, sizeof(bindaddr)) == -1)
{
    std::cout << "bind socket error." << std::endl;
    return -1;
}
```

再次重新编译程序，先启动一个客户端，此时的 TCP 连接状态如下：

```
[root@localhost testsocket]# lsof -i -Pn
COMMAND    PID USER    FD   TYPE DEVICE SIZE/OFF NODE NAME
server    1676 root    3u   IPv4  21933      0t0  TCP *:3000 (LISTEN)
server    1676 root    4u   IPv4  21934      0t0  TCP 127.0.0.1:3000->127.0.0.1:20000
(ESTABLISHED)
client    1678 root    3u   IPv4  21336      0t0  TCP 127.0.0.1:20000->127.0.0.1:3000
(ESTABLISHED)
```

通过上面的结果，我们发现 client 进程确实被使用 20000 端口连接到 server 进程。这时如果再开启一个 client 进程，则可以预想由于 20000 端口已被占用，新启动的 client 会由于调用 bind 函数出错而退出。实际验证一下：

```
[root@localhost testsocket]# ./client
bind socket error.

[root@localhost testsocket]#
```

结果确实和我们预想的一样。

在进行技术面试的时候，面试官有时会问 TCP 网络通信的客户端程序中的 socket 是否可以调用 bind 函数，相信读到这里，聪明的读者已经有答案了。

另外，在 Linux 的 nc 命令中有个-p 选项（字母 p 是小写），这个选项的作用就是使用 nc 命令模拟客户端程序时，可以使用指定的端口号连接到服务器程序上。还是以上面的服务器程序为例，这里不用 client 程序，改用 nc 命令模拟客户端。在 Shell 终端输入：

```
[root@localhost testsocket]# nc -v -p 9999 127.0.0.1 3000
Ncat: Version 6.40 ( http://nmap.org/ncat )
Ncat: Connected to 127.0.0.1:3000.
My name is zhangxf
My name is zhangxf
```

其中：-v 选项表示输出 nc 命令连接的详细信息，这里连接成功后，会输出 "Ncat: Connected to 127.0.0.1:3000"，提示已经连接到服务器的 3000 端口上了；-p 选项的参数值是 9999，表示我们要求 nc 命令在本地以端口号 9999 连接服务器。注意，不要与 3000 端口混淆，3000 是服务器的监听端口号，也就是我们连接的目标端口号；9999 是客户端使用的端口号。

用 lsof 命令来验证我们的 nc 命令是否确实以 9999 端口连接到 server 进程上了：

```
[root@localhost testsocket]# lsof -i -Pn
COMMAND    PID USER   FD   TYPE DEVICE SIZE/OFF NODE NAME
server    1676 root   3u   IPv4  21933      0t0  TCP *:3000 (LISTEN)
server    1676 root   7u   IPv4  22405      0t0  TCP 127.0.0.1:3000->127.0.0.1:9999 (ESTABLISHED)
nc        2005 root   3u   IPv4  22408      0t0  TCP 127.0.0.1:9999->127.0.0.1:3000 (ESTABLISHED)
```

结果确实与我们预期的一致。

4.5 select 函数的用法和原理

select 函数是网络通信编程中很常用的一个函数，我们应该熟练掌握它。虽然它是 BSD 标准之一的 socket 函数，但在 Linux 和 Windows 上，其行为表现还是有点区别的。先来看看 Linux 上的 select 函数。

4.5.1 Linux 上的 select 函数

select 函数用于检测在一组 socket 中是否有事件就绪。这里的事件就绪一般分为如下三类。

1）读事件就绪

◎ 在 socket 内核中，接收缓冲区中的字节数大于或等于低水位标记 SO_RCVLOWAT，此时调用 recv 或 read 函数可以无阻塞地读该文件描述符，并且返回值大于 0。

- ◎ TCP 连接的对端关闭连接,此时本端调用 recv 或 read 函数对 socket 进行读操作, recv 或 read 函数会返回 0 值。
- ◎ 在监听 socket 上有新的连接请求。
- ◎ 在 socket 上有未处理的错误。

2) 写事件就绪

- ◎ 在 socket 内核中,发送缓冲区中的可用字节数(发送缓冲区的空闲位置大小)大于或等于低水位标记 SO_SNDLOWAT 时,可以无阻塞地写,并且返回值大于 0。
- ◎ socket 的写操作被关闭(调用了 close 或 shutdown 函数)时,对一个写操作被关闭的 socket 进行写操作,会触发 SIGPIPE 信号。
- ◎ socket 使用非阻塞 connect 连接成功或失败时。

3) 异常事件就绪

在 socket 上收到带外数据,函数签名如下:

```
int select(int nfds,
       fd_set *readfds,
       fd_set *writefds,
       fd_set *exceptfds,
       struct timeval *timeout);
```

对其中的参数说明如下。

(1) nfds: Linux 上的 socket 也叫作 fd,将这个参数的值设置为所有需要使用 select 函数检测事件的 fd 中的最大值加 1。

(2) readfds: 需要监听可读事件的 fd 集合。

(3) writefds: 需要监听可写事件的 fd 集合。

(4) exceptfds: 需要监听异常事件的 fd 集合。

(5) timeout: 超时时间,即在这个参数设定的时间内检测这些 fd 的事件,超过这个时间后,select 函数将立即返回。这是一个 timeval 类型的结构体,其定义如下:

```
struct timeval
{
   long tv_sec;   /* 秒   */
   long tv_usec;  /* 微秒 */
};
```

select 函数的总超时时间是 timeout->tv_sec 和 timeout->tv_usec 的和,前者的时间单位是秒,后者的时间单位是微秒。

参数 readfds、writefds 和 exceptfds 的类型都是 fd_set,这是一个结构体信息,其定义位于/usr/include/sys/select.h 中:

```
/* fd_set 字段必须是一个 long 型数组  */
typedef long int   __fd_mask;

/* 某些版本的<linux/posix_types.h>文件定义了如下宏 */
#define __NFDBITS       (8 * (int) sizeof (__fd_mask))
```

```
#define __FD_ELT(d)      ((d) / __NFDBITS)
#define __FD_MASK(d)     ((__fd_mask) 1 << ((d) % __NFDBITS))
/* fd_set 结构用于 select 和 pselect 函数 */
typedef struct
{
    //笔者的 centOS 7.0 系统中的值：
    //__FD_SETSIZE = 1024
    //__NFDBITS = 64
    __fd_mask __fds_bits[__FD_SETSIZE / __NFDBITS];
#define __FDS_BITS(set) ((set)->__fds_bits)
} fd_set;

/* fd_set 结构体中文件描述符的最大数量 */
#define FD_SETSIZE          __FD_SETSIZE
```

假设未定义宏__USE_XOPEN，将上面的代码整理一下：

```
typedef struct
{
    long int __fds_bits[16];
} fd_set;
```

在将一个 fd 添加到 fd_set 这个集合中时需要使用 FD_SET 宏，其定义如下：

```
void FD_SET(int fd, fd_set *set);
```

其实现如下：

```
#define FD_SET(fd,fdsetp) __FD_SET(fd,fdsetp)
```

FD_SET 在内部又是通过宏__FD_SET 实现的，__FD_SET 的定义如下（位于/usr/include/bits/select.h 中）：

```
/*这里不使用 memset 函数来清零的原因是不想引入其他函数且数组长度并不大*/
#define __FD_ZERO(set) \
  do {                                                          \
    unsigned int __i;                                           \
    fd_set *__arr = (set);                                      \
    for (__i = 0; __i < sizeof (fd_set) / sizeof (__fd_mask); ++__i) \
      __FDS_BITS (__arr)[__i] = 0;                              \
  } while (0)


#define __FD_SET(d, set) \
  ((void) (__FDS_BITS (set)[__FD_ELT (d)] |= __FD_MASK (d)))
#define __FD_CLR(d, set) \
  ((void) (__FDS_BITS (set)[__FD_ELT (d)] &= ~__FD_MASK (d)))
#define __FD_ISSET(d, set) \
  ((__FDS_BITS (set)[__FD_ELT (d)] & __FD_MASK (d)) != 0)
```

重点看以下两行：

```
#define __FD_SET(d, set) \
  ((void) (__FDS_BITS (set)[__FD_ELT (d)] |= __FD_MASK (d)))
```

__FD_MASK 和__FD_ELT 宏在上面的代码中已经给出定义：

```
#define __FD_ELT(d)      ((d) / __NFDBITS)
#define __FD_MASK(d)     ((__fd_mask) 1 << ((d) % __NFDBITS))
```

fd_set 数组的定义如下：

```
typedef struct
{
   long int __fds_bits[16];   //可以看作128 bit 的数组
} fd_set;
```

__fds_bits 是 long int 类型的数组，long int 占 8 字节，每字节都有 8bit，每个 bit 都对应一个 fd 的事件状态，0 表示无事件，1 表示有事件，数组长度是 16。因此一共可以表示 $8 \times 8 \times 16 = 1024$ 个 fd 的状态，这是 select 函数支持的最大 fd 数量。

__FD_SET(d, set)的实际操作如下：

```
__FDS_BITS (set)[__FD_ELT (d)] |= __FD_MASK (d)
```

__FD_ELT(d)确定的是某个 fd（这里用 d 表示）在数组 __fds_bits 中的下标位置，计算方法是将 fd 与 __NFDBITS 求商（__NFDBITS 的值是 64，即 8×8）。

__FD_MASK 计算对应的 fd 在对应的 bit 位置的值，计算方法是先与 __NFDBITS 求余得到 n，然后执行 1 << n，即左移 n 位，然后将值设置到对应的 bit 上（|=操作）。

举个例子，假设现在 fd 的值是 57，那么 FD_SET(57, set)实际调用的是 __fds_bits[__FD_ELT(57)] |= __FD_MASK(57)，即 __fds_bits[57 / 64] |= (1 << (57 % 64))，__fds_bits[0] |=(0000 0010 0000 0000 0000 0000 0000 0000 0000 0000 0000 0000 0000 0000 0000 0000)（二进制），在数组下标为 0 的元素（64 bit）中第 57 个 bit 被置为 1（0~63 位）。

> 在 Linux 上，向 fd_set 集合中添加新的 fd 时，决定这个 fd 在 __fds_bits 数组的位置的实现使用的是位图法（bitmap）；在 Windows 上添加 fd 至 fd_set 的实现则是依次从数组的第 0 个位置开始向后递增。

也就是说，FD_SET 宏在本质上是在一个有 1024 个连续 bit（共计 64 字节）的内存的某个 bit 上设置一个标志。

同理，如果需要在 fd_set 中删除一个 fd，即将对应的 bit 置 0，则可以使用 FD_CLR，其定义如下：

```
void FD_CLR(int fd, fd_set *set);
```

如果需要将 fd_set 中所有的 fd 都清掉，即将所有 bit 都置 0，则可以使用宏 FD_ZERO：

```
void FD_ZERO(fd_set *set);
```

当 select 函数返回时，我们使用 FD_ISSET 宏判断在某个 fd 中是否有我们关心的事件。FD_ISSET 宏的定义如下：

```
int  FD_ISSET(int fd, fd_set *set);
```

FD_ISSET 宏在本质上就是检测对应的 bit 是否置位，实现如下：

```
#define __FD_ISSET(d, set) \
  ((__FDS_BITS (set)[__FD_ELT (d)] & __FD_MASK (d)) != 0)
```

这就是与 select 函数相关的几个宏的原理。示意图如下图所示。

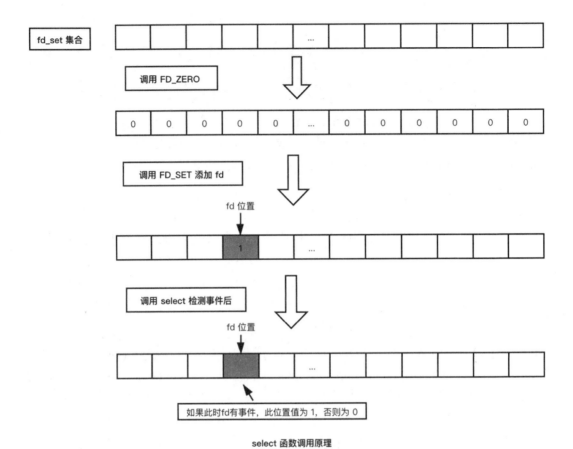

select 函数调用原理

来看一个具体示例：

```cpp
/**
 * 代码 4-5-1
 * select 函数示例，server 端，select_server.cpp
 * zhangyl 2018.12.24
 */
#include <sys/types.h>
#include <sys/socket.h>
#include <arpa/inet.h>
#include <unistd.h>
#include <iostream>
#include <string.h>
#include <sys/time.h>
#include <vector>
#include <errno.h>

//自定义代表无效 fd 的值
#define INVALID_FD -1

int main(int argc, char* argv[])
{
    //创建一个监听 socket
    int listenfd = socket(AF_INET, SOCK_STREAM, 0);
    if (listenfd == INVALID_FD)
    {
```

```cpp
        std::cout << "create listen socket error." << std::endl;
        return -1;
    }

    //初始化服务器地址
    struct sockaddr_in bindaddr;
    bindaddr.sin_family = AF_INET;
    bindaddr.sin_addr.s_addr = htonl(INADDR_ANY);
    bindaddr.sin_port = htons(3000);
    if (bind(listenfd, (struct sockaddr *)&bindaddr, sizeof(bindaddr)) == -1)
    {
        std::cout << "bind listen socket error." << std::endl;
        close(listenfd);
        return -1;
    }

    //启动监听
    if (listen(listenfd, SOMAXCONN) == -1)
    {
        std::cout << "listen error." << std::endl;
        close(listenfd);
        return -1;
    }

    //存储客户端 socket 的数组
    std::vector<int> clientfds;
    int maxfd;

    while (true)
    {
        fd_set readset;
        FD_ZERO(&readset);

        //将监听 socket 加入待检测的可读事件中
        FD_SET(listenfd, &readset);

        maxfd = listenfd;
        //将客户端 fd 加入待检测的可读事件中
        int clientfdslength = clientfds.size();
        for (int i = 0; i < clientfdslength; ++i)
        {
            if (clientfds[i] != INVALID_FD)
            {
                FD_SET(clientfds[i], &readset);

                if (maxfd < clientfds[i])
                    maxfd = clientfds[i];
            }
        }

        timeval tm;
        tm.tv_sec = 1;
        tm.tv_usec = 0;
        //暂且只检测可读事件，不检测可写和异常事件
        int ret = select(maxfd + 1, &readset, NULL, NULL, &tm);
        if (ret == -1)
        {
```

```cpp
                //出错，退出程序
                if (errno != EINTR)
                    break;
            }
            else if (ret == 0)
            {
                //select 函数超时，下次继续
                continue;
            }
            else
            {
                //检测到某个 socket 有事件
                if (FD_ISSET(listenfd, &readset))
                {
                    //监听 socket 的可读事件，表明有新的连接到来
                    struct sockaddr_in clientaddr;
                    socklen_t clientaddrlen = sizeof(clientaddr);
                    //4.接受客户端连接
                    int clientfd = accept(listenfd, (struct sockaddr *)&clientaddr, &clientaddrlen);
                    if (clientfd == INVALID_FD)
                    {
                        //接受连接出错，退出程序
                        break;
                    }

                    //只接受连接，不调用 recv 收取任何数据
                    std:: cout << "accept a client connection, fd: " << clientfd << std::endl;
                    clientfds.push_back(clientfd);
                }
                else
                {
                    //假设对端发来的数据长度不超过 63 个字符
                    char recvbuf[64];
                    int clientfdslength = clientfds.size();
                    for (int i = 0; i < clientfdslength; ++i)
                    {
                        if (clientfds[i] != INVALID_FD && FD_ISSET(clientfds[i], &readset))
                        {
                            memset(recvbuf, 0, sizeof(recvbuf));
                            //非监听 socket，接收数据
                            int length = recv(clientfds[i], recvbuf, 64, 0);
                            if (length <= 0)
                            {
                                //收取数据出错
                                std::cout << "recv data error, clientfd: " << clientfds[i] << std::endl;
                                close(clientfds[i]);
                                //不直接删除该元素，将该位置的元素标记为 INVALID_FD
                                clientfds[i] = INVALID_FD;
                                continue;
                            }

                            std::cout << "clientfd: " << clientfds[i] << ", recv data: " << recvbuf << std::endl;
                        }
```

```cpp
            }
        }
    }

    //关闭所有客户端 socket
    int clientfdslength = clientfds.size();
    for (int i = 0; i < clientfdslength; ++i)
    {
        if (clientfds[i] != INVALID_FD)
        {
            close(clientfds[i]);
        }
    }

    //关闭监听 socket
    close(listenfd);

    return 0;
}
```

编译并运行程序：

```
[root@localhost testsocket]# g++ -g -o select_server select_server.cpp
[root@localhost testsocket]# ./select_server
```

然后多开几个 Shell 窗口，这里不再编写客户端程序了，使用 Linux 上的 nc 指令模拟两个客户端。

在 Shell 窗口 1 中使用 nc 模拟的客户端 1 连接成功后发送字符串 hello123：

```
[root@localhost ~]# nc -v 127.0.0.1 3000
Ncat: Version 6.40 ( http://nmap.org/ncat )
Ncat: Connected to 127.0.0.1:3000.
hello123
```

在 Shell 窗口 2 中使用 nc 命令模拟的客户端 2 连接成功后发送字符串 helloworld：

```
[root@localhost ~]# nc -v 127.0.0.1 3000
Ncat: Version 6.40 ( http://nmap.org/ncat )
Ncat: Connected to 127.0.0.1:3000.
helloworld
```

此时服务端的输出结果如下：

```
[root@localhost chapter04]# ./select_server
accept a client connection, fd: 4
clientfd: 4, recv data: hello123

accept a client connection, fd: 5
```

注意，由于 nc 发送的数据是按换行符来区分的，每一个数据包默认的换行符都以\n 结束（当然，可以将-C 选项换成\r\n），所以服务器在收到数据后，显示出来的数据的每一行下面都有一个空白行。

当断开各个客户端连接时，服务端的 select 函数对各个客户端 fd 检测时，仍然会触发可读事件，此时对这些 fd 调用 recv 函数会返回 0（recv 函数返回 0，表明对端关闭了连接，这是一个很重要的知识点），服务端也关闭这些连接就可以了。

客户端断开连接后，服务端（select_server）的输出结果如下：

```
[root@myaliyun chapter04]# ./select_server
accept a client connection, fd: 4
accept a client connection, fd: 5
clientfd: 4, recv data: helloworld

clientfd: 5, recv data: hello123

recv data error, clientfd: 4
recv data error, clientfd: 5
```

代码 4-5-1 演示了一个简单的服务端程序实现的基本流程，代码虽然简单，但是非常具有典型性和代表性，而且同样适用于客户端网络通信，如果用于客户端的话，则只需用 select 检测连接 socket（即客户端用来连接服务器的 socket）就可以了。如果连接 socket 有可读事件，则调用 recv 函数接收数据，剩下的逻辑都是一样的。可以将代码 4-5-1 的逻辑绘制成如下所示的流程图。

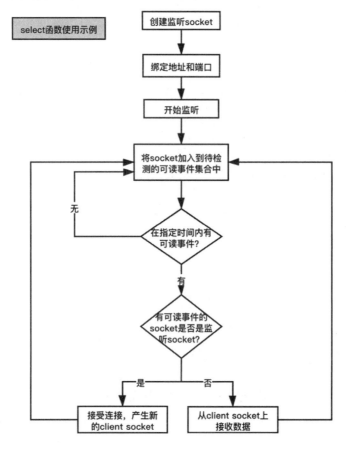

关于以上代码，在实际开发中有几个需要注意的事项，这里逐一进行说明。

（1）select 函数在调用前后可能会修改 readfds、writefds 和 exceptfds 这三个集合中的内容，所以如果想在下次调用 select 函数时复用这些 fd_set 变量，则要在下次调用前使用 FD_ZERO 将 fd_set 清零，然后调用 FD_SET 将需要检测事件的 fd 重新添加到 fd_set 中。

（2）select 函数也会修改 timeval 结构体的值，如果想复用这个变量，则必须给 timeval 变量重新设置值。在上面的例子中调用 select 函数一次后，变量 tv 的值也被修改，具体修改成多少，得看系统的表现。当然，这种特性不是跨平台的，在 Linux 上这样，在其他操作系统上却不一定这样（在 Windows 上就不会修改这个结构体的值），这一点在 Linux man 手册 select 函数的说明中说得很清楚。由于不同系统的实现可能不一样，man 手册建议将 select 函数修改 timeval 结构体的行为当作未定义的，即下次使用 select 函数复用这个变量时，要重新赋值。这是使用 select 函数时需要注意的第 2 个地方。

（3）select 函数的 timeval 结构体的 tv_sec 和 tv_usec 如果都被设置为 0，即检测事件的总时间被设置为 0，则其行为是 select 检测相关集合中的 fd，如果没有需要的事件，则立即返回。使用下面这段示例代码来验证一下：

```cpp
/**
 * 验证 select 函数的 timeval 参数被设置为 0 的行为, select_client_tv0.cpp
 * zhangyl 2018.12.25
 */
#include <sys/types.h>
#include <sys/socket.h>
#include <arpa/inet.h>
#include <unistd.h>
#include <iostream>
#include <string.h>
#include <errno.h>
#include <string.h>

#define SERVER_ADDRESS "127.0.0.1"
#define SERVER_PORT    3000

int main(int argc, char* argv[])
{
    //创建一个socket
    int clientfd = socket(AF_INET, SOCK_STREAM, 0);
    if (clientfd == -1)
    {
        std::cout << "create client socket error." << std::endl;
        return -1;
    }

    //连接服务器
    struct sockaddr_in serveraddr;
    serveraddr.sin_family = AF_INET;
    serveraddr.sin_addr.s_addr = inet_addr(SERVER_ADDRESS);
    serveraddr.sin_port = htons(SERVER_PORT);
    if (connect(clientfd, (struct sockaddr *)&serveraddr, sizeof(serveraddr)) == -1)
    {
```

```cpp
        std::cout << "connect socket error." << std::endl;
        close(clientfd);
        return -1;
    }

    int ret;
    while (true)
    {
        fd_set readset;
        FD_ZERO(&readset);
        //将监听 socket 加入待检测的可读事件中
        FD_SET(clientfd, &readset);
        timeval tm;
        tm.tv_sec = 0;
        tm.tv_usec = 0;

        //暂且只检测可读事件，不检测可写和异常事件
        ret = select(clientfd + 1, &readset, NULL, NULL, &tm);
        std::cout << "tm.tv_sec: " << tm.tv_sec << ", tm.tv_usec: " << tm.tv_usec << std::endl;
        if (ret == -1)
        {
            //除了被信号中断的情况，对其他情况都出错
            if (errno != EINTR)
                break;
        }
        else if (ret == 0)
        {
            //select 函数超时
            std::cout << "no event in specific time interval." << std::endl;
            continue;
        }
        else
        {
            if (FD_ISSET(clientfd, &readset))
            {
                //检测到可读事件
                char recvbuf[32];
                memset(recvbuf, 0, sizeof(recvbuf));
                //假设对端发数据的时候不超过 31 个字符
                int n = recv(clientfd, recvbuf, 32, 0);
                if (n < 0)
                {
                    //除了被信号中断的情况，对其他情况都出错
                    if (errno != EINTR)
                        break;
                }
                else if (n == 0)
                {
                    //对端关闭了连接
                    break;
                }
                else
                {
                    std::cout << "recv data: " << recvbuf << std::endl;
```

```
                }//end inner-if
            }
            else
            {
                std::cout << "other socket event." << std::endl;
            }//end middle-if
        }//end outer-if
    }//end while-loop


    //关闭socket
    close(clientfd);

    return 0;
}
```

执行结果确实如我们预期的，这里 select 函数只是简单地检测 clientfd，并不会等待固定的时间，然后立即返回。输出结果如下：

```
no event in specific time interval.
tm.tv_sec: 0, tm.tv_usec: 0
no event in specific time interval.
tm.tv_sec: 0, tm.tv_usec: 0
no event in specific time interval.
tm.tv_sec: 0, tm.tv_usec: 0
no event in specific time interval.
tm.tv_sec: 0, tm.tv_usec: 0
no event in specific time interval.
tm.tv_sec: 0, tm.tv_usec: 0
no event in specific time interval.
tm.tv_sec: 0, tm.tv_usec: 0
```

（4）如果将 select 函数的 timeval 参数设置 NULL，则 select 函数会一直阻塞下去，直到我们需要的事件触发。

我们将以上代码再修改一下：

```
/**
 * 验证select函数的时间参数被设置为NULL, select_client_tvnull.cpp
 * zhangyl 2018.12.25
 */
#include <sys/types.h>
#include <sys/socket.h>
#include <arpa/inet.h>
#include <unistd.h>
#include <iostream>
#include <string.h>
#include <errno.h>
#include <string.h>

#define SERVER_ADDRESS "127.0.0.1"
#define SERVER_PORT    3000

int main(int argc, char* argv[])
{
    //创建一个socket
```

```cpp
int clientfd = socket(AF_INET, SOCK_STREAM, 0);
if (clientfd == -1)
{
    std::cout << "create client socket error." << std::endl;
    return -1;
}

//连接服务器
struct sockaddr_in serveraddr;
serveraddr.sin_family = AF_INET;
serveraddr.sin_addr.s_addr = inet_addr(SERVER_ADDRESS);
serveraddr.sin_port = htons(SERVER_PORT);
if (connect(clientfd, (struct sockaddr *)&serveraddr, sizeof(serveraddr)) == -1)
{
    std::cout << "connect socket error." << std::endl;
    close(clientfd);
    return -1;
}

int ret;
while (true)
{
    fd_set readset;
    FD_ZERO(&readset);
    //将监听socket加入待检测的可读事件中
    FD_SET(clientfd, &readset);
    //timeval tm;
    //tm.tv_sec = 0;
    //tm.tv_usec = 0;

    //暂且只检测可读事件，不检测可写和异常事件
    ret = select(clientfd + 1, &readset, NULL, NULL, NULL);
    if (ret == -1)
    {
        //除了被信号中断的情况，对其他情况都出错
        if (errno != EINTR)
            break;
    }
    else if (ret == 0)
    {
        //select函数超时
        std::cout << "no event in specific time interval." << std::endl;
        continue;
    }
    else
    {
        if (FD_ISSET(clientfd, &readset))
        {
            //检测到可读事件
            char recvbuf[32];
            memset(recvbuf, 0, sizeof(recvbuf));
            //假设对端发数据的时候不超过31个字符
            int n = recv(clientfd, recvbuf, 32, 0);
            if (n < 0)
            {
```

```cpp
                //除了被信号中断的情况，对其他情况都出错
                if (errno != EINTR)
                    break;
            }
            else if (n == 0)
            {
                //对端关闭了连接
                break;
            }
            else
            {
                std::cout << "recv data: " << recvbuf << std::endl;
            }
        }
        else
        {
            std::cout << "other socket event." << std::endl;
        }
    }
    //关闭socket
    close(clientfd);

    return 0;
}
```

我们先在另一个 Shell 窗口中用 nc 命令模拟一个服务器，监听的 IP 地址和端口号是 0.0.0.0:3000：

```
[root@myaliyun ~]# nc -v -l 0.0.0.0 3000
Ncat: Version 6.40 ( http://nmap.org/ncat )
Ncat: Listening on 0.0.0.0:3000
```

然后回到原来的 Shell 窗口中编译以上 select_client_tvnull.cpp，并使用 gdb 运行程序，这次使用 gdb 运行程序的目的是在程序卡在某个位置时，可以使用 Ctrl+C 组合键把程序中断，看看程序阻塞在哪个函数调用处：

```
[root@myaliyun testsocket]# g++ -g -o select_client_tvnull select_client_tvnull.cpp
[root@myaliyun testsocket]# gdb select_client_tvnull
Reading symbols from /root/testsocket/select_client_tvnull...done.
(gdb) r
Starting program: /root/testsocket/select_client_tvnull
^C
Program received signal SIGINT, Interrupt.
0x00007ffff72e7783 in __select_nocancel () from /lib64/libc.so.6
Missing separate debuginfos, use: debuginfo-install glibc-2.17-196.el7_4.2.x86_64 libgcc-4.8.5-16.el7_4.1.x86_64 libstdc++-4.8.5-16.el7_4.1.x86_64
(gdb) bt
#0  0x00007ffff72e7783 in __select_nocancel () from /lib64/libc.so.6
#1  0x0000000000400c75 in main (argc=1, argv=0x7fffffffe5f8) at select_client_tvnull.cpp:51
(gdb) c
Continuing.
recv data: hello
```

```
^C
Program received signal SIGINT, Interrupt.
0x00007ffff72e7783 in __select_nocancel () from /lib64/libc.so.6
(gdb) c
Continuing.
recv data: world
```

如以上输出结果所示，我们使用 gdb 的 r 命令（run）将程序运行起来后，程序卡在某个地方，我们按 Ctrl+C 组合键（代码中的^C）中断程序后使用 bt 命令查看当前程序的调用堆栈，发现确实阻塞在 select 函数调用处；接着在服务端向客户端发送一个 hello 数据：

```
[root@myaliyun ~]# nc -v -l 0.0.0.0 3000
Ncat: Version 6.40 ( http://nmap.org/ncat )
Ncat: Listening on 0.0.0.0:3000
Ncat: Connection from 127.0.0.1.
Ncat: Connection from 127.0.0.1:55968.
hello
```

在客户端收到数据后，select 函数满足条件，立即返回，在将数据输出后继续进行下一轮 select 检测。我们使用 Ctrl + C 组合键将程序中断，发现程序又阻塞在 select 调用处；输入 c 命令（continue）让程序继续运行，再用服务端向客户端发送 world 字符串，select 函数再次返回并将数据打印出来，进入下一轮 select 检测，并继续在 select 处阻塞：

```
[root@myaliyun ~]# nc -v -l 0.0.0.0 3000
Ncat: Version 6.40 ( http://nmap.org/ncat )
Ncat: Listening on 0.0.0.0:3000
Ncat: Connection from 127.0.0.1.
Ncat: Connection from 127.0.0.1:55968.
hello
world
```

（5）在 Linux 上，select 函数的第 1 个参数必须被设置为需要检测事件所有 fd 中的最大值加 1。所以在上面的 select_server.cpp 中，每新产生一个 clientfd，都会与当前最大的 maxfd 做比较，如果大于当前 maxfd，则将 maxfd 更新为这个新的最大值。其最终目的是在 select 调用时作为第 1 个参数（加 1）传进去。

> 在 Windows 上，对 select 函数的第 1 个值传入任意值都可以，Windows 本身不使用这个值，只是为了兼容才保留这个参数，但是在实际开发中为了兼容跨平台代码，也会按惯例将这个值设置为最大 socket 加 1。

Linux select 函数的缺点也是显而易见的，如下所述。

（1）每次调用 select 函数时，都需要把 fd 集合从用户态复制到内核态，这个开销在 fd 较多时会很大，同时每次调用 select 函数都需要在内核中遍历传递进来的所有 fd，这个开销在 fd 较多时也很大。

（2）单个进程能够监视的文件描述符的数量存在最大限制，在 Linux 上一般为 1024，可以通过先修改宏定义然后重新编译内核来调整这一限制，但这样非常麻烦而且效率低下。

（3）select 函数在每次调用之前都要对传入的参数进行重新设定，这样做也比较麻烦。

（4）在 Linux 上，select 函数的实现原理是其底层使用了 poll 函数。

4.5.2 Windows 上的 select 函数

上面提到，在 Windows 上，select 函数结束后，不会修改其参数 timeval 的值。我们可以使用下面这段代码来验证：

```
01  bool Connect(const char* pServer, short nPort)
02  {
03      SOCKET hSocket = ::socket(AF_INET, SOCK_STREAM, 0);
04      if (hSocket == INVALID_SOCKET)
05          return false;
06
07      //将 socket 设置为非阻塞的
08      unsigned long on = 1;
09      if (::ioctlsocket(hSocket, FIONBIO, &on) == SOCKET_ERROR)
10          return false;
11
12      struct sockaddr_in addrSrv = { 0 };
13      struct hostent* pHostent = NULL;
14      unsigned int addr = 0;
15
16      if ((addrSrv.sin_addr.s_addr = inet_addr(pServer) == INADDR_NONE)
17      {
18          pHostent = ::gethostbyname(pServer);
19          if (!pHostent)
20              return false;
21          else
22              addrSrv.sin_addr.s_addr = *((unsigned long*)pHostent->h_addr);
23      }
24
25      addrSrv.sin_family = AF_INET;
26      addrSrv.sin_port = htons((nPort);
27      int ret = ::connect(hSocket, (struct sockaddr*)&addrSrv, sizeof(addrSrv));
28      if (ret == 0)
29          return true;
30
31      if (ret == SOCKET_ERROR && WSAGetLastError() != WSAEWOULDBLOCK)
32          return false;
33
34      fd_set writeset;
35      FD_ZERO(&writeset);
36      FD_SET(hSocket, &writeset);
37      struct timeval tm = { 3, 200 };
38      if (::select(hSocket + 1, NULL, &writeset, NULL, &tm) != 1)
39      {
40          printf("tm.tv_sec: %d, tm.tv_usec: %d\n", tm.tv_sec, tm.tv_usec);
41          return false;
42      }
```

```
43
44      printf("tm.tv_sec: %d, tm.tv_usec: %d\n", tm.tv_sec, tm.tv_usec);
45
46      return true;
47  }
```

在以上代码中，第 38 行调用了 select 函数，无论 select 是成功还是出错，我们都会打印出其参数的 tm 的值（第 40 和 44 行）。经测试验证，tm 结构体的两个成员值在 select 函数调用前后并没有发生改变。

> 虽然 Windows 并不会改变 select 的超时时间参数的值，但是为了满足代码的跨平台性，我们在实际开发中不应该依赖这种特性，而是在每次调用 select 函数前都重新设置超时时间参数。

4.6　socket 的阻塞模式和非阻塞模式

对 socket 在阻塞和非阻塞模式下各个 socket 函数的表现进行深入理解，是掌握网络编程的基本要求之一，也是重点和难点。

在阻塞和非阻塞模式下，我们常讨论的具有不同行为表现的 socket 函数一般有 connect、accept、send 和 recv。在 Linux 上对 socket 进行操作时也包括 write 函数和 read 函数。下面对 send 函数的讨论也适用于 write 函数，对 recv 函数的讨论也适用于 read 函数。

在正式讨论以上 4 个函数之前，先解释阻塞模式和非阻塞模式的概念。阻塞模式指当某个函数执行成功的条件当前不满足时，该函数会阻塞当前执行线程，程序执行流在超时时间到达或执行成功的条件满足后恢复继续执行。非阻塞模式则恰恰相反，即使某个函数执行成功的条件当前不能满足，该函数也不会阻塞当前执行线程，而是立即返回，继续执行程序流。

4.6.1　如何将 socket 设置为非阻塞模式

无论是 Windows 还是 Linux，默认创建的 socket 都是阻塞模式的。

在 Linux 上，我们可以使用 fcntl 函数或 ioctl 函数给创建的 socket 增加 O_NONBLOCK 标志来将 socket 设置为非阻塞模式。示例代码如下：

```
int oldSocketFlag = fcntl(sockfd, F_GETFL, 0);
int newSocketFlag = oldSocketFlag | O_NONBLOCK;
fcntl(sockfd, F_SETFL, newSocketFlag);
```

ioctl 函数与 fcntl 函数的使用方式基本一致，这里就不再给出示例代码了。

当然，Linux 上的 socket 函数也可以直接在创建时将 socket 设置为非阻塞模式，socket 函数签名如下：

```
int socket(int domain, int type, int protocol);
```

给 type 参数增加一个 SOCK_NONBLOCK 标志即可，例如：

```
int s = socket(AF_INET, SOCK_STREAM | SOCK_NONBLOCK, IPPROTO_TCP);
```

不仅如此，在 Linux 上利用 accept 函数返回的代表与客户端通信的 socket 也提供了一个扩展函数 accept4，直接将 accept 函数返回的 socket 设置为非阻塞的：

```
int accept(int sockfd, struct sockaddr *addr, socklen_t *addrlen);
int accept4(int sockfd, struct sockaddr *addr, socklen_t *addrlen, int flags);
```

只要将 accept4 函数的最后一个参数 flags 设置为 SOCK_NONBLOCK 即可。也就是说以下代码是等价的：

```
socklen_t addrlen = sizeof(clientaddr);
int clientfd = accept4(listenfd, &clientaddr, &addrlen, SOCK_NONBLOCK);
socklen_t addrlen = sizeof(clientaddr);
int clientfd = accept(listenfd, &clientaddr, &addrlen);
if (clientfd != -1)
{
    int oldSocketFlag = fcntl(clientfd, F_GETFL, 0);
    int newSocketFlag = oldSocketFlag | O_NONBLOCK;
    fcntl(clientfd, F_SETFL, newSocketFlag);
}
```

在 Windows 上可以调用 ioctlsocket 函数将 socket 设置为非阻塞模式，ioctlsocket 函数签名如下：

```
int ioctlsocket(SOCKET s, long cmd, u_long *argp);
```

将 cmd 参数设置为 FIONBIO，将 argp 设置为 0，即可将 socket 设置为阻塞模式，而将 argp 设置为非 0，即可将其设置为非阻塞模式。

示例如下：

```
//将 socket 设置为非阻塞模式
u_long argp = 1;
ioctlsocket(s, FIONBIO, &argp);

//将 socket 设置为阻塞模式
u_long argp = 0;
ioctlsocket(s, FIONBIO, &argp);
```

要使用 ioctlsocket 函数，则必须使用 Windows Vista 或 Windows Server 2003 及之后的版本。

在 Windows 上对一个 socket 调用了 WSAAsyncSelect 或 WSAEventSelect 函数后，再调用 ioctlsocket 函数将该 socket 设置为非阻塞模式会失败，我们必须先调用 WSAAsyncSelect 函数将 lEvent 参数设置为 0 或调用 WSAEventSelect 函数将 lNetworkEvents 参数设置为 0，这样会清除已经设置的 socket 相关标志位，接着调用 ioctlsocket 函数将该 socket 设置为阻塞模式才会成功。因为调用 WSAAsyncSelect 或 WSAEventSelect 函数时会自动将 socket 设置为非阻塞模式。

注意：无论是 Linux 的 fcntl 函数，还是 Windows 的 ioctlsocket 函数，都建议读者在实际编码中判断函数返回值以确定是否调用成功。

4.6.2　send 和 recv 函数在阻塞和非阻塞模式下的表现

send 和 recv 函数其实名不符实。send 函数在本质上并不是向网络上发送数据，而是将应用层发送缓冲区的数据拷贝到内核缓冲区中，至于数据什么时候会从网卡缓冲区中真正地发到网络中，要根据 TCP/IP 协议栈的行为来确定。如果 socket 设置了 TCP_NODELAY 选项（即禁用 nagel 算法），存放到内核缓冲区的数据就会被立即发出去；反之，一次放入内核缓冲区的数据包如果太小，则系统会在多个小的数据包凑成一个足够大的数据包后才会将数据发出去。

recv 函数在本质上并不是从网络上收取数据，而是将内核缓冲区中的数据拷贝到应用程序的缓冲区中。当然，在拷贝完成后会将内核缓冲区中的该部分数据移除。

可以用下面一张图来描述上述事实。

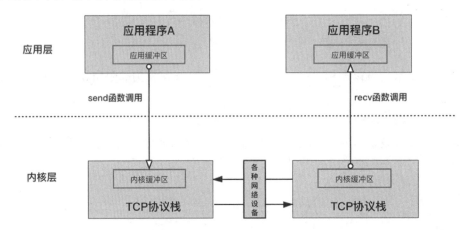

通过上图可以知道，不同的程序在进行网络通信时，发送的一方会将内核缓冲区的数据通过网络传输给接收方的内核缓冲区。在应用程序 A 与应用程序 B 建立 TCP 连接之后，假设应用程序 A 不断调用 send 函数，则数据会不断拷贝至对应的内核缓冲区中，如果应用程序 B 一直不调用 recv 函数，那么在应用程序 B 的内核缓冲区被填满后，应用程序 A 的内核缓冲区也会被填满，此时应用程序 A 继续调用 send 函数会是什么结果呢？具体的结果取决于该 socket 是否是阻塞模式，这里先给出结论：

（1）当 socket 是阻塞模式时，继续调用 send/recv 函数，程序会阻塞在 send/recv 调用处；

（2）当 socket 是非阻塞模式时，继续调用 send/recv 函数，send/recv 函数不会阻塞程序执行流，而是立即出错并返回，我们会得到一个相关的错误码，在 Linux 上该错误码为 EWOULDBLOCK 或 EAGAIN（这两个错误码的值相同），在 Windows 上该错误码为 WSAEWOULDBLOCK。

我们实际编写代码来验证以上两种情况。

1. socket 阻塞模式下 send 函数的表现

服务端代码（blocking_server.cpp）如下：

```cpp
/**
 * 验证阻塞模式下 send 函数的行为，server 端，blocking_server.cpp
 * zhangyl 2018.12.17
 */
#include <sys/types.h>
#include <sys/socket.h>
#include <arpa/inet.h>
#include <unistd.h>
#include <iostream>
#include <string.h>

int main(int argc, char* argv[])
{
    //1.创建一个监听 socket
    int listenfd = socket(AF_INET, SOCK_STREAM, 0);
    if (listenfd == -1)
    {
        std::cout << "create listen socket error." << std::endl;
        return -1;
    }

    //2.初始化服务器地址
    struct sockaddr_in bindaddr;
    bindaddr.sin_family = AF_INET;
    bindaddr.sin_addr.s_addr = htonl(INADDR_ANY);
    bindaddr.sin_port = htons(3000);
    if (bind(listenfd, (struct sockaddr *)&bindaddr, sizeof(bindaddr)) == -1)
    {
        std::cout << "bind listen socket error." << std::endl;
        close(listenfd);
        return -1;
    }

    //3.启动监听
    if (listen(listenfd, SOMAXCONN) == -1)
    {
        std::cout << "listen error." << std::endl;
        close(listenfd);
        return -1;
    }

    while (true)
    {
        struct sockaddr_in clientaddr;
        socklen_t clientaddrlen = sizeof(clientaddr);
        //4.接受客户端连接
        int clientfd = accept(listenfd, (struct sockaddr *)&clientaddr, &clientaddrlen);
        if (clientfd != -1)
        {
            //只接受连接，不调用 recv 收取任何数据
            std:: cout << "accept a client connection." << std::endl;
        }
    }
```

```cpp
    //5.关闭监听socket
    close(listenfd);

    return 0;
}
```

客户端代码（blocking_client.cpp）如下：

```cpp
/**
 * 验证阻塞模式下send函数的行为，client端，blocking_client.cpp
 * zhangyl 2018.12.17
 */
#include <sys/types.h>
#include <sys/socket.h>
#include <arpa/inet.h>
#include <unistd.h>
#include <iostream>
#include <string.h>

#define SERVER_ADDRESS  "127.0.0.1"
#define SERVER_PORT     3000
#define SEND_DATA       "helloworld"

int main(int argc, char* argv[])
{
    //1.创建一个socket
    int clientfd = socket(AF_INET, SOCK_STREAM, 0);
    if (clientfd == -1)
    {
        std::cout << "create client socket error." << std::endl;
        return -1;
    }

    //2.连接服务器
    struct sockaddr_in serveraddr;
    serveraddr.sin_family = AF_INET;
    serveraddr.sin_addr.s_addr = inet_addr(SERVER_ADDRESS);
    serveraddr.sin_port = htons(SERVER_PORT);
    if (connect(clientfd, (struct sockaddr *)&serveraddr, sizeof(serveraddr)) == -1)
    {
        std::cout << "connect socket error." << std::endl;
        close(clientfd);
        return -1;
    }

    //3.不断向服务器发送数据，一直到出错并退出循环
    int count = 0;
    while (true)
    {
        int ret = send(clientfd, SEND_DATA, strlen(SEND_DATA), 0);
        if (ret != strlen(SEND_DATA))
        {
            std::cout << "send data error." << std::endl;
            break;
        }
```

```
        else
        {
            count ++;
            std::cout << "send data successfully, count = " << count << std::endl;
        }
    }

    //4.关闭socket
    close(clientfd);

    return 0;
}
```

在 Shell 中分别编译这两个 cpp 文件，得到两个可执行程序 blocking_server 和 blocking_client：

```
g++ -g -o blocking_server blocking_server.cpp
g++ -g -o blocking_client blocking_client.cpp
```

我们先启动 blocking_server，用 gdb 启动 blocking_client，输入 run 命令让 blocking_client 运行。blocking_client 会不断向 blocking_server 发送 helloworld 字符串，在每次发送成功后，都会打印计数器 count 的值，计数器 count 的值会不断增加。在程序运行一段时间后，计数器 count 的值不再增加且程序不再有输出。操作过程及输出结果如下。

blocking_server 端：

```
[root@localhost testsocket]# ./blocking_server
accept a client connection.
[root@localhost testsocket]# gdb blocking_client
Reading symbols from /root/testsocket/blocking_client...done.
(gdb) run
//省略部分输出
send data successfully, count = 355384
send data successfully, count = 355385
send data successfully, count = 355386
send data successfully, count = 355387
send data successfully, count = 355388
send data successfully, count = 355389
send data successfully, count = 355390
```

此时程序不再有输出，说明我们的程序应该卡在某个地方，继续按 Ctrl+C 组合键让 gdb 中断，输入 bt 命令查看此时的调用堆栈，发现程序确实阻塞在 send 函数调用处：

```
^C
Program received signal SIGINT, Interrupt.
0x00007ffff72f130d in send () from /lib64/libc.so.6
(gdb) bt
#0  0x00007ffff72f130d in send () from /lib64/libc.so.6
#1  0x0000000000400b46 in main (argc=1, argv=0x7fffffffe598) at
blocking_client.cpp:41
(gdb)
```

以上示例验证了如果一端一直发送数据，对端应用层一直不收取数据（或收取数据的速度慢于发送的速度），则两端的内核缓冲区很快就会被填满，导致发送端调用 send 函数被阻塞。这里说的内核缓冲区其实有个专门的名字，即 TCP 窗口。也就是说，在 socket

阻塞模式下，send 函数在 TCP 窗口太小时会阻塞当前程序的执行流（即阻塞 send 函数所在线程的执行）。

另外，在上面的例子中，我们每次都发送一个"helloworld"（10 字节），一共发送了 355 390 次（每次测试的结果都略有不同），可以粗略算出 TCP 窗口大约等于 1.7M（10×355390/2）。

让我们再深入一点，利用 Linux tcpdump 工具查看这种情形下 TCP 窗口大小的动态变化。需要注意的是，在 Linux 上使用 tcpdump 命令需要有 root 权限。

我们开启 3 个 Shell 窗口，在第 1 个窗口先启动 blocking_server 进程，在第 2 个窗口用 tcpdump 抓取经过 TCP 3000 端口的数据包：

```
[root@localhost testsocket]# tcpdump -i any -nn -S 'tcp port 3000'
tcpdump: verbose output suppressed, use -v or -vv for full protocol decode
listening on any, link-type LINUX_SLL (Linux cooked), capture size 262144 bytes
```

接着在第 3 个 Shell 窗口启动 blocking_client。当 blocking_client 进程不再输出时，抓包的结果如下：

```
13:58:54.370662 IP 127.0.0.1.39102 > 127.0.0.1.3000: Flags [S], seq 317054582, win 43690, options [mss 65495,sackOK,TS val 2890857589 ecr 0,nop,wscale 7], length 0
13:58:54.370673 IP 127.0.0.1.3000 > 127.0.0.1.39102: Flags [S.], seq 1396473902, ack 317054583, win 43690, options [mss 65495,sackOK,TS val 2890857589 ecr 2890857589,nop,wscale 7], length 0
13:58:54.370684 IP 127.0.0.1.39102 > 127.0.0.1.3000: Flags [.], ack 1396473903, win 342, options [nop,nop,TS val 2890857589 ecr 2890857589], length 0
13:58:54.370699 IP 127.0.0.1.39102 > 127.0.0.1.3000: Flags [P.], seq 317054583:317054593, ack 1396473903, win 342, options [nop,nop,TS val 2890857589 ecr 2890857589], length 10
...省略部分输出...
13:59:37.020155 IP 127.0.0.1.39102 > 127.0.0.1.3000: Flags [.], ack 1396473903, win 342, options [nop,nop,TS val 2890900238 ecr 2890885390], length 0
13:59:37.020183 IP 127.0.0.1.3000 > 127.0.0.1.39102: Flags [.], ack 317972685, win 0, options [nop,nop,TS val 2890900238 ecr 2890870646], length 0
14:00:06.204170 IP 127.0.0.1.39102 > 127.0.0.1.3000: Flags [.], ack 1396473903, win 342, options [nop,nop,TS val 2890929421 ecr 2890900238], length 0
14:00:06.204214 IP 127.0.0.1.3000 > 127.0.0.1.39102: Flags [.], ack 317972685, win 0, options [nop,nop,TS val 2890929421 ecr 2890870646], length 0
```

抓取到的前三个数据包是 blocking_client 与 blocking_server 为建立连接而发起的 TCP 三次握手操作的数据包。

```
13:58:54.370662 IP 127.0.0.1.39102 > 127.0.0.1.3000: Flags [S], seq 317054582, win 43690, options [mss 65495,sackOK,TS val 2890857589 ecr 0,nop,wscale 7], length 0
13:58:54.370673 IP 127.0.0.1.3000 > 127.0.0.1.39102: Flags [S.], seq 1396473902, ack 317054583, win 43690, options [mss 65495,sackOK,TS val 2890857589 ecr 2890857589,nop,wscale 7], length 0
13:58:54.370684 IP 127.0.0.1.39102 > 127.0.0.1.3000: Flags [.], ack 1396473903, win 342, options [nop,nop,TS val 2890857589 ecr 2890857589], length 0
```

示意图如下所示。

第 4 章　网络编程重难点解析

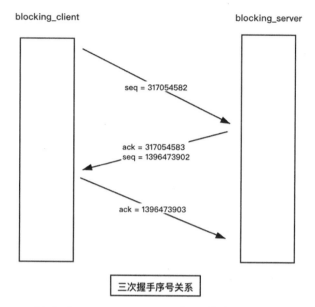

三次握手序号关系

在 blocking_client 每次向 blocking_server 发送数据后，blocking_server 都会应答 blocking_client，在每次应答的数据包中都会带上自己当前可用的 TCP 窗口大小（看上面的结果中 127.0.0.1.3000→127.0.0.1.39102 方向数据包的 win 字段大小变化），由于 TCP 流量控制和拥塞控制机制的存在，blocking_server 端的 TCP 窗口大小在短期内会慢慢增加，后面随着接收缓冲区中数据积压越来越多，TCP 窗口会慢慢变小，最终变为 0。

另外，tcpdump 显示对端的 TCP 窗口是 0 时，blocking_client 仍然可以继续发送一段时间的数据，此时的数据已经不是发往对端，而是逐渐填满到本端的内核发送缓冲区中，这也验证了 send 函数实际上是向内核缓冲区中拷贝数据这一行为。

2. socket 非阻塞模式下 send 函数的表现

再来验证非阻塞 socket 的 send 表现，服务端的代码不变，我们将 blocking_client.cpp 中的 socket 设置为非阻塞的，修改后的代码如下：

```
/**
 * 验证非阻塞模式下 send 函数的行为，client 端，nonblocking_client.cpp
 * zhangyl 2018.12.17
 */
#include <sys/types.h>
#include <sys/socket.h>
#include <arpa/inet.h>
#include <unistd.h>
#include <iostream>
#include <string.h>
#include <stdio.h>
#include <fcntl.h>
#include <errno.h>

#define SERVER_ADDRESS  "127.0.0.1"
#define SERVER_PORT     3000
#define SEND_DATA       "helloworld"
```

· 325 ·

```cpp
int main(int argc, char* argv[])
{
    //1.创建一个socket
    int clientfd = socket(AF_INET, SOCK_STREAM, 0);
    if (clientfd == -1)
    {
        std::cout << "create client socket error." << std::endl;
        return -1;
    }

    //2.连接服务器
    struct sockaddr_in serveraddr;
    serveraddr.sin_family = AF_INET;
    serveraddr.sin_addr.s_addr = inet_addr(SERVER_ADDRESS);
    serveraddr.sin_port = htons(SERVER_PORT);
    if (connect(clientfd, (struct sockaddr *)&serveraddr, sizeof(serveraddr)) == -1)
    {
        std::cout << "connect socket error." << std::endl;
        close(clientfd);
        return -1;
    }

    //连接成功以后,我们再将clientfd设置为非阻塞模式,
    //不能在创建时就设置,这样会影响到connect函数的行为
    int oldSocketFlag = fcntl(clientfd, F_GETFL, 0);
    int newSocketFlag = oldSocketFlag | O_NONBLOCK;
    if (fcntl(clientfd, F_SETFL, newSocketFlag) == -1)
    {
        close(clientfd);
        std::cout << "set socket to nonblock error." << std::endl;
        return -1;
    }

    //3.不断向服务器发送数据,或者出错退出
    int count = 0;
    while (true)
    {
        int ret = send(clientfd, SEND_DATA, strlen(SEND_DATA), 0);
        if (ret == -1)
        {
            //非阻塞模式下,send函数由于TCP窗口太小发不出去数据,错误码是EWOULDBLOCK
            if (errno == EWOULDBLOCK)
            {
                std::cout << "send data error as TCP Window size is too small." << std::endl;
                continue;
            }
            else if (errno == EINTR)
            {
                //如果被信号中断,则继续重试
                std::cout << "sending data interrupted by signal." << std::endl;
                continue;
            }
            else
```

```cpp
            {
                std::cout << "send data error." << std::endl;
                break;
            }
        }
        else if (ret == 0)
        {
            //对端关闭了连接，我们也关闭
            std::cout << "send data error." << std::endl;
            close(clientfd);
            break;
        }
        else
        {
            count ++;
            std::cout << "send data successfully, count = " << count << std::endl;
        }
    }

    //4.关闭 socket
    close(clientfd);

    return 0;
}
```

编译 nonblocking_client.cpp，得到可执行程序 nonblocking_client：

```
g++ -g -o nonblocking_client nonblocking_client.cpp
```

运行 nonblocking_client 一段时间后，对端和本端的 TCP 窗口已满，数据发送不出去，但是 send 函数不会阻塞，而是立即返回，返回值是-1（在 Windows 上返回 SOCKET_ERROR，这个宏的值也是-1），此时得到的错误码是 EWOULDBLOCK。执行结果如下：

```
(count 依此递增......)
send data successfully, count = 358654
send data successfully, count = 358655
send data successfully, count = 358656
send data successfully, count = 358657
send data successfully, count = 358658
send data error as TCP Window size is too small.
send data error as TCP Window size is too small.
send data error as TCP Window size is too small.
send data error as TCP Window size is too small.
send data error as TCP Window size is too small.
send data error as TCP Window size is too small.
send data error as TCP Window size is too small.
send data error as TCP Window size is too small.
send data error as TCP Window size is too small.
(更多的重复输出同上...)
```

3. socket 阻塞模式下 recv 函数的表现

在了解 send 函数的行为后，我们再来看一下阻塞模式下 recv 函数的表现。我们不需要修改服务端的代码，修改客户端的代码即可。如果服务端不向客户端发送数据，则此时客户端调用 recv 函数的执行流会阻塞在 recv 函数调用处。修改后，客户端的代码如下：

```cpp
/**
 * 验证阻塞模式下 recv 函数的行为，client 端，blocking_client_recv.cpp
 * zhangyl 2018.12.17
 */
#include <sys/types.h>
#include <sys/socket.h>
#include <arpa/inet.h>
#include <unistd.h>
#include <iostream>
#include <string.h>

#define SERVER_ADDRESS "127.0.0.1"
#define SERVER_PORT    3000
#define SEND_DATA      "helloworld"

int main(int argc, char* argv[])
{
    //1.创建一个 socket
    int clientfd = socket(AF_INET, SOCK_STREAM, 0);
    if (clientfd == -1)
    {
        std::cout << "create client socket error." << std::endl;
        return -1;
    }

    //2.连接服务器
    struct sockaddr_in serveraddr;
    serveraddr.sin_family = AF_INET;
    serveraddr.sin_addr.s_addr = inet_addr(SERVER_ADDRESS);
    serveraddr.sin_port = htons(SERVER_PORT);
    if (connect(clientfd, (struct sockaddr *)&serveraddr, sizeof(serveraddr)) == -1)
    {
        std::cout << "connect socket error." << std::endl;
        close(clientfd);
        return -1;
    }

    //3.直接调用 recv 函数，程序会阻塞在 recv 函数调用处
    char recvbuf[32] = {0};
    int ret = recv(clientfd, recvbuf, 32, 0);
    if (ret > 0)
    {
        std::cout << "recv successfully." << std::endl;
    }
    else
    {
        std::cout << "recv data error." << std::endl;
    }

    //4.关闭 socket
    close(clientfd);

    return 0;
}
```

编译 blocking_client_recv.cpp 并启动，我们发现程序既没有输出 recv 函数调用成功的信息，也没有输出 recv 函数调用失败的信息。我们将程序中断并使用 bt 命令查看此时的调用堆栈，发现程序确实阻塞在 recv 函数调用处：

```
[root@localhost testsocket]# g++ -g -o blocking_client_recv
blocking_client_recv.cpp
[root@localhost testsocket]# gdb blocking_client_recv
Reading symbols from /root/testsocket/blocking_client_recv...done.
(gdb) r
Starting program: /root/testsocket/blocking_client_recv
^C
Program received signal SIGINT, Interrupt.
0x00007ffff72f119d in recv () from /lib64/libc.so.6
Missing separate debuginfos, use: debuginfo-install glibc-2.17-196.el7_4.2.x86_64
libgcc-4.8.5-16.el7_4.2.x86_64 libstdc++-4.8.5-16.el7_4.2.x86_64
(gdb) bt
#0  0x00007ffff72f119d in recv () from /lib64/libc.so.6
#1  0x0000000000400b18 in main (argc=1, argv=0x7fffffffe588) at
blocking_client_recv.cpp:40
```

4. socket 非阻塞模式下 recv 函数的表现

在非阻塞模式下，如果当前无数据可读，则 recv 函数将立即返回，返回值为-1，错误码为 EWOULDBLOCK。将客户端的代码修改一下：

```cpp
/**
 * 验证非阻塞模式下 recv 函数的行为，client 端, nonblocking_client_recv.cpp
 * zhangyl 2018.12.17
 */
#include <sys/types.h>
#include <sys/socket.h>
#include <arpa/inet.h>
#include <unistd.h>
#include <iostream>
#include <string.h>
#include <stdio.h>
#include <fcntl.h>
#include <errno.h>

#define SERVER_ADDRESS "127.0.0.1"
#define SERVER_PORT    3000
#define SEND_DATA      "helloworld"

int main(int argc, char* argv[])
{
    //1.创建一个 socket
    int clientfd = socket(AF_INET, SOCK_STREAM, 0);
    if (clientfd == -1)
    {
        std::cout << "create client socket error." << std::endl;
        return -1;
    }

    //2.连接服务器
```

```cpp
struct sockaddr_in serveraddr;
serveraddr.sin_family = AF_INET;
serveraddr.sin_addr.s_addr = inet_addr(SERVER_ADDRESS);
serveraddr.sin_port = htons(SERVER_PORT);
if (connect(clientfd, (struct sockaddr *)&serveraddr, sizeof(serveraddr)) == -1)
{
    std::cout << "connect socket error." << std::endl;
    close(clientfd);
    return -1;
}

//连接成功以后，我们再将clientfd设置为非阻塞模式，
//不能在创建时就设置，这样会影响到connect函数的行为
int oldSocketFlag = fcntl(clientfd, F_GETFL, 0);
int newSocketFlag = oldSocketFlag | O_NONBLOCK;
if (fcntl(clientfd, F_SETFL, newSocketFlag) == -1)
{
    close(clientfd);
    std::cout << "set socket to nonblock error." << std::endl;
    return -1;
}

//直接调用recv函数，程序会阻塞在recv函数调用处
while (true)
{
    char recvbuf[32] = {0};
    //由于clientfd被设置成了非阻塞模式，所以无论是否有数据，recv函数都不会阻塞程序
    int ret = recv(clientfd, recvbuf, 32, 0);
    if (ret > 0)
    {
        //收到数据
        std::cout << "recv successfully." << std::endl;
    }
    else if (ret == 0)
    {
        //对端关闭连接
        std::cout << "peer close the socket." << std::endl;
        break;
    }
    else if (ret == -1)
    {
        if (errno == EWOULDBLOCK)
        {
            std::cout << "There is no data available now." << std::endl;
        }
        else if (errno == EINTR)
        {
            //如果被信号中断，则继续重试recv函数
            std::cout << "recv data interrupted by signal." << std::endl;
        }
        else
        {
            //真的出错了
            break;
        }
    }
```

```
        }
    }

    //3.关闭 socket
    close(clientfd);

    return 0;
}
```

执行结果与我们预期的一模一样，recv 函数在无数据可读的情况下并不会阻塞程序执行流，所以会一直有"There is no data available now."相关的输出：

```
There is no data available now.
There is no data available now.
There is no data available now.
There is no data available now.
There is no data available now.
There is no data available now.
There is no data available now.
There is no data available now.
There is no data available now.
There is no data available now.
```
（更多重复的输出省略）

4.6.3 非阻塞模式下 send 和 recv 函数的返回值总结

这里根据前面的讨论，总结 send 和 recv 函数各种返回值的含义，如下表所示。

返回值 n	返回值的含义
大于 0	成功发送（send）或接收（recv）n 字节
0	对端关闭连接
小于 0（-1）	出错、被信号中断、对端 TCP 窗口太小导致数据发送不出去或者当前网卡缓冲区已无数据可接收

这里逐一介绍这三种情况。

（1）返回值大于 0。当 send 和 recv 函数的返回值大于 0 时，表示发送或接收多少字节。需要注意的是，在这种情形下，我们一定要判断 send 函数的返回值是不是我们期望发送的字节数，而不是简单判断其返回值大于 0。举个例子：

```
int n = send(socket, buf, buf_length, 0);
if (n > 0)
{
    printf("send data successfully\n");
}
```

很多新手会写出以上代码，虽然返回值 n 大于 0，但在实际情况下，由于对端的 TCP 窗口可能因为缺少一部分字节就满了，所以 n 的值可能为(0,buf_length]。当 0 < n < buf_length 时，虽然此时 send 函数调用成功，但在业务上并不算正确，因为有部分数据并没有被发送出去。我们可能在一次测试中测不出 n 小于 buf_length 的情况，但不代表实际上不存在。所以，建议要么在返回值 n 等于 buf_length 时才认为正确，要么在一个循环中

调用 send 函数，如果数据一次性发送不完，则记录偏移量，下一次从偏移量处接着发送，直到全部发送完为止：

```
//不推荐的方式
int n = send(socket, buf, buf_length, 0);
if (n == buf_length)
{
    printf("send data successfully\n");
}

//推荐的方式：在一个循环里面根据偏移量发送数据
bool SendData(const char* buf, int buf_length)
{
    //已发送的字节数
    int sent_bytes = 0;
    int ret = 0;
    while (true)
    {
        ret = send(m_hSocket, buf + sent_bytes, buf_length - sent_bytes, 0);
        if (ret == -1)
        {
            if (errno == EWOULDBLOCK)
            {
                //严谨的做法是：如果发送不出去，则应该缓存尚未发出去的数据。
                break;
            }
            else if (errno == EINTR)
                continue;
            else
                return false;
        }
        else if (ret == 0)
        {
            return false;
        }

        sent_bytes += ret;
        if (sent_bytes == buf_length)
            break;
    }

    return true;
}
```

（2）返回值等于 0。在通常情况下，如果 send 或 recv 函数返回 0，我们就认为对端关闭了连接，我们这端也关闭连接即可，这是实际开发时最常见的处理逻辑。send 函数主动发送 0 字节时也会返回 0，这是一种特例，下一节会介绍。

（3）返回值小于 0。对于 send 或 recv 函数返回值小于 0 的情况（即返回-1），根据前面的讨论，此时并不表示 send 或者 recv 函数一定调用出错。这里以下表进行说明。

返回值和错误码	send 函数	recv 函数	操作系统
返回-1，错误码是 EWOUDBLOCK 或 EAGAIN	TCP 窗口太小，数据暂时发不出去	在当前内核缓冲区中无可读数据	Linux
返回-1，错误码是 EINTR	被信号中断，需要重试	被信号中断，需要重试	Linux
返回-1，错误码不是以上 3 种	出错	出错	Linux
返回-1，错误码是 WSAEWOUDBLOCK	TCP 窗口太小，数据暂时发不出去	在当前内核缓冲区中无可读数据	Windows
返回-1，错误码不是 WSAEWOUDBLOCK	出错	出错	Windows

注意：此表展现的是非阻塞模式下 socket 的 send 和 recv 返回值，对于阻塞模式下的 socket，如果返回值是-1（在 Windows 上即 SOCKET_ERROR），则一定表示出错。

4.6.4 阻塞与非阻塞 socket 的各自适用场景

阻塞的 socket 函数在调用 send、recv、connect、accept 等函数时，如果特定的条件不满足，就会阻塞其调用线程直至超时，非阻塞的 socket 恰恰相反。这并不意味着非阻塞模式比阻塞模式模式好，二者各有优缺点。

非阻塞模式一般用于需要支持高并发多 QPS 的场景（如服务器程序），但是正如前文所述，这种模式让程序的执行流和控制逻辑变得复杂；相反，阻塞模式逻辑简单，程序结构简单明了，常用于一些特殊场景中。非阻塞模式的使用非常普遍，这里不再举例，这里举两个可以使用阻塞模式的应用场景。

应用场景一：某程序需要临时发送一个文件，文件分段发送，每发送一段，对端都会给予一个响应，该程序可以单独开一个任务线程，在这个任务线程函数里面，使用先 send 后 recv 再 send 再 recv 的模式，每次 send 和 recv 都是阻塞模式的。

应用场景二：A 端与 B 端之间的通信只有问答模式，即 A 端每发送给 B 端一个请求，B 端必定会给 A 端一个响应，除此以外，B 端不会向 A 端推送任何数据，此时 A 端就可以采用阻塞模式，在每次 send 完请求后，都可以直接使用阻塞式的 recv 函数接收一定要有的应答包。

4.7 发送 0 字节数据的效果

4.6 节提到 send 或 recv 函数返回 0 表示对端关闭了连接。有人可能认为如果发送了 0 字节数据，那么 send 函数也会返回 0，对端会接收到 0 字节数据。真是这样吗？

我们通过一个例子来看看发送一个长度为 0 的数据，send 函数的返回值是什么，对端是否会接收到 0 字节数据。

server 端的代码如下：

```cpp
/**
 * 验证 recv 函数接收 0 字节数据时的行为，server 端，server_recv_zero_bytes.cpp
 * zhangyl 2018.12.17
 */
#include <sys/types.h>
#include <sys/socket.h>
#include <arpa/inet.h>
#include <unistd.h>
#include <iostream>
#include <string.h>
#include <vector>

int main(int argc, char* argv[])
{
    //1.创建一个监听 socket
    int listenfd = socket(AF_INET, SOCK_STREAM, 0);
    if (listenfd == -1)
    {
        std::cout << "create listen socket error." << std::endl;
        return -1;
    }

    //2.初始化服务器地址
    struct sockaddr_in bindaddr;
    bindaddr.sin_family = AF_INET;
    bindaddr.sin_addr.s_addr = htonl(INADDR_ANY);
    bindaddr.sin_port = htons(3000);
    if (bind(listenfd, (struct sockaddr *)&bindaddr, sizeof(bindaddr)) == -1)
    {
        std::cout << "bind listen socket error." << std::endl;
        close(listenfd);
        return -1;
    }

    //3.启动监听
    if (listen(listenfd, SOMAXCONN) == -1)
    {
        std::cout << "listen error." << std::endl;
        close(listenfd);
        return -1;
    }

    int clientfd;
    struct sockaddr_in clientaddr;
    socklen_t clientaddrlen = sizeof(clientaddr);
    //4.接受客户端连接
    clientfd = accept(listenfd, (struct sockaddr *)&clientaddr, &clientaddrlen);
    if (clientfd != -1)
    {
        while (true)
        {
            char recvBuf[32] = {0};
            //5.从客户端接收数据，客户端没有数据过来时会在 recv 函数处阻塞
            int ret = recv(clientfd, recvBuf, 32, 0);
            if (ret > 0)
```

```
            {
                std::cout << "recv data from client, data: " << recvBuf << std::endl;
            }
            else if (ret == 0)
            {
                //假设 recv 返回值为 0 时意味着收到了 0 字节数据
                std::cout << "recv 0 byte data." << std::endl;
                continue;
            }
            else
            {
                //出错
                std::cout << "recv data error." << std::endl;
                break;
            }
        }
    }

    //6.关闭客户端 socket
    close(clientfd);
    //7.关闭监听 socket
    close(listenfd);

    return 0;
}
```

以上代码的监听端口号是 3000，加粗代码行调用了 recv 函数，如果客户端一直没有数据，则程序会阻塞在这行。

client 端的代码如下：

```
/**
 * 验证非阻塞模式下 send 函数发送 0 字节数据时的行为，client 端
 * nonblocking_client_send_zero_bytes.cpp
 * zhangyl 2018.12.17
 */
#include <sys/types.h>
#include <sys/socket.h>
#include <arpa/inet.h>
#include <unistd.h>
#include <iostream>
#include <string.h>
#include <stdio.h>
#include <fcntl.h>
#include <errno.h>

#define SERVER_ADDRESS  "127.0.0.1"
#define SERVER_PORT     3000
#define SEND_DATA       ""

int main(int argc, char* argv[])
{
    //1.创建一个 socket
    int clientfd = socket(AF_INET, SOCK_STREAM, 0);
    if (clientfd == -1)
```

```cpp
    {
        std::cout << "create client socket error." << std::endl;
        return -1;
    }

    //2.连接服务器
    struct sockaddr_in serveraddr;
    serveraddr.sin_family = AF_INET;
    serveraddr.sin_addr.s_addr = inet_addr(SERVER_ADDRESS);
    serveraddr.sin_port = htons(SERVER_PORT);
    if (connect(clientfd, (struct sockaddr *)&serveraddr, sizeof(serveraddr)) == -1)
    {
        std::cout << "connect socket error." << std::endl;
        close(clientfd);
        return -1;
    }

    //连接成功以后,我们再将clientfd设置为非阻塞模式,
    //不能在创建时就设置,这样会影响到connect函数的行为
    int oldSocketFlag = fcntl(clientfd, F_GETFL, 0);
    int newSocketFlag = oldSocketFlag | O_NONBLOCK;
    if (fcntl(clientfd, F_SETFL, newSocketFlag) == -1)
    {
        close(clientfd);
        std::cout << "set socket to nonblock error." << std::endl;
        return -1;
    }

    //3.不断向服务器发送数据,或者出错退出
    int count = 0;
    while (true)
    {
        //发送0字节数据
        int ret = send(clientfd, SEND_DATA, 0, 0);
        if (ret == -1)
        {
            //在非阻塞模式下,send函数由于TCP窗口太小,发不出去数据,错误码是EWOULDBLOCK
            if (errno == EWOULDBLOCK)
            {
                std::cout << "send data error as TCP Window size is too small." << std::endl;
                continue;
            }
            else if (errno == EINTR)
            {
                //如果被信号中断,则继续重试
                std::cout << "sending data interrupted by signal." << std::endl;
                continue;
            }
            else
            {
                std::cout << "send data error." << std::endl;
                break;
            }
        }
```

```
        else if (ret == 0)
        {
            //发送了 0 字节数据
            std::cout << "send 0 byte data." << std::endl;
        }
        else
        {
            count ++;
            std::cout << "send data successfully, count = " << count << std::endl;
        }

        //每 3 秒发送一次
        sleep(3);
    }

    //5.关闭 socket
    close(clientfd);

    return 0;
}
```

先启动 server 端（可执行文件名为 server_recv_zero_bytes），我们接着使用 tcpdump 抓取经过端口 3000 的数据包，使用如下命令：

```
tcpdump -i any 'tcp port 3000'
```

最后启动 client 端（可执行文件名为 nonblocking_client_send_zero_bytes），client 端在连接 server 端成功后，每隔 3 秒调用 send 函数发送一段 0 字节数据，输出结果如下：

```
[root@myaliyun chapter04]# ./nonblocking_client_send_zero_bytes
send 0 byte data.
send 0 byte data.
send 0 byte data.
send 0 byte data.
send 0 byte data.
send 0 byte data.
send 0 byte data.
send 0 byte data.
send 0 byte data.
```

client 端确实是每隔 3 秒发送一次数据。此时使用 lsof -i -Pn 命令查看连接状态，发现连接状态也是正常的，如下图所示。

```
[root@myaliyun ~]# lsof -i -Pn | grep 3000
server_re 24951   root    3u  IPv4 8482110      0t0  TCP *:3000 (LISTEN)
server_re 24951   root    4u  IPv4 8482111      0t0  TCP 127.0.0.1:3000->127.0.0.1:38750 (ESTABLISHED)
nonblocki 24955   root    3u  IPv4 8482121      0t0  TCP 127.0.0.1:38750->127.0.0.1:3000 (ESTABLISHED)
```

在 tcpdump 抓包结果中，除了建立连接时的三次握手数据包，再无其他数据包，也就是说，send 函数发送 0 字节数据，此时 send 函数返回 0，但 client 端的操作系统协议栈并不会把这些数据发送出去：

```
[[root@myaliyun ~]# tcpdump -i any 'tcp port 3000'
tcpdump: verbose output suppressed, use -v or -vv for full protocol decode
listening on any, link-type LINUX_SLL (Linux cooked), capture size 262144 bytes
21:15:35.106740 IP localhost.38756 > localhost.hbci: Flags [S], seq 3145581735, win 43690, options [mss 65495,sackOK,TS val 145859887 ecr 0,nop,wscale 7], length 0
```

```
21:15:35.106751 IP localhost.hbci > localhost.38756: Flags [S.], seq 3377074905, ack
3145581736, win 43690, options [mss 65495,sackOK,TS val 145859887 ecr
145859887,nop,wscale 7], length 0
21:15:35.106762 IP localhost.38756 > localhost.hbci: Flags [.], ack 1, win 342,
options [nop,nop,TS val 145859887 ecr 145859887], length 0
```

因此，server 端也会一直没有输出，如果我们用的是 gdb 启动 server，则此时中断会发现，server 端由于没有数据，一直阻塞在 recv 函数调用处（server_recv_zero_bytes.cpp 文件中的加粗代码行），下图展示将 server 端用 gdb 中断下来的验证结果。

```
[root@myaliyun chapter04]#gdb server_recv_zero_bytes
GNU gdb (GDB) 8.3
Copyright (C) 2019 Free Software Foundation, Inc.
License GPLv3+: GNU GPL version 3 or later <http://gnu.org/licenses/gpl.html>
This is free software: you are free to change and redistribute it.
There is NO WARRANTY, to the extent permitted by law.
Type "show copying" and "show warranty" for details.
This GDB was configured as "x86_64-pc-linux-gnu".
Type "show configuration" for configuration details.
For bug reporting instructions, please see:
<http://www.gnu.org/software/gdb/bugs/>.
Find the GDB manual and other documentation resources online at:
    <http://www.gnu.org/software/gdb/documentation/>.

For help, type "help".
Type "apropos word" to search for commands related to "word"...
Reading symbols from server_recv_zero_bytes...
(gdb) r
Starting program: /root/mybooksources/chapter04/server_recv_zero_bytes
warning: File "/usr/local/lib64/libstdc++.so.6.0.24-gdb.py" auto-loading has been declined by
ir:$datadir/auto-load".
To enable execution of this file add
        add-auto-load-safe-path /usr/local/lib64/libstdc++.so.6.0.24-gdb.py
line to your configuration file "/root/.gdbinit".
To completely disable this security protection add
        set auto-load safe-path /
line to your configuration file "/root/.gdbinit".
For more information about this security protection see the
"Auto-loading safe path" section in the GDB manual.  E.g., run from the shell:
        info "(gdb)Auto-loading safe path"
^C
Program received signal SIGINT, Interrupt.
0x00007ffff7271e7d in recv () from /usr/lib64/libc.so.6
(gdb) bt
#0  0x00007ffff7271e7d in recv () from /usr/lib64/libc.so.6
#1  0x000000000040136f in main (argc=1, argv=0x7fffffffe2e8) at server_recv_zero_bytes.cpp:54
(gdb)
```

通过上面的测试，可以知道存在以下两种情形让 send 函数的返回值为 0：

（1）对端关闭连接时，我们正好尝试调用 send 函数发送数据；

（2）本端尝试调用 send 函数发送 0 字节数据。

而 recv 函数只有在对端关闭连接时才会返回 0，对端发送 0 字节数据，本端的 recv 函数是不会收到 0 字节数据的。

然而，这里的代码示例仅仅用于实验性讨论，发送一个 0 字节数据是没有任何意义的，希望读者在实际开发时避免写出可能调用 send 函数发送 0 字节数据的代码。

4.8　connect 函数在阻塞和非阻塞模式下的行为

当 socket 使用阻塞模式时，connect 函数会一直到有明确的结果才会返回（或连接成功或连接失败），如果服务器地址"较远"或者网络状况不好，连接速度较慢，则程序可能会在 connect 函数处阻塞好一会儿（如两三秒之久）。虽然这一般也不会对依赖于网络通信的程序造成什么影响，但在实际项目中，我们一般倾向于使用异步 connect 技术（非阻塞 connect），一般有如下步骤。

（1）创建 socket，将 socket 设置为非阻塞模式。

（2）调用 connect 函数，此时无论 connect 函数是否连接成功，都会立即返回；如果返回-1，则并不一定表示连接出错，如果此时错误码是 EINPROGRESS，则表示正在尝试连接。

（3）调用 select 函数，在指定的时间内判断该 socket 是否可写，如果可写，则说明连接成功，反之认为连接失败。

按上述流程编写代码如下：

```cpp
/**
 * 异步的 connect 写法，nonblocking_connect.cpp
 * zhangyl 2018.12.17
 */
#include <sys/types.h>
#include <sys/socket.h>
#include <arpa/inet.h>
#include <unistd.h>
#include <iostream>
#include <string.h>
#include <stdio.h>
#include <fcntl.h>
#include <errno.h>

#define SERVER_ADDRESS    "127.0.0.1"
#define SERVER_PORT       3000
#define SEND_DATA         "helloworld"

int main(int argc, char* argv[])
{
    //1.创建一个socket
    int clientfd = socket(AF_INET, SOCK_STREAM, 0);
    if (clientfd == -1)
    {
        std::cout << "create client socket error." << std::endl;
        return -1;
    }

    //将 clientfd 设置为非阻塞模式
    int oldSocketFlag = fcntl(clientfd, F_GETFL, 0);
    int newSocketFlag = oldSocketFlag | O_NONBLOCK;
    if (fcntl(clientfd, F_SETFL, newSocketFlag) == -1)
```

```cpp
    {
        close(clientfd);
        std::cout << "set socket to nonblock error." << std::endl;
        return -1;
    }

    //2.连接服务器
    struct sockaddr_in serveraddr;
    serveraddr.sin_family = AF_INET;
    serveraddr.sin_addr.s_addr = inet_addr(SERVER_ADDRESS);
    serveraddr.sin_port = htons(SERVER_PORT);
    for (;;)
    {
        int ret = connect(clientfd, (struct sockaddr *)&serveraddr, sizeof(serveraddr));
        if (ret == 0)
        {
            std::cout << "connect to server successfully." << std::endl;
            close(clientfd);
            return 0;
        }
        else if (ret == -1)
        {
            if (errno == EINTR)
            {
                //connect动作被信号中断，重试connect
                std::cout << "connecting interruptted by signal, try again." << std::endl;
                continue;
            }
            else if (errno == EINPROGRESS)
            {
                //连接正在尝试中
                break;
            }
            else
            {
                //真的出错了
                close(clientfd);
                return -1;
            }
        }
    }

    fd_set writeset;
    FD_ZERO(&writeset);
    FD_SET(clientfd, &writeset);
    struct timeval tv;
    tv.tv_sec = 3;
    tv.tv_usec = 0;
    //3.调用select函数判断socket是否可写
    if (select(clientfd + 1, NULL, &writeset, NULL, &tv) == 1)
    {
        std::cout << "[select] connect to server successfully." << std::endl;
    }
```

```
    else
    {
        std::cout << "[select] connect to server error." << std::endl;
    }

    close(clientfd);

    return 0;
}
```

为了区别到底是通过调用 connect 函数判断连接成功的,还是通过调用 select 函数判断连接成功的,我们在后者的输出内容中都加上了[select]标签以示区别。

先用 nc 命令启动一个服务端程序:

```
nc -v -l 0.0.0.0 3000
```

然后编译客户端程序并执行:

```
[root@localhost testsocket]# g++ -g -o nonblocking_connect nonblocking_connect.cpp
[root@localhost testsocket]# ./nonblocking_connect
[select] connect to server successfully.
```

把服务端程序关掉,再重新启动客户端,这时应该会连接失败,程序输出结果如下:

```
[root@localhost testsocket]# ./nonblocking_connect
[select] connect to server successfully.
```

奇怪?为什么连接不上也会得到一样的输出结果?程序出问题的原因如下。

(1)在 Windows 上,一个 socket 没有建立连接之前,我们使用 select 函数检测其是否可写,能得到正确的结果(不可写),连接成功后检测,会变为可写。所以,上述介绍的异步 connect 写法流程在 Windows 上是没有问题的。

(2)在 Linux 上一个 socket 没有建立连接之前,用 select 函数检测其是否可写,我们也会得到可写的结果,所以上述流程并不适用于 Linux。Linux 上正确的做法是,connect 之后,不仅要调用 select 检测是否可写,还要调用 getsockopt 检测此时 socket 是否出错,通过错误码来检测和确定是否连接上,错误码为 0 时表示连接上,反之表示未连接上。完整的代码如下:

```
/**
 * Linux上正确的异步connect写法, linux_nonblocking_connect.cpp
 * zhangyl 2018.12.17
 */
#include <sys/types.h>
#include <sys/socket.h>
#include <arpa/inet.h>
#include <unistd.h>
#include <iostream>
#include <string.h>
#include <stdio.h>
#include <fcntl.h>
#include <errno.h>
```

```cpp
#define SERVER_ADDRESS  "127.0.0.1"
#define SERVER_PORT     3000
#define SEND_DATA       "helloworld"

int main(int argc, char* argv[])
{
    //1.创建一个socket
    int clientfd = socket(AF_INET, SOCK_STREAM, 0);
    if (clientfd == -1)
    {
        std::cout << "create client socket error." << std::endl;
        return -1;
    }

    //将clientfd设置为非阻塞模式
    int oldSocketFlag = fcntl(clientfd, F_GETFL, 0);
    int newSocketFlag = oldSocketFlag | O_NONBLOCK;
    if (fcntl(clientfd, F_SETFL, newSocketFlag) == -1)
    {
        close(clientfd);
        std::cout << "set socket to nonblock error." << std::endl;
        return -1;
    }

    //2.连接服务器
    struct sockaddr_in serveraddr;
    serveraddr.sin_family = AF_INET;
    serveraddr.sin_addr.s_addr = inet_addr(SERVER_ADDRESS);
    serveraddr.sin_port = htons(SERVER_PORT);
    for (;;)
    {
        int ret = connect(clientfd, (struct sockaddr *)&serveraddr, sizeof(serveraddr));
        if (ret == 0)
        {
            std::cout << "connect to server successfully." << std::endl;
            close(clientfd);
            return 0;
        }
        else if (ret == -1)
        {
            if (errno == EINTR)
            {
                //connect动作被信号中断，重试connect
                std::cout << "connecting interruptted by signal, try again." << std::endl;
                continue;
            }
            else if (errno == EINPROGRESS)
            {
                //连接正在尝试中
                break;
            }
            else
            {
```

```cpp
            //真的出错了
            close(clientfd);
            return -1;
        }
    }
}

fd_set writeset;
FD_ZERO(&writeset);
FD_SET(clientfd, &writeset);
struct timeval tv;
tv.tv_sec = 3;
tv.tv_usec = 0;
//3.调用 select 函数判断 socket 是否可写
if (select(clientfd + 1, NULL, &writeset, NULL, &tv) != 1)
{
    std::cout << "[select] connect to server error." << std::endl;
    close(clientfd);
    return -1;
}

int err;
socklen_t len = static_cast<socklen_t>(sizeof err);
//4.调用 getsockopt 检测此时 socket 是否出错
if (::getsockopt(clientfd, SOL_SOCKET, SO_ERROR, &err, &len) < 0)
{
    close(clientfd);
    return -1;
}

if (err == 0)
    std::cout << "connect to server successfully." << std::endl;
else
    std::cout << "connect to server error." << std::endl;

close(clientfd);

return 0;
}
```

在 Linux 上，以上代码中的第 3 步（在注释 "3.调用 select 函数判断 socket 是否可写"处）可以使用 poll 函数代替 select 函数判断 socket 是否可写；在 Windows 上可以使用 WSAEventSelect 或 WSAAsyncSelect 函数判断连接是否成功。

4.9 连接时顺便接收第 1 组数据

在一些应用场景下，我们需要在网络通信双方建立连接成功后就把对端的第 1 组数据接收过来。

Linux 提供了 TCP_DEFER_ACCEPT 的 socket 选项，设置该选项之后只有连接建立成功且收到第 1 组对端数据时，accept 函数才会返回。以下是 libevent 网络库中该选项的用法：

```
//evutil.c
int
evutil_make_tcp_listen_socket_deferred(evutil_socket_t sock)
{
#if defined(EVENT__HAVE_NETINET_TCP_H) && defined(TCP_DEFER_ACCEPT)
    int one = 1;
    //TCP_DEFER_ACCEPT 选项让内核延迟调用 accept 函数，直到对端数据到来
    return setsockopt(sock, IPPROTO_TCP, TCP_DEFER_ACCEPT, &one,
        (ev_socklen_t)sizeof(one));
#endif
    return 0;
}
```

Windows 也提供了类似的功能，其扩展的函数 AcceptEx 用于接受连接后，可选择性地决定是否顺便收取第 1 组对端数据。其函数签名如下：

```
BOOL AcceptEx(
    SOCKET          sListenSocket,
    SOCKET          sAcceptSocket,
    PVOID           lpOutputBuffer,
    DWORD           dwReceiveDataLength,
    DWORD           dwLocalAddressLength,
    DWORD           dwRemoteAddressLength,
    LPDWORD         lpdwBytesReceived,
    LPOVERLAPPED    lpOverlapped
);
```

对其中的参数讲解如下。

（1）sListenSocket：是监听 socket。

（2）sAcceptSocket：对应新产生的 clientsocket。

（3）lpOutputBuffer：是一个输出缓冲区，其内容由三块构成，分别是收到的第 1 组数据和连接双方的地址（均为 sockaddr_in 结构）。

（4）dwReceiveDataLength：是 lpOutputBuffer 中用于存放第 1 组数据的缓冲区长度，如果将其设置为 0，则 AcceptEx 函数会立即返回，如果将其设置为大于 0，则 AcceptEx 函数会等待第 1 组数据收到后再返回。

（5）dwLocalAddressLength 和 dwRemoteAddressLength：是 lpOutputBuffer 中用于存放本端和对端地址结构的缓冲区长度，这两个参数的值必须大于或等于 16。

（6）lpdwBytesReceived：是实际收到的第 1 组数据的字节数。

（7）lpOverlapped：是一个 OVERLAPPED 结构的指针。

（8）返回值：在函数调用成功时返回 TRUE，在函数调用失败时返回 FALSE。

既然被动连接方可以在接受连接时顺便收取第 1 组数据，那么主动连接方也可以在发起连接时顺便发送第 1 组数据。Windows 提供了 ConnectEx 函数来达到此目的，其函数签名如下：

```
BOOL ConnectEx(
    SOCKET              s,
```

```
    const sockaddr*      name,
    int                  namelen,
    PVOID                lpSendBuffer,
    DWORD                dwSendDataLength,
    LPDWORD              lpdwBytesSent,
    LPOVERLAPPED         lpOverlapped
)
```

ConnectEx 函数除了具备大家熟悉的 connect 函数的功能，还可以通过参数 lpSendBuffer 和 dwSendDataLength 指定连接建立成功时发送的数据内容和长度。

和 AcceptEx 一样，我们不能直接使用函数名 ConnectEx 调用该函数，必须先通过调用 WSAIoctl 函数获取 ConnectEx 函数指针，再通过得到的函数指针调用 ConnectEx 函数。libevent 网络库中实现端口模块的代码演示了这一做法：

```
//event_iocp.c
static void *
get_extension_function(SOCKET s, const GUID *which_fn)
{
    void *ptr = NULL;
    DWORD bytes=0;
    //调用 WSAIoctl 获取相关函数指针
    WSAIoctl(s, SIO_GET_EXTENSION_FUNCTION_POINTER,
        (GUID*)which_fn, sizeof(*which_fn),
        &ptr, sizeof(ptr),
        &bytes, NULL, NULL);

    //这里不需要判断 WSAIoctl 是否调用成功
    //我们要么得到相关函数的有效指针，要么得到一个 NULL 指针
    return ptr;
}

static void
init_extension_functions(struct win32_extension_fns *ext)
{
    const GUID acceptex = WSAID_ACCEPTEX;
    const GUID connectex = WSAID_CONNECTEX;
    const GUID getacceptexsockaddrs = WSAID_GETACCEPTEXSOCKADDRS;
    SOCKET s = socket(AF_INET, SOCK_STREAM, 0);
    if (s == EVUTIL_INVALID_SOCKET)
        return;
    ext->AcceptEx = get_extension_function(s, &acceptex);
    //在 ext->ConnectEx 中存储了 ConnectEx 函数的指针
    ext->ConnectEx = get_extension_function(s, &connectex);
    ext->GetAcceptExSockaddrs = get_extension_function(s,
        &getacceptexsockaddrs);
    closesocket(s);

    extension_fns_initialized = 1;
}
```

4.10 如何获取当前 socket 对应的接收缓冲区中的可读数据量

本节讲解如何获取当前 socket 对应的接收缓冲区中的可读数据量。

4.10.1 分析

当一个非监听 socket 可读时，我们想知道其当前在接收缓冲区中已经有多少数据可读，类似于 Java JDK 中 java.io.InputStream.available 方法的功能：

```
//class InputStream;
//返回当前输入流中的可读字节数
//若该函数返回值大于 0，下次便可无阻塞地从该流中读取数据了
int available();
```

Windows 和 Linux 均提供了类似的功能。

在 Windows 上可以使用 ioctlsocket 这个 API 函数，函数签名如下：

```
int ioctlsocket(SOCKET s, long cmd, u_long* argp);
```

参数 s 是需要操作的 socket 句柄，参数 cmd 是对应的操作类型，参数 argp 存储操作后的结果。在函数调用成功时返回 0，在函数调用失败时返回非 0 值。

这个函数的功能非常强大，这里只讨论如何获取对应的 socket 接收缓冲区中的字节数，将 cmd 命令设置为 FIONREAD 即可。代码如下：

```
ulong bytesToRecv;
//clientsock 是需要操作的 socket 句柄
if (ioctlsocket(clientsock, FIONREAD, &bytesToRecv) == 0)
{
    //在函数调用成功后，bytesToRecv 的值即当前接收缓冲区中的数据的字节数
}
```

在 Linux 上可以使用 ioctl 函数，其函数签名如下：

```
#include <sys/ioctl.h>

int ioctl(int d, int request, ...);
```

其用法和返回值与 Windows 版本的 ioctlsocket 函数基本相同，这里不再赘述。

来看一个完整的例子：

```
/**
 * 演示如何获取当前 socket 对应的接收缓冲区中有多少数据可读，linux_ioctl.cpp
 * zhangyl 2019.11.12
 */
#include <sys/types.h>
#include <sys/socket.h>
#include <sys/ioctl.h>
#include <arpa/inet.h>
#include <unistd.h>
#include <fcntl.h>
#include <poll.h>
#include <iostream>
#include <string.h>
```

```cpp
#include <vector>
#include <errno.h>

//无效fd标记
#define INVALID_FD  -1

int main(int argc, char* argv[])
{
    //创建一个监听socket
    int listenfd = socket(AF_INET, SOCK_STREAM, 0);
    if (listenfd == INVALID_FD)
    {
        std::cout << "create listen socket error." << std::endl;
        return -1;
    }

    //将监听socket设置为非阻塞的
    int oldSocketFlag = fcntl(listenfd, F_GETFL, 0);
    int newSocketFlag = oldSocketFlag | O_NONBLOCK;
    if (fcntl(listenfd, F_SETFL, newSocketFlag) == -1)
    {
        close(listenfd);
        std::cout << "set listenfd to nonblock error." << std::endl;
        return -1;
    }

    //复用地址和端口号
    int on = 1;
    setsockopt(listenfd, SOL_SOCKET, SO_REUSEADDR, (char *)&on, sizeof(on));
    setsockopt(listenfd, SOL_SOCKET, SO_REUSEPORT, (char *)&on, sizeof(on));

    //初始化服务器地址
    struct sockaddr_in bindaddr;
    bindaddr.sin_family = AF_INET;
    bindaddr.sin_addr.s_addr = htonl(INADDR_ANY);
    bindaddr.sin_port = htons(3000);
    if (bind(listenfd, (struct sockaddr *)&bindaddr, sizeof(bindaddr)) == -1)
    {
        std::cout << "bind listen socket error." << std::endl;
        close(listenfd);
        return -1;
    }

    //启动监听
    if (listen(listenfd, SOMAXCONN) == -1)
    {
        std::cout << "listen error." << std::endl;
        close(listenfd);
        return -1;
    }

    std::vector<pollfd> fds;
    pollfd listen_fd_info;
    listen_fd_info.fd = listenfd;
    listen_fd_info.events = POLLIN;
```

```cpp
    listen_fd_info.revents = 0;
    fds.push_back(listen_fd_info);

    //是否存在无效的 fd 标志
    bool exist_invalid_fd;
    int n;
    while (true)
    {
        exist_invalid_fd = false;
        n = poll(&fds[0], fds.size(), 1000);
        if (n < 0)
        {
            //被信号中断
            if (errno == EINTR)
                continue;

            //出错，退出循环
            break;
        }
        else if (n == 0)
        {
            //超时，继续
            continue;
        }

        int size = fds.size();
        for (size_t i = 0; i < size; ++i)
        {
            //事件可读
            if (fds[i].revents & POLLIN)
            {
                if (fds[i].fd == listenfd)
                {
                    //监听 socket，接受新连接
                    struct sockaddr_in clientaddr;
                    socklen_t clientaddrlen = sizeof(clientaddr);
                    //接受客户端连接并将产生的 clientfd 加入 fds 集合中
                    int clientfd = accept(listenfd, (struct sockaddr *)&clientaddr, &clientaddrlen);
                    if (clientfd != -1)
                    {
                        //将客户端 socket 设置为非阻塞的
                        int oldSocketFlag = fcntl(clientfd, F_GETFL, 0);
                        int newSocketFlag = oldSocketFlag | O_NONBLOCK;
                        if (fcntl(clientfd, F_SETFL, newSocketFlag) == -1)
                        {
                            close(clientfd);
                            std::cout << "set clientfd to nonblock error." << std::endl;
                        }
                        else
                        {
                            struct pollfd client_fd_info;
                            client_fd_info.fd = clientfd;
                            client_fd_info.events = POLLIN;
                            client_fd_info.revents = 0;
```

```cpp
                    fds.push_back(client_fd_info);
                    std::cout << "new client accepted, clientfd: " << clientfd << std::endl;
                }
            }
        }
        else
        {
            //在 socket 可读时获取当前接收缓冲区中的字节数
            ulong bytesToRecv = 0;
            if (ioctl(fds[i].fd, FIONREAD, &bytesToRecv) == 0)
            {
                std::cout << "bytesToRecv: " << bytesToRecv << std::endl;
            }

            //普通 clientfd, 收取数据
            char buf[64] = { 0 };
            int m = recv(fds[i].fd, buf, 64, 0);
            if (m <= 0)
            {
                if (errno != EINTR && errno != EWOULDBLOCK)
                {
                    //出错或对端关闭了连接, 关闭对应的 clientfd, 并设置无效标志位
                    std::cout << "client disconnected, clientfd: " << fds[i].fd << std::endl;
                    close(fds[i].fd);
                    fds[i].fd = INVALID_FD;
                    exist_invalid_fd = true;
                }
            }
            else
            {
                std::cout << "recv from client: " << buf << ", clientfd: " << fds[i].fd << std::endl;
            }
        }
    }
    else if (fds[i].revents & POLLERR)
    {
        //TODO: 暂不处理
    }

}//end outer-for-loop

if (exist_invalid_fd)
{
    //统一清理无效的 fd
    for (std::vector<pollfd>::iterator iter = fds.begin(); iter != fds.end(); )
    {
        if (iter->fd == INVALID_FD)
            iter = fds.erase(iter);
        else
            ++iter;
    }
}
```

```
    }//end while-loop

    //关闭所有socket
    for (std::vector<pollfd>::iterator iter = fds.begin(); iter != fds.end(); ++ iter)
        close(iter->fd);

    return 0;
}
```

以上程序在3000端口开启了一个监听，使用poll函数检测在监听socket和clientsocket上是否有读事件。对于 clientsocket，当触发其可读事件（POLLIN）时表明有数据可读，我们调用ioctl函数获取当前socket接收缓冲区的字节数并将其打印出来（代码片段中的加粗行）。编译该程序并启动，然后使用nc命令模拟一个客户端进行测试。

客户端的输入结果如下：

```
[root@myserver ~]# nc -v 127.0.0.1 3000
Ncat: Version 6.40 ( http://nmap.org/ncat )
Ncat: Connected to 127.0.0.1:3000.
hello
world
xxxx
```

服务端的输出结果如下：

```
[zhangyl@mydevbox test]$ g++ -g -o linux_ioctl linux_ioctl.cpp
[zhangyl@mydevbox test]$ ./linux_ioctl
new client accepted, clientfd: 4
bytesToRecv: 6
recv from client: hello
, clientfd: 4
bytesToRecv: 6
recv from client: world
, clientfd: 4
bytesToRecv: 5
recv from client: xxxx
, clientfd: 4
```

需要注意的是，由于 nc 命令默认以换行符（\n）结束，因此无论是客户端还是服务端，输出后都多了一个空行，服务器每次收到的字符串数量（即 bytesToRecv 值）都是可见字符串部分加上一个换行符的长度，例如 hello\n 的长度是 6。

4.10.2 注意事项

这里有两个注意事项。

（1）对于以下代码，第 3 个参数 bytesToRecv 是一个输出参数，这对于大多数其他函数来说都意味着 bytesToRecv 可以不指定初始化值，因为函数在调用成功后会给该变量设置值：

```
ulong bytesToRecv = 0;
if (ioctl(fds[i].fd, FIONREAD, &bytesToRecv) == 0)
```

```
    {
        //代码省略
    }
```

但是对于 Linux 的 ioctl 函数是个例外，必须将 bytesToRecv 初始化为 0，才能在 ioctl 函数调用成功后得到正确的结果；Windows 的 ioctlsocket 函数则没有这个限定：

```
//对于 Windows，bytesToRecv 可以不进行初始化
ulong bytesToRecv;
if (ioctlsocket(clientsock, FIONREAD, &bytesToRecv) == 0)
{
}

//对于 Linux，bytesToRecv 必须初始化为 0，才能使用 ioctl 函数得到正确的结果
ulong bytesToRecv = 0;
if (ioctl(clientsock, FIONREAD, &bytesToRecv) == 0)
{
}
```

（2）有人可能认为在调用 recv 或 read 函数接收数据之前，可以调用 ioctlsocket 或 ioctl 函数获得数据的大小，然后根据大小分配缓冲区。伪代码如下：

```
ulong bytesToRecv = 0;
if (ioctl(clientsock, FIONREAD, &bytesToRecv) != 0)
{
    //出错，退出
    return;
}

//根据 bytesToRecv 分配缓冲区的大小
char* pRecvBuf = new char[bytesToRecv];
//调用 recv
int ret = recv(clientsock, pRecvBuf, bytesToRecv, 0);
```

以上代码逻辑其实是有问题的，因为我们在调用完 ioctlsocket 或 ioctl 函数后，在调用 recv 或 read 函数之前，可能接收缓冲区又新增了一段数据，导致实际调用 recv 可以收到的数据长度大于 bytesToRecv，因此建议读者不要基于这样的认知去做一些逻辑上的假设，以免编写出错误的逻辑。

实际的网络通信程序很少需要预先知道在接收缓冲区中有多少可读数据，一般根据实际的业务需求决定收取多少字节的数据。

4.11　Linux EINTR 错误码

在类 UNIX/Linux 中调用一些 socket 函数时（connect、send、recv、epoll_wait 等），除了在函数调用出错时会返回-1，这些函数可能被信号中断时也会返回-1，此时我们可以通过错误码 errno 判断是不是 EINTR，来确定是不是被信号中断。在实际编码时，请务必考虑到这种情况。这也是在上面的很多代码中要专门判断错误码是不是 EINTR 的原因，如果是，则说明被信号中断，我们需要再次调用该函数进行重试。千万不要一看到返回值是-1，就草草认定这些调用失败，进而做出错误的逻辑判断：

```
bool SendData(const char* buf , int buf_length)
{
    //已发送的字节数
    int sent_bytes = 0;
    int ret = 0;
    while (true)
    {
        ret = send(m_hSocket, buf + sent_bytes, buf_length - sent_bytes, 0);
        if (nRet == -1)
        {
            if (errno == EWOULDBLOCK)
            {
                //严谨的做法：这里如果发送不出去，则应该缓存尚未发出去的数据
                break;
            }
            else if (errno == EINTR)
                continue;
            else
                return false;
        }
        else if (nRet == 0)
        {
            return false;
        }

        sent_bytes += ret;
        if (sent_bytes == buf_length)
            break;
    }

    return true;
}
```

4.12 Linux SIGPIPE 信号

为了描述方便，以下将 TCP 的通信双方用 A 和 B 来代替。当 A 关闭连接时，若 B 继续向 A 发送数据，则根据 TCP 的规定，B 会收到 A 的一个 RST 报文应答，若 B 继续向这个服务器发送数据，系统就会产生一个 SIGPIPE 信号给该 B 进程，告诉它程这个连接已经断开了，不要再写了。系统对 SIGPIPE 信号的默认处理行为是让 B 进程退出。

操作系统对 SIGPIPE 信号的这种默认处理行为非常不友好，让我们来分析一下。下图是 TCP 通信四次挥手的示意图，TCP 通信是全双工的信道，可以看作两条单工信道，TCP 连接两端的两个端点各负责一条。当对端关闭时，虽然本意是关闭整个两条信道，但本端只收到 FIN 包。按照 TCP 规定的语义，表示对端只关闭了其所负责的那一条单工信道，虽然不再发送数据，但仍然可以继续接收数据。

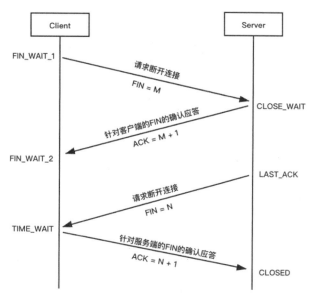

也就是说，因为 TCP 的限制，通信的一方无法获知对端的 socket 是调用了 close 还是调用了 shutdown：

```
int shutdown(int socket, int how);
```

shutdown 函数的参数 how 可以被设置为 SHUT_RD、SHUT_WR 或 SHUT_RDWR，分别表示关闭收通道、发通道或者同时关闭收发通道。

对一个已经收到 FIN 包的 socket 调用 read/recv 方法，如果接收缓冲区已空，则会返回 0，这就是常说的连接关闭状态。但第 1 次对其调用 write/send 方法时，如果发送缓冲区没问题，则发送成功（即 write/send 函数的返回值大于 0），但发送的报文会导致对端回应 RST 报文。因为上一次程序调用 write/send 时是正常的，所以再次尝试调用 write/send 函数时，会因为产生 SIGPIPE 信号而导致进程退出。

这种默认的行为对于我们开发程序造成一定的影响，尤其是后端服务程序。一般后端服务程序需要同时对多个客户端进行服务，不能因为与某个客户端的连接出了问题，而导致整个后端服务进程退出，从而不能继续为其他客户端服务。

为了避免这种情况出现，我们可以捕获 SIGPIPE 信号并对其进行处理或者忽略该信号，忽略该信号的代码如下：

```
signal(SIGPIPE, SIG_IGN);
```

这样设置后，第 2 次调用 write/send 方法时会返回 -1，同时，errno 错误码被置为 EPIPE，程序便能知道对端已经关闭。

4.13 Linux poll 函数的用法

poll 函数用于检测一组文件描述符（File Descriptor，简称 fd）上的可读可写和出错事件，其函数签名如下：

```
#include <poll.h>

int poll(struct pollfd* fds, nfds_t nfds, int timeout);
```

参数解释如下。

（1）fds：指向一个结构体数组首个元素的指针，每个数组元素都是一个 struct pollfd 结构，用于指定检测某个给定的 fd 的条件。

（2）nfds：参数 fds 结构体数组的长度。nfds_t 在本质上是 unsigned long int，其定义如下：

```
typedef unsigned long int nfds_t;
```

（3）timeout：表示 poll 函数的超时时间，单位为毫秒。

struct pollfd 结构体的定义如下：

```
struct pollfd {
    int       fd;       /* 待检测事件的 fd      */
    short     events;   /* 关心的事件组合       */
    short     revents;  /* 检测后得到的事件类型 */
};
```

struct pollfd 的 events 字段是由开发者设置的，告诉内核我们关注什么事件；而 revents 字段是 poll 函数返回时内核设置的，说明该 fd 发生了什么事件。events 和 revents 一般有如下表所示的取值。

事 件 宏	事件描述	是否可以作为输入（events）	是否可以作为输出（revents）
POLLIN	数据可读（包括普通数据&优先数据）	是	是
POLLOUT	数据可写（普通数据&优先数据）	是	是
POLLRDNORM	等同于 POLLIN	是	是
POLLRDBAND	优先级带数据可读（一般用于 Linux）	是	是
POLLPRI	高优先级数据可读，例如 TCP 带外数据	是	是
POLLWRNORM	等同于 POLLOUT	是	是
POLLWRBAND	优先级带数据可写	是	是
POLLRDHUP	TCP 连接被对端关闭，或者关闭了写操作，由 GNU 引入	是	是
POPPHUP	挂起	否	是
POLLERR	错误	否	是
POLLNVAL	文件描述符没有打开	否	是

poll 检测一组 fd 上的可读可写和出错事件的概念与前面介绍 select 的事件含义一样，这里不再赘述。poll 与 select 相比有如下优点：

（1）poll 不要求开发者计算最大文件描述符加 1 的大小；

（2）与 select 相比，poll 在处理大数量的文件描述符时速度更快；

（3）poll 没有最大连接数的限制，因为其存储 fd 的数组没有长度限制；

（4）在调用 poll 函数时，只需对参数进行一次设置就好了。

来看一个具体的例子：

```cpp
/**
 * 演示 poll 函数的用法，poll_server.cpp
 * zhangyl 2019.03.16
 */
#include <sys/types.h>
#include <sys/socket.h>
#include <arpa/inet.h>
#include <unistd.h>
#include <fcntl.h>
#include <poll.h>
#include <iostream>
#include <string.h>
#include <vector>
#include <errno.h>

//无效 fd 标记
#define INVALID_FD  -1

int main(int argc, char *argv[])
{
    //创建一个监听 socket
    int listenfd = socket(AF_INET, SOCK_STREAM, 0);
    if (listenfd == INVALID_FD)
    {
        std::cout << "create listen socket error." << std::endl;
        return -1;
    }

    //将监听 socket 设置为非阻塞的
    int oldSocketFlag = fcntl(listenfd, F_GETFL, 0);
    int newSocketFlag = oldSocketFlag | O_NONBLOCK;
    if (fcntl(listenfd, F_SETFL, newSocketFlag) == -1)
    {
        close(listenfd);
        std::cout << "set listenfd to nonblock error." << std::endl;
        return -1;
    }

    //复用地址和端口号
    int on = 1;
    setsockopt(listenfd, SOL_SOCKET, SO_REUSEADDR, (char *) &on, sizeof(on));
    setsockopt(listenfd, SOL_SOCKET, SO_REUSEPORT, (char *) &on, sizeof(on));

    //初始化服务器的地址
    struct sockaddr_in bindaddr;
    bindaddr.sin_family = AF_INET;
    bindaddr.sin_addr.s_addr = htonl(INADDR_ANY);
    bindaddr.sin_port = htons(3000);
    if (bind(listenfd, (struct sockaddr *) &bindaddr, sizeof(bindaddr)) == -1)
    {
        std::cout << "bind listen socket error." << std::endl;
        close(listenfd);
        return -1;
```

```cpp
    }

    //启动监听
    if (listen(listenfd, SOMAXCONN) == -1)
    {
        std::cout << "listen error." << std::endl;
        close(listenfd);
        return -1;
    }

    std::vector<pollfd> fds;
    pollfd listen_fd_info;
    listen_fd_info.fd = listenfd;
    listen_fd_info.events = POLLIN;
    listen_fd_info.revents = 0;
    fds.push_back(listen_fd_info);

    //是否存在无效的 fd 标志
    bool exist_invalid_fd;
    int n;
    while (true)
    {
        exist_invalid_fd = false;
        n = poll(&fds[0], fds.size(), 1000);
        if (n < 0)
        {
            //被信号中断
            if (errno == EINTR)
                continue;

            //出错，退出
            break;
        }
        else if (n == 0)
        {
            //超时，继续
            continue;
        }

        for (size_t i = 0; i < fds.size(); ++i)
        {
            //事件可读
            if (fds[i].revents & POLLIN)
            {
                if (fds[i].fd == listenfd)
                {
                    //监听 socket，接受新连接
                    struct sockaddr_in clientaddr;
                    socklen_t clientaddrlen = sizeof(clientaddr);
                    //接受客户端连接并将产生的 clientfd 加入 fds 集合中
                    int clientfd = accept(listenfd, (struct sockaddr *) &clientaddr, &clientaddrlen);
                    if (clientfd != -1)
                    {
                        //将客户端 socket 设置为非阻塞的
```

```cpp
                    int oldSocketFlag = fcntl(clientfd, F_GETFL, 0);
                    int newSocketFlag = oldSocketFlag | O_NONBLOCK;
                    if (fcntl(clientfd, F_SETFL, newSocketFlag) == -1)
                    {
                        close(clientfd);
                        std::cout << "set clientfd to nonblock error." << std::endl;
                    }
                    else
                    {
                        struct pollfd client_fd_info;
                        client_fd_info.fd = clientfd;
                        client_fd_info.events = POLLIN;
                        client_fd_info.revents = 0;
                        fds.push_back(client_fd_info);
                        std::cout << "new client accepted, clientfd: " << clientfd << std::endl;
                    }
                }
            }
            else
            {
                //普通 clientfd，收取数据
                char buf[64] = {0};
                int m = recv(fds[i].fd, buf, 64, 0);
                if (m <= 0)
                {
                    if (errno != EINTR && errno != EWOULDBLOCK)
                    {
                        //出错或对端关闭了连接，关闭对应的clientfd，并设置无效标志位
                        for (std::vector<pollfd>::iterator iter = fds.begin(); iter != fds.end(); ++iter)
                        {
                            if (iter->fd == fds[i].fd)
                            {
                                std::cout << "client disconnected, clientfd: " << fds[i].fd << std::endl;
                                close(fds[i].fd);
                                iter->fd = INVALID_FD;
                                exist_invalid_fd = true;
                                break;
                            }
                        }
                    }
                }
                else
                {
                    std::cout << "recv from client: " << buf << ", clientfd: " << fds[i].fd << std::endl;
                }
            }
        }
        else if (fds[i].revents & POLLERR)
        {
            //TODO: 暂且不处理
        }
```

```
        }//end outer-for-loop

        if (exist_invalid_fd)
        {
            //统一清理无效的 fd
            for (std::vector<pollfd>::iterator iter = fds.begin(); iter != fds.end();)
            {
                if (iter->fd == INVALID_FD)
                    iter = fds.erase(iter);
                else
                    ++iter;
            }
        }
    }//end while-loop

    //关闭所有 socket
    for (std::vector<pollfd>::iterator iter = fds.begin(); iter != fds.end(); ++iter)
        close(iter->fd);

    return 0;
}
```

编译以上程序生成 poll_server 并运行，然后使用 nc 命令模拟三个客户端，并向 poll_server 发送消息（这里三条消息分别是 "abcd" "1234" "helloworld"），然后断开与 poll_server 的连接。poll_server 的输出如下：

```
[root@myaliyun chapter04]# g++ -g -o poll_server poll_server.cpp
[root@myaliyun chapter04]# ./poll_server
new client accepted, clientfd: 4
recv from client: abcd
, clientfd: 4
new client accepted, clientfd: 5
recv from client: 1234
, clientfd: 5
new client accepted, clientfd: 6
recv from client: helloworld
, clientfd: 6
client disconnected, clientfd: 4
client disconnected, clientfd: 5
client disconnected, clientfd: 6
```

由于 nc 命令以\n 作为结束标志，所以 poll_server 收到客户端消息时显示分两行。

通过上面的示例代码，我们也能看出 poll 函数存在的一些缺点：

（1）在调用 poll 函数时，不管有没有意义，大量 fd 的数组在用户态和内核地址空间之间被整体复制；

（2）与 select 函数一样，poll 函数返回后，需要遍历 fd 集合来获取就绪的 fd，这样会使性能下降；

（3）同时连接的大量客户端在某一时刻可能只有很少的就绪状态，因此随着监视的描述符数量的增长，其效率也会线性下降。

与 select 函数实现非阻塞的 connect 原理一样，我们可以使用 poll 去实现，即通过 poll 检测 clientfd 在一定时间内是否可写，示例代码如下：

```cpp
/**
 * 在 Linux 上使用 poll 实现异步 connect
 * linux_nonblocking_connect_poll.cpp
 * zhangyl 2019.03.16
 */
#include <sys/types.h>
#include <sys/socket.h>
#include <arpa/inet.h>
#include <unistd.h>
#include <poll.h>
#include <iostream>
#include <string.h>
#include <stdio.h>
#include <fcntl.h>
#include <errno.h>

#define SERVER_ADDRESS  "127.0.0.1"
#define SERVER_PORT     3000
#define SEND_DATA       "helloworld"

int main(int argc, char* argv[])
{
    //创建一个 socket
    int clientfd = socket(AF_INET, SOCK_STREAM, 0);
    if (clientfd == -1)
    {
        std::cout << "create client socket error." << std::endl;
        return -1;
    }

    //将 clientfd 设置为非阻塞模式
    int oldSocketFlag = fcntl(clientfd, F_GETFL, 0);
    int newSocketFlag = oldSocketFlag | O_NONBLOCK;
    if (fcntl(clientfd, F_SETFL, newSocketFlag) == -1)
    {
        close(clientfd);
        std::cout << "set socket to nonblock error." << std::endl;
        return -1;
    }

    //连接服务器
    struct sockaddr_in serveraddr;
    serveraddr.sin_family = AF_INET;
    serveraddr.sin_addr.s_addr = inet_addr(SERVER_ADDRESS);
    serveraddr.sin_port = htons(SERVER_PORT);
    for (;;)
    {
        int ret = connect(clientfd, (struct sockaddr*)&serveraddr,
```

```cpp
            sizeof(serveraddr));
        if (ret == 0)
        {
            std::cout << "connect to server successfully." << std::endl;
            close(clientfd);
            return 0;
        }
        else if (ret == -1)
        {
            if (errno == EINTR)
            {
                //connect 动作被信号中断，重试 connect
                std::cout << "connecting interruptted by signal, try again." << std::endl;
                continue;
            }
            else if (errno == EINPROGRESS)
            {
                //连接正在尝试中
                break;
            }
            else
            {
                //真的出错了
                close(clientfd);
                return -1;
            }
        }
    }

    pollfd event;
    event.fd = clientfd;
    event.events = POLLOUT;
    int timeout = 3000;
    if (poll(&event, 1, timeout) != 1)
    {
        close(clientfd);
        std::cout << "[poll] connect to server error." << std::endl;
        return -1;
    }

    if (!(event.revents & POLLOUT))
    {
        close(clientfd);
        std::cout << "[POLLOUT] connect to server error." << std::endl;
        return -1;
    }

    int err;
    socklen_t len = static_cast<socklen_t>(sizeof err);
    if (::getsockopt(clientfd, SOL_SOCKET, SO_ERROR, &err, &len) < 0)
        return -1;

    if (err == 0)
        std::cout << "connect to server successfully." << std::endl;
```

```
    else
        std::cout << "connect to server error." << std::endl;

    //关闭socket
    close(clientfd);

    return 0;
}
```

运行效果与前面的 select 实现这个非阻塞的 connect 一样，这里不再呈现运行效果。

4.14 Linux epoll 模型

综合 select 和 poll 的一些优缺点，Linux 从内核 2.6 版本开始引入了更高效的 epoll 模型，本节详细介绍 epoll 模型。

4.14.1 基本用法

要想使用 epoll 模型，则必须先创建一个 epollfd，需要用到 epoll_create 函数：

```
#include <sys/epoll.h>

int epoll_create(int size);
```

参数 size 从 Linux 2.6.8 以后就不再使用了，但是必须为它设置一个大于 0 的值。若 epoll_create 函用调用成功，则返回一个非负值的 epollfd，否则返回-1。

有了 epollfd 之后，我们将需要检测事件的其他 fd 绑定到这个 epollfd 上，或者修改一个已经绑定上去的 fd 的事件类型，或者在不需要时将 fd 从 epollfd 上解绑，这都可以使用 epoll_ctl 函数完成：

```
int epoll_ctl(int epfd, int op, int fd, struct epoll_event* event);
```

对其中的参数说明如下。

（1）epfd：即上面提到的 epollfd。

（2）op：操作类型，取值有 EPOLL_CTL_ADD、EPOLL_CTL_MOD 和 EPOLL_CTL_DEL，分别表示在 epollfd 上添加、修改和移除 fd，当取值是 EPOLL_CTL_DEL 时，第 4 个参数 event 忽略不计，可以将其设置为 NULL。

（3）fd：即需要被操作的 fd。

（4）event：这是一个 epoll_event 结构体的地址。epoll_event 结构体的定义如下：

```
struct epoll_event
{
    uint32_t      events;  /* 需要检测的fd事件标志 */
    epoll_data_t  data;    /* 用户自定义的数据 */
};
```

epoll_event 结构体的 data 字段的类型是 epoll_data_t，我们可以利用这个字段设置一

个自定义数据,它在本质上是一个 Union 对象,在 64 位操作系统中大小是 8 字节,定义如下:

```
typedef union epoll_data
{
    void*        ptr;
    int          fd;
    uint32_t     u32;
    uint64_t     u64;
} epoll_data_t;
```

(5)函数返回值:epoll_ctl 若调用成功则返回 0,若调用失败则返回-1,我们可以通过 errno 错误码获取具体的错误原因。

创建 epollfd 后,设置好某个 fd 上需要检测的事件并将该 fd 绑定到 epollfd 上,就可以调用 epoll_wait 检测事件了。epoll_wait 函数签名如下:

```
int epoll_wait(int epfd, struct epoll_event* events, int maxevents, int timeout);
```

其参数的形式和 poll 函数很类似。参数 events 是一个 epoll_event 结构数组的首地址,它是一个输出参数,在函数调用成功后,在 events 中存放的是与就绪事件相关的 epoll_event 结构体数组;参数 maxevents 是数组元素的个数;timeout 是超时时间,单位是毫秒,如果将其设置为 0,则 epoll_wait 会立即返回。

epoll_wait 若调用成功,则会返回有事件的 fd 数量;若返回 0,则表示超时;若调用失败,则返回-1。

epoll_wait 的使用示例如下:

```
while (true)
{
    epoll_event epoll_events[1024];
    int n = epoll_wait(epollfd, epoll_events, 1024, 1000);
    if (n < 0)
    {
        //被信号中断
        if (errno == EINTR)
            continue;

        //出错,退出
        break;
    }
    else if (n == 0)
    {
        //超时,继续
        continue;
    }

    for (size_t i = 0; i < n; ++i)
    {
        if (epoll_events[i].events & EPOLLIN)
        {
            //处理可读事件
        }
```

```
        else if (epoll_events[i].events & EPOLLOUT)
        {
            //处理可写事件
        }
        else if (epoll_events[i].events & EPOLLERR)
        {
            //处理出错事件
        }
    }
}
```

4.14.2　epoll_wait 与 poll 函数的区别

通过对 poll 与 epoll_wait 函数的介绍可以发现：我们在 epoll_wait 函数调用完成后，可以通过参数 event 拿到所有有事件就绪的 fd（参数 event 仅仅是个输出参数）；而 poll 函数的事件集合参数（poll 函数的第 1 个参数）在调用前后数量都不会改变，只不过调用前通过 pollfd 结构体的 events 字段设置待检测的事件，调用后通过 pollfd 结构体的 revents 字段检测就绪的事件（参数 fds 既是入参也是出参）。

举个生活中的例子，某人不断给我们一些苹果，这些苹果有生有熟，调用 epoll_wait 函数相当于：先把苹果挨个投入 epoll 机器中（调用 epoll_ctl 函数）；然后调用 epoll_wait 函数加工，直接通过另一个袋子就能拿到所有熟苹果。调用 poll 函数相当于：先把收到的苹果装入一个袋子中，调用 poll 函数加工；在调用结束后拿到原来的袋子，在袋子中还是原来那么多苹果，只不过熟苹果被贴上了标签，我们需要挨个查看标签挑选熟苹果。

当然，这并不意味着 poll 函数的效率不如 epoll_wait 函数，一般在 fd 数量比较多但某段时间内就绪事件 fd 数量较少的情况下，epoll_wait 函数才会体现它的优势，也就是说在 socket 连接数量较大而活跃的连接较少时，epoll 模型更高效。

4.14.3　LT 模式和 ET 模式

与 poll 模式的事件宏相比，epoll 模式新增了一个事件宏 EPOLLET，即边缘触发模式（Edge Trigger，ET），我们称默认的模式为水平触发模式（Level Trigger，LT）。这两种模式的区别在于：

（1）对于水平触发模式，一个事件只要有，就会一直触发；

（2）对于边缘触发模式，在一个事件从无到有时才会触发。

这两个词汇来自电学术语，我们可以将 fd 上有数据的状态认为是高电平状态，将没有数据的状态认为是低电平状态，将 fd 可写状态认为是高电平状态，将 fd 不可写状态认为是低电平状态。那么水平模式的触发条件是处于高电平状态，而边缘模式的触发条件是新来的一次电信号将当前状态变为高电平状态。

水平模式的触发条件：①低电平→高电平；②处于高电平状态。

边缘模式的触发条件：低电平→高电平。

以 socket 的读事件为例，对于水平模式，只要在 socket 上有未读完的数据，就会一直产生 EPOLLIN 事件；而对于边缘模式，socket 上每新来一次数据就会触发一次，如果上一次触发后未将 socket 上的数据读完，也不会再触发，除非再新来一次数据。对于 socket 写事件，如果 socket 的 TCP 窗口一直不饱和，就会一直触发 EPOLLOUT 事件；而对于边缘模式，只会触发一次，除非 TCP 窗口由不饱和变成饱和再一次变成不饱和，才会再次触发 EPOLLOUT 事件。

socket 可读事件的水平模式触发条件：①socket 上无数据→socket 上有数据；②socket 处于有数据状态。

socket 可读事件的边缘模式触发条件：①socket 上无数据→socket 上有数据；②socket 又新来一次数据。

socket 可写事件的水平模式触发条件：①socket 可写→socket 不可写；②socket 不可写→ socket 可写。

socket 可写事件的边缘模式触发条件：socket 不可写→socket 可写。

也就是说，对于一个非阻塞 socket，如果使用 epoll 边缘模式检测数据是否可读，则触发可读事件后，一定要一次性地把 socket 上的数据收取干净。也就是说，一定要循环调用 recv 函数直到 recv 出错，错误码是 EWOULDBLOCK（EAGAIN 也一样，此时表示 socket 上的本次数据已经读完）；如果使用水平模式，则我们可以根据业务一次性地收取固定的字节数，或者到收完为止。在边缘模式下收取数据的代码示例如下：

```cpp
bool TcpSession::RecvEtMode()
{
    //每次只收取 256 字节
    char buff[256];
    while (true)
    {
        int nRecv = ::recv(m_clientfd, buff, 256, 0);
        if (nRecv == -1)
        {
            if (errno == EWOULDBLOCK)
                return true;
            else if (errno == EINTR)
                continue;

            return false;
        }
        //对端关闭了 socket
        else if (nRecv == 0)
            return false;

        m_inputBuffer.add(buff, (size_t)nRecv);
    }

    return true;
}
```

下面根据几个具体的例子比较 LT 模式与 ET 模式的区别。先测试 LT 模式与 ET 模式在处理读事件上的区别。代码如下：

```cpp
/**
 * 验证epoll的LT与ET模式的区别，epoll_server.cpp
 * zhangyl 2019.04.01
 */
#include<sys/types.h>
#include<sys/socket.h>
#include<arpa/inet.h>
#include<unistd.h>
#include<fcntl.h>
#include<sys/epoll.h>
#include<poll.h>
#include<iostream>
#include<string.h>
#include<vector>
#include<errno.h>
#include<iostream>

int main()
{
    //创建一个监听socket
    int listenfd = socket(AF_INET, SOCK_STREAM, 0);
    if (listenfd == -1)
    {
        std::cout << "create listen socket error" << std::endl;
        return -1;
    }

    //设置重用IP地址和端口号
    int on = 1;
    setsockopt(listenfd, SOL_SOCKET, SO_REUSEADDR, (char*)&on, sizeof(on));
    setsockopt(listenfd, SOL_SOCKET, SO_REUSEPORT, (char*)&on, sizeof(on));

    //将监听socker设置为非阻塞的
    int oldSocketFlag = fcntl(listenfd, F_GETFL, 0);
    int newSocketFlag = oldSocketFlag | O_NONBLOCK;
    if (fcntl(listenfd, F_SETFL, newSocketFlag) == -1)
    {
        close(listenfd);
        std::cout << "set listenfd to nonblock error" << std::endl;
        return -1;
    }

    //初始化服务器的地址
    struct sockaddr_in bindaddr;
    bindaddr.sin_family = AF_INET;
    bindaddr.sin_addr.s_addr = htonl(INADDR_ANY);
    bindaddr.sin_port = htons(3000);

    if (bind(listenfd, (struct sockaddr*)&bindaddr, sizeof(bindaddr)) == -1)
    {
        std::cout << "bind listen socker error." << std::endl;
        close(listenfd);
        return -1;
```

```cpp
    }

    //启动监听
    if (listen(listenfd, SOMAXCONN) == -1)
    {
        std::cout << "listen error." << std::endl;
        close(listenfd);
        return -1;
    }

    //创建epollfd
    int epollfd = epoll_create(1);
    if (epollfd == -1)
    {
        std::cout << "create epollfd error." << std::endl;
        close(listenfd);
        return -1;
    }

    epoll_event listen_fd_event;
    listen_fd_event.data.fd = listenfd;
    listen_fd_event.events = EPOLLIN;
    //若取消注释掉这一行，则使用ET模式
    //listen_fd_event.events |= EPOLLET; //代码第79行

    //将监听sokcet绑定到epollfd上
    if (epoll_ctl(epollfd, EPOLL_CTL_ADD, listenfd, &listen_fd_event) == -1)
    {
        std::cout << "epoll_ctl error" << std::endl;
        close(listenfd);
        return -1;
    }

    int n;
    while (true)
    {
        epoll_event epoll_events[1024];
        n = epoll_wait(epollfd, epoll_events, 1024, 1000);
        if (n < 0)
        {
            //被信号中断
            if (errno == EINTR)
                continue;

            //出错，退出
            break;
        }
        else if (n == 0)
        {
            //超时，继续
            continue;
        }

        for (size_t i = 0; i < n; ++i)
        {
```

```cpp
            //事件可读
            if (epoll_events[i].events & EPOLLIN)
            {
                if (epoll_events[i].data.fd == listenfd)
                {
                    //监听socket，接受新连接
                    struct sockaddr_in clientaddr;
                    socklen_t clientaddrlen = sizeof(clientaddr);
                    int clientfd = accept(listenfd, (struct sockaddr*)&clientaddr, &clientaddrlen);
                    if (clientfd != -1)
                    {
                        int oldSocketFlag = fcntl(clientfd, F_GETFL, 0);
                        int newSocketFlag = oldSocketFlag | O_NONBLOCK;
                        if (fcntl(clientfd, F_SETFL, newSocketFlag) == -1)
                        {
                            close(clientfd);
                            std::cout << "set clientfd to nonblocking error." << std::endl;
                        }
                        else
                        {
                            epoll_event client_fd_event;
                            client_fd_event.data.fd = clientfd;
                            client_fd_event.events = EPOLLIN;
                            //若取消注释这一行，则使用ET模式
                            //**client_fd_event.events |= EPOLLET**; //代码第135行
                            if (epoll_ctl(epollfd, EPOLL_CTL_ADD, clientfd, &client_fd_event) != -1)
                            {
                                std::cout << "new client accepted,clientfd: " << clientfd << std::endl;
                            }
                            else
                            {
                                std::cout << "add client fd to epollfd error" << std::endl;
                                close(clientfd);
                            }
                        }
                    }
                }
                else
                {
                    std::cout << "client fd: " << epoll_events[i].data.fd << " recv data." << std::endl;
                    //普通clientfd
                    char ch;
                    //每次只接收1字节
                    int m = recv(epoll_events[i].data.fd, &ch, 1, 0);
                    if (m == 0)
                    {
                        //对端关闭了连接，从epollfd上移除clientfd
                        if (epoll_ctl(epollfd, EPOLL_CTL_DEL, epoll_events[i].data.fd, NULL) != -1)
                        {
```

```cpp
                    std::cout << "client disconnected,clientfd:" << 
epoll_events[i].data.fd << std::endl;
                }
                close(epoll_events[i].data.fd);
            }
            else if (m < 0)
            {
                //出错
                if (errno != EWOULDBLOCK && errno != EINTR)
                {
                    if (epoll_ctl(epollfd, EPOLL_CTL_DEL, 
epoll_events[i].data.fd, NULL) != -1)
                    {
                        std::cout << "client disconnected,clientfd:" << 
epoll_events[i].data.fd << std::endl;
                    }
                    close(epoll_events[i].data.fd);
                }
                else
                {
                    //正常收到数据
                    std::cout << "recv from client:" << epoll_events[i].data.fd << 
", " << ch << std::endl;
                }
            }
        }
        else if (epoll_events[i].events & EPOLLERR)
        {
            //TODO：暂不处理
        }
    }
}

    close(listenfd);
    return 0;
}
```

我们先看看水平模式的行为，在以上代码片段中的加粗行都注释掉时，使用 LT 模式编译程序并运行：

```
[root@localhost testepoll]# g++ -g -o epoll_server epoll_server.cpp
[root@localhost testepoll]# ./epoll_server
```

然后另外开启一个 Shell 窗口，使用 nc 命令模拟一个客户端，在连接服务器成功后，向服务器发送一个消息 "abcef"：

```
[root@localhost ~]# nc -v 127.0.0.1 3000
Ncat: Version 7.50 ( https://nmap.org/ncat )
Ncat: Connected to 127.0.0.1:3000.
abcdef
```

此时服务端输出：

```
[root@localhost testepoll]# ./epoll_server
```

```
new client accepted,clientfd: 5
client fd: 5 recv data.
recv from client:5, a
client fd: 5 recv data.
recv from client:5, b
client fd: 5 recv data.
recv from client:5, c
client fd: 5 recv data.
recv from client:5, d
client fd: 5 recv data.
recv from client:5, e
client fd: 5 recv data.
recv from client:5, f
client fd: 5 recv data.
recv from client:5,
```

nc 命令实际发送了 a、b、c、d、e、f 和\n 这 7 个字符，由于服务端使用的是 LT 模式，每次接收一个字符，只要 socket 接收缓冲区中仍有数据可读，POLLIN 事件就会一直触发，所以服务器一共有 7 次输出，直到 socket 接收缓冲区没有数据时为止。

我们将 epoll_server.cpp 代码中加粗行的注释取消，使用 ET 模式再试一下，修改代码并重新编译，然后重新运行。再次使用 nc 命令模拟一个客户端连接后发送 "abcef"，服务器只会有一次输出，效果如下：

```
[root@localhost testepoll]# vi epoll_server.cpp
[root@localhost testepoll]# g++ -g -o epoll_server epoll_server.cpp
[root@localhost testepoll]# ./epoll_server
new client accepted,clientfd: 5
client fd: 5 recv data.
recv from client:5, a
```

由于使用了 ET 模式，所以只会触发一次 POLLIN 事件，如果此时没有新数据到来，就再也不会触发。所以，如果我们继续向服务器发送一条新数据如 123，则服务器将再次触发一次 POLLIN 事件，然后打印出字母 b。

nc 客户端的输出如下：

```
[root@localhost ~]# nc -v 127.0.0.1 3000
Ncat: Version 7.50 ( https://nmap.org/ncat )
Ncat: Connected to 127.0.0.1:3000.
abcdef
123
```

epoll_server 的输出如下：

```
[root@localhost testepoll]# vi epoll_server.cpp
[root@localhost testepoll]# g++ -g -o epoll_server epoll_server.cpp
[root@localhost testepoll]# ./epoll_server
new client accepted,clientfd: 5
client fd: 5 recv data.
recv from client:5, a
client fd: 5 recv data.
```

```
recv from client:5, b
```

所以，如果使用 ET 模式处理读事件，切记要将该次 socket 上的数据收完。再来测试 LT 模式与 ET 模式在处理写事件上的区别，修改以上代码如下：

```cpp
/**
 * 验证 epoll LT 与 ET 模式在处理写事件上的区别, epoll_server_write_event_lt.cpp
 * zhangyl 2019.04.01
 */
#include<sys/types.h>
#include<sys/socket.h>
#include<arpa/inet.h>
#include<unistd.h>
#include<fcntl.h>
#include<sys/epoll.h>
#include<poll.h>
#include<iostream>
#include<string.h>
#include<vector>
#include<errno.h>
#include<iostream>

int main()
{
    //创建一个监听 socket
    int listenfd = socket(AF_INET, SOCK_STREAM, 0);
    if (listenfd == -1)
    {
        std::cout << "create listen socket error" << std::endl;
        return -1;
    }

    //设置重用 IP 地址和端口号
    int on = 1;
    setsockopt(listenfd, SOL_SOCKET, SO_REUSEADDR, (char*)&on, sizeof(on));
    setsockopt(listenfd, SOL_SOCKET, SO_REUSEPORT, (char*)&on, sizeof(on));

    //将监听 socker 设置为非阻塞的
    int oldSocketFlag = fcntl(listenfd, F_GETFL, 0);
    int newSocketFlag = oldSocketFlag | O_NONBLOCK;
    if (fcntl(listenfd, F_SETFL, newSocketFlag) == -1)
    {
        close(listenfd);
        std::cout << "set listenfd to nonblock error" << std::endl;
        return -1;
    }

    //初始化服务器地址
    struct sockaddr_in bindaddr;
    bindaddr.sin_family = AF_INET;
    bindaddr.sin_addr.s_addr = htonl(INADDR_ANY);
    bindaddr.sin_port = htons(3000);

    if (bind(listenfd, (struct sockaddr*)&bindaddr, sizeof(bindaddr)) == -1)
```

```cpp
{
    std::cout << "bind listen socker error." << std::endl;
    close(listenfd);
    return -1;
}

//启动监听
if (listen(listenfd, SOMAXCONN) == -1)
{
    std::cout << "listen error." << std::endl;
    close(listenfd);
    return -1;
}


//创建 epollfd
int epollfd = epoll_create(1);
if (epollfd == -1)
{
    std::cout << "create epollfd error." << std::endl;
    close(listenfd);
    return -1;
}

epoll_event listen_fd_event;
listen_fd_event.data.fd = listenfd;
listen_fd_event.events = EPOLLIN;
//若取消注释这一行，则使用 ET 模式
//listen_fd_event.events |= EPOLLET;

//将监听 sokcet 绑定到 epollfd 上
if (epoll_ctl(epollfd, EPOLL_CTL_ADD, listenfd, &listen_fd_event) == -1)
{
    std::cout << "epoll_ctl error" << std::endl;
    close(listenfd);
    return -1;
}

int n;
while (true)
{
    epoll_event epoll_events[1024];
    n = epoll_wait(epollfd, epoll_events, 1024, 1000);
    if (n < 0)
    {
        //被信号中断
        if (errno == EINTR)
            continue;

        //出错，退出
        break;
    }
    else if (n == 0)
    {
        //超时，继续
```

```cpp
            continue;
        }

        for (size_t i = 0; i < n; ++i)
        {
            //事件可读
            if (epoll_events[i].events & EPOLLIN)
            {
                if (epoll_events[i].data.fd == listenfd)
                {
                    //监听socket，接受新连接
                    struct sockaddr_in clientaddr;
                    socklen_t clientaddrlen = sizeof(clientaddr);
                    int clientfd = accept(listenfd, (struct sockaddr*)&clientaddr, &clientaddrlen);
                    if (clientfd != -1)
                    {
                        int oldSocketFlag = fcntl(clientfd, F_GETFL, 0);
                        int newSocketFlag = oldSocketFlag | O_NONBLOCK;
                        if (fcntl(clientfd, F_SETFL, newSocketFlag) == -1)
                        {
                            close(clientfd);
                            std::cout << "set clientfd to nonblocking error." << std::endl;
                        }
                        else
                        {
                            epoll_event client_fd_event;
                            client_fd_event.data.fd = clientfd;
                            //同时监听新来的连接socket的读和写事件
                            client_fd_event.events = EPOLLIN | EPOLLOUT;
                            //取消注释这一行时，使用ET模式
                            //client_fd_event.events |= EPOLLET;
                            if (epoll_ctl(epollfd, EPOLL_CTL_ADD, clientfd, &client_fd_event) != -1)
                            {
                                std::cout << "new client accepted,clientfd: " << clientfd << std::endl;
                            }
                            else
                            {
                                std::cout << "add client fd to epollfd error" << std::endl;
                                close(clientfd);
                            }
                        }
                    }
                }
                else
                {
                    std::cout << "client fd: " << epoll_events[i].data.fd << " recv data." << std::endl;
                    //普通clientfd
                    char recvbuf[1024] = { 0 };
                    //每次只接收1字节
                    int m = recv(epoll_events[i].data.fd, recvbuf, 1024, 0);
```

```cpp
                if (m == 0)
                {
                    //对端关闭了连接，从 epollfd 上移除 clientfd
                    if (epoll_ctl(epollfd, EPOLL_CTL_DEL, epoll_events[i].data.fd, NULL) != -1)
                    {
                        std::cout << "client disconnected,clientfd:" << epoll_events[i].data.fd << std::endl;
                    }
                    close(epoll_events[i].data.fd);
                }
                else if (m < 0)
                {
                    //出错
                    if (errno != EWOULDBLOCK && errno != EINTR)
                    {
                        if (epoll_ctl(epollfd, EPOLL_CTL_DEL, epoll_events[i].data.fd, NULL) != -1)
                        {
                            std::cout << "client disconnected,clientfd:" << epoll_events[i].data.fd << std::endl;
                        }
                        close(epoll_events[i].data.fd);
                    }
                }
                else
                {
                    //正常收到数据
                    std::cout << "recv from client:" << epoll_events[i].data.fd << ", " << recvbuf << std::endl;
                }
            }
        }
        else if (epoll_events[i].events & EPOLLOUT)
        {
            //只处理客户端 fd 的可写事件
            if (epoll_events[i].data.fd != listenfd)
            {
                //打印结果
                std::cout << "EPOLLOUT triggered,clientfd: " << epoll_events[i].data.fd << std::endl;
            }
        }
        else if (epoll_events[i].events & EPOLLERR)
        {
            //TODO 暂不处理
        }
    }
}

close(listenfd);
return 0;
}
```

在以上代码中，我们对新来的连接 fd 同时注册读和写事件（加粗代码行），再次编译程序并运行：

```
[root@myaliyun chapter04]# g++ -g -o epoll_server_write_event_lt
epoll_server_write_event_lt.cpp
[root@myaliyun chapter04]# ./epoll_server_write_event_lt
```

然后使用 nc 命令模拟一个客户端去连接 epoll_server_write_event_lt：

```
[root@myaliyun ~]# nc -v 127.0.0.1 3000
Ncat: Version 6.40 ( http://nmap.org/ncat )
Ncat: Connected to 127.0.0.1:3000.
```

此时服务端（epoll_server_write_event_lt）会疯狂输出可写事件的触发消息：

```
new client accepted,clientfd: 5
EPOLLOUT triggered,clientfd: 5
EPOLLOUT triggered,clientfd: 5
EPOLLOUT triggered,clientfd: 5
EPOLLOUT triggered,clientfd: 5
EPOLLOUT triggered,clientfd: 5
EPOLLOUT triggered,clientfd: 5
...更多重复的输出省略...
```

之所以是这样，是因为我们注册了可写事件且使用的是 LT 模式。在 LT 模式下，由于这里的服务端对应的客户端 fd 一直是可写的，有写事件一直触发，所以会看到屏幕不断输出。

我们再将服务端与客户端建立连接时新建的 fd 设置为 ET 模式实验一下，修改代码如下：

```cpp
/**
 * 验证 epoll LT 与 ET 模式在处理写事件上的区别, epoll_server_write_event_et.cpp
 * zhangyl 2019.04.01
 */
#include<sys/types.h>
#include<sys/socket.h>
#include<arpa/inet.h>
#include<unistd.h>
#include<fcntl.h>
#include<sys/epoll.h>
#include<poll.h>
#include<iostream>
#include<string.h>
#include<vector>
#include<errno.h>
#include<iostream>

int main()
{
    //创建一个监听 socket
    int listenfd = socket(AF_INET, SOCK_STREAM, 0);
    if (listenfd == -1)
    {
        std::cout << "create listen socket error" << std::endl;
        return -1;
    }
```

```cpp
//设置重用 IP 地址和端口号
int on = 1;
setsockopt(listenfd, SOL_SOCKET, SO_REUSEADDR, (char*)&on, sizeof(on));
setsockopt(listenfd, SOL_SOCKET, SO_REUSEPORT, (char*)&on, sizeof(on));

//将监听 socker 设置为非阻塞的
int oldSocketFlag = fcntl(listenfd, F_GETFL, 0);
int newSocketFlag = oldSocketFlag | O_NONBLOCK;
if (fcntl(listenfd, F_SETFL, newSocketFlag) == -1)
{
    close(listenfd);
    std::cout << "set listenfd to nonblock error" << std::endl;
    return -1;
}

//初始化服务器的地址
struct sockaddr_in bindaddr;
bindaddr.sin_family = AF_INET;
bindaddr.sin_addr.s_addr = htonl(INADDR_ANY);
bindaddr.sin_port = htons(3000);

if (bind(listenfd, (struct sockaddr*) & bindaddr, sizeof(bindaddr)) == -1)
{
    std::cout << "bind listen socker error." << std::endl;
    close(listenfd);
    return -1;
}

//启动监听
if (listen(listenfd, SOMAXCONN) == -1)
{
    std::cout << "listen error." << std::endl;
    close(listenfd);
    return -1;
}


//创建 epollfd
int epollfd = epoll_create(1);
if (epollfd == -1)
{
    std::cout << "create epollfd error." << std::endl;
    close(listenfd);
    return -1;
}

epoll_event listen_fd_event;
listen_fd_event.data.fd = listenfd;
listen_fd_event.events = EPOLLIN;
//若取消注释这一行，则使用 ET 模式
//listen_fd_event.events |= EPOLLET;

//将监听 sokcet 绑定到 epollfd 上
if (epoll_ctl(epollfd, EPOLL_CTL_ADD, listenfd, &listen_fd_event) == -1
```

```cpp
        {
            std::cout << "epoll_ctl error" << std::endl;
            close(listenfd);
            return -1;
        }

        int n;
        while (true)
        {
            epoll_event epoll_events[1024];
            n = epoll_wait(epollfd, epoll_events, 1024, 1000);
            if (n < 0)
            {
                //被信号中断
                if (errno == EINTR)
                    continue;

                //出错,退出
                break;
            }
            else if (n == 0)
            {
                //超时,继续
                continue;
            }
            for (size_t i = 0; i < n; ++i)
            {
                //有读事件
                if (epoll_events[i].events & EPOLLIN)
                {
                    if (epoll_events[i].data.fd == listenfd)
                    {
                        //监听socket,接受新连接
                        struct sockaddr_in clientaddr;
                        socklen_t clientaddrlen = sizeof(clientaddr);
                        int clientfd = accept(listenfd, (struct sockaddr*) & clientaddr, &clientaddrlen);
                        if (clientfd != -1)
                        {
                            int oldSocketFlag = fcntl(clientfd, F_GETFL, 0);
                            int newSocketFlag = oldSocketFlag | O_NONBLOCK;
                            if (fcntl(clientfd, F_SETFL, newSocketFlag) == -1)
                            {
                                close(clientfd);
                                std::cout << "set clientfd to nonblocking error." << std::endl;
                            }
                            else
                            {
                                epoll_event client_fd_event;
                                client_fd_event.data.fd = clientfd;
                                //同时监听新来连接socket的读和写事件
                                client_fd_event.events = EPOLLIN | EPOLLOUT;//代码第133行
                                //若取消注释这一行,则使用ET模式
                                client_fd_event.events |= EPOLLET;
```

```cpp
                    if (epoll_ctl(epollfd, EPOLL_CTL_ADD, clientfd,
&client_fd_event) != -1)
                    {
                        std::cout << "new client accepted,clientfd: " << clientfd
<< std::endl;
                    }
                    else
                    {
                        std::cout << "add client fd to epollfd error" << std::endl;
                        close(clientfd);
                    }
                }
            }
            else
            {
                std::cout << "client fd: " << epoll_events[i].data.fd << " recv
data." << std::endl;
                //普通 clientfd
                char recvbuf[1024] = { 0 };
                //每次只接收 1 字节
                int m = recv(epoll_events[i].data.fd, recvbuf, 1024, 0);
                if (m == 0)
                {
                    //对端关闭了连接，从 epollfd 上移除 clientfd
                    if (epoll_ctl(epollfd, EPOLL_CTL_DEL, epoll_events[i].data.fd,
NULL) != -1)
                    {
                        std::cout << "client disconnected,clientfd:" <<
epoll_events[i].data.fd << std::endl;
                    }
                    close(epoll_events[i].data.fd);
                }
                else if (m < 0)
                {
                    //出错
                    if (errno != EWOULDBLOCK && errno != EINTR)
                    {
                        if (epoll_ctl(epollfd, EPOLL_CTL_DEL,
epoll_events[i].data.fd, NULL) != -1)
                        {
                            std::cout << "client disconnected,clientfd:" <<
epoll_events[i].data.fd << std::endl;
                        }
                        close(epoll_events[i].data.fd);
                    }
                }
                else
                {
                    //正常收到数据
                    std::cout << "recv from client:" << epoll_events[i].data.fd <<
", " << recvbuf << std::endl;

                    epoll_event client_fd_event;
                    client_fd_event.data.fd = epoll_events[i].data.fd;
```

```cpp
                    //再次给clientfd注册检测可写事件
                    client_fd_event.events = EPOLLIN | EPOLLOUT | EPOLLET;
                    //代码第133行
                    if (epoll_ctl(epollfd, EPOLL_CTL_MOD, epoll_events[i].data.fd, &client_fd_event) != -1)
                    {
                        std::cout << "epoll_ctl successfully, mode: EPOLL_CTL_MOD, clientfd:" << epoll_events[i].data.fd << std::endl;
                    }

                }
            }
        }
        else if (epoll_events[i].events & EPOLLOUT)
        {
            //只处理客户端fd的写事件
            if (epoll_events[i].data.fd != listenfd)
            {
                //打印结果
                std::cout << "EPOLLOUT triggered,clientfd: " << epoll_events[i].data.fd << std::endl;
            }
        }
        else if (epoll_events[i].events & EPOLLERR)
        {
            //TODO: 暂不处理
        }
    }
}

close(listenfd);
return 0;
}
```

在以上逻辑中，服务端在每次收到客户端消息时都会重新给客户端fd注册检测可写事件EPOLLOUT（加粗代码行），重新编译代码并启动服务（可执行文件名为epoll_server_write_event_et），再次使用nc命令模拟客户端给服务端发送几条消息（"msg1" "msg2" "msg3"）。

nc客户端的执行结果如下：

```
[root@myaliyun ~]# nc -v 127.0.0.1 3000
Ncat: Version 7.50 ( https://nmap.org/ncat )
Ncat: Connected to 127.0.0.1:3000.
msg1
msg2
msg3
```

epoll_server_write_event_et服务端的执行结果如下：

```
[root@myaliyun chapter04]# g++ -g -o epoll_server_write_event_et epoll_server_write_event_et.cpp
[root@myaliyun chapter04]# ./epoll_server_write_event_et
new client accepted,clientfd: 5
```

```
EPOLLOUT triggered,clientfd: 5
client fd: 5 recv data.
recv from client:5, msg1

epoll_ctl successfully, mode: EPOLL_CTL_MOD, clientfd:5
EPOLLOUT triggered,clientfd: 5
client fd: 5 recv data.
recv from client:5, msg2

epoll_ctl successfully, mode: EPOLL_CTL_MOD, clientfd:5
EPOLLOUT triggered,clientfd: 5
client fd: 5 recv data.
recv from client:5, msg3

epoll_ctl successfully, mode: EPOLL_CTL_MOD, clientfd:5
EPOLLOUT triggered,clientfd: 5
```

通过以上输出，我们可以发现，当使用 ET 模式下，即使服务端给客户端 fd 注册了检测可写事件，可写事件也不会一直触发，只会触发一次，触发完成后只有再次注册、检测可写事件时，可写事件才会继续触发。在 epoll_server_write_event_et 服务中是靠客户端来新消息驱动再次注册、检测可写事件的。也就是说，我们使用 ET 模式去处理可写事件，不必像 LT 模式那样为了避免不必要的可写事件重复触发，在可写事件触发后，如果不再需要，则应该立即移除对可写事件的注册。

这就意味着，使用 LT 模式时，如果我们的实现依赖于可写事件触发去发送数据，那么我们一定要在数据发送完成后移除检测可写事件，避免没有数据发送时无意义地触发。使用 ET 模式时，如果我们的实现也依赖于可写事件触发去发送数据，在可写事件触发后调用 send 函数（Linux 也可以使用 write 函数）发送数据，则如果数据本次不能全部发送完（对于非阻塞的 socket，此时 send 函数返回 -1，错误码为 EAGAIN 或 EWOULDBLOCK），则一定要继续注册、检测可写事件，否则我们剩余的数据就再也没有机会发送了，因为 ET 模式的可写事件再也不会被触发。

在目前主流的网络库中，发送数据的逻辑都不是上面所说的依赖于写事件触发，在写事件触发时去发送数据。这种做法不好，那好的做法是什么呢？在 7.6 节会详细介绍。

总结一下，如下所述。

（1）在 LT 模式下，读事件触发后可以按需收取想要的字节数，不用把本次接收的数据收取干净（即不用循环到 recv 或者 read 函数返回 -1，错误码为 EWOULDBLOCK 或 EAGAIN）；在 ET 模式下，读事件时必须把数据收取干净，因为我们不一定再有机会收取数据了，即使有机会，也可能因为没有及时处理上次没读完的数据，造成客户端响应延迟。

（2）在 LT 模式下，不需要写事件时一定要及时移除，避免不必要地触发且浪费 CPU 资源；在 ET 模式下，写事件触发后，如果还需要下一次的写事件触发来驱动任务（例如发送上次剩余的数据），则我们需要继续注册一次检测可写事件。

（3）LT 模式和 ET 模式各有优缺点，无所谓孰优孰劣。使用 LT 模式时，我们可以自由决定每次收取多少字节（对于普通 socket）或何时接收连接（对于监听 socket），但是可能会导致多次触发；使用 ET 模式时，我们必须每次都将数据收完（对于普通 socket）或立即调用 accept 接受连接（对于监听 socket），其优点是触发次数少。

我们一定要透彻地理解 epoll 模型的 LT 模式和 ET 模式在读写数据时的区别。因为，在现代互联网大环境下作为后台服务载体的主流操作系统是 Linux，而 epoll 模型是 Linux 上实现高性能服务网络模块的必备组件！只有理解了它们，我们才能编写出高性能的网络通信库乃至整个服务。

4.14.4 EPOLLONESHOT 选项

epoll 模型还有一个 EPOLLONESHOT 选项，如果某个 socket 注册了该标志，则其注册监听的事件（例如 EPOLLIN）在触发一次后再也不会触发，除非重新注册监听该事件类型。

我们通过一个实例来看一下 EPOLLONESHOT 选项的作用，代码如下：

```cpp
/**
 * 验证 epoll EPOLLONESHOT 选项, epoll_server_with_oneshot.cpp
 * zhangyl 2019.04.01
 */
#include<sys/types.h>
#include<sys/socket.h>
#include<arpa/inet.h>
#include<unistd.h>
#include<fcntl.h>
#include<sys/epoll.h>
#include<poll.h>
#include<iostream>
#include<string.h>
#include<vector>
#include<errno.h>
#include<iostream>

int main()
{
    //创建一个监听 socket
    int listenfd = socket(AF_INET, SOCK_STREAM, 0);
    if (listenfd == -1)
    {
        std::cout << "create listen socket error" << std::endl;
        return -1;
    }

    //设置重用的 IP 地址和端口号
    int on = 1;
    setsockopt(listenfd, SOL_SOCKET, SO_REUSEADDR, (char*)&on, sizeof(on));
    setsockopt(listenfd, SOL_SOCKET, SO_REUSEPORT, (char*)&on, sizeof(on));

    //将监听 socker 设置为非阻塞的
```

```cpp
    int oldSocketFlag = fcntl(listenfd, F_GETFL, 0);
    int newSocketFlag = oldSocketFlag | O_NONBLOCK;
    if (fcntl(listenfd, F_SETFL, newSocketFlag) == -1)
    {
        close(listenfd);
        std::cout << "set listenfd to nonblock error" << std::endl;
        return -1;
    }

    //初始化服务器的地址
    struct sockaddr_in bindaddr;
    bindaddr.sin_family = AF_INET;
    bindaddr.sin_addr.s_addr = htonl(INADDR_ANY);
    bindaddr.sin_port = htons(3000);

    if (bind(listenfd, (struct sockaddr*)&bindaddr, sizeof(bindaddr)) == -1)
    {
        std::cout << "bind listen socker error." << std::endl;
        close(listenfd);
        return -1;
    }

    //启动监听
    if (listen(listenfd, SOMAXCONN) == -1)
    {
        std::cout << "listen error." << std::endl;
        close(listenfd);
        return -1;
    }


    //创建 epollfd
    int epollfd = epoll_create(1);
    if (epollfd == -1)
    {
        std::cout << "create epollfd error." << std::endl;
        close(listenfd);
        return -1;
    }

    epoll_event listen_fd_event;
    listen_fd_event.data.fd = listenfd;
    listen_fd_event.events = EPOLLIN;

    //将监听 sokcet 绑定到 epollfd 上
    if (epoll_ctl(epollfd, EPOLL_CTL_ADD, listenfd, &listen_fd_event) == -1)
    {
        std::cout << "epoll_ctl error" << std::endl;
        close(listenfd);
        return -1;
    }

    int n;
    while (true)
    {
```

```cpp
        epoll_event epoll_events[1024];
        n = epoll_wait(epollfd, epoll_events, 1024, 1000);
        if (n < 0)
        {
            //被信号中断
            if (errno == EINTR)
                continue;

            //出错，退出
            break;
        }
        else if (n == 0)
        {
            //超时，继续
            continue;
        }
        for (size_t i = 0; i < n; ++i)
        {
            //事件可读
            if (epoll_events[i].events & EPOLLIN)
            {
                if (epoll_events[i].data.fd == listenfd)
                {
                    //监听socket，接受新连接
                    struct sockaddr_in clientaddr;
                    socklen_t clientaddrlen = sizeof(clientaddr);
                    int clientfd = accept(listenfd, (struct sockaddr*)&clientaddr, &clientaddrlen);
                    if (clientfd != -1)
                    {
                        int oldSocketFlag = fcntl(clientfd, F_GETFL, 0);
                        int newSocketFlag = oldSocketFlag | O_NONBLOCK;
                        if (fcntl(clientfd, F_SETFL, newSocketFlag) == -1)
                        {
                            close(clientfd);
                            std::cout << "set clientfd to nonblocking error." << std::endl;
                        }
                        else
                        {
                            epoll_event client_fd_event;
                            client_fd_event.data.fd = clientfd;
                            client_fd_event.events = EPOLLIN;
                            //为clientfd注册EPOLLONESHOT事件
                            client_fd_event.events |= EPOLLONESHOT;
                            if (epoll_ctl(epollfd, EPOLL_CTL_ADD, clientfd, &client_fd_event) != -1)
                            {
                                std::cout << "new client accepted,clientfd: " << clientfd << std::endl;
                            }
                            else
                            {
                                std::cout << "add client fd to epollfd error" << std::endl;
                                close(clientfd);
```

```cpp
                    }
                }
            }
            else
            {
                std::cout << "client fd: " << epoll_events[i].data.fd << " recv data." << std::endl;
                //普通 clientfd
                char ch;
                //每次只接收 1 字节
                int m = recv(epoll_events[i].data.fd, &ch, 1, 0);
                if (m == 0)
                {
                    //对端关闭了连接，从 epollfd 上移除 clientfd
                    if (epoll_ctl(epollfd, EPOLL_CTL_DEL, epoll_events[i].data.fd, NULL) != -1)
                    {
                        std::cout << "client disconnected,clientfd:" << epoll_events[i].data.fd << std::endl;
                    }
                    close(epoll_events[i].data.fd);
                }
                else if (m < 0)
                {
                    //出错
                    if (errno != EWOULDBLOCK && errno != EINTR)
                    {
                        if (epoll_ctl(epollfd, EPOLL_CTL_DEL, epoll_events[i].data.fd, NULL) != -1)
                        {
                            std::cout << "client disconnected,clientfd:" << epoll_events[i].data.fd << std::endl;
                        }
                        close(epoll_events[i].data.fd);
                    }
                }
                else
                {
                    //正常收到数据
                    std::cout << "recv from client:" << epoll_events[i].data.fd << ", " << ch << std::endl; //第 2 个加粗行
                }
            }
        }
        else if (epoll_events[i].events & POLLERR)
        {
            //TODO 暂不处理
        }
    }
}

close(listenfd);
return 0;
}
```

以上代码在第 1 个加粗行处为 clientfd 注册了 EPOLLONESHOT 事件，由于使用的是水平模式，所以 EPOLLIN 在每次触发后都只读取 1 个字符。我们使用 nc 命令模拟一个客户端，看下执行效果。nc 客户端的执行效果如下：

```
[root@myaliyun ~]# nc -v 127.0.0.1 3000
Ncat: Version 7.50 ( https://nmap.org/ncat )
Ncat: Connected to 127.0.0.1:3000.
hello
hello
```

服务端的执行效果如下：

```
[root@myaliyun chapter04]# g++ -g -o epoll_server_with_oneshot
epoll_server_with_oneshot.cpp
[root@myaliyun chapter04]# ./epoll_server_with_oneshot
new client accepted,clientfd: 5
client fd: 5 recv data.
recv from client:5, h
```

服务端每次只读取 1 个字符，由于注册了 EPOLLONESHOT 选项，所以在第 1 次 EPOLLIN 事件触发后，即使客户端再次发送"hello"字符串，服务端也不会触发 EPOLLIN 事件了。

将上面的代码稍微修改一下，在读取一个字符成功后，再次为这个 clientfd 注册 EPOLLIN 事件，在原代码第 2 个加粗行之后添加如下代码：

```cpp
//完整的代码位于 epoll_server_with_oneshot_2.cpp
//添加在原代码第 2 个加粗行之后
epoll_event client_fd_event;
client_fd_event.data.fd = epoll_events[i].data.fd;
client_fd_event.events = EPOLLIN;
//这里再次为clientfd注册EPOLLIN事件
if (epoll_ctl(epollfd, EPOLL_CTL_MOD, epoll_events[i].data.fd,
&client_fd_event) != -1)
{
    std::cout << "rearm EPOLLIN event to clientfd: " << epoll_events[i].data.fd << std::endl;
}
else
{
    //若epoll_ctl调用失败，则从epollfd上移除clientfd并关闭clientfd
    if (epoll_ctl(epollfd, EPOLL_CTL_DEL, epoll_events[i].data.fd, NULL) != -1)
    {
        std::cout << "remove clientfd from epoll fd successfully, clientfd:" << epoll_events[i].data.fd << std::endl;
    }
    close(epoll_events[i].data.fd);
}
```

重新编译程序并执行，nc 客户端连续发送两个"hello"字符串，服务端的输出结果如下：

```
[root@myaliyun chapter04]# g++ -g -o epoll_server_with_oneshot_2
```

```
epoll_server_with_oneshot_2.cpp
[root@myaliyun chapter04]# ./epoll_server_with_oneshot_2
new client accepted,clientfd: 5
client fd: 5 recv data.
recv from client:5, h
rearm EPOLLIN event to clientfd: 5
client fd: 5 recv data.
recv from client:5, e
rearm EPOLLIN event to clientfd: 5
client fd: 5 recv data.
recv from client:5, l
rearm EPOLLIN event to clientfd: 5
client fd: 5 recv data.
recv from client:5, l
rearm EPOLLIN event to clientfd: 5
client fd: 5 recv data.
recv from client:5, o
rearm EPOLLIN event to clientfd: 5
client fd: 5 recv data.
recv from client:5,

rearm EPOLLIN event to clientfd: 5
client fd: 5 recv data.
recv from client:5, h
rearm EPOLLIN event to clientfd: 5
client fd: 5 recv data.
recv from client:5, e
rearm EPOLLIN event to clientfd: 5
client fd: 5 recv data.
recv from client:5, l
rearm EPOLLIN event to clientfd: 5
client fd: 5 recv data.
recv from client:5, l
rearm EPOLLIN event to clientfd: 5
client fd: 5 recv data.
recv from client:5, o
rearm EPOLLIN event to clientfd: 5
client fd: 5 recv data.
recv from client:5,

rearm EPOLLIN event to clientfd: 5
```

这样每次就可以正常接收完整的数据包了。

在一些特殊的应用场景中，如果涉及多个线程同时处理某个 socket 上的事件，则为了避免数据乱序，我们不得不使用复杂的多线程同步机制；但是有了 EPOLLONESHOT 选项，我们就可以减少线程同步逻辑了。以 EPOLLIN 事件处理为例，多个线程同时从一个 socket 上读数据，可以使某个线程先处理，在该线程处理完之后再重新给该 socket 添加读事件，这样读事件再次触发时，就可以被其他线程继续处理了。这种做法在本质上还是保证同一个时刻只有一个线程在处理某个 socket 上的事件。当然，多个线程同时操作一个 socket 本来就是一种不好的表现，我们在实际开发时应该尽量避免。

4.15 高效的 readv 和 writev 函数

在实际开发中，高性能服务有一个原则：尽量减少系统调用。对一个文件描述符（fd，例如文件指针、套接字类型）的读或写，都是系统调用。我们有时需要将一个文件描述符对应的文件或套接字中的数据读到多个缓冲区中，或者将多个缓冲区中的数据同时写入一个文件描述符对应的文件或者套接字中。当然，我们可以通过多次调用 read 或者 write 函数，挨个操作每个缓冲区。Linux 提供了一系列 readv 和 writev 函数来完成上述工作，其函数签名分别如下：

```
#include <sys/uio.h>

ssize_t readv(int fd, const struct iovec *iov, int iovcnt);
ssize_t writev(int fd, const struct iovec *iov, int iovcnt);
ssize_t preadv(int fd, const struct iovec *iov, int iovcnt, off_t offset);
ssize_t pwritev(int fd, const struct iovec *iov, int iovcnt, off_t offset);
```

参数 fd 和 offset 的含义都是不言自明的。参数 iov 是一个 struct iovec 数组；参数 iovcnt 是该数组元素的个数；函数 readv 和 preadv 若调用成功，则返回总读取字节数；函数 writev 和 pwritev 若调用成功，则返回总写入字节数，函数调用失败时均返回-1。

struct iovec 的定义如下：

```
struct iovec {
    void  *iov_base;    /* 数据起始地址 */
    size_t iov_len;     /* 需要传输的字节数 */
};
```

字段 iov_base 是存放数据缓冲区的地址，字段 iov_len 是缓冲区的长度。

这里以 writev 函数为例来演示其用法。假设现在需要将 3 段数据同时写入一个 CSV 文件中，第 1 段数据是 CSV 的文件头信息，第 2 段和第 3 段数据是两只股票的信息，则示例代码如下：

```
char* headerInfo = "timestamp,symbol,side,volume,price\n";
char* firstSymbolInfo = "2014-12-12D06:53:11.127838000,600358,Buy,347,3.48\n";
char* secondSymbolInfo = "2014-12-12D06:53:11.127838000,600359,Buy,251,4.99\n";

struct iovec iov[3];
iov[0].iov_base = headerInfo;
iov[0].iov_len = strlen(headerInfo);
iov[1].iov_base = firstSymbolInfo;
iov[1].iov_len = strlen(firstSymbolInfo);
iov[2].iov_base = secondSymbolInfo;
iov[2].iov_len = strlen(secondSymbolInfo);

//fpCSV 为打开的 CSV 文件句柄
ssize_t nwritten = writev(fpCSV, iov, 2);
```

我们可能会在很多 C/C++ 网络库中见到 readv 和 writev 函数的身影，它们都用于提高程序的执行效率。

4.16 主机字节序和网络字节序

网络通信在本质上是不同的机器进行数据交换,数据的传输是通过字节流的方式进行传递,字节流的排序方式相对操作系统主机和网络传输来说,分为主机字节序和网络字节序。

4.16.1 主机字节序

不同的机器一般有不同的 CPU 型号,不同的 CPU 其字节序可能不一样。字节序指的是对于存储需要多字节(大于 1 字节)的整数来说,其每字节在不同的机器内存中存储的顺序就是主机字节序,一般分为以下两类。

(1) little-endian(LE,小端编码或小头编码)。对于一个整数值,如果使用小端字节序,则整数的高位被存储在内存地址高的位置,整数的低位被存储在内存地址低的位置(即高高低低),这种序列比较符合人的思维习惯。Intel x86 系列的系统使用的是小端编码方式。

(2) big-endian(BE,大端编码或大头编码)。对于一个整数值,如果使用大端字节序,则整数的高位被存储在内存地址低的位置,整数的低位被存储在内存地址高的位置(即高低低高),这是直观的字节序。Java 程序、Mac 机器上的程序一般是大端编码方式。

举个例子,对于内存中双字值 0x10203040(4 字节)的存储方式,如果使用小端编码,则其内存中的存储方式如下图所示。

0x10203040采用小端编码在内存中的存储示意图

如果使用大端编码存储 0x10203040,则内存中的存储如下图所示。

0x10203040采用大端编码在内存中的存储示意图

4.16.2 网络字节序

网络字节序是在 TCP/IP 中规定好的一种数据表示格式,它与具体的 CPU 类型、操作系统等无关,可以保证数据在不同的主机之间传输时能够被正确解释,采用 big-endian 排序方式。因此,为了不同的机器和系统可以正常交换数据,一般建议将需要传输的整型值转换为网络字节序,在前面的代码中使用端口时便将端口号数值从本地字节序转换为网络字节序:

```
//初始化服务器地址
struct sockaddr_in bindaddr;
bindaddr.sin_family = AF_INET;
bindaddr.sin_addr.s_addr = htonl(INADDR_ANY);
//将端口号3000转换为网络字节序
bindaddr.sin_port = htons(3000);
if (bind(listenfd, (struct sockaddr *)&bindaddr, sizeof(bindaddr)) == -1)
{
    std::cout << "bind listen socket error." << std::endl;
    return -1;
}
```

htons 函数即将一个 short 类型从本机字节序转换为网络字节序(Big Endian),我们可以这么记忆这个函数:host to net short => htons。与这个函数类似的还有一系列将整型转网络字节序的函数(以 Linux 为例):

```
#include <arpa/inet.h>

//host to net long
uint32_t htonl(uint32_t hostlong);
//host to net short
uint16_t htons(uint16_t hostshort);
```

与此相反,从网络上收到数据以后,如果需要将整数从网络字节序转换为本地字节序,则也有对应的系列函数:

```
#include <arpa/inet.h>

//net to host long
uint32_t ntohl(uint32_t netlong);
//net to host short
uint16_t ntohs(uint16_t netshort);
```

这类转换函数的实现原理也很简单,以将本地字节序转换为网络字节序为例,如果发现本机字节序就是网络字节序(即本机字节序就是大端编码),则什么也不做,反之将字节顺序互换。

那如何判断本机字节序是不是网络字节序呢?可以随意找一个 2 字节的十六进制数值测试一下,例如 0x1234,如果本机字节序是小端编码,则其值 12 被存储在高地址字节中,34 被存储在低地址字节中,这样强行把 0x1234 转换为 1 字节的 char 时,高字节被丢弃,剩下低字节值,就是 34;反之,如果本机字节序是大端编码,则在高地址字节中存储的是 34,在低地址字节中存储的是 12,当将其强转为 1 字节的 char 时,其值是 12,代码实现如下：

```cpp
//判断本机是否是网络字节序
bool isNetByteOrder()
{
    unsigned short mode = 0x1234;
    char* pmode = (char*)&mode;
    //如果将低字节放在低位，则是小端字节序列，即非网络字节序
    if (*pmode == 0x34)
        return false;
    return true;
}
```

我们在上面的基础上实现 htons 函数：

```cpp
uint16_t htons(uint16_t hostshort)
{
    //如果本机字节序已经是网络字节序，则直接返回
    if (isNetByteOrder())
        return hostshort;

    return ((uint16_t)(hostshort >> 8)) | ((uint16_t)((hostshort & 0x00ff) << 8));
}
```

其他函数的实现原理类似，读者可以自己实现一下，这里不再重复介绍。

4.16.3 操作系统提供的字节转换函数汇总

为了便于读者查阅，让我们对操作系统提供的这些函数进行汇总。

Windows：

```cpp
#include <winsock2.h>
/**
 * 将本机字节序转换为网络字节序
 */
htou_short htons(u_short hostshort);
u_long htonl(u_long hostlong);
unsigned __int64 htonll(unsigned __int64 Value);

/**
 * 将网络字节序转换为本机字节序
 */
u_short ntohs(u_short netshort);
u_long ntohl(u_long netlong);
unsigned __int64 ntohll(unsigned __int64 Value);
```

注意：在 Windows 上使用 socket 函数时需要先调用 WSAStartup 函数初始化 socket 库，但在使用这个系列的函数时不必调用 WSAStartup 函数。

Linux：

```cpp
#include <arpa/inet.h>

/**
 * 将本机字节序转换为网络字节序
```

```
*/
uint16_t htons(uint16_t hostshort);
uint32_t htonl(uint32_t hostlong);

/**
 * 将网络字节序转换为本机字节序
 */
uint16_t ntohs(uint16_t netshort);
uint32_t ntohl(uint32_t netlong);
```

对于 2 字节（short）和 4 字节（int），Linux 和 Windows 都提供了同名且函数签名形式一样的函数，我们在写跨平台代码时可直接调用。遗憾的是，在 Linux 上并没有提供与 Windows 有相同函数名和签名的对 8 字节（long long）的整数进行转换的函数。没关系，Linux 提供了另一批显式地将本节字节序与 Big Endian 或 Little Endian 互转的函数：

```
#include <endian.h>

uint16_t htobe16(uint16_t host_16bits);
uint16_t htole16(uint16_t host_16bits);
uint16_t be16toh(uint16_t big_endian_16bits);
uint16_t le16toh(uint16_t little_endian_16bits);

uint32_t htobe32(uint32_t host_32bits);
uint32_t htole32(uint32_t host_32bits);
uint32_t be32toh(uint32_t big_endian_32bits);
uint32_t le32toh(uint32_t little_endian_32bits);

uint64_t htobe64(uint64_t host_64bits);
uint64_t htole64(uint64_t host_64bits);
uint64_t be64toh(uint64_t big_endian_64bits);
uint64_t le64toh(uint64_t little_endian_64bits);
```

这一组函数使用起来非常方便，Windows 上的 htonll 函数对应 htobe64 函数，ntohll 函数对应 be64toh 函数。因此，我们在写跨平台兼容代码时可以这么写：

```
#ifdef WIN32
//在 Windows 上存在 ntohll 和 htonll，直接使用
#else
//在 Linux 上没有这两个函数，定义之
#define ntohll(x) be64toh(x)
#define htonll(x) htobe64(x)
#endif
```

4.17 域名解析 API 介绍

为了便于记忆，我们有时需要将自己的程序使用域名和端口号连接服务，在这种情况下需要使用 socket API gethostbyname 函数先把域名转换为 IP 地址，再使用 connect 函数连接。在 Linux 上，gethostbyname 函数签名如下：

```
#include <netdb.h>

struct hostent* gethostbyname(const char* name);
```

将域名转换为 IP 时，gethostbyname 函数将转换结果存放在一个 hostent 结构体中，将转换成功后的 IP 地址存放在 hostent 最后一个字段中。hostent 结构体类型的定义如下：

```
struct hostent
{
    char*   h_name;           /* official name of host */
    char**  h_aliases;        /* alias list */
    int     h_addrtype;       /* host address type */
    int     h_length;         /* length of address */
    char**  h_addr_list;      /* list of addresses */
}

#define h_addr h_addr_list[0]  /* for backward compatibility */
```

对其中的字段解释如下。

（1）h_name：地址的正式名称。

（2）h_aliases：地址的预备名称指针。

（3）h_addrtype：地址类型，通常是 AF_INE。

（4）h_length：地址的长度，以字节数为计量单位。

（5）h_addr_list：主机网络地址指针，网络字节顺序。其中，h_addr 是字段 h_addr_list 中的第一地址。

注意：虽然 h_addr_list[0]看起来是一个 char*类型，但它实际上是一个 uint32_t，这是 IP 地址的 32 bit 整数表示形式，如果需要将 32 bit 整数形式的 IP 地址转换为十进制点分法字符串，则只需再调用 inet_ntoa 函数即可。

来看一段示例代码：

```
#include <sys/types.h>
#include <sys/socket.h>
#include <netinet/in.h>
#include <arpa/inet.h>
#include <netdb.h>
#include <stdio.h>

//extern int h_errno;

bool connect_to_server(const char* server, short port)
{
    int hSocket = socket(AF_INET, SOCK_STREAM, 0);
    if (hSocket == -1)
        return false;

    struct sockaddr_in addrSrv = { 0 };
    struct hostent* pHostent = NULL;
    //unsigned int addr = 0;

    //如果传入的参数 server 的值是 somesite.com 这种域名域名形式，则 if 条件成立
    if (addrSrv.sin_addr.s_addr = inet_addr(server) == INADDR_NONE)
    {
```

```
        //接着调用gethostbyname将十进制点分法IP地址解析成4字节的整型数值
        pHostent = gethostbyname(server);
        if (pHostent == NULL)
            return false;

        //使用gethostbyname解析域名时可能会得到多个IP地址，常使用第1个IP地址
        addrSrv.sin_addr.s_addr = *((unsigned long*)pHostent->h_addr_list[0]);
    }

    addrSrv.sin_family = AF_INET;
    addrSrv.sin_port = htons(port);
    int ret = connect(hSocket, (struct sockaddr*)&addrSrv, sizeof(addrSrv));
    if (ret == -1)
        return false;

    return true;
}

int main()
{
    if (connect_to_server("baidu.com", 80))
        printf("connect successfully.\n");
    else
        printf("connect error.\n");

    return 0;
}
```

以上 connect_to_server 函数既支持直接传入域名，也支持传入 IP 地址：

```
connect_to_server("127.0.0.1", 8888);
connect_to_server("localhost", 8888);

connect_to_server("61.135.169.125", 80);
connect_to_server("baidu.com", 80);
```

实际上，使用 gethostbyname 函数时需要注意以下事项。

（1）gethostbyname 函数是不可重入函数，在 Linux 上建议使用 gethostbyname_r 函数替代它。

（2）gethostbyname 在解析域名时会阻塞当前执行线程，直到得到返回结果。

（3）在使用 gethostbyname 函数出错时，我们不能使用 errno 获取错误码信息（因此也不能使用 perror 函数打印错误信息），应该使用 h_errno 错误码（也可以调用 herror 函数打印错误信息）。herror 函数签名如下：

```
void herror(const char *s);
```

在新的 Linux 上，gethostbyname 和 gethostbyaddr 一样，已经被标记为废弃的，我们应该使用新的函数 getaddrinfo 去替代它们。Getaddrinfo 函数签名如下：

```
#include <sys/types.h>
#include <sys/socket.h>
#include <netdb.h>
```

```
int getaddrinfo(const char* node,
                const char* service,
                const struct addrinfo* hints,
                struct addrinfo** res);
```

若 getaddrinfo 函数调用成功，则返回 0，否则返回非 0 值，调用成功后将结果存储在参数 res 中。Addrinfo 函数结构体的定义如下：

```
struct addrinfo
{
    int              ai_flags;
    int              ai_family;
    int              ai_socktype;
    int              ai_protocol;
    socklen_t        ai_addrlen;
    struct sockaddr* ai_addr;
    char*            ai_canonname;
    struct addrinfo* ai_next;
};
```

如果我们不再需要 res 变量，则记得使用 freeaddrinfo 函数将其指向的资源释放：

```
void freeaddrinfo(struct addrinfo* res);
```

getaddrinfo 函数的使用示例如下：

```
struct addrinfo hints = {0};
hints.ai_flags = AI_CANONNAME;
hints.ai_family = family;
hints.ai_socktype = socktype;

struct addrinfo* res;
int n = getaddrinfo(host, service, &hints, &res);
if(n == 0)
{
    //调用成功，使用 res

    //释放 res 资源
    freeaddr(res);
}
```

getaddrinfo 函数同时支持对 IPv4 和 IPv6 地址的解析。在 redis-server 源码中就使用了这个函数去解析 IPv4 和 IPv6 的地址（位于 net.c 文件中）：

```
static int _redisContextConnectTcp(redisContext *c, const char *addr, int port,
                                   const struct timeval *timeout,
                                   const char *source_addr) {
    int s, rv, n;
    char _port[6];  /* strlen("65535"); */
    struct addrinfo hints, *servinfo, *bservinfo, *p, *b;
    int blocking = (c->flags & REDIS_BLOCK);
    int reuseaddr = (c->flags & REDIS_REUSEADDR);
    int reuses = 0;
    long timeout_msec = -1;

    servinfo = NULL;
    c->connection_type = REDIS_CONN_TCP;
```

```c
        c->tcp.port = port;

    if (c->tcp.host != addr) {
        if (c->tcp.host)
            free(c->tcp.host);

        c->tcp.host = strdup(addr);
    }

    if (timeout) {
        if (c->timeout != timeout) {
            if (c->timeout == NULL)
                c->timeout = malloc(sizeof(struct timeval));

            memcpy(c->timeout, timeout, sizeof(struct timeval));
        }
    } else {
        if (c->timeout)
            free(c->timeout);
        c->timeout = NULL;
    }

    if (redisContextTimeoutMsec(c, &timeout_msec) != REDIS_OK) {
        __redisSetError(c, REDIS_ERR_IO, "Invalid timeout specified");
        goto error;
    }

    if (source_addr == NULL) {
        free(c->tcp.source_addr);
        c->tcp.source_addr = NULL;
    } else if (c->tcp.source_addr != source_addr) {
        free(c->tcp.source_addr);
        c->tcp.source_addr = strdup(source_addr);
    }

    snprintf(_port, 6, "%d", port);
    memset(&hints,0,sizeof(hints));
    hints.ai_family = AF_INET;
    hints.ai_socktype = SOCK_STREAM;

    if ((rv = getaddrinfo(c->tcp.host,_port,&hints,&servinfo)) != 0) {
        hints.ai_family = AF_INET6;
        if ((rv = getaddrinfo(addr,_port,&hints,&servinfo)) != 0) {
            __redisSetError(c,REDIS_ERR_OTHER,gai_strerror(rv));
            return REDIS_ERR;
        }
    }
    for (p = servinfo; p != NULL; p = p->ai_next) {
addrretry:
        if ((s = socket(p->ai_family,p->ai_socktype,p->ai_protocol)) == -1)
            continue;

        c->fd = s;
        if (redisSetBlocking(c,0) != REDIS_OK)
            goto error;
```

```c
    if (c->tcp.source_addr) {
        int bound = 0;
        /* 利用 getaddrinfo 函数自动选择是使用 IPv4 还是 IPv6 地址 */
        if ((rv = getaddrinfo(c->tcp.source_addr, NULL, &hints, &bservinfo)) != 0) {
            char buf[128];
            snprintf(buf,sizeof(buf),"Can't get addr: %s",gai_strerror(rv));
            __redisSetError(c,REDIS_ERR_OTHER,buf);
            goto error;
        }

        if (reuseaddr) {
            n = 1;
            if (setsockopt(s, SOL_SOCKET, SO_REUSEADDR, (char*) &n,
                       sizeof(n)) < 0) {
                goto error;
            }
        }

        for (b = bservinfo; b != NULL; b = b->ai_next) {
            if (bind(s,b->ai_addr,b->ai_addrlen) != -1) {
                bound = 1;
                break;
            }
        }
        freeaddrinfo(bservinfo);
        if (!bound) {
            char buf[128];
            snprintf(buf,sizeof(buf),"Can't bind socket: %s",strerror(errno));
            __redisSetError(c,REDIS_ERR_OTHER,buf);
            goto error;
        }
    }
    if (connect(s,p->ai_addr,p->ai_addrlen) == -1) {
        if (errno == EHOSTUNREACH) {
            redisContextCloseFd(c);
            continue;
        } else if (errno == EINPROGRESS && !blocking) {
            /* This is ok. */
        } else if (errno == EADDRNOTAVAIL && reuseaddr) {
            if (++reuses >= REDIS_CONNECT_RETRIES) {
                goto error;
            } else {
                redisContextCloseFd(c);
                goto addrretry;
            }
        } else {
            if (redisContextWaitReady(c,timeout_msec) != REDIS_OK)
                goto error;
        }
    }
    if (blocking && redisSetBlocking(c,1) != REDIS_OK)
        goto error;
    if (redisSetTcpNoDelay(c) != REDIS_OK)
        goto error;
```

```
        c->flags |= REDIS_CONNECTED;
        rv = REDIS_OK;
        goto end;
    }
    if (p == NULL) {
        char buf[128];
        snprintf(buf,sizeof(buf),"Can't create socket: %s",strerror(errno));
        __redisSetError(c,REDIS_ERR_OTHER,buf);
        goto error;
    }

error:
    rv = REDIS_ERR;
end:
    freeaddrinfo(servinfo);
    return rv;  //Need to return REDIS_OK if alright
}
```

第 5 章

网络通信故障排查常用命令

进行后台开发工作时,有时需要排查和定位一些网络通信问题,本章将介绍一些常用的网络通信故障排查命令,这里以 Linux 为操作对象,使用 CentOS 7.0 进行演示。

5.1 ifconfig 命令

ifconfig 命令是查看当前系统的网卡和 IP 地址信息的常用命令。如果在机器上还没有安装 ifconfig 命令,则可以使用如下命令安装:

```
yum install net-tools
```

安装成功以后,就可以使用 ifconfig 查看机器的网卡信息了:

```
[root@localhost ~]# ifconfig
ens33: flags=4163<UP,BROADCAST,RUNNING,MULTICAST>  mtu 1500
        inet 192.168.206.140  netmask 255.255.255.0  broadcast 192.168.206.255
        inet6 fe80::1599:dcc4:b3e8:7fce  prefixlen 64  scopeid 0x20<link>
        ether 00:0c:29:ee:01:80  txqueuelen 1000  (Ethernet)
        RX packets 1005  bytes 77224 (75.4 KiB)
        RX errors 0  dropped 0  overruns 0  frame 0
        TX packets 349  bytes 47206 (46.0 KiB)
        TX errors 0  dropped 0 overruns 0  carrier 0  collisions 0

lo: flags=73<UP,LOOPBACK,RUNNING>  mtu 65536
        inet 127.0.0.1  netmask 255.0.0.0
        inet6 ::1  prefixlen 128  scopeid 0x10<host>
        loop  txqueuelen 1000  (Local Loopback)
        RX packets 4  bytes 352 (352.0 B)
        RX errors 0  dropped 0  overruns 0  frame 0
        TX packets 4  bytes 352 (352.0 B)
        TX errors 0  dropped 0 overruns 0  carrier 0  collisions 0
```

以上输出显示了本机当前激活的网卡列表,以及每个激活的网卡(网络设备)的 IPv4 地址、IPv6 地址、子网掩码、广播地址等信息,这里一共有两个网卡,分别是 ens33 和 lo。

我们可以使用-s 选项显示网卡信息的精简列表:

```
[root@localhost ~]# ifconfig -s
Iface      MTU     RX-OK RX-ERR RX-DRP RX-OVR     TX-OK TX-ERR TX-DRP TX-OVR Flg
ens33      1500     1086      0      0 0            379      0      0      0 BMRU
lo        65536        4      0      0 0              4      0      0      0 LRU
```

在默认情况下，ifconfig 命令只会显示激活的网卡信息，我们可以使用-a 选项显示所有（包括未激活）网卡信息。

使用 ifconfig up 命令可激活某个网卡，使用 ifconfig down 命令可禁用某个网卡，这两个命令的用法如下：

```
ifconfig 网卡名 up
ifconfig 网卡名 down
```

这里演示这两个命令的用法：

```
[root@localhost ~]# ifconfig
ens33: flags=4163<UP,BROADCAST,RUNNING,MULTICAST>  mtu 1500
        inet 192.168.206.140  netmask 255.255.255.0  broadcast 192.168.206.255
        inet6 fe80::1599:dcc4:b3e8:7fce  prefixlen 64  scopeid 0x20<link>
        ether 00:0c:29:ee:01:80  txqueuelen 1000  (Ethernet)
        RX packets 1398  bytes 114269 (111.5 KiB)
        RX errors 0  dropped 0  overruns 0  frame 0
        TX packets 601  bytes 97657 (95.3 KiB)
        TX errors 0  dropped 0  overruns 0  carrier 0  collisions 0

lo: flags=73<UP,LOOPBACK,RUNNING>  mtu 65536
        inet 127.0.0.1  netmask 255.0.0.0
        inet6 ::1  prefixlen 128  scopeid 0x10<host>
        loop  txqueuelen 1000  (Local Loopback)
        RX packets 12  bytes 1056 (1.0 KiB)
        RX errors 0  dropped 0  overruns 0  frame 0
        TX packets 12  bytes 1056 (1.0 KiB)
        TX errors 0  dropped 0  overruns 0  carrier 0  collisions 0

## 禁用网卡 lo 后默认只能看到一个网卡的信息
[root@localhost ~]# ifconfig lo down
[root@localhost ~]# ifconfig
ens33: flags=4163<UP,BROADCAST,RUNNING,MULTICAST>  mtu 1500
        inet 192.168.206.140  netmask 255.255.255.0  broadcast 192.168.206.255
        inet6 fe80::1599:dcc4:b3e8:7fce  prefixlen 64  scopeid 0x20<link>
        ether 00:0c:29:ee:01:80  txqueuelen 1000  (Ethernet)
        RX packets 1510  bytes 123232 (120.3 KiB)
        RX errors 0  dropped 0  overruns 0  frame 0
        TX packets 657  bytes 104751 (102.2 KiB)
        TX errors 0  dropped 0  overruns 0  carrier 0  collisions 0

[root@localhost ~]# ifconfig -a
ens33: flags=4163<UP,BROADCAST,RUNNING,MULTICAST>  mtu 1500
        inet 192.168.206.140  netmask 255.255.255.0  broadcast 192.168.206.255
        inet6 fe80::1599:dcc4:b3e8:7fce  prefixlen 64  scopeid 0x20<link>
        ether 00:0c:29:ee:01:80  txqueuelen 1000  (Ethernet)
        RX packets 1543  bytes 125948 (122.9 KiB)
        RX errors 0  dropped 0  overruns 0  frame 0
        TX packets 675  bytes 107251 (104.7 KiB)
        TX errors 0  dropped 0  overruns 0  carrier 0  collisions 0
```

```
lo: flags=8<LOOPBACK>  mtu 65536
        inet 127.0.0.1  netmask 255.0.0.0
        loop  txqueuelen 1000  (Local Loopback)
        RX packets 12  bytes 1056 (1.0 KiB)
        RX errors 0  dropped 0  overruns 0  frame 0
        TX packets 12  bytes 1056 (1.0 KiB)
        TX errors 0  dropped 0  overruns 0  carrier 0  collisions 0
## 再次启用网卡 lo
[root@localhost ~]# ifconfig lo up
[root@localhost ~]# ifconfig
ens33: flags=4163<UP,BROADCAST,RUNNING,MULTICAST>  mtu 1500
        inet 192.168.206.140  netmask 255.255.255.0  broadcast 192.168.206.255
        inet6 fe80::1599:dcc4:b3e8:7fce  prefixlen 64  scopeid 0x20<link>
        ether 00:0c:29:ee:01:80  txqueuelen 1000  (Ethernet)
        RX packets 1615  bytes 131924 (128.8 KiB)
        RX errors 0  dropped 0  overruns 0  frame 0
        TX packets 715  bytes 112423 (109.7 KiB)
        TX errors 0  dropped 0  overruns 0  carrier 0  collisions 0

lo: flags=73<UP,LOOPBACK,RUNNING>  mtu 65536
        inet 127.0.0.1  netmask 255.0.0.0
        inet6 ::1  prefixlen 128  scopeid 0x10<host>
        loop  txqueuelen 1000  (Local Loopback)
        RX packets 12  bytes 1056 (1.0 KiB)
        RX errors 0  dropped 0  overruns 0  frame 0
        TX packets 12  bytes 1056 (1.0 KiB)
        TX errors 0  dropped 0  overruns 0  carrier 0  collisions 0
```

ifconfig 命令还可以将一个 IP 地址绑定到某个网卡上,或将一个 IP 地址从某个网卡上解绑,命令如下:

```
# 将指定的 IP 地址绑定到某个网卡
ifconfig 网卡名 add IP 地址
# 从某个网卡上解绑指定的 IP 地址
ifconfig 网卡名 del IP 地址
```

这里通过实际操作进行演示:

```
[root@localhost ~]# ifconfig
ens33: flags=4163<UP,BROADCAST,RUNNING,MULTICAST>  mtu 1500
        inet 192.168.206.140  netmask 255.255.255.0  broadcast 192.168.206.255
        inet6 fe80::1599:dcc4:b3e8:7fce  prefixlen 64  scopeid 0x20<link>
        ether 00:0c:29:ee:01:80  txqueuelen 1000  (Ethernet)
        RX packets 1615  bytes 131924 (128.8 KiB)
        RX errors 0  dropped 0  overruns 0  frame 0
        TX packets 715  bytes 112423 (109.7 KiB)
        TX errors 0  dropped 0  overruns 0  carrier 0  collisions 0

lo: flags=73<UP,LOOPBACK,RUNNING>  mtu 65536
        inet 127.0.0.1  netmask 255.0.0.0
        inet6 ::1  prefixlen 128  scopeid 0x10<host>
        loop  txqueuelen 1000  (Local Loopback)
        RX packets 12  bytes 1056 (1.0 KiB)
        RX errors 0  dropped 0  overruns 0  frame 0
        TX packets 12  bytes 1056 (1.0 KiB)
```

```
        TX errors 0  dropped 0 overruns 0  carrier 0  collisions 0

[root@localhost ~]# ifconfig ens33 add 192.168.206.150
[root@localhost ~]# ifconfig
ens33: flags=4163<UP,BROADCAST,RUNNING,MULTICAST>  mtu 1500
        inet 192.168.206.140  netmask 255.255.255.0  broadcast 192.168.206.255
        inet6 fe80::1599:dcc4:b3e8:7fce  prefixlen 64  scopeid 0x20<link>
        ether 00:0c:29:ee:01:80  txqueuelen 1000  (Ethernet)
        RX packets 1804  bytes 145940 (142.5 KiB)
        RX errors 0  dropped 0  overruns 0  frame 0
        TX packets 781  bytes 119581 (116.7 KiB)
        TX errors 0  dropped 0 overruns 0  carrier 0  collisions 0

ens33:0: flags=4163<UP,BROADCAST,RUNNING,MULTICAST>  mtu 1500
        inet 192.168.206.150  netmask 255.255.255.0  broadcast 192.168.206.255
        ether 00:0c:29:ee:01:80  txqueuelen 1000  (Ethernet)

lo: flags=73<UP,LOOPBACK,RUNNING>  mtu 65536
        inet 127.0.0.1  netmask 255.0.0.0
        inet6 ::1  prefixlen 128  scopeid 0x10<host>
        loop  txqueuelen 1000  (Local Loopback)
        RX packets 12  bytes 1056 (1.0 KiB)
        RX errors 0  dropped 0  overruns 0  frame 0
        TX packets 12  bytes 1056 (1.0 KiB)
        TX errors 0  dropped 0 overruns 0  carrier 0  collisions 0
```

在以上操作中，网卡 ens33 绑定了 IP 地址 192.168.206.140，我们使用 ifconfig add 命令绑定了一个新的 IP 地址：192.168.206.150，现在就可以使用这个新的 IP 地址访问原来的网络了。

同理，如果要解绑这个 IP 地址，就可以按如下所示操作：

```
[root@localhost ~]# ifconfig ens33 del 192.168.206.150
[root@localhost ~]# ifconfig -a
ens33: flags=4163<UP,BROADCAST,RUNNING,MULTICAST>  mtu 1500
        inet 192.168.206.140  netmask 255.255.255.0  broadcast 192.168.206.255
        inet6 fe80::1599:dcc4:b3e8:7fce  prefixlen 64  scopeid 0x20<link>
        ether 00:0c:29:ee:01:80  txqueuelen 1000  (Ethernet)
        RX packets 2127  bytes 172321 (168.2 KiB)
        RX errors 0  dropped 0  overruns 0  frame 0
        TX packets 953  bytes 139954 (136.6 KiB)
        TX errors 0  dropped 0 overruns 0  carrier 0  collisions 0

lo: flags=73<UP,LOOPBACK,RUNNING>  mtu 65536
        inet 127.0.0.1  netmask 255.0.0.0
        inet6 ::1  prefixlen 128  scopeid 0x10<host>
        loop  txqueuelen 1000  (Local Loopback)
        RX packets 18  bytes 1560 (1.5 KiB)
        RX errors 0  dropped 0  overruns 0  frame 0
        TX packets 18  bytes 1560 (1.5 KiB)
        TX errors 0  dropped 0 overruns 0  carrier 0  collisions 0
```

Windows 上与 ifconfig 相对应的命令是 ipconfig，例如在 Windows 上查看所有网卡信息时可以使用 ipconfig /all。

5.2 ping 命令

Ping 命令是常用的命令之一，一般用于侦测本机到目标主机的网络是否畅通。其使用方法如下：

ping ip 地址

使用方法如下：

```
[root@localhost ~]# ping 120.55.94.78
PING 120.55.94.78 (120.55.94.78) 56(84) bytes of data.
64 bytes from 120.55.94.78: icmp_seq=1 ttl=128 time=11.0 ms
64 bytes from 120.55.94.78: icmp_seq=2 ttl=128 time=17.3 ms
64 bytes from 120.55.94.78: icmp_seq=3 ttl=128 time=16.2 ms
64 bytes from 120.55.94.78: icmp_seq=4 ttl=128 time=10.6 ms
64 bytes from 120.55.94.78: icmp_seq=5 ttl=128 time=10.2 ms
64 bytes from 120.55.94.78: icmp_seq=6 ttl=128 time=18.7 ms
64 bytes from 120.55.94.78: icmp_seq=7 ttl=128 time=15.8 ms
64 bytes from 120.55.94.78: icmp_seq=8 ttl=128 time=10.8 ms
64 bytes from 120.55.94.78: icmp_seq=9 ttl=128 time=10.8 ms
64 bytes from 120.55.94.78: icmp_seq=10 ttl=128 time=11.5 ms
```

上面的输出显示了目标网络可达。在 Linux 上，如果目标网络不可达，则使用 ping 命令会一直发送而无输出结果，直到按下 Ctrl + C 组合键中断才会有统计结果，效果如下：

```
[root@localhost ~]# ping 120.55.94.79
PING 120.55.94.79 (120.55.94.79) 56(84) bytes of data.
^C
--- 120.55.94.79 ping statistics ---
578 packets transmitted, 0 received, 100% packet loss, time 577119ms
```

在 Windows 上，这时使用 ping 命令会返回超时的包数量，而不是一直阻塞。在 Linux 上，这时使用 ping 命令会一直发送数据包，直到人工主动中断。在 Windows 机器上默认会发送 4 个数据包后停止，如果想一直发送而不停止，则可以使用-t 选项。

当然，ping 命令的目标也可以是一个域名，这样通过 ping 这个域名，就可以得到这个域名解析后的 IP 地址：

```
[root@localhost ~]# ping www.baidu.com
PING www.a.shifen.com (61.135.169.121) 56(84) bytes of data.
64 bytes from 61.135.169.121 (61.135.169.121): icmp_seq=1 ttl=128 time=30.3 ms
64 bytes from 61.135.169.121 (61.135.169.121): icmp_seq=2 ttl=128 time=28.8 ms
64 bytes from 61.135.169.121 (61.135.169.121): icmp_seq=3 ttl=128 time=29.0 ms
64 bytes from 61.135.169.121 (61.135.169.121): icmp_seq=4 ttl=128 time=31.9 ms
64 bytes from 61.135.169.121 (61.135.169.121): icmp_seq=5 ttl=128 time=28.8 ms
64 bytes from 61.135.169.121 (61.135.169.121): icmp_seq=6 ttl=128 time=27.8 ms
64 bytes from 61.135.169.121 (61.135.169.121): icmp_seq=7 ttl=128 time=29.0 ms
^C
--- www.a.shifen.com ping statistics ---
7 packets transmitted, 7 received, 0% packet loss, time 6011ms
rtt min/avg/max/mdev = 27.822/29.430/31.968/1.244 ms
```

这里得到 www.baidu.com 对应的 IP 地址是 61.135.169.121。

另外，ping 命令是通过发送 ICMP 数据包实现的。

5.3 telnet 命令

telnet 命令是最常用的网络调试命令之一。如果在自己的机器上还没有安装 telnet 命令，则可以使用如下命令安装：

```
yum install telnet
```

如果一个服务程序对外开启了监听服务，则我们可以使用 telnet ip port 连接上去，例如：

```
[root@localhost ~]# telnet 120.55.94.78 8888
Trying 120.55.94.78...
Connected to 120.55.94.78.
Escape character is '^]'.
```

如果不指定端口号，则 telnet 命令会使用默认的 23 号端口。

反过来说，可以通过 telnet 命令检测指定了 IP 地址和端口号的监听服务是否存在，比如可以使用它检测一个服务是否可以正常连接。又比如某次从某处得到一个代码下载地址，它是一个 svn 地址（svn://120.55.94.78/coderepo/someIM）。为了检测这个 svn 服务是否还能正常对外服务，可以先使用 ping 命令检测到达这个 IP 地址 120.55.94.78 的网络是否畅通：

```
[root@localhost ~]# ping 120.55.94.78
PING 120.55.94.78 (120.55.94.78) 56(84) bytes of data.
64 bytes from 120.55.94.78: icmp_seq=1 ttl=128 time=15.3 ms
64 bytes from 120.55.94.78: icmp_seq=2 ttl=128 time=14.3 ms
64 bytes from 120.55.94.78: icmp_seq=3 ttl=128 time=16.4 ms
64 bytes from 120.55.94.78: icmp_seq=4 ttl=128 time=16.1 ms
64 bytes from 120.55.94.78: icmp_seq=5 ttl=128 time=15.5 ms
^C
--- 120.55.94.78 ping statistics ---
5 packets transmitted, 5 received, 0% packet loss, time 4007ms
rtt min/avg/max/mdev = 14.343/15.568/16.443/0.723 ms
```

如果网络畅通，则再用 telnet 命令连接上去。由于 svn 服务器使用的默认端口是 3690，所以我们执行如下命令：

```
[root@localhost ~]# telnet 120.55.94.78 3690
Trying 120.55.94.78...
Connected to 120.55.94.78.
Escape character is '^]'.
( success ( 2 2 ) ( edit-pipeline svndiff1 absent-entries commit-revprops depth log-revprops atomic-revprops partial-replay ) ) )
```

如上所示，这个 svn 服务是正常开启和对外服务的。反之，如果 telnet 命令连接不上，则说明这个服务不能被外部网络正常连接，就没必要进一步尝试了。

同样，对于一个 Web 服务如 baidu.com，由于我们平时都可以通过输入"www.baidu.com"

访问百度的页面,Web 服务器默认的端口号是 80,所以我们使用"telnet www.baidu.com 80"应该也可以连接成功:

```
[root@localhost ~]# telnet www.baidu.com 80
Trying 115.239.211.112...
Connected to www.baidu.com.
Escape character is '^]'.
hello
HTTP/1.1 400 Bad Request

Connection closed by foreign host.
```

使用 telnet 命令连接上以后,随意发送一个 hello 消息,由于它是非法的 HTTP 请求,所以被服务器关闭了连接。

使用 telnet 命令不仅能连接某个服务器,还能与服务器交互,这通常用于操作一些接收纯文本数据的服务器程序,例如 FTP 服务、邮件服务等。为了演示如何使用 telnet 命令收发数据,这里模拟使用 telnet 命令向某个邮箱发送一封邮件,发送邮件时通常使用的是 SMTP,该协议默认使用的端口是 25。

假设我们的发件地址是 testformybook@163.com,收件地址是 balloonwj@qq.com。其中,发件地址是一个 163 邮箱。如果没有 163 邮箱,则可以申请一个,申请后进入邮箱,在设置页面获得网易邮箱 SMTP 服务的服务器地址。

我们得到的 SMTP 地址是 smtp.163.com,端口号是 25。同时,我们需要开启客户端授权,设置一个客户端授权码。

这里将授权码设置为 2019hhxxttxs。

早些年,很多邮件服务器都允许在其他客户端登录时只需输入正确的邮件服务器地址、用户名和密码。后来出于对安全的考虑,很多邮箱都采用了授权码机制,在其他第三方客户端登录该邮箱时需要输入授权码(不是密码),且需要用户主动打开允许第三方客户端登录的配置选项。

配置完成以后,我们现在就可以使用 telnet 命令连接 163 邮件服务器并发送邮件了。由于在登录过程中需要验证用户名和授权码,而且用户名和授权码必须使用 base64 编码之后的,所以我们先将用户名和授权码的 base64 码准备好,在用的时候直接复制过去。

原 文	base64 码
testformybook	dGVzdGZvcm15Ym9vaw==
2019hhxxttxs	MjAxOWhoeHh0dHhz

如果不知道 base64 编码的原理,则可以从网上搜索 base64 编解码工具。在整个演示过程需要用到如下 SMTP 命令。

命 令	含 义
helo	向 SMTP 服务器发送问候信息
auth login	请求登录验证
data	请求输入邮件正文

SMTP 是文本协议,每一个数据包都以 "\r\n" 结束(Windows 下的默认换行符)。演示一下操作过程:

```
[root@localhost ~]# telnet smtp.163.com 25
Trying 220.181.12.14...
```

```
Connected to smtp.163.com.
Escape character is '^]'.
220 163.com Anti-spam GT for Coremail System (163com[20141201])
helo 163.com
250 OK
auth login
334 dXNlcm5hbWU6
dGVzdGZvcm15Ym9vaw==
334 UGFzc3dvcmQ6
MjAxOWhoeHh0dHhz
235 Authentication successful
mail from: <testformybook@163.com>
250 Mail OK
rcpt to: <balloonwj@qq.com>
250 Mail OK
data
354 End data with <CR><LF>.<CR><LF>
from:testformybook@163.com
to: balloonwj@qq.com
subject: Test

Hello, this is a message from 163.
.
250 Mail OK queued as smtp10,DsCowADHAgQS1IBcwtExJA--.62308S2 1551946998
Connection closed by foreign host.
[root@localhost ~]#
```

这里分析以上操作过程。

（1）使用"telnet smtp.163.com 25"连接 163 邮件服务器，连接成功后，服务器向我们发送了一条欢迎消息：

`220 163.com Anti-spam GT for Coremail System (163com[20141201])\r\n`

（2）必须向服务器发送一条问候消息，使用"helo 163.com"，当然，"163.com"是问候的内容，可以随意填写，然后按回车键，最终组成的数据包内容如下：

`helo 163.com\r\n`

（3）服务器会回复一条状态码是 250 的消息，消息如下：

`250 OK\r\n`

（4）输入 auth login 命令请求验证，然后按回车键，实际发送给服务器的如下：

`auth login\r\n`

（5）服务器应答状态码 334，其中，"dXNlcm5hbWU6"是字符串"username:"的 base64 码：

`334 dXNlcm5hbWU6\r\n`

（6）输入我们的用户名 testformybook 的 base64 码，然后按回车键：

`dGVzdGZvcm15Ym9vaw==\r\n`

（7）服务器应答状态码 334，其中，"UGFzc3dvcmQ6"是字符串"Password:"的 base64 码。这里实际上要求我们输入的是前面介绍的授权码，而不是密码：

```
334 UGFzc3dvcmQ6\r\n
```

（8）输入"MjAxOWhoeHh0dHhz"，并按回车键：

```
MjAxOWhoeHh0dHhz\r\n
```

服务器提示我们授权成功（应答状态码235）：

```
235 Authentication successful\r\n
```

（9）输入邮件的发件地址和收件地址，服务器也会给我们相应的应答（此时应答状态码是250）：

```
mail from: <testformybook@163.com>\r\n
250 Mail OK\r\n
rcpt to: <balloonwj@qq.com>\r\n
250 Mail OK\r\n
```

（10）输入data命令设置邮件的主题、正文、对方收到邮件后显示的的发件人信息等：

```
data\r\n
354 End data with <CR><LF>.<CR><LF>
```

这时服务器应答状态码354，并且提示如果确定结束输入邮件正文，就先按回车键，再输入一个点"."，再接着按回车键，这样邮件就发送出去了。

（11）服务器应答状态码250：

```
250 Mail OK queued as smtp10,DsCowADHAgQS1IBcwtExJA--.62308S2 1551946998
```

如果想退出，则输入"quit"或"close"都可以。

最终，这封邮件就发出去了，到balloonwj@qq.com邮箱查看一下。

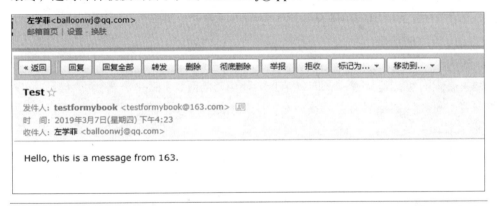

如果在实际实验时对端没有收到邮件，则请查看邮箱的垃圾箱或者反垃圾邮件设置，邮件有可能被反垃圾邮件机制拦截了。

在组装SMTP协议包时涉及很多状态码，常见的SMTP状态码介绍请参见6.12节。

由于我们使用的开发机器以Windows居多，所以在默认情况下，Windows的telnet命令是没有开启的。我们可以在"控制面板"→"程序"→"程序和功能"→"打开或关闭Windows功能"中打开Telnet功能，如下图所示。

5.4 netstat 命令

Windows 和 Linux 均提供了 netstat 命令，该命令经常用于查看网络连接状态。以 Linux 为例，netstat 命令的常见选项如下。

- -a：表示显示所有选项，不使用该选项时，netstat 默认不显示 LISTEN 相关选项。
- -t：表示仅显示 tcp 相关选项。
- -u：仅显示 udp 相关选项。
- -n：不显示别名，将能显示数字的全部转换为数字。
- -l：仅列出处于监听（listen）状态的服务。
- -p：显示建立相关链接的程序名。
- -r：显示路由信息、路由表。
- -e：显示扩展信息，例如 uid 等。
- -s：按各个协议进行统计（重要）。
- -c：每隔一个固定的时间执行该 netstat 命令。

这里详细介绍 -n 选项。在默认情况下，对于一些有别名的 IP 地址和端口号，netstat 命令会以其别名来显示，例如"127.0.0.1"会被显示为"localhost"，80 端口会被显示为"http"。如下所示：

```
[root@iZ238vnojlyZ ~]# netstat -at
Active Internet connections (servers and established)
Proto Recv-Q Send-Q Local Address           Foreign Address         State
tcp        0      0 0.0.0.0:svn             0.0.0.0:*               LISTEN
tcp        0      0 0.0.0.0:http            0.0.0.0:*               LISTEN
tcp        0      0 0.0.0.0:ssh             0.0.0.0:*               LISTEN
tcp        0      0 0.0.0.0:ddi-tcp-1       0.0.0.0:*               LISTEN
tcp        0      0 0.0.0.0:italk           0.0.0.0:*               LISTEN
tcp        0      0 0.0.0.0:dnp             0.0.0.0:*               LISTEN
tcp        0      0 localhost:32000         0.0.0.0:*               LISTEN
tcp        0      0 0.0.0.0:commtact-http   0.0.0.0:*               LISTEN
tcp        0    404 iZ238vnojlyZ:ssh        101.224.250.233:57844   ESTABLISHED
tcp        0      0 iZ238vnojlyZ:59520      100.100.45.131:http     ESTABLISHED
```

```
tcp        0      0 localhost:32000         localhost:31000         ESTABLISHED
tcp6       0      0 [::]:mysql              [::]:*                  LISTEN
tcp6       0      0 [::]:ftp                [::]:*                  LISTEN
tcp6       0      0 localhost:31000         localhost:32000         ESTABLISHED
```

加上 -n 选项看一下效果，可以看到所有 IP 地址和端口号都不再以别名形式显示了：

```
[root@iZ238vnojlyZ ~]# netstat -atn
Active Internet connections (servers and established)
Proto Recv-Q Send-Q Local Address           Foreign Address         State
tcp        0      0 0.0.0.0:3690            0.0.0.0:*               LISTEN
tcp        0      0 0.0.0.0:80              0.0.0.0:*               LISTEN
tcp        0      0 0.0.0.0:22              0.0.0.0:*               LISTEN
tcp        0      0 0.0.0.0:8888            0.0.0.0:*               LISTEN
tcp        0      0 0.0.0.0:12345           0.0.0.0:*               LISTEN
tcp        0      0 0.0.0.0:20000           0.0.0.0:*               LISTEN
tcp        0      0 127.0.0.1:32000         0.0.0.0:*               LISTEN
tcp        0      0 0.0.0.0:20002           0.0.0.0:*               LISTEN
tcp        0    404 120.55.94.78:22         101.224.250.233:57844   ESTABLISHED
tcp        0      0 10.117.203.175:59520    100.100.45.131:80       ESTABLISHED
tcp       15      0 127.0.0.1:32000         127.0.0.1:31000         ESTABLISHED
tcp6       0      0 :::3306                 :::*                    LISTEN
tcp6       0      0 :::21                   :::*                    LISTEN
tcp6       0      0 127.0.0.1:31000         127.0.0.1:32000         ESTABLISHED
```

在 Windows 上，除了可以使用 netstat 命令查看网络状态信息，还可以通过任务管理器打开资源监视器查看当前系统的各种网络连接状态（以下是 Win10 的截图）。

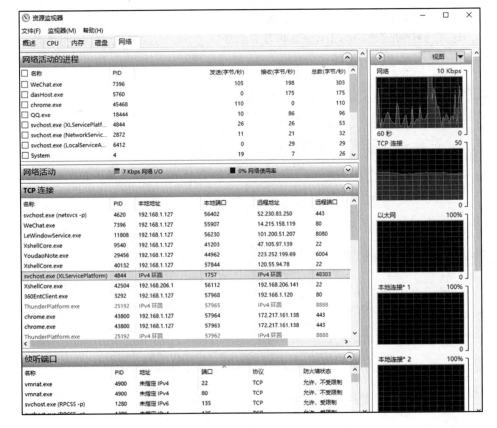

5.5 lsof 命令

lsof（list opened filedesciptor，列出已经打开的文件描述符）命令是 Linux 的扩展命令。在 Linux 上，所有与资源句柄相关的内容都可以被统一抽象为 fd（filedescriptor，文件描述符）。一个文件句柄就是一个 fd，一个 socket 对象也可以叫作 fd 等。

1. lsof 命令的基本用法

在默认情况下，系统是不存在这个命令的，我们需要安装它。使用如下命令安装：

```
yum install lsof
```

这个命令的使用效果如下图所示。在以下结果中列出了各个进程打开的 fd 类型。对于 Unix Socket，使用 lsof 命令会显示其详细的路径，打开的文件 fd 亦是如此。

```
[root@myaliyun ~]# lsof | more
COMMAND    PID  TID USER   FD      TYPE     DEVICE  SIZE/OFF     NODE NAME
systemd      1      root   cwd      DIR      253,1      4096        2 /
systemd      1      root   rtd      DIR      253,1      4096        2 /
systemd      1      root   txt      REG      253,1   1620416  1053844 /usr/lib/systemd/systemd
systemd      1      root   mem      REG      253,1     20112  1066432 /usr/lib64/libuuid.so.1.3.0
systemd      1      root   mem      REG      253,1    265624  1066433 /usr/lib64/libblkid.so.1.1.0
systemd      1      root   mem      REG      253,1     90248  1050434 /usr/lib64/libz.so.1.2.7
systemd      1      root   mem      REG      253,1    157424  1050447 /usr/lib64/liblzma.so.5.2.2
systemd      1      root   mem      REG      253,1     23968  1050682 /usr/lib64/libcap-ng.so.0.0.0
systemd      1      root   mem      REG      253,1     19896  1050665 /usr/lib64/libattr.so.1.1.0
systemd      1      root   mem      REG      253,1     19288  1058782 /usr/lib64/libdl-2.17.so
systemd      1      root   mem      REG      253,1    402384  1050423 /usr/lib64/libpcre.so.1.2.0
systemd      1      root   mem      REG      253,1   2151672  1049987 /usr/lib64/libc-2.17.so
systemd      1      root   mem      REG      253,1    141968  1050013 /usr/lib64/libpthread-2.17.so
systemd      1      root   mem      REG      253,1    747288  1075712 /usr/local/lib64/libgcc_s.so.1
systemd      1      root   mem      REG      253,1     43776  1058789 /usr/lib64/librt-2.17.so
systemd      1      root   mem      REG      253,1    277824  1066434 /usr/lib64/libmount.so.1.1.0
systemd      1      root   mem      REG      253,1     91848  1053492 /usr/lib64/libkmod.so.2.2.10
systemd      1      root   mem      REG      253,1    127184  1058739 /usr/lib64/libaudit.so.1.0.0
systemd      1      root   mem      REG      253,1     61672  1052185 /usr/lib64/libpam.so.0.83.1
systemd      1      root   mem      REG      253,1     20032  1050671 /usr/lib64/libcap.so.2.22
systemd      1      root   mem      REG      253,1    155784  1050421 /usr/lib64/libselinux.so.1
systemd      1      root   mem      REG      253,1    163400  1049976 /usr/lib64/ld-2.17.so
systemd      1      root    0u      CHR        1,3       0t0     5334 /dev/null
systemd      1      root    1u      CHR        1,3       0t0     5334 /dev/null
systemd      1      root    2u      CHR        1,3       0t0     5334 /dev/null
systemd      1      root    3u  a_inode       0,10         0     5330 [timerfd]
systemd      1      root    4u  a_inode       0,10         0     5330 [eventpoll]
systemd      1      root    5u  a_inode       0,10         0     5330 [signalfd]
```

使用 lsof 命令时需要注意以下三点。

（1）在默认情况下，lsof 命令的输出较多，我们可以使用 grep 命令过滤想要查看的进程打开的 fd 信息，例如：

```
lsof -i | grep myapp
```

或者使用 lsof -p pid 也能过滤指定的进程打开的 fd 信息（下图中的 "1723" 是 Nginx 进程的 PID）。

```
[root@myaliyun ~]# lsof -p 1723
COMMAND   PID  USER   FD   TYPE             DEVICE  SIZE/OFF    NODE NAME
nginx    1723  root  cwd    DIR              253,1      4096 1189484 /usr/local/nginx/sbin
nginx    1723  root  rtd    DIR              253,1      4096       2 /
nginx    1723  root  txt    REG              253,1   3830192 1204385 /usr/local/nginx/sbin/nginx
nginx    1723  root  mem    REG              253,1     61624 1058786 /usr/lib64/libnss_files-2.17.so
nginx    1723  root  mem    REG              253,1     11448 1049922 /usr/lib64/libfreebl3.so
nginx    1723  root  mem    REG              253,1   2151672 1049987 /usr/lib64/libc-2.17.so
nginx    1723  root  mem    REG              253,1     90248 1050434 /usr/lib64/libz.so.1.2.7
nginx    1723  root  mem    REG              253,1    402384 1050423 /usr/lib64/libpcre.so.1.2.0
nginx    1723  root  mem    REG              253,1     40664 1049991 /usr/lib64/libcrypt-2.17.so
nginx    1723  root  mem    REG              253,1    141968 1050013 /usr/lib64/libpthread-2.17.so
nginx    1723  root  mem    REG              253,1     19288 1058782 /usr/lib64/libdl-2.17.so
nginx    1723  root  mem    REG              253,1    163400 1049976 /usr/lib64/ld-2.17.so
nginx    1723  root  DEL    REG                0,4              26733 /dev/zero
nginx    1723  root   0u    CHR                1,3        0t0    5334 /dev/null
nginx    1723  root   1u    CHR                1,3        0t0    5334 /dev/null
nginx    1723  root   2w    REG              253,1  49749204 1446488 /usr/local/nginx/logs/error.log
nginx    1723  root   3u   unix 0xffff970038cc4800         0t0   26737 socket
nginx    1723  root   4w    REG              253,1 244264348 1446489 /usr/local/nginx/logs/access.log
nginx    1723  root   5w    REG              253,1  49749204 1446488 /usr/local/nginx/logs/error.log
nginx    1723  root   6u   IPv4              26732        0t0         TCP *:http (LISTEN)
nginx    1723  root   7u   unix 0xffff970038cc4c00         0t0   26738 socket
```

（2）使用 lsof 命令只能查看当前用户有权限查看的进程 fd 信息，对于其没有权限的进程，最右边一列会显示"Permission denied"，如下图所示。

```
[root@myaliyun ~]# su - zhangyl
Last login: Sun Jan 19 07:41:41 CST 2020 from 113.104.211.130 on pts/1
[zhangyl@myaliyun ~]$ lsof -p 1723
COMMAND   PID  USER   FD      TYPE DEVICE SIZE/OFF NODE NAME
nginx    1723  root  cwd   unknown                      /proc/1723/cwd (readlink: Permission denied)
nginx    1723  root  rtd   unknown                      /proc/1723/root (readlink: Permission denied)
nginx    1723  root  txt   unknown                      /proc/1723/exe (readlink: Permission denied)
nginx    1723  root NOFD                               /proc/1723/fd (opendir: Permission denied)
```

（3）lsof 命令第 1 栏进程名在显示时默认显示前 n 个字符，如果需要显示完整的进程名以方便过滤，则可以使用+c 选项。用法如下：

```
#最左侧的程序名最多显示15个字符
[root@myaliyun ~]$ lsof +c 15
```

若将值设置得太大，lsof 便不会采用我们设置的最大值，而是采用默认的最大值。

前面介绍了 socket 也是一种 fd，如果仅需显示系统的网络连接信息，则使用-i 选项即可，通过这个选项可以形象地显示系统当前的出入连接情况，操作和输出结果如下图所示。

```
[root@myaliyun ~]# lsof -i
COMMAND       PID   USER   FD   TYPE  DEVICE SIZE/OFF NODE NAME
sshd          455   root    3u  IPv4  8493547      0t0  TCP myaliyun:ssh->.:16498 (ESTABLISHED)
ntpd          522    ntp   16u  IPv4    12575      0t0  UDP *:ntp
ntpd          522    ntp   17u  IPv6    12576      0t0  UDP *:ntp
ntpd          522    ntp   18u  IPv4    12596      0t0  UDP localhost:ntp
ntpd          522    ntp   20u  IPv4    15777      0t0  UDP myaliyun:ntp
epoll_ser     681   root    3u  IPv4  8494198      0t0  TCP *:hbci (LISTEN)
epoll_ser     681   root    5u  IPv4  8494216      0t0  TCP localhost:hbci->localhost:38900 (ESTABLISHED)
nc            686   root    3u  IPv4  8494208      0t0  TCP localhost:38900->localhost:hbci (ESTABLISHED)
dhclient      743   root    6u  IPv4    13435      0t0  UDP *:bootpc
mysqld       1256  mysql   13u  IPv4    16416      0t0  TCP *:mysql (LISTEN)
sshd         1484   root    3u  IPv4  8494842      0t0  TCP myaliyun:ssh->.:17349 (ESTABLISHED)
sshd         1485   root    3u  IPv4    17012      0t0  TCP *:ssh (LISTEN)
nginx        1723   root    6u  IPv4    26732      0t0  TCP *:http (LISTEN)
nginx        1724   root    6u  IPv4    26732      0t0  TCP *:http (LISTEN)
sshd         2179   root    3u  IPv4  8502068      0t0  TCP myaliyun:ssh->.:16874 (ESTABLISHED)
sshd         2257   root    3u  IPv4  8502508      0t0  TCP myaliyun:ssh->.:16889 (ESTABLISHED)
AliYunDun   11511   root   22u  IPv4  8167008      0t0  TCP myaliyun:53006->100.100.30.25:http (ESTABLISHED)
route_ser   19613   root    5u  IPv4   833919      0t0  TCP *:trivnet1 (LISTEN)
file_serv   19644   root    6u  IPv4   834017      0t0  TCP *:asterix (LISTEN)
file_serv   19644   root    7u  IPv4   834018      0t0  TCP localhost:8601 (LISTEN)
```

看到上图中的连接方向了吧？

当然，和 netstat 命令一样，如果一个 IP 地址或者端口号存在别名，lsof-i 默认也会显示该 IP 地址或端口号的别名。使用-n 选项可以强制显示 IP 地址（不显示 IP 地址的别名），使用-P 选项可以强制显示端口号（不显示端口号的别名），对 IP 地址和端口号都不用别名显示的命令是 lsof-Pni。

2. 使用 lsof 命令恢复被删除的文件

这里演示使用 lsof 命令恢复被删除的文件的小技巧。为了避免出现意外并造成损失，建议在实验之前备份要删除的文件。

假设某个日志文件 fileserver.20201208142806.32470.log 被一个 fileserver 程序使用，则我们可以使用 lsof 命令进行验证：

```
[zhangyl@iZ238vnojlyZ logs]$ lsof | grep fileserve
```

下图中的输出证明 fileserver.20201208142806.32470.log 确实被进程 fileserver 使用，进程 ID 是 32470。

```
fileserve 32470 32476 fileserve    root  mem    REG      253,1   385384   33738980 /usr/lib64/ld-2.28.so
fileserve 32470 32476 fileserve    root   0u    CHR        1,3      0t0       8809 /dev/null
fileserve 32470 32476 fileserve    root   1u    CHR        1,3      0t0       8809 /dev/null
fileserve 32470 32476 fileserve    root   2u    CHR        1,3      0t0       8809 /dev/null
fileserve 32470 32476 fileserve    root   3u  a_inode     0,14        0       8803 [eventfd]
fileserve 32470 32476 fileserve    root   4u  a_inode     0,14        0       8803 [eventpoll]
fileserve 32470 32476 fileserve    root   5u    IPv4    58688291     0t0            TCP *:microsan (LISTEN)
fileserve 32470 32476 fileserve    root   6u    REG      253,1    744461   402752008 /home/zhangyl/flamingoserve
rXX/logs/fileserver.20201208142806.32470.log
fileserve 32470 32476 fileserve    root   7r    CHR        1,3      0t0       8809 /dev/null
fileserve 32470 32476 fileserve    root   8u  a_inode     0,14        0       8803 [eventfd]
fileserve 32470 32476 fileserve    root   9u  a_inode     0,14        0       8803 [eventpoll]
fileserve 32470 32476 fileserve    root  10u  a_inode     0,14        0       8803 [eventfd]
fileserve 32470 32476 fileserve    root  11u  a_inode     0,14        0       8803 [eventpoll]
fileserve 32470 32476 fileserve    root  12u  a_inode     0,14        0       8803 [eventfd]
fileserve 32470 32476 fileserve    root  13u  a_inode     0,14        0       8803 [eventpoll]
fileserve 32470 32476 fileserve    root  14u  a_inode     0,14        0       8803 [eventfd]
```

我们将该日志文件删除：

```
[zhangyl@iZ238vnojlyZ logs]$ rm -rf fileserver.20201208142806.32470.log
```

然后再次使用 lsof | grep fileserve 命令查看进程 fileserver 使用该文件的状态信息，如下图所示。此时在该进程中，该文件已经被标记为 deleted 状态了。

```
fileserve 32470 32477 fileserve    root  mem    REG      253,1   385384   33738980 /usr/lib64/ld-2.28.so
fileserve 32470 32477 fileserve    root   0u    CHR        1,3      0t0       8809 /dev/null
fileserve 32470 32477 fileserve    root   1u    CHR        1,3      0t0       8809 /dev/null
fileserve 32470 32477 fileserve    root   2u    CHR        1,3      0t0       8809 /dev/null
fileserve 32470 32477 fileserve    root   3u  a_inode     0,14        0       8803 [eventfd]
fileserve 32470 32477 fileserve    root   4u  a_inode     0,14        0       8803 [eventpoll]
fileserve 32470 32477 fileserve    root   5u    IPv4    58688291     0t0            TCP *:microsan (LISTEN)
fileserve 32470 32477 fileserve    root   6u    REG      253,1    744461   402752008 /home/zhangyl/flamingoserve
rXX/logs/fileserver.20201208142806.32470.log (deleted)
fileserve 32470 32477 fileserve    root   7r    CHR        1,3      0t0       8809 /dev/null
fileserve 32470 32477 fileserve    root   8u  a_inode     0,14        0       8803 [eventfd]
fileserve 32470 32477 fileserve    root   9u  a_inode     0,14        0       8803 [eventpoll]
fileserve 32470 32477 fileserve    root  10u  a_inode     0,14        0       8803 [eventfd]
```

我们进入 /proc/pid/fd/ 目录并使用 ll（ls -l 的别名）命令查看文件的状态，这里的 pid 要被换成相应进程的 ID，这里就是 32470：

```
[zhangyl@iZ238vnojlyZ 12976]$ cd /proc/32470/fd/
[zhangyl@iZ238vnojlyZ fd]$ ll
```

此时便找到了被删除的文件，如下图所示。

```
[root@iZbp14iz399acush5e8ok7Z fd]# cd /proc/32470/fd/
[root@iZbp14iz399acush5e8ok7Z fd]# ll
total 0
lrwx------ 1 root root 64 Feb 24 22:45 0 -> /dev/null
lrwx------ 1 root root 64 Feb 24 22:45 1 -> /dev/null
lrwx------ 1 root root 64 Feb 24 22:45 10 -> 'anon_inode:[eventfd]'
lrwx------ 1 root root 64 Feb 24 22:45 11 -> 'anon_inode:[eventpoll]'
lrwx------ 1 root root 64 Feb 24 22:45 12 -> 'anon_inode:[eventfd]'
lrwx------ 1 root root 64 Feb 24 22:45 13 -> 'anon_inode:[eventpoll]'
lrwx------ 1 root root 64 Feb 24 22:45 14 -> 'anon_inode:[eventfd]'
lrwx------ 1 root root 64 Feb 24 22:45 15 -> 'anon_inode:[eventpoll]'
lrwx------ 1 root root 64 Feb 24 22:45 16 -> 'anon_inode:[eventfd]'
lrwx------ 1 root root 64 Feb 24 22:45 17 -> 'anon_inode:[eventpoll]'
lrwx------ 1 root root 64 Feb 24 22:45 18 -> 'anon_inode:[eventfd]'
lrwx------ 1 root root 64 Feb 24 22:45 19 -> 'anon_inode:[eventpoll]'
lrwx------ 1 root root 64 Feb 24 22:45 2 -> /dev/null
lrwx------ 1 root root 64 Feb 24 22:45 3 -> 'anon_inode:[eventfd]'
lrwx------ 1 root root 64 Feb 24 22:45 4 -> 'anon_inode:[eventpoll]'
lrwx------ 1 root root 64 Feb 24 22:45 5 -> 'socket:[586882911]'
lrwx------ 1 root root 64 Feb 24 22:45 6 -> '/home/zhangyl/flamingoserverXX/logs/fileserver.20201208142806.32470.log (deleted)'
lr-x------ 1 root root 64 Feb 24 22:45 7 -> /dev/null
lrwx------ 1 root root 64 Feb 24 22:45 8 -> 'anon_inode:[eventfd]'
lrwx------ 1 root root 64 Feb 24 22:45 9 -> 'anon_inode:[eventpoll]'
```

右数第 3 列标号 6 那一行正好对应我们删除的文件，使用这个标号对文件进行恢复：

```
[zhangyl@iZ238vnojlyZ fd]$ cat 6 > /home/zhangyl/flamingoserver/logs/fileserver.20201208142806.32470.log
```

这样我们的文件就被恢复到/home/zhangyl/flamingoserver/logs/目录下了。

使用 lsof 命令恢复文件时需要注意：如果想成功恢复文件，则使用文件的进程必须处于存活状态，如果 fileserver 已经退出或被杀死，就无法用这种方式恢复；用这种方式恢复文件后，再次用 lsof 命令查看时，这个文件仍然是 deleted 状态。也就是说，虽然能够手工恢复，但是进程不会再使用这个文件了。对于这里的日志文件，fileserver 继续运行时也不会再向其中写入日志了。所以该命令一般只用于在某些特殊情况下删除一些重要数据文件时进行应急恢复。

5.6　nc 命令

nc 即 netcat 命令，在排查网络故障时非常有用，功能非常强大。在系统中默认是没有这个命令的，需要进行安装。安装方法如下：

```
yum install nc
```

nc 命令常用于模拟一个服务器程序被其他客户端连接，或者模拟一个客户端连接其他服务器，连接之后就可以进行数据收发。nc 命令默认使用 TCP，加上-u 选项后会使用 UDP。

下面介绍 nc 命令的常见用途。

（1）模拟一个服务器程序。我们使用-l 选项（单词 listen 的首字母）在某个 IP 地址和端口上开启一个监听服务，以便让其他客户端连接。通常为了显示更详细的信息，会带上-v 选项。示例如下：

```
[root@iZ238vnojlyZ ~]# nc -v -l 127.0.0.1 6000
Ncat: Version 6.40 ( http://nmap.org/ncat )
Ncat: Listening on 127.0.0.1:6000
```

以上命令启动了一个监听服务器，监听 IP 地址为 127.0.0.1，监听端口为 6000；如果机器可以被外网访问，则可以使用 0.0.0.0 这样的监听地址（也可以使用其他公网 IP 作为监听地址），示例如下：

```
[root@iZ238vnojlyZ ~]# nc -v -l 0.0.0.0 6000
Ncat: Version 6.40 ( http://nmap.org/ncat )
Ncat: Listening on 0.0.0.0:6000
```

（2）模拟一个客户端程序。使用 nc 命令模拟一个客户端程序时，我们不需要使用-l 选项，直接将其写上 IP 地址（或域名，nc 命令可以自动解析域名）和端口号即可，示例如下：

```
## 连接百度的 Web 服务器
[root@iZ238vnojlyZ ~]# nc -v www.baidu.com 80
Ncat: Version 6.40 ( http://nmap.org/ncat )
Ncat: Connected to 115.239.211.112:80.
```

该输出提示我们成功连接上百度的 Web 服务器。

我们知道，客户端连接服务器时一般都是操作系统随机分配一个可用的端口号连接到服务器，所以使用 nc 命令作为客户端时，可以使用-p 选项指定使用哪个端口号连接服务器，例如我们希望通过本地的 5555 端口连接百度的 Web 服务器时，则可以这么输入：

```
[root@iZ238vnojlyZ ~]# nc -v -p 5555 www.baidu.com 80
Ncat: Version 6.40 ( http://nmap.org/ncat )
Ncat: Connected to 115.239.211.112:80.
```

再开一个 Shell 窗口，使用在上文中介绍的 lsof 命令验证是否通过 5555 端口连接上了百度的 Web 服务器，其结果确实如我们所期望的：

```
[root@iZ238vnojlyZ ~]# lsof -Pni | grep nc
nc        32610    root    3u  IPv4 113369437     0t0  TCP
120.55.94.78:5555->115.239.211.112:80 (ESTABLISHED)
```

当然，使用 nc 命令与对端建立连接后，我们可以发送消息。

使用 nc -v -l 0.0.0.0 6000 命令模拟一个监听服务，再新建一个 Shell 窗口，使用 nc -v 127.0.0.1 6000 命令模拟一个客户端程序连接刚才的服务器；此时在客户端和服务器上就可以相互发送消息了，可以达到一个简化版的 IM 软件聊天效果。

服务端效果如下：

```
[root@myaliyun ~]# nc -v -l 0.0.0.0 6000
Ncat: Version 7.50 ( https://nmap.org/ncat )
Ncat: Listening on 0.0.0.0:6000
Ncat: Connection from 127.0.0.1.
Ncat: Connection from 127.0.0.1:56730.
What is your name?
zhangyuanlong
How are you?
fine, thank you and you?
I am fine too.Can you speak Chinese?
我会说中文，你呢？
我也会。
那太好了！
```

客户端效果如下：

```
[root@myaliyun ~]# nc -v 127.0.0.1 6000
Ncat: Version 7.50 ( https://nmap.org/ncat )
Ncat: Connected to 127.0.0.1:6000.
What is your name?
zhangyuanlong
How are you?
fine, thank you and you?
I am fine too.Can you speak Chinese?
我会说中文，你呢？
我也会。
那太好了！
```

> 如果使用 nc 命令发消息时不小心输入错误，则可以使用 Ctrl+Backspace 组合键删除输错的内容，直接使用 Backspace 键是无法删除的。

nc 命令默认会将 "\n" 作为每条消息的结束标志，如果指定了-C 选项，则使用 "\r\n" 作为消息结束的标志。

nc 命令不仅可以发送消息，也可以发送文件，这里进行演示。需要注意的是，接收文件方是服务端，发送文件方是客户端。接收文件方可以使用任意名称存储发送方的文件。

服务端（接收文件方）的命令如下：

```
nc -l IP地址 端口号 > 接收的文件名
```

客户端（发送文件方）的命令如下：

```
nc IP地址 端口号 < 发送的文件名
```

服务端的效果如下：

```
[root@myaliyun chapter05]# ll my_index.html
ls: cannot access my_index.html: No such file or directory
[root@myaliyun chapter05]# nc -v -l 12345 > my_index.html
Ncat: Version 7.50 ( https://nmap.org/ncat )
Ncat: Listening on :::12345
Ncat: Listening on 0.0.0.0:12345
Ncat: Connection from 127.0.0.1.
Ncat: Connection from 127.0.0.1:47094.
[root@myaliyun chapter05]# ll my_index.html
-rw-r--r-- 1 root root 0 Feb 25 00:54 my_index.html
```

客户端的效果如下：

```
[root@myaliyun ~]# ll | grep index.html
-rw-r--r-- 1 root    root          0 Feb 25 00:48 index.html
[root@myaliyun ~]# nc 127.0.0.1 12345 < index.html
```

在上述过程中，发送文件方（nc 客户端）发送了一个名为 index.html 的文件，接收文件方（nc 服务端）接收该文件并将其以 my_index.html 文件名存储到本地。

根据前面的介绍，若我们需要调试自己的服务器或者客户端程序，又不想自己开发相应的对端，就可以使用 nc 命令进行模拟。

当然，nc 命令非常强大，其功能远远超过本节介绍的这些，可以在 man 手册上获取关于 nc 命令的更多信息。

5.7 curl 命令

有读者应该用过图形化的 HTTP 请求模拟工具 PostMan，与 PostMan 一样，curl 命令是 Linux 和 Mac 机器上可以模拟发送 HTTP 请求的一个很常用的命令，可以使用如下命令进行安装：

```
yum install curl
```

curl 的基础用法如下：

```
curl 某个 URL
```

例如：

```
curl http://www.baidu.com
```

其默认行为是将目标页面的内容输出到 Shell 窗口：

```
[root@localhost ~]# curl http://www.baidu.com
<!DOCTYPE html>
<!--STATUS OK--><html> <head><meta http-equiv=content-type
content=text/html;charset=utf-8><meta http-equiv=X-UA-Compatible
content=IE=Edge><meta content=always name=referrer><link rel=stylesheet
type=text/css href=http://s1.bdstatic.com/r/www/cache/bdorz/baidu.min.css><title>
百度一下，你就知道</title></head> ...省略更多的输出结果...
```

我们也可以把页面保存到本地（等价于-o 选项），示例如下：

```
curl http://www.baidu.com > index.html
## 等价于
curl -o index.html http://www.baidu.com
```

HTTP 常用的请求方式是 GET 和 POST，我们可以使用-X 选项显式指定请求是采用 GET 方式还是 POST 方式（不指定时，采用默认的 GET 方式）：

```
curl -X GET http://www.baidu.com/index.php?s=9
```

如果采用 GET 方式，则 curl 提供了另一个专门的选项-G（或--get）进行设置。

如果采用 POST 方式，则除了需要使用-X 选项（或--request）指定请求方法，还需要使用-d 选项（或--data）指定 POST 的数据内容：

```
curl -X POST -d 'somepostdata' 'https://www.somesite.com/api/v1/chat'
```

我们有时需要在发送 HTTP 请求时设置增加一些头部信息，这时可以使用-H 选项（或--header）指定。如果有多个选项，则可以多次使用-H 选项来逐一设置，例如：

```
#模拟一个 HTML 表单上传文件功能，表单一共有两个字段 uploadVideo、title
#并设置了 Content-Type、Cookie 两个 HTTP 头
curl -X POST https://www.somesite.com/api/upload -F
"uploadVideo=@D:/videos/myvideo.mp4" -F "title=我的视频" -H "Content-Type:
multipart/form-data" -H "Cookie:
sessionKey=LSA2RSk-69392189-38e5-444f-8581-2e5c028ca2ag;
Device-Id=6905368379528347650"
```

当然，对一些常用的 HTTP 头字段，curl 提供了单独的选项。例如，对 user-agent 字段可以使用-A 选项（或--user-agent）进行设置；对 referer 字段可以使用-e 选项（或--referer）

进行设置，等等。

如果希望在 HTTP 的应答结果中包含 HTTP 头部信息，则可以使用-i 选项（或--include），演示如下：

```
[root@localhost ~]# curl -i http://www.baidu.com/
HTTP/1.1 200 OK
Accept-Ranges: bytes
Cache-Control: private, no-cache, no-store, proxy-revalidate, no-transform
Connection: Keep-Alive
Content-Length: 2381
Content-Type: text/html
Date: Sat, 09 Mar 2019 06:47:15 GMT
Etag: "588604c4-94d"
Last-Modified: Mon, 23 Jan 2017 13:27:32 GMT
Pragma: no-cache
Server: bfe/1.0.8.18
Set-Cookie: BDORZ=27315; max-age=86400; domain=.baidu.com; path=/

<!DOCTYPE html>
<!--STATUS OK--><html> <head><meta http-equiv=content-type
content=text/html;charset=utf-8><meta http-equiv=X-UA-Compatible
content=IE=Edge><meta content=always name=referrer>...省略更多的输出结果...
```

在某些情况下，如果只想显示 HTTP 应答的头部信息（不是 HTML 文档的头部），则可以使用-I 选项（或--head），演示如下：

```
[root@localhost ~]# curl -I http://www.baidu.com/
HTTP/1.1 200 OK
Accept-Ranges: bytes
Cache-Control: private, no-cache, no-store, proxy-revalidate, no-transform
Connection: Keep-Alive
Content-Length: 277
Content-Type: text/html
Date: Sat, 09 Mar 2019 06:42:16 GMT
Etag: "575e1f8a-115"
Last-Modified: Mon, 13 Jun 2016 02:50:50 GMT
Pragma: no-cache
Server: bfe/1.0.8.18
（这里有个换行）
```

注意：在以上输出结果中，最后一行的空行是刻意保留的，HTTP 头部的每一行都以 "\r\n" 结束，整个头部再以一个额外的 "\r\n" 结束，所以正好末尾有个空行。

如果需要使用代理发送 HTTP 请求，则可以使用-x 选项（或--proxy），使用形式如下：

```
curl -x <[protocol://][user:password@]proxyhost[:port]>
```

5.8　tcpdump 命令

tcpdump 命令是 Linux 提供的一个非常强大的抓包工具，熟练使用它，对我们排查网络问题非常有用。如果在机器上还没有安装该工具，则可以使用如下命令安装：
```
yum install tcpdump
```

如果要使用 tcpdump 命令，则必须有 sudo 权限。

tcpdump 命令常用的选项如下。

（1）-i：指定要捕获的目标网卡名，网卡名可以使用前面介绍的 ifconfig 命令获得；如果要抓取所有网卡上的包，则可以使用 any 关键字。代码如下：
```
## 抓取网卡 ens33 上的包
tcpdump -i ens33
## 抓取所有网卡上的包
tcpdump -i any
```

（2）-X：以 ASCII 和十六进制形式输出捕获的数据包内容，减去链路层的包头信息；-XX 以 ASCII 和十六进制形式输出捕获的数据包内容，包括链路层的包头信息。

（3）-n：不要将 IP 地址显示成别名；-nn 指不要将 IP 地址和端口显示成别名。

（4）-S：以绝对值显示包的 ISN 号（包序列号），默认以上一包的偏移量显示。

（5）-vv：显示详细的抓包数据；-vvv 指显示更详细的抓包数据。

（6）-w：将抓取的包的原始信息（不解析也不输出）写入文件中，后跟文件名。示例如下：
```
tcpdump -i any -w filename
```

（7）-r：从使用-w 选项保存的包文件中读取数据包信息。

除了可以使用选项，tcpdump 命令还支持各种数据包过滤的表达式，常见的形式如下：
```
## 仅显示经过 8888 端口的数据包（包括 tcp:8888 和 udp:8888）
tcpdump -i any 'port 8888'

## 仅显示经过 tcp:8888 端口的数据包
tcpdump -i any 'tcp port 8888'

## 仅显示源端口是 tcp:8888 的数据包
tcpdump -i any 'tcp src port 8888'

## 仅显示源端口是 tcp:8888 或目标端口是 udp:9999 的包
tcpdump -i any 'tcp src port 8888 or udp dst port 9999'

## 仅显示源地址是 127.0.0.1 且源端口是 tcp:9999 的包，以 ASCII 和十六进制显示详细的输出
## 不显示 IP 地址和端口号的别名
tcpdump -i any 'src host 127.0.0.1 and tcp src port 9999' -XX -nn -vv
```

下面通过三个具体的操作实例演示使用 tcpdump 命令抓包的过程。

1. 连接一个正常的监听端口

假设服务端的地址是 127.0.0.1:12345，则使用 nc 命令在一个 Shell 窗口创建一个服务器程序并在这个地址进行监听：
```
nc -v -l 127.0.0.1 12345
```

效果如下：

```
[root@myaliyun ~]# nc -v -l 127.0.0.1 12345
Ncat: Version 7.50 ( https://nmap.org/ncat )
Ncat: Listening on 127.0.0.1:12345
```

在另一个 Shell 窗口使用 tcpdump 命令对 12345 端口进行抓包：

```
tcpdump -i any 'port 12345' -XX -nn -vv
```

效果如下：

```
[root@myaliyun ~]# tcpdump -i any 'port 12345' -XX -nn -vv
tcpdump: listening on any, link-type LINUX_SLL (Linux cooked), capture size 262144
bytes
```

打开一个 Shell 窗口，使用 nc 命令创建一个客户端连接刚才用 nc 命令模拟的服务端：

```
[root@myaliyun ~]# nc -v 127.0.0.1 12345
Ncat: Version 7.50 ( https://nmap.org/ncat )
Ncat: Connected to 127.0.0.1:12345.
```

抓到的包如下图所示。

```
[root@myaliyun ~]# tcpdump -i any 'port 12345' -XX -nn -vv
tcpdump: listening on any, link-type LINUX_SLL (Linux cooked), capture size 262144 bytes
01:08:55.515597 IP (tos 0x0, ttl 64, id 57067, offset 0, flags [DF], proto TCP (6), length 60)
    127.0.0.1.47106 > 127.0.0.1.12345: Flags [S], cksum 0xfe30 (incorrect -> 0x06a5), seq 3453720739, win 4
3690, options [mss 65495,sackOK,TS val 159860296 ecr 0,nop,wscale 7], length 0
        0x0000:  0000 0304 0006 0000 0000 0000 0000 0800  ................
        0x0010:  4500 003c deeb 4000 4006 5dce 7f00 0001  E..<..@.@.]....
        0x0020:  7f00 0001 b802 3039 cddb 98a3 0000 0000  ......09......
        0x0030:  a002 aaaa fe30 0000 0204 ffd7 0402 080a  .....0..........
        0x0040:  0987 4648 0000 0000 0103 0307 0000 0000  ..FH............
        0x0050:  0000 0000 0000 0000 0000 0000            ............
01:08:55.515619 IP (tos 0x0, ttl 64, id 0, offset 0, flags [DF], proto TCP (6), length 60)
    127.0.0.1.12345 > 127.0.0.1.47106: Flags [S.], cksum 0xfe30 (incorrect -> 0x063d), seq 1759856546, ack
 3453720740, win 43690, options [mss 65495,sackOK,TS val 159860296 ecr 159860296,nop,wscale 7], length 0
        0x0000:  0000 0304 0006 0000 0000 0000 0000 0800  ................
        0x0010:  4500 003c 0000 4000 4006 3cba 7f00 0001  E..<..@.@.<....
        0x0020:  7f00 0001 3039 b802 68e5 47a2 cddb 98a4  ....09..h.G...
        0x0030:  a012 aaaa fe30 0000 0204 ffd7 0402 080a  .....0..........
        0x0040:  0987 4648 0987 4648 0103 0307 0000 0000  ..FH..FH........
        0x0050:  0000 0000 0000 0000 0000 0000            ............
01:08:55.515635 IP (tos 0x0, ttl 64, id 57068, offset 0, flags [DF], proto TCP (6), length 52)
    127.0.0.1.47106 > 127.0.0.1.12345: Flags [.], cksum 0xfe28 (incorrect -> 0xd881), seq 1, ack 1, win 342
, options [nop,nop,TS val 159860296 ecr 159860296], length 0
        0x0000:  0000 0304 0006 0000 0000 0000 0000 0800  ................
        0x0010:  4500 0034 deec 4000 4006 5dd5 7f00 0001  E..4..@.@.]....
        0x0020:  7f00 0001 b802 3039 98a4 68e5 47a3       ......09..h.G.
        0x0030:  8010 0156 fe28 0000 0101 080a 0987 4648  ...V.(........FH
        0x0040:  0987 4648 0000 0000 0000 0000 0000 0000  ..FH............
        0x0050:  0000 0000                                ....
```

由于我们没有在客户端和服务器之间发送任何消息，所以抓到的包其实是 TCP 连接的三次握手数据包：三次握手过程是客户端先向服务器发送一个 SYN；然后服务器应答一个 SYN + ACK，应答的序列号是递增 1 的，表示应答哪个请求，即从 3453720739 递增到 3453720740；接着客户端应答一个 ACK。这时，我们发现发包序列号和应答序列号都变成 1 了，这是因为 tcpdump 命令默认使用相对包序号，加上-S 选项后就变成绝对包序号了。

按 Ctrl+C 组合键中断 tcpdump 的抓包过程，并把用 nc 命令模拟的客户端和服务器程序停掉，然后在前面的 tcpdump 命令后面加上-S 选项后重新开启抓包，使用的命令如下：

```
tcpdump -i any 'port 12345' -XX -nn -vv -S
```

按顺序用 nc 命令再次启动模拟的服务端和客户端，得到的抓包结果如下图所示。

```
[root@myaliyun ~]# tcpdump -i any 'port 12345' -XX -nn -vv -S
tcpdump: listening on any, link-type LINUX_SLL (Linux cooked), capture size 262144 bytes
01:15:29.783846 IP (tos 0x0, ttl 64, id 56035, offset 0, flags [DF], proto TCP (6), length 60)
    127.0.0.1.47112 > 127.0.0.1.12345: Flags [S], cksum 0xfe30 (incorrect -> 0x2707), seq 2976747655, win 4
3690, options [mss 65495,sackOK,TS val 160254564 ecr 0,nop,wscale 7], length 0
        0x0000:  0000 0304 0006 0000 0000 0000 0000 0800  ................
        0x0010:  4500 003c dae3 4000 4006 61d6 7f00 0001  E..<..@.@.a.....
        0x0020:  7f00 0001 b808 3039 b16d 9087 0000 0000  ......09.m......
        0x0030:  a002 aaaa fe30 0000 0204 ffd7 0402 080a  .....0..........
        0x0040:  098d 4a64 0000 0000 0103 0307 0000 0000  ..Jd............
        0x0050:  0000 0000 0000 0000 0000 0000           ............
01:15:29.783861 IP (tos 0x0, ttl 64, id 0, offset 0, flags [DF], proto TCP (6), length 60)
    127.0.0.1.12345 > 127.0.0.1.47112: Flags [S.], cksum 0xfe30 (incorrect -> 0x4598), seq 49384059, ack 29
76747656, win 43690, options [mss 65495,sackOK,TS val 160254564 ecr 160254564,nop,wscale 7], length 0
        0x0000:  0000 0304 0006 0000 0000 0000 0000 0800  ................
        0x0010:  4500 003c 0000 4000 4006 3cba 7f00 0001  E..<..@.@.<.....
        0x0020:  7f00 0001 3039 b808 02f1 8a7b b16d 9088  ....09.....{.m..
        0x0030:  a012 aaaa fe30 0000 0204 ffd7 0402 080a  .....0..........
        0x0040:  098d 4a64 098d 4a64 0103 0307 0000 0000  ..Jd..Jd........
        0x0050:  0000 0000 0000 0000 0000 0000           ............
01:15:29.783874 IP (tos 0x0, ttl 64, id 56036, offset 0, flags [DF], proto TCP (6), length 52)
    127.0.0.1.47112 > 127.0.0.1.12345: Flags [.], cksum 0xfe28 (incorrect -> 0x17dd), seq 2976747656, ack 4
9384060, win 342, options [nop,nop,TS val 160254564 ecr 160254564], length 0
        0x0000:  0000 0304 0006 0000 0000 0000 0000 0800  ................
        0x0010:  4500 0034 dae4 4000 4006 61dd 7f00 0001  E..4..@.@.a.....
        0x0020:  7f00 0001 b808 3039 b16d 9088 02f1 8a7c  ......09.m.....|
        0x0030:  8010 0156 fe28 0000 0101 080a 098d 4a64  ...V.(........Jd
        0x0040:  098d 4a64 0000 0000 0000 0000 0000 0000  ..Jd............
        0x0050:  0000 0000                                ....
```

这次得到的包序号就是绝对序号了。

2. 连接一个不存在的监听端口

刚刚演示的实例是正常的 TCP 连接三次握手过程捕获到的数据包。假如连接的服务器 IP 地址存在，但监听端口号不存在，则看下 tcpdump 抓包的结果。除了在一个 Shell 窗口启动一个 tcpdump 抓包监测，在另一个 Shell 窗口用 nc 命令连接一个不存在的监听端口即可（以下代码尝试连接一个不存在的端口 1234）。

```
[root@myaliyun ~]# nc -v 127.0.0.1 1234
Ncat: Version 7.50 ( https://nmap.org/ncat )
Ncat: Connection refused.
```

tcpdump 抓包数据如下图所示。

```
[root@myaliyun ~]# tcpdump -i any 'port 1234' -XX -nn -vv -S
tcpdump: listening on any, link-type LINUX_SLL (Linux cooked), capture size 262144 bytes
01:18:34.239105 IP (tos 0x0, ttl 64, id 39205, offset 0, flags [DF], proto TCP (6), length 60)
    127.0.0.1.41182 > 127.0.0.1.1234: Flags [S], cksum 0xfe30 (incorrect -> 0x464f), seq 283542478, win 436
90, options [mss 65495,sackOK,TS val 160439019 ecr 0,nop,wscale 7], length 0
        0x0000:  0000 0304 0006 0000 0000 0000 0000 0800  ................
        0x0010:  4500 003c 9925 4000 4006 a394 7f00 0001  E..<.%@.@.......
        0x0020:  7f00 0001 a0de 04d2 10e6 83ce 0000 0000  ................
        0x0030:  a002 aaaa fe30 0000 0204 ffd7 0402 080a  .....0..........
        0x0040:  0990 1aeb 0000 0000 0103 0307 0000 0000  ................
        0x0050:  0000 0000 0000 0000 0000 0000           ............
01:18:34.239126 IP (tos 0x0, ttl 64, id 51942, offset 0, flags [DF], proto TCP (6), length 40)
    127.0.0.1.1234 > 127.0.0.1.41182: Flags [R.], cksum 0x7768 (correct), seq 0, ack 283542479, win 0, leng
th 0
        0x0000:  0000 0304 0006 0000 0000 0000 0000 0800  ................
        0x0010:  4500 0028 cae6 4000 4006 71e7 7f00 0001  E..(..@.@.q.....
        0x0020:  7f00 0001 04d2 a0de 0000 0000 10e6 83cf  ................
        0x0030:  5014 0000 7768 0000 0000 0000 0000 0000  P...wh..........
        0x0040:  0000 0000 0000 0000                      ........
```

这时客户端发送 SYN，目标主机由于并没有在 1234 这样的端口上开启监听，所以应答一个 ACK+RST，这个应答包会导致客户端的 connect 连接失败后返回。

3. 连接一个很遥远的 IP，或者网络繁忙

在实际应用中还存在一种情况，就是客户端访问一个很遥远的 IP 或者网络繁忙，服务器对客户端发送的 TCP 三次握手的网络 SYN 报文没有应答，这时会出现什么情况呢？

我们通过设置防火墙规则来模拟这种情况。使用 iptables -F 先将防火墙的已有规则都清理掉，然后给防火墙的 INPUT 链上增加一个规则：丢弃本地网卡 lo（也就是 127.0.0.1 这个回环地址）上的所有 SYN 包（关闭这个功能后再次执行 iptables –F 命令即可）：

```
iptables -F
iptables -I INPUT -p tcp --syn -i lo -j DROP
```

> 注意：如果对 CentOS 的防火墙 iptables 命令有兴趣，则可以使用 man iptables 在 man 手册中查看更详细的帮助。

在开启 tcpdump 抓包和设置防火墙规则之后，使用 nc 命令连接 127.0.0.1:12345 这个地址，nc 客户端在尝试连接一段时间后会超时失败（失败的原因是上面配置了丢弃所有 SYN 包，导致三次握手失败），nc 客户端执行结果如下：

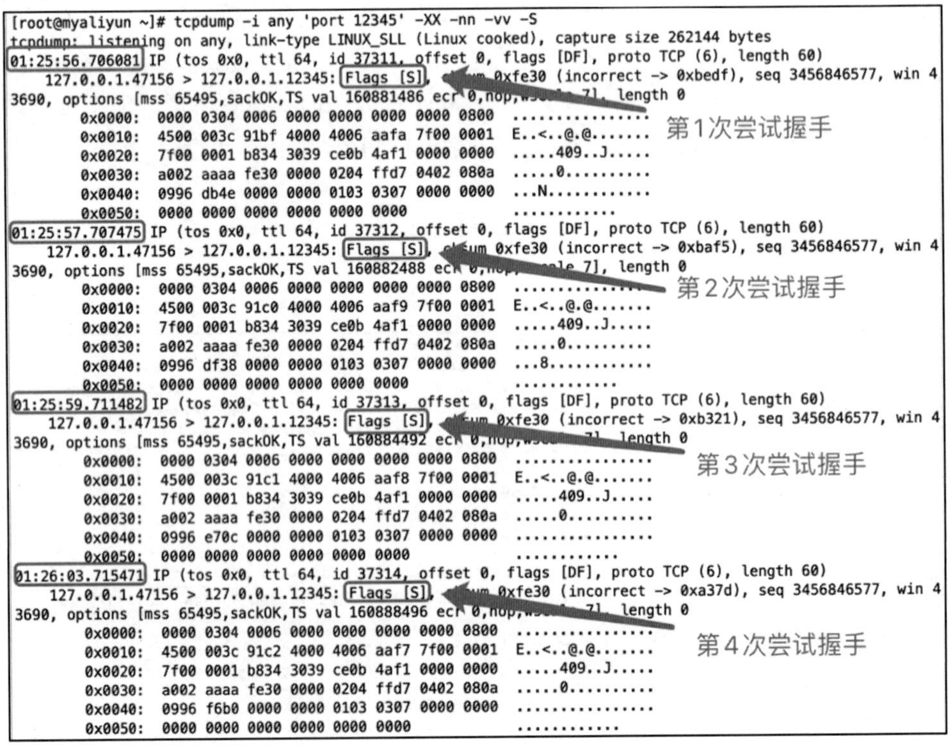

```
[root@myaliyun ~]# nc -v 127.0.0.1 12345
Ncat: Version 7.50 ( https://nmap.org/ncat )
Ncat: Connection timed out.
```

接着，我们得到 tcpdump 抓取的数据包，如下图所示。

通过抓包数据可以看到，如果本机连接不上某个服务器，则系统一共会重试多次，重试的时间间隔分别是 1、2、4、8、16 秒，依此类推，最后返回超时失败。这个重试次数在/proc/sys/net/ipv4/tcp_syn_retries 内核参数中进行设置，在笔者的测试机器上默认值为 6。

```
[root@myaliyun ~]# cat /proc/sys/net/ipv4/tcp_syn_retries
6
```

TCP 的四次挥手与三次握手基本类似，这里就不贴出 tcpdump 抓包的详情了，强烈建议不熟悉这块内容的读者实际练习一遍。

第 6 章

网络通信协议设计

在计算机技术体系中存在很多网络通信协议。通信协议实质上就是一段数据,通信双方事先约定好按规定的格式去编码和解码,最终达到传输消息的目的。TCP/IP 是目前各种计算机设备最常用的协议。当然,TCP/IP 也是一个协议簇,包含一组协议,其中靠近应用层且最常用的是 TCP 和 UDP。本章讲解设计应用层协议时需要理解的原理和注意事项,主要介绍面向连接的 TCP。

6.1 理解 TCP

TCP 是流式协议。流式协议,即协议的内容是流水一样的字节流,内容与内容之间没有明确的分界标志,需要我们人为地给这些内容划分边界。

举个例子,A 与 B 进行 TCP 通向,A 先后给 B 发送了两个大小分别是 100 字节和 200 字节的数据包,那么 B 是如何收到数据包的呢?B 可能先后收到 100、200 字节;也可能先后收到 50、250 字节;或者先后收到 100、100、100 字节;或者先后收到 20、20、60、100、50、50 字节……

不知道读者看出规律没有?规律就是 A 一共向 B 发送了 300 字节,B 可能以一次或者多次总数为 300 字节的任意形式收到。假设 A 向 B 发送了两个大小分别是 100 字节和 200 字节的数据包,则作为发送方的 A 来说,A 是知道如何划分这两个数据包的界限的,但是对于 B 来说,如果不人为规定将多少字节作为一个数据包,则 B 每次是不知道应该把收到的数据中的多少字节作为一个有效的数据包的,而规定每次把多少数据作为一个包就是协议格式需要定义的内容之一。

经常会有新手写出类似下面这样的代码。

发送端:

```
//省略创建socket、建立连接等部分不相关的逻辑
char buf[] = "the quick brown fox jumps over a lazy dog.";
int n = send(socket, buf, strlen(buf), 0);
//省略出错处理逻辑
```

接收端：

```
//省略创建 socket、建立连接等部分不相关的逻辑
char recvBuf[50] = { 0 };
int n = recv(socket, recvBuf, 50, 0);
//省略出错处理逻辑
printf("recvBuf: %s", recvBuf);
```

为了专注于问题本身的讨论，这里省略了建立连接和错误处理的逻辑。在以上代码中，发送端向接收端发送了一串字符"the quick brown fox jumps over a lazy dog."，接收端收到后将其打印了出来。

类似这样的代码在本机上一般会工作得很好，接收端也如期打印出期望的字符串，但是这样的代码在局域网或者公网环境下会出问题，即接收端打印的字符串可能并不完整；如果发送端连续多次发送字符串，则接收端打印的字符串可能不完整或者是乱码。打印的代码不完整的原因很好理解：对端某次收到的数据小于完整字符串的长度，recvBuf 数组开始被清空成\0，收到部分字符串后，该字符串的末尾仍然是\0，printf 函数寻找以\0 为结束标志的字符结束输出。乱码的原因：如果某次收入的数据不仅包含一个完整的字符串，还包含下一个字符串的部分内容，那么 recvBuf 数组将被填满，printf 函数输出时仍会寻找以\0 为结束标志的字符结束输出，这样读取的内存就越界了，一直找到\0 为止；而越界后的内存可能是一些不可读字符，显示乱码。

6.2 如何解决粘包问题

进行技术面试时，面试官经常会问："网络通信时，如何解决粘包、丢包或者包乱序问题？"这其实考察的就是网络基础知识。如果使用 TCP 进行通信，则在大多数场景下是不存在丢包和包乱序问题的。因为 TCP 通信是可靠的通信方式，TCP 栈通过序列号和包重传确认机制保证数据包的有序和一定被正确发送到目的地；如果使用 UDP 进行通信，且不允许少量丢包，就要自己在 UDP 的基础上实现类似 TCP 这种有序和可靠的传输机制了（例如 RTP、RUDP）。所以将该问题拆解后，就只剩下如何解决粘包的问题。

什么是粘包？粘包就是连续向对端发送两个或者两个以上的数据包，对端在一次收取中收到的数据包数量可能大于 1 个，当大于 1 个时，可能是几个（包括一个）包加上某个包的部分，或者干脆几个完整的包在一起。当然，也可能收到的数据只是一个包的部分，这种情况一般也叫作半包。

粘包示意图如下图所示。

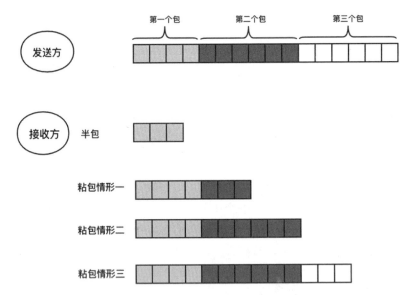

无论是半包问题还是粘包问题，因为 TCP 是流式数据格式，所以其解决思路还是从收到的数据中把包与包的边界区分出来。如何区分呢？一般有以下三种方法。

（1）固定包长的数据包。固定包长，即每个协议包的长度都是固定的。假如我们规定每个协议包的大小都是 64 字节，每收满 64 字节，就取出来解析（如果不够，就先存起来），则这种通信协议的格式简单但灵活性差。如果包的内容长度小于指定的字节数，对剩余的空间就需要填充特殊的信息，例如\0（如果不填充特殊的内容，那么如何区分包里面的正常内容与填充信息呢）；如果包的内容超过指定的字节数，又得分包分片，则需要增加额外的处理逻辑——在发送端进行分包分片，在接收端重新组装包片。

（2）以指定的字符（串）为包的结束标志。这种协议包比较常见，即在字节流中遇到特殊的符号值时就认为到一个包的末尾了。例如 FTP 或 SMTP，在一个命令或者一段数据后面加上\r\n（即 CRLF）表示一个包的结束。对端收到数据后，每遇到一个 "\r\n"，就把之前的数据当作一个数据包。这种协议一般用于一些包含各种命令控制的应用中，其不足之处就是如果协议数据包的内容部分需要使用包结束标志字符，就需要对这些字符做转码或者转义操作，以免被接收方错误地当成包结束标志而误解析。

（3）包头+包体格式。这种格式的包一般分为两部分，即包头和包体，包头是固定大小的，且包头必须包含一个字段来说明接下来的包体有多大。例如：

```
struct msg_header
{
    int32_t bodySize;
    int32_t cmd;
};
```

就是一个典型的包头格式，bodySize 指定了这个包的包体是多大。由于包头的大小是固定的（这里是 size(int32_t) + sizeof(int32_t) = 8 字节），所以对端先收取包头大小的字节内容（当然，如果不够，则还是将其先缓存起来，直到收够为止），然后解析包头，根据包头中指定的包体大小收取包体,等包体收够了,就组装成一个完整的包来处理。在某些实现中,包头中的 bodySize 可能被另一个叫作 packageSize 的字段代替,这个字段用于表示整个包的大小(即包头加上包体的大小),这时,我们只要用 packageSize 减去包头大小(这里是 sizeof(msg_header))就能算出包体的大小,原理同上。

> 在使用大多数网络库时,我们通常需要根据协议的格式自己对数据包分界和解析,一般的网络库不提供这种功能是因为需要支持不同的协议。由于协议的不确定性,网络库无法预先提供具体的解包代码。当然,这不是绝对的,也有一些网络库提供了这种功能。在 Java Netty 网络框架中提供了 FixedLengthFrameDecoder 类处理长度是定长的协议包,提供了 DelimiterBasedFrameDecoder 类处理将特殊字符作为结束符的协议包,提供了 ByteToMessageDecoder 类处理自定义格式的协议包(可用来处理包头+包体这种格式的数据包)。然而,在继承 ByteToMessageDecoder 的子类中,我们需要根据自己的协议的具体格式重写 decode 方法对数据包进行解包。

6.3 解包与处理

在理解前面介绍的数据包的三种格式后,这里介绍应该如何处理前述三种格式的数据包。其处理流程都是一样的,这里以包头+包体这种格式的数据包来说明。流程图如下图所示。

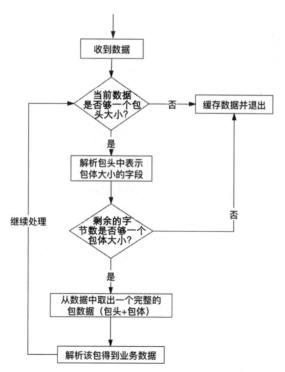

假设我们的包头格式如下:

```cpp
//强制1字节对齐
#pragma pack(push, 1)
//协议头
struct msg_header
{
    int32_t bodysize;    //包体大小
};
#pragma pack(pop)
```

那么上面的流程实现代码如下:

```cpp
//包的最大字节数限制为10MB
#define MAX_PACKAGE_SIZE    10 * 1024 * 1024

void ChatSession::OnRead(const std::shared_ptr<TcpConnection>& conn, Buffer* pBuffer, Timestamp receivTime)
{
    while (true)
    {
        if (pBuffer->readableBytes() < (size_t)sizeof(msg_header))
        {
            //不够一个包头大小,直接退出
            return;
        }

        //取包头信息
        msg_header header;
        memcpy(&header, pBuffer->peek(), sizeof(msg_header));

        //包头有错误,立即关闭连接
        if (header.bodysize <= 0 || header.bodysize > MAX_PACKAGE_SIZE)
        {
            //客户端发送非法数据包,服务器主动关闭它
            LOGE("Illegal package, bodysize: %lld, close TcpConnection, client: %s", header.bodysize, conn->peerAddress().toIpPort().c_str());
            conn->forceClose();
            return;
        }

        //收到的数据不够一个完整的包
        if (pBuffer->readableBytes() < (size_t)header.bodysize + sizeof(msg_header))
            return;

        pBuffer->retrieve(sizeof(msg_header));
        //inbuf用来存放当前要处理的包
        std::string inbuf;
        inbuf.append(pBuffer->peek(), header.bodysize);
        pBuffer->retrieve(header.bodysize);
        //解包和业务处理
        if (!Process(conn, inbuf.c_str(), inbuf.length()))
        {
            //客户端发送非法数据包,服务器主动关闭它
            LOGE("Process package error, close TcpConnection, client: %s", conn->peerAddress().toIpPort().c_str());
```

```
            conn->forceClose();
            return;
        }
    }//end while-loop
}
```

以上代码中的操作与流程图所示是一致的。pBuffer 是一个自定义的接收缓冲区，这部分代码已经将收到的数据放入这个缓冲区中，所以判断当前已收取的字节数时只需使用这个对象的相应方法即可。在以上代码中有些细节需要强调。

（1）取包头时应该拷贝一份数据包头大小的数据出来，而不是从缓冲区 pBuffer 中直接将数据取出来（将取出来的数据从 pBuffer 中移除），这是因为若接下来根据包头中的字段得到包体大小时，剩余的数据不够一个包体的大小，则我们还得把这个包头数据放回缓冲区中。为了避免这种不必要的操作，只有缓冲区中的数据大小够整个包的大小（代码中的 header.bodysize + sizeof(msg)）时，我们才需要把整个包大小的数据从缓冲区中移除。

（2）通过包头得到包体的大小时，我们一定要对 bodysize 的数值进行校验，这里要求 bodysize 必须大于 0 且不大于 10×1024×1024（即 10MB）。当然，我们可以根据实际情况决定 bodysize 的上下限（大小是 0 字节的包在某些业务场景下是允许的）。记住，一定要判断这个上下限，因为假设这是一个非法的客户端发来的数据，其 bodysize 设置了一个较大的数值，例如 1×1024×1024×1024（即 1GB），则我们的逻辑会让我们一直缓存该客户端发来的数据，服务器内存将很快被耗尽，操作系统在检测到我们的进程占用的内存达到一定阈值时会杀死我们的进程，导致服务不能再正常对外服务。如果判断了 bodysize 字段是否满足自己设置的上下限，则对于非法的 bodysize，直接关闭这路连接即可。这也是服务的一种自我保护措施，避免因为非法数据包带来的损失。还有另外一种情况，bodysize 也可能不是预期的合理值，即因为网络环境差或者某次数据解析逻辑错误，导致后续的数据错位，把不该当包头数据的数据当作了包头，这时解析出来的 bodysize 也可能不是合理的值；同样，在这种情形下也会被这段检验逻辑检测到，最终关闭连接。

（3）注意，整个判断包头、包体及处理包的逻辑都被放在一个 while 循环里，这是必要的。如果没有这个 while 循环，则当我们一次性收到多个包时，只会处理一个，下次接着处理就需要等到新的一批数据来临时再次触发这个逻辑。这样造成的结果就是，对端向我们发送了多个请求，我们最多只能应答一个，后面的应答得等到对端再次向我们发送数据才能进行。这就是对粘包逻辑的正确处理。

以上逻辑和代码是基本的粘包和半包处理机制，也就是技术上的解包处理逻辑。在理解它的基础之上，我们可以给解包拓展很多功能，例如再给我们的协议包增加一个支持压缩的功能，我们的包头会变成如下所示：

```
#pragma pack(push, 1)
//协议头
struct msg_header
{
    char     compressflag;     //压缩标志，如果为1，则启用压缩，反之不启用压缩
    int32_t  originsize;       //包体压缩前的大小
    int32_t  compresssize;     //包体压缩后的大小
```

```
    char          reserved[16];      //保留字段,用于将来拓展
};
#pragma pack(pop)
```

修改后的代码如下:

```cpp
//包的最大字节数限制为10MB
#define MAX_PACKAGE_SIZE    10 * 1024 * 1024

void ChatSession::OnRead(const std::shared_ptr<TcpConnection>& conn, Buffer* pBuffer, Timestamp receivTime)
{
    while (true)
    {
        if (pBuffer->readableBytes() < (size_t)sizeof(msg_header))
        {
            //不够一个包头大小,直接退出
            return;
        }

        //取包头信息
        msg_header header;
        memcpy(&header, pBuffer->peek(), sizeof(msg_header));

        //数据包被压缩过
        if (header.compressflag == PACKAGE_COMPRESSED)
        {
            //包头有错误,立即关闭连接
            if (header.compresssize <= 0 || header.compresssize > MAX_PACKAGE_SIZE ||
                header.originsize <= 0 || header.originsize > MAX_PACKAGE_SIZE)
            {
                //客户端发送非法数据包,服务器主动关闭它
                LOGE("Illegal package, compresssize: %lld, originsize: %lld, close TcpConnection, client: %s", header.compresssize, header.originsize, conn->peerAddress().toIpPort().c_str());
                conn->forceClose();
                return;
            }

            //收到的数据不够一个完整的包
            if (pBuffer->readableBytes() < (size_t)header.compresssize + sizeof(msg_header))
                return;

            pBuffer->retrieve(sizeof(msg_header));
            std::string inbuf;
            inbuf.append(pBuffer->peek(), header.compresssize);
            pBuffer->retrieve(header.compresssize);
            std::string destbuf;
            if (!ZlibUtil::UncompressBuf(inbuf, destbuf, header.originsize))
            {
                LOGE("uncompress error, client: %s", conn->peerAddress().toIpPort().c_str());
                conn->forceClose();
                return;
```

```cpp
            }
            //业务逻辑处理
            if (!Process(conn, destbuf.c_str(), destbuf.length()))
            {
                //客户端发送非法数据包，服务器主动关闭它
                LOGE("Process error, close TcpConnection, client: %s",
conn->peerAddress().toIpPort().c_str());
                conn->forceClose();
                return;
            }
        }
        //数据包未压缩
        else
        {
            //包头有错误，立即关闭连接
            if (header.originsize <= 0 || header.originsize > MAX_PACKAGE_SIZE)
            {
                //客户端发送非法数据包，服务器主动关闭它
                LOGE("Illegal package, compresssize: %lld, originsize: %lld, close TcpConnection, client: %s", header.compresssize, header.originsize,
conn->peerAddress().toIpPort().c_str());
                conn->forceClose();
                return;
            }

            //收到的数据不够一个完整的包
            if (pBuffer->readableBytes() < (size_t)header.originsize + sizeof(msg_header))
                return;

            pBuffer->retrieve(sizeof(msg_header));
            std::string inbuf;
            inbuf.append(pBuffer->peek(), header.originsize);
            pBuffer->retrieve(header.originsize);
            //业务逻辑处理
            if (!Process(conn, inbuf.c_str(), inbuf.length()))
            {
                //客户端发送非法数据包，服务器主动关闭它
                LOGE("Process error, close TcpConnection, client: %s",
conn->peerAddress().toIpPort().c_str());
                conn->forceClose();
                return;
            }
        }//end else

    }//end while-loop
}
```

这段代码先根据包头的压缩标志字段判断包体是否有压缩，如果有压缩，则取出包体大小解压缩，解压缩后的数据才是真正的业务数据。整个程序的执行流程如下图所示。

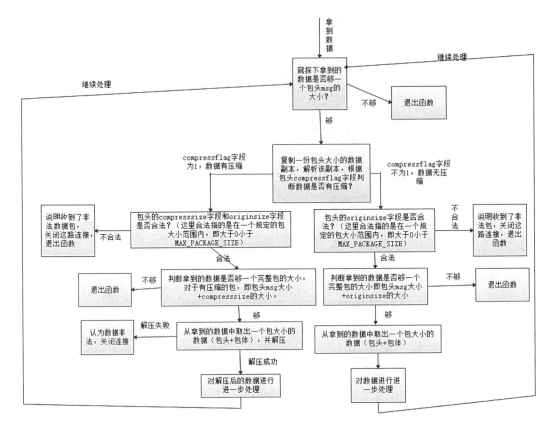

在以上代码中,变量 pBuffer 是当前连接的接收缓冲区,关于如何设计接收缓冲区,将在 7.7 节详细介绍。

6.4 从 struct 到 TLV

计算机世界存在各种格式的网络通信协议,不同的应用场景使用的网络通信协议可能不同,对于一名合格的后端开发者来说,根据应用场景设计合理的网络通信协议是对其的基本要求。那么在设计网络通信协议时,我们需要考虑哪些因素呢?

6.4.1 协议的演化

假设现在 A 与 B 之间要传输一个关于用户信息的数据包,则可以将该数据包格式定义成如下形式:

```
#pragma pack(push, 1)
struct userinfo
{
    //命令号
    int32_t cmd;
    //用户性别
    char    gender;
    //用户昵称
```

```
    char    name[8];
};
#pragma pack(pop)
```

相信很多读者都定义过这样的协议，其数据结构简单明了，对端只要直接拷贝数据并按顺序解析各个字段就可以了。但是，需求总在变化，例如随着业务的发展，我们需要在这个结构中增加一个字段表示用户的年龄，则可将协议结构修改如下：

```
#pragma pack(push, 1)
struct userinfo
{
    //命令号
    int32_t cmd;
    //用户性别
    char    gender;
    //用户昵称
    char    name[8];
    //用户年龄
    int32_t age;
};
#pragma pack(pop)
```

在这种情况下，在技术上直接增加一个字段并不能完全解决问题，因为新修改的协议格式在旧的客户端上无法兼容（旧的客户端已经分发出去），这时我们升级服务端的协议格式，会导致旧的客户端无法使用。所以在最初设计协议时，需要增加一个版本号字段，针对不同的版本做不同的处理：

```
/**
 * 旧的协议，版本号是1
 */
#pragma pack(push, 1)
struct userinfo
{
    //版本号
    short   version;
    //命令号
    int32_t cmd;
    //用户性别
    char    gender;
    //用户昵称
    char    name[8];
};
#pragma pack(pop)

/**
 * 新的协议，版本号是2
 */
#pragma pack(push, 1)
struct userinfo
{
    //版本号
    short   version;
    //命令号
    int32_t cmd;
```

```
    //用户性别
    char    gender;
    //用户昵称
    char    name[8];
    //用户年龄
    int32_t age;
};
#pragma pack(pop)
```

可以用以下伪代码来兼容新旧协议：

```
//从包中读取一个 short 型字段
short version = <从包中读取一个 short 型字段>;
if (version == 1)
{
    //当作旧的协议格式进行处理
}
else if (version == 2)
{
    //当作新的协议格式进行处理
}
```

以上是兼容旧版协议的常见做法，这样也存在一个问题，如果我们的业务需求变化快，则可能需要经常调整协议字段（增、删、改），这样我们的版本号数量会较多，代码会变成类似下面这种形式：

```
//从包中读取一个 short 型字段
short version = <从包中读取一个 short 型字段>;
if (version == 版本号1)
{
    //对版本号1的格式进行处理
}
else if (version == 版本号2)
{
    //对版本号2的格式进行处理
}
else if (version == 版本号3)
{
    //对版本号3的格式进行处理
}
else if (version == 版本号4)
{
    //对版本号4的格式进行处理
}
else if (version == 版本号5)
{
    //对版本号5的格式进行处理
}
...省略更多内容...
```

以上只考虑了协议顶层结构，没有考虑更复杂的嵌套结构，不管怎样，这样的代码会变得越来越难以维护。

这里只是为了说明问题,在实际开发中,建议读者在设计协议时尽量考虑周全,避免反复修改协议的结构。

前述协议格式还存在另一个问题:对于 name 字段,其长度为 8 字节,这种定长的字段,长度大小不具有伸缩性:若太长,则在很多情况下都用不完,会造成内存和网络带宽的浪费;若太短,则在某些情况下不够用。那么有没有方法来解决呢?

方法是有的:对于字符串类型的字段,我们可以在该字段前面加一个表示字符串长度(length)的标志,那么上面的协议在内存中的状态可以表示成如下图所示。

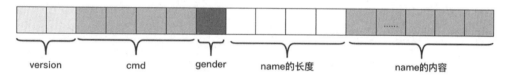

这种方法解决了定义字符串类型时太长浪费、太短不够用的问题,但是没有解决修改协议(如新增字段)需要兼容众多旧版本的问题,对于后面的问题,可以通过在每个字段前面都加一个 type 类型来解决。我们可以使用一个 char 类型来表示常用的类型,规定如下表所示。

类　　型	Type 值	类型描述
bool	0	布尔值
char	1	char 型
int16	2	16 位整型
int32	3	32 位整型
int64	4	64 位整形
string	5	字符串或二进制序列
list	6	列表
map	7	map
……		

那么对于以上协议,其内存格式变成如下图所示。

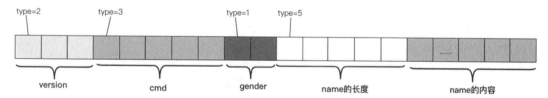

这样,每个字段的类型就是自解释了,这就是 TLV(Type Length Value)。对这种格式的协议,我们可以方便地增删改字段类型,程序在解析时会根据每个字段的 type 得到字段的类型。

在实际开发中,TLV 类型虽然易于扩展,但也存在以下缺点。

（1）TLV 格式因为每个字段都增加了一个 type 类型，所以占用的空间增大。

（2）我们在解析字段时需要额外增加一些判断 type 的逻辑，去判断字段的类型并做相应的处理，即：

```
//读取第 1 字节得到 type
if (type == Type::BOOL)
{
    //bool 型处理
}
else if (type == Type::CHAR)
{
    //char 型处理
}
else if (type == Type::SHORT)
{
    //short 型处理
}
...省略更多的类型...
```

如上面的代码所示，对每个字段都需要有这样的逻辑判断，这样的编码方式是非常麻烦的。

（3）即使我们知道了每个字段的技术类型（相对于业务来说），每个字段的业务含义也仍然需要我们制定文档格式，也就是说，TLV 格式只是做到了在技术上自解释。

所以在实际开发中，完全遵循 TLV 格式的协议并不多，尤其是针对一些整型类型的字段来说，例如整型字段一旦知道其类型，长度就固定下来了，比如 short 类型占 2 字节，int32 类型占 4 字节，因此不必浪费一段空间存储其长度信息。

TLV 格式还可以嵌套，如下图所示。

有的项目在 TLV 格式的基础上扩展了一种 TTLV 格式的协议，即 Tag-Type-Length-Value，在每个字段前面都增加了一个 Tag 类型，这时 Type 表示数据类型，Tag 的含义由协议双方协定。

6.4.2 协议的分类

根据协议的内容是否是文本格式（即人为可读格式），我们将协议分为文本协议和二进制协议，像 HTTP 的包头部分和 FTP 等都是典型的文本协议的例子。

6.4.3 协议设计工具

虽然 TLV 很简单，但是每做一套新的协议都要从头编解码、调试，而写编解码存在大量的复制粘贴过程，容易出错。

因此出现了一种叫作 IDL（Interface Description Language）的语言规范，它是一种描述语言，也是一种中间语言。IDL 将协议的使用类型规范化，提供跨语言特性。我们可以定义一个描述协议格式的 IDL 文件，然后通过 IDL 工具分析 IDL 文件，就可以生成各种语言版本的协议代码。Google Protobuf 库自带的工具 protoc 就是一个 IDL 工具。

6.5 整型数值的压缩

在实际设计协议时，整型数值（如 int32、int64）在协议字段中出现的频率非常高。以上面介绍的 TLV 格式为例，L 代表每个字段的长度，假设用一个 int32 类型表示，int32 占 4 字节，则对于无符号的 int32 类型来说，其可表示的范围为 0～4294967295。在实际应用中不会用到太长的字段值，因此可以根据字段实际的 length 值使用 1～n 字节表示这个 int32 的值。

在实际处理中，一字节（Byte）共有 8 位（bit），该字节的最高位用来作为标志位，说明一个整型数值是否到此字节结束，如果某字节的最高位为 0，则表示该整型值的内容到此字节结束；最高位为 1，表示下一字节仍然是该整型值的内容。说得有点抽象，来看一个具体的例子。假设在一串字节流中存在如下二进制数字表示某个整型值：

```
第 1 字节       第 2 字节       第 3 字节       第 4 字节
10111011       11110000       01110000       11110111   ...其他省略...
```

其中，第 1 字节是 10111011，其最高位是 1，说明其下一字节仍然属于表示该整型的序列；第 2 字节是 11110000，最高位仍然是 1；第 3 字节是 01110000，最高位是 0，表示这个整数的字节序列到此结束。假定我们压缩时的顺序是低位内容在内存地址较小的位置，高位内容在内存地址较大的位置，则将每字节的标志位（最高位）去掉后，其值如下：

```
第 3 字节   第 2 字节   第 1 字节
1110000    1110000    0111011    => 11100 00111000 00111011
11100 00111000 00111011 转换成十进制是 1849403。
```

使用上述技巧进行压缩的整型，由于一字节只使用了低 7 位（最高位为标志位，一般称之为"字节前导位"），一个 int32 的整型数值共 4 字节（4 × 8 = 32）位，对其使用上述技巧压缩后，其长度可能是 1～5 字节。在实际的协议中基本上很少使用超过 3 字节以上的长度，因此这种压缩还是比较实用的（节省空间）。

有了上面的分析，对于一个无符号 int32 的整型数值的压缩算法如下（节选自 POCO C++库，代码格式略有调整）：

```
01 //poco-master\Foundation\src\BinaryWriter.cpp
02 //将一个 uint32 压缩成 1～5 字节的算法
03 void BinaryWriter::write7BitEncoded(UInt32 value)
04 {
05     do
06     {
07         unsigned char c = (unsigned char) (value & 0x7F);
08         value >>= 7;
09         if (value)
```

```
10              c |= 0x80;
11
12          _ostr.write((const char*) &c, 1);
13      }
14      while (value);
15  }
```

以上代码对一个 uint32_t 整型 value 从低到高每次取 7bit，判断 value 的值在去掉 7bit 后是否有剩余（非 0 则说明有剩余，例如代码第 8 和 9 行），如果有剩余，则将当前字节最高 bit（标志位）设置为 1，这样在得到一字节的值后，将其放入字节流容器 _ostr 中，字节流容器的类型只要有连续的内存存储序列即可，例如 std::string。

假设现在 value 的值是十进制 125678，其二进制是 1 1110 1010 1110 1110，来看一下上述函数的执行过程。

第 1 次循环：十六进制 0x7F 的二进制为 0111 1111，执行如下代码

```
unsigned char c = (unsigned char) (value & 0x7F);
```

后，c = 110（十进制），二进制是 0110 1110；接着将 value 右移 7 bit，看看还有没有剩余（与 0 判断），此时 value 变为 981（十进制），对应二进制 11 1101 0101，代码第 9 行 if 条件为真，说明一字节表示不了这个数值，给算出的字节 c 最高位 bit 设置标志值 1（与 0x80 做或运算，0x80 的二进制是 1000 0000，代码第 10 行），得到第 1 字节值 238（十进制），对应二进制 1110 1110。

第 2 次循环：c 开始等于 85（十进制），执行代码第 7、8 行后，发现 value 的值仍有剩余，再次在该字节的高位设置标志 1，得到第 2 字节值 213（十进制）。

第 3 次循环：c 开始等于 7，执行代码第 7、8 行后，发现 value 的值已经没有剩余，得到第 3 字节值 7，然后退出循环。

程序的执行过程如下图所示。

	1111 0101 0111 0111 0	
第1次循环	1 1 1 1 0 1 0 1 1 1 0 1 1 1 0	110 & 0x7F
第2次循环	1 1 1 1 0 1 0 1	85 & 0x7F
第3次循环	1 1 0	7

理解整型的压缩算法后，就很容易弄明白其对应的解压缩算法了，代码如下（同样节选自 POCO C++ 库，代码格式略有调整）：

```
//poco-master\Foundation\src\BinaryReader.cpp
//将一字节流中连续的1~5字节还原成一个uint32整型
void BinaryReader::read7BitEncoded(UInt32& value)
{
    char c;
    value = 0;
    int s = 0;
    do
    {
        c = 0;
        _istr.read(&c, 1);
        UInt32 x = (c & 0x7F);
        x <<= s;
        value += x;
        s += 7;
    }
    while (c & 0x80);
}
```

以上代码从字节流容器_istr中挨个读取1字节，将当前字节与0x7F进行与运算，以取得该字节的低7位内容（代码第12行），然后将字节的内容与0x80进行与运算，以判断该字节的最高位是否为1，进而确定下一字节是不是也属于整型值的内容。

同样，对于uint64的整型数值，我们可以将其压缩成1~10字节的字节数组，其压缩和解压缩算法与uint32的整型数值一样，这里不再贴出。

6.6 设计通信协议时的注意事项

通过学习前面的章节，相信读者对协议设计有一定的了解了。本节讲解设计协议时的一些注意事项。

6.6.1 字节对齐

在前面讨论的协议示例中有一组成对的#pragma XX 指令，其中，#pragma pack(push, n)用于告诉编译器将接下来的所有结构体（这里就是userinfo 协议）的每一个字段都按n字节对齐，这里n为1，所以按一字节对齐，即去除所有padding字节。这样做是为了使内存更加紧凑及节省存储空间。不再需要这个对齐功能后，应该使用#pragma pack(pop)让编译器恢复之前的对齐方式：

```
#pragma pack(push, 1)
struct userinfo
{
    //版本号
    short    version;
    //命令号
    int32_t  cmd;
    //用户性别
    char     gender;
    //用户昵称
```

```
    char        name[8];
    //用户年龄
    int32_t age;
};
#pragma pack(pop)
```

> 注意：#pragma pack(push, n)与#pragma pack(pop)一定要成对使用，如果漏掉其中任何一个，则编译出来的代码可能会出现很多奇怪的运行结果。

6.6.2 显式地指定整型字段的长度

对于一个 int 型字段，在其作为协议传输时，我们应该显式地指定该类型的长度，也就是说，应该使用 int32_t、int64_t 这样的类型代替 int、long。之所以这么做，是因为在不同字长的机器上，默认的 int 和 long 的长度可能不一样，例如 long 型在 32 位操作系统上的长度是 4 字节，在 64 位机器上的长度是 8 字节。如果不显式指定这种整形的长度，则可能因为不同机器字长的不同，导致协议解析出错或者产生错误的结果。

6.6.3 涉及浮点数时要考虑精度问题

计算机表示浮点数时存在精度取舍不准确的问题，例如对于 1.000000，有的计算机可能会得到 0.999999。在某些应用中，如果这个浮点数的业务单位较大（例如表示金额，单位为亿），就会造成很大的影响。因此为了避免不同的机器解析得到不同的结果，建议在网络传输时将浮点数值放大相应的倍数，变成整数或者转换为字符串来传输。

6.6.4 大小端问题

第 4 章详细介绍了大小端问题（即主机字节序和网络字节序），在设计协议格式时，如果在协议中存在整型字段，则建议使用同一字节序。通常的做法是在进行网络传输时将所有整型都转换为网络字节序（大端编码，Big Endian），避免不同的机器在解析时因为大小端问题而得到不同的整型值。

当然，不一定非要转换为网络字节序，如果明确地知道通信双方使用的是相同的字节序，则也可以不转换。

6.6.5 协议与自动升级功能

对于一个商业产品，发布出去的客户端一般通过客户端的自动升级功能获得更新（iOS App 除外，苹果公司要求所有 App 都必须在其 App Store 上更新版本，禁止热更新）。在客户端与服务器通信的所有协议格式中，自动升级协议是最重要的一个，无论版本如何迭代，一定要保证自动升级协议的新旧兼容，这样做有如下原因。

（1）如果新服务器不能兼容旧客户端中的自动升级协议，那么旧客户端用户将无法升级成新版本，这样的产品相当于把自己"阉割"了。对于不少产品，不通过自动升级而让众多用户去官网下载新版本是很难做到的，这种决策可能会导致大量用户流失。

(2)对于测试不完善或者快速迭代中的产品,只要保证自动升级功能正常,旧版本的任何 Bug 和瑕疵都可以通过升级新版本解决。这对于想投放市场试水但又可能设计不充分的产品尤其重要。自动升级功能一般是将当前版本的版本号与服务端新版本的版本号进行比较,如果二者之间存在一个大版本号的差别(如 1.0.0 与 2.0.0),即有重大功能更新,则应该强制客户端更新下载最新版本。如果只是一个小版本号的更新(如 1.0.0 与 1.1.0),则可以让用户选择是否更新。当然,如果是新版本修正了前一个版本中严重影响使用的 Bug,则也应当强制用户更新。

6.7 包分片

包分片指的是应用层对包的拆分。当一个包的数据较大,超过一个包的最长长度时,我们需要对包进行分片。有人可能会有疑问:分成多个包就行了,为什么要对包进行分片?在实际应用中一般会根据业务需求对包的类型进行编号,例如使用一个 wCmd 表示业务号,但某些业务类型某次携带的数据可能较大,超过了单个包的最大长度,这时需要将该数据拆分成多个包片,但其业务号隶属于同一个包,这就是包分片。

理解包分片的原理后,设计包分片的功能就很简单了,这里提供了两种思路。

(1)设置分片标志。在包头部分设置一个字段来表示当前包是否属于某个大包的分片,分片标志字段一般有 4 种取值类型:无分片标志、包的第 1 个分片标志、包的最后一个分片标志、第 1 个分片与最后一个分片之间的包分片标志。

(2)在每个包分片的包头部分都有该包的总分片数量和当前分片编号。对于 TCP 来说,由于其数据传输本身是有序的,所以对于多个分片,只要我们的一端按顺序依次发送,另外一端收包时就一定会按发送的顺序收到,我们也不用考虑包分片的顺序问题。

来看一个具体的包分片的例子。假设现在有如下协议头定义:

```
//与客户端交互协议包头
#pragma pack(push, 1)
typedef struct tagNtPkgHead
{
    unsigned char    bStartFlag;      //协议包起始标志 0xFF
    unsigned char    bVer;            //版本号
    unsigned char    bEncryptFlag;    //加密标志(如果不加密,则为 0)
    unsigned char    bFrag;           //是否有包分片(1: 有包分片; 0: 无包分片)
    unsigned short   wLen;            //总包长
    unsigned short   wCmd;            //命令号
    unsigned short   wSeq;            //包的序列号,业务使用
    unsigned short   wCrc;            //Crc16 校验码
    unsigned int     dwSID;           //会话 ID
    unsigned short   wTotal;          //有包分片时的分片总数
    unsigned short   wCurSeq;         //有包分片时的分片序号,从 0 开始,无分片时也为 0
} NtPkgHead, *PNtPkgHead;
#pragma pack(pop)
```

对端处理包分片时的逻辑伪代码如下：

```cpp
UINT CSocketClient::RecvDataThreadProc(LPVOID lpParam)
{
    LOG_NORMAL("Start recv data thread.");
    DWORD               dwWaitResult;
    std::string         strPkg;
    //临时存放一个完整的包数据的变量
    std::string         strTotalPkg;
    unsigned short      uPkgLen = 0;
    unsigned int        uBodyLen = 0;
    unsigned int        uTotalPkgLen = 0;
    unsigned int        uCmd = 0;
    NtPkgHead           pkgHead;
    unsigned short      uTotal = 0;
    //记录上一次的包分片序号，包分片序号从 0 开始
    unsigned short      uCurSeq = 0;
    int                 nWaitTimeout = 1;

    CSocketClient*      pSocketClient = (CSocketClient*)lpParam;

    while (!m_bExit)
    {
        //检测是否有数据
        if (!pSocketClient->CheckReceivedData())
        {
            //休眠 10 毫秒
            Sleep(10);
            continue;
        }

        //接收数据，并放入 pSocketClient->m_strRecvBuf 中
        if (!pSocketClient->Recv())
        {
            LOG_ERROR("Recv data error");

            //收数据出错，清空接收缓冲区，可以做一些关闭连接、重连等动作
            pSocketClient->m_strRecvBuf.clear();

            Reconnect();
            continue;
        }

        //一定要放在一个循环里解包，因为在当前缓冲区中可能存在多个数据包
        while (true)
        {
            //判断当前收到的数据是否够一个包头大小
            if (pSocketClient->m_strRecvBuf.length() < sizeof(NtPkgHead))
                break;

            memset(&pkgHead, 0, sizeof(pkgHead));
            memcpy_s(&pkgHead, sizeof(pkgHead), pSocketClient->m_strRecvBuf.c_str(), sizeof(pkgHead));
```

```cpp
    //对包消息头进行检验
    if (!CheckPkgHead(&pkgHead))
    {
        //如果包头检验不通过，则缓冲区里面的数据已经是脏数据了，直接清空
        //可以做一些关闭连接并重连的动作
        LOG_ERROR("Check package head error, discard data %d bytes",
(int)pSocketClient->m_strRecvBuf.length());

        pSocketClient->m_strRecvBuf.clear();

        Reconnect();
        break;
    }

    //判断当前数据是否够一个整包的大小
    uPkgLen = ntohs(pkgHead.wLen);
    if (pSocketClient->m_strRecvBuf.length() < uPkgLen)
        break;

    strPkg.clear();
    strPkg.append(pSocketClient->m_strRecvBuf.c_str(), uPkgLen);

    //从接收缓冲区中移除已经处理的数据部分
    pSocketClient->m_strRecvBuf.erase(0, uPkgLen);

    uTotal = ::ntohs(pkgHead.wTotal);
    uCurSeq = ::ntohs(pkgHead.wCurSeq);
    //无分片或者是第1个分片
    if (uCurSeq == 0)
    {
        strTotalPkg.clear();
        uTotalPkgLen = 0;
    }

    uBodyLen = uPkgLen - sizeof(NtPkgHead);
    uTotalPkgLen += uBodyLen;
    strTotalPkg.append(strPkg.data() + sizeof(NtPkgHead), uBodyLen);

    //如果无分包或者是包的最后一个分片，则将组装后的包发送出去
    if (uTotal == 0 || (uTotal != 0 && uTotal == uCurSeq + 1))
    {
        uCmd = ::ntohs(pkgHead.wCmd);

        //ProxyPackage 是解析出来的业务包定义
        ProxyPackage proxyPackage;
        //复制业务号
        proxyPackage.nCmd = uCmd;
        //复制包长度
        proxyPackage.nLength = uTotalPkgLen;
        //复制包体内容
        proxyPackage.pszJson = new char[uTotalPkgLen];
```

```
                memset(proxyPackage.pszJson, 0, uTotalPkgLen * sizeof(char));
                memcpy_s(proxyPackage.pszJson, uTotalPkgLen, strTotalPkg.c_str(),
strTotalPkg.length());

                //将一个完整的包交给业务处理
                pSocketClient->m_pNetProxy->AddPackage((const char*)&proxyPackage,
sizeof(proxyPackage));
            }
        }//end inner-while-loop
    }//end outer-while-loop


    LOG_NORMAL("Exit recv data thread.");

    return 0;
}
```

以上代码在一个网络数据收取线程中，先检测是否有可读数据，如果有可读数据，则从 socket 上读取该数据并存入接收缓冲区 pSocketClient->m_strRecvBuf 中，然后判断收到的数据是否够一个包头大小（sizeof(NtPkgHead)），如果不够，则退出当前循环，等待后续数据的到来；如果够，则对包头数据进行校验，然后从包头中得到整包的大小（ntohs(pkgHead.wLen)，这里表示整包大小的字段 wLen 使用了网络字节序，我们调用 ntohs 函数可以得到本机字节序)。接着判断收到的数据是否够一个整包的大小，如果不够，则退出当前循环等待后续数据的到来；如果够，则根据记录当前包分片序号的变量 uCurSeq（uCurSeq = ::ntohs(pkgHead.wCurSeq)）确定该包是否是某个分片。uCurSeq 等于 0，说明此次是从一个新的包片或完整的包开始的；从接收缓冲区中将当前包片或者完整包的数据放入变量 strTotalPkg 中存储起来（注意，pkgHead.wTotal 和 pkgHead.wCurSeq 均使用了网络字节序，需要转换成本地字节序）。接着，根据包头字段 pkgHead.wTotal 和 pkgHead.wCurSeq 转换成本机字节序的值，判断这是否是一个完整的包（当 uTotal == 0 时）或者是最后一个包分片（当 uTotal != 0 && uTotal == uCurSeq + 1 时），此时 strTotalPkg 存放的就是一个完整的包数据了，接着将其拷贝出来（这里是拷贝到 ProxyPackage 结构中）进行业务逻辑处理。如果当前包片只是一个大包的中间包片，则继续进行下一轮数据的处理。在 strTotalPkg 中存放的数据在到达一个完整的包时，会在业务处理后、下一轮循环存入新的包片数据前清空（代码第 81 行）。

对以上流程可用下图表示。

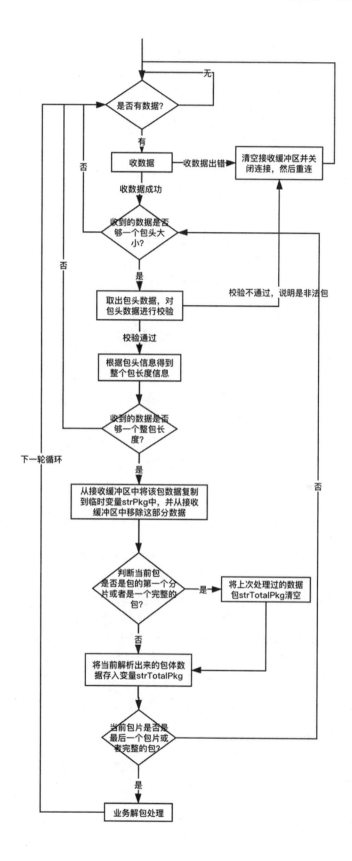

6.8 XML 与 JSON 格式的协议

XML 和 JSON 这两种格式由于其良好的自我解释性，是开发中使用非常广泛的两种数据格式。一个 XML 格式的示例如下：

```
<?xml version="1.0" encoding="utf-8"?>
<User>
    <userid>1001</userid>
    <username>xiaofang</username>
    <nickname>小方</nickname>
    <facetype>0</facetype>
    <gender>1</gender>
    <birthday>19900101</birthday>
    <signature>生活需要很多的力气呀。xx</signature>
    <clienttype>1</clienttype>
</User>
```

一个 JSON 格式的示例如下：

```
{
    "userid": 1001,
    "username": "xiaofang",
    "nickname": "小方",
    "facetype": 0,
    "gender": 1,
    "birthday": 19900101,
    "signature": "生活需要很多的力气呀。xx",
    "clienttype": 1
}
```

那么 XML 和 JSON 格式可以单独作为网络通信协议吗？当然可以，但是单独使用 XML 或者 JSON 格式作为网络通信协议的服务非常少，其原因是在给数据包分界得到一个个完整的包时非常不便。

无论是 XML 格式还是 JSON 格式，如果单独作为协议，则一般由于业务数据内容的不同，每个包的长度都不一样，在 TCP 流式数据中只能采取固定结束符的方式来分割。对于不确定长度的 XML 和 JSON 字符串，在频繁进行数据交换的网络通信程序中，每次解包前都得遍历一次 XML 或 JSON 字符串以寻找特定的包结束符。例如，对于上面 XML 格式的示例，就是在流式数据中寻找</User>字符串以作为一个 xml 格式的包结束符；对于上面的 json 格式，寻找右大括号（}）作为 JSON 格式的包结束符，这种寻找特定标记的方式容易造成误判，还需要在此基础上加上其他限定标记。例如，对 JSON 字符串的最后一个节点可以加一个特殊字段标志作为结束标记。以上述 JSON 字符串为例，在其末尾增加一个 endFlag 标志后变成如下形式：

```
{
    "userid": 1001,
    "username": "xiaofang",
    "nickname": "小方",
    "facetype": 0,
    "gender": 1,
    "birthday": 19900101,
```

```
    "signature": "生活需要很多的力气呀。xx",
    "clienttype": 1,
    "endFlag": 0
}
```

这样当该 JSON 字符串作为一个数据包时，在判断包结束标志时可以通过寻找 "endFlag": 0 加一个 "}" 这样的字符串作为包分界符号。但这种方法很不灵活，JSON 字符串在某些系统或库中被解析时，各个字段的位置顺序可能会被调整，也就是说，像 "endFlag": 0 这样的字段可能会被调整到 JSON 字符串的非末尾位置；另外，在 JSON 字符串被格式化后，某些字段值后面会被追加\n 或\r\n 这样的换行符，给程序寻找指定的包结束符带来困扰。

所以在通常情况下，XML 或者 JSON 格式不会被单独作为协议格式，而是作为某个协议的一部分出现，例如如下格式：

```
struct msg
{
    //在消息头 header 中说明整个包的大小，减去 header、cmd 和 seq 的大小就是 buf 的长度
    //即 XML 或者 JSON 的长度
    msgheader  header;
    int32_t    cmd;
    int32_t    seq;
    //buf 是一个字符串，可以是 XML 或者 JSON 格式
    char*      buf;
};
```

> 业界也有使用 XML 格式的协议，例如 XMPP，可通过 XMPP 官网了解。

6.9 一个自定义协议示例

本节结合前面的理论知识，演示如何自定义一个灵活的通信协议类。笔者在开源即时通信软件 Flamingo 的源码中自定义了一个协议格式，这个协议也分为包头和包体两部分，其中包头的定义如下：

```
#pragma pack(push, 1)
//协议头
struct chat_msg_header
{
    char       compressflag;      //压缩标志，如果为 1，则启用压缩，反之不启用压缩
    int32_t    originsize;        //包体压缩前的大小
    int32_t    compresssize;      //包体压缩后的大小
    char       reserved[16];      //保留字段，用于将来扩展
};
#pragma pack(pop)
```

包体的内容长度无论是否设置了 compressflag 压缩标志，最后的实际长度都是 originsize，得到了包体的内容后，我们就可以按通信两端规定好的协议逐个解析业务字段了。

假设是一个聊天内容协议，发送方的示例代码如下：

```
//发送方组装包体的格式
std::string outbuf;
net::BinaryStreamWriter writeStream(&outbuf);
//消息类型，msgType
writeStream.WriteInt32(msg_type_chat);
//消息序号, seq
writeStream.WriteInt32(m_seq);
//发送者 ID, senderId
writeStream.Write(senderId);
//消息内容，chatMsg
writeStream.WriteString(chatMsg);
//接受者 ID, receiverId
writeStream.WriteInt32(receiverId);
writeStream.Flush();
```

在以上代码开始处定义了一个自动扩展的字符串缓冲区 outbuf（这里使用了 std::string），然后依次写入字段信息（如下表所示）。

字段标号	字段名	类型/字节数量	说 明
字段 1	msgType	int32/4 字节	消息类型
字段 2	seq	int32/4 字节	消息序号
字段 3	senderId	int32/4 字节	发送者 ID
字段 4	chatMsg	string/长度由消息内容定	聊天消息，可以定义为一个 JSON 字符串
字段 5	receiverId	int32/4 字节	接收者 ID

写入上述几个字段后，消息体的结构如下图所示。

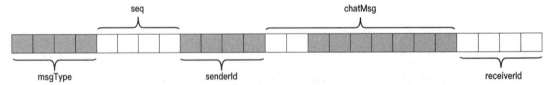

在上图中有一个小细节，即对于 string 类型的消息，在写入实际的字符串内容之前会先写入这个字符串的长度，writeStream.WriteString(chatMsg)函数的实现如下：

```
//WriteString 实际上调用了 WriteCString 方法
bool BinaryStreamWriter::WriteString(const string& str)
{
    return WriteCString(str.c_str(), str.length());
}

bool BinaryStreamWriter::WriteCString(const char* str, size_t len)
{
    std::string buf;
    write7BitEncoded(len, buf);

    m_data->append(buf);
    m_data->append(str, len);
    return true;
}
```

这里的 m_data 是前面介绍的 outbuf 的指针，也就是说，使用一个 std::string 存放二进制流时，BinaryStreamWriter::WriteCString 方法会先将字符串的长度写入流中，再写入字符串本身的内容，对于字符串的长度，会根据其长度值压缩成 1~5 字节：

```
//将一个4字节的整型数值压缩成1~5字节
void write7BitEncoded(uint32_t value, std::string& buf)
{
    do
    {
        unsigned char c = (unsigned char)(value & 0x7F);
        value >>= 7;
        if (value)
            c |= 0x80;

        buf.append(1, c);
    } while (value);
}
```

写入上述 5 个字段后会调用 writeStream.Flush 方法，该方法的实现如下：

```
void BinaryStreamWriter::Flush()
{
    char* ptr = &(*m_data)[0];
    unsigned int ulen = htonl(m_data->length());
    memcpy(ptr, &ulen, sizeof(ulen));
}
```

其作用是在流的前 4 字节处存放流数据的长度，存储长度使用的是网络字节序，这 4 字节在创建 BinaryStreamWriter 对象时被预留出来：

```
std::string outbuf;
net::BinaryStreamWriter writeStream(&outbuf);
```

以上代码调用了 BinaryStreamWriter 的构造函数：

```
enum
{
    //4字节头长度
    BINARY_PACKLEN_LEN_2 = 4,
    CHECKSUM_LEN = 2,
};

BinaryStreamWriter::BinaryStreamWriter(string* data) :
    m_data(data)
{
    m_data->clear();
    char str[BINARY_PACKLEN_LEN_2 + CHECKSUM_LEN];
    m_data->append(str, sizeof(str));
}
```

实际上，在 m_data 指向的流起始处一共预留了 6 字节，前 4 字节存放将来整个流数据的长度（网络字节序），后两字节存放数据的校验和（checksum，这里未使用）。

因此调用 writeStream.Flush 方法后，流对象的结构如下图所示，图中 4 字节的 streamLength 表示包体长度。

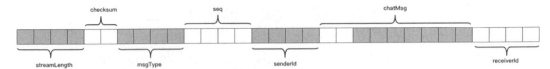

这个二进制流虽然在这里的含义是包体部分,但是已经可以做到自我分界和解析了。在简单的业务中,我们先读取 4 字节的 streamLength,然后根据 streamLength 转换成本机字节序后的长度来获取实际的内容长度。但是这样不够方便,所以这里没有在这个流的基础上继续扩展,而是选择在这个流的前面再加一个包头定义(即以上代码中的 chat_msg_header struct)。chat_msg_header struct(包头)和这里的流(包体)组装成一个完整的包:

```cpp
//p 即包体流的指针
void TcpSession::sendPackage(const char* p, int32_t length)
{
    string srcbuf(p, length);
    string destbuf;
    //按需压缩
    if (m_bNeedCompress)
    {
        if (!ZlibUtil::compressBuf(srcbuf, destbuf))
        {
            LOGE("compress buf error");
            return;
        }
    }


    string strPackageData;
    chat_msg_header header;
    if (m_bNeedCompress)
    {
        //设置压缩标志
        header.compressflag = PACKAGE_COMPRESSED;
        //设置压缩后的包体大小
        header.compresssize = destbuf.length();
    }
    else
        header.compressflag = PACKAGE_UNCOMPRESSED;

    //设置压缩前的包体大小
    header.originsize = length;

    //插入真正的包头
    strPackageData.append((const char*)&header, sizeof(header));
    strPackageData.append(destbuf);

    //将整个包发到网络上
    conn->send(strPackageData);
}
```

实际上，我们可以基于 BinaryStreamWriter 这个流对象进行扩展，不用单独再定义一个 chat_msg_header 结构体作为包头。

BinaryStreamWriter 对浮点数的处理，是先将浮点数按一定的精度转换成字符串，然后将字符串写入流中：

```
//isNULL 参数表示可以写入一个 double 类型的占位符
bool BinaryStreamWriter::WriteDouble(double value, bool isNULL)
{
    char doublestr[128];
    if (isNULL == false)
    {
        sprintf(doublestr, "%f", value);
        WriteCString(doublestr, strlen(doublestr));
    }
    else
        WriteCString(doublestr, 0);

    return true;
}
```

以上便是对这个自定义协议的装包过程，在解包过程中实现了一个 BinaryStreamReader 类，该类的操作对象是去除了包头的 chat_msg_header 结构后拿到的包体流。上述聊天协议的解包代码如下：

```
bool CRecvMsgThread::HandleMessage(const std::string& strMsg)
{
    //strMsg 是包体流，如何得到包体流在 6.3 节已经介绍过了
    net::BinaryStreamReader readStream(strMsg.c_str(), strMsg.length());

    //读取消息类型
    int32_t msgType;
    if (!readStream.ReadInt32(msgType))
    {
        return false;
    }

    //读取消息序列号
    if (!readStream.ReadInt32(m_seq))
    {
        return false;
    }

    //根据消息类型做处理
    switch (msgType)
    {
        //聊天消息
        case msg_type_chat:
        {
            //从流中读取发送者 ID
            int32_t senderId;
            if (!readStream.ReadInt32(senderId))
            {
```

```cpp
            break;
        }

        //从流中读取聊天消息本身
        std::string chatMsg;
        size_t chatMsgLength;
        if (!readStream.ReadString(&chatMsg, 0, chatMsgLength))
        {
            return false;
        }

        //从流中读取接收者 ID
        int32_t receiverId;
        if (!readStream.ReadInt32(receiverId))
        {
            break;
        }

        //对聊天消息进行处理
        HandleChatMessage(senderId, receiverId, data);
    }
    break;

    //对其他消息进行处理
}//end switch

return false;
}
```

以上代码根据写入的流的字段类型顺序依次读出相应的字段值，在拿到各个字段的值后就可以进行相应的业务处理了。

BinaryStreamReader 和 BinaryStreamWriter 类的完整实现如下。

ProtocolStream.h（文件位置为 flamingoserver/net/ProtocolStream.h）：

```cpp
/**
 * 一个强大的协议类, protocolstream.h
 * zhangyl 2017.05.27
 */

#ifndef __PROTOCOL_STREAM_H__
#define __PROTOCOL_STREAM_H__

#include <stdlib.h>
#include <sys/types.h>
#include <string>
#include <sstream>
#include <stdint.h>

//二进制协议的打包解包类，内部服务器之间的通信统一采用这些类
namespace net
{
    enum
    {
```

```cpp
    TEXT_PACKLEN_LEN = 4,
    TEXT_PACKAGE_MAXLEN = 0xffff,
    BINARY_PACKLEN_LEN = 2,
    BINARY_PACKAGE_MAXLEN = 0xffff,

    TEXT_PACKLEN_LEN_2 = 6,
    TEXT_PACKAGE_MAXLEN_2 = 0xffffff,

    BINARY_PACKLEN_LEN_2 = 4,                    //4字节头长度
    BINARY_PACKAGE_MAXLEN_2 = 0x10000000,  //包最大长度是256MB，足够用

    CHECKSUM_LEN = 2,
};

//计算校验和
unsigned short checksum(const unsigned short* buffer, int size);
//将一个4字节的整型数值压缩成1～5字节
void write7BitEncoded(uint32_t value, std::string& buf);
//将一个8字节的整型值编码成1～10字节
void write7BitEncoded(uint64_t value, std::string& buf);

//将一个1～5字节的字符数组值还原成4字节的整型值
void read7BitEncoded(const char* buf, uint32_t len, uint32_t& value);
//将一个1～10字节的值还原成4字节的整型值
void read7BitEncoded(const char* buf, uint32_t len, uint64_t& value);

class BinaryStreamReader final
{
public:
    BinaryStreamReader(const char* ptr, size_t len);
    ~BinaryStreamReader() = default;

    virtual const char* GetData() const;
    virtual size_t GetSize() const;
    bool IsEmpty() const;
    bool ReadString(std::string* str, size_t maxlen, size_t& outlen);
    bool ReadCString(char* str, size_t strlen, size_t& len);
    bool ReadCCString(const char** str, size_t maxlen, size_t& outlen);
    bool ReadInt32(int32_t& i);
    bool ReadInt64(int64_t& i);
    bool ReadShort(short& i);
    bool ReadChar(char& c);
    size_t ReadAll(char* szBuffer, size_t iLen) const;
    bool IsEnd() const;
    const char* GetCurrent() const { return cur; }

public:
    bool ReadLength(size_t& len);
    bool ReadLengthWithoutOffset(size_t& headlen, size_t& outlen);

private:
    BinaryStreamReader(const BinaryStreamReader&) = delete;
    BinaryStreamReader& operator=(const BinaryStreamReader&) = delete;

private:
```

```cpp
        const char*  const m_ptr;
        const size_t       m_len;
        const char*        m_cur;
    };

    class BinaryStreamWriter final
    {
    public:
        BinaryStreamWriter(std::string* data);
        ~BinaryStreamWriter() = default;

        virtual const char* GetData() const;
        virtual size_t GetSize() const;
        bool WriteCString(const char* str, size_t len);
        bool WriteString(const std::string& str);
        bool WriteDouble(double value, bool isNULL = false);
        bool WriteInt64(int64_t value, bool isNULL = false);
        bool WriteInt32(int32_t i, bool isNULL = false);
        bool WriteShort(short i, bool isNULL = false);
        bool WriteChar(char c, bool isNULL = false);
        size_t GetCurrentPos() const { return m_data->length(); }
        void Flush();
        void Clear();

    private:
        BinaryStreamWriter(const BinaryStreamWriter&) = delete;
        BinaryStreamWriter& operator=(const BinaryStreamWriter&) = delete;

    private:
        std::string* m_data;
    };

}//end namespace

#endif //!__PROTOCOL_STREAM_H__
```

ProtocolStream.cpp 文件的实现如下（文件位置为 flamingoserver/net/ProtocolStream.cpp ）：

```cpp
#ifndef _WIN32
#include <arpa/inet.h>
#else
#include <Winsock2.h>
#pragma comment(lib, "Ws2_32.lib")
#endif

#include "ProtocolStream.h"
#include <string.h>
#include <stdio.h>
#include <sys/types.h>
#include <cassert>
#include <algorithm>
#include <stdio.h>

using namespace std;
```

```cpp
namespace net
{
    //计算校验和
    unsigned short checksum(const unsigned short* buffer, int size)
    {
        unsigned int cksum = 0;
        while (size > 1)
        {
            cksum += *buffer++;
            size -= sizeof(unsigned short);
        }
        if (size)
        {
            cksum += *(unsigned char*)buffer;
        }
        //将32位数转换成16位数
        while (cksum >> 16)
            cksum = (cksum >> 16) + (cksum & 0xffff);

        return (unsigned short)(~cksum);
    }

    //将一个4字节的整型数值压缩成1~5字节
    void write7BitEncoded(uint32_t value, std::string& buf)
    {
        do
        {
            unsigned char c = (unsigned char)(value & 0x7F);
            value >>= 7;
            if (value)
                c |= 0x80;

            buf.append(1, c);
        } while (value);
    }

    //将一个8字节的整型值编码成1~10字节
    void write7BitEncoded(uint64_t value, std::string& buf)
    {
        do
        {
            unsigned char c = (unsigned char)(value & 0x7F);
            value >>= 7;
            if (value)
                c |= 0x80;

            buf.append(1, c);
        } while (value);
    }

    //将一个1~5字节的字符数组值还原成4字节的整型值
    void read7BitEncoded(const char* buf, uint32_t len, uint32_t& value)
    {
        char c;
```

```cpp
        value = 0;
        int bitCount = 0;
        int index = 0;
        do
        {
            c = buf[index];
            uint32_t x = (c & 0x7F);
            x <<= bitCount;
            value += x;
            bitCount += 7;
            ++index;
        } while (c & 0x80);
    }

    //将一个1~10字节的值还原成4字节的整型值
    void read7BitEncoded(const char* buf, uint32_t len, uint64_t& value)
    {
        char c;
        value = 0;
        int bitCount = 0;
        int index = 0;
        do
        {
            c = buf[index];
            uint64_t x = (c & 0x7F);
            x <<= bitCount;
            value += x;
            bitCount += 7;
            ++index;
        } while (c & 0x80);
    }

    BinaryStreamReader::BinaryStreamReader(const char* ptr, size_t len)
        : m_ptr(ptr), m_len(len), m_cur(ptr)
    {
        m_cur += BINARY_PACKLEN_LEN_2 + CHECKSUM_LEN;
    }

    bool BinaryStreamReader::IsEmpty() const
    {
        return m_len <= BINARY_PACKLEN_LEN_2;
    }

    size_t BinaryStreamReader::GetSize() const
    {
        return m_len;
    }

    bool BinaryStreamReader::ReadCString(char* str, size_t strlen, /* out */ size_t& outlen)
    {
        size_t fieldlen;
        size_t headlen;
        if (!ReadLengthWithoutOffset(headlen, fieldlen)) {
            return false;
```

```cpp
        //用户缓冲区的内容长度不够
        if (fieldlen > strlen) {
            return false;
        }

        //偏移到数据的位置
        m_cur += headlen;
        if (m_cur + fieldlen > m_ptr + m_len)
        {
            outlen = 0;
            return false;
        }
        memcpy(str, m_cur, fieldlen);
        outlen = fieldlen;
        m_cur += outlen;
        return true;
    }

    bool BinaryStreamReader::ReadString(string* str, size_t maxlen, size_t& outlen)
    {
        size_t headlen;
        size_t fieldlen;
        if (!ReadLengthWithoutOffset(headlen, fieldlen)) {
            return false;
        }

        //用户缓冲区的内容长度不够
        if (maxlen != 0 && fieldlen > maxlen) {
            return false;
        }

        //偏移到数据的位置
        m_cur += headlen;
        if (m_cur + fieldlen > m_ptr + m_len)
        {
            outlen = 0;
            return false;
        }
        str->assign(m_cur, fieldlen);
        outlen = fieldlen;
        m_cur += outlen;
        return true;
    }

    bool BinaryStreamReader::ReadCCString(const char** str, size_t maxlen, size_t& outlen)
    {
        size_t headlen;
        size_t fieldlen;
        if (!ReadLengthWithoutOffset(headlen, fieldlen)) {
            return false;
        }
        //用户缓冲区的内容长度不够
        if (maxlen != 0 && fieldlen > maxlen) {
```

```cpp
        return false;
    }

    //偏移到数据的位置
    m_cur += headlen;
    if (m_cur + fieldlen > m_ptr + m_len)
    {
        outlen = 0;
        return false;
    }
    *str = m_cur;
    outlen = fieldlen;
    m_cur += outlen;
    return true;
}

bool BinaryStreamReader::ReadInt32(int32_t& i)
{
    const int VALUE_SIZE = sizeof(int32_t);

    if (m_cur + VALUE_SIZE > m_ptr + m_len)
        return false;

    memcpy(&i, m_cur, VALUE_SIZE);
    i = ntohl(i);

    cur += VALUE_SIZE;

    return true;
}

bool BinaryStreamReader::ReadInt64(int64_t& i)
{
    char int64str[128];
    size_t length;
    if (!ReadCString(int64str, 128, length))
        return false;

    i = atoll(int64str);

    return true;
}

bool BinaryStreamReader::ReadShort(short& i)
{
    const int VALUE_SIZE = sizeof(short);

    if (m_cur + VALUE_SIZE > m_ptr + m_len) {
        return false;
    }

    memcpy(&i, m_cur, VALUE_SIZE);
    i = ntohs(i);

    m_cur += VALUE_SIZE;
```

```cpp
    return true;
}

bool BinaryStreamReader::ReadChar(char& c)
{
    const int VALUE_SIZE = sizeof(char);

    if (m_cur + VALUE_SIZE > m_ptr + m_len) {
        return false;
    }

    memcpy(&c, m_cur, VALUE_SIZE);
    m_cur += VALUE_SIZE;

    return true;
}

bool BinaryStreamReader::ReadLength(size_t& outlen)
{
    size_t headlen;
    if (!ReadLengthWithoutOffset(headlen, outlen)) {
        return false;
    }

    m_cur += headlen;
    return true;
}

bool BinaryStreamReader::ReadLengthWithoutOffset(size_t& headlen, size_t& outlen)
{
    headlen = 0;
    const char* temp = cur;
    char buf[5];
    for (size_t i = 0; i < sizeof(buf); i++)
    {
        memcpy(buf + i, temp, sizeof(char));
        temp++;
        headlen++;

        if ((buf[i] & 0x80) == 0x00)
            break;
    }
    if (m_cur + headlen > m_ptr + m_len)
        return false;

    unsigned int value;
    read7BitEncoded(buf, headlen, value);
    outlen = value;

    return true;
}

bool BinaryStreamReader::IsEnd() const
```

```cpp
{
    return m_cur == m_ptr + m_len;
}

const char* BinaryStreamReader::GetData() const
{
    return m_ptr;
}

size_t BinaryStreamReader::ReadAll(char* szBuffer, size_t iLen) const
{
    size_t iRealLen = min(iLen, m_len);
    memcpy(szBuffer, m_ptr, iRealLen);
    return iRealLen;
}

//===========class BinaryStreamWriter implementation============//
BinaryStreamWriter::BinaryStreamWriter(string* data) :
    m_data(data)
{
    m_data->clear();
    char str[BINARY_PACKLEN_LEN_2 + CHECKSUM_LEN];
    m_data->append(str, sizeof(str));
}

bool BinaryStreamWriter::WriteCString(const char* str, size_t len)
{
    std::string buf;
    write7BitEncoded(len, buf);
    m_data->append(buf);
    m_data->append(str, len);
    return true;
}

bool BinaryStreamWriter::WriteString(const string& str)
{
    return WriteCString(str.c_str(), str.length());
}

const char* BinaryStreamWriter::GetData() const
{
    return m_data->data();
}

size_t BinaryStreamWriter::GetSize() const
{
    return m_data->length();
}

bool BinaryStreamWriter::WriteInt32(int32_t i, bool isNULL)
{
    int32_t i2 = 999999999;
    if (isNULL == false)
        i2 = htonl(i);
    m_data->append((char*)& i2, sizeof(i2));
```

```cpp
        return true;
    }

    bool BinaryStreamWriter::WriteInt64(int64_t value, bool isNULL)
    {
        char int64str[128];
        if (isNULL == false)
        {
#ifndef _WIN32
            sprintf(int64str, "%ld", value);
#else
            sprintf(int64str, "%lld", value);
#endif
            WriteCString(int64str, strlen(int64str));
        }
        else
            WriteCString(int64str, 0);
        return true;
    }

    bool BinaryStreamWriter::WriteShort(short i, bool isNULL)
    {
        short i2 = 0;
        if (isNULL == false)
            i2 = htons(i);
        m_data->append((char*)& i2, sizeof(i2));
        return true;
    }

    bool BinaryStreamWriter::WriteChar(char c, bool isNULL)
    {
        char c2 = 0;
        if (isNULL == false)
            c2 = c;
        (*m_data) += c2;
        return true;
    }

    bool BinaryStreamWriter::WriteDouble(double value, bool isNULL)
    {
        char  doublestr[128];
        if (isNULL == false)
        {
            sprintf(doublestr, "%f", value);
            WriteCString(doublestr, strlen(doublestr));
        }
        else
            WriteCString(doublestr, 0);
        return true;
    }

    void BinaryStreamWriter::Flush()
    {
        char* ptr = &(*m_data)[0];
        unsigned int ulen = htonl(m_data->length());
```

```
        memcpy(ptr, &ulen, sizeof(ulen));
    }

    void BinaryStreamWriter::Clear()
    {
        m_data->clear();
        char str[BINARY_PACKLEN_LEN_2 + CHECKSUM_LEN];
        m_data->append(str, sizeof(str));
    }
}//end namespace
```

该协议类提供了对 char、short、int32、int64、string 等常用的字段类型的读写功能，功能非常强大，免去了定义各种结构体的麻烦。但是从业务上讲，在实际开发中，每个字段的含义及读写字段的顺序都需要通信双方提前协商好。

6.10 理解 HTTP

HTTP（Hypertext Transfer Protocol，超级文本传输协议）是用途最广泛的网络协议之一，但不少开发者只停留在会用工具或库发送 HTTP 请求阶段。笔者面试过不少求职者，问及 HTTP 的具体格式时，很多求职者都不能回答清楚，有人甚至将 HTML 文档页面的头部<head>标签当作 HTTP 数据包的包头。另外，大多数开发者都知道 HTTP GET 和 POST 方法，也可以利用一些现成的库或框架发送 GET 或 POST 请求，但是对 GET 或 POST 方法的数据在协议包中的存放位置一无所知，更不用说服务器如何识别并解析这些数据了。

需要用到 HTTP 服务器时，很多人只能依靠 Apache、Nginx 这样现成的 HTTP Web Server，对如何实现一个 HTTP 服务器无从下手。如果在实际应用场景中，服务端需要支持一些简单 HTTP 请求，例如某个服务需要提供一个 HTTP 格式的健康检查接口，则使用 Apache、Nginx 等 HTTP 服务器程序实在太"重"，不如自己实现一个简单的 HTTP 服务。

6.10.1 HTTP 格式介绍

HTTP 是建立在 TCP 之上的应用层协议，其格式如下：

```
GET 或 POST 请求的 URL 路径（一般是去掉域名的路径）    HTTP 协议版本号\r\n
字段 1 名：字段 1 值\r\n
字段 2 名：字段 2 值\r\n
...
字段 n 名 ：字段 n 值\r\n
\r\n
HTTP 包体内容
```

如上所示，HTTP 由包头（第 1~6 行）和包体（第 7 行以后）两部分组成，包头与包体之间使用一个\r\n 分割，包头的每一行均以\r\n 结束，在包头结束时再添加一个\r\n（空行）表示包头结束。也就是说，HTTP 在大多数情况下都是文本形式的明文格式，这也是 Hypertext Transfer Protocol 中 text（文本）的名称含义。

由于 HTTP 包头的每一行都以\r\n 结束，所以 HTTP 包头以\r\n\r\n 结束，我们在用程序解析 HTTP 格式的数据时可以通过\r\n\r\n 界定包头的结束位置和包体的起始位置。

也就是说，HTTP 也分为 head（包头）和 body（包体）两部分，注意，这里的 head 和 body 不是指 HTML 文档中的<head>和<body>标签。实际上，HTML 文档中的内容（当然也包括其中的<head>和<body>标签）仅是 HTTP 包 body（包体）的一部分。

```
1  <!DOCTYPE html>          <head>标记包含的内容是 HTML 文档的头，不是
2  <html>                    HTTP协议包的头，它属于HTTP协议包包体的一部分。
3  <head>
4  <title>Welcome to nginx!</title>
5  <style>
6      body {
7          width: 35em;
8          margin: 0 auto;
9          font-family: Tahoma, Verdana, Arial, sans-serif;
10     }
11 </style>                  <body>标记包含的内容也属于HTTP
12 </head>                   协议包包体的一部分。
13 <body>
14 <h1>Welcome to nginx!</h1>
15 <p>If you see this page, the nginx web server is successfully installed and
16 working. Further configuration is required.</p>
17
18 <p>For online documentation and support please refer to
19 <a href="http://nginx.org/">nginx.org</a>.<br/>
20 Commercial support is available at
21 <a href="http://nginx.com/">nginx.com</a>.</p>
22
23 <p><em>Thank you for using nginx.</em></p>
24 </body>
25 </html>
26
```

6.10.2　GET 与 POST 方法

HTTP 请求的方法有 GET、POST、HEAD、PUT、DELETE 等，其中 GET 和 POST 是我们用得最多的方法。下面以一个具体的例子说明使用 GET 方法时的 HTTP 包格式。

假设我们在浏览器中请求访问"http://www.hootina.org/index_2013.php"，则这是一个典型的 GET 方法，浏览器为我们组装的 HTTP 数据包格式如下：

```
GET /index_2013.php HTTP/1.1\r\n
Host: www.hootina.org\r\n
Connection: keep-alive\r\n
Upgrade-Insecure-Requests: 1\r\n
User-Agent: Mozilla/5.0 (Windows NT 6.1; Win64; x64) AppleWebKit/537.36 (KHTML, like Gecko) Chrome/65.0.3325.146 Safari/537.36\r\n
Accept: text/html,application/xhtml+xml,application/xml;q=0.9,image/webp,image/apng,*/*;q=0.8\r\n
Accept-Encoding: gzip, deflate\r\n
Accept-Language: zh-CN,zh;q=0.9,en;q=0.8\r\n
\r\n
```

打开浏览器的调试窗口，可以看到上面的 HTTP 包信息。例如，对于 Chrome 浏览器，在页面上右键选择"检查"菜单即可，快捷键为 F12。

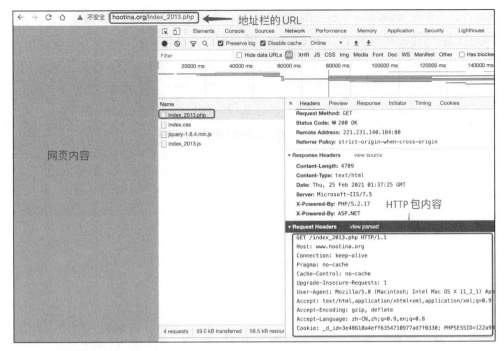

上面这个请求只有包头,没有包体,HTTP 的包体部分不是必需的,也就是说,GET 请求一般没有包体部分。

如果 GET 请求带参数,那么参数一般被附加在请求的 URL 后面,参数与参数之间使用&分割,例如请求 http://www.hootina.org/index_2013.php?param1=value1&param2=value2&param3=value3 有三个参数 param1、param2 和 param3,其对应的参数值分别是 value1、value2、value3。

这个请求组装的 HTTP 报文格式如下:

```
GET /index_2013.php?param1=value1&param2=value2&param3=value3 HTTP/1.1\r\n
Host: www.hootina.org\r\n
Connection: keep-alive\r\n
Upgrade-Insecure-Requests: 1\r\n
User-Agent: Mozilla/5.0 (Windows NT 6.1; Win64; x64) AppleWebKit/537.36 (KHTML, like Gecko) Chrome/65.0.3325.146 Safari/537.36\r\n
Accept: text/html,application/xhtml+xml,application/xml;q=0.9,image/webp,image/apng,*/*;q=0.8\r\n
Accept-Encoding: gzip, deflate\r\n
Accept-Language: zh-CN,zh;q=0.9,en;q=0.8\r\n
\r\n
```

由于浏览器对 URL 的长度最大值有限制,因此放在 URL 后面的 GET 参数数量和长度也是有限制的,不同浏览器的最大长度限制值不一样。

再来看看 POST 方法。POST 方法请求的数据被放在 HTTP 包的什么位置呢?这里以登录 12306 网站时输入用户名、密码和选择正确的图片验证码后单击"登录"按钮为例进行说明,这是一个典型的 HTTP POST 请求。

浏览器以 POST 方式组装了 HTTP 包发送给我们的用户名、密码和图片验证码等信息，组装的 HTTP 包内容格式如下：

```
POST /passport/web/login HTTP/1.1\r\n
Host: kyfw.12306.cn\r\n
Connection: keep-alive\r\n
Content-Length: 55\r\n
Accept: application/json, text/javascript, */*; q=0.01\r\n
Origin: https://kyfw.12306.cn\r\n
X-Requested-With: XMLHttpRequest\r\n
User-Agent: Mozilla/5.0 (Windows NT 6.1; Win64; x64) AppleWebKit/537.36 (KHTML, like Gecko) Chrome/65.0.3325.146 Safari/537.36\r\n
Content-Type: application/x-www-form-urlencoded; charset=UTF-8\r\n
Referer: https://kyfw.12306.cn/otn/login/init\r\n
Accept-Encoding: gzip, deflate, br\r\n
Accept-Language: zh-CN,zh;q=0.9,en;q=0.8\r\n
Cookie: _passport_session=0b2cc5b86eb74bcc976bfa9dfef3e8a20712;
_passport_ct=18d19b0930954d76b8057c732ce4cdcat8137;
route=6f50b51faa11b987e576cdb301e545c4; RAIL_EXPIRATION=1526718782244;
RAIL_DEVICEID=QuRAhOyIWv9lwWEhkq03x5Yl_livKZxx7gW6_-52oTZQda1c4zmVWxdw5Zk79xSDFH
e9LJ57F8luYOFp_yahxDXQAOmEV8U1VgXavacuM2UPCFy3knfn42yTsJM3EYOy-hwpsP-jTb2OXevJj5
acf40XsvsPDcM7; BIGipServerpool_passport=300745226.50215.0000;
BIGipServerotn=1257243146.38945.0000;
BIGipServerpassport=1005060362.50215.0000\r\n
\r\n
username=balloonwj%40qq.com&password=iloveyou&appid=otn
```

其中，username=balloonwj%40qq.com&password=iloveyou&appid=otn 就是我们的 POST 请求携带的数据，但是需要注意以下事项。

（1）用户名是 balloonwj@qq.com，在包体里面变成 balloonwj%40qq.com，其中%40 是@符号的 16 进制转码形式。浏览器会对部分 URL 或者包体部分的部分字符做 16 进制转码。

（2）这里有 3 个变量，分别是 username、password 和 appid，它们之间使用&分割，但这不意味着 POST 请求传递多个数据时必须使用&分割。只不过这里是浏览器的 HTML

表单（在浏览器中输入用户名和密码的文本框是 HTML 表单的一种）分割多个变量采用的默认方式而已。读者可以根据自己的需求灵活组织 POST 方法携带的数据格式，只要收发端协商好格式即可。例如分割多个变量时也可以采用如下三种方法。

方法一：

```
username=balloonwj%40qq.com|password=iloveyou|appid=otn
```

方法二：

```
username:balloonwj%40qq.com\r\n
password:iloveyou\r\n
appid:otn\r\n
```

方法三：

```
username,password,appid=balloonwj%40qq.com,iloveyou,otn
```

POST 方法请求的数据被放在 HTTP 包体中（\r\n\r\n 标志之后）。如下图所示，由于 HTTP 是基于 TCP 的，而 TCP 是流式协议，所以包头部分可以通过多出的\r\n 分界，但对端如何知道包体部分有多长呢？这是协议本身要解决的问题。目前一般有两种方式，第 1 种方式是采用 HTTP chunk；第 2 种方式是在包头中设置一个 Content-Length 字段（字段名不区分大小写），这个字段的值标识了包体的长度。

下图中的 Content-Length 值为 55，这是数据 username=balloonwj%40qq.com&password=iloveyou&appid=otn 字符串的长度，服务器在收到一个数据包后，会先从包头中解析出这个字段的值，再根据这个值读取相应长度的数据作为 HTTP 的包体数据。

注意：本节介绍的技术仅用于个人学习研究使用，请勿用于破坏或攻击包括 12306 在内的任何站点，违者自负法律责任。

GET 与 POST 请求的安全性比较

了解 HTTP 的 GET 和 POST 方法之后，有的开发者可能会有这样一种认识，即 POST 请求比 GET 请求要安全，其理由是：GET 请求的数据直接暴露在 URL 后面，而 POST 请求的数据在 HTTP 包的包体里面。这种认识其实是有些偏颇的，因为就算是 POST 请求，仍然可以通过打开浏览器的调试窗口查看，或者进行抓包分析，所以理论上说对于这两种类型的请求方式，数据被破解的难度是相同的。因此在实际开发中，无论是对 POST 请求还是对 GET 请求，我们都会对一些关键性的信息进行混淆和加密。

在对安全要求更高的应用中，例如交易支付，会将 SSL 与 HTTP 结合起来，即 HTTPS。HTTPS 在 HTTP 的基础上通过传输加密和身份认证保证了传输过程的安全性，SSL（Secure Socket Layer，安全套接字层）即在 TCP 层与 HTTP 层再加入一个 SSL 层。HTTPS 使用了一个不同于 HTTP 的默认端口（443 端口）及一个加密/身份验证层用于互联网上对安全敏感的通信。

6.10.3　HTTP chunk 编码

HTTP 在传输过程中如果包体过大，比如使用 HTTP 上传一个大文件，或者传输动态产生的内容给对端时，传输方无法预先知道传输的内容有多大，这时就可以使用 HTTP chunk 编码技术了。

HTTP chunk 编码的技术原理是将整个 HTTP 包体分成多个小块，每一块都有自己的字段来说明自身的长度，对端收到这些块后，去除说明的部分，将多个小块合并在一起得到完整的包体内容。传输方在 HTTP 包头中设置了 Transfer-Encoding:chunked 来告诉对端这个数据是分块传输的（代替 Content-Length 字段）。

分块传输的编码格式如下：

```
[chunkSize][\r\n][chunkData][\r\n][chunkSize][\r\n][chunkData][\r\n][chunkSize=0][\r\n][\r\n]
```

在这个编码格式中使用了若干个 chunk，每个 chunk 都为 [chunkSize][\r\n][chunkData][\r\n]，最后以一个 chunkSize 为 0 且没有 chunkData 的 chunk 结束。每个 chunk 都由两部分组成，第 1 部分是该 chunk 的长度（chunkSize），第 2 部分是指定长度的内容，每部分都用\r\n 隔开。需要注意的是，最后一个长度为 0 的 chunk 只有 chunkSize，没有 chunkData，因此其格式变成了[chunkSize=0][\r\n][\r\n]。

chunkSize 以十六进制的 ASCII 码表示每字节，例如某个 chunkSize 部分一共有两字节，第 1 字节的值是 35（十六进制），第 2 字节的值是 36（十六进制），十六进制的 35 和十六进制的 36 在 ASCII 码表中分别表示阿拉伯数字 5 和 6，因此被 HTTP 解释为十六进制的 56，十六进制的 56 对应十进制的 86，后面紧跟\r\n（0d 0a），再接着是连续的 86 字节的 chunkData。chunk 数据以 0 长度的 chunk 块结束，也就是 30 0d 0a 0d 0a，十六进制的 30 对应阿拉伯数字 0。这个 chunk 的结构如下图所示。

| 0x35 | 0x36 | 0x0d | 0x0a | 长度为0x56的chunkData | 0x0d | 0x0a | 0x30 | 0x0d | 0x0a | 0x0d | 0x0a |

在介绍了 HTTP chunk 编码格式后，对端对 chunk 格式的解压缩也很容易了。首先，对端要在收到的 HTTP 头部找到 Transfer-Encoding 字段，并且其值是 chunked，说明这个 HTTP 数据包是使用 chunk 技术编码的，接下来按格式对分块进行解析就可以了。

（1）找到 HTTP 包体开始的地方（HTTP 头部\r\n\r\n 的下一个位置）和接下来一个\r\n 中间的部分，这是第 1 个 chunkSize 的内容。

（2）假设第 1 个 chunkSize 的长度是 n 字节，则对照 ASCII 码表将这 n 字节依次转换成数字，然后将这些数字拼凑起来当成十六进制数值，再转换成十进制，这就是接下来chunkData 的长度。

（3）跳过\r\n，获取下一个数据块的 chunkSize 和 chunkData，直到遇到一个 chunkSize 为 0 的数据块。

（4）将各个数据块的 chunkData 按顺序拼接，以得到这个 HTTP 数据包的完整包体。

理顺上面的逻辑后，用代码实现就比较简单了，这里不展示具体的例子了。

6.10.4　HTTP 客户端的编码实现

如果能理解前面讲解的 HTTP 格式，就可以自己通过代码组装 HTTP 报文来发送 HTTP 请求了，这也是各种 HTTP 工具和库模拟 HTTP 请求的基本原理。

举个例子，我们要请求"http://www.hootina.org/index_2013.php"这个 URL，则先取出 URL 中的域名部分，即 hootina.org，然后通过 socket API gethostbyname 函数得到 hootina.org 这个域名对应的 IP 地址，因为在这个 URL 中没有显式指定请求的端口号，所以使用 HTTP 的默认端口号 80。有了 IP 和端口号之后，我们使用 socket API connect 函数连接服务器，然后根据上面介绍的格式组装成 HTTP 包，并利用 socket API send 函数将组装的协议包发出去，如果服务器有应答，就可以使用 socket API recv 函数接收数据，然后按 HTTP 格式进行解包（分为解析包头和包体两个步骤）。

6.10.5　HTTP 服务端的实现

这里简化一些问题，假设客户端发送的请求都是 GET 请求，则在客户端发来 HTTP 请求之后，我们拿到 HTTP 包就可以做相应的处理。这里以 Flamingo 服务器实现一个支持 HTTP 格式的注册请求为例，假设用户在浏览器中输入了以下内容，就可以实现一个注册功能：

```
http://101.37.25.166:12345/register.do?p={"username": "13917043329", "nickname": "balloon", "password": "123"}
```

这里的 HTTP 使用的是 12345 端口，而不是默认的 80 端口。如何监听 12345 端口属于网络编程基础知识，这里不再赘述。收到数据后按如下逻辑进行处理：

```
//代码片段 6-10-5
void HttpSession::OnRead(const std::shared_ptr<TcpConnection>& conn, Buffer* pBuffer, Timestamp receiveTime)
{
    string inbuf;
    //先把所有数据都取出来
    inbuf.append(pBuffer->peek(), pBuffer->readableBytes());
    //因为一个 HTTP 包头的数据至少有\r\n\r\n，所以大于 4 个字符
    //若小于或等于 4 个字符，则说明数据未收完，退出，等待网络底层接着收取
    if (inbuf.length() <= 4)
        return;

    //我们收到的 GET 请求数据包的格式一般如下：
```

```cpp
    /*
    GET
/register.do?p={%22username%22:%20%2213917043329%22,%20%22nickname%22:%20%22ball
oon%22,%20%22password%22:%20%22123%22} HTTP/1.1\r\n
    Host: 120.55.94.78:12345\r\n
    Connection: keep-alive\r\n
    Upgrade-Insecure-Requests: 1\r\n
    User-Agent: Mozilla/5.0 (Windows NT 6.1; Win64; x64) AppleWebKit/537.36 (KHTML,
like Gecko) Chrome/65.0.3325.146 Safari/537.36\r\n
    Accept-Encoding: gzip, deflate\r\n
    Accept-Language: zh-CN, zh; q=0.9, en; q=0.8\r\n
    \r\n
    */
    //检查是否以\r\n\r\n结束，如果不是，则说明包头不完整，退出
    string end = inbuf.substr(inbuf.length() - 4);
    if (end != "\r\n\r\n")
        return;

    //以\r\n分割每一行
    std::vector<string> lines;
    StringUtil::Split(inbuf, lines, "\r\n");
    if (lines.size() < 1 || lines[0].empty())
    {
        conn->forceClose();
        return;
    }

    std::vector<string> chunk;
    StringUtil::Split(lines[0], chunk, " ");
    //在chunk中至少有3个字符串：GET、URL和HTTP版本号
    if (chunk.size() < 3)
    {
        conn->forceClose();
        return;
    }

    LOG_INFO << "url: " << chunk[1] << " from " << conn->peerAddress().toIpPort();
    //inbuf =
/register.do?p={%22username%22:%20%2213917043329%22,%20%22nickname%22:%20%22ball
oon%22,%20%22password%22:%20%22123%22}
    std::vector<string> part;
    //通过"?"分割成前后两端，前面是URL，后面是参数
    StringUtil::Split(chunk[1], part, "?");
    //在chunk中至少有三个字符串：GET、URL和HTTP版本号
    if (part.size() < 2)
    {
        conn->forceClose();
        return;
    }

    string url = part[0];
    string param = part[1].substr(2);

    if (!Process(conn, url, param))
    {
```

```
        LOG_ERROR << "handle http request error, from:" <<
conn->peerAddress().toIpPort() << ", request: " << pBuffer->retrieveAllAsString();
    }

    //短连接，处理完后关闭连接
    conn->forceClose();
}
```

在以上代码中，在接收到的字节流中必须存在\r\n\r\n 标志（即至少有一个 HTTP 包头部分），然后利用\r\n 分割得到每一行，其中第 1 行的数据如下：

```
GET /register.do?p={%22username%22:%20%2213917043329%22,%20%22nickname%22:%20%22balloon%22,%20%22password%22:%20%22123%22} HTTP/1.1
```

> 注意：这里为了便于说明问题，并没有处理对端使用 HTTP chunk 格式传输数据包的情形。

其中，"%22" 是"（双引号）的 URL 转码形式，"%20" 是空格的 URL 转码形式。然后我们根据空格将第 1 行的数据分成 3 段，第 2 段就是网址 URL 和参数：

```
/register.do?p={%22username%22:%20%2213917043329%22,%20%22nickname%22:%20%22balloon%22,%20%22password%22:%20%22123%22}
```

然后根据网址与参数之间的问号将第 2 段数据再分成两段，第 1 段是 URL，第 2 段是 GET 参数的内容，根据 URL 匹配网址：

```
bool HttpSession::Process(const std::shared_ptr<TcpConnection>& conn, const
std::string& url, const std::string& param)
{
    if (url.empty())
        return false;

    if (url == "/register.do")
        OnRegisterResponse(param, conn);
    else if (url == "/login.do")
        OnLoginResponse(param, conn);
    else
        return false;

    return true;
}
```

如果是注册请求，则会走注册处理逻辑：

```
void HttpSession::OnRegisterResponse(const std::string& data, const
std::shared_ptr<TcpConnection>& conn)
{
    string retData;
    string decodeData;
    URLEncodeUtil::Decode(data, decodeData);
    BussinessLogic::RegisterUser(decodeData, conn, false, retData);
    if (!retData.empty())
    {
        std::string response;
        URLEncodeUtil::Encode(retData, response);
```

```
        MakeupResponse(retData, response);
        conn->send(response);

        LOG_INFO << "Response to client: cmd=msg_type_register" << ", data=" << retData
<< conn->peerAddress().toIpPort();;
    }
}
```

注册结果被放在 retData 中，为了发送给客户端，我们对应答结果中的特殊字符（如双引号）进行转码，例如如下返回结果：

```
{"code":0, "msg":"ok"}
```

则会被转码如下：

```
{%22code%22:0,%20%22msg%22:%22ok%22}
```

将数据组装成 HTTP 发送给客户端。发送给客户端的应答协议与 HTTP 请求协议有一点点差别，就是将请求的 URL 路径转换成 HTTP 响应码，例如 200 表示应答正常返回、404 表示页面不存在。应答协议的格式如下：

```
GET 或 POST 响应码 HTTP 协议版本号\r\n
字段 1 名：字段 1 值\r\n
字段 2 名：字段 2 值\r\n
     ...
字段 n 名 ：字段 n 值\r\n
\r\n
HTTP 协议包体内容
```

以下是一个实际的 HTTP 响应包的例子：

```
HTTP/1.1 200 OK\r\n
Content-Type: text/html\r\n
Content-Length:42\r\n
\r\n
{%22code%22:%200,%20%22msg%22:%20%22ok%22}
```

注意，包头中的 Content-Length 字段值必须被设置成包体（即字符串{%22code%22:%200,%20%22msg%22:%20%22ok%22}）的长度，这里是 42。

浏览器会得到如下图所示的应答结果。

```
← → C ⌂  ▲ 不安全 | 101.37.25.166:12345/register.do?p={"use
{
    code: 0,
    msg: "ok"
}
```

需要注意的是，HTTP 请求一般是短连接，即在一次请求完数据后连接会断开，这里也实现了这个功能，即代码片段 6-10-5 中加粗行的实现是无论一个 HTTP 请求是否成功，服务器在处理完后都会立即关闭连接：

```
conn->forceClose();
```

当然，在以上实现代码中还存在一些没处理好的地方。例如，以上代码在处理收到的数据时，没有考虑到如下情形：即对不满足一个 HTTP 包头时的处理，如果某个客户端（不是使用浏览器）通过程序模拟了一个连接请求，但是迟迟不发送含有\r\n\r\n 的数据，这路连接就会被一直占用。因此，我们可以判断收到的数据长度，防止别有用心的客户端向我们的服务器乱发数据。我们假定每个 HTTP 请求的最大数据包长度是 2048，如果用户发送的数据不含有\r\n\r\n 且累积超过 2048 个，就认为连接非法，将其断开。将代码修改成如下形式：

```cpp
void HttpSession::OnRead(const std::shared_ptr<TcpConnection>& conn, Buffer* pBuffer, Timestamp receivTime)
{
    //LOG_INFO << "Recv a http request from " << conn->peerAddress().toIpPort();

    string inbuf;
    //先把所有数据都取出来
    inbuf.append(pBuffer->peek(), pBuffer->readableBytes());
    //因为一个 HTTP 包头的数据至少有\r\n\r\n，所以大于 4 个字符
    //若小于或等于 4 个字符，则说明数据未收完，退出，等待网络底层接着收取
    if (inbuf.length() <= 4)
        return;

    //我们收到的 GET 请求数据包的一般格式如下:
    /*
    GET /register.do?p={%22username%22:%20%2213917043329%22,%20%22nickname%22:%20%22balloon%22,%20%22password%22:%20%22123%22} HTTP/1.1\r\n
    Host: 120.55.94.78:12345\r\n
    Connection: keep-alive\r\n
    Upgrade-Insecure-Requests: 1\r\n
    User-Agent: Mozilla/5.0 (Windows NT 6.1; Win64; x64) AppleWebKit/537.36 (KHTML, like Gecko) Chrome/65.0.3325.146 Safari/537.36\r\n
    Accept-Encoding: gzip, deflate\r\n
    Accept-Language: zh-CN, zh; q=0.9, en; q=0.8\r\n
    \r\n
    */

    string end = inbuf.substr(inbuf.length() - 4);
    if (end != "\r\n\r\n")
    {
        if (inbuf.length() >= MAX_URL_LENGTH)
        {
            //超过 2048 个字符，且不含有\r\n\r\n，我们认为是非法请求
            conn->forceClose();
        }

        //检查是否以\r\n\r\n 结束，如果不是且长度不超过 2048 个字符，
        //则说明包头不完整，退出
        return;
    }

    //找到完整的包头
    //以\r\n 分割每一行
    std::vector<string> lines;
    StringUtil::Split(inbuf, lines, "\r\n");
    if (lines.size() < 1 || lines[0].empty())
```

```
{
    conn->forceClose();
    return;
}

//后续代码与代码片段 6-10-5 一样，此处为了节约篇幅，省略之
}
```

以上代码逻辑只能解决客户端发送非法数据的情况，如果一个客户端连上来也不给我们发送任何数据，这段逻辑就无能为力了。如果不断有客户端这么做，就会浪费我们大量的连接资源，所以还需要一个定时器去定时检测有哪些 HTTP 连接超过一定时间没向我们发送数据，找到后将连接断开。这又涉及如何设计服务器定时器的问题。

> 在实际开发中设计通用的 Web 服务器时，需要对对端发送过来的 HTTP 数据做的校验远比在这个例子中考虑的情形要多。这里只是为了说明问题方便，仅列举了少数情形。大多数 Web 服务器在收到请求时都可能会从磁盘加载某个文件的内容返回给客户端，其基本原理也是利用文件的内容组装成 HTTP 应答包。

6.10.6　HTTP 与长连接

在大多数开发者的认知中，一提到 HTTP 和 HTTP 请求，就必然要和短连接关联起来，这种认识是不正确的。尽管目前大多数 Web 服务器实现的 HTTP 连接基本上都是短连接，但可以在 HTTP 头中设置 keepalive 字段，该字段用于建议连接的双方使用长连接进行通信，而不是对每次请求都建立新的连接，但这只是一个建议选项，不少 Web 服务器并不遵循 keepalive 字段的建议。

相反，在实际商业项目中有这样一种场景会使用长连接的 HTTP：在一些业务需要对外保密的企业中，一般禁止内部员工通过除浏览器外的任何客户端访问外网，企业内部会安装一些安全软件，这类软件会过滤掉除 HTTP 格式外的所有数据包。为了应对这种网络场景，可以使用 HTTP 长连接访问外网，这是很多客户端支持 HTTP 代理的原因。

6.10.7　libcurl

Libcurl 是一个被广泛使用的跨平台的用于发送 HTTP 请求的第三方 C/C++库，其基本用法如下。

（1）使用 curl_easy_init 函数初始化一个 CURL 对象：

```
CURL* curl = curl_easy_init();
```

（2）调用 curl_easy_setopt 函数为 CURL 对象设置相关选项，例如请求的 URL 地址、方法、最大超时时间，以及是否需要在请求结果中保留 HTTP 头信息、将请求结果放在哪里等：

```
//设置请求的 URL
curl_easy_setopt(curl, CURLOPT_URL, url);
//若设置 CURLOPT_POST 这个选项值，则使用 POST 方法进行请求，否则为 GET 方法
```

```
curl_easy_setopt(curl, CURLOPT_POST, 1L);
//将请求结果放到 response 中, response 的类型是 std::string
curl_easy_setopt(curl, CURLOPT_WRITEDATA, (void*)&response);
//设置连接超时时间
curl_easy_setopt(curl, CURLOPT_CONNECTTIMEOUT, connTimeout);
//设置读取数据的超时时间
curl_easy_setopt(curl, CURLOPT_TIMEOUT, readTimeout);
//在请求结果中保留应答的 HTTP 头信息
curl_easy_setopt(curl, CURLOPT_HEADER, 1L);
```

（3）调用 curl_easy_perform 函数发送实际的 HTTP 请求，并得到一个类型为 CURLcode 的返回值，如果返回值为 CURLE_OK，则说明请求成功：

```
CURLcode res = curl_easy_perform(curl);
```

（4）无论调用成功与否，都调用 curl_easy_cleanup 函数释放 curl_easy_init 函数分配的 CURL 对象：

```
curl_easy_cleanup(curl);
```

这里基于 libcurl 做了一个简单的封装，实现如下。

CurlClient.h 文件的内容如下：

```
/**
 * 封装 curl 的 HTTP 库, CurlClient.h
 * zhangyl 2019.08.27
 */

#ifndef __CURL_CLIENT_H__
#define __CURL_CLIENT_H__

#include <string>

#include "curl.h"

class CCurlClient final
{
public:
    CCurlClient();
    ~CCurlClient();

    CCurlClient(const CCurlClient& rhs) = delete;
    CCurlClient& operator=(const CCurlClient& rhs) = delete;

    /**
     * 初始化 libcurl
     * 非线程安全函数, 建议在程序初始化时调用一次, 以免因多线程调用 curl_easy_init 而崩溃
     */
    static void init();
    /**
     * 反初始化 libcurl
     * 非线程安全函数, 建议在程序退出时调用一次
     */
    static void uninit();
```

```
    /** 发送 HTTP GET 请求
     * @param URL 请求的网址
     * @param headers 随请求发送的自定义 HTTP 头信息, 多个自定义头之间使用\r\n 分割, 最后一个
以\r\n 结束, 若无自定义 HTTP 头信息, 则将其设置为 nullptr
     * @param response 如果请求成功, 则存储 HTTP 请求结果 (注意: response 在函数调用中是追加
模式, 也就是说如果上一次 response 的值不清空, 则调用这个函数时会追加, 而不是覆盖)
     * @param autoRedirect 请求得到 HTTP 3xx 的状态码是否自动重定向至新的 URL
     * @param bReserveHeaders 请求的结果中 (存储于 response), 是否保留 HTTP 头信息
     * @param connTimeout 连接超时时间, 单位为秒 (对于某些 HTTP URI 资源不好的, 若总是返回超
时, 则可以将该参数设置得大一点)
     * @param readTimeout 读取数据超时时间, 单位为秒 (对于某些 HTTP URI 资源不好的, 若总是返
回超时, 则可以将该参数设置得大一点)
     */
    bool get(const char* url, const char* headers, std::string& response, bool
autoRedirect = false, bool reserveHeaders = false, int64_t connTimeout = 1L, int64_t
readTimeout = 5L);

    /** 发送 HTTP POST 请求
     * @param url 请求的网址
     * @param headers 随请求发送的自定义 HTTP 头信息, 多个自定义头之间使用\r\n 分割, 最后一个
以\r\n 结束, 若无自定义 HTTP 头信息, 则将其设置为 nullptr
     * @param postParams post 参数内容
     * @param response 如果请求成功, 则存储 HTTP 请求的结果 (注意: response 在函数调用中是追
加模式, 也就是说如果上一次 response 的值不清空, 则调用这个函数时会追加, 而不是覆盖)
     * @param autoRedirect 请求得到 HTTP 3xx 的状态码是否自动重定向至新的 URL
     * @param bReserveHeaders 请求的结果中 (存储于 response) 是否保留 HTTP 头信息
     * @param connTimeout 连接超时时间, 单位为秒 (对于某些 HTTP URI 资源不好的, 若总是返回超
时, 则可以将该参数设置得大一点)
     * @param readTimeout 读取数据超时时间, 单位为秒 (对于某些 HTTP URI 资源不好的, 若总是返
回超时, 则可以将该参数设置得大一点)
     */
    bool post(const char* url, const char* headers, const char* postParams,
std::string& response, bool autoRedirect = false, bool reserveHeaders = false,
int64_t connTimeout = 1L, int64_t readTimeout = 5L);


private:
    static bool  m_bGlobalInit;
};

#endif //!__CURL_CLIENT_H__
```

CurlClient.cpp 文件的内容如下:

```
/**
 * 封装 curl 的 HTTP 库, CurlClient.cpp
 * zhangyl 2019.08.27
 */

#include "CurlClient.h"
#include <iostream>
```

```cpp
//有数据响应时的回调函数
size_t reqReply(void* ptr, size_t size, size_t nmemb, void* stream)
{
    std::string* str = (std::string*)stream;
    (*str).append((char*)ptr, size * nmemb);
    return size * nmemb;
}

bool CCurlClient::m_bGlobalInit = false;

CCurlClient::CCurlClient()
{
}

CCurlClient::~CCurlClient()
{
}

void CCurlClient::init()
{
    if (!m_bGlobalInit)
    {
        curl_global_init(CURL_GLOBAL_ALL);
        m_bGlobalInit = true;
    }
}

void CCurlClient::uninit()
{
    if (m_bGlobalInit)
    {
        curl_global_cleanup();
        m_bGlobalInit = false;
    }
}

//HTTP GET
bool CCurlClient::get(const char* url, const char* headers, std::string& response,
                      bool autoRedirect/* = false*/, bool reserveHeaders/* = false*/,
                      int64_t connTimeout/* = 1L*/, int64_t readTimeout/* = 5L*/)
{
    //初始化CURL对象
    CURL* curl = curl_easy_init();
    if (curl == nullptr)
        return false;

    //设置请求的URL
    curl_easy_setopt(curl, CURLOPT_URL, url);
    //不使用HTTPS,如果需要使用HTTPS,则将最后一个参数设置为true
    curl_easy_setopt(curl, CURLOPT_SSL_VERIFYPEER, false);
    curl_easy_setopt(curl, CURLOPT_SSL_VERIFYHOST, false);
    curl_easy_setopt(curl, CURLOPT_VERBOSE, 0L);
```

```cpp
    curl_easy_setopt(curl, CURLOPT_READFUNCTION, NULL);
    curl_easy_setopt(curl, CURLOPT_WRITEFUNCTION, reqReply);
    curl_easy_setopt(curl, CURLOPT_WRITEDATA, (void*)&response);
    //禁用 SIGALRM+sigsetjmp/siglongjmp 的超时机制
    //采用其他超时机制，因为该机制修改了一个全局变量，所以在多线程下可能出现问题
    curl_easy_setopt(curl, CURLOPT_NOSIGNAL, 1L);
    curl_easy_setopt(curl, CURLOPT_CONNECTTIMEOUT, connTimeout);
    curl_easy_setopt(curl, CURLOPT_TIMEOUT, readTimeout);

    //遇到 HTTP 3xx 状态码是否自动重定位
    if (autoRedirect)
        curl_easy_setopt(curl, CURLOPT_FOLLOWLOCATION, 1L);
    else
        curl_easy_setopt(curl, CURLOPT_FOLLOWLOCATION, 0L);

    if (reserveHeaders)
        curl_easy_setopt(curl, CURLOPT_HEADER, 1L);
    else
        curl_easy_setopt(curl, CURLOPT_HEADER, 0L);

    //添加自定义 HTTP 头信息
    if (headers != nullptr)
    {
        struct curl_slist* chunk = NULL;
        chunk = curl_slist_append(chunk, headers);
        curl_easy_setopt(curl, CURLOPT_HTTPHEADER, chunk);
    }

    //发起 HTTP 请求
    CURLcode res = curl_easy_perform(curl);

    //释放 CURL 对象
    curl_easy_cleanup(curl);

    return res == CURLcode::CURLE_OK;
}

//http POST
bool CCurlClient::post(const char* url, const char* headers, const char* postParams,
std::string& response,
                bool autoRedirect/* = false*/, bool reserveHeaders/* = false*/,
int64_t connTimeout/* = 1L*/, int64_t readTimeout/* = 5L*/)
{
    //初始化 CURL 对象
    CURL* curl = curl_easy_init();
    if (curl == nullptr)
        return false;

    //设置请求方式为 post
    curl_easy_setopt(curl, CURLOPT_POST, 1L);
    //设置请求的 URL
    curl_easy_setopt(curl, CURLOPT_URL, url);
```

```cpp
//设置 POST 请求的参数
curl_easy_setopt(curl, CURLOPT_POSTFIELDS, postParams);
//不启用 HTTPS
curl_easy_setopt(curl, CURLOPT_SSL_VERIFYPEER, false);
curl_easy_setopt(curl, CURLOPT_SSL_VERIFYHOST, false);
//设置相关回调函数
curl_easy_setopt(curl, CURLOPT_VERBOSE, 0L);
curl_easy_setopt(curl, CURLOPT_READFUNCTION, NULL);
curl_easy_setopt(curl, CURLOPT_WRITEFUNCTION, reqReply);
curl_easy_setopt(curl, CURLOPT_WRITEDATA, (void*)&response);
//禁用 SIGALRM+sigsetjmp/siglongjmp 的超时机制,
//采用其他超时机制,因为该机制修改了一个全局变量,所以在多线程下可能会出问题
curl_easy_setopt(curl, CURLOPT_NOSIGNAL, 1L);
curl_easy_setopt(curl, CURLOPT_CONNECTTIMEOUT, connTimeout);
curl_easy_setopt(curl, CURLOPT_TIMEOUT, readTimeout);

//遇到 HTTP 3xx 状态码是否自动重定位
if (autoRedirect)
    curl_easy_setopt(curl, CURLOPT_FOLLOWLOCATION, 1L);
else
    curl_easy_setopt(curl, CURLOPT_FOLLOWLOCATION, 0L);

if (reserveHeaders)
    curl_easy_setopt(curl, CURLOPT_HEADER, 1L);
else
    curl_easy_setopt(curl, CURLOPT_HEADER, 0L);

//添加自定义头信息
if (headers != nullptr)
{
    struct curl_slist* chunk = NULL;
    chunk = curl_slist_append(chunk, headers);
    curl_easy_setopt(curl, CURLOPT_HTTPHEADER, chunk);
}

//发起 HTTP 请求
CURLcode res = curl_easy_perform(curl);

//释放 CURL 对象
curl_easy_cleanup(curl);

return res == CURLcode::CURLE_OK;
}
```

在编写以上代码时有如下几个注意事项。

（1）curl_easy_init 函数在第 1 次调用时,其内部会调用 curl_global_init(CURL_GLOBAL_ALL)函数, curl_global_init 函数会初始化 libcurl 的一些内部全局状态,这是非线程安全的。也就是说,如果多个线程同时首次调用 curl_easy_init 函数,则可能会出现线程安全问题。例如对于程序不知原因地在 curl_easy_init 函数中崩溃,可以将 curl_global_init 函数调用单独提取出来封装成一个 init 方法,我们的程序在调用这个库的

CCurlClient::get/post 方法之前应该调用且只调用一次 CCurlClient::init 方法。

（2）以下两行代码分别设置了 HTTP 请求的连接超时时间和读取数据的超时时间，对于一些 URL 资源地址网络不很畅通的 Web 服务器，应该将这两个超时值稍微设置得长一点，而不是使用这里给的默认值，否则可能会因为超时时间太短而无法得到期望的结果：

```
curl_easy_setopt(curl, CURLOPT_CONNECTTIMEOUT, connTimeout);
curl_easy_setopt(curl, CURLOPT_TIMEOUT, readTimeout);
```

（3）对于某些 HTTP 请求的结果，资源所在的 Web 服务器会返回一个以 3xx 开头的重定向 HTTP 状态码，用于提示资源已被挪到其他地址，可以使用重定向提示的地址访问所要的资源。对于这种情况，我们需要将 CCurlClient::get/post 方法的 autoRedirect 参数值设置为 true，以告诉 libcurl 自动重定向至新的 URL 来获取数据。

CCurlClient 类的用法示例如下：

```
std::string httpResponse;
CCurlClient curlClient;
std::ostringstream headers;
headers << "CLIENT-ID: " << m_strClientID << "\r\n" << "ACCOUNT-ID: " << m_strAccountID << "\r\n";
if (!curlClient.get(someURL, headers.str().c_str(), httpResponse))
{
    //TODO: 请求失败的操作
    return;
}

//请求成功后，应答结果被存储在 httpResponse 中
//TODO: 对 httpResponse 做进一步操作
```

6.10.8　Restful 接口与 Java Spring MVC

建议提供 HTTP 接口的 Web 服务都遵循 Restful 设计风格，这是一种将 HTTP URL 的路径定义为可对资源类型和特性进行描述的规范。

Spring MVC 框架提供的 Controller 很容易提供一个对外访问的 HTTP 接口（当然，实际提供此功能的是该框架自带的 Tomcat 服务器）。一个 Controller 可以按如下方式提供一个 HTTP 接口：

```
@GetMapping(value = "/symbol")
public String fixKlineItem(@RequestParam(value = "symbol") String symbol,
                @RequestParam(value = "type") String type) {
    //TODO: 对 HTTP 请求的一些处理工作
    return "{\"symbol\": \"600053\"}";
}
```

以上 Rest 接口提供了一个对外访问的 HTTP 接口，可以通过如下 URL 访问：

```
http://somehost.com/symbo?symbol=600053&type=1
```

这个接口需要提供两个参数，分别是 symbol 和 type。

采用这种方式开发对外的 HTTP 接口很方便，不需要手动处理接收过来的 HTTP 请求协议包，但这并不意味着这些工作不存在，这些工作已经由 Tomcat 和 Spring MVC 框架

替开发者做了，开发者可以专注于业务逻辑的开发。但需要注意，虽然我们不再需要手动解析 HTTP 包，但还是要弄明白框架的工作原理和流程。

6.11 SMTP、POP3 与邮件客户端

电子邮件服务是现代互联网的基础服务之一，熟练掌握邮件相关的协议以及邮件服务器和客户端的工作原理，是对一个合格的后端开发工程师的基本要求之一。本节将详细介绍与邮件相关的协议，并基于这些协议来介绍邮件服务器与客户端的基本实现原理。

6.11.1 邮件协议简介

与邮件收发有关的协议有 POP3、SMTP 和 IMAP 等。

（1）POP3（Post Office Protocol 3，邮局协议的第 3 个版本）规定了怎样将个人计算机连接到 Internet 的邮件服务器和下载电子邮件的电子协议，是因特网电子邮件的第 1 个离线协议标准，允许用户从服务器上把邮件存储到本地主机（即自己的计算机）上，同时删除保存在邮件服务器上的邮件。POP3 服务器是遵循 POP3 协议的邮件接收服务器。

（2）SMTP（Simple Mail Transfer Protocol，简单邮件传输协议）是一组用于从源地址到目的地址传输邮件的规范，帮助每台计算机在发送或中转邮件时都找到下一个目的地。SMTP 服务器是遵循协议 SMTP 的邮件发送服务器。

（3）IMAP（Internet Mail Access Protocol，交互式邮件存取协议）是与 POP3 协议类似的邮件访问标准协议之一。不同的是，开启了 IMAP 后，在电子邮件客户端收取的邮件仍然保留在服务器上，同时客户端上的操作都会被反馈到服务器上，例如删除邮件、标记已读等时，服务器上的邮件也会做相应的动作。所以无论是从浏览器登录邮箱还是从客户端软件登录邮箱，看到的邮件及状态都是一致的，而 POP3 对邮件的操作只会在本地邮件客户端上起作用。

如果需要自己编写相关的邮件收发客户端,则需要登录对应的邮件服务器开启相应的 POP3、SMTP、IMAP 服务。以 163 邮箱为例：登录 163 邮箱，单击页面正上方的"设置"菜单项，再单击左侧的"POP3/SMTP/IMAP"菜单项，其中"开启 SMTP 服务"选项是系统默认勾选的。勾选图中另外两个选项，单击"确定"按钮，即可开启成功。不勾选图中两个选项时，单击"确定"按钮，即可关闭成功。

网易 163 免费邮箱相关的服务器信息如下图所示。

服务器名称	服务器地址	SSL协议端口号	非SSL协议端口号
IMAP	imap.163.com	993	143
SMTP	smtp.163.com	465/994	25
POP3	pop.163.com	995	110

POP3、SMTP、IMAP 是前面介绍的使用指定字符（串）作为包结束标志的典型例子。接下来以 SMTP 和 POP3 为例进行讲解。

6.11.2 SMTP

SMTP 用于发送邮件，格式如下：

```
关键字 自定义内容\r\n
```

如上所示，对自定义内容根据关键字的类型选择是否设置，采用 SMTP 的客户端常用的关键字如下：

```
//连接邮件服务器后登录服务器之前向服务器发送的问候信息
HELO 自定义问候语\r\n

//请求登录邮件服务器
AUTH LOGIN\r\n
base64 形式的用户名\r\n
base64 形式的密码\r\n

//设置发件人的邮箱地址
MAIL FROM:发件人地址\r\n

//设置收件人地址，每次发送时都可设置一个收件人地址，如果有多个收件人地址，就要分别设置对应的次数
rcpt to:收件人地址\r\n

//发送邮件正文的开始标志
DATA\r\n
//发送邮件正文，注意邮件正文以.\r\n结束
邮件正文\r\n.\r\n

//登出服务器
QUIT\r\n
```

采用了 SMTP 的邮件服务器，其常用的关键字是各种应答码，在应答码后面可以带上自定义信息，并以\r\n结束，格式如下：

```
应答码 自定义消息\r\n
```

常用的应答码含义如下。

- ◎ 211：帮助返回系统状态。
- ◎ 214：帮助信息。
- ◎ 220：服务准备就绪。
- ◎ 221：关闭连接。
- ◎ 235：用户验证成功。
- ◎ 250：请求操作就绪。
- ◎ 251：用户不在本地，转寄到其他路径。
- ◎ 334：等待用户输入验证信息。
- ◎ 354：开始邮件输入。
- ◎ 421：服务不可用。
- ◎ 450：操作未执行，邮箱忙。
- ◎ 451：操作中止，本地错误。

- ◎ 452：操作未执行，存储空间不足。
- ◎ 500：命令不可识别或语言错误。
- ◎ 501：参数语法错误。
- ◎ 502：命令不支持（未实现）。
- ◎ 503：命令顺序错误。
- ◎ 504：命令参数不支持。
- ◎ 550：操作未执行，邮箱不可用。
- ◎ 551：非本地用户。
- ◎ 552：因存储空间不足而中止。
- ◎ 553：操作未执行，邮箱名不正确。
- ◎ 554：传输失败。

更多的 SMTP 细节可以参考相应的 RFC 文档。

下面看一个具体的使用 SMTP 发送邮件的代码示例，假设现在要实现一个邮件报警系统，则根据上文的介绍，实现一个 SmtpSocket 类来综合常用邮件的功能。

SmtpSocket.h：

```cpp
/**
 * 发送邮件类，SmtpSocket.h
 * zhangyl 2019.05.11
 */
#pragma once

#include <string>
#include <vector>
#include "Platform.h"

class SmtpSocket final
{
public:
    static bool sendMail(const std::string& server, short port, const std::string& from, const std::string& fromPassword,
                         const std::vector<std::string>& to, const std::string& subject, const std::string& mailData);

public:
    SmtpSocket(void);
    ~SmtpSocket(void);

    bool isConnected() const { return m_hSocket; }
    bool connect(const char* pszUrl, short port = 25);
    bool logon(const char* pszUser, const char* pszPassword);
    bool setMailFrom(const char* pszFrom);
    bool setMailTo(const std::vector<std::string>& sendTo);
    bool send(const std::string& subject, const std::string& mailData);

    void closeConnection();
    //退出
    void quit();
```

```cpp
private:
    //验证从服务器返回的前三位代码和传递进来的参数是否一样
    bool checkResponse(const char* recvCode);
private:
    bool                        m_bConnected;
    SOCKET                      m_hSocket;
    std::string                 m_strUser;
    std::string                 m_strPassword;
    std::string                 m_strFrom;
    std::vector<std::string>    m_strTo;
};
```

SmtpSocket.cpp：

```cpp
#include "SmtpSocket.h"
#include <sstream>
#include <time.h>
#include <string.h>
#include "Base64Util.h"
#include "Platform.h"
bool SmtpSocket::sendMail(const std::string& server, short port, const std::string& from, const std::string& fromPassword,
                    const std::vector<std::string>& to, const std::string& subject, const std::string& mailData)
{
    size_t atSymbolPos = from.find_first_of("@");
    if (atSymbolPos == std::string::npos)
        return false;

    std::string strUser = from.substr(0, atSymbolPos);

    SmtpSocket smtpSocket;
    //smtp.163.com 25
    if (!smtpSocket.connect(server.c_str(), port))
        return false;

    //testformybook 2019hhxxttxs
    if (!smtpSocket.logon(strUser.c_str(), fromPassword.c_str()))
        return false;

    //testformybook@163.com
    if (!smtpSocket.setMailFrom(from.c_str()))
        return false;

    if (!smtpSocket.setMailTo(to))
        return false;

    if (!smtpSocket.send(subject, mailData))
        return false;

    return true;
}
```

```cpp
SmtpSocket::SmtpSocket() : m_bConnected(false), m_hSocket(-1)
{
}

SmtpSocket::~SmtpSocket()
{
    quit();
}

bool SmtpSocket::checkResponse(const char* recvCode)
{
    char recvBuffer[1024] = { 0 };
    long lResult = 0;
    lResult = recv(m_hSocket, recvBuffer, 1024, 0);
    if (lResult == SOCKET_ERROR || lResult < 3)
        return false;

    return  recvCode[0] == recvBuffer[0] && \
            recvCode[1] == recvBuffer[1] && \
            recvCode[2] == recvBuffer[2] ? true : false;
}

void SmtpSocket::quit()
{
    if (m_hSocket < 0)
        return;

    //退出
    if (::send(m_hSocket, "QUIT\r\n", strlen("QUIT\r\n"), 0) == SOCKET_ERROR)
    {
        closeConnection();
        return;
    }

    if (!checkResponse("221"))
        return;
}

bool SmtpSocket::logon(const char* pszUser, const char* pszPassword)
{
    if (m_hSocket < 0)
        return false;

    //发送"AUTH LOGIN"
    if (::send(m_hSocket, "AUTH LOGIN\r\n", strlen("AUTH LOGIN\r\n"), 0) == SOCKET_ERROR)
        return false;

    if (!checkResponse("334"))
        return false;

    //发送经base64编码的用户名
    char szUserEncoded[64] = { 0 };
    Base64Util::encode(szUserEncoded, pszUser, strlen(pszUser), '=', 64);
    strncat(szUserEncoded, "\r\n", 64);
```

```cpp
    if (::send(m_hSocket, szUserEncoded, strlen(szUserEncoded), 0) == SOCKET_ERROR)
        return false;

    if (!checkResponse("334"))
        return false;

    //发送经 base64 编码的密码
    //验证密码
    char szPwdEncoded[64] = { 0 };
    Base64Util::encode(szPwdEncoded, pszPassword, strlen(pszPassword), '=', 64);
    strncat(szPwdEncoded, "\r\n", 64);
    if (::send(m_hSocket, szPwdEncoded, strlen(szPwdEncoded), 0) == SOCKET_ERROR)
        return false;

    if (!checkResponse("235"))
        return false;

    m_strUser = pszUser;
    m_strPassword = pszPassword;

    return true;
}

void SmtpSocket::closeConnection()
{
    if (m_hSocket >= 0)
    {
        closesocket(m_hSocket);
        m_hSocket = -1;
        m_bConnected = false;
    }
}

bool SmtpSocket::connect(const char* pszUrl, short port/* = 25*/)
{
    struct sockaddr_in server = { 0 };
    struct hostent* pHostent = NULL;
    unsigned int addr = 0;

    closeConnection();
    m_hSocket = socket(AF_INET, SOCK_STREAM, IPPROTO_TCP);

    if (m_hSocket < 0)
        return false;

    long tmSend(15 * 1000L), tmRecv(15 * 1000L), noDelay(1);
    setsockopt(m_hSocket, IPPROTO_TCP, TCP_NODELAY, (char*)& noDelay, sizeof(long));
    setsockopt(m_hSocket, SOL_SOCKET, SO_SNDTIMEO, (char*)& tmSend, sizeof(long));
    setsockopt(m_hSocket, SOL_SOCKET, SO_RCVTIMEO, (char*)& tmRecv, sizeof(long));

    if (inet_addr(pszUrl) == INADDR_NONE)
    {
        pHostent = gethostbyname(pszUrl);
    }
    else
    {
```

```cpp
        addr = inet_addr(pszUrl);
        pHostent = gethostbyaddr((char*)& addr, sizeof(addr), AF_INET);
    }

    if (!pHostent)
        return false;

    server.sin_family = AF_INET;
    server.sin_port = htons((u_short)port);
    server.sin_addr.s_addr = *((unsigned long*)pHostent->h_addr);
    if (::connect(m_hSocket, (struct sockaddr*) & server, sizeof(server)) == SOCKET_ERROR)
        return false;

    if (!checkResponse("220"))
        return false;

    //向服务器发送"HELO"+服务器名
    //string strTmp="HELO "+SmtpAddr+"\r\n";
    char szSend[256] = { 0 };
    snprintf(szSend, sizeof(szSend), "HELO %s\r\n", pszUrl);
    if (::send(m_hSocket, szSend, strlen(szSend), 0) == SOCKET_ERROR)
        return false;

    if (!checkResponse("250"))
        return false;

    m_bConnected = true;

    return true;
}

bool SmtpSocket::setMailFrom(const char* pszFrom)
{
    if (m_hSocket < 0)
        return false;

    char szSend[256] = { 0 };
    snprintf(szSend, sizeof(szSend), "MAIL FROM:<%s>\r\n", pszFrom);
    if (::send(m_hSocket, szSend, strlen(szSend), 0) == SOCKET_ERROR)
        return false;

    if (!checkResponse("250"))
        return false;

    m_strFrom = pszFrom;

    return true;
}

bool SmtpSocket::setMailTo(const std::vector<std::string>& sendTo)
{
    if (m_hSocket < 0)
        return false;

    char szSend[256] = { 0 };
```

```cpp
    for (const auto& iter : sendTo)
    {
        snprintf(szSend, sizeof(szSend), "rcpt to: <%s>\r\n", iter.c_str());
        if (::send(m_hSocket, szSend, strlen(szSend), 0) == SOCKET_ERROR)
            return false;

        if (!checkResponse("250"))
            return false;
    }

    m_strTo = sendTo;

    return true;
}

bool SmtpSocket::send(const std::string& subject, const std::string& mailData)
{
    if (m_hSocket < 0)
        return false;

    std::ostringstream osContent;

    //注意:邮件正文的内容与其他附属字样之间一定要空一行
    osContent << "Date: " << time(nullptr) << "\r\n";
    osContent << "from: " << m_strFrom << "\r\n";
    osContent << "to: ";
    for (const auto& iter : m_strTo)
    {
        osContent << iter << ";";
    }
    osContent << "\r\n";
    osContent << "subject: " << subject << "\r\n";
    osContent << "Content-Type: text/plain; charset=UTF-8\r\n";
    osContent << "Content-Transfer-Encoding: quoted-printable\r\n\r\n";
    osContent << mailData << "\r\n.\r\n";

    std::string data = osContent.str();
    const char* lpSendBuffer = data.c_str();

    //发送"DATA\r\n"
    if (::send(m_hSocket, "DATA\r\n", strlen("DATA\r\n"), 0) == SOCKET_ERROR)
        return false;

    if (!checkResponse("354"))
        return false;

    long dwSend = 0;
    long dwOffset = 0;
    long lTotal = data.length();
    long lResult = 0;
    const long SEND_MAX_SIZE = 1024 * 100000;
    while ((long)dwOffset < lTotal)
    {
        if (lTotal - dwOffset > SEND_MAX_SIZE)
```

```
            dwSend = SEND_MAX_SIZE;
        else
            dwSend = lTotal - dwOffset;

        lResult = ::send(m_hSocket, lpSendBuffer + dwOffset, dwSend, 0);
        if (lResult == SOCKET_ERROR)
            return false;

        dwOffset += lResult;
    }

    if (!checkResponse("250"))
        return false;

    return true;
}
```

然后使用另外一个类 MailMonitor 对 SmtpSocket 对象的功能进行高层抽象。

MailMonitor.h：

```
/**
 * 邮件监控线程，MailMonitor.h
 * zhangyl 2019.05.11
 */
#pragma once

#include <string>
#include <vector>
#include <list>
#include <memory>
#include <mutex>
#include <condition_variable>
#include <thread>

struct MailItem
{
    std::string subject;
    std::string content;
};

class MailMonitor final
{
public:
    static MailMonitor& getInstance();

private:
    MailMonitor() = default;
    ~MailMonitor() = default;
    MailMonitor(const MailMonitor & rhs) = delete;
    MailMonitor& operator=(const MailMonitor & rhs) = delete;

public:
    bool initMonitorMailInfo(const std::string& servername, const std::string& mailserver, short mailport, const std::string& mailfrom, const std::string& mailfromPassword, const std::string& mailto);
    void uninit();
```

```cpp
    void wait();

    void run();

    bool alert(const std::string& subject, const std::string& content);
private:
    void alertThread();

    void split(const std::string& str, std::vector<std::string>& v, const char* delimiter = "|");

private:
    //用于标识是哪一台服务器发送的邮件
    std::string                 m_strMailName;
    std::string                 m_strMailServer;
    short                       m_nMailPort;
    std::string                 m_strFrom;
    std::string                 m_strFromPassword;
    std::vector<std::string>    m_strMailTo;
    //待写入的日志
    std::list<MailItem>         m_listMailItemsToSend;
    std::shared_ptr<std::thread> m_spMailAlertThread;
    std::mutex                  m_mutexAlert;
    std::condition_variable     m_cvAlert;
    //退出标志
    bool                        m_bExit;
    //运行标志
    bool                        m_bRunning;
};
```

MailMonitor.cpp：

```cpp
#include "MailMonitor.h"
#include <functional>
#include <sstream>
#include <iostream>
#include <string.h>
#include "SmtpSocket.h"

MailMonitor& MailMonitor::getInstance()
{
    static MailMonitor instance;
    return instance;
}

bool MailMonitor::initMonitorMailInfo(const std::string& servername, const std::string& mailserver, short mailport, const std::string& mailfrom, const std::string& mailfromPassword, const std::string& mailto)
{
    m_strMailName = servername;

    m_strMailServer = mailserver;
    m_nMailPort = mailport;
    m_strFrom = mailfrom;
    m_strFromPassword = mailfromPassword;
```

```cpp
    split(mailto, m_strMailTo, ";");

    std::ostringstream osSubject;
    osSubject << "[" << m_strMailName << "]";

    SmtpSocket::sendMail(m_strMailServer, m_nMailPort, m_strFrom,
m_strFromPassword, m_strMailTo, osSubject.str(), "You have started Mail Alert
System.");

    return true;
}

void MailMonitor::uninit()
{
    m_bExit = true;

    m_cvAlert.notify_one();

    if (m_spMailAlertThread->joinable())
        m_spMailAlertThread->join();
}

void MailMonitor::wait()
{
    if (m_spMailAlertThread->joinable())
        m_spMailAlertThread->join();
}

void MailMonitor::run()
{
    m_spMailAlertThread.reset(new std::thread(std::bind(&MailMonitor::alertThread,
this)));
}

void MailMonitor::alertThread()
{
    m_bRunning = true;

    while (true)
    {
        MailItem mailItem;
        {
            std::unique_lock<std::mutex> guard(m_mutexAlert);
            while (m_listMailItemsToSend.empty())
            {
                if (m_bExit)
                    return;

                m_cvAlert.wait(guard);
            }

            mailItem = m_listMailItemsToSend.front();
            m_listMailItemsToSend.pop_front();
        }
```

```cpp
        std::ostringstream osSubject;
        osSubject << "[" << m_strMailName << "]" << mailItem.subject;
        SmtpSocket::sendMail(m_strMailServer, m_nMailPort, m_strFrom,
m_strFromPassword, m_strMailTo, osSubject.str(), mailItem.content);
    }//end outer-while-loop

    m_bRunning = false;
}

bool MailMonitor::alert(const std::string& subject, const std::string& content)
{
    if (m_strMailServer.empty() || m_nMailPort < 0 || m_strFrom.empty() ||
m_strFromPassword.empty() || m_strMailTo.empty())
        return false;

    MailItem mailItem;
    mailItem.subject = subject;
    mailItem.content = content;

    {
        std::lock_guard<std::mutex> lock_guard(m_mutexAlert);
        m_listMailItemsToSend.push_back(mailItem);
        m_cvAlert.notify_one();
    }

    return true;
}

void MailMonitor::split(const std::string& str, std::vector<std::string>& v, const 
char* delimiter/* = "|"*/)
{
    if (delimiter == NULL || str.empty())
        return;

    std::string buf(str);
    size_t pos = std::string::npos;
    std::string substr;
    int delimiterlength = strlen(delimiter);
    while (true)
    {
        pos = buf.find(delimiter);
        if (pos != std::string::npos)
        {
            substr = buf.substr(0, pos);
            if (!substr.empty())
                v.push_back(substr);

            buf = buf.substr(pos + delimiterlength);
        }
        else
        {
            if (!buf.empty())
                v.push_back(buf);
            break;
        }
    }
}
```

}

我们在 main 函数中模拟生成一条新的报警邮件。

main.cpp：

```cpp
/**
 *  邮件报警 demo
 *  zhangyl 2020.04.09
 **/
#include <iostream>
#include <stdlib.h>
#include "Platform.h"
#include "MailMonitor.h"

//Winsock 网络库初始化
#ifdef WIN32
NetworkInitializer windowsNetworkInitializer;
#endif

#ifndef WIN32
void prog_exit(int signo)
{
    std::cout << "program recv signal [" << signo << "] to exit." << std::endl;

    //停止邮件发送服务
    MailMonitor::getInstance().uninit();
}

#endif

const std::string servername = "MailAlertSysem";
const std::string mailserver = "smtp.163.com";
const short mailport = 25;
const std::string mailuser = "testformybook@163.com";
const std::string mailpassword = "2019hhxxttxs";
const std::string mailto = "balloonwj@qq.com;analogous_love@qq.com";

int main(int argc, char* argv[])
{
#ifndef WIN32
    //设置信号处理
    signal(SIGCHLD, SIG_DFL);
    signal(SIGPIPE, SIG_IGN);
    signal(SIGINT, prog_exit);
    signal(SIGTERM, prog_exit);
#endif

    bool bInitSuccess = MailMonitor::getInstance().initMonitorMailInfo(servername,
mailserver, mailport, mailuser, mailpassword, mailto);
    if (bInitSuccess)
        MailMonitor::getInstance().run();

    const std::string subject = "Alert Mail";
    const std::string content = "This is an alert mail from " + mailuser;
    MailMonitor::getInstance().alert(subject, content);
```

```
    //等待邮件报警线程退出
    MailMonitor::getInstance().wait();

    return 0;
}
```

以上代码使用了 163 邮箱账号 testformybook@163.com 向 QQ 邮箱账户 balloonwj@qq.com 和 analogous_love@qq.com 分别发送了邮件，发送邮件的函数是 MailMonitor::alert()，实际发送邮件的函数是 SmtpSocket::send()。

在 Windows 或者 Linux 上编译运行程序时，我们的两个邮箱都会分别收到两封邮件，如下图所示。

产生第 1 封邮件的原因是我们在 main 函数中调用 MailMonitor::getInstance().initMonitorMailInfo 函数初始化邮箱服务器名、地址、端口号、用户名和密码时，MailMonitor::initMonitorMailInfo 函数内部会调用 SmtpSocket::sendMail 函数发送一封邮件，通知指定的联系人邮件报警系统已经启动：

```
bool MailMonitor::initMonitorMailInfo(const std::string& servername, const
std::string& mailserver, short mailport, const std::string& mailfrom, const
std::string& mailfromPassword, const std::string& mailto)
{
    //无关代码省略

    SmtpSocket::sendMail(m_strMailServer, m_nMailPort, m_strFrom,
m_strFromPassword, m_strMailTo, osSubject.str(), "You have started Mail Alert
System.");

    return true;
}
```

产生第 2 封邮件的原因是我们在 main 函数中主动调用了产生报警邮件的函数：

```
const std::string subject = "Alert Mail";
const std::string content = "This is an alert mail from " + mailuser;
MailMonitor::getInstance().alert(subject, content);
```

这里以第 1 封邮件为例来说明整个邮件发送过程中，我们的程序（客户端）与 163 邮件服务器之间协议数据的交换内容，核心的邮件发送功能在 SmtpSocket::sendMail 函数中实现：

```cpp
bool SmtpSocket::sendMail(const std::string& server, short port, const std::string&
from, const std::string& fromPassword,
                    const std::vector<std::string>& to, const std::string&
subject, const std::string& mailData)
{
    size_t atSymbolPos = from.find_first_of("@");
    if (atSymbolPos == std::string::npos)
        return false;

    std::string strUser = from.substr(0, atSymbolPos);

    SmtpSocket smtpSocket;
    //smtp.163.com 25
    if (!smtpSocket.connect(server.c_str(), port))
        return false;

    //testformybook 2019hhxxttxs
    if (!smtpSocket.logon(strUser.c_str(), fromPassword.c_str()))
        return false;

    //testformybook@163.com
    if (!smtpSocket.setMailFrom(from.c_str()))
        return false;

    if (!smtpSocket.setMailTo(to))
        return false;

    if (!smtpSocket.send(subject, mailData))
        return false;

    return true;
}
```

这个函数先创建 socket，再使用指定的地址和端口号连接服务器（smtpSocket.connect 函数内部额外做了一步将域名解析成 IP 地址的工作），连接成功后开始和服务端进行数据交换：

```
client: 尝试连接服务器
client: 连接成功
server: 220\r\n

client: HELO 自定义问候语\r\n
server: 250\r\n

client: AUTH LOGIN\r\n
server: 334\r\n

client: base64 编码后的用户名\r\n
server: 334\r\n

client: base64 编码后的密码\r\n
server: 235\r\n

client: MAIL FROM:<发件人地址>\r\n
server: 250\r\n
```

```
client: rcpt to:<收件人地址 1>\r\n
server: 250\r\n

client: rcpt to:<收件人地址 2>\r\n
server: 250\r\n

client: DATA\r\n
server: 354\r\n

client: 邮件正文\r\n.\r\n
server: 250\r\n

client: QUIT\r\n
server: 221\r\n
```

上述过程如下图所示。

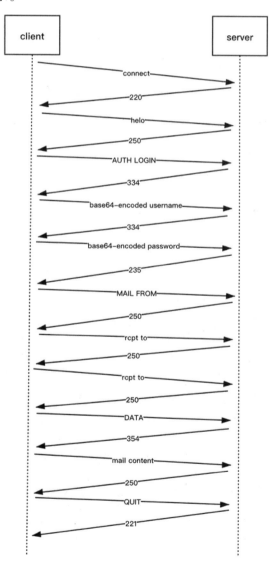

邮件最终发送出去了。这里模拟了客户端使用 SMTP 向服务端发送邮件，我们自己实现服务端接收客户端发送的邮件请求也是一样的道理。这就是 SMTP 的格式。SMTP 是以特定标记作为分隔符的协议格式的典型。

> 我们可以在 Windows 或者 Linux 主机上测试上述程序。为了避免在网络上产生大量垃圾邮件，阿里云等云主机服务商默认禁止将数据发往其他服务器 25 号端口。所以在阿里云等云主机上测试时，需要申请解除该限制，或者将邮件服务器默认使用的 25 号端口改为其他端口（一般改为 465 端口）。

前面介绍了 SMTP 常用的协议命令，SMTP 支持的完整命令列表可以参考 rfc5321 文档。

6.11.3　POP3

再来看看 POP3：

```
client: 尝试连接邮箱 POP 服务器，连接成功
server: +OK Welcome to coremail Mail Pop3 Server
(163coms[10774b260cc7a37d26d71b52404dcf5cs])\r\n

client: USER 用户名\r\n
server: +OK core mail

client: PASS 密码\r\n
server: +OK 202 message(s) [3441786 byte(s)]\r\n

client: LIST\r\n
server: +OK 5 30284\r\n1 8284\r\n2 11032\r\n3 2989\r\n4 3871\r\n5 4108\r\n.\r\n

client: RETR 100\r\n
server:
+OK 4108 octets\r\n
Received: from sonic310-21.consmr.mail.gq1.yahoo.com (unknown [98.137.69.147])
 by mx29 (Coremail) with SMTP id T8CowABHlztmml5erAoHAQ--.23443S3;
Wed, 04 Mar 2020 01:56:57 +0800 (CST)\r\n
DKIM-Signature: v=1; a=rsa-sha256; c=relaxed/relaxed; d=yahoo.com; s=s2048;
 t=1583258213; bh=ABL3sF+YL/syl+mwknwxiAlvKPRNYq4AYTujNrPA86g=;
 h=Date:From:Reply-To:Subject:References:From:Subject;\r\n
 b=OrAQTs0GJnA
...省略部分内容...
6Mzu2lmr07WwMCE7wgqwOSWRnYNCz2rWcLmXA_TVDtdJ85
bHZ79FY6Vs5pGJjp.7YgDnVqysBp95w--\r\n
Received: from sonic.gate.mail.ne1.yahoo.com by sonic310.consmr.mail.gq1.yahoo.com
 with HTTP; Tue, 3 Mar 2020 17:56:53 +0000\r\n
Date: Tue, 3 Mar 2020 17:56:49 +0000 (UTC)\r\n
From: Peter Edward Copley <noodlelife@yahoo.com>\r\n
Reply-To: Peter Edward Copley <pshun3592@gmail.com>\r\n
Message-ID: <729348196.5391236.1583258209467@mail.yahoo.com>\r\n
Subject: Re:Hello\r\n
MIME-Version: 1.0\r\n
```

```
Content-Type: multipart/alternative; \r\n
boundary="----=_Part_5391235_1821490954.1583258209466"\r\n
References: <729348196.5391236.1583258209467.ref@mail.yahoo.com>\r\n
X-Mailer: WebService/1.1.15302 YMailNorrin Mozilla/5.0 (Windows NT 10.0; Win64; x64)
AppleWebKit/537.36 (KHTML, like Gecko) Chrome/80.0.3987.122 Safari/537.36\r\n
X-CM-TRANSID:T8CowABHlztmml5erAoHAQ--.23443S3\r\n
Authentication-Results: mx29; spf=pass smtp.mail=noodlelife@yahoo.com;\r\n
dkim=pass header.i=@yahoo.com\r\n
X-Coremail-Antispam: 1Uf129KBjDUn29KB7ZKAUJUUUUU529EdanIXcx71UUUUU7v73
VFW2AGmfu7bjvjm3AaLaJ3UbIYCTnIWIevJa73UjIFyTuYvjxU-NtIUUUUU\r\n
\r\n
------=_Part_5391235_1821490954.1583258209466\r\n
.\r\n

client: QUIT\r\n
server: +OK core mail\r\n
```

上述过程如下图所示。

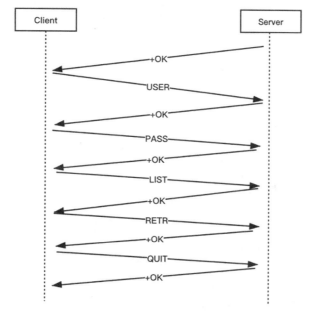

我们在收取邮件正文之后，就可以根据邮件正文中的各种 tag 解析邮件的内容，以得到邮件的 MessageID、收件人、发件人、邮件主题、正文和附件。注意，附件的内容也会被拆成特定的编码格式放在邮件中。邮件正文按 boundary 分成多个块，例如上文中的 boundary="----=_Part_5391235_1821490954.1583258209466"。

来看一个具体的例子。

上述邮件的主题是"测试邮件",内容是纯文本"这是一封测试邮件,含有两个附件。"其中还有两个附件,一个是 self.jpg 图片,一个是 test.docx 文档。将邮件下载后得到的邮件原文如下:

```
+OK 93763 octets
Received: from qq.com (unknown [183.3.226.165])
        by mx27 (Coremail) with SMTP id TcCowABHJo+dMqReI+72Bg--.18000S3;
        Sat, 25 Apr 2020 20:52:45 +0800 (CST)
DKIM-Signature: v=1; a=rsa-sha256; c=relaxed/relaxed; d=qq.com; s=s201512;
        t=1587819165; bh=RLNDml5+GusG7KQTgkjeS/Mpn1m/LmqBUaz6Nmo6ukY=;
        h=From:To:Subject:Mime-Version:Date:Message-ID;
        b=K3sJK+aPQ9zHu1GUvKckofm3cfocpze10XBp9FufVVVYS423myQnFWMaREpGGbeaS
         vrCGdjawcfhXpkvGZnhOkJZrtut1er5zWZRkmsDnqvoekRURXKt3wWyOv5WUuSPHZI
         NzGjMQbtYmbWjFla7zs1Cg81UQKRtg1s5KxWwGVQ=
X-QQ-FEAT: CPmoSFXLZ/TSSc3nxNJn8bUc57myjtkH8mxkmSC9/G9nP1mNDXcYVAAERmmiE
        038rlXj8w6qkTmh1317bdJp9MqMMEUSgpJC5DulJn4k6WCURo4NEYDiuUQK/J+YfUQnpETt
        w4aQYpj6nKAIqKgorGGK0zy6oQWavfOgssyvSU15d6wqlw904x6aZhS3KAUAM4+eGitBRk9
        fxUEABnV/opGuLtZ/fex+UsUAVgXFbTZPoYjhxoM4ZKJsDEJ38x/9QHR1FymBebmAvNzzbB
        JT45M4OYwynKE/mrFR1FPSeXA=
X-QQ-SSF: 00010000000000F0000000000000000Z
X-HAS-ATTACH: no
X-QQ-BUSINESS-ORIGIN: 2
X-Originating-IP: 255.21.142.175
X-QQ-STYLE:
X-QQ-mid: webmail504t1587819163t7387219
From: "=?gb18030?B?1/PRp7fG?=" <balloonwj@qq.com>
To: "=?gb18030?B?dGVzdGZvcm15Ym9vaw==?=" <testformybook@163.com>
Subject: =?gb18030?B?suLK1NPKvP4=?=
Mime-Version: 1.0
Content-Type: multipart/mixed;
        boundary="----=_NextPart_5EA4329B_0FBAC2B8_51634C9D"
Content-Transfer-Encoding: 8Bit
Date: Sat, 25 Apr 2020 20:52:43 +0800
X-Priority: 3
Message-ID: <tencent_855A7727508F28D762951979338305E06B08@qq.com>
```

```
X-QQ-MIME: TCMime 1.0 by Tencent
X-Mailer: QQMail 2.x
X-QQ-Mailer: QQMail 2.x
X-QQ-SENDSIZE: 520
Received: from qq.com (unknown [127.0.0.1])
    by smtp.qq.com (ESMTP) with SMTP
    id ; Sat, 25 Apr 2020 20:52:44 +0800 (CST)
Feedback-ID: webmail:qq.com:bgweb:bgweb16
X-CM-TRANSID:TcCowABHJo+dMqReI+72Bg--.18000S3
Authentication-Results: mx27; spf=pass smtp.mail=balloonwj@qq.com; dki
    m=pass header.i=@qq.com
X-Coremail-Antispam: 1Uf129KBjDUn29KB7ZKAUJUUUUU529EdanIXcx71UUUUU7v73
    VFW2AGmfu7bjvjm3AaLaJ3UbIYCTnIWIevJa73UjIFyTuYvjxU-LIDUUUUU

This is a multi-part message in MIME format.

------=_NextPart_5EA4329B_0FBAC2B8_51634C9D
Content-Type: multipart/alternative;
    boundary="----=_NextPart_5EA4329B_0FBAC2B8_71508FA9";


------=_NextPart_5EA4329B_0FBAC2B8_71508FA9
Content-Type: text/plain;
    charset="gb18030"
Content-Transfer-Encoding: base64

1eLKx9K7t+Ky4srU08q8/qOsuqzT0MG9uPa4vbz+oaM=

------=_NextPart_5EA4329B_0FBAC2B8_71508FA9
Content-Type: text/html;
    charset="gb18030"
Content-Transfer-Encoding: base64

PG1ldGEgaHR0cC11cXVpdj0iQ29udGVudC1UeXBlIiBjb250ZW50PSJ0ZXh0L2h0bWw7IGNo
YXJzZXQ9R0IxODAzMCI+PGRpdj7V4srH0ru34rLiytTTyrz+o6y6rNPQwb249ri9vP6hozwv
ZGl2Pg==

------=_NextPart_5EA4329B_0FBAC2B8_71508FA9--

------=_NextPart_5EA4329B_0FBAC2B8_51634C9D
Content-Type: application/octet-stream;
    charset="gb18030";
    name="self.jpg"
Content-Disposition: attachment; filename="self.jpg"
Content-Transfer-Encoding: base64

/9j/4AAQSkZJRgABAQEAYABgAAD/2wBDAAgGBgcGBQgHBwcJCQgKDBQNDAsLDBkSEw8UHRof
（限于篇幅，这里省去了部分内容）
RzNNWJfvT1Kz+LvDctsBHqW0p0Bik/8AiKrr4v0HH/H+P+/Mn+FFFZz3Gf/Z

------=_NextPart_5EA4329B_0FBAC2B8_51634C9D
Content-Type: application/octet-stream;
    charset="gb18030";
    name="test.docx"
Content-Disposition: attachment; filename="test.docx"
Content-Transfer-Encoding: base64
```

```
UEsDBBQABgAIAAAAIQCshlBXjgEAAMAFAAATAAgCW0NvbnRlbnRfVHlwZXNdLnhtbCCiBAIo
（限于篇幅，这里省去了部分内容）
AQAAxQIAABAAAAAAAAAAAAAAAAAAcb8AAGRvY1Byb3BzL2FwcC54bWxQSwUGAAAAAA0ADQBM
AwAAF8IAAAAA

------=_NextPart_5EA4329B_0FBAC2B8_51634C9D--
.
```

如何解析这样的邮件格式呢？

这封邮件的内容主要由两部分组成，第 1 部分是 OK 关键字，第 2 部分是邮件的内容，邮件的内容以点+\r\n 结束。其中，邮件内容前面的一部分是一个 tag 和 tag 值，我们可以从这些 tag 中得到邮件的 MessageID、收件人姓名和地址、发件人姓名和地址、邮件主题，例如：

```
From: "=?gb18030?B?1/PRp7fG?=" <balloonwj@qq.com>
To: "=?gb18030?B?dGVzdGZvcm15Ym9vaw==?=" <testformybook@163.com>
Subject: =?gb18030?B?suLK1NPKvP4=?=
Date: Sat, 25 Apr 2020 20:52:43 +0800
Message-ID: <tencent_855A7727508F28D762951979338305E06B08@qq.com>
```

其中，邮件的收发人姓名（From 和 To）使用了 base64 编码，我们对其使用 base64 解码即可还原其内容：

```
Content-Type: multipart/mixed; 说明邮件由多个部分组成。
```

我们先根据 boundary="----=_NextPart_5EA4329B_0FBAC2B8_71508FA9";中指定的 ----=_NextPart_5EA4329B_0FBAC2B8_71508FA9 分隔符得到除邮件附件内容外的邮件正文内容，一共有两段。

正文段一：

```
Content-Type: text/plain;
    charset="gb18030"
Content-Transfer-Encoding: base64

1eLKx9K7t+Ky4srU08q8/qOsuqzT0MG9uPa4vbz+oaM=
```

这段内容为纯文本格式（text/plain），使用 base64 编码，字符集格式为 gb18030，解码之后得到正文，如下图所示。

正文段二：

```
Content-Type: text/html;
    charset="gb18030"
Content-Transfer-Encoding: base64

PG11dGEgaHR0cC11cXVpdj0iQ29udGVudC1UeXBlIiBjb250ZW50PSJ0ZXh0L2h0bWw7IGNo
YXJzZXQ9R0IxODAzMCI+PGRpdj7V4srH0ru34rLiytTTyrz+o6y6rNPQwb249ri9vP6hozwv
ZGl2Pg==
```

这段内容为富文本格式（text/html），采用 base64 编码，字符集格式为 gb18030，解码之后得到正文，即邮件中带超级链接的英语广告，是 163 邮件服务器自动将其插入邮件正文中的，如下图所示。

接下来就是两个附件的内容了，使用的编码格式也是 base64。我们对其使用 base64 解码，将其还原成 ASCII 字节流后作为文件的内容，再取 tag 中附件的文件名生成对应的文件，即可还原附件。

6.11.4　邮件客户端

除了上面说的三种协议，有的邮件还使用了 Exchange 协议，可以参考微软官方关于 Exchange 协议的说明文档。

理解上述邮件协议之后，我们就可以编写自己的邮件客户端了，而且可以自由定制邮件的展示功能（例如在收到的邮件内部插入自定义的英语广告）。

6.12　WebSocket 协议

WebSocket 协议是用于解决 HTTP 通信的无状态、短连接（通常是）和服务端无法主动向客户端推送数据等问题而开发的新型协议，其通信基础也基于 TCP。由于较旧的浏览器可能不支持 WebSocket 协议，所以使用 WebSocket 协议的通信双方在进行 TCP 三次握手之后，还要额外地进行一次握手，这次参与握手的通信双方的报文格式是基于 HTTP 改造的。

6.12.1　WebSocket 协议的握手过程

本节介绍 WebSocket 协议通信中的最后一次握手过程。最后一次握手开始后，一方向另外一方发送了一个 HTTP 格式的报文，这个报文的格式大致如下：

```
GET /realtime HTTP/1.1\r\n
Host: 127.0.0.1:9989\r\n
Connection: Upgrade\r\n
Pragma: no-cache\r\n
Cache-Control: no-cache\r\n
User-Agent: Mozilla/5.0 (Windows NT 10.0; Win64; x64)\r\n
Upgrade: websocket\r\n
Origin: http://xyz.com\r\n
Sec-WebSocket-Version: 13\r\n
Accept-Encoding: gzip, deflate, br\r\n
Accept-Language: zh-CN,zh;q=0.9,en;q=0.8\r\n
Sec-WebSocket-Key: IqcAWodjyPDJuhGgZwkpKg==\r\n
Sec-WebSocket-Extensions: permessage-deflate; client_max_window_bits\r\n
\r\n
```

对这个格式有如下要求：

（1）握手必须是一个有效的 HTTP 请求；

（2）请求的方法必须是 GET，且 HTTP 版本必须是 1.1；

（3）请求必须包含 Host 字段信息；

（4）请求必须包含 Upgrade 字段信息，值必须为 websocket；

（5）请求必须包含 Connection 字段信息，值必须为 Upgrade；

（6）请求必须包含 Sec-WebSocket-Key 字段，该字段值是客户端的标识编码成 base64 格式后的内容；

（7）请求必须包含 Sec-WebSocket-Version 字段信息，值必须是 13；

（8）请求必须包含 Origin 字段；

（9）请求可能包含 Sec-WebSocket-Protocol 字段来规定子协议；

（10）请求可能包含 Sec-WebSocket-Extensions 字段来规定协议的扩展；

（11）请求可能包含其他字段如 cookie 等。

对端在收到该数据包后如果支持 WebSocket 协议，则会回复一个 HTTP 格式的应答，这个应答报文的格式大致如下：

```
HTTP/1.1 101 Switching Protocols\r\n
Upgrade: websocket\r\n
Connection: Upgrade\r\n
Sec-WebSocket-Accept: 5wC5L6joP6tl31zpj9OlCNv9Jy4=\r\n
\r\n
```

上面列出了应答报文中必须包含的几个字段和对应的值,即 Upgrade、Connection、Sec-WebSocket-Accept,注意:第 1 行必须是 HTTP/1.1 101 Switching Protocols\r\n。

对于 Sec-WebSocket-Accept 字段,其值是根据对端传过来的 Sec-WebSocket-Key 的值经过一定的算法计算出来的,这样应答的双方才能匹配。算法流程如下:

(1)将 Sec-WebSocket-Key 的值与固定的字符串 "258EAFA5-E914-47DA-95CA-C5AB0DC85B11" 拼接;

(2)将拼接后的字符串进行 SHA-1 处理,然后将结果进行 base64 编码。

算法公式如下:

```
mask   = "258EAFA5-E914-47DA-95CA-C5AB0DC85B11";   //这是算法中要用到的固定字符串
accept = base64( sha1( Sec-WebSocket-Key + mask ) );
```

这里用 C++实现了该算法:

```cpp
namespace uWS {

struct WebSocketHandshake {
    template <int N, typename T>
    struct static_for {
        void operator()(uint32_t *a, uint32_t *b) {
            static_for<N - 1, T>()(a, b);
            T::template f<N - 1>(a, b);
        }
    };

    template <typename T>
    struct static_for<0, T> {
        void operator()(uint32_t *a, uint32_t *hash) {}
    };

    template <int state>
    struct Sha1Loop {
        static inline uint32_t rol(uint32_t value, size_t bits) {return (value << bits) | (value >> (32 - bits));}
        static inline uint32_t blk(uint32_t b[16], size_t i) {
            return rol(b[(i + 13) & 15] ^ b[(i + 8) & 15] ^ b[(i + 2) & 15] ^ b[i], 1);
        }

        template <int i>
        static inline void f(uint32_t *a, uint32_t *b) {
            switch (state) {
            case 1:
                a[i % 5] += ((a[(3 + i) % 5] & (a[(2 + i) % 5] ^ a[(1 + i) % 5])) ^ a[(1 + i) % 5]) + b[i] + 0x5a827999 + rol(a[(4 + i) % 5], 5);
                a[(3 + i) % 5] = rol(a[(3 + i) % 5], 30);
                break;
            case 2:
                b[i] = blk(b, i);
                a[(1 + i) % 5] += ((a[(4 + i) % 5] & (a[(3 + i) % 5] ^ a[(2 + i) % 5])) ^ a[(2 + i) % 5]) + b[i] + 0x5a827999 + rol(a[(5 + i) % 5], 5);
```

```cpp
                    a[(4 + i) % 5] = rol(a[(4 + i) % 5], 30);
                    break;
                case 3:
                    b[(i + 4) % 16] = blk(b, (i + 4) % 16);
                    a[i % 5] += (a[(3 + i) % 5] ^ a[(2 + i) % 5] ^ a[(1 + i) % 5]) + b[(i + 4) % 16] + 0x6ed9eba1 + rol(a[(4 + i) % 5], 5);
                    a[(3 + i) % 5] = rol(a[(3 + i) % 5], 30);
                    break;
                case 4:
                    b[(i + 8) % 16] = blk(b, (i + 8) % 16);
                    a[i % 5] += (((a[(3 + i) % 5] | a[(2 + i) % 5]) & a[(1 + i) % 5]) | (a[(3 + i) % 5] & a[(2 + i) % 5])) + b[(i + 8) % 16] + 0x8f1bbcdc + rol(a[(4 + i) % 5], 5);
                    a[(3 + i) % 5] = rol(a[(3 + i) % 5], 30);
                    break;
                case 5:
                    b[(i + 12) % 16] = blk(b, (i + 12) % 16);
                    a[i % 5] += (a[(3 + i) % 5] ^ a[(2 + i) % 5] ^ a[(1 + i) % 5]) + b[(i + 12) % 16] + 0xca62c1d6 + rol(a[(4 + i) % 5], 5);
                    a[(3 + i) % 5] = rol(a[(3 + i) % 5], 30);
                    break;
                case 6:
                    b[i] += a[4 - i];
            }
        }
    };

    //sha1函数的实现
    static inline void sha1(uint32_t hash[5], uint32_t b[16]) {
        //为了节省篇幅，该实现参见随书代码
    }

    //base64编码函数
    static inline void base64(unsigned char *src, char *dst) {
        //为了节省篇幅，该实现参见随书代码
    }

public:
    /**
     * 生成Sec-WebSocket-Accept算法
     * @param input 对端传过来的Sec-WebSocket-Key值
     * @param output 存放生成的Sec-WebSocket-Accept值
     */
    static inline void generate(const char input[24], char output[28]) {
        uint32_t b_output[5] = {
            0x67452301, 0xefcdab89, 0x98badcfe, 0x10325476, 0xc3d2e1f0
        };
        uint32_t b_input[16] = {
            0, 0, 0, 0, 0, 0, 0x32353845, 0x41464135, 0x2d453931, 0x342d3437,
0x44412d39,
            0x3543412d, 0x43354142, 0x30444338, 0x35423131, 0x80000000
        };

        for (int i = 0; i < 6; i++) {
            b_input[i] = (input[4 * i + 3] & 0xff) | (input[4 * i + 2] & 0xff) << 8
```

```
        | (input[4 * i + 1] & 0xff) << 16 | (input[4 * i + 0] & 0xff) << 24;
    }
    sha1(b_output, b_input);
    uint32_t last_b[16] = {0, 0, 0, 0, 0, 0, 0, 0, 0, 0, 0, 0, 0, 0, 0, 480};
    sha1(b_output, last_b);
    for (int i = 0; i < 5; i++) {
        uint32_t tmp = b_output[i];
        char *bytes = (char *) &b_output[i];
        bytes[3] = tmp & 0xff;
        bytes[2] = (tmp >> 8) & 0xff;
        bytes[1] = (tmp >> 16) & 0xff;
        bytes[0] = (tmp >> 24) & 0xff;
    }
    base64((unsigned char *) b_output, output);
};
```

握手完成之后，通信双方就可以保持连接并相互发送数据了。

6.12.2　WebSocket 协议的格式

WebSocket 协议除了上文提到的握手过程中使用的是 HTTP 数据格式，之后的通信双方使用的是另一种自定义格式。每个 WebSocket 数据包都被称为一个 Frame（帧），其格式如下：

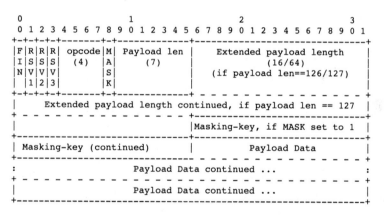

下面逐一介绍上文中各字段的含义。

第 1 字节（上图中第 1 个数字 0 至第 1 个数字 7）的内容如下。

（1）FIN 标志：占第 1 字节中的第 1 位（bit），即第 1 字节的最高位（1 字节含有 8 位（bit）），该标志置 0 时表示当前包未结束，后续有该包的分片；置 1 时表示当前包已结束，后续无该包的分片。我们在解包时如果发现该标志为 0，则需要将当前包的包体数据（即 Payload Data）缓存起来，与后续的包分片组装在一起才是一个完整的包体数据。

（2）RSV1、RSV2、RSV3：每个各占 1 位，一共 3 位，这 3 位是保留字段（默认都是 0），可以用它们作为通信双方协商好的一些特殊标志。

（3）opCode：操作类型，占 4 位。目前的操作类型及其取值如下：

```cpp
//4 bits
enum OpCode
{
    //表示后续还有新的 Frame
    CONTINUATION_FRAME      = 0x0,
    //包体是文本类型的 Frame
    TEXT_FRAME              = 0x1,
    //包体是二进制类型的 Frame
    BINARY_FRAME            = 0x2,
    //保留值
    RESERVED1               = 0x3,
    RESERVED2               = 0x4,
    RESERVED3               = 0x5,
    RESERVED4               = 0x6,
    RESERVED5               = 0x7,
    //建议对端关闭的 Frame
    CLOSE                   = 0x8,
    //心跳包中的 ping Frame
    PING                    = 0x9,
    //心跳包中的 pong Frame
    PONG                    = 0xA,
    //保留值
    RESERVED6               = 0xB,
    RESERVED7               = 0xC,
    RESERVED8               = 0xD,
    RESERVED9               = 0xE,
    RESERVED10              = 0xF
};
```

第 2 字节（上图中第 1 个数字 8 至第 2 个数字 5）的内容如下。

（1）MASK 标志：占 1 位（bit），该标志为 1 时，表明该 Frame 在包体长度字段后面携带 4 字节的 masking-key 信息，为 0 时则没有 masking-key 信息。

（2）Payload len：占 7 位，表示包体的长度信息。由于 Payload len 值使用了 1 字节的低 7 位（7bit），因此其能表示的长度范围是 0～127，其中 126 和 127 被当作特殊标志使用。当该字段值是 0～125 时，表示跟在 masking-key 字段后面的就是包体内容长度；当该值是 126 时，接下来两字节的内容表示跟在 masking-key 字段后面的包体内容长度（即图中的 Extended Payload Length）。由于两字节表示的最大无符号整数是 0xFFFF（十进制是 65535，因此编译器提供了一个宏 UINT16_MAX 来表示这个值）。如果包体长度超过 65535，包长度就记录不下了，此时应该将 Payload len 设置为 127，以使用更多的字节数表示包体长度。当 Payload len 是 127 时，接下来用 8 字节的内容表示跟在 masking-key 字段后面的包体内容长度（Extended Payload Length）。总结起来，若 Payload len 在 0～125 之间，则 Extended Payload Len 不存在，即占 0 字节；若 Payload len 等于 126，则 Extended Payload Length 占两字节；若 Payload len 等于 127，则 Extended Payload Length 占 8 字节。另外，若 Payload len 等于 125 或 126，则接下来存储实际包长信息分别使用两字节或 8 字节，其值必须被转换为网络字节序（Big Endian）。

（3）Masking-key：如果前面的 mask 标志被设置成 1，则该字段存在，占 4 字节；反之，在 Frame 中不存在存储 Masking-key 字段的字节。另外，客户端（主动发起握手请求的一方）向服务端（被动接受握手的另一方）发送的 frame 信息（包信息）中的 mask 标志必须是 1，而服务端向客户端发送的 frame 信息中的 mask 标志是 0，客户端发送给服务端的数据帧中存在 4 字节的 masking-key，而服务端发送给客户端的数据帧中不存在 Masking-key 信息（笔者在 Websocket 协议的 RFC 文档中并没有看到有这种强行规定，在研究了一些 Websocket 库的实现后发现，此结论并不一定成立，客户端发送的数据也可能没有被设置 mask 标志）。如果存在 Masking-key 信息，则数据帧中的数据（图中的 Payload Data）都是与 Masking-key 运算后的内容。无论是将原始数据与 Masking-key 运算后得到传输的数据，还是将传输的数据还原成原始数据，其算法都如下。

假设：original-octet-i 为原始数据的第 i 字节；transformed-octet-i 为转换后的数据的第 i 字节；j 为 i mod 4 的结果；masking-key-octet-j 为 mask key 的第 j 字节。original-octet-i 与 masking-key-octet-j 异或后，得到 transformed-octet-i：

```
j = i MOD 4
transformed-octet-i = original-octet-i XOR masking-key-octet-j
```

用 C++实现该算法：

```cpp
/**
 * @param src 函数调用前是原始需要传输的数据，函数调用后是 mask 或者 unmask 后的内容
 * @param maskingKey 4 字节
 */
void maskAndUnmaskData(std::string& src, const char* maskingKey)
{
    char j;
    for (size_t n = 0; n < src.length(); ++n)
    {
        j = n % 4;
        src[n] = src[n] ^ maskingKey[j];
    }
}
```

举个例子，假设有一个客户端发送给服务器的数据包，mask=1，即存在 4 字节的 masking-key，则当包体数据长度为 0～125 时，该包的结构如下，包头总计 6 字节：

```
第 1 字节第 0 位      => FIN
第 1 字节第 1～3 位   => RSV1 + RSV2 + RSV3
第 1 字节第 4～7 位   => opcode
第 2 字节第 0 位      => mask（等于 1）
第 2 字节第 1～7 位   => 包体长度
第 3～6 字节          =>  masking-key
第 7 字节及以后       =>  包体内容
```

当包体数据长度大于 125 且小于或等于 UINT16_MAX 时，该包的结构如下，包头总共 8 字节：

```
第 1 字节第 0 位      => FIN
第 1 字节第 1～3 位   => RSV1 + RSV2 + RSV3
第 1 字节第 4～7 位   => opcode
第 2 字节第 0 位      => mask（等于 1）
第 2 字节第 1～7 位   => 开启扩展包头长度标志，值为 126
```

第 3~4 字节　　　　　=>　包头长度
第 5~8 字节　　　　　=>　masking-key
第 9 字节及以后　　　=>　包体内容

当包体数据长度大于 UINT16_MAX 时，该包的结构如下，包头总共 14 字节：

第 1 字节第 0 位　　　=>　FIN
第 1 字节第 1~3 位　　=>　RSV1 + RSV2 + RSV3
第 1 字节第 4~7 位　　=>　opcode
第 2 字节第 0 位　　　=>　mask（等于 1）
第 2 字节第 1~7 位　　=>　开启扩展包头长度标志，值为 127
第 3~10 字节　　　　　=>　包头长度
第 11~14 字节　　　　 =>　masking-key
第 15 字节及以后　　　=>　包体内容

由于存储包体的长度使用了 8 字节存储（无符号），因此最大包体长度是 0xFFFFFFFFFFFFFFFF，这是一个非常大的数字，但在实际开发中用不到这么长的包体，且当包体超过一定值时就应该分包（分片）了。

所以，分包的逻辑很简单：假设将一个包分成 3 个包片，那么应该将第 1 个和第 2 个包片第 1 字节的第 1 位 FIN 设置为 0，将 OpCode 设置为 CONTINUATION_FRAME（也是 0）；将第 3 个包片 FIN 设置为 1，表示该包至此结束，将 OpCode 设置为想要的类型（如 TEXT_FRAME、BINARY_FRAME 等）。对端收到该包时，如果发现标志 FIN = 0 或 OpCode = 0，则将该包包体的数据暂存起来，直到收到 FIN = 1、OpCode ≠ 0 的包，将该包的数据与前面收到的数据放在一起，组装成一个完整的业务数据。示例代码如下：

```
//某次解包后得到包体 payloadData，根据 FIN 标志判断，
//如果 FIN = true，则说明一个完整的业务数据包已经收完整了
//调用 processPackage 函数处理该业务数据
//否则暂存于 m_strParsedData 中
//每次处理完一个完整的业务包数据，就将暂存区 m_strParsedData 中的数据清空
if (FIN)
{
    m_strParsedData.append(payloadData);
    processPackage(m_strParsedData);
    m_strParsedData.clear();
}
else
{
    m_strParsedData.append(payloadData);
}
```

6.12.3　WebSocket 协议的压缩格式

WebSocket 协议对于包体也支持压缩，是否需要开启压缩，需要通信双方在握手时进行协商。再看一下握手时主动发起一方的包内容：

```
GET /realtime HTTP/1.1\r\n
Host: 127.0.0.1:9989\r\n
Connection: Upgrade\r\n
Pragma: no-cache\r\n
Cache-Control: no-cache\r\n
User-Agent: Mozilla/5.0 (Windows NT 10.0; Win64; x64)\r\n
```

```
Upgrade: websocket\r\n
Origin: http://xyz.com\r\n
Sec-WebSocket-Version: 13\r\n
Accept-Encoding: gzip, deflate, br\r\n
Accept-Language: zh-CN,zh;q=0.9,en;q=0.8\r\n
Sec-WebSocket-Key: IqcAWodjyPDJuhGgZwkpKg==\r\n
Sec-WebSocket-Extensions: permessage-deflate; client_max_window_bits\r\n
\r\n
```

在该包中，在 Sec-WebSocket-Extensions 字段中有一个值 permessage-deflate，这是 Websocket 数据包是否支持压缩功能的标志。如果发起方支持压缩功能，则在发起握手的数据包中添加该标志，若对端收到后也支持压缩功能，则在应答的包中也设置该标志；如果没有该标志，则表示不支持压缩。例如：

```
HTTP/1.1 101 Switching Protocols\r\n
Upgrade: websocket\r\n
Connection: Upgrade\r\n
Sec-WebSocket-Accept: 5wC5L6joP6tl31zpj9OlCNv9Jy4=\r\n
Sec-WebSocket-Extensions: permessage-deflate; client_no_context_takeover
\r\n
```

如果双方都支持压缩，则此后通信包中的包体部分都是经过压缩的，反之是未经过压缩的。在解包得到包体（即 Payload Data）后，如果握手时有压缩标志并且对方也回复了支持压缩，则需要对该包体进行解压缩；同理，在发送数据之前组装产生 WebSocket 包时，需要先对包体（即 Payload Data）进行压缩。

收包后解压缩的示例代码如下：

```cpp
bool MyWebSocketSession::processPackage(const std::string& data)
{
    std::string out;
    //m_bClientCompressed 在握手时确定是否支持压缩
    if (m_bClientCompressed)
    {
        //解压缩
        if (!ZlibUtil::inflate(data, out))
        {
            LOGE("uncompress failed, dataLength: %d", data.length());
            return false;
        }
    }
    else
        out = data;

    //如果不需要解压缩，则 out=data，反之 out 是解压缩后的数据
    LOGI("receid data: %s", out.c_str());


    return Process(out);
}
```

对包进行压缩的算法如下：

```cpp
size_t dataLength = data.length();
std::string destbuf;
if (m_bClientCompressed)
{
    //按需压缩
    if (!ZlibUtil::deflate(data, destbuf))
    {
        LOGE("compress buf error, data: %s", data.c_str());
        return;
    }
}
else
    destbuf = data;

LOGI("destbuf.length(): %d", destbuf.length());
```

压缩和解压缩算法即 gzip 压缩算法。压缩和解压缩的函数为 zlib 库的 deflate（压缩）和 inflate（解压缩）函数，具体内容请参见随书配套代码。

注意，在使用 zlib 的 deflate 函数进行压缩时，压缩完毕后要将压缩后的字节流末尾多余的 4 字节删掉，这是因为 deflate 函数在压缩后会将内容为 00 00 ff ff 的特殊标志放入压缩后的缓冲区中，这个标志不是我们需要的正文内容。

6.12.4　WebSocket 协议装包与解包示例

这里以服务端发送给客户端的装包代码为例，根据上文所述，服务端发包时不需要设置 mask 标志，因此包中不需要填充 4 字节的 Masking-key，所以也不需要对包体内容进行 mask 运算（以下代码没有为大包进行分片）：

```cpp
void MyWebSocketSession::send(const std::string& data, int32_t opcode/* = MyOpCode::TEXT_FRAME*/, bool compress/* = false*/)
{
    //data 是待发送的业务数据
    size_t dataLength = data.length();
    std::string destbuf;
    //按需压缩
    if (m_bClientCompressed && dataLength > 0)
    {
        if (!ZlibUtil::deflate(data, destbuf))
        {
            LOGE("compress buf error, data: %s", data.c_str());
            return;
        }
    }
    else
        destbuf = data;

    LOGI("destbuf.length(): %d", destbuf.length());

    dataLength = destbuf.length();

    char firstTwoBytes[2] = { 0 };
    //设置分片标志 FIN
```

```cpp
    firstTwoBytes[0] |= 0x80;

    //设置 opcode
    firstTwoBytes[0] |= opcode;

    const char compressFlag = 0x40;
    if (m_bClientCompressed)
        firstTwoBytes[0] |= compressFlag;

    //mask = 0;
    //实际发送的数据包
    std::string actualSendData;

    //包体长度小于 126,不使用扩展的包体长度字节
    if (dataLength < 126)
    {
        firstTwoBytes[1] = dataLength;
        actualSendData.append(firstTwoBytes, 2);
    }
    //包体长度大于或等于 126 且小于 UINT16_MAX,使用两字节的扩展包体长度
    else if (dataLength <= UINT16_MAX)    //两字节的无符号整数最大数的值(65535)
    {
        firstTwoBytes[1] = 126;
        char extendedPlayloadLength[2] = { 0 };
        //转换为网络字节序
        uint16_t tmp = ::htons(dataLength);
        memcpy(&extendedPlayloadLength, &tmp, 2);
        actualSendData.append(firstTwoBytes, 2);
        actualSendData.append(extendedPlayloadLength, 2);
    }
    //包体长度大于 UINT16_MAX,使用 8 字节的扩展包体长度
    else
    {
        firstTwoBytes[1] = 127;
        char extendedPlayloadLength[8] = { 0 };
        //转换为网络字节序
        uint64_t tmp = ::htonll((uint64_t)dataLength);
        memcpy(&extendedPlayloadLength, &tmp, 8);
        actualSendData.append(firstTwoBytes, 2);
        actualSendData.append(extendedPlayloadLength, 8);
    }

    //actualSendData 是实际组包后的内容
    actualSendData.append(destbuf);

    //发送到网络上
    sendPackage(actualSendData.c_str(), actualSendData.length());
}
```

服务端收到客户端的数据包时,解包过程稍微复杂一点:根据客户端传送过来的数据包是否设置了 mask 标志,决定是否必须取出 4 字节的 masking-key,然后使用它们对包体内容进行还原,得到包体后还需要根据是否有压缩标志以决定是否解压缩,然后根据 FIN 标志确定是将包体数据当作一个完整的业务数据,还是先将其暂存起来等收完整后再处理:

```cpp
bool MyWebSocketSession::decodePackage(Buffer* pBuffer, const
std::shared_ptr<TcpConnection>& conn)
{
    //readableBytesCount 是当前收到的数据长度
    size_t readableBytesCount = pBuffer->readableBytes();

    const int32_t TWO_FLAG_BYTES = 2;

    //最大包头长度
    const int32_t MAX_HEADER_LENGTH = 14;
    char pBytes[MAX_HEADER_LENGTH] = {0};
    //在已经收到的数据大于最大包长时,仅拷贝可能是包头的最大部分
    if (readableBytesCount > MAX_HEADER_LENGTH)
        memcpy(pBytes, pBuffer->peek(), MAX_HEADER_LENGTH * sizeof(char));
    else
        memcpy(pBytes, pBuffer->peek(), readableBytesCount * sizeof(char));

    //检测是否有 FIN 标志
    bool FIN = (pBytes[0] & 0x80);
    //TODO: 这里就不校验了,因为服务器和未知的客户端之间无约定
    //bool RSV1, RSV2, RSV3;
    //取第 1 字节的低 4 位获取数据类型
    int32_t opcode = (int32_t)(pBytes[0] & 0xF);

    //取第 2 字节的最高位,理论上客户端发给服务端的这个字段必须被设置为 1
    bool mask = ((pBytes[1] & 0x80));

    int32_t headerSize = 0;
    int64_t bodyLength = 0;
    //按 mask 标志加上 4 字节的 masking-key 长度
    if (mask)
        headerSize += 4;

    //取第 2 字节的低 7 位,即得到 payload length
    int32_t payloadLength = (int32_t)(pBytes[1] & 0x7F);
    if (payloadLength <= 0 && payloadLength > 127)
    {
        LOGE("invalid payload length, payloadLength: %d, client: %s", payloadLength,
conn->peerAddress().toIpPort().c_str());
        return false;
    }

    if (payloadLength > 0 && payloadLength <= 125)
    {
        headerSize += TWO_FLAG_BYTES;
        bodyLength = payloadLength;
    }
    else if (payloadLength == 126)
    {
        headerSize += TWO_FLAG_BYTES;
        headerSize += sizeof(short);

        if ((int32_t)readableBytesCount < headerSize)
            return true;

        short tmp;
        memcpy(&tmp, &pBytes[2], 2);
```

```cpp
        int32_t extendedPayloadLength = ::ntohs(tmp);
        bodyLength = extendedPayloadLength;
        //包体长度不符合要求
        if (bodyLength < 126 || bodyLength > UINT16_MAX)
        {
            LOGE("illegal extendedPayloadLength, extendedPayloadLength: %d, client: %s", bodyLength, conn->peerAddress().toIpPort().c_str());
            return false;
        }
    }
    else if (payloadLength == 127)
    {
        headerSize += TWO_FLAG_BYTES;
        headerSize += sizeof(uint64_t);

        //包体长度不够
        if ((int32_t)readableBytesCount < headerSize)
            return true;

        int64_t tmp;
        memcpy(&tmp, &pBytes[2], 8);
        int64_t extendedPayloadLength = ::ntohll(tmp);
        bodyLength = extendedPayloadLength;
        //包体长度不符合要求
        if (bodyLength <= UINT16_MAX)
        {
            LOGE("illegal extendedPayloadLength, extendedPayloadLength: %lld, client: %s", bodyLength, conn->peerAddress().toIpPort().c_str());
            return false;
        }
    }

    if ((int32_t)readableBytesCount < headerSize + bodyLength)
        return true;

    //取出包头
    pBuffer->retrieve(headerSize);
    std::string payloadData(pBuffer->peek(), bodyLength);
    //取出包体
    pBuffer->retrieve(bodyLength);

    if (mask)
    {
        char maskingKey[4] = { 0 };
        //headerSize - 4 即 masking-key 的位置
        memcpy(maskingKey, pBytes + headerSize - 4, 4);
        //对包体数据进行 unmask 还原
        unmaskData(payloadData, maskingKey);
    }

    if (FIN)
    {
        //最后一个分片，与之前的合并（如果有的话）后进行处理
        m_strParsedData.append(payloadData);
        //包处理出错
        if (!processPackage(m_strParsedData, (MyOpCode)opcode, conn))
            return false;
```

```
        m_strParsedData.clear();
    }
    else
    {
        //非最后一个分片，先缓存起来
        m_strParsedData.append(payloadData);
    }

    return true;
}
```

6.12.5 解析握手协议

这里以服务端处理客户端主动发送过来的握手协议为准，在代码中检测了上文中介绍的几个必需字段和值，同时获取了客户端是否支持压缩的标志，如果所有检测都能通过，则将其组装成应答协议包，根据自己是否支持压缩来带上压缩标志：

```
//websocket 握手逻辑处理
bool MyWebSocketSession::handleHandshake(const std::string& data, const std::shared_ptr<TcpConnection>& conn)
{
    std::vector<std::string> vecHttpHeaders;
    //按\r\n拆分成每一行
    StringUtil::Split(data, vecHttpHeaders, "\r\n");
    //至少有3行
    if (vecHttpHeaders.size() < 3)
        return false;

    std::vector<std::string> v;
    size_t vecLength = vecHttpHeaders.size();
    for (size_t i = 0; i < vecLength; ++i)
    {
        //第1行获得参数名称和协议版本号
        if (i == 0)
        {
            if (!parseHttpPath(vecHttpHeaders[i]))
                return false;
        }
        else
        {
            //解析头标志
            v.clear();
            StringUtil::Cut(vecHttpHeaders[i], v, ":");
            if (v.size() < 2)
                return false;

            StringUtil::trim(v[1]);
            m_mapHttpHeaders[v[0]] = v[1];
        }
    }

    //请求必须包含 Upgrade: websocket 头，值必须为 websocket
    auto target = m_mapHttpHeaders.find("Connection");
    if (target == m_mapHttpHeaders.end() || target->second != "Upgrade")
```

```cpp
        return false;

    //请求必须包含 Connection: Upgrade 头，值必须为 Upgrade
    target = m_mapHttpHeaders.find("Upgrade");
    if (target == m_mapHttpHeaders.end() || target->second != "websocket")
        return false;

    //请求必须包含 Host 头
    target = m_mapHttpHeaders.find("Host");
    if (target == m_mapHttpHeaders.end() || target->second.empty())
        return false;

    //请求必须包含 Origin 头
    target = m_mapHttpHeaders.find("Origin");
    if (target == m_mapHttpHeaders.end() || target->second.empty())
        return false;

    target = m_mapHttpHeaders.find("User-Agent");
    if (target != m_mapHttpHeaders.end())
    {
        m_strUserAgent = target->second;
    }

    //检测是否支持压缩
    target = m_mapHttpHeaders.find("Sec-WebSocket-Extensions");
    if (target != m_mapHttpHeaders.end())
    {
        std::vector<std::string> vecExtensions;
        StringUtil::Split(target->second, vecExtensions, ";");

        for (const auto& iter : vecExtensions)
        {
            if (iter == "permessage-deflate")
            {
                m_bClientCompressed = true;
                break;
            }
        }
    }

    target = m_mapHttpHeaders.find("Sec-WebSocket-Key");
    if (target == m_mapHttpHeaders.end() || target->second.empty())
        return false;

    char secWebSocketAccept[29] = {};
    balloon::WebSocketHandshake::generate(target->second.c_str(),
secWebSocketAccept);
    std::string response;
    makeUpgradeResponse(secWebSocketAccept, response);
    conn->send(response);

    m_bUpdateToWebSocket = true;


    return true;
}
```

```cpp
//处理websocket第1次请求的HTTP数据包
bool MyWebSocketSession::parseHttpPath(const std::string& str)
{
    std::vector<std::string> vecTags;
    StringUtil::Split(str, vecTags, " ");
    if (vecTags.size() != 3)
        return false;

    //TODO: 应该改为不区分大小写的比较
    if (vecTags[0] != "GET")
        return false;

    std::vector<std::string> vecPathAndParams;
    StringUtil::Split(vecTags[1], vecPathAndParams, "?");
    //至少有一个路径参数
    if (vecPathAndParams.empty())
        return false;

    m_strURL = vecPathAndParams[0];
    if (vecPathAndParams.size() >= 2)
        m_strParams = vecPathAndParams[1];

    //WebSocket协议版本号必须为1.1
    if (vecTags[2] != "HTTP/1.1")
        return false;

    return true;
}

//将HTTP升级成Websocket协议,并响应客户端
void MyWebSocketSession::makeGradeResponse(const char* secWebSocketAccept,
std::string& response)
{
    response = "HTTP/1.1 101 Switching Protocols\r\n"
               "Content-Length: 0\r\n"
               "Upgrade: websocket\r\n"
               "Sec-Websocket-Accept: ";
    response += secWebSocketAccept;
    response +="\r\n"
               "Server: WebsocketServer 1.0.0\r\n"
               "Connection: Upgrade\r\n"
               "Sec-WebSocket-Version: 13\r\n";
    if (m_bClientCompressed)
        response += "Sec-WebSocket-Extensions: permessage-deflate; client_no_context_takeover\r\n";
    response += "Host: 127.0.0.1:9988\r\n";
    //时间可被改成动态的
    response +="Date: Wed, 21 Jun 2017 03:29:14 GMT\r\n"
               "\r\n";
}
```

第 7 章

单个服务的基本结构

我们通常希望服务器程序能满足高性能、高并发的要求,那什么是高性能、高并发呢?从技术角度来讲,服务器程序的"高性能"指服务器程序能流畅、低延迟地应答客户端的各类请求;"高并发"指服务器程序可以在同一时间支持较多的客户端连接和数据请求。如果一个服务器程序只能简单地接受 n 个客户端连接请求(n 可能很大),但不能同时流畅地处理这些客户端请求,就算不上高性能、高并发。

如何将单个服务的性能做到最优呢?一个服务的性能不仅取决于服务所在机器的硬件配置(内存、CPU、磁盘、网络带宽等),还取决于软件层面上服务器程序的逻辑结构设计。本章将从软件层面讨论如何设计一个良好的服务器程序。

7.1 网络通信组件的效率问题

一个服务器程序要对外提供服务,就需要与外部程序通信,这些外部程序往往是分散在不同机器上的不同进程(即客户端),一般采用网络通信(一般用得最多的是 socket 通信)。因此网络通信组件是服务器程序的基础组件,其设计得好坏直接影响服务器对外服务的能力。

7.1.1 高效网络通信框架的设计原则

业务的不同导致网络通信框架在一些细节上的设计略有不同,但网络通信框架的设计方法大多是通用的、有规律可循的,本节会讨论几个建议遵循的设计原则。

1. 尽量少等待

目前,流行的网络通信框架虽然有很多,例如 libevent、libuv、Boost Asio 等,但实现这些网络通信框架的技术手段往往大同小异。一个好的网络通信框架至少要解决以下 7 个问题。

(1)如何检测有新的客户端连接到来?

（2）如何接受客户端的连接请求？

（3）如何检测客户端是否有数据发送过来？

（4）如何收取客户端发送的数据？

（5）如何检测异常的客户端连接？检测到之后，如何处理？

（6）如何向客户端发送数据？

（7）如何在客户端发送完数据后关闭连接？

有网络编程基础的读者，都能解决上面的一些问题。例如：对第 1、3 个问题，使用 I/O Multiplexing（之后简称"I/O 复用"）技术的 select、poll、epoll 等相关套接字函数；对第 2 个问题，使用 socket API accept 函数；对第 4 个问题，使用 recv 函数；对第 6 个问题，使用 send 函数。的确如此，这些基础的 socket API 构成了网络通信的基础，无论网络通信框架设计得如何巧妙，都是利用这些基础的 socket API 构建出来的。

所以如何巧妙组织这些基础的 socket API，成为解决问题的关键。服务器的高性能、高并发实际上只是一个技术实现所达到的效果而已，不管怎样，从程序设计的角度来说，高性能、高并发服务只是一个或一组程序。一般来说，能尽量满足尽量少等待原则的程序就是高性能的（高效的）。这里说的"高性能"不是指"忙得忙死、闲得闲死"，而是都可以闲着，但只要有"活儿干"，尽量一起干以达到工作效率最优。

这里举几个工作效率不是最优的网络通信的例子：

◎ 默认情况下，在 recv 函数没有数据时，线程会阻塞在 recv 函数调用处；

◎ 默认情况下，如果 TCP 窗口不是足够大，则数据无法发出，send 函数也会阻塞当前调用线程；

◎ 默认情况下，connect 函数发起连接时会有一定时长的阻塞；

◎ 向对端发送一段数据，接着使用 recv 函数接收对端的应答数据，如果对端一直不应答，当前调用线程就会阻塞在 recv 函数调用处。

以上例子所展现的都不是高效网络编程方式，因为它们都不满足尽量少等待原则。在网络通信中，有些等待不是必需的。那么，有没有一种方法，使上述过程不需要等待？最好是不仅不需要等待，而且在网络操作完成时能通知我们，利用等待的时间让程序做其他事情。答案是肯定的，使用 I/O 复用技术即可。

2. 尽量减少做无用功的时间

目前 Windows 支持的 I/O 复用技术有 select、WSAAsyncSelect、WSAEventSelect 和 IOCP（完成端口），Linux 支持的 I/O 复用技术有 select、poll 和 epoll 模型。前面已经详细介绍过这些 I/O 复用函数的用法，这里讨论一些深层次的内容。以上列举的 I/O 复用函数可以分为以下两个级别。

◎ 第 1 级别：select 和 poll。

◎ 第 2 级别：WSAAsyncSelect、WSAEventSelect、IOCP、epoll。

这种划分级别的依据是什么呢？

先来分析第 1 级别的函数。select 和 poll 函数在本质上还是在一定时间内主动查询在一组 socket 句柄（一个或是多个）上是否有网络事件（可读事件、可写事件或出错事件等），也就是说，我们必须每隔一段时间就主动做这些检测操作。如果在这段时间内检测到一些网络事件，检测操作消耗的时间就没有白费；如果在这段时间内没有事件呢？那就相当于做了无用功，是对系统资源的一种浪费。原因是，假设一个服务器有多个连接，在 CPU 时间片有限的情况下，我们花费了一定时间检测一部分 socket 的网络事件，却发现什么事件都没有，而我们在这段时间内可能有其他事情需要处理，那我们为什么要花时间去做这种检测呢？把这个时间用在我们需要做的事情上不是更好吗？所以对于服务器网络通信组件来说，要想高效，就应该尽量避免花时间主动查询一些 socket 是否有网络事件，而是等到这些 socket 有网络事件时让系统主动通知我们，我们再去处理。这就是第 2 级别的函数做的事情。

第 2 级别的函数相当于变主动查询为被动通知，即网络事件发生时，系统会通知我们处理。只不过第 2 级别的函数通知我们的方式各不相同，WSAAsyncSelect 函数利用 Windows 窗口消息队列的事件机制将通知发给我们设置的窗口过程函数，IOCP 模型利用 GetQueuedCompletionStatus 函数从挂起状态唤醒并返回，epoll 模型利用 epoll_wait 函数返回就绪事件。

例如，connect 函数发起连接时，如果将连接 socket 设置为非阻塞模式，程序就不需要等待 connect 函数的返回结果，可以立即返回；等连接完成之后，WSAAsyncSelect 函数会产生 FD_CONNECT 事件，告知我们连接是否成功，epoll 模型会生成 EPOLLOUT 事件通知我们。又例如，socket 有数据可读时，WSAAsyncSelect 函数会产生 FD_READ 事件，epoll 模型会产生 EPOLLIN 事件，等等。

总之，对网络通信组件的性能有高要求时，尽量不要主动查询各个 socket 事件，而是等待操作系统通知我们。

基于上面的讨论，这里提出第 2 个原则：尽量减少做无用功的时间。在服务程序资源足够的情况下，这样做可能体现不出什么优势，但是对于有大量的任务要处理、需要支持高并发服务的情况，这样做优势很明显。

> 通过上面的分析，读者应该明白了对于高性能的服务，同样是 I/O 复用 API，为什么不使用 select、poll 函数了。

另外，使用 I/O 复用 API，如果某个 socket 失效，就应及时从 I/O 复用 API 上移除该 socket，否则可能造成死循环或者浪费 CPU 检测周期的问题。

3. 检测网络事件的高效做法

根据上文介绍的两个原则，我们在高性能服务器设计中一般将 socket 设置成非阻塞模式，利用 I/O 复用函数检测各个 socket 上的事件（读、写、出错等事件）。当然，阻塞的 socket 通信模式并非一无是处。

下面回答本章开头提出的 7 个问题。

对于第 1、2 个问题，在默认情况下，如果没有新的客户端连接请求，则对监听 socket（调用 bind 和 listen 函数的 socket）调用 accept 函数会阻塞调用线程，使用 I/O 复用函数以后，如果 epoll_wait 函数检测到监听 socket 有 EPOLLIN 事件，或者 WSAAsyncSelect 函数检测到有 FD_ACCEPT 事件，就表明此时有新连接到来，再调用 accept 函数就不会阻塞调用线程了。

对于第 3、4 个问题，调用 accept 函数返回的新 socket 也应该被设置成非阻塞模式，而且应该在 epoll_wait 或 WSAAsyncSelect 函数报告这个 socket 有可读事件时收取数据，这样才不会做无用功。那么一次性收取多少数据合适呢？可以根据自己的实际需求来决定，甚至可以在一个循环中反复调用 recv（或 read）函数。对于非阻塞模式的 socket，如果没有数据可读，则 recv（或 read）函数会立即返回（返回值是-1），此时得到的错误码 EWOULDBLOCK（或 EAGAIN）表明当前已经没有数据了。代码示例如下：

```cpp
bool CMySocket::Recv()
{
    int nRet = 0;
    while (true)
    {
        char buff[512];
        nRet = ::recv(m_hSocket, buff, 512, 0);
        if (nRet == SOCKET_ERROR)
        {
            //调用 recv 函数，直到错误码是 WSAEWOULDBLOCK
            if (errno == EWOULDBLOCK)
                break;
            else
                return false;
        }
        else if (nRet < 1)
            return false;

        m_strRecvBuf.append(buff, nRet);
    }

    return true;
}
```

对于第 5 个问题，同样，若 I/O 复用函数（如 epoll_wait、WSAAsyncSelect）收到异常事件（如 EPOLLERR）或关闭事件（如 FD_CLOSE）的通知，我们就知道有异常产生了，对异常的处理一般是关闭相应的 socket。另外，如果 send/recv（或 read/write）函数操作某个 socket 时返回 0，则一般可以认为对端关闭了连接，对于本端，此时这路连接也没有存在的必要了，我们可以关闭本端对应的 socket。需要说明的是，TCP 连接是状态机，I/O 复用函数一般无法检测出两个端点之间路由错误导致的链路问题，所以对于这种情形，我们需要通过定时器结合心跳包来检测。定时器和心跳包相关的内容将在 9.2 节详细介绍。

对于第 6 个问题，向客户端发送数据比收取数据稍微麻烦一点，也是需要讲究技巧的。对于 epoll 模型的水平触发模式（Level Trigger），我们首先不能像检测读事件一样一开始就注册检测写事件标志，因为一旦注册了检测写事件标志，则在一般情况下，只要对端正常收取数据，对应的 socket 就通常是可写的，这会导致频繁触发写事件通知。但并不是每次有写事件触发时都有数据要发送，所以正确的做法是：在 epoll 模型水平触发模式下，如果有数据要发送，则先调用 send 或 write 函数尝试直接发送；如果发送不了或者只发送出去部分数据，则将剩余的数据先缓存起来（需要一个缓冲区来存放剩余的数据，即"发送缓冲区"），再为该 socket 注册检测写事件标志，等下次写事件触发时再发送剩余的数据；如果剩下的数据还是不能完全发送完，则继续等待下一次写事件触发通知，在下一次写事件触发通知后，继续发送数据。如此反复，直到所有数据都发送出去为止。一旦所有数据都发送出去了，就及时为 socket 移除检测写事件标志，避免再次触发无用的写事件通知。

上面提到的发送缓冲区不仅用来存放本次没有发送完的数据，还用来存放发送过程中上层传来的需要发送的新数据。为了保证顺序，新数据应该被追加在当前剩余数据的后面。每次有写事件触发时，我们都应该按数据的先后依次从发送缓冲区中取出再发送，即"先来的数据先发送，后来的数据后发送"。7.6 节会详细介绍这部分逻辑和具体的代码实现。

对于第 7 个问题，比较难处理，因为这里的"发送完"不一定是真的发送完，发送数据用的 send 或 write 函数即使返回成功，也只能说明向操作系统网络协议栈里面写入数据成功，并不代表数据被成功发送到网络。至于最后操作系统协议栈中的数据能否被发送出去及何时被发送出去，很难判断，发送出去后对方是否收到，就更难判断了。所以，我们目前只能简单地认为 send 或者 write 函数返回我们期望的字节数时，就算数据发送完成。在这种情形下发送"完"数据后，就可以关闭连接了。当然，我们也可以使用 shutdown 函数达到"半关闭"效果（即只关闭 socket 的发送或接收通道）。

> socket 有个 linger 选项，可以设置某个 socket 在关闭时，剩下的数据最多可以逗留的时间。如果在逗留的时间内数据还不能完全发送出去，那剩余的数据就真的被操作系统丢弃了。

7.1.2　连接的被动关闭与主动关闭

在实际应用中，连接的被动关闭指我们检测到了连接的异常事件（例如，触发 EPOLLERR 事件、send/recv 函数返回 0 使对端关闭连接），这时这路连接已经没有存在的必要了，需要被迫关闭连接。而主动关闭连接指我们主动调用 close/closesocket 函数关闭 socket 来关闭某个连接，例如客户端向服务端发送了非法数据（如网络攻击中的一些刺探性数据包），这时服务端出于安全考虑，需要主动关闭与该客户端的连接。

7.1.3　长连接和短连接

网络通信双方之间的连接根据保持状态分为长连接和短连接，长连接指长时间保持通信双方的连接状态，是相对于短连接而言的。

长连接操作通常为：连接→数据传输→保持连接→数据传输→保持连接→……→关闭连接，在没有数据交换时需要定时发送保活心跳包，以维持连接状态。长连接主要用于通信双方频繁通信的场景中，缺点是通信双方要增加相应的逻辑维护连接的状态，连接信息本身也需要一定的系统消耗；优点是可以进行实时数据交换。

短连接操作通常为：连接→数据传输→关闭连接，在没有数据交换时直接关闭连接即可。短连接一般用于数据传输完成后即可关闭或者对通信双方的状态信息实时性要求不高的场景中。例如 Web 服务器一般使用短连接与浏览器通信，在 Web 服务器将页面信息发送给浏览器后即可关闭连接，在需要时可再次建立连接。短连接的优点是通信双方无须长时间维护连接状态信息，可节省连接资源；缺点是如果传输数据的频率较高，则可能需要频繁建立和关闭连接，也无法实时推送消息。

> "Web 服务器一般使用短连接与浏览器通信"严格来讲是不准确的，在一些 HTTP 通信中，通信双方可能会接受对端 HTTP 包头中 keepalive 选项的建议，在多次通信之间保持连接状态。
>
> 另外，虽然在大多数场景下 HTTP 通信双方都使用短链接，不支持服务端的消息推送，但这正在改变，最新的 HTTP 2.0 协议标准就支持服务端的推送。

7.2 原始的服务器结构

单个服务器的结构是随着业务需求的升级而不断演进的。原始的服务器结构，是创建好监听 socket，在一个循环里面接受新的连接并产生对应的客户端 fd，然后利用这个客户端 fd 与客户端通信（收发数据）。

在一个循环里面接受新的连接并产生对应的客户端 fd，然后利用这个客户端 fd 与客户端通信的示例代码如下：

```
/**
 * TCP 服务器通信的基本模型
 * zhangyl 2018.12.13
 */
#include <sys/types.h>
#include <sys/socket.h>
#include <arpa/inet.h>
#include <unistd.h>
#include <iostream>
#include <string.h>

int main()
{
    //1.创建一个监听 socket
    int listenfd = socket(AF_INET, SOCK_STREAM, 0);
    if (listenfd == -1)
```

```cpp
    {
        std::cout << "create listen socket error." << std::endl;
        return -1;
    }

    //2.初始化服务器地址
    struct sockaddr_in bindaddr;
    bindaddr.sin_family = AF_INET;
    bindaddr.sin_addr.s_addr = htonl(INADDR_ANY);
    bindaddr.sin_port = htons(3000);
    if (bind(listenfd, (struct sockaddr *)&bindaddr, sizeof(bindaddr)) == -1)
    {
        std::cout << "bind listen socket error." << std::endl;
        return -1;
    }

    //3.启动监听
    if (listen(listenfd, SOMAXCONN) == -1)
    {
        std::cout << "listen error." << std::endl;
        return -1;
    }

    while (true)
    {
        struct sockaddr_in clientaddr;
        socklen_t clientaddrlen = sizeof(clientaddr);
        //4.接受客户端连接
        int clientfd = accept(listenfd, (struct sockaddr *)&clientaddr, &clientaddrlen);
        if (clientfd != -1)
        {
            char recvBuf[32] = {0};
            //5.从客户端接收数据
            int ret = recv(clientfd, recvBuf, 32, 0);
            if (ret > 0)
            {
                std::cout << "recv data from client, data: " << recvBuf << std::endl;
                //6.将收到的数据原封不动地发送给客户端
                ret = send(clientfd, recvBuf, strlen(recvBuf), 0);
                if (ret != strlen(recvBuf))
                    std::cout << "send data error." << std::endl;

                std::cout << "send data to client successfully, data: " << recvBuf << std::endl;
            }
            else
            {
                std::cout << "recv data error." << std::endl;
            }

            close(clientfd);
        }
    }
```

```
    //7.关闭监听 socket
    close(listenfd);

    return 0;
}
```

将以上代码抽出主干部分并整理成如下伪代码：

```
int main()
{
    //1.初始化阶段

    while (true)
    {
        //2.利用 accept 函数接受连接，产生客户端 fd

        //3.利用步骤 2 中的 fd 与某个客户端通信
    }

    //4.资源清理

    return 0;
}
```

在以上流程中，程序在每轮循环中都只能处理一个客户端连接请求，要处理下一个客户端连接请求，就必须等当前操作完成后进入下一轮循环才行。采用这种结构的缺点很明显：不支持并发，更不支持高并发。

7.3　一个连接对应一个线程模型

因为原始的服务器结构不支持并发，所以随着计算机引入多线程模型，软件开发者想出了另一种服务器结构，即为每一个客户端连接都创建一个线程，这样多个线程就可以并行执行，在每个独立的线程中为对应的客户端提供服务。示例代码如下：

```
//监听线程
UINT WINAPI MyMainThread(LPVOID lPvoid)
{
    UINT        nThreadID = 0;
    SOCKET      sListenSocket = (SOCKET)lPvoid;
    SOCKET      sClientSocket = 0;
    while (1)
    {
        //等待客户连接
        sockaddr_in clientAddr = { 0 };
        int clientAddrLength = sizeof(clientAddr);
        if ((sClientSocket = accept(sListenSocket, (struct sockaddr*)&clientAddr, &clientAddrLength)) == INVALID_SOCKET)
            break;

        LOG_NORMAL("New client connected: %s:%d", inet_ntoa(clientAddr.sin_addr), ntohs(clientAddr.sin_port));
```

```
        //启动客户签到线程
        _beginthreadex(NULL, 0, MyChildThread, (LPVOID)sClientSocket, 0, &nThreadID);
    }

    closesocket(sListenSocket);
    return 0;
}
//接收连接线程
UINT WINAPI MyChildThread(LPVOID lPvoid)
{
    SOCKET sClientSocket = (SOCKET)lPvoid;
    CLIENTITEM clientItem = { 0 };
    int nCmd = HandleClientMain(sClientSocket, &clientItem);

    LOG_NORMAL("Client cmd = %d", nCmd);
    if (nCmd == -1)
        closesocket(sClientSocket);
    else if (nCmd == CONN_MAIN)
        LoginTrans(sClientSocket, &clientItem);
    else
        InterTrans(sClientSocket, &clientItem, nCmd);
    return 0;
}
```

如以上代码所示，在某个线程 MyMainThread 中（可以是主线程，也可以是非主线程）调用 accept 函数接受客户端连接，在连接建立后，为每个新连接都创建一个工作线程（MyChildThread）。当然，为了让工作线程正常处理所负责的连接上的来往数据，这里利用线程函数参数将 socket 句柄传给工作线程（注意以上代码中创建线程的_beginthreadex 函数的第 4 个参数 sClientSocket）。

显然，这种一个连接一个线程的做法也不支持高并发，当连接数达到一定量时，会创建很多 MyChildThread 线程，CPU 在线程之间的切换也是一笔不小的开销，CPU 时间片最后都浪费在各个线程之间的切换上了，严重影响程序的执行效率。

7.4 Reactor 模式

目前主流的网络通信库，从 C/C++的 libevent 到 Java 的 Netty 再到 Python 的 Twisted 等，使用的都是 Reactor 模式（译为"反应器模式"或"反射器模式"）。Reactor 模式是一种事件处理设计模式，在 I/O 请求到达后，服务处理程序使用 I/O 复用技术同步地将这些请求派发给相关的请求处理程序。流程图如下。

从流程图来看，该设计模式很简单，但它的设计思想并不简单。它解决了计算机世界

中普遍存在的一个问题：请求太多，资源太少。也就是说，一个对外服务的程序，其接受的各种输入输出请求的数量可能是非常多的，但由于服务程序处理能力有限，其处理这些请求的资源数量（即图中的处理程序）是有限的。所以，上图中输入输出的请求数量之和一般远远超过处理程序的数量，多路复用器（I/O Demultiplexer）将这些数量众多的输入输出请求分发给有限的处理程序。

所以一个 Reactor 模式结构一般包含以下模块：
◎ 资源请求事件（Resource Request）；
◎ 多路复用器与事件分发器（I/O Demultiplexer & Event Dispatcher）；
◎ 事件处理器（EventHandler）。

这里以目前大多数饭店的运营模式为例来说明 Reactor 模式。饭店一般由某个服务员负责服务某几桌客户，顾客有需求时（点菜、结账等）可以告诉服务员，由服务员把这些需求转给其他相关人员（将点菜需求转给厨房工作人员，将结账需求转给收银工作人员）。如此操作，即使饭店顾客爆满，靠几个服务员也能有条不紊地运转整个饭店。

这里将上述例子对应到具体的服务器程序技术上。以 socket 的读写为例，输入输出请求指 socket 上有数据可读或者需要往 socket 上写入数据，而 I/O 多路复用器对应操作系统的相关 I/O 复用函数。在 Windows 上有 select 技术（函数），在 Linux 上有 select 函数、poll 函数、epol 模型。使用这些 I/O 复用函数后，Reactor 模式对应的流程图如下图所示。

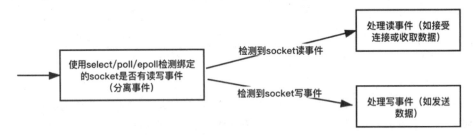

7.5　one thread one loop 思想

基于 7.4 节介绍的 Reactor 模式，这里引出 one thread one loop（一个线程对应一个循环）思想。

7.5.1　one thread one loop 程序的基本结构

one thread one loop 中的线程指的是网络线程，也就是说，在每个网络线程函数里面都有一个循环（loop），每个循环的流程都如下所示，因此每个网络线程都做着同样的事情。这个循环流程如下：

```
//线程函数
void* thread_func(void* thread_arg)
{
    //这里做一些需要的初始化工作
```

```
while (线程退出标志)
{
    //步骤一：利用 select/poll/epoll 等 I/O 复用技术分离出读写事件

    //步骤二：处理读事件或写事件

    //步骤三：做其他事情
}

//这里做一些清理工作
}
```

关键部分是线程函数里面的 while 循环部分，步骤一利用 I/O 复用技术分离出 socket 读写事件；步骤二处理读事件和写事件，需要注意的是，在 Linux 上除了 socket 对象，其他类型的对象（如管道、文件句柄）也可以被挂到 I/O 复用函数上。

这里暂时先讨论 socket 对象，以处理读事件为例进行说明。

◎ 对于监听 socket，我们认为处理监听 socket 的读事件一般就是接受新连接等操作，我们不仅可以接受新连接，还可以多做一点事情，例如将 accept 函数返回的客户端 socket 绑定到 I/O 复用函数上，以及创建连接对象等。

◎ 对于普通的 socket，在不考虑出错的情况下，处理普通 socket 的读事件实际上就是调用 recv 或者 read 函数收取数据，或者进一步对收到的数据进行解包，并做一些业务逻辑处理。举个例子，假设收到的数据经解包后是登录请求，则可以接着处理这些登录请求，并应答客户端。在这种场景下，处理读事件实际上包括收数据、解包、处理数据、应答客户端这 4 个步骤。

处理 socket 的写事件，一般是做发送数据操作。

在步骤三中提到要做其他事情，其他事情有哪些？这就要具体问题具体对待了。例如，我们可以将步骤二中的解包操作或验证包数据有效性的操作放到步骤三中。毕竟从程序设计结构的角度来看，检测 socket 网络事件（步骤一）和收发数据属于网络通信层的工作，而解包和处理包数据属于业务层的工作。对于一个设计良好的结构，这两个工作应该是分离的。当然，在步骤三中还可以做更多的事情，很快就会介绍到。

7.5.2 线程的分工

根据上文介绍的线程函数中的循环结构，服务端为了能流畅处理多个客户端连接请求，一般在某个线程 A 里面调用 accept 函数产生新的客户端连接并生成相应的 socket，然后将这些新连接的 socket 传递给另外数个工作线程 B1、B2、B3、B4 等，这些工作线程负责处理这些新连接上的网络 I/O 事件（即收发数据），这些工作线程还同时处理系统中的另一些事务。在实际开发中，线程 A 往往对应进程的主线程，为了叙述方便，在下文中除了特别强调，否则将 A 线程称为主线程，将 B1、B2、B3、B4 等称为工作线程。

将工作线程的代码框架用伪代码表示如下：

```
while (!m_bQuitFlag)
{
    epoll_or_select_func();

    handle_io_events();

    handle_other_things();
}
```

在 epoll_or_select_func()中通过 select/poll/epoll 等 I/O 复用函数去检测 socket 上的网络事件，若检测到这些事件，则下一步调用 handle_io_events()处理这些事件（一般是收发数据），在做完之后可能还要处理其他任务，这时调用 handle_other_things()即可。

主线程（线程 A）和工作线程（线程 B1、B2、B3、B4 等）分工策略的优点如下。

（1）线程 A 只需处理新来的连接即可，不用处理普通 socket 的网络 I/O 事件。如果在线程 A 里面既处理新连接又处理网络 I/O 事件，则可能由于线程忙于处理网络 I/O 事件，无法及时处理新的客户端连接请求。

（2）线程 A 接受连接产生新的 socket，我们可以根据一定的负载均衡策略，将这些 socket 分配给各个工作线程。round robin（轮询策略）是其中一种很简便、常用的策略，该策略在不考虑中途有连接断开的情况下将 A 线程产生的新 socket 依次分配给各个工作线程，将第 1 个分配给 B1，将再来的一个分配给 B2，再将一个分配给 B3……如此反复。线程 A 会记录各个工作线程上的 socket 数量，这样可以平衡资源利用率，避免一些工作线程"忙死"而另一些工作线程"闲死"的情况出现。

（3）在工作线程不满负载的情况下，可以让工作线程处理其他任务。例如现在有 4 个工作线程，但只有 3 个连接，工作线程 B4 就可以在 handle_other_things()中处理其他任务。

在上述 while 循环里面，epoll_or_select_func()中的 poll、select、epoll_wait 等函数一般设置了一个超时时间。如果将超时时间设置为 0，那么在没有任何网络 I/O 事件和其他任务需要处理的情况下，这些工作线程实际上会空转，白白浪费了 CPU 时间片。如果将超时时间设置为大于 0，则在没有网络 I/O 事件但 handle_other_things()中有其他任务需要处理时，poll、select、epoll_wait 等函数会在挂起指定时间后才能返回，导致 handle_other_things()中的任务不能及时执行。那这里该怎么设置 I/O 复用函数的超时时间呢？

其实解决问题的思路不在于超时时间的设置，我们想达到的效果是：如果没有网络 I/O 事件和其他任务要处理，那么这些工作线程最好直接挂起而不是空转；如果有其他任务要处理，那么这些工作线程要能立刻处理这些任务，而不是在 poll、select、epoll_wait 等函数处挂起指定时间后才处理其他任务。

解决方案是，我们仍然会给 poll、select、epoll_wait 等函数设置一定的超时时间（大于 0），但对于 handle_other_things 函数的执行，会采取一种特殊的唤醒策略。以 Linux 的 epoll_wait 函数为例，不管在 epollfd 上有没有 socket fd，我们都会为 epollfd 挂载一个特殊的 fd，这个 fd 被称为 wakeup fd（唤醒 fd）。当我们有其他任务需要立即处理时（也就是需要 handle_other_thing()立刻执行时），向这个唤醒 fd 上随便写入 1 字节的内容，这个唤醒 fd 就立即变成可读的了，epoll_wait 函数会立即被唤醒并返回，接下来就可以马上执行 handle_other_thing 函数了，这样其他任务就可以得到立即处理；反之，在没有其他任务也没有网络 I/O 事件时，epoll_or_select_func 函数就挂在那里什么也不做。poll、select 函数的处理策略与 epoll_wait 相同。

7.5.3 唤醒机制的实现

本节讲解唤醒机制的实现。

1. 唤醒机制在 Linux 上的实现

唤醒机制在 Linux 上的实现方式有以下几种。

1）使用管道 fd（pipe）

创建一个管道，将管道的一端（管道 fd 中的一个）绑定到 epollfd 上，需要唤醒时，向管道的另一端（管道 fd 中的另一个）写入 1 字节，工作线程会被立即唤醒。创建管道的相关 API 函数如下：

```
#include <unistd.h>
#include <fcntl.h>

int pipe(int pipefd[2]);

int pipe2(int pipefd[2], int flags);
```

2）使用 Linux 2.6 新增的 eventfd

eventfd 的使用方法与管道 fd 一样，将调用 eventfd 函数返回的 eventfd 绑定到 epollfd 上，需要唤醒时，向这个 eventfd 上写入一字节，I/O 复用函数会被立即唤醒。创建 eventfd 函数签名如下：

```
#include <sys/eventfd.h>

int eventfd(unsigned int initval, int flags);
```

3）使用 socketpair

socketpair 是一对相互连接的 socket，相当于服务端和客户端的两个端点，每一端都可以读写数据，向其中一端写入数据后，就可以从另一端读取数据了。创建 socketpair 的函数签名如下：

```
#include <sys/types.h>
#include <sys/socket.h>

int socketpair(int domain, int type, int protocol, int sv[2]);
```

调用这个函数返回的两个文件描述符是 sv[0]和 sv[1]，即 socketpair 函数的第 4 个参数是一个输出参数，向 sv[0]、sv[1]中的任何一个 fd 写入字节后，就可以从另一个 fd 中读取写入的字节了。我们将读取方的 fd 绑定到 epollfd 上，在需要时向写入方的 fd 写入 1 字节，工作线程就会立即被唤醒。

> 注意：与创建普通 socket 稍微不同的是，使用 socketpair 函数创建 socketpair 时，必须将第 1 个参数 domain 设置为 AF_UNIX。

2. 唤醒 socket 在 Windows 上的实现

在 Windows 上使用 select 函数作为 I/O 复用函数时，由于 Windows 的 select 函数只支持绑定套接字这种类型的句柄，因此在 Windows 上一般只能模仿 Linux 创建 socketpair 的思路，即手动创建两个 socket；然后调用 connect、accept 函数建立一个连接，相当于将其中一个 socket 作为客户端 socket（用来调用 connect 函数的 socket），去连接某个监听 socket，将另外一个 socket 作为监听端接受连接后（调用 accept 函数）返回的 socket；然后将读取数据的那一端的 socket 绑定到 select 函数上并检测其可读事件。当然，这种创建唤醒 socket 的做法同样适用于 Linux。

> 如下图所示，在以上一段文字中共有 3 个 socket：一端调用 connect 函数发起连接请求的 socket（记为 socketA）；另一端的监听 socket（记为 socketB）；利用 socketB 调用 accept 函数返回的 socket（记为 socketC）。其中的 socketA 和 socketC 就是唤醒 socket（在 Linux 上称之为唤醒 fd）。

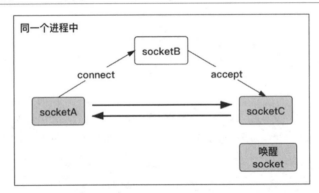

创建唤醒 fd（socket）的具体实现如下：

```
//创建唤醒 fd
bool EventLoop::createWakeupfd()
{
#ifdef WIN32
    m_wakeupFdListen = sockets::createOrDie();
    m_wakeupFdSend = sockets::createOrDie();

    //在 Windows 上需要创建一对 socket
    struct sockaddr_in bindaddr;
    bindaddr.sin_family = AF_INET;
    bindaddr.sin_addr.s_addr = htonl(INADDR_LOOPBACK);
    //将 port 设为 0，然后进行 bind,
    bindaddr.sin_port = 0;
    sockets::setReuseAddr(m_wakeupFdListen, true);
```

```cpp
    sockets::bindOrDie(m_wakeupFdListen, bindaddr);
    sockets::listenOrDie(m_wakeupFdListen);

    struct sockaddr_in serveraddr;
    int serveraddrlen = sizeof(serveraddr);
    //接着通过 getsockname 获取 port，这可以满足获取随机端口的需求
    if (getsockname(m_wakeupFdListen, (sockaddr*)&serveraddr, &serveraddrlen) < 0)
    {
        //让程序挂掉
        LOGF("Unable to bind address info, EventLoop: 0x%x", this);
        return false;
    }

    int useport = ntohs(serveraddr.sin_port);
    LOGD("wakeup fd use port: %d", useport);

    //serveraddr.sin_family = AF_INET;
    //serveraddr.sin_addr.s_addr = inet_addr("127.0.0.1");
    //serveraddr.sin_port = htons(INNER_WAKEUP_LISTEN_PORT);
    if (::connect(m_wakeupFdSend, (struct sockaddr*)&serveraddr, sizeof(serveraddr)) < 0)
    {
        //让程序挂掉
        LOGF("Unable to connect to wakeup peer, EventLoop: 0x%x", this);
        return false;
    }

    struct sockaddr_in clientaddr;
    socklen_t clientaddrlen = sizeof(clientaddr);
    m_wakeupFdRecv = ::accept(m_wakeupFdListen, (struct sockaddr*)&clientaddr, &clientaddrlen);
    if (m_wakeupFdRecv < 0)
    {
        //让程序挂掉
        LOGF("Unable to accept wakeup peer, EventLoop: 0x%x", this);
        return false;
    }

    sockets::setNonBlockAndCloseOnExec(m_wakeupFdSend);
    sockets::setNonBlockAndCloseOnExec(m_wakeupFdRecv);

#else
    //在 Linux 上有一个 eventfd 就够了，可以实现读写
    m_wakeupFd = ::eventfd(0, EFD_NONBLOCK | EFD_CLOEXEC);
    if (m_wakeupFd < 0)
    {
        //让程序挂掉
        LOGF("Unable to create wakeup eventfd, EventLoop: 0x%x", this);
        return false;
    }

#endif
```

```
        return true;
}
```

在以上代码中有一个实现细节需要注意：在 Windows 上，作为服务端的一方，创建一个监听 socket（代码中的 m_wakeupFdListen）后，需要调用 bind 函数绑定特定的 IP 和端口号，这里不要使用一个固定的端口号。一旦使用固定的端口号，服务在某个机器上运行时就可能会因为端口号已经被占用，导致 bind 函数调用失败，无法创建出唤醒 socket，进而影响程序的功能。唤醒 fd 是程序的关键，所以在以上代码中，createWakeupfd 函数中的任意一步操作失败，都会导致唤醒 fd 无法创建，主动打印一条 Fatal 级别的日志并让程序退出，这表明关键性的初始化步骤无法完成。

所以，我们将端口号设置为 0（以上代码中的第 1 个加粗行），根据 4.4 节对 bind 函数的介绍，操作系统会为我们分配一个可用的端口号。但主动连接的一方调用 connect 函数时需要指定明确的 IP 地址和端口号，这时调用 getsockname 函数获取操作系统为 bind 函数分配的端口号就可以了（以上代码中的第 2 个加粗行）。整个流程图如下图所示。

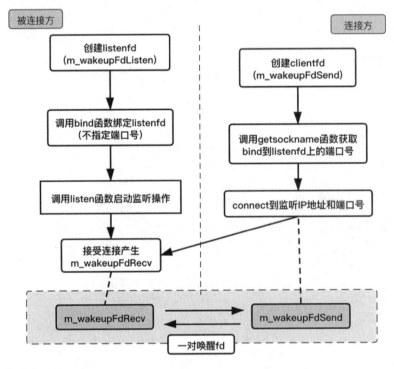

唤醒函数的实现如下：

```
bool EventLoop::wakeup()
{
    //变量 one 为随便写入的数据，只要让唤醒 fd 触发读事件即可，数据内容随意
    uint64_t one = 1;
#ifdef WIN32
    int32_t n = sockets::write(m_wakeupFdSend, &one, sizeof(one));
#else
    int32_t n = sockets::write(m_wakeupFd, &one, sizeof(one));
#endif
```

```cpp
        if (n != sizeof one)
        {
#ifdef WIN32
            DWORD error = ::WSAGetLastError();
            LOGSYSE("EventLoop::wakeup() writes %d  bytes instead of 8, fd: %d, error: %d",
n, m_wakeupFdSend, (int32_t)error);
#else
            int error = errno;
            LOGSYSE("EventLoop::wakeup() writes %d  bytes instead of 8, fd: %d, error: %d, errorinfo: %s", n, m_wakeupFd, error, strerror(error));
#endif

            return false;
        }

        return true;
}
```

无论使用哪种类型的 fd 作为唤醒 fd，一定要在唤醒后及时将唤醒 fd 中的数据读出来，即消耗掉这个 fd 中的数据，否则 fd 会因为有数据不断触发读事件，失去唤醒的作用。

从唤醒 fd 中读取数据：

```cpp
bool EventLoop::handleRead()
{
    uint64_t one = 1;
#ifdef WIN32
    int32_t n = sockets::read(m_wakeupFdRecv, &one, sizeof(one));
#else
    int32_t n = sockets::read(m_wakeupFd, &one, sizeof(one));
#endif

    if (n != sizeof one)
    {
#ifdef WIN32
        DWORD error = ::WSAGetLastError();
        LOGSYSE("EventLoop::wakeup() read %d  bytes instead of 8, fd: %d, error: %d",
n, m_wakeupFdRecv, (int32_t)error);
#else
        int error = errno;
        LOGSYSE("EventLoop::wakeup() read %d  bytes instead of 8, fd: %d, error: %d, errorinfo: %s", n, m_wakeupFd, error, strerror(error));
#endif
        return false;
    }

    return true;
}
```

可以在唤醒 fd 的读事件触发后调用唤醒 fd 读取数据函数 EventLoop::handleRead()。

7.5.4　handle_other_things 方法的实现逻辑

在了解唤醒机制之后，我们来看一下 handle_other_things 方法的实现逻辑。handle_other_things 方法可以被设计成从其他任务集合中取出任务来执行的方法：

```
void EventLoop::handle_other_things()
{
    std::vector<OtherThingFunctor> otherThingFunctors;
    m_callingPendingFunctors = true;

    //这对大括号用于减少锁 m_mutex 的范围
    {
        std::unique_lock<std::mutex> lock(m_mutex);
        otherThingFunctors.swap(m_pendingOtherThingFunctors);
    }

    thingCount := otherThingFunctors.size()
    for (size_t i = 0; i < thingCount; ++i)
    {
        otherThingFunctors[i]();
    }

    m_callingPendingFunctors = false;
}
```

m_pendingOtherThingFunctors 是一个类成员变量，为一个存放其他任务的集合，这里使用了 std::vector，工作线程本身会从这个容器中取出任务来执行，我们将任务封装成一个个函数对象（OtherThingFunctor），从容器中取出后直接执行即可。这里使用了一个特殊的小技巧：为了减小锁对象 m_mutex（也是成员变量）加锁的范围，并提高程序执行效率，这里使用了一个局部变量 otherThingFunctors 将成员变量 m_pendingOtherThingFunctors 中的数据倒换进这个局部变量中。

添加 "other_things" 时，可以在任意线程中添加，也就是说，可以在网络线程之外的线程中添加任务，因此可能涉及多个线程同时操作 m_pendingOtherThingFunctors 对象，所以需要对其进行加锁保护。添加 other_things 的代码如下：

```
void EventLoop::queueInLoop(const Functor& cb)
{
    //这对大括号用于减少锁 m_mutex 的范围
    {
        std::unique_lock<std::mutex> lock(m_mutex);
        m_pendingOtherThingFunctors.push_back(cb);
    }

    //如果在其他线程调用了这个函数，
    //则立即尝试唤醒 handle_other_things 方法所在的线程
    if (!isInLoopThread() || m_callingPendingFunctors)
    {
        //wakeup()即上文介绍的 EventLoop::wakeup 方法
        wakeup();
```

```
    }
}
```

最后，在某些业务场景中，other_things 可能有多种类型，因此可以存在多个 handle_other_things 方法，程序结构演变如下：

```
while (!m_bQuitFlag)
{
    epoll_or_select_func();

    handle_io_events();

    handle_other_things1();

    handle_other_things2();

    handle_other_things3();

    //根据实际需要，可以有更多的handle_other_things()
}
```

另外，handle_other_things()系列的方法可以在 one thread one loop 结构中的 while 循环内部的任意位置，不一定非要放在 handle_io_events 函数后面，我们有时也称 handle_other_things()为钩子函数（Hook Functions），例如：

```
while (!m_bQuitFlag)
{
    handle_other_things1();

    epoll_or_select_func();

    handle_other_things2();

    handle_io_events();

    handle_other_things3();

    //根据实际需要，可以有更多的handle_other_things()
}
```

7.5.5 带定时器的程序结构

如果想在 one thread one loop 结构的线程中做一些定时任务，则需要定时器功能，我们可以在线程循环执行流中加上检测和处理定时器事件的逻辑，通常将其放在程序循环执行流的第 1 步。加上定时器逻辑后，程序结构如下：

```
while (!m_bQuitFlag)
{
    check_and_handle_timers();

    epoll_or_select_func();

    handle_io_events();
```

```
    handle_other_things();
}
```

这里需要注意的是，在 epoll_or_select_func()中使用的 I/O 复用函数的超时时间要尽量不大于 check_and_handle_timers()中所有定时器的最小时间间隔，以免定时器的逻辑处理延迟较大。关于定时器逻辑的具体实现将在 7.9 节详细介绍。

7.5.6　one thread one loop 的效率保障

在整个 loop 结构中，为了保证各个步骤高效执行，除了 epoll_or_select_func 步骤中的 I/O 复用函数可能会造成等待，在任何其他步骤中都不能有阻塞整个流程或者耗时的操作。如果业务决定了在定时器逻辑（对应 check_and_handle_timers 函数）、读写事件处理逻辑（对应 handle_io_events 函数）或其他自定义逻辑（对应 handle_other_things 函数）中有耗时的操作，就需要再开新的业务线程去处理这些耗时的操作，I/O 线程（loop 所在的线程）本身不能处理耗时的操作。业务线程在处理耗时操作完毕后，可以将处理结果或者数据通过特定方式返回给 I/O 线程，7.10 节会详细介绍具体做法。

7.6　收发数据的正确做法

在网络通信中，我们可能既要通过 socket 发送数据，也要通过 socket 收取数据。那么，一般的网络通信框架是如何收发数据的呢？注意，这里讨论的是基于先使用 I/O 复用函数（select、poll、epoll 等）判断 socket 读写事件再来收发数据，其他情形比较简单，不再赘述。

以服务端为例，服务端接受客户端连接后，会产生一个与客户端连接对应的 socket（在 Linux 上也叫作 fd，为了叙述方便，下文称之为 clientfd），我们可以通过这个 clientfd 收取从客户端发来的数据，也可以通过这个 clientfd 将数据发往客户端。但是收与发在操作流程上是有明显区别的。

7.6.1　如何收取数据

对于收取数据，在接受连接并得到 clientfd 后，我们会将该 clientfd 绑定到相应的 I/O 复用函数上并监听其可读事件。不同 I/O 复用函数的可读事件标志不一样，例如对于 poll 模型，可读标志是 POLLIN；对于 epoll 模型，可读事件标志是 EPOLLIN。在可读事件触发后，我们调用 recv 函数从 clientfd 上收取数据（这里不考虑出错情况），根据不同的业务需求，我们可能会收取部分数据或一次性收完所有数据。我们会将收取到的数据放入接收缓冲区，然后做解包操作。这就是收取数据的全部流程。对于使用 epoll 的 LT 模式（水平触发模式），我们可以在每次读事件触发后都只收取部分数据；但对于 ET 模式（边缘触发模式），我们必须在每次读事件触发后都将 fd 上的数据全部收完。

收完数据的标志是 recv 或 read 函数返回-1，错误码 errno 等于 EWOULDBLOCK（EAGAIN，这两个宏的值一样）。

这就是收取数据的正确做法，流程如下图所示。

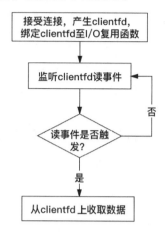

7.6.2 如何发送数据

对于发送数据，使用 epoll 的 ET 模式这种场景除外，使用 epoll 的 LT 模式或者其他 I/O 复用函数时，我们通常都不会一开始就为 clientfd 注册监听可写事件的标志，这是因为只要对端正常收取数据，就一般不会出现因 TCP 窗口太小导致本端 send 或 write 函数无法写成功的问题。因此在大多数情况下，本端的 clientfd 都是可写的，如果注册了监听可写事件标志，则会导致可写事件通知一直触发，而此时不一定有数据需要发送。所以，如果有数据需要发送，则一般都是先调用 send 或者 write 函数直接发送，如果在发送过程中 send 函数或 write 函数返回-1，并且错误码是 EWOULDBLOCK（或 EAGAIN），则表明此时 TCP 窗口已经太小，数据已经无法再发送。如果仍然有剩下的部分数据未发送，我们才会为 clientfd 注册监听可写事件标志，并将剩余的数据存入自定义的 socket 发送缓冲区（应用层设计的 socket 发送缓冲区）中，等到写事件触发后再接着将发送缓冲区中剩余的数据发送出去。如果仍然有部分数据不能发送出去，则继续注册监听可写事件标志，当已经无数据需要发送时，应该立即移除对写事件的监听。这是目前主流网络库的做法。

如果在监听写事件期间，业务层又产生了新的数据需要发送，我们就需要将这些新来的数据放到刚才剩余的待发数据的后面，即下次写事件触发后，先发送旧的数据再接着发送后来的新数据。

发送数据的操作流程图如下图所示。

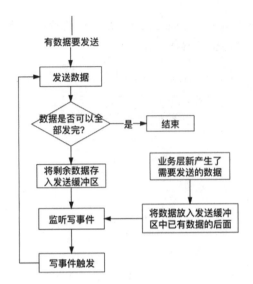

上述逻辑的实现代码如下。

直接尝试发送消息的处理逻辑如下:

```cpp
/**
 *@param data 待发送的数据
 *@param len 待发送的数据长度
 */
void TcpConnection::sendMessage(const void* data, size_t len)
{
    int32_t nwrote = 0;
    size_t remaining = len;
    bool faultError = false;

    //当前未监听可写事件,且在发送缓冲区中没有遗留数据
    if (!m_channel->isWriting() && m_outputBuffer.readableBytes() == 0)
    {
        //直接发送数据
        nwrote = sockets::write(m_channel->fd(), data, len);
        if (nwrote >= 0)
        {
            remaining = len - nwrote;
        }
        else //nwrote < 0
        {
            nwrote = 0;
            //错误码不等于EWOULDBLOCK,说明发送出错
            if (errno != EWOULDBLOCK)
            {
                if (errno == EPIPE || errno == ECONNRESET)
                {
                    faultError = true;
                }
            }
        }
    }
```

```cpp
    //发送未出错且还有剩余字节未发送出去
    if (!faultError && remaining > 0)
    {
        //将剩余部分加入发送缓冲区中
        m_outputBuffer.append(static_cast<const char*>(data) + nwrote, remaining);
        if (!m_channel->isWriting())
        {
            //注册可写事件
            m_channel->enableWriting();
        }
    }
}
```

当数据不能一次性全部发送出去时，为相应的 client 注册监听写事件标志。写事件触发后的处理逻辑如下：

```cpp
//可写事件触发后会调用 handleWrite 函数
void TcpConnection::handleWrite()
{
    //将发送缓冲区中的数据发送出去
    int32_t n = sockets::write(m_channel->fd(), m_outputBuffer.peek(), m_outputBuffer.readableBytes());
    if (n > 0)
    {
        //发送多少，就从发送缓冲区中移除多少
        m_outputBuffer.retrieve(n);
        //如果在发送缓冲区中已经没有剩余，则移除监听可写事件
        if (m_outputBuffer.readableBytes() == 0)
        {
            //移除监听可写事件
            m_channel->disableWriting();
        }
    }
    else
    {
        //发数据出错处理
        handleClose();
    }
}
```

使用 epoll LT 模式注册监听一次写事件后，在写事件触发时会尝试继续发送数据，如果此时数据还不能全部发送完，就不用再次注册监听写事件标志了，因为上次注册的监听写事件标志继续有效；对于 epoll ET 模式，先注册监听写事件标志，写事件触发后，会尝试继续发送数据，如果此时数据还不能全部发送完，则需要再次注册监听写事件标志，以便让写事件下次再触发（给予再次发送数据的机会）。当然，这只是理论情况，在实际开发中，对于一段数据反复发送都不能完全发送完的场景（例如对端先不收，后面每隔很长时间再收 1 字节），我们就可以设置一个最大发送次数或最大发送总时间限制，超过这些限制后，我们可以认为对端出了问题，应该立即清空发送缓冲区并关闭相应的连接。

这里总结检测一个 fd 的读写事件的区别。

◎ 使用 select、poll 或 epoll LT 模式时，可以直接为待检测 fd 注册监听读事件标志。

◎ 使用 select、poll 或 epoll LT 模式时，不要直接为待检测 fd 注册监听可写事件标志，应该先尝试发送数据，若 TCP 窗口太小发不出去，则再为待检测 fd 注册监听可写事件标志，一旦数据发送完，就应该立即移除监听写事件标志。
◎ 使用 epoll ET 模式时，如果需要发送数据，则每次都要为 fd 注册监听写事件标志。

> 对于监听 fd，一般只监听其读事件，监听 fd 没有写事件处理逻辑。

7.6.3 不要多个线程同时利用一个 socket 收（发）数据

TCP 通信是全双工的，收取数据和发送数据是独立的，所以利用同一个 socket 收取数据和发送数据不会相互影响。这里建议不要多个线程同时利用一个 socket 收取数据（或发送数据），指的不是说收取数据和发送数据必须被放在同一个线程里面进行，相反，在实际开发中，有不少应用对同一个 socket 收取数据使用一个线程，发送数据使用另一个线程，但是需要额外做一些工作以同步收发两个线程中 socket 的出错状态。

但是一定不要多个线程同时使用一个 socket 发送数据，或者多个线程同时使用一个 socket 收取数据。

TCP 数据是有序的，以发送数据为例，如果多个线程同时对一个 socket 调用 send 函数，最终对端收到的数据顺序就无法保证了。例如现在有 3 个线程分别要发送 A、B、C 三个数据块，对端期望收到的顺序是 A、B、C，但由于发送端使用了 3 个线程发送，所以对端收到的数据顺序不一定是 A、B、C，除非使用一定的线程同步策略让三个线程按 A、B、C 的顺序发送数据，但这种同步策略是非常麻烦的，而且没有必要。与其写这样的逻辑，还不如将其放在一个线程中操作。

同理，对于收取数据，多个线程会同时调用 recv 函数，每个线程都可能收取部分数据，那么最终按什么顺序将这些数据还原成发送端发送的数据顺序呢？

不仅是 socket，管道亦如此，不建议多个线程同时在同一个管道上进行读操作（或写操作）。

有读者可能会问："我们平时所说的多线程上传或者下载文件，不是多线程同时对一个文件内容做读写吗？"这是不一样的，多线程上传或下载文件的原理是将文件按一定的大小切割成不同的内容块，然后开启多个连接，每个线程都操作一个连接对指定编号的文件内容块进行读写操作，在各个线程都完工后，按内容编号将文件重新组织起来。这在本质上并不是多个线程同时操作一个 socket，而是每个线程都只操作属于自己的文件块。

7.7 发送、接收缓冲区的设计要点

上一节介绍了收发数据的正确做法，本节介绍在发送和收取数据的过程中分别用到的发送缓冲区和接收缓冲区。

7.7.1 为什么需要发送缓冲区和接收缓冲区

网络层在发送数据的过程中，由于 TCP 窗口太小，会导致数据无法发送出去，而上层可能不断产生新的数据，此时就需要将数据先存储起来，以便等 socket 可写时再次发送，这个存储数据的地方就叫作发送缓冲区。

对于接收缓冲区也是一样的道理，在收到数据后，我们可以直接对其进行解包，但是这样做并不好，有三个理由。

- 理由一：除去一些通用协议格式（例如 HTTP），大多数业务使用的都是自定义协议格式，也就是说对一个数据包里面数据格式的解读应该是业务层应该做的。不同的业务一般有不同的协议格式，协议格式与具体的业务有关，网络通信层一般不知道也不需要知道上层协议数据的具体格式。为了让网络层更加通用，网络通信层应该与业务层解耦。
- 理由二：即使知道协议格式，由于 TCP 是流式协议，某一次收到的数据长度也不一定够一个完整的包大小，此时需要一个地方将这些不完整的数据先缓存起来，以便等数据足够一个包大小时再处理。
- 理由三：即便接收到的数据足够一个包，但出于一些特殊的业务逻辑要求，我们仍需将收到的数据暂时缓存起来，等满足一定条件时再取出来处理。

鉴于以上理由，我们的网络层确实需要一个接收缓冲区，将收取的数据按需存放在该缓冲区里面，交由专门的业务线程或者业务逻辑从接收缓冲区中取出数据，并解包处理业务。

7.7.2 如何设计发送缓冲区和接收缓冲区

根据前面的描述，无论是发送缓冲区还是接收缓冲区，一般都建议将其设计成一个内存连续的存储容器。

> 当然，我们也可以将发送缓冲区和接受缓冲区设计成不连续的内存，例如链表结构，每个链表的节点都是一个存储数据的内存块。这种设计在存取数据时相对麻烦。

在通常情况下，发送缓冲区和接收缓冲区根据功能至少需要提供两类接口，即存数据的接口和取数据的接口。对于发送缓冲区，由于上层交给网络层的数据是有序的，所以若某次需要发送的数据未发完，则其剩余的数据一定排在后续产生的数据前面；对于接收缓冲区，由于其不断从 socket 上读取数据，所以后面读到的数据一定排在前面读到的数据后面。

另外，应该将发送缓冲区和接收缓冲区的容量设置成多大呢？预分配的内存太小时可能不够用，太大时可能造成浪费，所以答案是像 std::string、vector 一样，设计出一个容量可以动态增加的结构，按需分配，容量不够时可以扩展容量。

既然是用于收发数据的缓冲区，所以我们可能需要向其中写入或者从其中读取各种数据类型，例如 char、short、int32、int64、string 等，这是我们在设计缓冲区对象时需要考虑的情形，也就是说，需要缓冲区结构提供写入和读取这些数据类型的接口。

对于接收缓冲区，我们可能需要从接收缓冲区中寻找特殊的标志，例如若某个业务的数据包以 "\n" 为结束标志，我们就需要在其中寻找 "\n" 标志以确定缓冲区中的数据长度是否至少足够一个包的长度。

下图演示了一种缓冲区设计结构，这种结构的缓冲区在创建时会自动分配一块固定大小的内存，如下图所示。

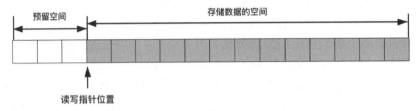

在这个结构中，内存是连续的，而且由预留空间和存储数据的空间组成。其中，预留空间可以做一些特殊用途使用，例如存储一些元数据信息等，预留空间的大小可以被设置为 0，它不是必需的；存储数据的空间由于同时提供读数据和写数据功能，所以分别使用一个读指针和写指针来标明读写位置。当缓冲区是空的时，读写指针的位置相同。在写入一段数据后，该缓冲区的结构如下图所示。

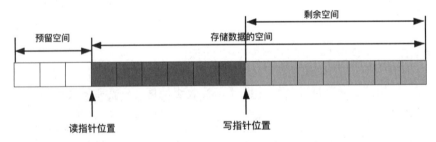

此时该缓冲区中已存储数据的范围是读指针位置～写指针位置的区间，下一次需要读取数据时，应该从缓冲区的读指针位置开始读，最大可以读取的数据长度等于写指针位置减去读指针位置的值。假设读取了 n 字节，n 小于最大可读取数据的长度，则该缓冲区结构如下图所示。

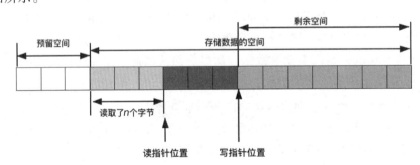

下一次写数据时会从写指针的位置继续写。为了更清楚地说明问题，假设预留的空间大小为 prependableBytes，读指针使用 readerIndex，写指针使用 writerIndex，总缓冲区长度为 size，现在需要写入 m 字节，则会有以下几种情形。

◎ 当 $m \leqslant$ size−writerIndex 时（即将写入的数据长度小于或等于写指针位置到缓冲区结束位置），可以直接在 writerIndex 位置写入。

◎ 当 $m >$ size−writerIndex 时，从 writerIndex 位置一直到缓冲区结束位置的内容已经不够写入了，可以对当前缓冲区未充分利用的内存进行整理。哪里有未充分使用的内存呢？为了充分利用空间，我们不会一开始就扩容，而是把未利用的空间充分利用，方法是：从预留空间结束位置到读指针之前的位置，该空间的数据已经被读取，先将读指针位置挪到预留空间结束处（即起始位置），然后将原来读指针之后的数据也挪到起始位置，最后将写指针向前挪到数据结束的位置。挪动后的缓冲区结构如下图所示。

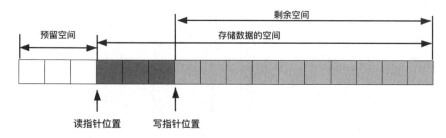

挪动之后，如果剩余空间足够写入 m 字节的数据，则在新的 writerIndex 位置写入；如果挪动之后，剩余的空间仍然不够写入 m 字节，则需要重新扩展缓冲区，即新建一个更大的缓冲区，将现有的缓冲区结构和数据复制过去。扩容示意图如下所示。

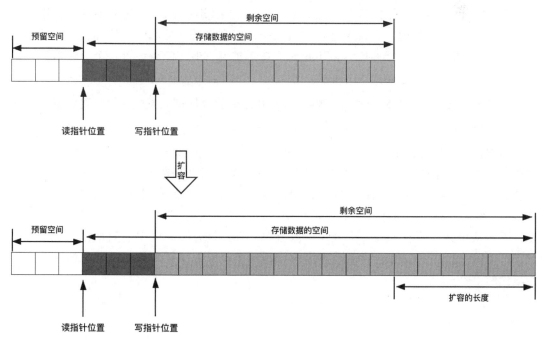

用代码实现上述逻辑并不难,我们甚至可以基于 stl 提供的 std::string、std::vector 等现成的类来实现,这里就不再贴出具体的实现代码了。

关于接收缓冲区和发送缓冲区,需要另外强调几点。

◎ 对于服务端程序来说,由于需要同时服务多个客户端,而每一路连接都会有一个接收缓冲区和发送缓冲区(per-socket buffer),所以不同连接从 socket 上读取出来但还没有被业务处理的数据,会被放在自己对应的接收缓冲区中;因 TCP 窗口太小暂时发不出去的数据,会被存放在自己所属连接的发送缓冲区中。服务进程、网络通信组件、连接对象和发送缓冲区、接收缓冲区的关系示意图如下图所示。

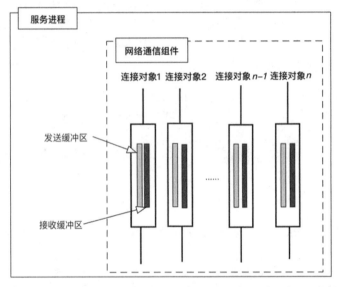

◎ 缓冲区的容量上限一般是有限制的,尤其对于服务程序,由于需要支持多个连接,所以为了节约内存,每个连接的接收缓冲区和发送缓冲区的初始容量一般都不会设置得很大。在缓冲区容量不足时应按需扩展,但一定要设置一个上限值,且这个上限值一般不会太大。

对于发送缓冲区,如果较长时间内发送缓冲区中的数据都未发送出去,我们就可以认为这路连接出问题了,可以将该缓冲区回收并关闭连接。

同理,对于接收缓冲区,如果接收缓冲区中的数据一直滞留甚至积压,我们就要好好检查自己处理数据的逻辑是否有问题,看看为何不能及时处理数据。当然,在实际开发中还会遇到这样一种情况:对端短时间内向服务端发送了大量数据,此时我们需要做一个限制策略,例如 3 秒内某路连接接收缓冲区中的数据已经达到 30MB,这时可以设置一个标志,不再从该路连接的 socket 上继续读取数据,直到接收缓冲区中的数据被处理掉一部分,再清除该标志,以便继续从该连接的 socket 上收取数据。这是一种常见的对单个连接限流的策略。

7.7.3 服务端发送数据时对端一直不接收的问题

这类问题一般出现在跨部门尤其是与外部开发人员合作时。假设现在有这样一种情形：我们的服务器对外提供服务，已经规定好协议格式，客户端由外部人员开发。

我们的服务端在向对端（客户端）发送数据时，对端可能因为一些问题（可能是逻辑 bug 或者其他一些问题）一直不从 socket 上收取数据，但服务端可能会定期产生一些数据发送给客户端，在发送一段时间后，由于 TCP 窗口太小，导致数据发送不出去，这样待发送的数据会在服务端相应连接的发送缓冲区中积压很久。如果我们不做任何处理，则服务很快会因为内存耗尽而被操作系统杀死。

对于这种情况，一般建议从以下两方面增加防御措施。

（1）为每路连接的发送缓冲区大小设置上限（如 2MB，可以根据单个数据包的大小和业务情况来定），设置方法在上文中已经介绍过了。当某路连接上的数据发送不出去时，在将数据存入发送缓冲区前，先判断缓冲区的最大剩余空间（包括允许扩容后的容量），如果剩余空间已经小于我们要放入的数据大小，也就是说缓冲区中的数据大小超过了我们规定的缓冲区容量上限，我们就认为该连接出了问题，关闭该路连接并回收相应的资源（如清空缓冲区、回收套接字资源等）即可。示例代码如下：

```
//m_outputBuffer 为发送缓冲区对象
size_t remainingLen = m_outputBuffer.remainingBytes();
//如果加入缓冲区中的数据长度超出了发送缓冲区的最大剩余量
if (remainingLen < dataToAppend.length())
{
    //关闭连接
    forceClose();
    return;
}

m_outputBuffer.append(static_cast<const char*>(dataToAppend.c_str()),
dataToAppend.length());
```

（2）如果因为一些原因（可能是逻辑 bug）导致一部分数据已经被积压在发送缓冲区中一段时间了，此后服务端未产生新的待发送的数据，此时发送缓冲区的数据容量未超过缓冲区上限。对此种情形若不做任何处理，发送缓冲区的数据就会一直积压，白白浪费系统资源。对于这种情形，我们一般会设置一个定时器，每隔一段时间（如 6 秒）检查一下在各路连接的发送缓冲区中是否还有数据未发送出去，也就是说，如果一个连接超过一定时间还存在未发送出去的数据，我们也认为该路连接出现问题，需要关闭该路连接并回收相应的资源（如清空缓冲区、回收套接字资源等）。示例代码如下：

```
//每 3 秒检测一次
const int SESSION_CHECK_INTERVAL = 6000;
SetTimer(SESSION_CHECK_TIMER_ID, SESSION_CHECK_INTERVAL);

void CSessionManager::OnTimer()
{
    for (auto iter = m_mapSession.begin(); iter != m_mapSession.end(); ++iter)
    {
```

```
        if (!CheckSession(iter->value))
        {
            //关闭session，回收相关的资源
            iter->value->ForceClose();

            iter = m_mapSession.erase(iter);
        }
    }
}

//检测在相应连接的发送缓冲区中是否还有未发送的数据
void CSessionManager::CheckSession(CSession* pSession)
{
    return pSession->GetConnection().m_outputBuffer.IsEmpty();
}
```

以上代码每隔 6 秒就会检测所有 Session 对应的 Connection 对象，如果发现发送缓冲区非空，则说明该连接的发送缓冲区中的数据已驻留 6 秒，将关闭该连接并清理资源。

7.8 网络库的分层设计

"对于计算机科学领域中的任何问题，都可以通过增加一个间接的中间层来解决"这句话几乎概括了计算机软件体系结构的设计要点。计算机软件体系结构从上到下都是按照严格的层次结构设计的，不仅整个体系如此，体系里面的每个组件如 OS 本身、很多应用程序、软件系统甚至很多硬件结构也如此。

7.8.1 网络库设计中的各个层

常见的网络通信库根据功能也可以分成很多层，根据离业务的远近从上到下依次是 Session 层、Connection 层、Channel 层、Socket 层，其中 Session 层属于业务层，Connection 层、Channel 层、Socket 层属于技术层，示意图如下。

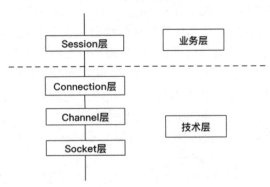

下面依次介绍各层的作用。

1. Session 层

Session 层处于顶层，在设计上不属于网络框架本身，用于记录各种业务状态数据和处理各种业务逻辑。在业务逻辑处理完毕后，如果需要进行网络通信，则依赖 Connection 层进行数据收发。例如一个 IM 服务的 Session 类可能有如下接口和成员数据：

```cpp
typedef std::shared_ptr<TcpConnection> TcpConnectionPtr;
typedef  const TcpConnectionPtr& CTcpConnectionPtrR;
class ChatSession
{
public:
    ChatSession(CTcpConnectionPtrR conn, int sessionid);
    virtual ~ChatSession();

    int32_t GetSessionId()
    {
        return m_id;
    }

    int32_t GetUserId()
    {
        return m_userinfo.userid;
    }

    std::string GetUsername()
    {
        return m_userinfo.username;
    }

    int32_t GetClientType()
    {
        return m_userinfo.clienttype;
    }

    int32_t GetUserStatus()
    {
        return m_userinfo.status;
    }

    int32_t GetUserClientType()
    {
        return m_userinfo.clienttype;
    }

    void SendUserStatusChangeMsg(int32_t userid, int type, int status = 0);
private:
    //各个业务逻辑的处理方法
    bool Process(CTcpConnectionPtrR conn, const char* inbuf, size_t buflength);

    void OnHeartbeatResponse(CTcpConnectionPtrR conn);
    void OnRegisterResponse(const std::string& data, CTcpConnectionPtrR conn);
    void OnLoginResponse(const std::string& data, CTcpConnectionPtrR conn);
    void OnGetFriendListResponse(CTcpConnectionPtrR conn);
    void OnFindUserResponse(const std::string& data, CTcpConnectionPtrR conn);
    void OnChangeUserStatusResponse(const std::string& data, CTcpConnectionPtrR conn);
```

```cpp
    TcpConnectionPtr GetConnectionPtr()
    {
        if (m_tmpConn.expired())
            return NULL;

        return m_tmpConn.lock();
    }
    //调用下层Connection层发送数据的方法
    void Send(int32_t cmd, int32_t seq, const std::string& data);
    void Send(int32_t cmd, int32_t seq, const char* data, int32_t dataLength);
    void Send(const std::string& p);
    void Send(const char* p, int32_t length);
private:
    int32_t             m_id;                   //session id
    OnlineUserInfo      m_userinfo;             //该Session对应的用户信息
    int32_t             m_seq;                  //当前Session数据包的序列号
    bool                m_isLogin;              //当前Session对应的用户是否已登录
    //引用下层Connection层的成员变量
    //但不管理TcpConnection对象的生命周期
    std::weak_ptr<TcpConnection>      m_tmpConn;
};
```

在以上代码中除了业务状态数据和业务接口，还有一个send系列的函数，这个函数依赖Connection对象进行数据收发。但是Session对象并不拥有Connection对象，也就是说Session对象不控制Connection对象的生命周期。这是因为虽然Session对象的主动销毁（如收到非法的客户端数据并关闭Session对象）会引起Connection对象的销毁，但Connection对象本身也可能因为网络出错等原因被销毁，进而引起Session对象被销毁。因此，在上述类接口描述中，ChatSession类使用了一个std::weak_ptr来引用TCPConnection对象。这是需要注意的地方。

2. Connection层

Connection层是技术层的顶层，每一路客户端连接都对应一个Connection对象，该层一般用于记录连接的各种状态信息。常见的状态信息有连接状态、数据收发缓冲区信息、数据流量信息、本端和对端的地址和端口号信息等，同时提供对各种网络事件的处理接口，这些接口或被本层自己使用，或被Session层使用。Connection持有一个Channel对象，而且掌管Channel对象的生命周期。

一个Connection对象可以提供的接口和记录的数据状态如下：

```cpp
class TcpConnection
{
public:
    TcpConnection(EventLoop* loop,
            const string& name,
            int sockfd,
            const InetAddress& localAddr,
            const InetAddress& peerAddr);
    ~TcpConnection();
```

```cpp
    const InetAddress& localAddress() const { return m_localAddr;}
    const InetAddress& peerAddress() const { return m_peerAddr; }
    bool connected() const { return m_state == kConnected; }

    void send(const void* message, int len);
    void send(const string& message);
    void send(Buffer* message);
    void shutdown();

    void forceClose();

    void setConnectionCallback(const ConnectionCallback& cb);
    void setMessageCallback(const MessageCallback& cb);
    void setCloseCallback(const CloseCallback& cb);
    void setErrorCallback(const ErrorCallback& cb);

    Buffer* getInputBuffer();
    Buffer* getOutputBuffer();

private:
    enum StateE { kDisconnected, kConnecting, kConnected, kDisconnecting };
    void handleRead(Timestamp receiveTime);
    void handleWrite();
    void handleClose();
    void handleError();
    void sendInLoop(const string& message);
    void sendInLoop(const void* message, size_t len);
    void shutdownInLoop();
    void forceCloseInLoop();
    void setState(StateE s) { m_state = s; }

  private:
    //连接状态信息
    StateE                    m_state;
    //引用 Channel 对象
    std::shared_ptr<Channel>  m_spChannel;
    //本端的地址信息
    const InetAddress         m_localAddr;
    //对端的地址信息
    const InetAddress         m_peerAddr;

    ConnectionCallback        m_connectionCallback;
    MessageCallback           m_messageCallback;
    CloseCallback             m_closeCallback;
    ErrorCallback             m_errorCallback;

    //接收缓冲区
    Buffer                    m_inputBuffer;
    //发送缓冲区
    Buffer                    m_outputBuffer;
    //流量统计类
    CFlowStatistics           m_flowStatistics;
};
```

3. Channel 层

Channel 层一般持有一个 socket 句柄，是实际进行数据收发的地方，因而一个 Channel 对象会记录当前需要监听的各种网络事件（读写和出错事件）的状态，同时提供对这些事件状态的查询和增删改接口。在部分网络库的实现中，Channel 对象管理着 socket 对象的生命周期，因此 Channel 对象需要提供创建和关闭 socket 对象的接口；而在另外一些网络库的实现中由 Connection 对象直接管理 socket 对象的生命周期，也就是说没有 Channel 层。所以，Channel 层不是必需的。

由于 TCP 收发数据是全双工的（收发走独立的通道，互不影响），所以收发逻辑一般不会有依赖关系，但收发操作一般会被放在同一个线程中进行，这样做的目的是防止在收发过程中改变 socket 状态时，对另一个操作产生影响。假设收发操作分别使用一个线程，在一个线程中收数据时因出错而关闭了连接，但另一个线程可能正在发送数据，则这样会出问题。

一个 Channel 对象提供的函数接口和状态数据如下：

```
class Channel
{
public:
    Channel(EventLoop* loop, int fd);
    ~Channel();

    void handleEvent(Timestamp receiveTime);

    int fd() const;
    int events() const;
    void setRevents(int revt);
    void addRevents(int revt);
    void removeEvents();
    bool isNoneEvent() const;

    bool enableReading();
    bool disableReading();
    bool enableWriting();
    bool disableWriting();
    bool disableAll();

    bool isWriting() const;

private:
    const int    m_fd;        //当前需要检测的事件
    int          m_events;    //处理后的事件
    int          m_revents;
```

4. Socket 层

严格来说，并不存在 Socket 层，这一层通常只是对常用的 socket 函数进行封装，例如屏蔽不同操作系统操作 socket 函数的差异性来实现跨平台，方便上层使用。

> 如果存在 Channel 层，则 Socket 层的上层就是 Channel 层；如果不存在 Channel 层，则 Socket 层的上层就是 Connection 层。

Socket 层也不是必需的，因此很多网络库都没有 Socket 层。下面是某 Socket 层对常用 socket 函数的功能进行一层简单封装的接口示例：

```
namespace sockets
{
#ifdef WIN32
#else
    typedef int SOCKET;
#endif

    SOCKET createOrDie();
    SOCKET createNonblockingOrDie();

    void setNonBlockAndCloseOnExec(SOCKET sockfd);

    void setReuseAddr(SOCKET sockfd, bool on);
    void setReusePort(SOCKET sockfd, bool on);

    int connect(SOCKET sockfd, const struct sockaddr_in& addr);
    void bindOrDie(SOCKET sockfd, const struct sockaddr_in& addr);
    void listenOrDie(SOCKET sockfd);
    int accept(SOCKET sockfd, struct sockaddr_in* addr);
    int32_t read(SOCKET sockfd, void *buf, int32_t count);
#ifndef WIN32
    ssize_t readv(SOCKET sockfd, const struct iovec *iov, int iovcnt);
#endif
    int32_t write(SOCKET sockfd, const void *buf, int32_t count);
    void close(SOCKET sockfd);
    void shutdownWrite(SOCKET sockfd);

    void toIpPort(char* buf, size_t size, const struct sockaddr_in& addr);
    void toIp(char* buf, size_t size, const struct sockaddr_in& addr);
    void fromIpPort(const char* ip, uint16_t port, struct sockaddr_in* addr);

    int getSocketError(SOCKET sockfd);

    struct sockaddr_in getLocalAddr(SOCKET sockfd);
    struct sockaddr_in getPeerAddr(SOCKET sockfd);
}
```

在实际开发中，有的服务在设计网络通信模块时会将 Connection 层与 Channel 层合并成一层，当然，这取决于业务的复杂程度。所以在某些服务代码中只看到 Connection 对象或者 Channel 对象时，请不要觉得奇怪。

另外，对于服务端程序，抛开业务本身，从技术层面上来说，我们需要一个 Server 对象（如 TcpServer）来集中管理多个 Connection 对象，这也是网络库自身需要处理好的部分。一个 TcpServer 对象可能需要提供如下函数接口和状态数据：

```cpp
class TcpServer
{
public:
    typedef std::function<void(EventLoop*)> ThreadInitCallback;
    enum Option
    {
        kNoReusePort,
        kReusePort,
    };

    TcpServer(EventLoop* loop,
              const InetAddress& listenAddr,
              const std::string& nameArg,
              Option option = kReusePort);
    ~TcpServer();

    void addConnection(int sockfd, const InetAddress& peerAddr);
    void removeConnection(const TcpConnection& conn);

    typedef std::map<string, TcpConnectionPtr> ConnectionMap;

private:
    int             m_nextConnId;
    ConnectionMap   m_connections;
};
```

对于客户端程序来说，同样可以设计出一个 TCPClient 对象来管理各个 Connector（连接器对象）。

Session 对象虽然与 Connection 对象一一对应，但在业务层（网络通信框架之外）需要有专门的类来管理这些 Session 对象的生命周期，我们一般把这个专门的类称为 SessionManager 或者 SessionFactory。

7.8.2 将 Session 进一步分层

不同的服务，其业务可能千差万别，在实际开发中，我们可以根据业务场景将 Session 层进一步拆分成多个层，使每一层都专注于自己的业务逻辑。例如，假设现在有一个需要支持聊天消息压缩的即时通信服务，我们可以将 Session 划分为三个层，从上到下依次是 ChatSession、CompressionSession 和 TcpSession。ChatSession 负责处理聊天业务本身，CompressSession 负责数据的解压缩，TcpSession 负责将数据加工成网络层需要的格式或者将网络层发送的数据还原成业务需要的格式（如数据装包和解包），示意图如下。

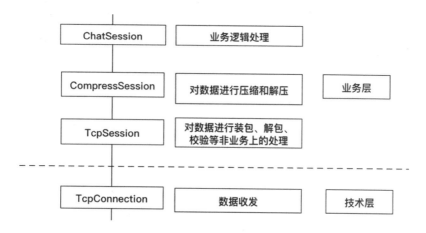

7.8.3 连接信息与 EventLoop/Thread 的对应关系

综合各层对象，一个 socket（fd）只对应一个 Channel 对象、一个 Connection 对象及一个 Session 对象，这组对象构成了一路连接信息（技术加业务上的）。结合前面介绍的 one thread one loop 思想，每一路连接信息都只能属于一个 loop，也就是说只属于某个线程；但是反过来，一个 loop 或者一个线程可以同时拥有多个连接信息，这就保证了我们只会在同一个线程里面处理特定的 socket 收发事件。

7.9 后端服务中的定时器设计

定时器模块是后端服务常用的功能之一，用于周期性地执行某些任务的场景中。设计定时器模块的方法很多，但关键是定时器的效率问题。让我们先从最简单的开始。

7.9.1 最简单的定时器

一个最简单的定时器功能可以按如下思路实现：

```
void* threadFunc(void* arg)
{
    while (m_bRunning)
    {
        //休眠 3 秒
        sleep(3000);

        //检测所有会话的心跳
        checkSessionHeartbeat();
    }
    return NULL;
}
```

以上代码在一个独立的线程中每隔 3 秒对所有 Session 做一次心跳检测。这是一个非常简单的实现逻辑，有些读者可能会觉得这样做有点"简单粗暴"。其实，这段代码来源于一个真实的商业项目，至今仍然工作得很好。在一些特殊场景下，我们确实可以按这种思路来实现定时器，只不过可能将 sleep 函数换成一些可以设置超时或等待时间的、让线程挂起或等待的函数（如 select、poll 等）。

但是上述实现定时器的方法毕竟适用场景太少，也不能与本章介绍的 one thread one loop 结构相结合，one thread one loop 结构中的定时器才是本节的重点。前面介绍的定时器实现只是"热身"，现在让我们正式开始。

7.9.2 定时器设计的基本思路

根据实际的场景需求，我们的定时器对象一般需要一个唯一标识、过期时间、重复次数、定时器到期时触发的动作，因此一个定时器对象可以被设计成如下结构：

```
typedef std::function<void()> TimerCallback;

//定时器对象
class Timer
{
    Timer() ;
    ~Timer();

    void run()
    {
        callback();
    }

    //其他实现在下文中会逐步完善...

private:
    int64_t         m_id;              //定时器的 ID，唯一标识一个定时器
    time_t          m_expiredTime;     //定时器的到期时间
    int32_t         m_repeatedTimes;   //定时器重复触发的次数
    TimerCallback   m_callback;        //定时器触发后的回调函数
};
```

注意，正如前面章节所强调的，在定时器回调函数 m_callback 中不能有耗时或者阻塞线程的操作，如果存在这种操作，则为了不影响整个循环流的执行，需要将这些耗时或者阻塞的操作移到其他线程中。

在 7.5 节中提到，使用定时器的 one thread one loop 结构如下：

```
while (!m_bQuitFlag)
{
    check_and_handle_timers();

    epoll_or_select_func();

    handle_io_events();

    handle_other_things();
}
```

我们在 check_and_handle_timers 函数中对各个定时器对象进行了处理（检测是否到

期，如果到期，则调用相应的定时器函数完成定时任务）。先从最简单的情形开始讨论，将定时器对象放在一个 std::list 对象中：

```
//m_listTimers 可以是 EventLoop 的成员变量
std::list<Timer*> m_listTimers;

void EventLoop::check_and_handle_timers()
{
    for (auto& timer : m_listTimers)
    {
        //判断定时器是否到期
        if (timer->isExpired())
        {
            timer->run();
        }
    }
}
```

为了方便管理所有定时器对象，我们可以专门新建一个 TimerManager 类对定时器对象进行管理，该对象提供了增加、移除和判断定时器是否到期等的接口：

```
class TimerManager
{
public:
    TimerManager() = default;
    ~TimerManager() = default;

    /** 添加定时器
     * @param repeatedTimes 定时器重复次数，设置为-1 时表示一直重复下去
     * @param interval      触发间隔
     * @param timerCallback 定时器触发时的回调函数
     * @return 返回创建成功的定时器 ID
     */
    int64_t addTimer(int32_t repeatedTimes, int64_t interval, const TimerCallback& timerCallback);

    /** 移除指定 ID 的定时器
     * @param timerId 待移除的定时器 ID
     * @return 成功移除定时器时返回 true，反之返回 false
     */
    bool removeTimer(int64_t timerId);

    /** 检测定时器是否到期，如果到期，则触发定时器函数
     */
    void checkAndHandleTimers();
private:
    std::list<Timer*> m_listTimers;
};
```

check_and_handle_timers 函数的实现如下：

```
void EventLoop::check_and_handle_timers()
{
    //m_timerManager 可以是 EventLoop 的成员变量
    m_timerManager.checkAndHandleTimers();
```

addTimer、removeTimer、checkAndHandleTimers 的实现如下：

```cpp
int64_t TimerManager::addTimer(int32_t repeatedTimes, int64_t interval, const TimerCallback& timerCallback)
{
    Timer* pTimer = new Timer(repeatedTimes, interval, timerCallback);
    m_listTimers.push_back(pTimer);
    return pTimer->getId();
}

bool TimerManager::removeTimer(int64_t timerId)
{
    for (auto iter = m_listTimers.begin(); iter != m_listTimers.end(); ++iter)
    {
        if ((*iter)->getId() == timerId)
        {
            m_listTimers.erase(iter);
            return true;
        }
    }

    return false;
}

void TimerManager::checkAndHandleTimers()
{
    Timer* deletedTimer;
    for (auto iter = m_listTimers.begin(); iter != m_listTimers.end(); )
    {
        if ((*iter)->isExpired())
        {
            (*iter)->run();

            if ((*iter)->getRepeatedTimes() == 0)
            {
                //定时器不需要重复，从集合中移除该对象
                deletedTimer = *iter;
                iter = m_listTimers.erase(iter);
                delete deletedTimer;
                continue;
            }
            else
            {
                ++iter;
            }
        }
    }
}
```

在向 addTimer 函数传递必要的参数后创建一个 Timer 对象，并返回唯一标识该定时器对象的 ID，后续就可以通过定时器 ID 操作这个定时器对象了。

这里的定时器 ID 使用了一个单调递增的 int64_t 类型的整数，我们也可以使用其他类

型,例如 uid,只要能唯一区分每个定时器对象即可。当然,在这里的设计逻辑中,可能存在多个线程多个 EventLoop,每个 EventLoop 都含有一个 m_timerManager 对象,但我们希望所有定时器 ID 都能够全局唯一,所以这里每次生成定时器 ID 时都使用了一个整型原子变量的 ID 基数,将它设置为 Timer 对象的静态成员变量,在每次需要生成新的定时器 ID 时都将其递增 1 即可。这里利用 C++ 11 的 std::mutex 对 s_initialId 进行保护:

```
//Timer.h
class Timer
{
public:
    Timer::Timer(int32_t repeatedTimes, int64_t interval, const TimerCallback&
timerCallback);
    ~Timer() {}

    bool isExpired();

    void run()
    {
        callback();
    }

    //其他无关代码省略

public:
    static int64_t generateId();        //生成一个唯一 ID

private:
    static int64_t         s_initialId{0};  //定时器 ID 的基准值,初始值为 0
    static std::mutex      s_mutex{};       //保护 s_initialId 的互斥体对象
};

//Timer.cpp
int64_t Timer::generateId()
{
    int64_t tmpId;
    s_mutex.lock();
    ++s_initialId;
    tmpId = s_initialId;
    s_mutex.unlock();

    return tmpId;
}

Timer::Timer(int32_t repeatedTimes, int64_t interval, const TimerCallback&
timerCallback)
{
    m_repeatedTimes = repeatedTimes;
    m_interval = interval;

    //当前时间加上触发间隔得到下一次的过期时间
    m_expiredTime = (int64_t)time(nullptr) + interval;
    m_callback = timerCallback;
    m_id = Timer::generateId();
```

}

定时器的下一次过期时间 m_expiredTime 是添加定时器的时间点加上触发间隔 interval，参见以上代码片段中的加粗行，也就是说，这里使用绝对时间点作为定时器的过期时间，读者在自己的实现中也可以使用相对时间间隔。

在这里的实现中，定时器还有个表示触发次数的变量：m_repeatedCount，m_repeatedCount 为-1 时表示不限制触发次数（即一直触发），m_repeatedCount 大于 0 时表示触发 m_repeatedCount 次，每触发一次，m_repeatedCount 就递减 1，一直到 m_repeatedCount 等于 0 时将其从定时器集合中移除：

```
void TimerManager::checkAndHandleTimers()
{
    Timer* deletedTimer;
    for (auto iter = m_listTimers.begin(); iter != m_listTimers.end(); )
    {
        if ((*iter)->isExpired())
        {
            //执行定时器事件
            (*iter)->run();

            if ((*iter)->getRepeatedTimes() == 0)
            {
                //定时器不需要重复触发从集合中移除该对象
                deletedTimer = *iter;
                iter = m_listTimers.erase(iter);
                delete deletedTimer;
                continue;
            }
            else
            {
                ++iter;
            }
        }
    }
}
```

在以上代码中先遍历定时器对象集合，然后调用 Timer::isExpired 函数判断当前定时器对象是否到期，该函数的实现如下，实现很简单，即用定时器的到期时间与当前系统时间做比较：

```
bool Timer::isExpired()
{
    int64_t now = time(nullptr);
    return now >= m_expiredTime;
}
```

如果一个定时器已经到期，则执行定时器 Timer::run()，该函数不仅调用定时器回调函数，还更新定时器对象的状态信息（如触发次数和下一次触发的时间点）：

```
void Timer::run()
{
    m_callback();
```

```
    if (m_repeatedTimes >= 1)
    {
        --m_repeatedTimes;
    }

    //计算下一次的触发时间
    m_expiredTime += m_interval;
}
```

除了可以在定时器触发次数变为 0 时将定时器对象从定时器列表中移除，也可以调用 removeTimer 函数主动从定时器列表中移除一个定时器对象：

```
bool TimerManager::removeTimer(int64_t timerId)
{
    for (auto iter = m_listTimers.begin(); iter != m_listTimers.end(); ++iter)
    {
        if ((*iter)->getId() == timerId)
        {
            m_listTimers.erase(iter);
            return true;
        }
    }

    return false;
}
```

removeTimer 函数通过定时器 ID 成功移除一个定时器对象时返回 true，否则返回 false。

完整的代码如下。

Timer.h：

```
#ifndef __TIMER_H__
#define __TIMER_H__

#include <functional>

typedef std::function<void()> TimerCallback;

class Timer
{
public:
    /**
     * @param repeatedTimes    定时器重复次数，设置为-1时表示一直重复下去
     * @param interval         下一次触发的时间间隔
     * @param timerCallback    定时器触发后的回调函数
     */
    Timer(int32_t repeatedTimes, int64_t interval, const TimerCallback&
timerCallback);
    ~Timer();

    int64_t getId()
    {
        return m_id;
    }
```

```cpp
    bool isExpired();

    int32_t getRepeatedTimes()
    {
        return m_repeatedTimes;
    }

    void run();

    //暂且省略其他实现
public:
    //生成一个唯一ID
    static int64_t generateId();

private:
    int64_t             m_id;                //定时器的ID,唯一标识一个定时器
    time_t              m_expiredTime;       //定时器的到期时间
    int32_t             m_repeatedTimes;     //定时器重复触发的次数
    TimerCallback       m_callback;          //定时器触发后的回调函数
    int64_t             m_interval;          //触发时间间隔

    static int64_t      s_initialId{0};      //定时器的ID基准值,初始值为0
    static std::mutex   s_mutex{};           //保护s_initialId的互斥体对象
};
#endif //!__TIMER_H__
```

Timer.cpp:

```cpp
#include "Timer.h"
#include <time.h>

int64_t Timer::generateId()
{
    int64_t tmpId;
    s_mutex.lock();
    ++s_initialId;
    tmpId = s_initialId;
    s_mutex.unlock();

    return tmpId;
}

Timer::Timer(int32_t repeatedTimes, int64_t interval, const TimerCallback&
timerCallback)
{
    m_repeatedTimes = repeatedTimes;
    m_interval = interval;

    //当前时间加上触发间隔得到下一次的过期时间
    m_expiredTime = (int64_t)time(nullptr) + interval;

    m_callback = timerCallback;

    //生成唯一ID
```

```cpp
    m_id = Timer::generateId();
}

bool Timer::isExpired() const
{
    int64_t now = time(nullptr);
    return now >= m_expiredTime;
}

void Timer::run()
{
    m_callback();

    if (m_repeatedTimes >= 1)
    {
        --m_repeatedTimes;
    }

    m_expiredTime += m_interval;
}
```

TimerManager.h:

```cpp
#ifndef __TIMER_MANAGER_H__
#define __TIMER_MANAGER_H__

#include <stdint.h>
#include <list>

#include "Timer.h"

void defaultTimerCallback()
{
}

class TimerManager
{
public:
    TimerManager() = default;
    ~TimerManager() = default;

    /** 添加定时器
     * @param repeatedCount    重复次数
     * @param interval         触发间隔
     * @param timerCallback    定时器回调函数
     * @return                 返回创建成功的定时器 ID
     */
    int64_t addTimer(int32_t repeatedCount, int64_t interval, const TimerCallback&
timerCallback);

    /** 移除指定 ID 的定时器
     * @param timerId  待移除的定时器 ID
     * @return 成功移除时定时器返回 true, 否则返回 false
     */
    bool removeTimer(int64_t timerId);
```

```cpp
    /** 检测定时器是否到期，如果到期，则触发定时器函数
    */
    void checkAndHandleTimers();

private:
    std::list<Timer*> m_listTimers;
};
#endif //!__TIMER_MANAGER_H__
```

TimerManager.cpp：

```cpp
#include "TimerManager.h"

int64_t TimerManager::addTimer(int32_t repeatedCount, int64_t interval, const TimerCallback& timerCallback)
{
    Timer* pTimer = new Timer(repeatedCount, interval, timerCallback);
    m_listTimers.push_back(pTimer);
    return pTimer->getId();
}

bool TimerManager::removeTimer(int64_t timerId)
{
    for (auto iter = m_listTimers.begin(); iter != m_listTimers.end(); ++iter)
    {
        if ((*iter)->getId() == timerId)
        {
            m_listTimers.erase(iter);
            return true;
        }
    }

    return false;
}

void TimerManager::checkAndHandleTimers()
{
    Timer* deletedTimer;
    for (auto iter = m_listTimers.begin(); iter != m_listTimers.end(); )
    {
        if ((*iter)->isExpired())
        {
            (*iter)->run();

            if ((*iter)->getRepeatedTimes() == 0)
            {
                //在定时器不需要触发时，从集合中移除该对象
                deletedTimer = *iter;
                iter = m_listTimers.erase(iter);
                delete deletedTimer;
                continue;
            }
            else
```

```
            {
                ++iter;
            }
        }
    }
}
```

以上就是定时器的基本设计思路，我们一定要明白在这个流程中一个定时器对象具有哪些属性，以及如何管理定时器对象。当然，这里自顶向下一共有三层，分别是 EventLoop、TimerManager、Timer，其中 TimerManager 对象不是必需的，在一些设计中直接用 EventLoop 封装的相应方法对 Timer 对象进行管理。

理解 one thread one loop 中定时器的设计思路之后，我们来看看上述定时器实现中的性能问题。

7.9.3 定时器逻辑的性能优化

上述定时器实现存在严重的性能问题，即每次检测定时器对象是否触发时都要遍历整个定时器集合，移除定时器对象时也需要遍历整个定时器集合。我们其实可以将定时器对象按过期时间从小到大排序，这样检测定时器对象时，只要从最小过期时间开始检测即可，一旦找到过期时间大于当前时间的定时器对象，就不需要继续检测后面的定时器对象了。

1. 定时器对象集合的数据结构优化一

我们可以在每次将定时器对象添加到集合时都自动进行排序，如果仍然使用 std::list 作为定时器集合，则可以给 std::list 自定义一个排序函数（从小到大排序），代码实现如下：

```
//Timer.h
class Timer
{
public:
    //省略无关代码

    int64_t getExpiredTime() const
    {
        return m_expiredTime;
    }
};
//TimerManager.h
struct TimerCompare
{
    bool operator() (const Timer* lhs, const Timer* rhs)
    {
        return lhs->getExpiredTime() < rhs->getExpiredTime();
    }
}
```

每次添加定时器时都调用自定义排序函数对象 TimerCompare（以下代码片段中的加粗行）：

```cpp
int64_t TimerManager::addTimer(int32_t repeatedCount, int64_t interval, const
TimerCallback& timerCallback)
{
    Timer* pTimer = new Timer(repeatedCount, interval, timerCallback);
    m_listTimers.push_back(pTimer);
    //对定时器对象按过期时间从小到大排序
    m_listTimers.sort(TimerCompare());
    return pTimer->getId();
}
```

将定时器对象按过期时间从小到大排好序后,检测各个定时器对象是否触发时不用再遍历整个定时器对象集合,只要从过期时间最小的定时器对象开始检测,一直找到过期时间大于当前系统时间的定时器对象就可以停止检测了,实现逻辑如下:

```cpp
void TimerManager::checkAndHandleTimers()
{
    //在遍历过程中是否调整了部分定时器的过期时间
    bool adjusted = false;
    Timer* deletedTimer;
    for (auto iter = m_listTimers.begin(); iter != m_listTimers.end(); )
    {
        if ((*iter)->isExpired())
        {
            //在 run 函数中执行完定时器任务后会对定时器的过期时间进行调整
            (*iter)->run();

            if ((*iter)->getRepeatedTimes() == 0)
            {
                //在定时器不需要再触发时从集合中移除该对象
                deletedTimer = *iter;
                iter = m_listTimers.erase(iter);
                delete deletedTimer;
                continue;
            }
            else
            {
                ++iter;
                //对集合中有定时器调整了过期时间的情况进行标记
                adjusted = true;
            }
        }
        else
        {
            //若找到大于当前系统时间的定时器对象,则无须继续检查,退出循环
            break;
        }// end if
    }// end for-loop

    //由于调整了部分定时器的过期时间,所以需要重新排序
    if (adjusted)
        m_listTimers.sort(TimerCompare());
}
```

在以上代码中有个细节需要注意：假设现在的系统时刻是 now，定时器集合中定时器的过期时间从小到大依次为 t1、t2、t3、t4、t5……tn，假设 t4 < now < t5，即此刻 t1、t2、t3、t4 对应的定时器会触发，触发后，会从 t1、t2、t3、t4 中减去相应的时间间隔，更新后的 t1′、t2′、t3′、t4′就不一定小于 t5 ~ tn 了，因此需要再次对定时器集合进行排序。但是存在一种情形：t1 ~ t5 触发后对应的触发次数正好变为 0，因此需要从定时器列表中移除它们，在这种情形下就不需要对定时器列表进行排序了。因此以上代码使用了一个 adjusted 变量记录是否有过期时间被更新且未被从列表中移除的定时器对象，如果有，则之后再次对定时器集合进行排序。

上述设计虽然解决了定时器遍历效率低下的问题，但是无法解决移除一个定时器时仍然需要遍历的问题，使用链表结构的 std::list 插入非常方便，但定位某个具体元素的效率较低。我们可以将 std::list 转换成 std::map，当然，我们仍然需要对 std::map 中的定时器对象按过期时间进行自定义排序。

2. 定时器对象集合的数据结构优化二

为了提高定时器的效率，我们一般采用两种常用的方法：时间轮和时间堆。

时间轮的基本思想是将现在时刻 t 加上一个时间间隔 interval，以 interval 为步长，将各个定时器对象的过期时间按步长分布在不同的时间槽（time slot）中，当在一个时间槽中出现多个定时器对象时，这些定时器对象按加入槽的顺序串成链表，时间轮的示意图如下所示。

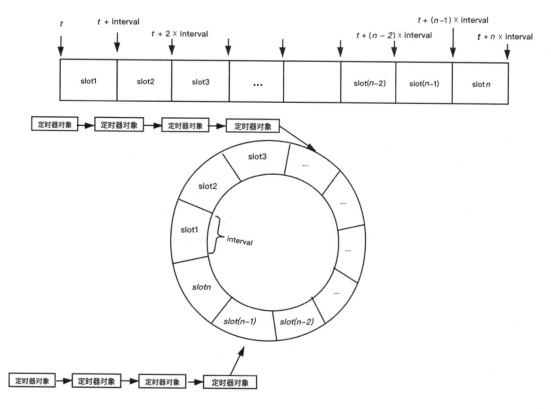

因为每个时间槽的时间间隔都是一定的，因此对时间轮中的定时器对象的检测会有两种方法。

（1）在每次检测时都判断当前系统时间处于哪个时间槽中，比该槽序号小的槽中的所有定时器都已到期，执行对应的定时器函数之后，移除不需要重新触发的定时器，或重新计算需要下一次触发的定时器对象的时间并重新计算，将其移到新的时间槽中。这适用于 one loop one thread 结构。

（2）在每次检测时都假设当前时间与之前相比跳动了一个时间轮的间隔，适用场景较少，不适用于 one loop one thread 结构。

时间轮实际上是将一个链表按时间分组，这虽然提高了一些效率，但还是存在问题，尤其是某个时间槽对应的链表较长时。

时间堆指利用数据结构中小根堆（Min Heap）到期时间的大小来组织定时器对象，如下图所示，图中小根堆的各个节点都代表一个定时器对象，它们按过期时间从小到大排列。使用小根堆在管理定时器对象和执行效率上都要优于前面方案中的 std::list 和 std::map，这是目前一些主流网络库中涉及定时器部分的实现，例如 Libevent。笔者在实际项目中会使用 stl 提供的优先队列即 std::priority_queue 作为定时器的实现，使用 std::priority_queue 的排序方式是从小到大，这是因为 std::priority 从小到大排序时，其内部实现的数据结构也是小根堆。

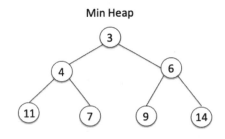

7.9.4　对时间的缓存

在使用定时器功能时，免不了要使用获取操作系统时间的函数，而在大多数操作系统上，获取系统时间的函数都属于系统调用，一次系统调用相对于 one thread one loop 结构中的其他逻辑来说可能耗时更多。因此为了提高效率，在一些对时间要求精度不是特别高的情况下，我们可能会缓存一些时间，在较近的下次如果需要系统时间，则可以使用上次缓存的时间，而不是再次调用获取系统时间的函数。目前有不少网络库和商业服务在定时器逻辑这一块都使用了这一策略。

上述逻辑的伪代码如下：

```
while (!m_bQuitFlag)
{
    //这里第1次获取系统时间，并缓存它
    get_system_time_and_cache();
```

```
    //利用上一步获取的系统时间进行一些耗时短的操作
    do_something_quickly_with_system_timer();

    //这里可以不用再次获取系统时间,而是将第 1 步缓存的时间作为当前系统时间
    use_cached_time_to_check_and_handle_timers();

    epoll_or_select_func();

    handle_io_events();

    handle_other_things();
}
```

最后总结一下:定时器的实现原理和逻辑并不复杂,关键点是如何为定时器对象集合设计出高效的数据结构,使每次从定时器集合中增加、删除、修改和遍历定时器对象时都更高效。另外,为了进一步提高定时器逻辑的执行效率,在某些场景下,我们可以利用上次缓存的系统时间来避免再一次调用获取时间的系统 API 的开销。

定时器的设计还有其他一些需要考虑的问题,例如定时器逻辑如何解决服务器机器时间被人为提前或者延后,以及定时器事件的时间精度等,可以自行研究和解决。

> 定时器逻辑如何解决服务器机器时间被人为提前或者延后的问题,在 Redis 服务端的源码中有优雅的实现。

7.10 处理业务数据时是否一定要单独开线程

如前文所述,一个 loop 的主要结构一般如下:

```
while (!m_bQuitFlag)
{
    epoll_or_select_func();

    handle_io_events();

    handle_other_things();
}
```

对于一些业务逻辑处理较简单、不会太耗时的应用来说,handle_io_events 方法除了可以用来收发数据,也可以直接用来做业务逻辑处理。在这种情形下,handle_io_events 方法的逻辑结构如下:

```
void handle_io_events()
{
    //收发数据
    recv_or_send_data();

    //解包并处理数据
    decode_packages_and_process();
}
```

其中，在 recv_or_send_data 方法中调用 send/recv API 进行网络数据收发操作。以收数据为例，收完数据并将其存入接收缓冲区后，接下来进行解包处理，然后进行业务处理。比如对于一个登录数据包的处理，其业务一般是验证登录的账户和密码是否正确并记录其登录行为等。从程序函数调用堆栈来看，这些业务处理逻辑其实是直接在网络收发数据线程中处理的，即网络线程调用 handle_io_events 方法，在 handle_io_events 方法中调用 decode_packages_and_process 方法，在 decode_packages_and_process 方法中做具体的业务逻辑处理，示意图如下。

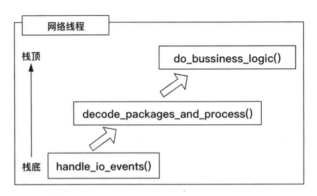

为了让网络层与业务层脱耦，在网络层中通常会提供一些回调函数的接口，我们可以将这些回调函数指向具体的业务处理函数。以 libevent 网络库的用法为例：

```
int main(int argc, char **argv)
{
    struct event_base *base;
    struct evconnlistener *listener;
    struct event *signal_event;

    struct sockaddr_in sin;

    base = event_base_new();

    memset(&sin, 0, sizeof(sin));
    sin.sin_family = AF_INET;
    sin.sin_port = htons(PORT);

    //listener_cb 是我们自定义的回调函数
    listener = evconnlistener_new_bind(base, listener_cb, (void *)base,
        LEV_OPT_REUSEABLE|LEV_OPT_CLOSE_ON_FREE, -1,
        (struct sockaddr*)&sin,
        sizeof(sin));

    if (!listener) {
        fprintf(stderr, "Could not create a listener!\n");
        return 1;
    }

    //signal_cb 是我们自定义的回调函数
    signal_event = evsignal_new(base, SIGINT, signal_cb, (void *)base);
```

```
if (!signal_event || event_add(signal_event, NULL)<0) {
    fprintf(stderr, "Could not create/add a signal event!\n");
    return 1;
}

//启动 loop
event_base_dispatch(base);

evconnlistener_free(listener);
event_free(signal_event);
event_base_free(base);

printf("done\n");
return 0;
}
```

以上代码根据 libevent 自带的 helloworld 示例修改而来，其中 listener_cb 和 signal_cb 是自定义的回调函数，有相应的事件触发后，libevent 的事件循环会调用我们设置的回调，在这些回调函数中，我们可以编写自己的业务逻辑代码。

这种基本的服务器结构如下图所示。

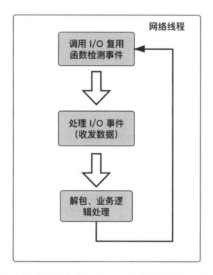

以上示意图展示了这个结构的基本逻辑，在这结构的基础上可以延伸出很多变体。在解包、业务逻辑处理部分（位于 handle_io_events() 的 decode_packages_and_process 方法中），如果业务逻辑处理过程比较耗时（例如，从数据库取大量数据、写文件），就会导致网络线程在这些操作上停留时间很长，很久以后才能执行下一次循环，影响网络事件的检测和数据收发的及时性，导致整个程序的效率低下。

因此对于这种情形，我们需要将业务处理逻辑单独拆分出来交给另外的业务工作线程处理，业务工作线程可以是单独一个线程，也可以是含有一组线程的线程池，业务数据从网络线程组流向业务线程组。这样的程序结构如下图所示。

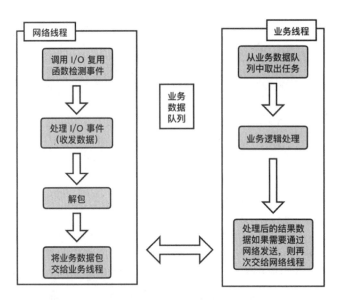

在上图中，对于网络线程将业务数据包交给业务线程的流程，可以使用一个共享的业务数据队列来实现，此时网络线程是生产者，业务线程从业务数据队列中取出任务去处理，业务线程是消费者。业务线程在处理完成后如需将结果数据发送出去，则再将数据交给网络线程。这时，经过处理的数据会从业务线程再次流向网络线程，那么如何将数据从业务线程交给网络线程呢？以发送数据为例，一般有三种方法。

（1）直接调用相应的发送数据的方法，如果我们的网络线程本身也会调用这些发送数据的方法，那么此时网络线程和业务线程可能同时对发送数据的方法进行调用，相当于多个线程同时调用 socket send 函数，这样可能导致同一个连接上的数据顺序有问题。此时应该利用锁机制，保证同一时刻只有一个线程可以调用发送数据的方法。这里给出一段伪代码，假设 TcpConnection 对象表示某路连接，则无论是网络线程还是业务线程，在处理完数据后需要发送数据时都会调用 TcpConnection::sendData 方法，TcpConnection::sendData 方法的实现如下：

```
void TcpConnection::sendData(const std::string& data)
{
    //加锁，保证同一时刻只有一个线程调用接下来的socket send 函数
    std::lock_guard<std::mutex> scoped_lock(m_mutexForConnection);
    //这里调用 socket send 函数
}
```

但是，该方法在设计上存在不足之处，即发送数据应该是网络层的事情，不能由其他模块（这里指业务线程）越俎代庖。

（2）在存在定时器结构的情况下，网络线程结构如下：

```
while (!m_bQuitFlag)
{
    check_and_handle_timers();

    epoll_or_select_func();
```

```
    handle_io_events();
}
```

业务线程可以将需要发送的数据放入另一个共享区域中（例如相应的 TcpConnection 对象的一个成员变量中），定时器定时将其从这个共享区域中取出来，再发送出去。该方法的优点是网络线程做了它该做的事情，缺点是需要添加定时器，程序逻辑变得复杂，而且定时器每隔一段时间才会触发，发送的数据可能会有一定的延迟。

（3）利用线程执行流中的 handle_other_things 方法，我们再来看看前面介绍的基本结构：

```
while (!m_bQuitFlag)
{
    epoll_or_select_func();

    handle_io_events();

    handle_other_things();
}
```

7.5 节讲解 one thread one loop 思想时介绍了 handle_other_things 函数可以做一些"其他事情"，这个函数可以在需要执行时通过唤醒机制唤醒。业务线程将数据放入某个共享区域中作为"其他事情"，然后在 handle_other_things 函数中执行数据的发送。如果读到这里时仍能思路清晰，则说明读者已经大致明白一个不错的服务器框架的实现方式了。上面介绍的服务器结构是目前主流的、基于 Reactor 模式的服务程序（如 libevent、libuv 网络库）的通用结构。下面再深入讨论一下。

在实际应用中，很多程序的业务逻辑处理其实是不耗时的，也就是说，这些业务逻辑的处理速度很快。由于 CPU 核数有限，所以在线程数量超过 CPU 数量时，各个线程（网络线程和业务线程）也不是真正地并行执行，即使开了一组业务线程也不一定真正地并发执行。在这种业务逻辑处理不耗时的情况下，可以在网络线程里面直接处理业务逻辑，这样并不会影响网络线程的执行效率。

前面讲到，在 handle_io_events 方法中会直接处理业务逻辑，如果在处理业务逻辑时会产生新的任务，则可以将这些新任务作为"其他事情"投递给网络线程，最终的网络线程的 handle_other_things 方法会处理这些新任务。此时的服务器程序结构如下图所示。

再次强调：利用 handle_io_events 或 handle_other_things 方法处理业务逻辑，仅适用于业务逻辑中不会有耗时操作的场景，如果在 handle_io_events 方法或 handle_other_things 方法中有耗时操作，则还需要单独开业务线程。虽然线程数量超过 CPU 数量时，各个线程不会真正并行，但那是操作系统线程调度的事情，做应用层开发时不必关心这一点。

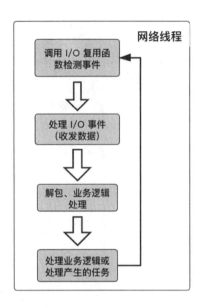

7.11 非侵入式结构与侵入式结构

在通常情况下，一个后端服务器的结构是无法固定下来的，因为我们不知道在服务的技术框架上面会搭建什么业务，结构随着业务的不同而不同。如果是这样的话，我们是不是就无法讨论和抽象出一个可以适用于大多数业务场景的服务程序结构了呢？其实未必，大多数服务都会与其他服务或客户端进行网络通信，那么这个服务一定包含网络通信模块。这样的话，网络通信模块将是不同业务的服务都有的一部分。以此为支点，我们尝试继续探究一套适用于大多数业务的通用程序结构。

在假设网络通信框架结构确定的情况下，根据通信数据从网络框架中流入流出的情况，我们将服务程序的结构分为非侵入式结构和侵入式结构两种。

7.11.1 非侵入式结构

非侵入式结构更简单一点，我们先来讨论它。非侵入式，指的是一个服务中的所有通信或业务数据都在网络通信框架内部流动，也就是说没有外部数据源注入网络通信模块或从网络通信模块中流出。举个例子，一个 IM 服务程序，在通常情况下，无论是单聊消息还是群聊消息，其核心业务本身的数据流都是在网络通信模块内部流动的。单聊时，A 用户向 B 用户发送一条消息，实际上消息流是从 A 用户的 Connection 对象传递到 B 用户的 Connection 对象上的，然后通过 B 的 Connection 对象的发送方法将数据发送出去。群聊也一样，数据从一个用户的 Connection 对象同时传递给其他多个用户的 Connection 对象。无论是哪种情况，这些 Connection 对象都是网络通信模块的内部结构。

消息在内部流动示意图

7.11.2 侵入式结构

如果有外部消息流入网络通信模块或从网络通信模块流出，就相当于有外部消息"侵入"网络通信结构，我们把这种服务器结构称为侵入式服务结构，示意图如下所示。

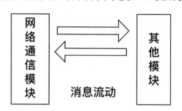

侵入式结构示意图

侵入式服务器的结构除网络通信组件外，其他组件的结构设计可以多种多样。来看两种通用结构。

◎ 通用结构一：业务线程（或称数据源线程）将数据处理后交给网络通信组件发送。
◎ 通用结构二：网络解包后需要将任务交给专门的业务线程处理，处理完后需要再次通过网络通信组件发送出去。

通用结构一其实是通用结构二的后半部分，因此这里重点讨论通用结构二。

在 one thread one loop 思想下，每个网络线程的基本结构都如下：

```
while (!m_bQuitFlag)
{
    epoll_or_select_func();

    handle_io_events();

    handle_other_things();
}
```

若 handle_io_events 收完网络数据后解包，则由于解包后得到的任务处理逻辑比较耗时，所以需要把这些任务交给专门的业务线程处理。业务线程可以是一组工作的消费者线程，我们可以将这些任务放在某个队列中。这里网络组件的线程（网络线程）是生产者，业务工作线程是消费者，我们可以使用互斥体、临界区（CriticalSection，Windows 系统特有）或条件变量等技术协调生产者和消费者，也就是说会涉及一个公共队列系统。这是一种常用的实现，数据会从网络组件流向业务组件。

接下来，如果业务组件需要对处理后的数据做网络通信操作，则此时如何将处理后的数据由业务组件交给网络组件呢？一般有两种方法，下面会详细讲解这两种方法。

1. 方法一

直接通过某些标识（如业务对应的 socket fd、sessionID 等）找到这些数据在网络组件中对应的 session，利用这些 session 将数据直接发送出去。

例如，在某个场景下，业务组件在处理完数据后需要将这些数据发送给所有用户，示例代码如下：

```cpp
void WebSocketSessionManager::pushDataToAll(const std::string& dataToPush)
{
    std::lock_guard<std::mutex> scoped_lock(m_mutexForSession);
    for (auto& session : m_mapSessions)
        session.second->pushSomeData(dataToPush);
}
```

在以上代码中，业务组件只要拿到网路组件管理所有 session 的对象 WebSocketSessionManager 即可，然后利用 WebSocketSessionManager 对象的 pushDataToAll 方法将数据发送给所有用户。dataToPush 是需要发送给所有用户的数据，因此在 pushDataToAll 方法中遍历所有 Session 对象（Session 对象被记录在 m_mapSessions 中）并通过 Session 对象挨个发送 dataToPush。

如果业务组件处理后的数据是发送给某个用户的，则示例代码如下：

```cpp
bool WebSocketSessionManager::pushDataToSingle(const std::string& accountID, const std::string& dataToPush)
{
    std::lock_guard<std::mutex> scoped_lock(m_mutexForSession);
    for (auto& session : m_mapSessions)
    {
        if (session.second->isAccountIDMatched(accountID))
        {
            session.second->pushSomeData(dataToPush);
            return true;
        }
    }

    return false;
}
```

以上代码中的业务组件根据数据中的 accountID 定位到网络组件中的具体 session，然后将数据发送给该 session 对应的用户。

这是业务组件处理完数据后向网络线程传送数据的常用方法，但存在如下两个缺点。

1）缺点一

这里从调用关系来看，实际上是业务线程调用网络线程相关的接口函数发送数据的，也就是说，其本质上是业务组件直接发起的网络操作。如果按功能来划分，则发送数据应该属于网络线程的功能，业务线程不应该发送数据。由于 Session 对象属于网络线程（网络线程管理着 Session 对象的生命周期），而这里的业务线程直接操作了 Session 对象，因此在以上示例代码中使用了 mutex（成员变量 m_mutexForSession）在相应的发送函数中对 Session 对象集合 m_mapSessions 进行保护。

这种方法的示意图如下。

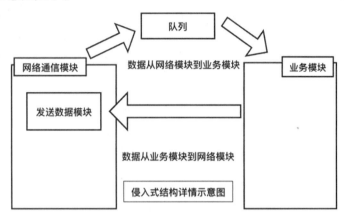

虽然这种做法不太合理，但实际上很多服务程序都在这么做：当业务组件调用这些发送方法时，通过 mutex 将这些 session 锁定。然而这样做存在一个效率问题，这里还是以上面向所有用户发送数据的示例来说明这个问题：

```
void WebSocketSessionManager::pushDataToAll(const std::string& dataToPush)
{
    std::lock_guard<std::mutex> scoped_lock(m_mutexForSession);
    for (auto& session : m_mapSessions)
        session.second->pushSomeData(dataToPush);
}
```

这段代码实际上调用了每个 session 对象的 pushSomeData 方法（加粗代码行），如果 Session 对象的 pushSomeData 方法耗时较长（耗时较长是相对来说的，在实际开发中要避免让这个函数耗时过长），则由于记录 Session 对象的 m_mapSessions 对象此时正被业务模块（业务线程）使用，所以如果网络线程想修改 m_mapSessions 对象，就必须等业务线程调用 WebSocketSessionManager::pushDataToAll 函数结束，这可能会影响网络线程的执行效率。因此有些开发者会这么设计：

```
void WebSocketSessionManager::pushDataToAll(const std::string& dataToPush)
{
    std::map<int64_t, BusinessSession*>   mapLocalSessions;

    {
        std::lock_guard<std::mutex> scoped_lock(m_mutexForSession);
        //从 m_mapSessions 拷贝 session 对象指针去 mapLocalSessions
        mapLocalSessions = m_mapSessions;
    }

    //这里使用 mapLocalSessions，让网络线程继续操作 m_mapSessions
    for (auto& session : mapLocalSessions)
        session.second->pushSomeData(dataToPush);
}
```

以上代码使用了一个临时变量 mapLocalSessions 将在 m_mapSessions 中记录的 Session 指针复制一份出来，这样的话，m_mutexForSession 锁保护的范围就大幅度减小了，业务线程可以尽快释放对 m_mapSessions 的占用，网络线程可以很快地使用 m_mapSessions。

这个看似很不错的设计方案，却存在着严重的错误：记录在 m_mapSessions 和 mapLocalSessions 中的各个 Session 指针都指向一个对象，倘若此时某个连接断开，网络线程就会销毁 m_mapSessions 中相应的 Session 对象，然而业务线程还可能拿着这个 Session 对象的指针继续操作（这里是发数据），此时这个指针已经是一个野指针了，所以会导致程序崩溃。

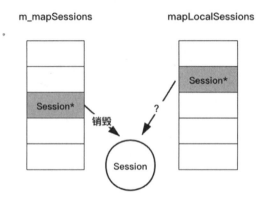

有的读者会说，在 mapLocalSessions 中记录的 Session 对象不要使用原始指针，可以使用一个智能指针，但该智能指针不管理该 Session 的生命周期，如下所示：

```
std::map<int64_t, std::weak_ptr<BusinessSession>> mapLocalSessions;
```

在决定是否使用一个 Session 对象前，这样做可以及时发现该 Session 对象是否有效，但若业务线程正在使用某个 Session 对象，进入 session.second->pushSomeData 函数后，Session 被网络线程回收了，则业务线程访问该 Session 对象的任何成员变量时都会因访问非法内存而导致程序崩溃。

所以，这是不正确的设计，在这种场景下千万不要使用这种减小锁范围的技术。为保证性能，应该尽量让 session.second->pushSomeData 函数的实现逻辑运行起来更快一点。

2）缺点二

方法一在一些场景中存在严重错误，假设我们的服务存在如下两条信息流。

◎ 信息流一：业务组件产生的数据需要发送。
◎ 信息流二：网络线程本身与客户端或者下游服务交互后产生的数据也要发送。

这两类数据如果被发送给同一个连接，但这两类数据有一定的顺序要求，则这样会很糟糕：因为在这样的设计中，业务线程会间接使用某个 Session 发送数据，网络线程会直接使用同一个 Session 发送数据，相当于多个线程同时调用 send 函数在同一个 socket 上发送数据，这样的话，每个单独的数据包不一定会出错，但是多个数据包之间的顺序就可能不正确了。

笔者曾在开发一套交易系统的行情推送服务时遇到过这样的问题，如下图所示，在客户端订阅了某类行情信息后，行情推送服务的网络模块会通过一个内部的 RPC 请求服务去拉取一个全量数据 t 并推送给客户端，同时行情推送服务的业务模块会从 RocketMQ 中取出增量数据 t1、d2、t3、d4 等（t 系列增量数据用于改造全量数据 t），然后侵入网络组件发送给客户端。客户端在收到增量数据后会以全量数据为基础，对全量数据进行增删改操作，也就是说在用户未收到全量数据之前，会丢弃其收到的增量数据（业务场景决定客户端无法在收到全量数据之前对增量数据进行缓存）。但在被丢弃的增量数据中可能有一部分属于这个全量数据，客户端在收到全量数据后再利用后续新收到的增量数据去改造全量数据，这样改造后的全量数据就可能不正确了。

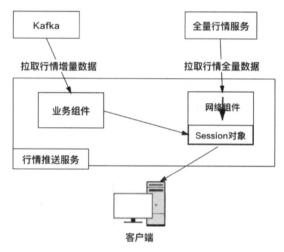

这就是方法一不适用的场景，即侵入网络组件的其他组件产生的数据有多个源，且多个源有顺序要求。反过来说，如果侵入网络组件产生的数据源只有一个，或者有多个数据源但数据源之间的数据没有顺序依赖，则这种设计是适用的。

方法一不适用的场景示意图如下。

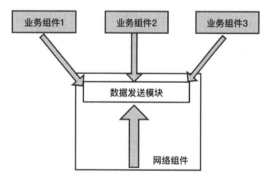

那么对于有多个数据源且数据源之间的数据有顺序依赖的情况，有没有办法继续使用方法一呢？有：可以在多个数据源处理完数据后，交给一个专门将数据源排序的组件，再由排序组件统一调用网络组件的数据发送模块。需要注意的是，也要将网络组件内部产生的需要发送的数据交给排序组件。示意图如下。

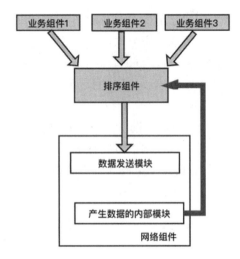

对这种场景可以举一个例子，我们在做交易系统的行情推送服务时，由于推送的数据有多个来源，有的来源于 RocketMQ 模块，有的来源于管理后台接口，有的来源于网络通信模块内部，所以需要将这三类数据按一定的顺序发送给用户。笔者就是采用以上示意图展示的结构来设计的，对其中的排序组件使用了一个队列，让来自不同数据源的数据按一定顺序进入队列中，排序组件从队列中挨个取出排好序的数据，接着调用网络通信组件的数据发送模块进行发送。

2. 方法二

方法一是在业务组件里面直接调用网络组件的，有点越俎代庖。方法二是将业务组件需要发送的数据交给网络组件自己去发送的，常用的实现方法是将对应的数据加入数据所属的那个连接的网络线程中。再来看看这个结构：

```
while (!m_bQuitFlag)
{
    epoll_or_select_func();

    handle_io_events();

    handle_other_things();
}
```

可以使用另一个队列，业务组件将数据交给这个队列，然后告知对应的网络组件中的线程需要收取任务并执行。这个逻辑在前面介绍过了，即利用唤醒机制执行 handle_other_things 函数。

这里给出一种实现，业务组件调用 EventLoop::runInLoop 方法将数据交给队列，EventLoop::runInLoop 方法的实现如下：

```
void EventLoop::runInLoop(const Functor& taskCallback)
{
    if (isInLoopThread())
    {
        taskCallback();
    }
```

```
    else
    {
        queueInLoop(taskCallback);
    }
}
```

以上代码中的 taskCallback 是需要执行的任务,由于业务线程和网络线程不是同一个线程,因此会执行 EventLoop::queueInLoop 方法,这样任务就被放到 EventLoop 的成员变量 m_pendingFunctors 容器中了,然后调用唤醒函数 wakeup。EventLoop::queueInLoop 方法的实现如下:

```
void EventLoop::queueInLoop(const Functor& taskCallback)
{
    {
        std::unique_lock<std::mutex> lock(m_mutex);
        m_pendingFunctors.push_back(taskCallback);
    }

    if (!isInLoopThread() || m_doingOtherThings)
    {
        wakeup();
    }
}
```

唤醒后的线程执行 handle_other_things 函数,该函数从 m_pendingFunctors 中取出任务进行执行:

```
void EventLoop::handle_other_things()
{
    std::vector<Functor> functors;
    m_doingOtherThings = true;

    {
        std::unique_lock<std::mutex> lock(m_mutex);
        functors.swap(m_pendingFunctors);
    }

    size_t size = functors.size();
    for (size_t i = 0; i < size; ++i)
    {
        functors[i]();
    }

    m_doingOtherThings = false;
}
```

通过这个流程,可以让网络组件本身去发送业务组件交给它的数据。

希望读者能深刻理解侵入式服务结构和非侵入式服务结构的特点和细节,以及侵入式服务结构中网络组件与业务组件交换数据的两种方法,根据实际业务设计出高质量的服务框架。

7.12 带有网络通信模块的服务器的经典结构

为了行文方便,下面将监听 socket 称为 listenfd,将由调用 accept 函数返回的 socket 称为 clientfd。

7.12.1 为何要将 listenfd 设置成非阻塞模式

我们知道,如果需要使用 I/O 复用函数统一管理各个 fd,则需要将 clientfd 设置成非阻塞模式,那么 listenfd 一定要被设置成非阻塞模式吗?答案是不一定——只要不用 I/O 复用函数去管理 listenfd 就可以了。如果 listenfd 不被设置成非阻塞模式,那么 accept 函数在没有新连接时就会阻塞。

1. 结构一:listenfd 为阻塞模式,为 listenfd 独立分配一个接受连接线程

有很多服务程序结构确实采用了阻塞模式的 listenfd,为了不让 accept 函数在没有连接时因阻塞对程序的其他逻辑执行流造成影响,我们通常将 accept 函数放在一个独立的线程中。这个线程的伪代码如下:

```
//接受连接线程
void* accept_thread_func(void* param)
{
    //可以在这里做一些初始化工作

    while (退出标志)
    {
        struct sockaddr_in clientaddr;
        socklen_t clientaddrlen = sizeof(clientaddr);
        //没有连接时,线程会被阻塞在 accept 函数处
        int clientfd = accept(listenfd, (struct sockaddr *)&clientaddr, &clientaddrlen);
        if (clientfd == -1)
        {
            //出错了,可以在此做一些清理资源动作,例如关闭 listenfd
            break;
        }

        //将 clientfd 交给其他 I/O 线程的 I/O 复用函数
        //由于跨线程操作,所以需要使用锁对公共操作的资源进行保护
    }
}
```

其他 I/O 线程的结构依旧是利用 I/O 复用函数处理 clientfd 的 one thread one loop 结构,这里以 epoll_wait 为例:

```
//其他 I/O 线程
void* io_thread_func(void* param)
{
    //可以在这里做一些初始化工作

    while (退出标志)
```

```
{
    epoll_event epoll_events[1024];
    //所有 clientfd 都被挂载到 epollfd，由 epoll_wait 统一检测读写事件
    n = epoll_wait(epollfd, epoll_events, 1024, 1000);

    //在 epoll_wait 返回时处理对应 clientfd 上的读写事件

    //其他一些操作
}
}
```

当然，这里的 I/O 线程可以存在多个，结构示意图如下。

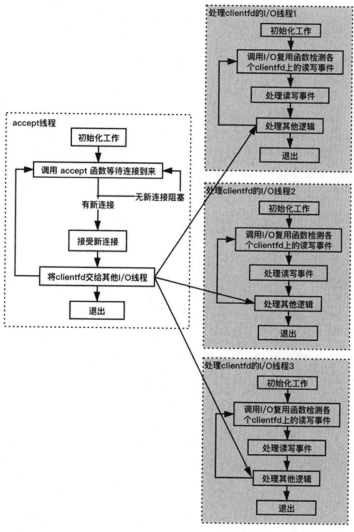

将 clientfd 从 accept_thread_func 交给 io_thread_func 的方法也有很多，这里以使用一个互斥锁进行实现为例：

```
//存储 accept 函数产生的 clientfd 的多线程共享变量
std::vector<int>    g_vecClientfds;
//保护 g_vecClientfds 的互斥体
```

```cpp
std::mutex            g_clientfdMutex;
//接受连接线程
void* accept_thread_func(void* param)
{
    //可以在这里做一些初始化工作

    while (退出标志)
    {
        struct sockaddr_in clientaddr;
        socklen_t clientaddrlen = sizeof(clientaddr);
        //没有连接时，线程会被阻塞在 accept 函数处
        int clientfd = accept(listenfd, (struct sockaddr *)&clientaddr, &clientaddrlen);
        if (clientfd == -1)
        {
            //出错了，可以在此做一些清理资源动作，例如关闭 listenfd
            break;
        }

        //将 clientfd 交给其他 I/O 线程的 I/O 复用函数
        //由于是跨线程操作，所以需要一些锁对公共操作的资源进行保护
        std::lock_guard<std::mutex> scopedLock(g_clientfdMutex);
        g_vecClientfds.push_back(clientfd);
    }
}

//其他 I/O 线程
void* io_thread_func(void* param)
{
    //可以在这里做一些初始化工作

    while (退出标志)
    {
        epoll_event epoll_events[1024];
        //将所有 clientfd 都挂载到 epollfd，由 epoll_wait 统一检测读写事件
        n = epoll_wait(epollfd, epoll_events, 1024, 1000);

        //在 epoll_wait 返回时处理对应 clientfd 的读写事件

        //其他一些操作

        //从共享变量 g_vecClientfds 中取出新的 clientfd
        retrieveNewClientfds(epollfd);
    }
}

void retrieveNewClientfds(int epollfd)
{
    std::lock_guard<std::mutex> scopedLock(g_clientfdMutex);
    if (!g_vecClientfds.empty())
    {
        //遍历 g_vecClientfds 取出各个 fd，设置 fd 挂载到所在线程的 epollfd 上
        //在全部取出后，清空 g_vecClientfds
```

```
        g_vecClientfds.clear();
    }
}
```

注意，在以上代码中由于要求 clientfd 是非阻塞模式的，所以将设置 clientfd 为非阻塞模式的逻辑放在 accept_thread_func 或 io_thread_func 中均可。

以上代码有点效率问题：某个时刻 accept_thread_func 向 g_vecClientfds 中添加了一个 clientfd，如果此时 io_thread_func 函数正阻塞在 epoll_wait 处，我们就要唤醒 epoll_wait，7.5 节已介绍如何设计这个唤醒逻辑，这里不再重复。

2. 结构二：listenfd 为阻塞模式，使用同一个 one thread one loop 结构处理 listenfd 的事件

单独为 listenfd 分配一个线程毕竟是对资源的一种浪费，有读者可能想到这样一种方案：listenfd 虽然被设置为阻塞模式，但可以将 listenfd 挂载到某个 loop 的 epollfd 上，在 epoll_wait 返回且 listenfd 上有读事件时，调用 accept 函数就不会阻塞了。伪代码如下：

```
void* io_thread_func(void* param)
{
    //可以在这里做一些初始化工作

    while (退出标志)
    {
        epoll_event epoll_events[1024];
        //listenfd 和 clientfd 都被挂载到 epollfd 中，由 epoll_wait 统一检测读写事件
        n = epoll_wait(epollfd, epoll_events, 1024, 1000);

        if (如果在 listenfd 上有事件)
        {
            //此时调用 accept 函数就不会阻塞
            int clientfd = accept(listenfd, ...);

            //对 clientfd 做进一步处理
        }

        //其他一些操作
    }
}
```

如以上代码所示，在这种情况下确实可以将 listenfd 设置成阻塞模式，调用 accept 函数时也不会造成流程阻塞。但这样的设计存在严重的效率问题：在每一轮循环中一次只能接受一个连接（每次循环都仅调用了一次 accept 函数），如果客户端的连接数较多，则这种处理速度可能跟不上客户端的连接请求速度，所以要在一个循环里面处理 accept 函数，但实际情形是我们无法确定下一轮调用 accept 函数时，在 backlog 队列中是否还有新连接，如果没有，则由于 listenfd 是阻塞模式，所以调用 accept 函数会阻塞。

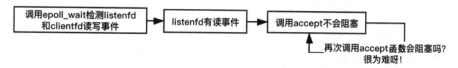

3. 结构三：listenfd 为非阻塞模式，使用同一个 one thread one loop 结构处理 listenfd 的事件

将 listenfd 设置为非阻塞模式后，我们就不会有以上窘境了。伪代码如下：

```
void* io_thread_func(void* param)
{
    //可以在这里做一些初始化工作

    while (退出标志)
    {
        epoll_event epoll_events[1024];
        //listenfd 和 clientfd 都被挂载到 epollfd，由 epoll_wait 统一检测读写事件
        n = epoll_wait(epollfd, epoll_events, 1024, 1000);

        if (如果在 listenfd 上有事件)
        {
            while (true)
            {
                //此时调用 accept 函数就不会阻塞
                int clientfd = accept(listenfd, ...);
                if (clientfd == -1)
                {
                    //错误码是 EWOULDBLOCK，说明此时已经没有新连接了
                    //可以退出内层的 while 循环
                    if (errno == EWOULDBLOCK)
                        break;
                    //被信号中断时重新调用一次 accept 函数即可
                    else if (errno == EINTR)
                        continue;
                    else
                    {
                        //对其他情况都认为出错
                        //做一次错误处理逻辑
                    }
                } else {
                    //正常接受连接
                    //对 clientfd 做进一步处理
                }//end inner-if
            }//end inner-while-loop

        }//end outer-if

        //其他一些操作
    }//end outer-while-loop
}
```

将 listenfd 设置成非阻塞模式还有一个好处：我们可以自己定义在一次 listenfd 读事件处理中最多接受多少连接，该逻辑也很容易实现，将以上代码中内层 while 循环的判断条件从 true 改成判定特定的次数即可：

```
void* io_thread_func(void* param)
{
    //可以在这里做一些初始化工作
```

```
    //每次处理的最大连接数量
    const int MAX_ACCEPTS_PER_CALL = 200;
    //当前数量
    int currentAccept;

    while (退出标志)
    {
        epoll_event epoll_events[1024];
        //将 listenfd 和 clientfd 都挂载到 epollfd 上，由 epoll_wait 统一检测读写事件
        n = epoll_wait(epollfd, epoll_events, 1024, 1000);

        if (listenfd 上有事件)
        {
            currentAccept = 0;
            while (currentAccept <= MAX_ACCEPTS_PER_CALL)
            {
                //此时调用 accept 函数不会阻塞
                int clientfd = accept(listenfd, ...);
                if (clientfd == -1)
                {
                    //错误码是 EWOULDBLOCK，说明此时已经没有新连接了
                    //可以退出内层的 while 循环了
                    if (errno == EWOULDBLOCK)
                        break;
                    //被信号中断时重新调用一次 accept 函数即可
                    else if (errno == EINTR)
                        continue;
                    else
                    {
                        //对其他情况认为出错
                        //做一次错误处理逻辑
                    }
                } else {
                    //累加处理数量
                    ++currentAccept;
                    //正常接受连接
                    //对 clientfd 做进一步处理
                }//end inner-if
            }//end inner-while-loop

        }//end outer-if

        //其他操作
    }//end outer-while-loop
}
```

这是一段比较常用的逻辑，以 redis-server 源码中的使用为例：

```
//https://github.com/balloonwj/redis-6.0.3/blob/master/src/networking.c
//networking.c 文件第 971 行
void acceptTcpHandler(aeEventLoop *el, int fd, void *privdata, int mask) {
    //MAX_ACCEPTS_PER_CALL 在 Redis 中是 1000
    int cport, cfd, max = MAX_ACCEPTS_PER_CALL;
    char cip[NET_IP_STR_LEN];
```

```
//每次最多处理 max 个连接数量
while(max--) {
    cfd = anetTcpAccept(server.neterr, fd, cip, sizeof(cip), &cport);
    if (cfd == ANET_ERR) {
        //若未到每次处理新连接的最大数时无新连接待接收,则直接 while 循环
        if (errno != EWOULDBLOCK)
            serverLog(LL_WARNING,
                "Accepting client connection: %s", server.neterr);
        return;
    }
    serverLog(LL_VERBOSE,"Accepted %s:%d", cip, cport);
    acceptCommonHandler(connCreateAcceptedSocket(cfd),0,cip);
}
```

7.12.2 基于 one thread one loop 结构的经典服务器结构

理解 listenfd 为什么被建议设置成非阻塞模式后,我们将 listenfd 挂载到某个 loop 所属的 epollfd 上与 clientfd 统一处理就没有疑问了。接下来进一步讨论这一结构。

1. listenfd 单独使用一个 loop,clientfd 被分配至其他 loop

这在实际商业服务器中是比较常用的一个结构,listenfd 被单独挂载到一个线程 loop 的 epollfd 上(这个线程一般是主线程),为了表述方便,我们将这个线程称为主线程,将对应的 loop 称为主 loop。产生新的 clientfd 并将其按一定的策略挂载到其他线程 loop 的 epollfd 上,我们将这些线程称为工作线程,将对应的 loop 称为工作 loop。

例如使用轮询策略(round robin),我们可以将 clientfd 均匀地分配给其他工作线程,如下图所示。

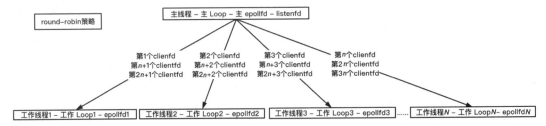

对轮询策略可以做一些优化:将 clientfd 挂载到各个工作 loop 上之后,由于连接断开时,工作 loop 会移除连接对应的 clientfd,所以在一段时间后,各个工作 loop 上的 clientfd 数量都可能不一样,会出现数量差别很大的极端情况,因此主 loop 在分配新产生的 clientfd 时可以先查询各个工作 loop 上当前实际的 clientfd 数量,把当前新产生的 clientfd 分配给持有 clientfd 最少的工作 loop,如下图所示。

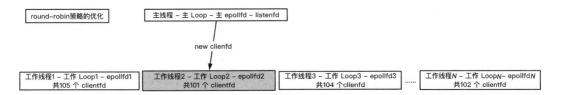

当然，我们也可以根据一定的策略比重分配 clientfd。假设现在有 4 个工作线程（对应 4 个工作 loop），其分配比重为 1:4:4:1，在程序运行一段时间后没有断开连接的情况，则这 4 个工作 loop 上的 clientfd 数量比例在理论上应该也是 1:4:4:1。

2. listenfd 不单独使用一个 loop，将所有 clientfd 都按一定策略分配给各个 loop

对于一些建立和断开连接操作不是很频繁的场景，实际上没必要让 listenfd 单独使用一个线程，因为如果在这种场景下让 listenfd 单独使用一个 Loop，这个线程在大多数情况下就可能处于空闲状态，而负责 clientfd 的其他线程可能比较忙碌，例如对于用户量较大的即时通信服务器、实时对战类型的游戏服务器，接受连接并不是高频操作，连接上的数据收发操作才是高频操作。如果采用 listenfd 单独使用一个线程的策略，则不仅浪费资源，效率也不高，所以应该让 listenfd 所在的线程也参与 clientfd 读写事件的处理。

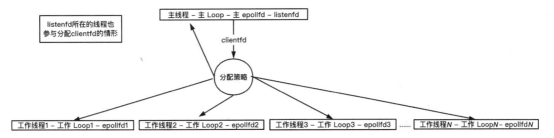

Redis 6.0 引入了多线程 I/O，在多线程启用状态下，主线程既参与处理 listenfd 逻辑，也参与分配 clientfd 和处理 clientfd 的读写事件，这是 listenfd 不单独使用一个 loop 且所有 clientfd 都按一定策略分配给各个 loop 的典型案例。下一章会详细介绍这些内容。

3. listenfd 和所有 clientfd 均使用一个 Loop

这是上述情形的特例，一般用于整个 loop 都是高效的内存操作的情形，例如 Redis-server 的 I/O 线程情形，即单线程 I/O 情形。

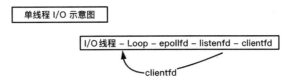

事实上，本章介绍的 one thread one loop 思想及 clientfd 在多个 loop 之间的分配策略同样适用于多进程模型，例如 Nginx，只不过在像 Nginx 这样的服务中使用单个进程来接受和处理连接后，主线程变成了主进程（Master Process），工作线程变成了工作进程（Worker Process），原来在同一个进程地址空间中可直接

将一个 clientfd 数据投递给其他线程的方式，变成了利用进程通信技术将 clientfd 从主进程传递给工作进程。不管怎样，无论是主进程还是工作进程中的主线程结构、loop 结构，这些进程中的主线程都是 one thread one loop 结构。

7.12.3 服务器的性能瓶颈

7.11 节讨论了一个服务程序的基本结构，从线程维度来看，可以分为网路线程和业务处理线程，其中网络线程的执行逻辑一般比较固定，业务线程的执行逻辑则随着业务的不同而不同。

在机器物理资源有限的情况下，我们假定某个服务线程数量也是有限的。为了合理分配线程资源，让程序性能最大化，我们需要找到程序的性能瓶颈在哪里。按照业务类型的不同，我们一般将服务器程序归为两类：I/O 密集型和计算密集型。

（1）I/O 密集型指在程序业务上没有复杂的计算或者耗时的业务逻辑要处理，在大多数情况下是频繁的网络收发操作，这类服务有 IM 服务、交易系统中的行情推送服务、实时对战游戏的服务等。

（2）计算密集型指在程序业务逻辑中存在耗时的计算，这类服务有数据处理服务、调度服务等。

如果服务是 I/O 密集型的，我们就需要将线程数量向网络通信组件倾斜，反过来，如果服务是计算密集型的，我们就应该将线程数量向业务模块倾斜。假设现在的总线程数量是 10 个，那么对于 I/O 密集型服务，我们的网络线程数量可以被设置为大于 5，而业务线程数量小于 5；反过来，对于计算密集型服务，我们可以将网络线程数量设置为小于 5，业务线程数量大于 5，具体数量可以根据网络通信逻辑与业务处理逻辑在整个服务中的资源占用比例决定，谁的占用资源率大，倾斜给谁的线程数量就应该越多。

第 8 章
Redis 网络通信模块源码分析

8.1 调试 Redis 环境与准备

上一章介绍了 one thread one loop 思想，Redis 的网络通信结构就是以这一思想为蓝本的，本节将以 Redis 为例进一步讲解这一思想。本节假设读者不清楚 Redis 网络通信层的结构，会以 Redis 源码分析与 gdb 调试相结合的方式，探究 Redis 并逐步理解 Redis 网络通信模块的结构。

先来介绍一下笔者探究 Redis 的环境：在 Mac 机器上使用 VSCode 阅读和分析 Redis 源码，在 Linux 机器上使用 gdb 调试 Redis。笔者的相关工具如下。

◎ Linux 调试机器：CentOS 7.0，gdb 8.3。
◎ 代码阅读机器：macOS Catalina，VSCode 1.45.1。
◎ Redis 版本：6.0.3（写作本书时的 Redis 最新版本）。

不同版本的 Redis 源码可能存在与本章内容中的行号不完全匹配的情况，为了阅读本章内容时与源码相匹配，建议读者从 Redis 官方下载 redis-6.0.3 版本。对于代码阅读、调试工具及相关操作系统，不一定必须与笔者的保持一致。

8.1.1 Redis 源码编译与启动

下载 redis-6.0.3 源码后解压缩，并在 CentOS 系统中编译出带有调试符号的可执行文件。

进入 redis-6.0.3/src 目录，其中的 redis-server 和 redis-cli 文件是我们需要调试的。先用 gdb 关联 redis-server 文件，并通过 set args 命令设置 redis-server 的配置文件 redis.conf 的路径，然后启动 redis-server。操作如下：

```
[root@myaliyun ~]# cd redis-6.0.3/src
[root@myaliyun src]# gdb redis-server
GNU gdb (GDB) 8.3
Reading symbols from redis-server...
(gdb) set args "../redis.conf"
```

```
(gdb) run
Starting program: /root/redis-6.0.3/src/redis-server "../redis.conf"
[Thread debugging using libthread_db enabled]
Using host libthread_db library "/usr/lib64/libthread_db.so.1".
7470:C 10 Jun 2020 20:32:19.625 # oO0OoO0OoO0Oo Redis is starting oO0OoO0OoO0Oo
7470:C 10 Jun 2020 20:32:19.625 # Redis version=6.0.3, bits=64, commit=00000000,
modified=0, pid=7470, just started
7470:C 10 Jun 2020 20:32:19.625 # Configuration loaded
                _._
           _.-``__ ''-._
      _.-``    `.  `_.  ''-._           Redis 6.0.3 (00000000/0) 64 bit
  .-`` .-```.  ```\/    _.,_ ''-._
 (    '      ,       .-`  | `,    )     Running in standalone mode
 |`-._`-...-` __...-.``-._|'` _.-'|     Port: 6379
 |    `-._   `._    /     _.-'    |     PID: 7470
  `-._    `-._  `-./  _.-'    _.-'
 |`-._`-._    `-.__.-'    _.-'_.-'|
 |    `-._`-._        _.-'_.-'    |     http://redis.io
  `-._    `-._`-.__.-'_.-'    _.-'
 |`-._`-._    `-.__.-'    _.-'_.-'|
 |    `-._`-._        _.-'_.-'    |
  `-._    `-._`-.__.-'_.-'    _.-'
      `-._    `-.__.-'    _.-'
          `-._        _.-'
              `-.__.-'

7470:M 10 Jun 2020 20:32:19.627 # WARNING: The TCP backlog setting of 511 cannot be
enforced because /proc/sys/net/core/somaxconn is set to the lower value of 128.
7470:M 10 Jun 2020 20:32:19.627 # Server initialized
7470:M 10 Jun 2020 20:32:19.627 # WARNING overcommit_memory is set to 0! Background
save may fail under low memory condition. To fix this issue add 'vm.overcommit_memory
= 1' to /etc/sysctl.conf and then reboot or run the command 'sysctl
vm.overcommit_memory=1' for this to take effect.
7470:M 10 Jun 2020 20:32:19.627 # WARNING you have Transparent Huge Pages (THP) support
enabled in your kernel. This will create latency and memory usage issues with Redis.
To fix this issue run the command 'echo never >
/sys/kernel/mm/transparent_hugepage/enabled' as root, and add it to your
/etc/rc.local in order to retain the setting after a reboot. Redis must be restarted
after THP is disabled.
[New Thread 0x7ffff0bb9700 (LWP 7475)]
[New Thread 0x7ffff03b8700 (LWP 7476)]
[New Thread 0x7fffefbb7700 (LWP 7477)]
[New Thread 0x7fffef3b6700 (LWP 7478)]
[New Thread 0x7fffeebb5700 (LWP 7479)]
[New Thread 0x7fffee3b4700 (LWP 7480)]
7470:M 10 Jun 2020 20:32:19.634 * Ready to accept connections
```

以上所示便是 redis-server 启动成功后的画面。

我们再开一个 Shell 窗口，再次进入 redis-6.0.3/src 目录，然后使用 gdb 启动 Redis 客户端 redis-cli：

```
[root@myaliyun src]# gdb redis-cli
GNU gdb (GDB) 8.3
Reading symbols from redis-cli...
```

```
(gdb) r
Starting program: /root/redis-6.0.3/src/redis-cli
[Thread debugging using libthread_db enabled]
Using host libthread_db library "/usr/lib64/libthread_db.so.1".
127.0.0.1:6379>
```

以上所示便是 redis-cli 启动成功后的画面。

8.1.2 通信示例与术语约定

由于本章主要讲解 Redis 的网络通信模块，并不关心 Redis 的其他内容，因此使用了一个简单的通信实例进行说明，即通过 redis-cli 产生一个 key 为 "hello"、value 为 "world" 的 Redis key-value 数据来探究 Redis 的网络通信模块。示例如下：

```
127.0.0.1:6379> set hello world
OK
127.0.0.1:6379>
```

为了方便行文，先约定几个技术术语。

- listenfd：监听 fd，用于绑定 IP 和端口号，并调用 listen 函数启动监听操作的 socket 对象。
- clientfd：客户端 fd，由服务端调用 accept 函数接受连接返回的、对应某路客户端连接的 socket。
- connfd：创建后用于调用 connect 函数连接服务器的 socket。

在 Linux 上，socket 也是一种文件描述符（File Descriptor），因此也称之为 fd。

8.2 探究 redis-server 端的网络通信模块

这里先研究 redis-server 端的网络通信模块。

8.2.1 监听 fd 的初始化工作

应用层编写网络通信程序的流程一般如下。

（1）服务端创建监听 socket。

（2）将监听 socket 绑定到需要的 IP 地址和端口上（调用 Socket API bind 函数）。

（3）启动监听（调用 socket API listen 函数）。

（4）等待客户端连接的到来，调用 Socket API accept 函数接受客户端连接，并产生一个与该客户端对应的客户端 socket（即 clientfd）。

（5）处理客户端 socket 上的网络数据收发，在必要时关闭该 socket。

这里先来探究以上流程中的第 1~3 步在 redis-server 中是如何实现的。redis-server 默认对外使用的端口号是 6379，我们可以使用这个端口号作为探究的线索。

笔者这里使用的是 VSCode 的搜索功能，读者使用的其他 IDE 一般都会提供源码搜索功能。

我们在 Redis 源码文件中进行全局搜索，找到调用 Socket API bind 函数的位置。经过过滤和筛选后，我们确定 bind 函数位于 anet.c 文件的 anetListen 函数中：

```c
//anet.c 文件第 454 行
static int anetListen(char *err, int s, struct sockaddr *sa, socklen_t len, int backlog)
{
    if (bind(s,sa,len) == -1) {
        anetSetError(err, "bind: %s", strerror(errno));
        close(s);
        return ANET_ERR;
    }

    if (listen(s, backlog) == -1) {
        anetSetError(err, "listen: %s", strerror(errno));
        close(s);
        return ANET_ERR;
    }
    return ANET_OK;
}
```

先使用 gdb 将 redis-server 启动起来，然后按 Ctrl+C 组合键将运行中的 redis-server 中断，接着用 gdb 的 b 命令在 anetListen 函数上加个断点，重新运行 redis-server，操作如下：

```
7470:M 10 Jun 2020 20:32:19.634 * Ready to accept connections
^C
Thread 1 "redis-server" received signal SIGINT, Interrupt.
0x00007ffff71e2603 in epoll_wait () from /usr/lib64/libc.so.6
(gdb) b anetListen
Breakpoint 1 at 0x42aab2: file anet.c, line 455.
(gdb) r
The program being debugged has been started already.
Start it from the beginning? (y or n) y
Starting program: /root/redis-6.0.3/src/redis-server "../redis.conf"
[Thread debugging using libthread_db enabled]
Using host libthread_db library "/usr/lib64/libthread_db.so.1".
7770:C 10 Jun 2020 20:38:11.791 # oO0OoO0OoO0Oo Redis is starting oO0OoO0OoO0Oo
7770:C 10 Jun 2020 20:38:11.792 # Redis version=6.0.3, bits=64, commit=00000000, modified=0, pid=7770, just started
7770:C 10 Jun 2020 20:38:11.792 # Configuration loaded

Breakpoint 1, anetListen (err=0x568c48 <server+680> "", s=6, sa=0x6174c0, len=16, backlog=511) at anet.c:455
455         if (bind(s,sa,len) == -1) {
(gdb)
```

当程序中断在 anetListen 函数时，使用 bt 命令查看此时的调用堆栈：

```
(gdb) bt
#0  anetListen (err=0x568c48 <server+680> "", s=6, sa=0x6174c0, len=16, backlog=511) at anet.c:455
#1  0x000000000042ad26 in _anetTcpServer (err=0x568c48 <server+680> "", port=6379,
```

```
bindaddr=0x5debf0 "127.0.0.1", af=2, backlog=511) at anet.c:501
#2  0x000000000042adc2 in anetTcpServer (err=0x568c48 <server+680> "", port=6379,
bindaddr=0x5debf0 "127.0.0.1", backlog=511) at anet.c:519
#3  0x0000000000430db3 in listenToPort (port=6379, fds=0x568b14 <server+372>,
count=0x568b54 <server+436>) at server.c:2680
#4  0x00000000004312ef in initServer () at server.c:2792
#5  0x000000000043712a in main (argc=2, argv=0x7fffffffe308) at server.c:5128
```

通过上述堆栈并结合堆栈#2处显示的6379端口号信息,可以确认这是我们要找的逻辑,并且这个逻辑是在主线程中发起的,因为从堆栈输出上看,底层堆栈是main函数(#5)。

看下堆栈#1处的代码,发现anetListen函数在anet.c文件第501行被调用(以下代码片段中的加粗行):

```
//anet.c 文件第 479 行
static int _anetTcpServer(char *err, int port, char *bindaddr, int af, int backlog)
{
    int s = -1, rv;
    char _port[6];  /* strlen("65535") */
    struct addrinfo hints, *servinfo, *p;

    snprintf(_port,6,"%d",port);
    memset(&hints,0,sizeof(hints));
    hints.ai_family = af;
    hints.ai_socktype = SOCK_STREAM;
    hints.ai_flags = AI_PASSIVE;    /* No effect if bindaddr != NULL */

    if ((rv = getaddrinfo(bindaddr,_port,&hints,&servinfo)) != 0) {
        anetSetError(err, "%s", gai_strerror(rv));
        return ANET_ERR;
    }
    for (p = servinfo; p != NULL; p = p->ai_next) {
        if ((s = socket(p->ai_family,p->ai_socktype,p->ai_protocol)) == -1)
            continue;

        if (af == AF_INET6 && anetV6Only(err,s) == ANET_ERR) goto error;
        if (anetSetReuseAddr(err,s) == ANET_ERR) goto error;
        //这一行对应 anet.c 文件第 501 行
        if (anetListen(err,s,p->ai_addr,p->ai_addrlen,backlog) == ANET_ERR) s =
ANET_ERR;
        goto end;
    }
    if (p == NULL) {
        anetSetError(err, "unable to bind socket, errno: %d", errno);
        goto error;
    }

error:
    if (s != -1) close(s);
    s = ANET_ERR;
end:
    freeaddrinfo(servinfo);
    return s;
}
```

使用 f 命令将堆栈切换至#1，输入 info args 命令就可以查看传给 anetListen 函数的参数：

```
(gdb) f 1
#1  0x000000000042ad26 in _anetTcpServer (err=0x568c48 <server+680> "", port=6379,
bindaddr=0x5debf0 "127.0.0.1", af=2, backlog=511) at anet.c:501
501             if (anetListen(err,s,p->ai_addr,p->ai_addrlen,backlog) == ANET_ERR)
s = ANET_ERR;
(gdb) info args
err = 0x568c48 <server+680> ""
port = 6379
bindaddr = 0x5debf0 "127.0.0.1"
af = 2
backlog = 511
(gdb)
```

在以上 _anetTcpServer 函数中使用系统 API getaddrinfo 函数来解析得到当前主机的 IP 地址和端口信息。这里没有选择使用 gethostbyname 这个 API，是因为 gethostbyname 函数仅能解析 IPv4 相关的主机信息，而 getaddrinfo 函数可同时解析与 IPv4 和 IPv6 相关的主机信息。getaddrinfo 函数签名如下：

```
int getaddrinfo(const char *node, const char *service,
                const struct addrinfo *hints,
                struct addrinfo **res);
```

读者可以自行在 Linux man 手册上查看这个函数的具体用法，这里就不再介绍了。通常在服务端调用 getaddrinfo 函数之前，会将 hints 参数的 ai_flags 设置为 AI_PASSIVE 标志，用于 bind；主机名 node 参数通常被设置为 NULL，用于让函数返回通配地址[::]。当然，如果是客户端调用 getaddrinfo，则 hints 参数的 ai_flags 一般不设置 AI_PASSIVE 标志，但是主机名 node 参数和服务名 service 参数应该不能被设置为 NULL。

解析完协议信息后，利用得到的主机信息（地址簇、IP 地址、协议类型）创建监听 socket，并为该 socket 开启 reuseAddr 选项，然后调用 anetListen 函数，在该函数中利用监听 socket 先 bind 后 listen。

至此，redis-server 就可以在 6379 端口上接受客户端连接了。

8.2.2 接受客户端连接

同样的方法，要研究 redis-server 是如何接受客户端连接的，只要在源码文件中搜索 Socket API accept 函数即可。

经过定位，我们最终找到了 anetGenericAccept 函数（位于 anet.c 文件第 545 行），在该函数中调用了 accept API：

```
//anet.c 文件第 545 行
static int anetGenericAccept(char *err, int s, struct sockaddr *sa, socklen_t *len)
{
    int fd;
    while(1) {
        fd = accept(s,sa,len);
```

```
        if (fd == -1) {
            if (errno == EINTR)
                continue;
            else {
                anetSetError(err, "accept: %s", strerror(errno));
                return ANET_ERR;
            }
        }
        break;
    }
    return fd;
}
```

我们先用 b 命令在 anetGenericAccept 函数处加个断点，然后重新运行 redis-server。redis-server 运行起来后，gdb 都没有触发该断点，这是因为此时并没有任何客户端连接 redis-server。我们新打开一个 Shell 窗口，在该窗口中进入 redis-6.0.3/src 目录启动 redis-cli，redis-cli 会尝试连接 redis-server。启动 redis-cli 后，触发了刚才加的断点，我们使用 bt 命令查看此时的调用堆栈：

```
[New Thread 0x7ffff0bb9700 (LWP 8454)]
[New Thread 0x7ffff03b8700 (LWP 8455)]
[New Thread 0x7fffefbb7700 (LWP 8456)]
[New Thread 0x7fffef3b6700 (LWP 8457)]
[New Thread 0x7fffeebb5700 (LWP 8458)]
[New Thread 0x7fffee3b4700 (LWP 8459)]
8451:M 10 Jun 2020 20:52:04.584 * Ready to accept connections

Thread 1 "redis-server" hit Breakpoint 2, anetGenericAccept (err=0x568c48
<server+680> "", s=6, sa=0x7fffffffe000, len=0x7fffffffdffc) at anet.c:548
548             fd = accept(s,sa,len);
(gdb) bt
#0  anetGenericAccept (err=0x568c48 <server+680> "", s=6, sa=0x7fffffffe000,
len=0x7fffffffdffc) at anet.c:548
#1  0x000000000042afa8 in anetTcpAccept (err=0x568c48 <server+680> "", s=6,
ip=0x7fffffffe0d0 "@\341\377\377\377\177", ip_len=46, port=0x7fffffffe104) at
anet.c:566
#2  0x0000000000442540 in acceptTcpHandler (el=0x5e5830, fd=6, privdata=0x0, mask=1)
at networking.c:979
#3  0x00000000004298bf in aeProcessEvents (eventLoop=0x5e5830, flags=27) at
ae.c:479
#4  0x0000000000429ab6 in aeMain (eventLoop=0x5e5830) at ae.c:539
#5  0x00000000004372bb in main (argc=2, argv=0x7fffffffe308) at server.c:5175
(gdb)
```

分析上述调用堆栈后，我们梳理一下这里的调用流程。在 main 函数的 initServer 函数中先创建监听 socket、绑定地址，然后开启监听，接着调用 aeMain 函数（ae.c 文件第 536 行）启动一个循环来不断地处理"事件"，循环代码如下：

```
//ae.c 文件第 536 行
void aeMain(aeEventLoop *eventLoop) {
    eventLoop->stop = 0;
    while (!eventLoop->stop) {
        aeProcessEvents(eventLoop, AE_ALL_EVENTS|
                        AE_CALL_BEFORE_SLEEP|
```

```
                        AE_CALL_AFTER_SLEEP);
    }
}
```

循环的退出条件是 eventLoop->stop 为 1。事件处理函数 aeProcessEvents 的代码如下（ae.c 文件第 386 行）：

```
//代码片段 8-2-2
//为了节约篇幅，以下代码片段省略部分代码和注释
//ae.c 文件第 386 行
int aeProcessEvents(aeEventLoop *eventLoop, int flags)
{
    int processed = 0, numevents;
    //如果设置检测任何事件的标志，则直接返回
    if (!(flags & AE_TIME_EVENTS) && !(flags & AE_FILE_EVENTS)) return 0;
    if (eventLoop->maxfd != -1 ||
        ((flags & AE_TIME_EVENTS) && !(flags & AE_DONT_WAIT))) {
        int j;
        aeTimeEvent *shortest = NULL;
        struct timeval tv, *tvp;
        //查找最近快到期的定时器事件
        if (flags & AE_TIME_EVENTS && !(flags & AE_DONT_WAIT))
            shortest = aeSearchNearestTimer(eventLoop);
        if (shortest) {
            long now_sec, now_ms;

            aeGetTime(&now_sec, &now_ms);
            tvp = &tv;

            long long ms =
                (shortest->when_sec - now_sec)*1000 +
                shortest->when_ms - now_ms;

            if (ms > 0) {
                tvp->tv_sec = ms/1000;
                tvp->tv_usec = (ms % 1000)*1000;
            } else {
                tvp->tv_sec = 0;
                tvp->tv_usec = 0;
            }
        } else {
            if (flags & AE_DONT_WAIT) {
                tv.tv_sec = tv.tv_usec = 0;
                tvp = &tv;
            } else {
                /* 将 tvp 设置为 NULL 将会导致接下来调用 I/O 复用函数没有事件时会一直阻塞 */
                tvp = NULL;
            }
        }

        if (eventLoop->flags & AE_DONT_WAIT) {
            tv.tv_sec = tv.tv_usec = 0;
            tvp = &tv;
        }
```

```c
        if (eventLoop->beforesleep != NULL && flags & AE_CALL_BEFORE_SLEEP)
            eventLoop->beforesleep(eventLoop);

        //将 tvp 传给 aeApiPoll 函数作为 I/O 复用函数的超时时间参数
        numevents = aeApiPoll(eventLoop, tvp);

        if (eventLoop->aftersleep != NULL && flags & AE_CALL_AFTER_SLEEP)
            eventLoop->aftersleep(eventLoop);

        for (j = 0; j < numevents; j++) {
            aeFileEvent *fe = &eventLoop->events[eventLoop->fired[j].fd];
            int mask = eventLoop->fired[j].mask;
            int fd = eventLoop->fired[j].fd;
            int fired = 0;

            int invert = fe->mask & AE_BARRIER;

            if (!invert && fe->mask & mask & AE_READABLE) {
                fe->rfileProc(eventLoop,fd,fe->clientData,mask);
                fired++;
                fe = &eventLoop->events[fd];
            }

            if (fe->mask & mask & AE_WRITABLE) {
                if (!fired || fe->wfileProc != fe->rfileProc) {
                    fe->wfileProc(eventLoop,fd,fe->clientData,mask);
                    fired++;
                }
            }

            if (invert) {
                fe = &eventLoop->events[fd];
                if ((fe->mask & mask & AE_READABLE) &&
                    (!fired || fe->wfileProc != fe->rfileProc))
                {
                    fe->rfileProc(eventLoop,fd,fe->clientData,mask);
                    fired++;
                }
            }

            processed++;
        }
    }
    /* 检测定时器事件 */
    if (flags & AE_TIME_EVENTS)
        processed += processTimeEvents(eventLoop);

    return processed;
}
```

该段代码先通过 flags 变量判断是否设置了事件检测标志，如果设置了事件检测标志，则接着判断是否设置了定时器事件检测标志。如果设置了定时器事件检测标志（AE_TIME_EVENTS），则寻找最近要到期的定时器（ae.c 文件第 272 行）：

```c
/* Search the first timer to fire.
 * This operation is useful to know how many time the select can be
 * put in sleep without to delay any event.
 * If there are no timers NULL is returned.
 *
 * Note that's O(N) since time events are unsorted.
 * Possible optimizations (not needed by Redis so far, but...):
 * 1) Insert the event in order, so that the nearest is just the head.
 *    Much better but still insertion or deletion of timers is O(N).
 * 2) Use a skiplist to have this operation as O(1) and insertion as O(log(N)).
 */
//ae.c 文件第 272 行
static aeTimeEvent *aeSearchNearestTimer(aeEventLoop *eventLoop)
{
    aeTimeEvent *te = eventLoop->timeEventHead;
    aeTimeEvent *nearest = NULL;

    while(te) {
        if (!nearest || te->when_sec < nearest->when_sec ||
            (te->when_sec == nearest->when_sec &&
             te->when_ms < nearest->when_ms))
            nearest = te;
        te = te->next;
    }
    return nearest;
}
```

以上代码有详细的注释，也非常好理解。代码注释告诉我们，由于这里的定时器集合使用了链表，是无序的，所以在寻找到期的定时器时需要遍历整个链表，算法复杂度是 $O(n)$。同时，在注释中也"暗示"了将来 Redis 在这块的优化方向，即将这个链表按定时器对象的到期时间从小到大排序，这样链表的头部就是我们需要的最近时间点的定时器对象，算法复杂度是 $O(1)$。或者使用 Redis 中的 skiplist，算法复杂度是 $O(\log(N))$。

接着 redis-server 获取当前系统时间（ae.c 文件第 406、408 行，代码片段 8-2-2 中的加粗行）：

```c
//ae.c 文件第 406 行
long now_sec, now_ms;
//ae.c 文件第 408 行
aeGetTime(&now_sec, &now_ms);
```

先将最早要到期的定时器时间减去当前系统时间获得一个时间间隔 tvp，然后将 tvp 作为接下来 I/O 复用函数的超时时间参数（通过 aeApiPoll 函数传入）。这样做的好处是对最近到期的定时器事件的处理不会因为与接下来的 I/O 复用函数挂起时间不一致而受到太大影响。反过来说，I/O 复用函数的超时时间要么大于 tvp，要么小于 tvp，如果 I/O 复用函数的挂起时间大于最近到期的定时器事件的时间（tvp），则会造成定时器事件延期执行；如果 I/O 复用函数的挂起时间小于最近到期的定时器事件的时间（tvp），则可能造成对定时器事件是否到期做了多次不必要的检测。我们将这段逻辑从代码片段 8-2-2 中提取出来：

```
long now_sec, now_ms;

aeGetTime(&now_sec, &now_ms);
tvp = &tv;

long long ms =
    (shortest->when_sec - now_sec)*1000 +
    shortest->when_ms - now_ms;

if (ms > 0) {
    tvp->tv_sec = ms/1000;
    tvp->tv_usec = (ms % 1000)*1000;
} else {
    tvp->tv_sec = 0;
    tvp->tv_usec = 0;
}
```

如果发现最近的定时器事件的到期时间已经比当前系统时间小（ms > 0 的 else 分支，以上代码中的加粗行），则说明这个定时器事件需要立即执行，接下来的 I/O 复用函数的等待时间被设置为 0，即调用 I/O 复用函数检测一次 I/O 事件后立即返回（将 I/O 函数超时时间设置为 0 会让 I/O 复用函数不进行任何等待而直接返回）。

这是 Redis 实现带有定时器逻辑的 one thread one loop 结构时的一个小技巧。

Redis 对 one thread one loop 结构中各个函数的顺序略有改造：

```
while (!m_bQuitFlag)
{
    //这里对定时器事件的检测是为了给下一步的 I/O 复用函数设置一个合理的等待时间
    check_timers();

    epoll_or_select_func();

    handle_io_events();

    check_handle_timers();
}
```

代码片段 8-2-2 中时间间隔 tvp 变量接下来作为 numevents = aeApiPoll(eventLoop, tvp); 调用的参数之一（上文已经对 tvp 做了介绍），aeApiPoll 函数在 Linux 系统中实际调用的是 epoll_wait 函数。Redis 在不同的操作系统上使用不同的 I/O 复用函数，在 Windows 系统中使用 select 函数；在 Mac 系统中使用 kqueue 函数。这里以在 Linux 系统中使用 epoll_wait 函数调用为例来说明（ae_epoll.c 文件第 108 行）：

```
//ae_epoll.c 文件第108 行
static int aeApiPoll(aeEventLoop *eventLoop, struct timeval *tvp) {
    aeApiState *state = eventLoop->apidata;
    int retval, numevents = 0;

    retval = epoll_wait(state->epfd,state->events,eventLoop->setsize,
            tvp ? (tvp->tv_sec*1000 + tvp->tv_usec/1000) : -1);
    if (retval > 0) {
        int j;
```

```
            numevents = retval;
            for (j = 0; j < numevents; j++) {
                int mask = 0;
                struct epoll_event *e = state->events+j;

                if (e->events & EPOLLIN) mask |= AE_READABLE;
                if (e->events & EPOLLOUT) mask |= AE_WRITABLE;
                if (e->events & EPOLLERR) mask |= AE_WRITABLE|AE_READABLE;
                if (e->events & EPOLLHUP) mask |= AE_WRITABLE|AE_READABLE;
                eventLoop->fired[j].fd = e->data.fd;
                eventLoop->fired[j].mask = mask;
            }
        }
        return numevents;
}
```

epoll_wait 函数签名如下：

```
int epoll_wait(int epfd, struct epoll_event *events, int maxevents, int timeout);
```

最后一个参数 timeout 的设置非常有讲究，如果传入进来的 tvp 值是 NULL，则根据上文的分析，说明没有定时器事件，将等待时间设置为-1，这会让 epoll_wait 无限期地挂起，直到有事件产生时才会被唤醒。代码逻辑如下：

```
//ae.c 文件第 405 行
if (shortest) {
    //省略代码
} else {

    if (flags & AE_DONT_WAIT) {
        tv.tv_sec = tv.tv_usec = 0;
        tvp = &tv;
    } else {
        tvp = NULL;
    }
}
```

epoll_wait 函数挂起的好处是不浪费 CPU 时间片；反之，如前文所述，将 timeout 设置成最近到期的定时器事件间隔（定时器过期时间减去当前系统时间），将 epoll_wait 函数的等待时间设置为最近到期的定时器时间间隔，可以及时唤醒 epoll_wait 函数，这样程序流可以尽快处理到期的定时器事件。定时器事件的处理逻辑将在 8.2.11 节介绍。

epoll_wait 函数处理的所有 fd 信息（包括监听 fd 和普通客户端 fd）及其对应的事件信息都被记录在事件循环对象 aeEventLoop 中（ae.h 文件第 100 行）：

```
/* loop 对象的状态信息 */
typedef struct aeEventLoop {
    int maxfd;
    int setsize;
    long long timeEventNextId;
    time_t lastTime;
    aeFileEvent *events;
    aeFiredEvent *fired;
```

```
    aeTimeEvent *timeEventHead;
    int stop;
    void *apidata;
    aeBeforeSleepProc *beforesleep;
    aeBeforeSleepProc *aftersleep;
    int flags;
} aeEventLoop;
```

aeEventLoop 对象就是 one thread one loop 结构中的 loop 对象，其中，events 字段记录了注册到该 loop 上的所有 fd（客户端 fd、监听 fd）；fired 字段记录 I/O 复用函数返回后所有有读写事件的 fd；apidata 字段存储与特定 I/O 复用函数相关联的信息，例如 epoll_wait 函数需要用到的 epollfd。当某个 fd 上有事件触发时，就从 events 字段中找到该 fd，并把事件类型掩码（aeFileEvent 结构的 mask 字段）一起记录到 aeEventLoop 的 fired 字段中。这里先介绍接受客户端连接的流程，再介绍 epoll_wait 函数使用的 epollfd 是在何时何处创建的，以及监听 fd 和客户端 fd 是如何挂载到 epollfd 上的。

从 I/O 复用函数中得到 fd 读写事件后，接下来就要处理这些事件了。在主循环 aeProcessEvents 函数中，先从 aeEventLoop 对象的 fired 字段（这是一个数组）取出有事件的 fd，然后根据事件类型（读事件和写事件）分别对这些事件进行处理：

```
//以下代码片段位于 aeProcessEvents 函数中
//aeProcessEvents 函数位于 ae.c 文件第 453 行
for (j = 0; j < numevents; j++) {
    aeFileEvent *fe = &eventLoop->events[eventLoop->fired[j].fd];
    int mask = eventLoop->fired[j].mask;
    int fd = eventLoop->fired[j].fd;
    int fired = 0; /* Number of events fired for current fd. */

    //处理读事件
    if (!invert && fe->mask & mask & AE_READABLE) {
        fe->rfileProc(eventLoop,fd,fe->clientData,mask);
        fired++;
        fe = &eventLoop->events[fd]; /* Refresh in case of resize. */
    }

    //处理写事件
    if (fe->mask & mask & AE_WRITABLE) {
        if (!fired || fe->wfileProc != fe->rfileProc) {
            fe->wfileProc(eventLoop,fd,fe->clientData,mask);
            fired++;
        }
    }

    //省略部分代码

    processed++;
}
```

读事件处理函数 rfileProc 和写事件处理函数 wfileProc 都是 aeFileProc 函数指针类型，aeFileProc 类型的定义如下：

```
typedef void aeFileProc(struct aeEventLoop *eventLoop, int fd, void *clientData, int
mask);
```

rfileProc 和 wfileProc 是事件类型结构体 aeFileEvent 的成员字段，在产生新的 fd 时设置。aeFileEvent 结构的定义如下：

```
/* File event 结构* /
typedef struct aeFileEvent {
    int mask; /* 取值是 AE_(READABLE|WRITABLE)的标志之一 */
    aeFileProc *rfileProc;
    aeFileProc *wfileProc;
    void *clientData;
} aeFileEvent;
```

8.2.3 epollfd 的创建

搜索创建 epollfd 的 epoll_create 函数名，最终在 ae_epoll.c 文件中找到调用 epoll_create 函数的 aeApiCreate 函数：

```
//ae_epoll.c 文件第 39 行
static int aeApiCreate(aeEventLoop *eventLoop) {
    aeApiState *state = zmalloc(sizeof(aeApiState));

    if (!state) return -1;
    state->events = zmalloc(sizeof(struct epoll_event)*eventLoop->setsize);
    if (!state->events) {
        zfree(state);
        return -1;
    }
    state->epfd = epoll_create(1024);
    if (state->epfd == -1) {
        zfree(state->events);
        zfree(state);
        return -1;
    }
    eventLoop->apidata = state;
    return 0;
}
```

如以上代码所示，在调用 epoll_create 创建好 epollfd 后（得到 state->epfd），将整个 state 对象一起记录到前文说的 loop 对象（aeEventLoop）的 apidata 字段中。接下来如果在程序中需要将某个 fd 挂载到该 epollfd 上（EPOLL_CTL_ADD），或者从该 epollfd 上卸载 fd（EPOLL_CTL_DEL）、修改事件（EPOLL_CTL_MOD），就可以通过 aeEventLoop->apidata->state->epfd 引用到该 epollfd。Redis 是一个 C 程序，对于 C++程序，epollfd 一般作为 loop 对象的成员变量，其原理是一样的。

先使用 gdb 的 b 命令在 aeApiCreate 函数上加个断点，然后使用 run 命令重新运行 redis-server，在断点触发后，使用 bt 命令查看此时的调用堆栈，可以发现 epollfd 的创建函数 aeApiCreate 也是在上文介绍的 initServer 函数中调用的。操作过程如下：

```
[root@myaliyun src]# gdb redis-server
Reading symbols from redis-server...
(gdb) set args "../redis.conf"
(gdb) b aeApiCreate
Breakpoint 1 at 0x428741: file ae_epoll.c, line 40.
(gdb) r
Starting program: /root/redis-6.0.3/src/redis-server "../redis.conf"
[Thread debugging using libthread_db enabled]
Using host libthread_db library "/usr/lib64/libthread_db.so.1".
17279:C 10 Jun 2020 23:51:19.382 # oO0OoO0OoO0Oo Redis is starting oO0OoO0OoO0Oo
17279:C 10 Jun 2020 23:51:19.382 # Redis version=6.0.3, bits=64, commit=00000000,
modified=0, pid=17279, just started
17279:C 10 Jun 2020 23:51:19.382 # Configuration loaded

Breakpoint 1, aeApiCreate (eventLoop=0x5e5420) at ae_epoll.c:40
40          aeApiState *state = zmalloc(sizeof(aeApiState));
(gdb) bt
#0  aeApiCreate (eventLoop=0x5e5420) at ae_epoll.c:40
#1  0x0000000000428c74 in aeCreateEventLoop (setsize=10128) at ae.c:80
#2  0x0000000000431266 in initServer () at server.c:2781
#3  0x000000000043712a in main (argc=2, argv=0x7fffffffe308) at server.c:5128
(gdb)
```

如上面的演示所示，在 aeCreateEventLoop 函数中不仅调用 aeApiCreate 函数创建了 epollfd，也创建了整个事件循环需要的 loop 对象——aeEventLoop，并把 aeEventLoop 对象记录在 Redis 的一个全局变量 server 的 el 字段中。initServer 函数中的相关代码如下：

```
void initServer(void) {
    //省略其他代码

    //server.c 文件第 2781 行
    server.el = aeCreateEventLoop(server.maxclients+CONFIG_FDSET_INCR);

    //省略其他代码
}
```

全局变量 server 是一个结构体类型，其定义如下：

```
//server.c 文件第 72 行
struct redisServer server; /* Redis 的全局状态定义 */
//server.h 文件第 1030 行
struct redisServer {

    //省略部分字段
    aeEventLoop *el;
    unsigned int lruclock;
    //太长了，省略部分字段
}
```

8.2.4 监听 fd 与客户端 fd 是如何挂载到 epollfd 上的

要把监听 fd 挂载到 epollfd 上，就需要调用 epoll_ctl 这个系统 API。全文搜索一下 epoll_ctl 函数名，可在 ae_epoll.c 文件中找到 aeApiAddEvent 函数，在 aeApiAddEvent 函

数中调用了 epoll_ctl 函数。aeApiAddEvent 函数的定义如下：

```c
//ae_epoll.c 文件第 73 行
static int aeApiAddEvent(aeEventLoop *eventLoop, int fd, int mask) {
    aeApiState *state = eventLoop->apidata;
    struct epoll_event ee = {0};
    //如果一个 fd 已经监控了某些类型的事件，我们就需要对该 fd 进行 EPOLL_CTL_MOD 操作；
    //否则对该 fd 进行 EPOLL_CTL_ADD 操作
    int op = eventLoop->events[fd].mask == AE_NONE ?
            EPOLL_CTL_ADD : EPOLL_CTL_MOD;

    ee.events = 0;
    mask |= eventLoop->events[fd].mask; /* Merge old events */
    if (mask & AE_READABLE) ee.events |= EPOLLIN;
    if (mask & AE_WRITABLE) ee.events |= EPOLLOUT;
    ee.data.fd = fd;
    if (epoll_ctl(state->epfd,op,fd,&ee) == -1) return -1;
    return 0;
}
```

以上代码逻辑的主要功能是：当需要把一个 fd 绑定到 epollfd 上去时，先要通过 eventLoop（aeEventLoop 类型）的 events 字段中记录的 fd 和事件标志确认此 fd 是否被挂载到 epollfd 上，如果 fd 已被挂载，则使用 epoll_ctl 更改 fd 的事件标志（使用 EPOLL_CTL_MOD flag），如果没有被挂载，则将 fd 挂载到 epollfd 上（使用 EPOLL_CTL_ADD flag）并设置 fd 的事件掩码。

我们在调用 aeApiAddEvent 函数处添加一个断点，再重启 redis-server，断点触发后的调用堆栈如下：

```
(gdb) b aeApiAddEvent
Breakpoint 1 at 0x4288a0: file ae_epoll.c, line 74.
(gdb) r
Starting program: /root/redis-6.0.3/src/redis-server "../redis.conf"
[Thread debugging using libthread_db enabled]
Using host libthread_db library "/usr/lib64/libthread_db.so.1".
18432:C 11 Jun 2020 00:12:32.192 # oO0OoO0OoO0Oo Redis is starting oO0OoO0OoO0Oo
18432:C 11 Jun 2020 00:12:32.192 # Redis version=6.0.3, bits=64, commit=00000000,
modified=0, pid=18432, just started
18432:C 11 Jun 2020 00:12:32.192 # Configuration loaded

Breakpoint 1, aeApiAddEvent (eventLoop=0x5e5830, fd=6, mask=1) at ae_epoll.c:74
74          aeApiState *state = eventLoop->apidata;
(gdb) bt
#0  aeApiAddEvent (eventLoop=0x5e5830, fd=6, mask=1) at ae_epoll.c:74
#1  0x0000000000428f1e in aeCreateFileEvent (eventLoop=0x5e5830, fd=6, mask=1,
proc=0x4424ff <acceptTcpHandler>, clientData=0x0) at ae.c:162
#2  0x000000000043184e in initServer () at server.c:2886
#3  0x000000000043712a in main (argc=2, argv=0x7fffffffe308) at server.c:5128
(gdb)
```

从以上调用堆栈可以看出，aeApiAddEvent 函数首次在 initServer 函数中被调用。结合上文分析的监听 fd 的创建过程，去掉无关代码，抽出主脉络得到 initServer 函数的伪代码如下：

```
void initServer(void) {

    //记录程序进程 ID
    server.pid = getpid();

    //创建程序的 aeEventLoop 对象和 epfd 对象
    server.el = aeCreateEventLoop(server.maxclients+CONFIG_FDSET_INCR);

    //创建 Redis 的定时器,用于执行定时任务 cron
    aeCreateTimeEvent(server.el, 1, serverCron, NULL, NULL) == AE_ERR

    //将监听 fd 绑定到 epfd 上
     aeCreateFileEvent(server.el, server.ipfd[j], AE_READABLE,
acceptTcpHandler,NULL) == AE_ERR

    //创建一个管道,用于在需要时唤醒 epoll_wait 挂起的整个 EventLoop

    aeCreateFileEvent(server.el, server.module_blocked_pipe[0], AE_READABLE,
moduleBlockedClientPipeReadable,NULL) == AE_ERR)
}
```

> 注意:这里所说的 initServer 函数的"主脉络"是指我们关心的与网络通信功能相关的主脉络,不代表这个函数中的其他代码不重要。

如何验证上述断点处挂载到 epollfd 上的 fd 就是监听 fd 呢?这很容易,在创建监听 fd 时,我们先使用 gdb 将这个 fd 的值打印出来,然后在上述断点触发时再次打印挂载到 epollfd 上的 fd 值,如果这两个 fd 值相等,就说明上述断点处的 fd 就是监听 fd。例如,笔者的计算机某次运行时,监听 fd 的值是 11,如下图(调试工具实际用的是 cgdb)所示。

```
491     if ((rv = getaddrinfo(bindaddr,_port,&hints,&servinfo)) != 0) {
492         anetSetError(err, "%s", gai_strerror(rv));
493         return ANET_ERR;
494     }
495     for (p = servinfo; p != NULL; p = p->ai_next) {
496         if ((5)= socket(p->ai_family,p->ai_socktype,p->ai_protocol)) == -1)
497             continue;
498
499  >      if (af == AF_INET6 && anetV6Only(err,s) == ANET_ERR) goto error;
500         if (anetSetReuseAddr(err,s) == ANET_ERR) goto error;
501         if (anetListen(err,s,p->ai_addr,p->ai_addrlen,backlog) == ANET_ERR) s = ANET_E
502         goto end;
503     }
504     if (p == NULL) {
505         anetSetError(err, "unable to bind socket, errno: %d", errno);
506         goto error;
507     }
/root/redis-6.0.3/src/anet.c
(gdb) r
Starting program: /root/redis-6.0.3/src/redis-server
[Thread debugging using libthread_db enabled]
Using host libthread_db library "/usr/lib64/libthread_db.so.1".
2691:C 25 Feb 2021 10:32:14.775 # o000o000o000o Redis is starting o000o000o000o
2691:C 25 Feb 2021 10:32:14.775 # Redis version=6.0.3, bits=64, commit=00000000, modified=0
, pid=2691, just started
2691:C 25 Feb 2021 10:32:14.775 # Warning: no config file specified, using the default conf
ig. In order to specify a config file use /root/redis-6.0.3/src/redis-server /path/to/redis
.conf

Breakpoint 1, _anetTcpServer (err=0x568c48 <server+680> "", port=6379, bindaddr=0x0, af=10,
  backlog=511) at anet.c:499
(gdb) p s            ← 输出侦听 fd 的值
$1 = 11
(gdb)
```

继续触发上述断点，确认此时挂载到 epfd 上的 fd 值。

如下图所示，添加在 aeApiEvent 处的断点触发时传入的 fd 值也是 11，这说明此时挂载到 epollfd 上的 fd 是监听 fd。当然，在将监听 fd 挂载到 epollfd 的同时，也设置了只关注监听 fd 的读事件，并设置读事件处理函数为 acceptTcpHandler。

```
66          aeApiState *state = eventLoop->apidata;
67
68          close(state->epfd);
69          zfree(state->events);
70          zfree(state);
71     }
72
73     static int aeApiAddEvent(aeEventLoop *eventLoop, int fd, int mask) {
74  ->     aeApiState *state = eventLoop->apidata;
75         struct epoll_event ee = {0}; /* avoid valgrind warning */
76         /* If the fd was already monitored for some event, we need a MOD
77          * operation. Otherwise we need an ADD operation. */
78         int op = eventLoop->events[fd].mask == AE_NONE ?
79                 EPOLL_CTL_ADD : EPOLL_CTL_MOD;
80
81         ee.events = 0;
82         mask |= eventLoop->events[fd].mask; /* Merge old events */
/root/redis-6.0.3/src/ae_epoll.c
(gdb) p s
$1 = 11
(gdb) b aeApiAddEvent        ← 添加断点

(gdb) c
Continuing.

Breakpoint 1, _anetTcpServer (err=0x568c48 <server+680> "", port=6379, bindaddr=0x0, af=2,
backlog=511) at anet.c:499
(gdb) c
Continuing.

Breakpoint 2, aeApiAddEvent (eventLoop=0x5e5750, fd=11, mask=1) at ae_epoll.c:74
(gdb) p fd          ← 断点触发时的 fd 值
$2 = 11
(gdb)
```

通常情况下，对于监听 fd，我们一般只需关注其读事件，当监听 fd 有读事件时，说明有新的连接到来。调用 aeCreateFileEvent 函数挂载监听 fd 至 epollfd，并设置关注其读事件、设置回调函数为 acceptTcpHandler，代码如下：

```
//server.c 文件第 2886 行
if (aeCreateFileEvent(server.el, server.ipfd[j], AE_READABLE,
        acceptTcpHandler,NULL) == AE_ERR)
{
    serverPanic("Unrecoverable error creating server.ipfd file event.");
}
```

acceptTcpHandler 函数就是监听 fd 的读事件回调函数 rfileProc 的实际指向，还记得前文中介绍的 fd 读事件处理函数 rfileProc 和写事件处理函数 wfileProc 吗？

在 acceptTcpHandler 函数中调用了 anetTcpAccept 函数，在 anetTcpAccept 函数中又调用了 anetGenericAccept 函数。我们先用 gdb 在 anetGenericAccept 函数调用处设置一个断点，然后重启 redis-server，再新开一个 Shell 窗口启动一个 redis-cli 客户端，gdb 会在 anetGenericAccept 调用处中断，此时的调用堆栈可以反映上述调用关系（acceptTcpHandler

→anetTcpAccept→anetGenericAccept，箭头方向是调用方向）：

```
Thread 1 "redis-server" hit Breakpoint 1, anetGenericAccept (err=0x568c48
<server+680> "", s=7, sa=0x7fffffffe000, len=0x7fffffffdffc) at anet.c:548
548             fd = accept(s,sa,len);
(gdb) l
543     }
544
545     static int anetGenericAccept(char *err, int s, struct sockaddr *sa, socklen_t
*len) {
546         int fd;
547         while(1) {
548             fd = accept(s,sa,len);
549             if (fd == -1) {
550                 if (errno == EINTR)
551                     continue;
552                 else {
(gdb) bt
#0  anetGenericAccept (err=0x568c48 <server+680> "", s=7, sa=0x7fffffffe000,
len=0x7fffffffdffc) at anet.c:548
#1  0x000000000042afa8 in anetTcpAccept (err=0x568c48 <server+680> "", s=7,
ip=0x7fffffffe0d0 "@\341\377\377\377\177", ip_len=46, port=0x7fffffffe104) at
anet.c:566
#2  0x0000000000442540 in acceptTcpHandler (el=0x5e5760, fd=7, privdata=0x0, mask=1)
at networking.c:979
#3  0x00000000004298bf in aeProcessEvents (eventLoop=0x5e5760, flags=27) at
ae.c:479
#4  0x0000000000429ab6 in aeMain (eventLoop=0x5e5760) at ae.c:539
#5  0x00000000004372bb in main (argc=1, argv=0x7fffffffe308) at server.c:5175
(gdb)
```

redis-server 在 acceptTcpHandler 函数中成功接受新连接后，产生客户端 cfd（即 clientfd），接着调用 acceptCommonHandler 函数，并在传递第 1 个参数时调用 connCreateAcceptedSocket 函数创建与每个 cfd 对应的 connection 对象：

```
//代码片段 8-2-4
//networking.c 文件第 971 行
void acceptTcpHandler(aeEventLoop *el, int fd, void *privdata, int mask) {
    int cport, cfd, max = MAX_ACCEPTS_PER_CALL;
    char cip[NET_IP_STR_LEN];
    UNUSED(el);
    UNUSED(mask);
    UNUSED(privdata);

    while(max--) {
        cfd = anetTcpAccept(server.neterr, fd, cip, sizeof(cip), &cport);
        if (cfd == ANET_ERR) {
            if (errno != EWOULDBLOCK)
                serverLog(LL_WARNING,
                    "Accepting client connection: %s", server.neterr);
            return;
        }
        serverLog(LL_VERBOSE,"Accepted %s:%d", cip, cport);
        //在传递第 1 个参数时调用 connCreateAcceptedSocket 创建 connection 对象
```

```
    acceptCommonHandler(connCreateAcceptedSocket(cfd),0,cip);
    }
}
```

connCreateAcceptedSocket 函数内部调用 connCreateSocket 函数，connCreateSocket 函数是实际创建 connection 对象的函数。connCreateAcceptedSocket 函数的定义如下：

```
//connection.c 文件第 91 行
connection *connCreateAcceptedSocket(int fd) {
    //connCreateSocket 是实际创建 connection 对象的函数
    connection *conn = connCreateSocket();
    conn->fd = fd;
    conn->state = CONN_STATE_ACCEPTING;
    return conn;
}
```

connCreateSocket 函数的定义如下：

```
//connection.c 文件第 71 行
connection *connCreateSocket() {
    connection *conn = zcalloc(sizeof(connection));
    conn->type = &CT_Socket;
    conn->fd = -1;

    return conn;
}
```

connection 对象代表一个连接对象，其定义如下：

```
//connection.h 文件第 69 行
struct connection {
    ConnectionType *type;
    ConnectionState state;
    short int flags;
    short int refs;
    int last_errno;
    void *private_data;
    ConnectionCallbackFunc conn_handler;
    ConnectionCallbackFunc write_handler;
    ConnectionCallbackFunc read_handler;
    int fd;
};
```

connection 对象的 fd 字段记录了上述所说的 cfd。请注意 connection 对象的 type 字段，在这个字段中记录了相应的 fd 发生读写等事件时的各种回调函数和用于设置回调函数的函数，字段 type 的类型是 ConnectionType*。ConnectionType 类型的定义如下：

```
//connection.h 文件第 53 行
typedef struct ConnectionType {
    void (*ae_handler)(struct aeEventLoop *el, int fd, void *clientData, int mask);
    int (*connect)(struct connection *conn, const char *addr, int port, const char *source_addr, ConnectionCallbackFunc connect_handler);
    //监听 fd 不使用这个回调
    int (*write)(struct connection *conn, const void *data, size_t data_len);
    //cfd 会设置这个字段
    int (*read)(struct connection *conn, void *buf, size_t buf_len);
```

```
    void (*close)(struct connection *conn);
    int (*accept)(struct connection *conn, ConnectionCallbackFunc accept_handler);
    int (*set_write_handler)(struct connection *conn, ConnectionCallbackFunc
handler, int barrier);
    int (*set_read_handler)(struct connection *conn, ConnectionCallbackFunc
handler);
    const char *(*get_last_error)(struct connection *conn);
    int (*blocking_connect)(struct connection *conn, const char *addr, int port, long
long timeout);
    ssize_t (*sync_write)(struct connection *conn, char *ptr, ssize_t size, long long
timeout);
    ssize_t (*sync_read)(struct connection *conn, char *ptr, ssize_t size, long long
timeout);
    ssize_t (*sync_readline)(struct connection *conn, char *ptr, ssize_t size, long
long timeout);
} ConnectionType;
```

在 ConnectionType 对象的各个字段中，代表回调函数有 ae_handler、connect、write、read、close、accept 字段，用于设置回调函数的字段有 set_write_handler、set_read_handler、get_last_error、blocking_connect、sync_write、sync_read、sync_readline 字段。

用于设置回调函数的函数的各个字段在 fd 创建（无论是 listenfd 还是 clientfd）、调用 connCreateSocket 函数创建 connection 对象时统一指定。如何指定呢？在 connCreateSocket 函数中有这么一行：

```
//connection.c 文件第 71 行
connection *connCreateSocket() {
    connection *conn = zcalloc(sizeof(connection));
    //这一行设置回调函数的函数
    conn->type = &CT_Socket;
    conn->fd = -1;

    return conn;
}
```

上文已经介绍了 conn->type 的类型为 ConnectionType*，以上代码中 conn->type 指向一个全局变量 CT_Socket，CT_Socket 的定义如下：

```
//connection.c 文件第 329 行
ConnectionType CT_Socket = {
    .ae_handler = connSocketEventHandler,
    .close = connSocketClose,
    .write = connSocketWrite,
    .read = connSocketRead,
    .accept = connSocketAccept,
    .connect = connSocketConnect,
    .set_write_handler = connSocketSetWriteHandler,
    .set_read_handler = connSocketSetReadHandler,
    .get_last_error = connSocketGetLastError,
    .blocking_connect = connSocketBlockingConnect,
    .sync_write = connSocketSyncWrite,
    .sync_read = connSocketSyncRead,
    .sync_readline = connSocketSyncReadLine
};
```

如此设置之后，clientfd 对应的 connection 对象的 type 字段的各个回调字段就被设置好了，例如 ae_handler 字段实际指向 connSocketEventHandler 函数，set_read_handler 实际指向 connSocketSetReadHandler 函数。

需要注意的是，不同类型的 fd 仅使用 ConnectionType 这个结构中的某几个回调函数，例如对于监听 fd 就不使用 write 回调函数，对于 clientfd（cfd），在初始状态下我们只关心其读事件，因此只要设置 read 字段就可以了。在哪里设置呢？继续往下看。

创建好的 connection 对象会被当作 acceptCommonHandler 函数的第 1 个参数（代码片段 8-2-4 中的加粗行）。acceptCommonHandler 函数的定义如下：

```
//networking.c 文件第 915 行
static void acceptCommonHandler(connection *conn, int flags, char *ip) {

    //省略部分代码

    /* 创建 connection 和 client 对象 */
    if ((c = createClient(conn)) == NULL) {
        char conninfo[100];
        serverLog(LL_WARNING,
            "Error registering fd event for the new client: %s (conn: %s)",
            connGetLastError(conn),
            connGetInfo(conn, conninfo, sizeof(conninfo)));
        connClose(conn);
        return;
    }

    //省略部分代码
}
```

在 acceptCommonHandler 函数中调用 createClient 函数（以上代码中的加粗行），createClient 函数用于创建 client 对象，其定义如下：

```
//networking.c 文件第 88 行
client *createClient(connection *conn) {
    //省略部分代码

    connNonBlock(conn);
    connEnableTcpNoDelay(conn);

    connKeepAlive(conn,server.tcpkeepalive);
    connSetReadHandler(conn, readQueryFromClient);

    //省略部分代码
}
```

以上代码的工作步骤如下：

（1）将客户端 fd 设置成非阻塞的（逻辑在 connNonBlock 函数中）；

（2）将该 fd 挂载到 epollfd 上，同时记录到整个程序的 aeEventLoop 对象上（逻辑在 connSetReadHandler 中，下文很快会介绍）；

（3）调用 connSetReadHandler 函数设置处理该 fd（上文中的 cfd）读事件的回调函数为 readQueryFromClient。

connSetReadHandler 函数实际调用的是上文介绍的 connection 对象的 type 字段的 set_read_handler 字段指向的回调函数。connSetReadHandler 函数的实现如下：

```
//connection.h 文件第 161 行
//第 2 个参数 func 的值是 readQueryFromClient，并一路往下传递
static inline int connSetReadHandler(connection *conn, ConnectionCallbackFunc func)
{
    return conn->type->set_read_handler(conn, func);
}
```

conn->type->set_read_handler 实际指向 connSocketSetReadHandler 函数，在 connSocketSetReadHandler 函数中将 clientfd 对应的 connection 对象的 read_handler 设置为 readQueryFromClient。connSocketSetReadHandler 函数的定义如下：

```
//connection.c 文件第 225 行
static int connSocketSetReadHandler(connection *conn, ConnectionCallbackFunc func)
{
    if (func == conn->read_handler) return C_OK;

    //func 的值是 readQueryFromClient
    //因此这里将 conn->read_handler 设置为 readQueryFromClient 函数
    conn->read_handler = func;
    if (!conn->read_handler)
        aeDeleteFileEvent(server.el,conn->fd,AE_READABLE);
    else
        if (aeCreateFileEvent(server.el,conn->fd,
                    AE_READABLE,conn->type->ae_handler,conn) == AE_ERR) return C_ERR;
    return C_OK;
}
```

在 connSocketSetReadHandler 函数中，当 clientfd 首次创建时，调用 aeCreateFileEvent 函数将 clientfd 挂载到 epollfd 上，并设置关注其读事件和设置读事件的处理函数（即上文中三个步骤中的步骤 2），读事件触发后的回调函数是 conn->type->ae_handler，aeCreateFileEvent 对 clientfd 与 listenfd 在此处的处理逻辑是一样的。

conn->type->ae_handler 实际指向 connSocketEventHandler 函数。connSocketEventHandler 函数的定义如下：

```
//connection.c 文件第 241 行
static void connSocketEventHandler(struct aeEventLoop *el, int fd, void *clientData,
int mask)
{
    //省略部分代码

    int call_read = (mask & AE_READABLE) && conn->read_handler;

    if (!invert && call_read) {
        //conn->read_handler 实际指向 readQueryFromClient
        if (!callHandler(conn, conn->read_handler)) return;
    }
```

```
    //省略部分代码
}
```

在 connSocketEventHandler 函数中,对读事件的处理最终调用的是 callHandler(conn, conn->read_handler)函数。callHandler 函数的实现如下:

```
//connhelpers.h 文件第 77 行
static inline int callHandler(connection *conn, ConnectionCallbackFunc handler) {
    connIncrRefs(conn);
    //实际指向关系
    //handler(conn) => conn->read_handler(conn) => readQueryFromClient(conn)
    if (handler) handler(conn);
    connDecrRefs(conn);
    if (conn->flags & CONN_FLAG_CLOSE_SCHEDULED) {
        if (!connHasRefs(conn)) connClose(conn);
        return 0;
    }
    return 1;
}
```

由于 conn->read_handler 实际指向的是 readQueryFromClient,因此最终会调用 readQueryFromClient 函数。

上述通过回调指针在各个函数之间传递很容易让人迷惑,我们可以先在 readQueryFromClient 函数入口处设置一个断点,然后使用 redis-cli 发送一条指令给 redis-server,断点触发时的调用堆栈如下:

```
Thread 1 "redis-server" hit Breakpoint 1, readQueryFromClient (conn=0x629510) at networking.c:1891
1891         client *c = connGetPrivateData(conn);
(gdb) bt
#0  readQueryFromClient (conn=0x629510) at networking.c:1891
#1  0x00000000004dc455 in callHandler (conn=0x629510, handler=0x444809 <readQueryFromClient>) at connhelpers.h:79
#2  0x00000000004dcae2 in connSocketEventHandler (el=0x5e5750, fd=8, clientData=0x629510, mask=1) at connection.c:281
#3  0x00000000004298bf in aeProcessEvents (eventLoop=0x5e5750, flags=27) at ae.c:479
#4  0x0000000000429ab6 in aeMain (eventLoop=0x5e5750) at ae.c:539
#5  0x00000000004372bb in main (argc=1, argv=0x7fffffffe318) at server.c:5175
(gdb)
```

我们将从监听 fd 的初始化到客户端 fd 的读事件回调函数被调用这一过程绘制成如下流程图。

第 8 章 Redis 网络通信模块源码分析

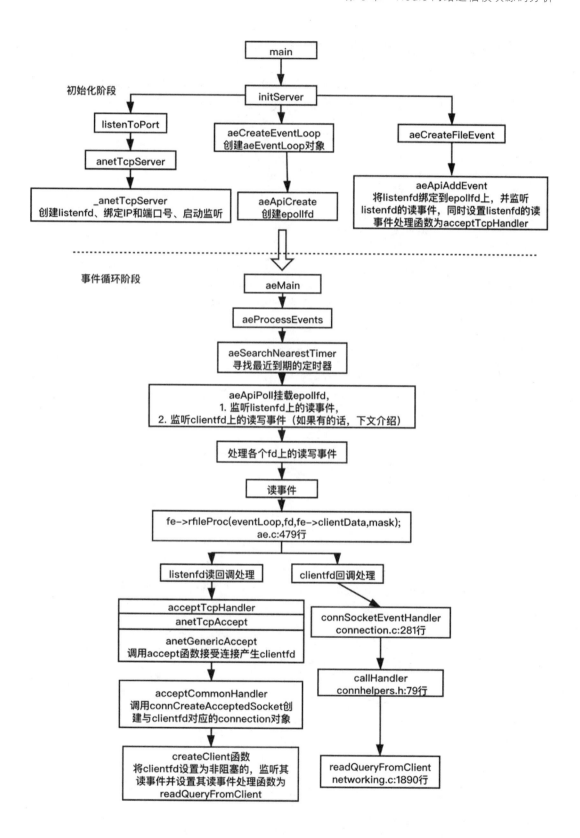

8.2.5　readQueryFromClient 函数

readQueryFromClient 函数的主要逻辑是从 clientfd 上收取数据并将其解包成相应的 Redis 命令加以处理。readQueryFromClient 函数的实现如下：

```
//networking.c 文件第 1890 行
void readQueryFromClient(connection *conn) {
    //省略部分代码
    int nread, readlen;
    size_t qblen;

    //这里的逻辑用于 Redis 多线程 I/O
    if (postponeClientRead(c)) return;

    //每次最大收取的字节数 PROTO_IOBUF_LEN=1024×16
    readlen = PROTO_IOBUF_LEN;

    //省略部分代码

    //计算当前接收缓冲区已有数据的偏移量
    qblen = sdslen(c->querybuf);
    //扩大缓冲区数量至 readlen
    c->querybuf = sdsMakeRoomFor(c->querybuf, readlen);
    //从 clientfd 上读取数据
    nread = connRead(c->conn, c->querybuf+qblen, readlen);

    //省略部分代码

    //接收数据后更新接收缓冲区中的数据长度
    sdsIncrLen(c->querybuf,nread);

    //省略部分代码

    //处理接收缓冲区中的数据
    processInputBuffer(c);
}
```

readQueryFromClient 函数在准备好合适的接收缓冲区后，会调用 connRead 函数进行数据读取操作。connRead 函数的实现如下：

```
//connection.h 文件第 147 行
static inline int connRead(connection *conn, void *buf, size_t buf_len) {
    return conn->type->read(conn, buf, buf_len);
}
```

通过以上代码可以知道，connRead 函数内部实际调用了 conn->type->read 指向的函数，conn->type->read 是一个函数指针，实际指向 connSocketRead 函数，在 connSocketRead 函数中程序会调用系统 read 函数进行数据收取，代码逻辑如下：

```
//connection.c 文件第 173 行
static int connSocketRead(connection *conn, void *buf, size_t buf_len) {
    int ret = read(conn->fd, buf, buf_len);
    if (!ret) {
        conn->state = CONN_STATE_CLOSED;
    } else if (ret < 0 && errno != EAGAIN) {
```

```
        conn->last_errno = errno;
        conn->state = CONN_STATE_ERROR;
    }

    return ret;
}
```

调用系统 read 函数收取的数据被记录在 client 对象的 querybuf 中（这是 clientfd 的接收缓冲区），收取数据后接着调用 processInputBuffer 函数对数据进行解包，得到客户端请求的 Redis 命令，然后调用 processCommandAndResetClient 函数处理这些命令。processInputBuffer 函数中的关键代码如下：

```
//networking.c 文件第 1807 行
void processInputBuffer(client *c) {
    //省略部分代码

    /* 我们终于准备开始执行 Redis 命令了 */
    if (processCommandAndResetClient(c) == C_ERR) {

        return;
    }

    //省略部分代码
}
```

在 processCommandAndResetClient 函数内部又调用了 processCommand 函数，processCommand 是实际处理客户端命令的函数，processCommandAndResetClient 函数的实现如下：

```
//networking.c 文件第 1789 行
int processCommandAndResetClient(client *c) {
    int deadclient = 0;
    server.current_client = c;
    if (processCommand(c) == C_OK) {
        commandProcessed(c);
    }

    //省略部分代码
}
```

Redis 解包的具体过程要结合 Redis 通信协议来介绍，将在 8.4 节详细介绍。

8.2.6 如何处理可写事件

processCommand 函数在处理完命令之后生成给客户端的应答数据，执行这个逻辑的函数是 addReply，addReply 函数的定义如下：

```
//networking.c 文件第 306 行
void addReply(client *c, robj *obj) {
    if (prepareClientToWrite(c) != C_OK) return;

    if (sdsEncodedObject(obj)) {
        if (_addReplyToBuffer(c,obj->ptr,sdslen(obj->ptr)) != C_OK)
            _addReplyProtoToList(c,obj->ptr,sdslen(obj->ptr));
```

```
    } else if (obj->encoding == OBJ_ENCODING_INT) {

        char buf[32];
        size_t len = ll2string(buf,sizeof(buf),(long)obj->ptr);
        if (_addReplyToBuffer(c,buf,len) != C_OK)
            _addReplyProtoToList(c,buf,len);
    } else {
        serverPanic("Wrong obj->encoding in addReply()");
    }
}
```

addReply 函数内部先调用 prepareClientToWrite 函数，prepareClientToWrite 函数内部会检测在当前客户端的发送缓冲区中是否存在未发送数据，如果存在，则说明当前 socket 处于不可写状态，然后为该 socket 注册监听写事件并设置写事件回调函数。prepareClientToWrite 函数的实现如下：

```
//networking.c 文件第 220 行
int prepareClientToWrite(client *c) {
    //省略部分代码

    //检测在当前客户端的发送缓冲区中是否存在未发送数据
    //如果存在，则将数据附加至未发送数据的尾部并注册监听写事件，设置写事件回调函数
    if (!clientHasPendingReplies(c)) clientInstallWriteHandler(c);

    //省略部分代码
}
```

addReply 函数接着调用 _addReplyToBuffer 函数，_addReplyToBuffer 函数的实现如下：

```
//networking.c 文件第 247 行
int _addReplyToBuffer(client *c, const char *s, size_t len) {
    size_t available = sizeof(c->buf)-c->bufpos;

    if (c->flags & CLIENT_CLOSE_AFTER_REPLY) return C_OK;

    if (listLength(c->reply) > 0) return C_ERR;

    if (len > available) return C_ERR;

    memcpy(c->buf+c->bufpos,s,len);
    c->bufpos+=len;
    return C_OK;
}
```

_addReplyToBuffer 函数将待发送的数据放入发送缓冲区（c->buf，即 client 对象的 buf 字段）中，如果在发送缓冲区已经有数据了，则本次的数据会被放置在上一次数据的尾部。

前面介绍了 redis-server 如何处理读事件，整个流程就是注册读事件回调函数，在回调函数中调用系统 read 函数收取数据，然后解析数据得到 Redis 命令，处理这些命令后接着将应答数据包放到 client 对象的发送缓冲区 buf 字段中。现在待发送的数据已被存入发送缓冲区了，那么这些数据是何时被发出去的呢？

还记得在前面章节中介绍的 while 事件循环吗？再来回顾这个 while 事件循环的代码：

```
//ae.c 文件第 536 行
```

第 8 章 Redis 网络通信模块源码分析

```
void aeMain(aeEventLoop *eventLoop) {
    eventLoop->stop = 0;
    while (!eventLoop->stop) {
        if (eventLoop->beforesleep != NULL)
            eventLoop->beforesleep(eventLoop);
        aeProcessEvents(eventLoop, AE_ALL_EVENTS|AE_CALL_AFTER_SLEEP);
    }
}
```

在以上代码中，先判断 eventLoop->beforesleep 对象是否已设置（加粗代码行），eventLoop->beforesleep 是一个回调函数，在 redis-server 初始化时就已经设置好了。设置逻辑如下：

```
//代码片段 8-2-6
//ae.c 文件第 549 行
void aeSetBeforeSleepProc(aeEventLoop *eventLoop, aeBeforeSleepProc *beforesleep)
{
    eventLoop->beforesleep = beforesleep;
}
```

我们先在 aeSetBeforeSleepProc 这个函数入口处设置一个断点，然后重启 redis-server 来验证在何处设置的 eventLoop->beforesleep 回调。操作如下：

```
(gdb) b aeSetBeforeSleepProc
Breakpoint 1 at 0x429ada: file ae.c, line 550.
(gdb) r
Starting program: /root/redis-6.0.3/src/redis-server
[Thread debugging using libthread_db enabled]
Using host libthread_db library "/usr/lib64/libthread_db.so.1".
31514:C 21 Jun 2020 15:57:56.254 # oO0OoO0OoO0Oo Redis is starting oO0OoO0OoO0Oo
31514:C 21 Jun 2020 15:57:56.254 # Redis version=6.0.3, bits=64, commit=00000000, modified=0, pid=31514, just started
31514:C 21 Jun 2020 15:57:56.254 # Warning: no config file specified, using the default config. In order to specify a config file use /root/redis-6.0.3/src/redis-server /path/to/redis.conf

Breakpoint 1, aeSetBeforeSleepProc (eventLoop=0x5e5750, beforesleep=0x42f866 <beforeSleep>) at ae.c:550
550         eventLoop->beforesleep = beforesleep;
(gdb) bt
#0  aeSetBeforeSleepProc (eventLoop=0x5e5750, beforesleep=0x42f866 <beforeSleep>) at ae.c:550
#1  0x00000000004319ac in initServer () at server.c:2916
#2  0x000000000043712a in main (argc=1, argv=0x7fffffffe318) at server.c:5128
(gdb)
```

使用 f 1 命令切换到堆栈#1，并输入 l 命令（list 命令的简写）显示断点附近的代码：

```
(gdb) f 1
#1  0x00000000004319ac in initServer () at server.c:2916
2916        aeSetBeforeSleepProc(server.el,beforeSleep);
(gdb) l
2911             "blocked clients subsystem.");
2912        }
2913
2914        /* Register before and after sleep handlers (note this needs to be done
```

```
2915            * before loading persistence since it is used by
processEventsWhileBlocked. */
2916           aeSetBeforeSleepProc(server.el,beforeSleep);
2917           aeSetAfterSleepProc(server.el,afterSleep);
2918
2919           /* Open the AOF file if needed. */
2920           if (server.aof_state == AOF_ON) {
(gdb)
```

在 server.c 第 2916 行将 eventLoop->beforesleep 回调设置成 beforeSleep 函数。因此每一轮循环都会调用这个 beforeSleep 函数。aeSetBeforeSleepProc 函数的第 1 个参数 server.el 就是前面介绍过的 aeEventLoop 对象（代码片段 8-2-6），在这个 beforeSleep 函数中调用了 handleClientsWithPendingWritesUsingThreads 函数（位于文件 server.c 中）：

```
//server.c 文件第 2106 行
void beforeSleep(struct aeEventLoop *eventLoop) {
    //省略部分代码

    //server.c 文件第 2186 行
    handleClientsWithPendingWritesUsingThreads();

    //省略部分代码
}
```

handleClientsWithPendingWritesUsingThreads 函数的逻辑是把记录在每个 client 的发送缓冲区中的数据发送出去。具体看一下发送逻辑（位于 networking.c 文件中）：

```
//networking.c 文件第 3031 行
int handleClientsWithPendingWritesUsingThreads(void) {
    //省略部分代码

    if (server.io_threads_num == 1 || stopThreadedIOIfNeeded()) {
        return handleClientsWithPendingWrites();
    }

    /* 按需启动新的 I/O 线程 */
    if (!io_threads_active) startThreadedIO();

    //省略部分代码
}
```

在 Redis 6.0 版本之前，Redis 的网络 I/O 线程仅有一个线程，即单线程做 I/O 操作，从 Redis 6.0 之后引入了多线程 I/O，在 handleClientsWithPendingWritesUsingThreads 函数中有这样一段逻辑——如果在 redis.conf 配置文件中多线程网络 I/O 未开启或者客户端连接数较少，则直接调用 handleClientsWithPendingWrites 函数，反之，根据在 redis.conf 配置文件中配置的多线程数量来开启新的网络 I/O 线程（启动新的网络 I/O 线程时使用 startThreadedIO 函数）。

先来看看不开启多线程网络 I/O 的情形，在这种情形下会调用 handleClientsWithPendingWrites 函数，该函数的逻辑如下：

```
//networking 文件第 1398 行
int handleClientsWithPendingWrites(void) {
```

```
    listIter li;
    listNode *ln;
    int processed = listLength(server.clients_pending_write);

    listRewind(server.clients_pending_write,&li);
    while((ln = listNext(&li))) {
        client *c = listNodeValue(ln);
        c->flags &= ~CLIENT_PENDING_WRITE;
        listDelNode(server.clients_pending_write,ln);

        if (c->flags & CLIENT_PROTECTED) continue;

        if (writeToClient(c,0) == C_ERR) continue;

        if (clientHasPendingReplies(c)) {
            int ae_barrier = 0;

            if (server.aof_state == AOF_ON &&
                server.aof_fsync == AOF_FSYNC_ALWAYS)
            {
                ae_barrier = 1;
            }
            if (connSetWriteHandlerWithBarrier(c->conn, sendReplyToClient,
ae_barrier) == C_ERR) {
                freeClientAsync(c);
            }
        }
    }
    return processed;
}
```

handleClientsWithPendingWrites 函数先从全局变量 server 对象（前面已介绍过这个对象）的 clients_pending_write 字段（存储 client 对象的链表）中挨个取出有数据要发送的 client 对象，然后调用 writeToClient 函数尝试将在 client 的发送缓冲区中存储的应答数据发出去。writeToClient 函数的逻辑如下：

```
//networking.c 文件第 1292 行
int writeToClient(client *c, int handler_installed) {
    ssize_t nwritten = 0, totwritten = 0;
    size_t objlen;
    clientReplyBlock *o;

    while(clientHasPendingReplies(c)) {
        if (c->bufpos > 0) {
            nwritten = connWrite(c->conn,c->buf+c->sentlen,c->bufpos-c->sentlen);
            if (nwritten <= 0) break;
            c->sentlen += nwritten;
            totwritten += nwritten;

            if ((int)c->sentlen == c->bufpos) {
                c->bufpos = 0;
                c->sentlen = 0;
            }
        } else {
            o = listNodeValue(listFirst(c->reply));
            objlen = o->used;
```

```c
            if (objlen == 0) {
                c->reply_bytes -= o->size;
                listDelNode(c->reply,listFirst(c->reply));
                continue;
            }

            nwritten = connWrite(c->conn, o->buf + c->sentlen, objlen - c->sentlen);
            if (nwritten <= 0) break;
            c->sentlen += nwritten;
            totwritten += nwritten;

            if (c->sentlen == objlen) {
                c->reply_bytes -= o->size;
                listDelNode(c->reply,listFirst(c->reply));
                c->sentlen = 0;

                if (listLength(c->reply) == 0)
                    serverAssert(c->reply_bytes == 0);
            }
        }

        if (totwritten > NET_MAX_WRITES_PER_EVENT &&
            (server.maxmemory == 0 ||
             zmalloc_used_memory() < server.maxmemory) &&
            !(c->flags & CLIENT_SLAVE)) break;
    }
    server.stat_net_output_bytes += totwritten;
    if (nwritten == -1) {
        if (connGetState(c->conn) == CONN_STATE_CONNECTED) {
            nwritten = 0;
        } else {
            serverLog(LL_VERBOSE,
                "Error writing to client: %s", connGetLastError(c->conn));
            freeClientAsync(c);
            return C_ERR;
        }
    }
    if (totwritten > 0) {

        if (!(c->flags & CLIENT_MASTER)) c->lastinteraction = server.unixtime;
    }
    if (!clientHasPendingReplies(c)) {
        c->sentlen = 0;

        if (handler_installed) connSetWriteHandler(c->conn, NULL);

        /* Close connection after entire reply has been sent. */
        if (c->flags & CLIENT_CLOSE_AFTER_REPLY) {
            freeClientAsync(c);
            return C_ERR;
        }
    }
    return C_OK;
}
```

writeToClient 函数先把存储在 client 对象的 buf 字段中的数据发出去，如果发送出错的话就释放这个 client 对象。如果数据能够全部发送完，则发送完毕之后会移除相应 fd 上对写事件的监听（前提是该 fd 添加了对写事件的监听）；如果当前 client 对象设置了 CLIENT_CLOSE_AFTER_REPLY 标志，则在发送完数据后会立即释放这个 client 对象。

实际发送数据的函数是 connWrite，这个函数内部调用最初设置的写回调函数：

```
//connection.h 文件第 135 行
static inline int connWrite(connection *conn, const void *data, size_t data_len) {
    return conn->type->write(conn, data, data_len);
}
```

如前文所分析的，conn->type->write 实际指向 connSocketWrite 函数。connSocketWrite 函数的实现如下：

```
//代码片段 8-2-6-2
//connection.c 文件第 163 行
static int connSocketWrite(connection *conn, const void *data, size_t data_len) {
    int ret = write(conn->fd, data, data_len);
    if (ret < 0 && errno != EAGAIN) {
        conn->last_errno = errno;
        conn->state = CONN_STATE_ERROR;
    }

    return ret;
}
```

当然，可能存在这样一种情况：由于网络或者客户端的原因，redis-server 发给某个客户端的数据一直发送不出去，或者只有部分数据可以发出去（例如：服务端向客户端发数据，客户端的应用层一直不从 TCP 内核缓冲区中读取数据，这样在服务器发送一段时间的数据后，客户端内核缓冲区变满，服务端再发数据就会发不出去了。由于 fd 是非阻塞的，这时服务器调用 send 或者 write 函数会直接返回，返回值是 -1，错误码是 EAGAIN，如代码片段 8-2-6-2 中的加粗行所示）。不管是哪种情况，数据一次发不完，这时就需要监听写事件了。在 handleClientsWithPendingWrites 函数中有如下代码：

```
//networking.c 文件第 1398 行
int handleClientsWithPendingWrites(void) {
    //省略部分代码

        if (writeToClient(c,0) == C_ERR) continue;

        if (clientHasPendingReplies(c)) {
            //省略部分代码

            if (connSetWriteHandlerWithBarrier(c->conn, sendReplyToClient,
ae_barrier) == C_ERR) {
                freeClientAsync(c);
            }
        }

    //省略部分代码
}
```

实际做注册监听写事件工作的函数是 connSetWriteHandlerWithBarrier，该函数的实现如下：

```
//connection.h 文件第 170 行
static inline int connSetWriteHandlerWithBarrier(connection *conn,
ConnectionCallbackFunc func, int barrier) {
    return conn->type->set_write_handler(conn, func, barrier);
}
```

在以上代码中，如前文所述，conn->type->set_write_handler 实际指向 connSocketSetWriteHandler 函数。connSocketSetWriteHandler 函数的逻辑如下：

```
//connection.c 文件第 206 行
static int connSocketSetWriteHandler(connection *conn, ConnectionCallbackFunc func,
int barrier) {
    if (func == conn->write_handler) return C_OK;

    conn->write_handler = func;
    if (barrier)
        conn->flags |= CONN_FLAG_WRITE_BARRIER;
    else
        conn->flags &= ~CONN_FLAG_WRITE_BARRIER;
    if (!conn->write_handler)
        aeDeleteFileEvent(server.el,conn->fd,AE_WRITABLE);
    else
        //在这一行注册可写事件
        if (aeCreateFileEvent(server.el,conn->fd,AE_WRITABLE,
                conn->type->ae_handler,conn) == AE_ERR) return C_ERR;
    return C_OK;
}
```

需要注册监听可写事件时，执行 aeCreateFileEvent 函数；反之移除对写事件的监听，执行 aeDeleteFileEvent 函数，移除对写事件的监听的情形是在写事件触发后数据全部发送完毕。

这里写事件（标志是 AE_WRITABLE）触发的回调函数是 sendReplyToClient。也就是说，当下一次写事件触发时，调用的就是 sendReplyToClient 函数。可以猜到，sendReplyToClient 函数发送数据的逻辑和前面介绍的 writeToClient 函数的逻辑一模一样：

```
//networking.c 文件第 1389 行
void sendReplyToClient(connection *conn) {
    client *c = connGetPrivateData(conn);
    writeToClient(c,1);
}
```

至此，redis-server 发送数据的逻辑也理清楚了。这里简单做个总结，如下所述。

（1）如果有数据要发送给某个 client，而不是一上来就注册写事件、等触发可写事件再发送，则通常的做法是，在应答数据产生的地方直接发送。如果因为对端 TCP 窗口太小而导致数据发送不完，则需要将剩余的数据存储至发送缓冲区中并注册监听写事件，等下次写事件触发后再尝试发送，一直到数据全部发送完毕再移除可写事件。

（2）redis-server 数据的发送逻辑与本总结第 1 条稍微有点差别，差别在于将数据发送的时机放到 loop 的某个时间点上（这里是在 ProcessEvents 之前）。

8.2.7　Redis 6.0 多线程网络 I/O

接下来介绍网络 I/O 使用多个线程的情形，要使用多个线程，必须先开启这个功能。

在 Redis 6.0 中默认是不开启多线程网络 I/O 的，可以通过修改 redis.conf 配置文件中的相关配置项来打开。在 redis.conf 文件中开启多线程网络 I/O 的配置项如下：

```
# So for instance if you have a four cores boxes, try to use 2 or 3 I/O
# threads, if you have a 8 cores, try to use 6 threads. In order to
# enable I/O threads use the following configuration directive:
#
# io-threads 4
#
# Setting io-threads to 1 will just use the main thread as usually.
# When I/O threads are enabled, we only use threads for writes, that is
# to thread the write(2) syscall and transfer the client buffers to the
# socket. However it is also possible to enable threading of reads and
# protocol parsing using the following configuration directive,
# by setting it to yes:
#
# io-threads-do-reads no
#
```

将 io-threads 配置项前面的 "#" 去掉，将其设置成期望的线程数量（这里默认数量是 4），将 io-threads-do-reads 配置前面的 "#" 也去掉，将其值改为 yes（默认是 no）。

修改这两个配置项后，我们先使用 gdb 命令 set args "../redis.conf"为 redis-server 设置命令行参数，然后重启 redis-server 如下：

```
(gdb) set args "../redis.conf"
(gdb) r
The program being debugged has been started already.
Start it from the beginning? (y or n) y
Starting program: /root/redis-6.0.3/src/redis-server "../redis.conf"
[Thread debugging using libthread_db enabled]
Using host libthread_db library "/usr/lib64/libthread_db.so.1".
```

在 redis-server 运行起来之后，按 Ctrl+C 组合键将程序中断，使用 info threads 命令查看此时的线程信息：

```
(gdb) info threads
  Id   Target Id                                         Frame
* 1    Thread 0x7ffff7feb740 (LWP 11992) "redis-server"  0x00007ffff71e2603 in
epoll_wait () from /usr/lib64/libc.so.6
  2    Thread 0x7ffff0bb9700 (LWP 11993) "bio_close_file" 0x00007ffff74bc965 in
pthread_cond_wait@@GLIBC_2.3.2 () from /usr/lib64/libpthread.so.0
  3    Thread 0x7ffff03b8700 (LWP 11994) "bio_aof_fsync" 0x00007ffff74bc965 in
pthread_cond_wait@@GLIBC_2.3.2 () from /usr/lib64/libpthread.so.0
  4    Thread 0x7fffefbb7700 (LWP 11995) "bio_lazy_free" 0x00007ffff74bc965 in
pthread_cond_wait@@GLIBC_2.3.2 () from /usr/lib64/libpthread.so.0
  5    Thread 0x7fffef3b6700 (LWP 11996) "io_thd_1"      0x00007ffff74bf4ed in
__lll_lock_wait () from /usr/lib64/libpthread.so.0
  6    Thread 0x7fffeebb5700 (LWP 11997) "io_thd_2"      0x00007ffff74bf4ed in
__lll_lock_wait () from /usr/lib64/libpthread.so.0
  7    Thread 0x7fffee3b4700 (LWP 11998) "io_thd_3"      0x00007ffff74bf4ed in
```

```
__lll_lock_wait () from /usr/lib64/libpthread.so.0
(gdb)
```

如上面的输出所示,与未开启多线程网络 I/O 的情形相比,redis-server 多了名为 io_thd_1、io_thd_2、io_thd_3 的线程,加上主线程(线程名为 redis-server 的线程)一共是 4 个 I/O 线程(在配置文件中,io-threads = 4)。我们重点看下这三个 I/O 工作线程,由于这三个工作线程的逻辑是一样的,所以以 io_thd_1 为例来介绍。先使用 thread 5(5 是 io_thd_1 线程的线程编号)命令切换到 io_thd_1 线程,然后使用 bt 命令查看这个线程的调用堆栈:

```
(gdb) bt
#0  0x00007ffff74bf4ed in __lll_lock_wait () from /usr/lib64/libpthread.so.0
#1  0x00007ffff74badcb in _L_lock_883 () from /usr/lib64/libpthread.so.0
#2  0x00007ffff74bac98 in pthread_mutex_lock () from /usr/lib64/libpthread.so.0
#3  0x0000000000447907 in IOThreadMain (myid=0x1) at networking.c:2921
#4  0x00007ffff74b8dd5 in start_thread () from /usr/lib64/libpthread.so.0
#5  0x00007ffff71e202d in clone () from /usr/lib64/libc.so.6
```

堆栈#3 处的代码如下:

```c
//networking.c 文件第 2903 行
void *IOThreadMain(void *myid) {

    long id = (unsigned long)myid;
    char thdname[16];

    snprintf(thdname, sizeof(thdname), "io_thd_%ld", id);
    //设置线程的名称
    redis_set_thread_title(thdname);
    redisSetCpuAffinity(server.server_cpulist);

    while(1) {
        /* 等待启动 */
        for (int j = 0; j < 1000000; j++) {
            if (io_threads_pending[id] != 0) break;
        }

        /* 让主线程可以停掉当前线程 */
        if (io_threads_pending[id] == 0) {
            pthread_mutex_lock(&io_threads_mutex[id]);
            pthread_mutex_unlock(&io_threads_mutex[id]);
            continue;
        }

        serverAssert(io_threads_pending[id] != 0);

        if (tio_debug) printf("[%ld] %d to handle\n", id,
(int)listLength(io_threads_list[id]));

        listIter li;
        listNode *ln;
        listRewind(io_threads_list[id],&li);
        while((ln = listNext(&li))) {
            client *c = listNodeValue(ln);
```

```
            if (io_threads_op == IO_THREADS_OP_WRITE) {
                writeToClient(c,0);
            } else if (io_threads_op == IO_THREADS_OP_READ) {
                readQueryFromClient(c->conn);
            } else {
                serverPanic("io_threads_op value is unknown");
            }
        }
        listEmpty(io_threads_list[id]);
        io_threads_pending[id] = 0;

        if (tio_debug) printf("[%ld] Done\n", id);
    }
}
```

IOThreadMain 函数是工作线程的线程函数，主要逻辑是一些初始化工作和一个 while 循环，初始化工作的主要逻辑是设置线程的名称。这就是在 gdb 中看到工作线程的线程名为 io_thd_1、io_thd_2、io_thd_3 的原因。

工作线程的线程 id（严格来说，其实这里应该叫作线程序号）是主线程创建工作线程时通过线程参数传递过来的，线程 id 从 1 开始，序号 0 保留给主线程。

主线程在 main 函数中调用 InitServerLast 函数，InitServerLast 函数又调用了 initThreadedIO 函数，在 initThreadedIO 函数中 redis-server 会根据配置文件中的 io-threads 配置项创建对应数量的 I/O 工作线程数量（由于主线程也是 I/O 线程，所以实际的工作线程数量是 io-threads-1）。我们先在 initThreadedIO 函数调用处加个断点，然后重启 gdb，就可以看到对应的调用关系和相应的代码位置：

```
Thread 1 "redis-server" hit Breakpoint 2, initThreadedIO () at networking.c:2954
2954        io_threads_active = 0;   //省略了此处注释
(gdb) bt
#0  initThreadedIO () at networking.c:2954
#1  0x0000000000431aa8 in InitServerLast () at server.c:2954
#2  0x0000000000437195 in main (argc=2, argv=0x7fffffffe308) at server.c:5142
(gdb)
```

initThreadedIO 函数的实现如下：

```
//代码片段 8-2-7-1
//networking.c 文件第 2953 行
void initThreadedIO(void) {
    io_threads_active = 0;

    if (server.io_threads_num == 1) return;
    //最大允许的 I/O 工作线程数量是 IO_THREADS_MAX_NUM=128
    if (server.io_threads_num > IO_THREADS_MAX_NUM) {
        serverLog(LL_WARNING,"Fatal: too many I/O threads configured. "
                             "The maximum number is %d.", IO_THREADS_MAX_NUM);
        exit(1);
    }

    for (int i = 0; i < server.io_threads_num; i++) {

        io_threads_list[i] = listCreate();
```

```
        //编号为 0 时是主线程
        if (i == 0) continue;

        pthread_t tid;
        //初始化 io_threads_mutex[i]对象
        pthread_mutex_init(&io_threads_mutex[i],NULL);
        io_threads_pending[i] = 0;
        //锁定 io_threads_mutex[i]
        pthread_mutex_lock(&io_threads_mutex[i]);
        //创建对应的 I/O 工作线程
        if (pthread_create(&tid,NULL,IOThreadMain,(void*)(long)i)!=0) {
            serverLog(LL_WARNING,"Fatal: Can't initialize IO thread.");
            exit(1);
        }
        io_threads[i] = tid;
    }
}
```

通过以上代码段，我们可以得到两个结论：①redis-server 最大允许的 I/O 工作线程数量为 128 个（IO_THREADS_MAX_NUM 宏的值）；②序号为 0 的线程是主线程，因此实际的工作线程数量是 io-threads −1。

创建新的 I/O 线程之前，redis-server 会为每个 I/O 线程都创建一个链表 io_threads_list[i]，这个链表用于存储代表客户端的 client 对象，这些链表被存储在全局数组 io_threads_list 中，与线程序号一一对应；redis-server 同时创建相应数量的整型变量（unsigned long）并将其存储于另一个全局数组 io_threads_pending 中，同样与线程序号一一对应，这些整型变量和另一组 Linux 互斥体对象（存储在 io_threads_mutex 数组中）一起让主线程控制工作线程的启动与停止。主线程控制相应的工作线程启动与停止的逻辑如下：

（1）将 io_threads_pending[i]设置为 0；

（2）在代码片段 8-2-7-1 的 for 循环中初始化 io_threads_mutex[i]对象后，立刻调用 pthread_mutex_lock(&io_threads_mutex[i])将这些互斥体锁定（代码片段 8-2-7-1 中的加粗行）；

（3）创建对应的 I/O 工作线程，在这些 I/O 工作线程的线程函数 IOThreadMain 中有如下代码：

```
//代码片段 8-2-7-2
//networking.c 文件第 2903 行
//IOThreadMain 是 I/O 工作线程的线程函数
void *IOThreadMain(void *myid) {
    //省略部分代码

    while(1) {

        for (int j = 0; j < 1000000; j++) {
            if (io_threads_pending[id] != 0) break;
        }

        /* 让主线程有机会可以停止这些 I/O 工作线程 */
        if (io_threads_pending[id] == 0) {
```

```
        pthread_mutex_lock(&io_threads_mutex[id]);
        pthread_mutex_unlock(&io_threads_mutex[id]);
        continue;
    }

    //省略部分代码

    //处理链表逻辑
    listIter li;
    listNode *ln;
    listRewind(io_threads_list[id],&li);
    while((ln = listNext(&li))) {
        client *c = listNodeValue(ln);
        if (io_threads_op == IO_THREADS_OP_WRITE) {
            writeToClient(c,0);
        } else if (io_threads_op == IO_THREADS_OP_READ) {
            readQueryFromClient(c->conn);
        } else {
            serverPanic("io_threads_op value is unknown");
        }
    }
    listEmpty(io_threads_list[id]);
    io_threads_pending[id] = 0;
    }
}
```

在以上代码中，I/O 工作线程在执行 pthread_mutex_lock(&io_threads_ mutex[id])行（以上加粗代码行）时，由于 io_threads_mutex[id]这个互斥体已经被主线程加锁了，因此 I/O 工作线程被阻塞在这一行。如果想让 I/O 工作线程停止阻塞，即启用这些 I/O 工作线程，则可以调用 startThreadedIO 函数。startThreadedIO 函数的实现如下：

```
//networking.c 文件第 2985 行
void startThreadedIO(void) {
    //省略部分代码
    for (int j = 1; j < server.io_threads_num; j++)
        pthread_mutex_unlock(&io_threads_mutex[j]);
    io_threads_active = 1;
}
```

startThreadedIO 函数会对相应的互斥体对象 io_threads_mutex[id]进行解锁，同时设置启用 I/O 工作线程的标志变量 io_threads_active，在下文中会介绍这个标志变量。

有读者可能会注意到：在代码片段 8-2-7-2 中，即使解锁 io_threads_mutex[id]互斥体（代码片段 8-2-7-2 中的加粗行），在执行代码中的 continue 语句之后，下一轮循环由于 io_threads_pending[id]仍然为 0，循环会继续加锁、解锁再执行 continue 语句，仍然无法进入由 client 对象组成的链表对象的处理逻辑中。因此，除了解锁 io_threads_ mutex[id]互斥体，还必须将 io_threads_pending[id]设置为非 0 值，才能执行 I/O 工作线程中处理链表的逻辑。那么 io_threads_pending[id]是在什么地方被设置成非 0 值的呢？

答案是在 beforeSleep 函数（位于主线程中）中分别调用了 handleClientsWithPendingReadsUsingThreads 函数和 handleClientsWithPendingWritesUsingThreads 函数，这两个函数

分别对应读和写的情况。beforeSleep 函数的实现如下：

```
//server.c 文件第 2106 行
void beforeSleep(struct aeEventLoop *eventLoop) {
    //省略部分代码

    //处理有读操作的 clients
    handleClientsWithPendingReadsUsingThreads();

    //省略部分代码

    //处理有写操作的 clients
    handleClientsWithPendingWritesUsingThreads();

    //省略部分代码
}
```

先来看看读的情况，handleClientsWithPendingReadsUsingThreads 函数的实现如下：

```
//代码片段 8-2-7-3
//networking 文件第 3126 行
int handleClientsWithPendingReadsUsingThreads(void) {
    if (!io_threads_active || !server.io_threads_do_reads) return 0;
    int processed = listLength(server.clients_pending_read);
    if (processed == 0) return 0;

    if (tio_debug) printf("%d TOTAL READ pending clients\n", processed);

    listIter li;
    listNode *ln;
    listRewind(server.clients_pending_read,&li);
    int item_id = 0;
    //主线程给工作线程分配 client 对象的策略
    while((ln = listNext(&li))) {
        client *c = listNodeValue(ln);
        int target_id = item_id % server.io_threads_num;
        listAddNodeTail(io_threads_list[target_id],c);
        item_id++;
    }

    //networking.c 3147 行
    io_threads_op = IO_THREADS_OP_READ;
    for (int j = 1; j < server.io_threads_num; j++) {
        int count = listLength(io_threads_list[j]);
        io_threads_pending[j] = count;
    }

    //networking.c 3153 行
    //主线程处理分配给自己的 client 对象
    listRewind(io_threads_list[0],&li);
    while((ln = listNext(&li))) {
        client *c = listNodeValue(ln);
        readQueryFromClient(c->conn);
    }
    listEmpty(io_threads_list[0]);

    //主线程正在利用一个 while 循环等待 I/O 工作
```

```
//线程完成对自己 client 对象的处理
//============================
while(1) {
    unsigned long pending = 0;
    for (int j = 1; j < server.io_threads_num; j++)
        pending += io_threads_pending[j];
    //在 pending 等于 0 时退出这个 while 循环
    if (pending == 0) break;
}
if (tio_debug) printf("I/O READ All threads finshed\n");
//============================

while(listLength(server.clients_pending_read)) {
    ln = listFirst(server.clients_pending_read);
    client *c = listNodeValue(ln);
    c->flags &= ~CLIENT_PENDING_READ;
    listDelNode(server.clients_pending_read,ln);

    if (c->flags & CLIENT_PENDING_COMMAND) {
        c->flags &= ~CLIENT_PENDING_COMMAND;
        if (processCommandAndResetClient(c) == C_ERR) {

            continue;
        }
    }
    processInputBuffer(c);
}
return processed;
}
```

代码片段 8-2-7-3 先通过 io_threads_active 和 server.io_threads_do_reads 两个标志判断是否开启 I/O 工作线程，如果没有开启，则直接退出该函数，并在主线程中执行所有 I/O 操作。

如果开启了 I/O 线程，则第 1 个 while 循环中的逻辑（第 1 个加粗代码行）是主线程给 I/O 工作线程分配 client 对象的策略。这里的策略也很简单，即使用 Round-Robin 策略（轮询策略）将当前处理序号与线程数量求余，并分别将对应的 client 对象放入相应的线程（包括主线程）存储 client 对象的链表中。通俗地说，假设现在加上主线程一共有 4 个 I/O 线程，则将第 0 个 client 对象分配给主线程，将第 1 个 client 对象分配给 1 号工作线程，将第 2 个 client 对象分配给 2 号工作线程，将第 3 个 client 对象分配给 3 号工作线程，将第 4 个 client 对象再次分配给主线程，将第 5 个 client 对象再次分配给 1 号工作线程，将第 6 个 client 对象再次分配给 2 号工作线程，以此类推。

在分配好 client 对象到相应的 I/O 工作线程的链表中后，redis-server 设置与这些 I/O 工作线程相对应的 io_threads_pending[j]变量值为非 0 值，这里实际设置给 io_threads_pending[j]变量的值是对应 I/O 工作线程的链表长度，这是因为在 client 对象少于 I/O 工作线程数量的情况下，如果某个 I/O 工作线程的链表长度为 0，就没必要唤醒该 I/O 工作线程。这段逻辑对应代码片段 8-2-7-3 中的斜体部分。

由于主线程自己也参与分配 client 对象，因此主线程给 I/O 工作线程分配好相应的 client 对象并设置唤醒标志（io_threads_pending[j]）后，主线程接下来需要处理自己被分配到的 client 对象。这段逻辑对应代码片段 8-2-7-3 中的第 2 个加粗代码行。

主线程处理分配给自己的 client 对象的方法：先从自己的链表（io_threads_list[0]）中挨个取出 client 对象，然后调用 readQueryFromClient 读取数据和解包。主线程在处理完链表中的 client 对象后，会将自己的链表清空。

同样的道理，I/O 工作线程在处理自己的链表时与主线程的操作相同：

```
//networking.c 文件第 2903 行
//I/O 工作线程函数
void *IOThreadMain(void *myid) {
    //省略部分代码

    while(1) {
        //省略部分代码

        if (io_threads_pending[id] == 0) {
            pthread_mutex_lock(&io_threads_mutex[id]);
            pthread_mutex_unlock(&io_threads_mutex[id]);
            continue;
        }

        //省略部分代码

        listIter li;
        listNode *ln;
        listRewind(io_threads_list[id],&li);
        while((ln = listNext(&li))) {
            client *c = listNodeValue(ln);
            if (io_threads_op == IO_THREADS_OP_WRITE) {
                writeToClient(c,0);
            } else if (io_threads_op == IO_THREADS_OP_READ) {
                readQueryFromClient(c->conn);
            } else {
                serverPanic("io_threads_op value is unknown");
            }
        }
        //处理完成后将清空自己的链表
        listEmpty(io_threads_list[id]);
        //重置状态标志值
        io_threads_pending[id] = 0;

        //省略部分代码
    }
}
```

在以上代码中，I/O 工作线程在处理完自己链表中的 client 对象后也会清空链表并重置 io_threads_pending[id]标志，而此时主线程正在利用一个 while 循环等待 I/O 工作线程完成对自己 client 对象的处理。由于每个 I/O 工作线程在处理完自己链表中的 client 对象后都会将自己的 io_threads_pending[id]标志重置为 0，所以主线程中 for 循环累加的 pending 值最终会变为 0，退出 while 循环。这一逻辑对应代码片段 8-2-7-3 中以 "//=====" 标记的范围。

以上就是在 Redis 6.0 版本中利用 I/O 工作线程处理读事件的逻辑。然而事情还没有结束，如果仔细研究源码会发现以下两个问题。

问题一

要想让主线程在 handleClientsWithPendingReadsUsingThreads 函数中给 I/O 工作线程分配需要做读操作的 client 对象，就必须满足 io_threads_active 和 server.io_threads_do_reads 这两个标志同时是非 0 值。代码如下：

```
//networking.c 文件第 3126 行
int handleClientsWithPendingReadsUsingThreads(void) {
    //io_threads_active 和 server.io_threads_do_reads 必须同时是非 0 值
    if (!io_threads_active || !server.io_threads_do_reads) return 0;

    //省略部分代码
}
```

对 server.io_threads_do_reads 的值只要在 redis.conf 中配置就可以了，但是 io_threads_active 标志一开始就被设置为 0（以下代码片段中的加粗行），逻辑如下：

```
//networking.c 文件第 2953 行
void initThreadedIO(void) {
    io_threads_active = 0;

    //省略部分代码
}
```

在所有代码文件中都搜索过后，我们发现 io_threads_active 标志只会在 startThreadedIO 函数中被设置为非 0 值：

```
//networking.c 文件第 2985 行
void startThreadedIO(void) {
    //省略部分代码

    for (int j = 1; j < server.io_threads_num; j++)
        pthread_mutex_unlock(&io_threads_mutex[j]);
    io_threads_active = 1;
}
```

但是在 handleClientsWithPendingReadsUsingThreads 函数中并没有调用 startThreadedIO 函数，这是问题一。

问题二

根据 I/O 工作线程中的逻辑，I/O 工作线程必须同时满足两个条件才能执行处理自己链表中的 client 对象的逻辑：

条件一：io_threads_pending[id] 不等于 0。这个条件在 handleClientsWithPending-ReadsUsingThreads 给 I/O 工作线程分配 client 对象后就已经满足。

条件二：I/O 工作线程在创建后，被阻塞在 pthread_mutex_lock (&io_threads_ mutex[id]) 处（以下代码片段的加粗行），必须解锁 io_threads_mutex[id] 这个互斥体才能让线程继续往下运行。这个条件暂时不满足，代码逻辑如下：

```
//I/O 工作线程函数
//networking.c 文件第 2903 行
void *IOThreadMain(void *myid) {
    //省略部分代码

    while(1) {
        //省略部分代码

        if (io_threads_pending[id] == 0) {
            //工作线程在创建后被阻塞在这里
            pthread_mutex_lock(&io_threads_mutex[id]);
            pthread_mutex_unlock(&io_threads_mutex[id]);
            continue;
        }

        //省略部分代码

        listIter li;
        listNode *ln;
        listRewind(io_threads_list[id],&li);
        while((ln = listNext(&li))) {
            client *c = listNodeValue(ln);
            if (io_threads_op == IO_THREADS_OP_WRITE) {
                writeToClient(c,0);
            } else if (io_threads_op == IO_THREADS_OP_READ) {
                readQueryFromClient(c->conn);
            } else {
                serverPanic("io_threads_op value is unknown");
            }
        }
        listEmpty(io_threads_list[id]);
        io_threads_pending[id] = 0;

        //省略部分代码
    }
}
```

结合前面的所有分析，主线程只有调用了 startThreadedIO 函数先解锁 io_threads_mutex[id]并设置 io_threads_active 标志为非 0 值，在 handleClientsWithPendingReadsUsingThreads 函数中才会分配 client 对象给 I/O 工作线程并设置 io_threads_pending[id]为非 0 值，这样 I/O 工作线程才能处理分配给自己的 client 对象。所以大的前提是 startThreadedIO 函数被调用。但是通过分析流程可以发现，主线程的事件循环第 1 次调用 beforeSleep 函数，beforeSleep 函数第 1 次调用 handleClientsWithPendingReadsUsingThreads 函数时 startThreadedIO 未被调用，也就是说，主线程第 1 次是不会使用 I/O 工作线程处理 client 对象上的读操作的。那么 startThreadedIO 函数到底是在哪里被调用的呢？

答案是在 beforeSleep 函数中除了调用了 handleClientsWithPendingReadsUsingThreads 处理读操作，还调用了另一个函数 handleClientsWithPendingWritesUsingThreads 用于处理写操作，代码逻辑如下：

```
//server.c 文件第 2106 行
void beforeSleep(struct aeEventLoop *eventLoop) {
    //省略部分代码
```

```
    //处理 client 的读操作
    handleClientsWithPendingReadsUsingThreads();

    //省略部分代码

    //处理 client 的写操作
    handleClientsWithPendingWritesUsingThreads();

    //省略部分代码
}
```

handleClientsWithPendingWritesUsingThreads 函数的实现如下:

```
//代码片段 8-2-7-4
//networking.c 文件第 3031 行
int handleClientsWithPendingWritesUsingThreads(void) {
    int processed = listLength(server.clients_pending_write);
    if (processed == 0) return 0;

    //逻辑一
    //networking.c 文件第 3037 行
    if (server.io_threads_num == 1 || stopThreadedIOIfNeeded()) {
        //如果是单线程模式或者停止 I/O 工作线程, 则执行流在此处直接退出
        return handleClientsWithPendingWrites();
    }

    //逻辑二
    //根据 io_threads_active 标志值启动 I/O 工作线程
    if (!io_threads_active) startThreadedIO();

    if (tio_debug) printf("%d TOTAL WRITE pending clients\n", processed);

    //逻辑三
    // (1) 使用 Round-Robin 策略将各个 fd 待处理的读事件分配给各个工作线程
    listIter li;
    listNode *ln;
    listRewind(server.clients_pending_write,&li);
    int item_id = 0;
    while((ln = listNext(&li))) {
        client *c = listNodeValue(ln);
        c->flags &= ~CLIENT_PENDING_WRITE;
        int target_id = item_id % server.io_threads_num;
        listAddNodeTail(io_threads_list[target_id],c);
        item_id++;
    }

    // (2) 设置 io_threads_pending[id]标志值以激活工作线程
    io_threads_op = IO_THREADS_OP_WRITE;
    for (int j = 1; j < server.io_threads_num; j++) {
        int count = listLength(io_threads_list[j]);
        io_threads_pending[j] = count;
    }

    // (3) 主线程处理分配给自己的链表中的 client 对象的写事件
    listRewind(io_threads_list[0],&li);
    while((ln = listNext(&li))) {
```

```
        client *c = listNodeValue(ln);
        writeToClient(c,0);
    }
    listEmpty(io_threads_list[0]);

    // (4) 等待工作线程处理完对应的链表中的所有写事件
    while(1) {
        unsigned long pending = 0;
        for (int j = 1; j < server.io_threads_num; j++)
            pending += io_threads_pending[j];
        if (pending == 0) break;
    }
    if (tio_debug) printf("I/O WRITE All threads finshed\n");

    // (5) 在所有写事件处理完毕后，如果仍有一些 client 存在未发送完的数据，
    //则继续为这些 client 注册写事件
    listRewind(server.clients_pending_write,&li);
    while((ln = listNext(&li))) {
        client *c = listNodeValue(ln);

        if (clientHasPendingReplies(c) &&
                connSetWriteHandler(c->conn, sendReplyToClient) == AE_ERR)
        {
            freeClientAsync(c);
        }
    }
    listEmpty(server.clients_pending_write);
    return processed;
}
```

handleClientsWithPendingWritesUsingThreads 有几段关键逻辑，我们挨个看一下。

1. 逻辑一

按需停止 I/O 线程，在 I/O 线程停止的情况下直接调用 handleClientsWithPendingWrites 函数处理客户端可写事件。

在 stopThreadedIOIfNeeded 函数中按需停止 I/O 工作线程，逻辑如下：

```
//networking.c 文件第 3017 行
int stopThreadedIOIfNeeded(void) {
    int pending = listLength(server.clients_pending_write);

    if (server.io_threads_num == 1) return 1;

    if (pending < (server.io_threads_num*2)) {
        if (io_threads_active) stopThreadedIO();
        return 1;
    } else {
        return 0;
    }
}
```

在以上代码中给出了需要停止 I/O 工作线程的条件：即在需要处理写操作的客户端数量少于 I/O 线程数的两倍，例如 I/O 线程数量为 4（包括主线程），而当前需要处理写操作的客户端数量少于 8 的时候，停止 I/O 工作线程。停止 I/O 工作线程的方法是调用

stopThreadedIO 函数，在该函数中会给 io_threads_mutex[id]加锁，并将 io_threads_active 标志设置为 0。stopThreadedIO 函数的实现如下：

```
//networking.c 文件第 2994 行
void stopThreadedIO(void) {

    handleClientsWithPendingReadsUsingThreads();

    //省略部分代码

    for (int j = 1; j < server.io_threads_num; j++)
        pthread_mutex_lock(&io_threads_mutex[j]);
    io_threads_active = 0;
}
```

在停止 I/O 工作线程之前必须调用 handleClientsWithPendingReadsUsingThreads 函数（以上加粗代码行），这是需要注意的地方，因为之前分配给各个 I/O 工作线程的待处理读事件可能还没来得及处理，所以必须在停止 I/O 线程之前主动调用一次 handleClientsWithPendingReadsUsingThreads 函数，将这些读事件处理掉。

2. 逻辑二

如果逻辑一中的 if 条件不满足，handleClientsWithPendingWritesUsingThreads 函数就不会提前 return，而是接着根据 io_threads_active 标志值启动 I/O 工作线程。

根据上面的分析，我们可以总结出启动 I/O 工作线程需要同时满足的三个条件：

◎ 在配置文件中配置了工作线程数量大于 1；
◎ 待处理写事件的客户端数量达到所有 I/O 线程数量的两倍或更多；
◎ I/O 工作线程还没处于激活状态（全局变量 io_threads_active 记录该状态）。

3. 逻辑三

handleClientsWithPendingWritesUsingThreads 函数（主线程中）接下来的几个逻辑就和 handleClientsWithPendingReadsUsingThreads 函数中的一样了：

（1）使用 Round-Robin 策略将有读操作的 client 依次分配给各个 I/O 工作线程；
（2）设置 io_threads_pending[id]标志值，以激活 I/O 工作线程；
（3）主线程处理分配给自己的 client 对象的写操作；
（4）等待工作线程处理完对应的链表中所有的写操作；
（5）在所有写操作都处理完毕之后，如果仍有一些 client 存在未发送完的数据，则继续为这些 client 注册写事件。

笔者已经在代码片段 8-2-7-4 中通过注释标记出这些步骤相应的代码段。

无论是读操作还是写操作，I/O 工作线程在处理完自己链表中的 client 对象后都会立刻挂起自己，之后主线程根据具体情形决定是否再次唤醒它们。

从本质上来看，在 Redis 6.0 中新增的多线程 I/O 处理逻辑只是在 client 较多的情况下，利用多个 I/O 工作线程加快读写操作的处理速度，在 I/O 工作线程处理这些 client 期间，主线程可以处理分配给自己的 client 对象，如果主线程比 I/O 工作线程早处理完分配给自己的 client 对象，则主线程必须继续等待，直到所有 I/O 工作线程处理 client 完毕。

我们将 Redis 6.0 多线程网络 I/O 处理逻辑绘制成如下流程图。

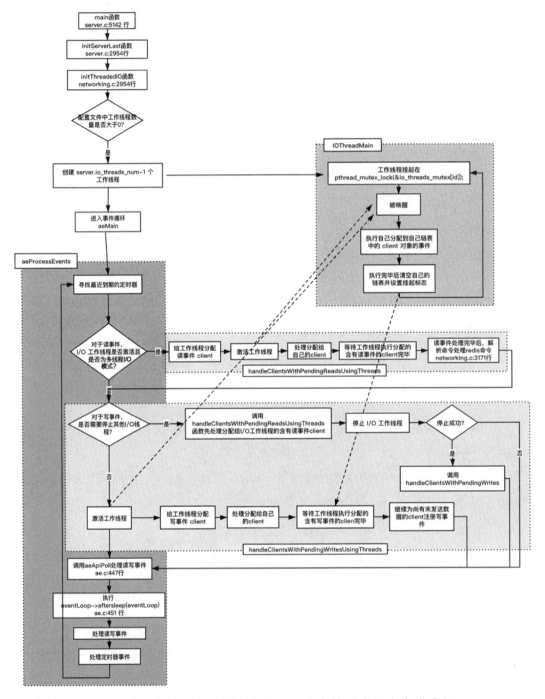

图中 aftersleep 函数和处理定时器事件的逻辑将在接下来的内容中介绍。

8.2.8　Redis 对客户端的管理

redis-server 使用了一个 client 结构去管理连接上来的客户端对象，client 结构的定义如下：

```
//server.h 文件第 768 行
typedef struct client {
    //标记客户端的唯一自增 ID
    uint64_t id;
    //连接对象
    connection *conn;
    int resp;
    redisDb *db;
    robj *name;
    //接收缓冲区
    sds querybuf;
    //接收缓冲区的读指针位置
    size_t qb_pos;

    //省略部分字段
} client;
```

client 对象与底层网络通信相关的信息，例如 clientfd、连接状态、读写事件回调函数，被记录在 client->conn 字段中，client->conn 字段类型是 connection*。connection 结构的定义如下：

```
//connection.h 文件第 69 行
struct connection {
    ConnectionType *type;
    ConnectionState state;
    short int flags;
    //connection 对象的引用计数，
    //用于标记是否需要释放该 connection 对象，8.2.9 节会介绍
    short int refs;
    int last_errno;
    //指向自己的 client 对象
    void *private_data;
    ConnectionCallbackFunc conn_handler;
    ConnectionCallbackFunc write_handler;
    ConnectionCallbackFunc read_handler;
    int fd;
};
```

为了方便 connection 对象引用自己的 client 对象，在 connection 对象的 private_data 字段中记录了 client 对象的指针，这样 connection 对象和 client 对象就可以方便地相互引用了，示意图如下。

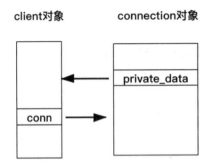

redis-server 先创建 connection 对象，再利用 connection 对象创建 client 对象，创建 client 对象的函数 createClient 的实现如下：

```
//networking.c 文件第 88 行
client *createClient(connection *conn) {
   client *c = zmalloc(sizeof(client));

   //省略部分代码

   //networking.c 文件第 101 行
   //在 connSetPrivateData 函数中让 conn-> private_data 指向 client 对象
   connSetPrivateData(conn, c);

   //省略部分代码

   //client 对象在全局链表中的节点指针
   listNode *client_list_node;

   //省略部分代码
}

//connection.c 文件第 128 行
void connSetPrivateData(connection *conn, void *data) {
   conn->private_data = data;
}
```

createClient 函数在接受连接时的 acceptCommonHandler 函数中被调用，调用代码如下：

```
//networking.c 文件第 915 行
static void acceptCommonHandler(connection *conn, int flags, char *ip) {
   client *c;

   //省略部分代码
   //redis-server 可接受的最大连接数是 server.maxclients
   if (listLength(server.clients) >= server.maxclients) {
      char *err = "-ERR max number of clients reached\r\n";

      if (connWrite(conn,err,strlen(err)) == -1) {

      }
      server.stat_rejected_conn++;
      connClose(conn);
      return;
```

```
        }

        /* 在这里创建 client 和 connection 对象 */
        if ((c = createClient(conn)) == NULL) {
            //省略部分代码

            return;
        }

        //省略部分代码
}
```

以上代码中的加粗行展示了 redis-server 可接受的最大连接数是 server.maxclients，我们可以在 redis.conf 配置文件中修改这个值：

```
# redis.conf 文件第 808 行
# maxclients 10000
```

server.maxclients 的默认值是 10000，当连接数超过这个值时，redis-server 会给新连接返回一个错误消息 "-ERR max number of clients reached\r\n"，并关闭这个连接。

在 createClient 函数中创建并初始化好 client 对象后，redis-server 通过调用 linkClient 函数将 client 对象追加到 server.clients 这个全局链表中，逻辑如下：

```
//networking.c 文件第 88 行
client *createClient(connection *conn) {
    client *c = zmalloc(sizeof(client));

    //省略部分代码

    //networking.c 文件第 167 行
    if (conn) linkClient(c);

    //省略部分代码

    return c;
}
```

linkClient 函数的实现如下：

```
//networking.c 文件第 78 行
void linkClient(client *c) {
    listAddNodeTail(server.clients,c);
    //使用 client->client_list_node 字段记录 client 对象在链表中的节点指针，
    //这样调用 unlinkClient 函数从链表中移除这个节点时，直接释放该节点的内存即可，
    //不需要遍历链表
    c->client_list_node = listLast(server.clients);
    uint64_t id = htonu64(c->id);
    raxInsert(server.clients_index,(unsigned char*)&id,sizeof(id),c,NULL);
}
```

server.clients 是存储 client 对象的链表，为 list 结构体类型，这是一个双向链表，定义如下：

```
//adlist.h 文件第 47 行
typedef struct list {
```

```
    listNode *head;
    listNode *tail;
    void *(*dup)(void *ptr);
    void (*free)(void *ptr);
    int (*match)(void *ptr, void *key);
    unsigned long len;
} list;
```

链表中每个节点的类型都是 listNode，其定义如下：

```
//adlist.h 文件第 36 行
typedef struct listNode {
    struct listNode *prev;
    struct listNode *next;
    void *value;
} listNode;
```

client 对象定义了一个字段 client_list_node 来记录自己在链表 server.clients 中的节点的指针，这样如果需要从该链表中删除该 client 对象节点，则使用 client->client_list_node 字段就可以很方便地做删除操作，即先将当前节点的上一个节点的 next 字段指向当前节点的下一个节点，将当前节点的下一个节点的 prev 字段指向当前节点的上一个节点（双向链表），然后释放当前节点占用的内存即可，示意图如下所示。

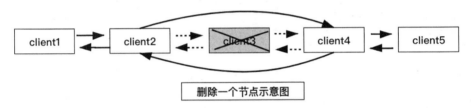

删除一个节点示意图

从链表中删除节点的逻辑如下：

```
//从 server.clients 中删除某个 client 节点
listDelNode(server.clients,c->client_list_node);

//adlist.c 文件第 167 行
void listDelNode(list *list, listNode *node)
{
    //将当前节点的上一个节点的 next 字段指向当前节点的下一个节点
    if (node->prev)
        node->prev->next = node->next;
    else
        list->head = node->next;
    //将当前节点的下一个节点的 prev 字段指向当前节点的上一个节点
    if (node->next)
        node->next->prev = node->prev;
    else
        list->tail = node->prev;
    //释放节点本身的内存
    if (list->free) list->free(node->value);
    zfree(node);
    //链表长度减 1
    list->len--;
}
```

此外，每个 client 对象都有一个 id 字段，类型是 uint64_t，这个字段用于唯一标识一个 client 对象，这个字段的值利用全局整型变量 server.next_client_id 生成，server.next_client_id 全局整型变量在每次产生新的 client 对象时都递增 1，server.next_client_id 从 0 开始递增：

```c
//networking.c 文件第 88 行
client *createClient(connection *conn) {
    client *c = zmalloc(sizeof(client));

    //省略部分代码

    //networking.c 文件第 105 行
    //递增 1，产生新的 client_id 并分配给新客户端
    uint64_t client_id = ++server.next_client_id;
    c->id = client_id;

    //省略部分代码
}
```

client 对象的 id 字段值最终被转换成网络字节序作为 server.clients_index 对象中的 key 去索引一个 client 对象，逻辑如下：

```c
//networking.c 文件第 78 行
void linkClient(client *c) {

    uint64_t id = htonu64(c->id);
    raxInsert(server.clients_index,(unsigned char*)&id,sizeof(id),c,NULL);
}
```

server.clients_index 对象类型是 Redis 自定义的一个特殊数据结构 rax，由于 rax 内容与本章主题无关，所以这里就不介绍了，有兴趣的读者可以研究一下（rax 结构体的定义位于 rax.h 文件第 98 行）。

在事件循环中，某个 fd 的读或写事件触发时，Redis 是如何通过 fd 找到对应的 client 对象的呢？

答案是在全局对象 aeEventLoop 中定义了两个数组指针字段 events 和 fired：

```c
//ae.h 文件第 100 行
typedef struct aeEventLoop {
    int maxfd;
    int setsize;
    long long timeEventNextId;
    time_t lastTime;
    aeFileEvent *events; /* 注册的事件 */
    aeFiredEvent *fired; /* 触发的事件 */
    aeTimeEvent *timeEventHead;
    int stop;
    void *apidata;
    aeBeforeSleepProc *beforesleep;
    aeBeforeSleepProc *aftersleep;
    int flags;
} aeEventLoop;
```

events 和 fired 这两个指针在程序启动时均被设置成指向一个很大的数组,逻辑如下:

```
//代码片段 8-2-8
//ae.c 文件第 63 行
aeEventLoop *aeCreateEventLoop(int setsize) {
    aeEventLoop *eventLoop;
    int i;

    if ((eventLoop = zmalloc(sizeof(*eventLoop))) == NULL) goto err;
    //注意这两行
    eventLoop->events = zmalloc(sizeof(aeFileEvent)*setsize);
    eventLoop->fired = zmalloc(sizeof(aeFiredEvent)*setsize);
    if (eventLoop->events == NULL || eventLoop->fired == NULL) goto err;
    eventLoop->setsize = setsize;
    eventLoop->lastTime = time(NULL);
    eventLoop->timeEventHead = NULL;
    eventLoop->timeEventNextId = 0;
    eventLoop->stop = 0;
    eventLoop->maxfd = -1;
    eventLoop->beforesleep = NULL;
    eventLoop->aftersleep = NULL;
    eventLoop->flags = 0;
    //注意这一行
    if (aeApiCreate(eventLoop) == -1) goto err;

    for (i = 0; i < setsize; i++)
        eventLoop->events[i].mask = AE_NONE;
    return eventLoop;

err:
    if (eventLoop) {
        zfree(eventLoop->events);
        zfree(eventLoop->fired);
        zfree(eventLoop);
    }
    return NULL;
}
```

在代码片段 8-2-8 中调用了 aeApiCreate 函数,并使用 eventLoop 作为参数(代码片段 8-2-8 中的第 3 个加粗行),在 aeApiCreate 函数中创建了与 eventLoop->events 和 eventLoop->fired 这两个数组长度相同的另一个数组 state->events 来记录注册到各个 fd 上的事件信息,eventLoop->apidata 又指向 state 对象,逻辑如下:

```
//ae_epoll.c 文件第 39 行
static int aeApiCreate(aeEventLoop *eventLoop) {
    aeApiState *state = zmalloc(sizeof(aeApiState));

    if (!state) return -1;
    state->events = zmalloc(sizeof(struct epoll_event)*eventLoop->setsize);
    if (!state->events) {
        zfree(state);
        return -1;
    }
    state->epfd = epoll_create(1024);
    if (state->epfd == -1) {
```

```
        zfree(state->events);
        zfree(state);
        return -1;
    }
    eventLoop->apidata = state;
    return 0;
}
```

因此我们可以通过 eventLoop->apidata->events[fd] 获知某个 fd 注册了哪些事件，eventLoop->apidata 的类型是 aeApiState。aeApiState 的定义如下：

```
//ae_epoll.c 文件第 34 行
typedef struct aeApiState {
    int epfd;
    //大小为 eventLoop->setsize 的数组指针
    struct epoll_event *events;
} aeApiState;
```

eventLoop 的 events 指针指向的数组的元素类型是 aeFileEvent，eventLoop 的 fired 指针指向的数组的元素类型是 aeFiredEvent。aeFileEvent 和 aeFiredEvent 这两个类型的定义如下：

```
//ae.h 文件第 72 行
typedef struct aeFileEvent {
    int mask; /* AE_(READABLE|WRITABLE|BARRIER)中的一种 */
    aeFileProc *rfileProc;
    aeFileProc *wfileProc;
    void *clientData;
} aeFileEvent;

//ae.h 文件第 94 行
typedef struct aeFiredEvent {
    int fd;
    int mask;
} aeFiredEvent;
```

由于 events 和 fired 这两个数组足够大，而代表客户端的 clientfd 的类型是一个 int，且 clientfd 在同一时间内不会重复，因此 redis-server 将 clientfd 作为数组的下标，而通过数组下标对应的位置值（aeFileEvent*类型）可以得到该 clientfd 注册的事件标志（mask 字段）、读写事件回调函数（rfileProc 和 wfileProc 字段），aeFileEvent->clientData 字段则在注册事件时被设置为指向 clientfd 对应的 connection 对象。connection 对象的 private_data 字段记录了对应的 client 对象的指针。以注册一个 clientfd 的读事件为例，在读事件触发后，redis-server 调用 connSocketSetReadHandler 函数进行处理。connSocketSetReadHandler 函数的实现如下：

```
//connection.c 文件第 225 行
static int connSocketSetReadHandler(connection *conn, ConnectionCallbackFunc func)
{
    if (func == conn->read_handler) return C_OK;

    conn->read_handler = func;
    if (!conn->read_handler)
        aeDeleteFileEvent(server.el,conn->fd,AE_READABLE);
```

```
    else
        //注意这一行
        if (aeCreateFileEvent(server.el,conn->fd,
              AE_READABLE,conn->type->ae_handler,conn) == AE_ERR)
            return C_ERR;
    return C_OK;
}
```

在 connSocketSetReadHandler 函数中调用了 aeCreateFileEvent 函数为 clientfd（这里使用 conn->fd 表示）设置事件掩码和回调函数。aeCreateFileEvent 函数的实现如下：

```
//ae.c 文件第 153 行
int aeCreateFileEvent(aeEventLoop *eventLoop, int fd, int mask,
        aeFileProc *proc, void *clientData)
{
    if (fd >= eventLoop->setsize) {
        errno = ERANGE;
        return AE_ERR;
    }

    //这一行将 fd 作为数组下标与对应的事件信息绑定
    aeFileEvent *fe = &eventLoop->events[fd];

    if (aeApiAddEvent(eventLoop, fd, mask) == -1)
        return AE_ERR;

    //设置掩码和读写事件回调函数
    fe->mask |= mask;
    if (mask & AE_READABLE) fe->rfileProc = proc;
    if (mask & AE_WRITABLE) fe->wfileProc = proc;
    fe->clientData = clientData;
    if (fd > eventLoop->maxfd)
        eventLoop->maxfd = fd;
    return AE_OK;
}
```

至此，fd、connection、client 对象及 fd 对应的事件类型和回调函数的关系就确定了。我们以读事件为例来梳理一下这个流程。对于 Linux 系统，在事件循环中 epoll_wait 函数返回后，执行逻辑如下：

```
//ae_epoll.c 文件第 108 行
static int aeApiPoll(aeEventLoop *eventLoop, struct timeval *tvp) {
    aeApiState *state = eventLoop->apidata;
    int retval, numevents = 0;

    retval = epoll_wait(state->epfd,state->events,eventLoop->setsize,
            tvp ? (tvp->tv_sec*1000 + tvp->tv_usec/1000) : -1);
    if (retval > 0) {
        int j;
        //epoll_wait 函数返回后的逻辑
        numevents = retval;
        for (j = 0; j < numevents; j++) {
            int mask = 0;
            struct epoll_event *e = state->events+j;

            if (e->events & EPOLLIN) mask |= AE_READABLE;
```

```
            if (e->events & EPOLLOUT) mask |= AE_WRITABLE;
            if (e->events & EPOLLERR) mask |= AE_WRITABLE|AE_READABLE;
            if (e->events & EPOLLHUP) mask |= AE_WRITABLE|AE_READABLE;
            eventLoop->fired[j].fd = e->data.fd;
            eventLoop->fired[j].mask = mask;
        }
    }
    return numevents;
}
```

在以上代码中，redis-server 设置 epoll_wait 函数的第 2 个参数使用一个很大的值：eventLoop->setsize，这个值默认是 10128，等于最大连接数 server.maxclients + CONFIG_FDSET_INCR。epoll_wait 函数在返回后，将本次触发事件的序号（变量 j）作为数组下标，将 fd、事件掩码记录到 eventLoop->fired 数组对应的位置上。

aeApiPoll 函数在调用返回后，遍历 eventLoop->fired 数组，取出其中有效的数组元素，得到有事件的 fd 和事件掩码 mask，然后以 fd 为下标从 eventLoop->events 数组中取出 aeFileEvent 对象，这样通过 aeFileEvent 对象就能得到对应的 fd 的读写事件回调函数和 connection 对象了（通过 fe->clientData 字段获取 connection 对象），逻辑如下：

```
//ae.c 文件第 386 行
int aeProcessEvents(aeEventLoop *eventLoop, int flags)
{
    //省略部分代码

    numevents = aeApiPoll(eventLoop, tvp);

    //省略部分代码

    //遍历 eventLoop->fired 数组
    for (j = 0; j < numevents; j++) {
        //通过 fd 得到 aeFileEvent 对象
        aeFileEvent *fe = &eventLoop->events[eventLoop->fired[j].fd];
        //得到事件掩码 mask 和 fd
        int mask = eventLoop->fired[j].mask;
        int fd = eventLoop->fired[j].fd;
        int fired = 0;

        int invert = fe->mask & AE_BARRIER;

        //得到读写事件回调函数 rfileProc 和 wfileProc
        if (!invert && fe->mask & mask & AE_READABLE) {
            fe->rfileProc(eventLoop,fd,fe->clientData,mask);
            fired++;
            fe = &eventLoop->events[fd];
        }

        if (fe->mask & mask & AE_WRITABLE) {
            if (!fired || fe->wfileProc != fe->rfileProc) {
                fe->wfileProc(eventLoop,fd,fe->clientData,mask);
                fired++;
            }
        }
```

```
            if (invert) {
                fe = &eventLoop->events[fd];
                if ((fe->mask & mask & AE_READABLE) &&
                    (!fired || fe->wfileProc != fe->rfileProc))
                {
                    fe->rfileProc(eventLoop,fd,fe->clientData,mask);
                    fired++;
                }
            }

            processed++;
        }
    }

//省略部分代码
}
```

以上逻辑过程和相应的数据结构可以用如下流程图表示。

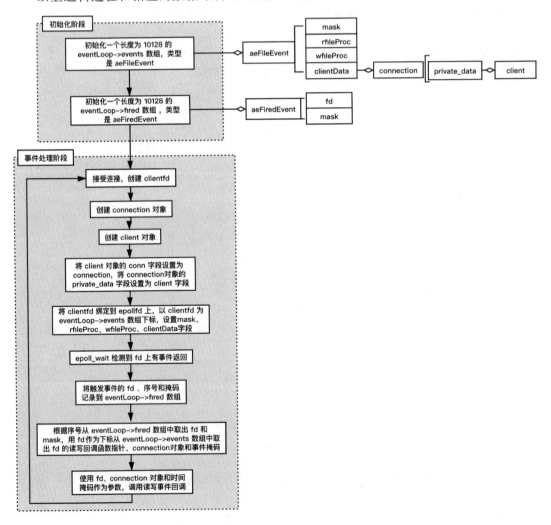

根据 8.2.4 节的分析，客户端 fd（clientfd）的读事件回调函数是 readQueryFromClient，因此 clientfd 的读事件回调实际是调用 readQueryFromClient 函数，在调用 readQueryFromClient 函数时，通过函数第 1 个参数的 connection 对象的 private_data 字段得到 client 对象，由于 client 对象含有各种信息字段，所以这样就可以进行相关读逻辑处理了。相关代码如下：

```
//networking.c 文件第 1890 行
void readQueryFromClient(connection *conn) {
    //通过 connection 对象获取 client 对象
    client *c = connGetPrivateData(conn);

    //省略部分代码
}

//connection.c 文件第 133 行
void *connGetPrivateData(connection *conn) {
    return conn->private_data;
}
```

同理，处理 clientfd 的写事件的回调函数，最终调用的是 sendReplyToClient 函数，在 sendReplyToClient 函数中也通过 connection 对象得到 client 对象，然后进一步操作：

```
//networking.c 文件第 1389 行
void sendReplyToClient(connection *conn) {
    //通过 connection 对象获取 client 对象
    client *c = connGetPrivateData(conn);
    writeToClient(c,1);
}
```

前面介绍了所有 client 对象都会被放在 server.clients 这个双向链表中，server 对象还有另外两个与 server.clients 类型相同的字段 clients_pending_read 和 clients_pending_write，这两个字段分别用于存放有读事件和写事件的 client 对象。在开启 I/O 工作线程时，主线程会在某个 fd 触发读或写事件时将 fd 对应的 client 对象放入 clients_pending_read 或 clients_pending_write 字段中。以读事件为例，在 readQueryFromClient 函数中有如下调用：

```
//networking.c 文件第 1890 行
void readQueryFromClient(connection *conn) {
    client *c = connGetPrivateData(conn);

    //postponeClientRead 函数判断是否开启了 I/O 工作线程
    //如果是，则将 client 对象放入 clients_pending_read 链表中后返回
    if (postponeClientRead(c)) return;

    //省略部分代码
}
```

如果含有读事件的 client 对象被放入 clients_pending_read 链表中，那么 readQueryFromClient 函数就不会继续往下执行了（如以上代码片段中的加粗行所示），因为接下来对 fd 数据读取和解析的工作将交由 I/O 工作线程来做。postponeClientRead 函数的逻辑如下：

```
//networking.c 文件第 3106 行
int postponeClientRead(client *c) {
```

```
//满足if条件后将client对象加入server.clients_pending_read链表中
if (io_threads_active &&
    server.io_threads_do_reads &&
    !ProcessingEventsWhileBlocked &&
    !(c->flags & (CLIENT_MASTER|CLIENT_SLAVE|CLIENT_PENDING_READ)))
{
    c->flags |= CLIENT_PENDING_READ;
    listAddNodeHead(server.clients_pending_read,c);
    return 1;
} else {
    return 0;
}
```

接着进行下一轮循环，主线程在 handleClientsWithPendingReadsUsingThreads 函数中使用 Round-Robin 策略将 server.clients_pending_read 中的 client 对象分配给各个 I/O 工作线程，这个逻辑在 8.2.7 节已经介绍过了。

同样的逻辑也适用于写事件处理和 clients_pending_write 链表处理，这里不再赘述。

8.2.9 客户端断开流程

redis-server 断开客户端连接主要有以下两种情形。

1. 情形一、连接数到达上限，断开连接

当连接数达到 redis-server 的连接数上限时，redis-server 就会给客户端发送一条错误信息，然后关闭连接。关闭连接的函数是 connClose。connClose 函数的定义如下：

```
//connection.h 文件第 174 行
static inline void connClose(connection *conn) {
    conn->type->close(conn);
}
```

在以上代码中，conn->type->close 是个回调指针，conn->type->close 实际指向 connSocketClose 函数。connSocketClose 函数的实现如下：

```
//代码片段 8-2-9-1
//connection.c 文件第 144 行
static void connSocketClose(connection *conn) {
    if (conn->fd != -1) {
        //尝试移除 fd 上的读写事件
        aeDeleteFileEvent(server.el,conn->fd,AE_READABLE);
        aeDeleteFileEvent(server.el,conn->fd,AE_WRITABLE);
        close(conn->fd);
        conn->fd = -1;
    }

    //如果 conn.refs 的引用计数大于 0，
    //则为 conn 设置延迟关闭标志 CONN_FLAG_CLOSE_SCHEDULED
    if (connHasRefs(conn)) {
        conn->flags |= CONN_FLAG_CLOSE_SCHEDULED;
        return;
    }
```

```
    zfree(conn);
}
```

情形一只创建了 connection 对象而没有创建 client 对象,因此只需在关闭 fd 后释放 connection 对象即可。让我们再回顾一下 acceptCommonHandler 函数中的逻辑,在达到最大连接上限时,给新连接上来的客户端发送一条 err 信息,然后调用 connClose 函数关闭连接,逻辑如下:

```
//networking.c 文件第 915 行
static void acceptCommonHandler(connection *conn, int flags, char *ip) {
    //省略部分代码

    if (listLength(server.clients) >= server.maxclients) {
        char *err = "-ERR max number of clients reached\r\n";

        //省略部分代码

        //将 err 信息发送给客户端
        if (connWrite(conn,err,strlen(err)) == -1) {
            /* 仅仅是为了消除编译器警告信息 */
        }
        server.stat_rejected_conn++;

        //在这里调用 connClose 函数,在 connClose 函数中会关闭 fd
        connClose(conn);
        return;
    }

    //省略部分代码
}
```

每次需要释放 connection 对象时都不是立即销毁 connection 对象,而是先判断 connection 对象的引用计数是否为 0,如果为 0 才销毁它,这个引用计数值被记录在 connection 对象的 refs 字段中。创建 connection 对象使用的是 zcalloc 函数,这个函数会把 connection 对象占用的内存清零,所以 connection 对象的 refs 字段也会被一并设置为 0,这是 refs 的初始值。refs 这个引用计数字段不是为了实现类似智能指针一样自动销毁自己的功能,而是结合 connection 对象的 flag 标志(该 flag 只支持 CONN_FLAG_CLOSE_SCHEDULED 标志),在合适的时机去销毁 connection 对象。这样说可能比较难以理解,这里详细分析一下,先看看销毁 connection 对象的逻辑,代码如下:

```
//connhelpers.h 文件第 77 行
static inline int callHandler(connection *conn, ConnectionCallbackFunc handler) {
    //递增 conn.refs 的值
    connIncrRefs(conn);
    if (handler) handler(conn);
    //递减 conn.refs 的值
    connDecrRefs(conn);
    if (conn->flags & CONN_FLAG_CLOSE_SCHEDULED) {
        if (!connHasRefs(conn)) connClose(conn);
        return 0;
    }
```

```
    return 1;
}
```

设计 callHandler 函数有两个目的：①让某些逻辑可以复用，减少重复的逻辑；②让第 1 个参数 handler 中的业务逻辑全部执行完再关闭连接。

callHandler 的第 2 个参数 handler 是一个自定义函数指针，由于无法保证在这个自定义函数中不会调用 connSocketClose 函数造成 handler 流程没走完就提前关闭连接，因此在调用 handler 之前，先递增 connection 对象的 refs 字段值。而关闭连接的函数 connSocketClose 内部会判断 connection 对象的 refs 字段是否大于 0，如果大于 0，则设置 connection 对象的 flag 字段值为延迟关闭标志 CONN_FLAG_CLOSE_SCHEDULED，等 handler 指向的函数执行完毕后再递减 connection 对象的 refs 值，接着判断 connection 对象的 flag 是否设置了 CONN_FLAG_CLOSE_SCHEDULED，如果设置了该标志，则说明在前面的逻辑中要求关闭这个 connection，此时再关闭这个 connection。

这个逻辑有点绕，可以绘制成如下流程图。

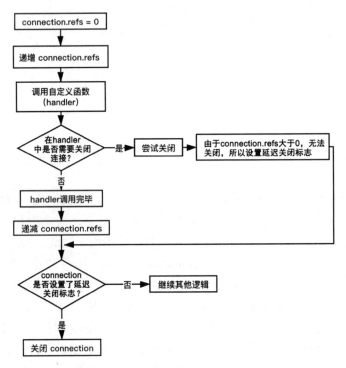

redis-server 在 connSocketClose 函数中调用 aeDeleteFileEvent 函数从 epollfd 上解绑 clientfd、移除 clientfd 的读写事件并关闭 clientfd。aeDeleteFileEvent 函数的实现如下：

```
//代码片段中 8-2-9-2
//ae.c 文件第 173 行
void aeDeleteFileEvent(aeEventLoop *eventLoop, int fd, int mask)
{
    if (fd >= eventLoop->setsize) return;
    aeFileEvent *fe = &eventLoop->events[fd];
    //按需移除事件
```

```
    if (fe->mask == AE_NONE) return;

    if (mask & AE_WRITABLE) mask |= AE_BARRIER;
    //从 epollfd 上解绑 clientfd、移除 clientfd 的读写事件
    aeApiDelEvent(eventLoop, fd, mask);
    fe->mask = fe->mask & (~mask);
    //重新计算 maxfd
    if (fd == eventLoop->maxfd && fe->mask == AE_NONE) {
        //更新 maxfd
        int j;

        //倒序遍历 eventLoop->events 数组
        //找到 mask 不是 AE_NONE 的 fd, 即新的 maxfd
        for (j = eventLoop->maxfd-1; j >= 0; j--)
            if (eventLoop->events[j].mask != AE_NONE) break;
        eventLoop->maxfd = j;
    }
}
```

对于 aeDeleteFileEvent 函数，有以下三个注意事项。

◎ 虽然 aeDeleteFileEvent 函数的上层调用（connSocketClose 函数）会同时尝试移除一个 fd 上的读事件和写事件，但 aeDeleteFileEvent 函数内部会根据自己记录的事件掩码（fe->mask）去移除已注册的事件，言下之意是某个 fd 并不一定同时注册了读写事件。这段逻辑对应代码片段 8-2-9-1 中的斜体代码块。

◎ 如果要移除某个 fd 上已经注册的所有事件，则只需将 eventLoop->events 数组中以 fd 为下标的值（值类型是 aeFileEvent）的 mask 字段值设置为 0（用 AE_NONE 表示），而不需要实际销毁对应的 aeFileEvent 对象。在这种情况下，redis-server 就会将这个 fd 从 epollfd 上解绑并关闭该 fd，但 eventLoop->events 数组在该位置处的 aeFileEvent 对象可以继续保留，以方便下一次系统接收新的连接时产生相同值的 fd 设置到数组这个位置时复用。

对于 Linux 系统，fd 的值是 int 类型，某次新建的一个 fd 的值是 n，关闭该 fd 后，再次新建一个 fd，操作系统可能还会将 n 这个值分配给这个新建的 fd。Windows 系统的套接字的 SOCKET 句柄值也是如此。

◎ eventLoop 对象的 maxfd 字段记录当前所有 fd 中值最大的一个，当有 fd 被关闭时，需要根据被关闭的 fd 值来决定是否需要重新计算这个 maxfd，计算逻辑是先看下这个被关闭的 fd 是否等于当前 maxfd：如果否，则说明不是当前具有最大值的 fd 被关闭，无须重新计算 maxfd；如果是，就需要重新在剩余的 fd 中找一个值最大的作为新的 maxfd。只需从 maxfd-1 开始倒序遍历 eventLoop->events 数组找到 mask 不是 AE_NONE 的 fd（mask 不是 AE_NONE 的 fd 才是有效 fd）。假设遍历 eventLoop->event 数组时最终找到符合要求的下标是 j，那么 j 就是剩下所有有效 fd 中最大的值，接着将 maxfd 值更新为 j 即可。寻找新的 maxfd 的示意图如下。

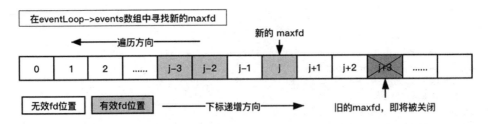

为什么需要记录所有 fd 中的最大值 maxfd 呢？因为 redis-server 可以使用不同的操作系统函数（上文介绍的是 epoll_wait 函数）作为 I/O 复用函数，当在 Linux 系统上使用 select 函数作为 I/O 复用函数时，select 函数的第 1 个参数要求是需要监视的 fd 集合中的最大值加 1。redis-server 使用 select 函数作为 I/O 复用函数的代码如下：

```c
//ae_select.c 文件第 77 行
static int aeApiPoll(aeEventLoop *eventLoop, struct timeval *tvp) {
    //省略部分代码

    //注意，select 函数的第 1 个参数是 eventLoop->maxfd+1
    retval = select(eventLoop->maxfd+1,
            &state->_rfds,&state->_wfds,NULL,tvp);

    //省略部分代码
}
```

2. 情形二、处理读写事件时出错或者 redis-server 主动关闭连接

以读事件的处理为例，根据前文所述，读事件的回调函数为 readQueryFromClient 函数，收取数据失败时调用 freeClientAsync 函数。以这种情形为例，代码逻辑如下：

```c
//networking.c 文件第 1890 行
void readQueryFromClient(connection *conn) {
    //省略部分代码

    nread = connRead(c->conn, c->querybuf+qblen, readlen);
    if (nread == -1) {
        if (connGetState(conn) == CONN_STATE_CONNECTED) {
            return;
        } else {
            serverLog(LL_VERBOSE, "Reading from client: %s",connGetLastError(c->conn));
            //conn 状态不对
            freeClientAsync(c);
            return;
        }
    } else if (nread == 0) {
        serverLog(LL_VERBOSE, "Client closed connection");
        //对端关闭了连接
        freeClientAsync(c);
        return;
    }

    //省略部分代码
}
```

freeClientAsync 函数将 client 对象放入 server.clients_to_close 队列中，不是立即释放 client 对象而是将其放入待释放队列中，这就是这个函数名中"Async"一词的含义。freeClientAsync 函数的实现如下：

```
//networking.c 文件第 1241 行
void freeClientAsync(client *c) {
    //省略部分代码

    if (server.io_threads_num == 1) {

        listAddNodeTail(server.clients_to_close,c);
        return;
    }
    static pthread_mutex_t async_free_queue_mutex = PTHREAD_MUTEX_INITIALIZER;
    pthread_mutex_lock(&async_free_queue_mutex);
    listAddNodeTail(server.clients_to_close,c);
    pthread_mutex_unlock(&async_free_queue_mutex);
}
```

在以上代码中，如果 redis-server 没有启动 I/O 工作线程，则对 server.clients_to_close 队列的生产（存入 client 对象）和消费（取出 client 对象）都位于主线程中，因此不需要加锁保护，freeClientAsync 函数直接将 client 对象放入 server.clients_to_close 队列；如果启动了 I/O 工作线程，则需要使用互斥体 async_free_queue_mutex 对 server.clients_to_close 队列进行保护，因为主线程和 I/O 工作线程都会调用 readQueryFromClient 函数，readQueryFromClient 函数又调用 freeClientAsync 函数，因此在主线程和工作线程中都可能将 client 对象放入 server.clients_to_close 队列（生产者是主线程和 I/O 工作线程）。经研究发现，接下来从该队列中取出 client 对象去做实际回收 client 的操作也位于主线程中（消费者是主线程）。那消费者的逻辑到底在哪里呢？

实际释放 client 对象的操作位于 beforeSleep 函数中（该函数在主线程中被调用），这里重点关注这个函数的另一段逻辑，代码如下：

```
//server.c 文件第 2106 行
void beforeSleep(struct aeEventLoop *eventLoop) {
    //省略部分代码

    //server.c 文件第 2189 行
    freeClientsInAsyncFreeQueue();

    //省略部分代码
}
```

freeClientsInAsyncFreeQueue 函数的定义如下：

```
//networking.c 文件第 1262 行
int freeClientsInAsyncFreeQueue(void) {
    int freed = listLength(server.clients_to_close);
    while (listLength(server.clients_to_close)) {
        listNode *ln = listFirst(server.clients_to_close);
        client *c = listNodeValue(ln);

        c->flags &= ~CLIENT_CLOSE_ASAP;
```

```
        freeClient(c);
        listDelNode(server.clients_to_close,ln);
    }
    return freed;
}
```

以上代码遍历 server.clients_to_close 队列并挨个取出 client 对象，然后调用 freeClient 函数释放它。freeClient 函数的实现如下：

```
//networking.c 文件 1117 行
void freeClient(client *c) {
    //省略部分代码

    sdsfree(c->querybuf);
    sdsfree(c->pending_querybuf);
    c->querybuf = NULL;

    //省略部分代码

    //在这个函数中释放 client 对象
    unlinkClient(c);

    //省略部分代码

    //释放 client 对象
    zfree(c)
}
```

如以上代码片段所示，在 freeClient 函数中先释放 client 对象关联的缓冲区对象 c->querybuf、c->pending_querybuf，然后调用 unlinkClient 函数从相关链表中移除 client 对象，最后调用 zfree 函数释放 client 对象。

unlinkClient 函数的作用是从相关链表中移除 client 对象（这是函数名中"unlink"一词的含义），并调用 connClose 函数释放相应的 client 对象：

```
//networking.c 文件第 1051 行
void unlinkClient(client *c) {
    //省略部分代码

    if (c->conn) {
        //从相关链表中移除 client 对象
        if (c->client_list_node) {
            uint64_t id = htonu64(c->id);
            raxRemove(server.clients_index,(unsigned char*)&id,sizeof(id),NULL);
            listDelNode(server.clients,c->client_list_node);
            c->client_list_node = NULL;
        }

        //省略部分代码

        //释放 connection 对象
        connClose(c->conn);
        c->conn = NULL;
    }
```

```
    //省略部分代码
}
```

unlinkClient 函数调用 connClose 函数释放相应的 connection 对象,释放 connection 对象同时意味着释放连接对应的底层 fd 和取消注册读写事件,这些逻辑在情形一中介绍 connClose 函数时已经详细介绍过了。

8.2.10 Redis 中收发缓冲区的设计

本节讲解 Redis 中的收发缓冲区是如何设计的,先来看看接收缓冲区的设计。

1. 接收缓冲区

Redis 中的接收缓冲区对象是 client 对象的 querybuf 字段,其类型是 sds,sds 在本质上是 char*:

```
//sds.h 文件第 43 行
typedef char *sds;
```

如果发送缓冲区的类型只是 char*,那么如何记录缓冲区的长度和当前读写位置呢?

client.querybuf 对象在 client 对象创建时调用 sdsempty 函数初始化,逻辑如下:

```
//networking.c 文件第 88 行
client *createClient(connection *conn) {
    client *c = zmalloc(sizeof(client));

    //省略部分代码

    c->querybuf = sdsempty();

    //省略部分代码
}
```

sdsempty 函数内部仅仅调用了 sdnewlen 函数:

```
//sds.c 文件第 149 行
sds sdsempty(void) {
    return sdsnewlen("",0);
}
```

sdsnewlen 函数的实现如下:

```
//代码片段 8-2-10
//sds.c 文件第 89 行
sds sdsnewlen(const void *init, size_t initlen) {
    void *sh;
    sds s;
    //这里根据 initlen 得到 type 值
    char type = sdsReqType(initlen);

    //为了减少内存重分配次数,当 initlen=0 时,
    //本应该使用 SDS_TYPE_5 类型,却使用了 SDS_TYPE_8 类型
    if (type == SDS_TYPE_5 && initlen == 0) type = SDS_TYPE_8;
    int hdrlen = sdsHdrSize(type);
    unsigned char *fp;
```

```c
    sh = s_malloc(hdrlen+initlen+1);
    if (sh == NULL) return NULL;
    if (init==SDS_NOINIT)
        init = NULL;
    else if (!init)
        memset(sh, 0, hdrlen+initlen+1);
    s = (char*)sh+hdrlen;
    fp = ((unsigned char*)s)-1;
    switch(type) {
        case SDS_TYPE_5: {
            *fp = type | (initlen << SDS_TYPE_BITS);
            break;
        }
        case SDS_TYPE_8: {
            SDS_HDR_VAR(8,s);
            sh->len = initlen;
            sh->alloc = initlen;
            *fp = type;
            break;
        }
        case SDS_TYPE_16: {
            SDS_HDR_VAR(16,s);
            sh->len = initlen;
            sh->alloc = initlen;
            *fp = type;
            break;
        }
        case SDS_TYPE_32: {
            SDS_HDR_VAR(32,s);
            sh->len = initlen;
            sh->alloc = initlen;
            *fp = type;
            break;
        }
        case SDS_TYPE_64: {
            SDS_HDR_VAR(64,s);
            sh->len = initlen;
            sh->alloc = initlen;
            *fp = type;
            break;
        }
    }
    if (initlen && init)
        memcpy(s, init, initlen);
    s[initlen] = '\0';
    return s;
}
```

在 sdsnewlen 函数中根据传入的第 2 个参数 initlen 的值（初始化长度）将 sds 对象分为如下类型（逻辑位于 sdsReqType 函数中）：

（1）当 initlen < 32 时，type = SDS_TYPE_5；

（2）当 32 ≤ initlen < 256 时，type = SDS_TYPE_8；

(3)当 256≤ initlen < 64k 时，type = SDS_TYPE_16；

(4)当 64k≤ initlen < 4G 时，type = SDS_TYPE_32；

(5)当 initlen≥4G 时，type = SDS_TYPE_64。

对各个 type 类型的定义如下：

```
//sds.h 文件第 76 行
#define SDS_TYPE_5  0
#define SDS_TYPE_8  1
#define SDS_TYPE_16 2
#define SDS_TYPE_32 3
#define SDS_TYPE_64 4
```

sds 对象的内存结构如下图所示。

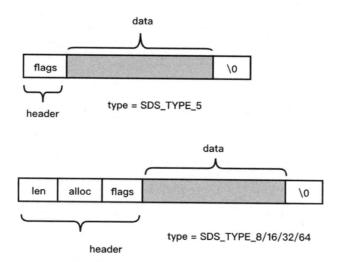

sds 在内存开始处存储的是数据的"头部信息"（或者叫元数据信息），头部信息记录了这段内存数据的一些状态信息。当 type=SDS_TYPE_5 时，头部信息只有一个 flags 字段；当 type=SDS_TYPE_8/16/32/64 时，头部有 len、alloc、flags 三个字段。在初始状态下，len 和 alloc 都记录着实际给存储数据分配的长度，flags 字段记录着 type 类型的信息，sds 所在内存最后一字节被填充为\0（ASCII 的 0 值）。这些结构的定义如下：

```
//sds.h 文件第 45 行
typedef char *sds;

struct __attribute__ ((__packed__)) sdshdr5 {
    unsigned char flags;
    char buf[];
};
struct __attribute__ ((__packed__)) sdshdr8 {
    uint8_t len;
    uint8_t alloc;
    unsigned char flags;
    char buf[];
};
```

```
struct __attribute__ ((__packed__)) sdshdr16 {
    uint16_t len;
    uint16_t alloc;
    unsigned char flags;
    char buf[];
};
struct __attribute__ ((__packed__)) sdshdr32 {
    uint32_t len;
    uint32_t alloc;
    unsigned char flags;
    char buf[];
};
struct __attribute__ ((__packed__)) sdshdr64 {
    uint64_t len;
    uint64_t alloc;
    unsigned char flags;
    char buf[];
};
```

type = SDS_TYPE_5 的内存结构在 Redis 中从未被用到,这里列出来只是为了体现完整性。

根据上面的分析,初始状态下 client.querybuf 的内存结构如下图所示。

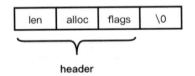

上图中的 len、alloc、flags 字段各占 1 字节,没有 data 区域,len 和 alloc 的值均为 0,type 类型为 SDS_TYPE_8。

考虑到初始化一个空 sds 类型时,这个 sds 大概率是用于之后向其中存入数据的,因此为了减少后面存入数据时内存扩展需要重新分配内存的次数,当 initlen=0 时,本应使用 SDS_TYPE_5 类型,却使用了 SDS_TYPE_8 类型(代码片段 8-2-10 中的加粗行展示了这一逻辑)。

当利用 client.querybuf 存储数据时,redis-server 执行以下逻辑:

```
//networking.c 文件第 1890 行
void readQueryFromClient(connection *conn) {
    client *c = connGetPrivateData(conn);
    int nread, readlen;
    size_t qblen;

    //省略部分代码

    readlen = PROTO_IOBUF_LEN;

    //省略部分代码

    qblen = sdslen(c->querybuf);
```

```
    if (c->querybuf_peak < qblen) c->querybuf_peak = qblen;
    c->querybuf = sdsMakeRoomFor(c->querybuf, readlen);
    nread = connRead(c->conn, c->querybuf+qblen, readlen);

    //省略部分代码
}
```

以上代码中的加粗行计算当前接收缓冲区中的可用空间大小，实际上就是取 sds 结构的 len 字段的值。connRead 尝试一次最大读取 PROTO_IOBUF_LEN 字节的数据（PROTO_IOBUF_LEN = 16k），因此需要先调用 sdsMakeRoomFor 函数为 client.query 扩展空间，如何扩展呢？且看 sdsMakeRoomFor 函数的实现：

```
//sds.c 文件第 204 行
sds sdsMakeRoomFor(sds s, size_t addlen) {
    void *sh, *newsh;
    size_t avail = sdsavail(s);
    size_t len, newlen;
    char type, oldtype = s[-1] & SDS_TYPE_MASK;
    int hdrlen;

    /* 如果剩余空间足够，就直接返回 */
    if (avail >= addlen) return s;

    len = sdslen(s);
    sh = (char*)s-sdsHdrSize(oldtype);
    newlen = (len+addlen);
    if (newlen < SDS_MAX_PREALLOC)
        newlen *= 2;
    else
        newlen += SDS_MAX_PREALLOC;

    type = sdsReqType(newlen);

    if (type == SDS_TYPE_5) type = SDS_TYPE_8;

    hdrlen = sdsHdrSize(type);
    if (oldtype==type) {
        newsh = s_realloc(sh, hdrlen+newlen+1);
        if (newsh == NULL) return NULL;
        s = (char*)newsh+hdrlen;
    } else {

        newsh = s_malloc(hdrlen+newlen+1);
        if (newsh == NULL) return NULL;
        memcpy((char*)newsh+hdrlen, s, len+1);
        s_free(sh);
        s = (char*)newsh+hdrlen;
        s[-1] = type;
        sdssetlen(s, len);
    }
    sdssetalloc(s, newlen);
    return s;
}
```

在以上代码中，扩展 sds 对象的空间时必须保留已存在于 sds 对象中的数据，先计算待扩展的缓冲区中的剩余容量 avail，sds 的剩余容量 avail 等于 alloc 减去 len。如果当前剩余空间 avail 大于或等于需要新增的大小 addlen，则没必要扩展空间，直接返回即可。反之需要扩展的最终大小是当前已存储数据的长度 len 加上新增的数据长度 addlen，即以上代码中的 newlen。redis-server 在这里使用了这样一个内存大小扩展策略：当 newlen < SDS_MAX_PREALLOC 时，newlen 变成其自身的 2 倍（即 newlen×2），反之当 newlen≥SDS_MAX_PREALLOC 时，newlen 加上 SDS_MAX_PREALLOC 作为新的 newlen 值。

接着根据 newlen 重新计算内部的 type 类型，重新分配内存生成新的 sds 结构，然后重新设置新的 sds 对象的 header 区域的 len、alloc 和 flags 字段值，并将旧的 sds 对象的 data 区域的数据拷贝至新的 sds 对象的 data 区域，然后释放旧的 sds 对象的内存。这就是 redis-server 中扩展自定义接收缓冲区的方案。

调用收取数据的函数 connRead 成功后，假设实际收取数据的长度为 nread，此时我们需要调用 sdsIncrLen 函数更新 client.querybuf 对象的内部数据状态，则调用逻辑如下：

```
//networking.c 文件第 1940 行
sdsIncrLen(c->querybuf,nread);
```

由于上一步中的 sds 内部结构已扩展了足够多的空间，因此 sdsIncrLen 函数只需更新 sds 内部已用内存长度 len 的值就可以了，即将 len 的值更新成 len 加上 nread 后的值。

虽然 sds 的内部结构能解决内存扩展问题，但是我们把 sds 结构作为接收缓冲区用于解包时从 client.querybuf 中读取一部分数据后，会造成从 header 的结束位置到读取位置之间的数据为空，因此我们必须记录上一次读取的位置。示意图如下。

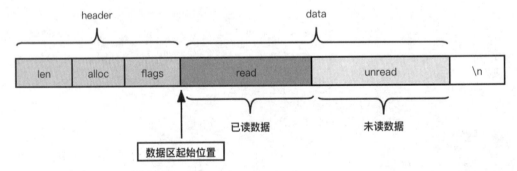

如上图所示，对 client.querybuf 中的数据进行读取时，若出现已经读取一部分且还剩余一部分的情形，则必须记录当前读取的位置，否则下一次就不知道从什么位置开始读取了，当前已读取位置被记录在 client 对象的 qb_pos（"qb" 是 "querybuf" 的缩写）中，我们在每次读取一部分数据后只更新 qb_pos 字段即可，下一次读取时从 qb_pos 记录的位置开始读。

2. 发送缓冲区

接收缓冲区使用 sds 结构，与这一结构相比，redis-server 发送缓冲区的结构比较简单，

client 对象有以下 3 个字段：

```
//server.h 文件第 768 行
typedef struct client {
    //省略其他无关字段

    size_t sentlen;

    //省略其他无关字段

    int bufpos;
    //发送缓冲区
    char buf[PROTO_REPLY_CHUNK_BYTES];
} client;
```

发送缓冲区的 buf 字段是一个大小为 16KB 的固定长度的数组（PROTO_REPLY_CHUNK_BYTES=16k），bufpos 字段记录存储的数据长度，sentlen 记录每次发送了多少，当发送缓冲区中的部分数据已发送但仍剩余部分数据时，其内存结构如下图所示。

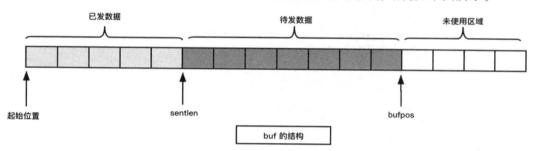

buf 的结构

总的来说，redis-server 的接收缓冲区和发送缓冲区的设计和使用方法非常典型、简便，建议读者认真体会。

8.2.11 定时器逻辑

本节深入讨论 redis-server 中的定时器是如何设计的。8.2.2 节介绍了在事件处理函数中如何寻找最近到期的定时器对象，这里就这个问题继续讨论。在 aeProcessEvents 函数的结尾处有这样一段代码：

```
//ae.c 文件第 386 行
int aeProcessEvents(aeEventLoop *eventLoop, int flags)
{
    //省略部分代码

    //ae.c 文件第 507 行
    if (flags & AE_TIME_EVENTS)
        processed += processTimeEvents(eventLoop);

    return processed;
}
```

如果存在定时器事件，则调用 processTimeEvents 函数进行处理，处理逻辑如下：

```
//ae.c 文件第 287 行
/* 处理定时器事件 */
```

```c
static int processTimeEvents(aeEventLoop *eventLoop) {
    int processed = 0;
    aeTimeEvent *te, *prev;
    long long maxId;
    time_t now = time(NULL);

    //省略部分代码

    if (now < eventLoop->lastTime) {
        te = eventLoop->timeEventHead;
        while(te) {
            te->when_sec = 0;
            te = te->next;
        }
    }

    //记录每次的定时器处理时间
    eventLoop->lastTime = now;

    prev = NULL;
    //定时器对象被存储于 eventLoop->timeEventHead 链表中，
    //因此这里的 te 是链表的第 1 个头节点
    te = eventLoop->timeEventHead;
    maxId = eventLoop->timeEventNextId-1;
    while(te) {
        long now_sec, now_ms;
        long long id;

        if (te->id == AE_DELETED_EVENT_ID) {
            //删除一个定时器节点逻辑
            aeTimeEvent *next = te->next;
            if (prev == NULL)
                eventLoop->timeEventHead = te->next;
            else
                prev->next = te->next;
            if (te->finalizerProc)
                te->finalizerProc(eventLoop, te->clientData);
            zfree(te);
            te = next;
            continue;
        }

        if (te->id > maxId) {
            te = te->next;
            continue;
        }
        aeGetTime(&now_sec, &now_ms);
        //判断定时器对象是否到期
        if (now_sec > te->when_sec ||
            (now_sec == te->when_sec && now_ms >= te->when_ms))
        {
            int retval;

            id = te->id;
```

```
            retval = te->timeProc(eventLoop, id, te->clientData);
            processed++;
            if (retval != AE_NOMORE) {
                aeAddMillisecondsToNow(retval,&te->when_sec,&te->when_ms);
            } else {
                te->id = AE_DELETED_EVENT_ID;
            }
        }
        prev = te;
        te = te->next;
    }
    return processed;
}
```

这段代码的核心逻辑是遍历 eventLoop->timeEventHead 链表中的每个定时器对象，如果某个定时器对象已被标记为删除状态（定时器对象的 id 字段是否等于 AE_DELETED_EVENT_ID 标志），则从 eventLoop-> timeEventHead 链表中移除该定时器对象，即让被移除的定时器对象的上一个节点的 next 指针指向下一个节点，对应以上代码片段中的加粗行。

如果定时器对象未被标记为删除状态，则接着判断该定时器对象是否到期。判断到期的逻辑是与当前系统时间比较，如果定时器已经到期，则调用定时器对象的回调函数 timeProc 进行处理。这段逻辑对应以上代码段中的斜体部分。

这段代码没有什么特别需要注意的地方，但其中考虑了一种特殊场景，就是假设人为地将当前计算机时间调到了未来某个时刻，再调回来，这样就会出现 now（当前时间）小于 eventLoop->lastTime（eventLoop->lastTime 记录了上一次处理定时器对象的时间）。出现这种情况怎么办呢？Redis 作者的做法是遍历该定时器对象链表，将该链表中所有定时器对象的到期时间都设置为 0，这些定时器就会立即得到处理了。Redis 作者在代码注释中解释了这种场景的处理策略：

```
//ae.c 文件第 294~301 行
/* If the system clock is moved to the future, and then set back to the
 * right value, time events may be delayed in a random way. Often this
 * means that scheduled operations will not be performed soon enough.
 *
 * Here we try to detect system clock skews, and force all the time
 * events to be processed ASAP when this happens: the idea is that
 * processing events earlier is less dangerous than delaying them
 * indefinitely, and practice suggests it is. */
```

注释中的 "ASAP" 是英文 "As Soon As Possible"（尽快）的缩写。

那么 redis-server 中到底在哪些地方使用了定时器呢？我们可以在 Redis 源码中搜索定时器创建函数 aeCreateTimeEvent，在 initServer 函数中有这么一行：

```
//server.c 文件第 2878 行
if (aeCreateTimeEvent(server.el, 1, serverCron, NULL, NULL) == AE_ERR) {
    serverPanic("Can't create event loop timers.");
    exit(1);
}
```

在以上代码中，定时器的回调函数是 serverCron。经常使用 Linux 系统的读者应该熟悉"Cron"一词，它是 Linux 的一种定时任务机制，在 redis-server 中用于 Cron 任务。

8.2.12 钩子函数

在通常情况下，在一个 EventLoop 中除了有定时器、I/O 复用和 I/O 事件处理逻辑，还可以根据需求自定义一些逻辑处理函数（对应第 7 章中介绍的 handle_other_things 逻辑），在 redis-server 中这类逻辑处理函数被称为"钩子函数"。钩子函数可以位于 Loop 的任何位置，前面介绍的 beforeSleep 函数就是事件处理之前的自定义钩子函数（位于定时器对象检测逻辑之前）。

在 redis-server 中，I/O 复用函数调用与 I/O 事件处理逻辑之间也有一个自定义钩子函数 aftersleep，代码如下：

```c
//networking.c 文件第 386 行
int aeProcessEvents(aeEventLoop *eventLoop, int flags)
{
    //无关代码省略
    //I/O 复用函数调用
    numevents = aeApiPoll(eventLoop, tvp);

    //aftersleep 钩子函数处理逻辑
    if (eventLoop->aftersleep != NULL && flags & AE_CALL_AFTER_SLEEP)
        eventLoop->aftersleep(eventLoop);

    //I/O 事件处理逻辑
    for (j = 0; j < numevents; j++) {
        //无关代码省略
    }
}
```

eventLoop->aftersleep 钩子函数在 main 函数中被设置，逻辑如下：

```c
//server.c 文件第 4969 行
int main(int argc, char **argv) {
    //无关代码省略

    //设置 eventLoop->beforesleep 钩子（server.c 文件第 2916 行）
    aeSetBeforeSleepProc(server.el,beforeSleep);
    //设置 eventLoop->aftersleep 钩子（server.c 文件第 2917 行）
    aeSetAfterSleepProc(server.el,afterSleep);

    return 0;
}
```

8.2.13 redis-server 端网络通信模块小结

通过前面的讲解，我们用一张图来概括 redis-server 端的网络通信模型，如下图所示。

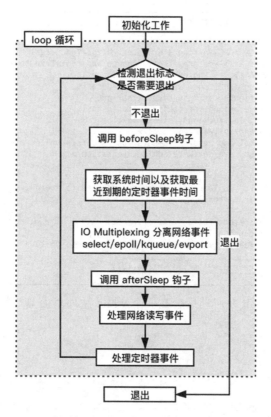

如上图所示，redis-server 是利用 one loop one thread 思想实现的典型程序结构，也是目前主流的网络通信结构的代表。在 Redis 6.0 之前的版本中，或者在 Redis 6.0 及后续版本中但没启动额外的 I/O 工作线程的情况下，Redis 在主线程中处理自定义的钩子函数、调用 I/O 复用函数、处理各种 I/O 事件和定时器事件，所以我们通常称 Redis 的网络通信模型是单线程的。

8.3　探究 redis-cli 端的网络通信模型

这里接着探究 Redis 源码自带的客户端 redis-cli 的网络通信模块。

我们使用 gdb 让 redis-cli 运行起来，这里原本打算使用 Ctrl+C 组合键让程序中断下来查看在 redis-cli 运行时有几个线程，但是经实验之后发现，按 Ctrl+C 组合键并不能让程序中断，反而让 redis-cli 进程退出。该操作如下图所示。

```
[root@myaliyun src]# gdb redis-cli
GNU gdb (GDB) 8.3
Copyright (C) 2019 Free Software Foundation, Inc.
License GPLv3+: GNU GPL version 3 or later <http://gnu.org/licenses/gpl.html>
This is free software: you are free to change and redistribute it.
There is NO WARRANTY, to the extent permitted by law.
Type "show copying" and "show warranty" for details.
This GDB was configured as "x86_64-pc-linux-gnu".
Type "show configuration" for configuration details.
For bug reporting instructions, please see:
<http://www.gnu.org/software/gdb/bugs/>.
Find the GDB manual and other documentation resources online at:
    <http://www.gnu.org/software/gdb/documentation/>.

For help, type "help".
Type "apropos word" to search for commands related to "word"...
Reading symbols from redis-cli...
(gdb) r                                            直接按Ctrl+C会导致
Starting program: /root/redis-6.0.3/src/redis-cli  redis-cli退出
[Thread debugging using libthread_db enabled]
Using host libthread_db library "/usr/lib64/libthread_db.so.1".
127.0.0.1:6379>
[Inferior 1 (process 3837) exited normally]
(gdb)
```

退出的原因是 redis-cli 在启动后会在一个 while 循环里面等待用户输入，这个逻辑位于 reply 函数中：

```
//redis-cli.c 文件第1909行
static void repl(void) {
    //省略部分代码

    //redis-cli.c 文件第1939行
    while((line = linenoise(context ? config.prompt : "not connected> ")) != NULL)
{
        //省略部分代码
    }
    exit(0);
}
```

以上代码表明，如果用户的输入不符合预期，例如按下了 Ctrl+C 组合键，使 while 条件不成立导致 while 循环结束，就会退出主进程。所以这种让被 gdb 调试的 redis-cli 中断的方法行不通，我们换种方法来试一下：先直接运行 redis-cli，然后查看 redis-cli 的进程 ID，接着使用 gdb attach 进程 ID 命令将 gdb attach 到 redis-cli 进程上，操作如下：

```
[root@myaliyun src]# ps -ef | grep redis-cli
root      27559 25850  0 14:35 pts/2    00:00:00 ./redis-cli
root      27813 26804  0 14:39 pts/0    00:00:00 grep --color=auto redis-cli
[root@myaliyun src]# gdb attach 27559
...省略部分输出...
Attaching to process 27559
Reading symbols from /root/redis-6.0.3/src/redis-cli...
...省略部分输出...
(gdb) info threads
  Id   Target Id                                         Frame
* 1    Thread 0x7f3d3c6cf740 (LWP 27559) "redis-cli"  0x00007f3d3bba16e0 in
__read_nocancel () from /usr/lib64/libpthread.so.0
(gdb)
```

在 gdb attach 到 redis-cli 进程后，使用 info threads 命令查看线程数量。通过上面的输出，我们发现 redis-cli 只有一个线程，即主线程。既然只有一个线程，那么我们可以断定 redis-cli 发给 redis-server 的命令肯定都是同步的，这里同步的意思是 redis-cli 发送命令后会一直等待 redis-server 的应答或者直到应答超时返回。

在 redis-cli 的 main 函数中有这样一段代码：

```
//redis-cli.c 文件第 8091 行
if (argc == 0 && !config.eval) {
    signal(SIGPIPE, SIG_IGN);

    cliConnect(0);
    repl();
}
```

其中，cliConnect(0)的实现如下：

```
//redis-cli.c 文件第 859 行
static int cliConnect(int force) {
    if (context == NULL || force) {
        if (context != NULL) {
            redisFree(context);
        }

        if (config.hostsocket == NULL) {
            context = redisConnect(config.hostip,config.hostport);
        } else {
            context = redisConnectUnix(config.hostsocket);
        }

        if (context->err) {
            fprintf(stderr,"Could not connect to Redis at ");
            if (config.hostsocket == NULL)
fprintf(stderr,"%s:%d: %s\n",config.hostip,config.hostport,context->errstr);
            else
                fprintf(stderr,"%s: %s\n",config.hostsocket,context->errstr);
            redisFree(context);
            context = NULL;
            return REDIS_ERR;
        }


        anetKeepAlive(NULL, context->fd, REDIS_CLI_KEEPALIVE_INTERVAL);

        if (cliAuth() != REDIS_OK)
            return REDIS_ERR;
        if (cliSelect() != REDIS_OK)
            return REDIS_ERR;
    }
    return REDIS_OK;
}
```

cliConnect 函数做的工作可以分为三步：①context = redisConnect(config.hostip,config.hostport)；②cliAuth()；③cliSelect()。

我们先来看下步骤 1 的 redisConnect 函数，redisConnect 函数实际又调用了 redisConnectWithOptions 函数，后者又调用 redisContextConnectBindTcp 函数，redisContextConnectBindTcp 函数内部又调用了_redisContextConnectTcp 函数。为了更清楚地表达这里的函数调用关系，我们在_redisContextConnectTcp 函数调用处加个断点，然后使用 run 命令重启 redis-cli，在断点触发后输入 bt 命令查看此时的调用堆栈。操作过程如下：

```
(gdb) b _redisContextConnectTcp
Breakpoint 2 at 0x42e7c2: file net.c, line 342.
(gdb) r
The program being debugged has been started already.
Start it from the beginning? (y or n) y
Starting program: /root/redis-6.0.3/src/redis-cli
[Thread debugging using libthread_db enabled]
Using host libthread_db library "/usr/lib64/libthread_db.so.1".

Breakpoint 2, _redisContextConnectTcp (c=0x44e050, addr=0x44e011 "127.0.0.1",
port=6379, timeout=0x0, source_addr=0x0) at net.c:342
342            int blocking = (c->flags & REDIS_BLOCK);
(gdb) bt
#0  _redisContextConnectTcp (c=0x44e050, addr=0x44e011 "127.0.0.1", port=6379,
timeout=0x0, source_addr=0x0) at net.c:342
#1  0x000000000042ef17 in redisContextConnectBindTcp (c=0x44e050, addr=0x44e011
"127.0.0.1", port=6379, timeout=0x0, source_addr=0x0) at net.c:513
#2  0x0000000000426202 in redisConnectWithOptions (options=0x7fffffffe1a0) at
hiredis.c:767
#3  0x0000000000426317 in redisConnect (ip=0x44e011 "127.0.0.1", port=6379) at
hiredis.c:792
#4  0x000000000040c585 in cliConnect (flags=0) at redis-cli.c:866
#5  0x00000000004213f8 in main (argc=0, argv=0x7fffffffe320) at redis-cli.c:8098
(gdb)
```

_redisContextConnectTcp 函数是实际连接 redis-server 的地方，先调用操作系统函数 getaddrinfo 解析传入的 IP 地址和端口号（这里使用的 IP 地址和端口号分别是 127.0.0.1 和 6379）；然后创建 socket（连接 socket），并将 socket 设置成非阻塞模式；接着调用 Socket API connect 函数，由于连接 socket 是非阻塞模式的，所以 connect 函数会立即返回-1；接着调用 redisContextWaitReady 函数，在 redisContextWaitReady 函数中调用 poll 函数检测连接 socket 是否可写（写事件标志为 POLLOUT），如果连接 socket 可写，则表示连接 redis-server 成功。

由于_redisContextConnectTcp 函数的实现代码较多，所以这里去掉了一些无关代码，整理出关键逻辑后的代码片段如下：

```
//net.c 文件第 335 行
static int _redisContextConnectTcp(redisContext *c, const char *addr, int port,
                                   const struct timeval *timeout,
                                   const char *source_addr) {
    //省略部分代码

    //解析传入的 IP 地址和端口号
```

```
    rv = getaddrinfo(c->tcp.host,_port,&hints,&servinfo)) != 0

    //创建连接socket
    s = socket(p->ai_family,p->ai_socktype,p->ai_protocol)) == -1

    //将连接socket设置为非阻塞模式
    redisSetBlocking(c,0) != REDIS_OK

    //连接redis-server
    connect(s,p->ai_addr,p->ai_addrlen)

    //检测连接socket是否可写
    redisContextWaitReady(c,timeout_msec) != REDIS_OK

    return rv;
}
```

redisContextWaitReady 函数的实现如下:

```
//net.c文件第241行
static int redisContextWaitReady(redisContext *c, long msec) {
    struct pollfd   wfd[1];

    wfd[0].fd     = c->fd;
    wfd[0].events = POLLOUT;

    if (errno == EINPROGRESS) {
        int res;

        if ((res = poll(wfd, 1, msec)) == -1) {
            __redisSetErrorFromErrno(c, REDIS_ERR_IO, "poll(2)");
            redisContextCloseFd(c);
            return REDIS_ERR;
        } else if (res == 0) {
            errno = ETIMEDOUT;
            __redisSetErrorFromErrno(c,REDIS_ERR_IO,NULL);
            redisContextCloseFd(c);
            return REDIS_ERR;
        }

        if (redisCheckSocketError(c) != REDIS_OK)
            return REDIS_ERR;

        return REDIS_OK;
    }

    __redisSetErrorFromErrno(c,REDIS_ERR_IO,NULL);
    redisContextCloseFd(c);
    return REDIS_ERR;
}
```

使用 b redisContextWaitReady 增加一个断点,然后使用 run 命令重新运行 redis-cli,程序会停在我们设置的断点处,然后使用 bt 命令得到当前调用堆栈,整个操作过程如下:

```
(gdb) b redisContextWaitReady
Breakpoint 4 at 0x42e460: file net.c, line 244.
```

```
(gdb) r
The program being debugged has been started already.
Start it from the beginning? (y or n) y
Starting program: /root/redis-6.0.3/src/redis-cli
[Thread debugging using libthread_db enabled]
Using host libthread_db library "/usr/lib64/libthread_db.so.1".

Breakpoint 4, redisContextWaitReady (c=0x44e050, msec=-1) at net.c:244
244         wfd[0].fd    = c->fd;
(gdb) bt
#0  redisContextWaitReady (c=0x44e050, msec=-1) at net.c:244
#1  0x000000000042edad in _redisContextConnectTcp (c=0x44e050, addr=0x44e011
"127.0.0.1", port=6379, timeout=0x0, source_addr=0x0) at net.c:475
#2  0x000000000042ef17 in redisContextConnectBindTcp (c=0x44e050, addr=0x44e011
"127.0.0.1", port=6379, timeout=0x0, source_addr=0x0) at net.c:513
#3  0x0000000000426202 in redisConnectWithOptions (options=0x7fffffffe1a0) at
hiredis.c:767
#4  0x0000000000426317 in redisConnect (ip=0x44e011 "127.0.0.1", port=6379) at
hiredis.c:792
#5  0x000000000040c585 in cliConnect (flags=0) at redis-cli.c:866
#6  0x00000000004213f8 in main (argc=0, argv=0x7fffffffe320) at redis-cli.c:8098
(gdb)
```

redis-cli 连接 redis-server 成功以后，redis-cli 接着会调用上文中提到的 cliAuth 和 cliSelect 函数，这两个函数会根据 config.auth 和 config.dbnum 参数的配置来决定是否向 redis-server 发送相关命令。由于这里没有配置这两个参数，所以这两个函数实际上什么也不做。cliSelect 函数对应的逻辑如下：

```
583     static int cliSelect(void) {
(gdb) n
585         if (config.dbnum == 0) return REDIS_OK;
(gdb) p config.dbnum
$11 = 0
```

接着调用 repl 函数，在 repl 函数中有一个 while 循环，不断地从用户的输入中获取信息并解析、处理，逻辑如下：

```
//redis-cli.c 文件第 1909 行
static void repl(void) {
    //省略无关代码
    while((line = linenoise(context ? config.prompt : "not connected> ")) != NULL)
{
        if (line[0] != '\0') {
            argv = cliSplitArgs(line,&argc);
            //将用户的命令保存到历史记录文件中
            if (history) linenoiseHistoryAdd(line);
            if (historyfile) linenoiseHistorySave(historyfile);

            if (argv == NULL) {
                printf("Invalid argument(s)\n");
                linenoiseFree(line);
                continue;
            } else if (argc > 0) {
                if (strcasecmp(argv[0],"quit") == 0 ||
                    strcasecmp(argv[0],"exit") == 0)
```

```c
        {
            exit(0);
        } else if (argv[0][0] == ':') {
            cliSetPreferences(argv,argc,1);
            continue;
        } else if (strcasecmp(argv[0],"restart") == 0) {
            if (config.eval) {
                config.eval_ldb = 1;
                config.output = OUTPUT_RAW;
                return;
            } else {
                printf("Use 'restart' only in Lua debugging mode.");
            }
        } else if (argc == 3 && !strcasecmp(argv[0],"connect")) {
            sdsfree(config.hostip);
            config.hostip = sdsnew(argv[1]);
            config.hostport = atoi(argv[2]);
            cliRefreshPrompt();
            cliConnect(1);
        } else if (argc == 1 && !strcasecmp(argv[0],"clear")) {
            linenoiseClearScreen();
        } else {
            long long start_time = mstime(), elapsed;
            int repeat, skipargs = 0;
            char *endptr;

            repeat = strtol(argv[0], &endptr, 10);
            if (argc > 1 && *endptr == '\0' && repeat) {
                skipargs = 1;
            } else {
                repeat = 1;
            }

            //对远端命令的处理
            issueCommandRepeat(argc-skipargs, argv+skipargs, repeat);


            if (config.eval_ldb_end) {
                config.eval_ldb_end = 0;
                cliReadReply(0);
                printf("\n(Lua debugging session ended%s)\n\n",
                    config.eval_ldb_sync ? "" :
                    " -- dataset changes rolled back");
            }

            elapsed = mstime()-start_time;
            if (elapsed >= 500 &&
                config.output == OUTPUT_STANDARD)
            {
                printf("(%.2fs)\n",(double)elapsed/1000);
            }
        }
    }

    sdsfreesplitres(argv,argc);
```

```
        }
        linenoiseFree(line);
    }
    exit(0);
}
```

以上代码的逻辑是得到用户输入的一行命令后，先将其保存到历史记录中（以便下一次按键盘上的上下箭头键再次获取），然后校验命令的合法性，如果是本地命令（不需要发送给 redis-server 的命令，例如 quit、exit），则直接执行；如果是远端命令，则调用 issueCommandRepeat 函数发送给服务端。issueCommandRepeat 函数的逻辑如下：

```
//redis-cli.c 文件第 1820 行
static int issueCommandRepeat(int argc, char **argv, long repeat) {
    while (1) {
        config.cluster_reissue_command = 0;
        if (cliSendCommand(argc,argv,repeat) != REDIS_OK) {
            //如果是网络问题，则重连后再次尝试发送
            cliConnect(1);

            if (cliSendCommand(argc,argv,repeat) != REDIS_OK) {
                //重连成功之后仍然发送出错，只好输出错误
                cliPrintContextError();
                return REDIS_ERR;
            }
        }

        if (config.cluster_mode && config.cluster_reissue_command) {
            cliConnect(1);
        } else {
            break;
        }
    }
    return REDIS_OK;
}
```

在 issueCommandRepeat 函数中实际发送命令的函数是 cliSendCommand，在 cliSendCommand 函数中又调用了 cliReadReply 函数，后者又调用了 redisGetReply 函数，在 redisGetReply 函数中又调用了 redisBufferWrite 函数，在 redisBufferWrite 函数中，redis-cli 最终调用系统 API write 将用户输入的命令发送了出去。

redisBufferWrite 函数的实现如下：

```
//hiredis.c 文件第 903 行
int redisBufferWrite(redisContext *c, int *done) {
    int nwritten;

    if (c->err)
        return REDIS_ERR;

    if (sdslen(c->obuf) > 0) {
        nwritten = write(c->fd,c->obuf,sdslen(c->obuf));
        if (nwritten == -1) {
            if ((errno == EAGAIN && !(c->flags & REDIS_BLOCK)) || (errno == EINTR))
```

```
{
        } else {
            __redisSetError(c,REDIS_ERR_IO,NULL);
            return REDIS_ERR;
        }
    } else if (nwritten > 0) {
        if (nwritten == (signed)sdslen(c->obuf)) {
            sdsfree(c->obuf);
            c->obuf = sdsempty();
        } else {
            sdsrange(c->obuf,nwritten,-1);
        }
    }
}
if (done != NULL) *done = (sdslen(c->obuf) == 0);
return REDIS_OK;
}
```

使用 b redisBufferWrite 命令增加一个断点,然后使用 run 命令重新运行 redis-cli,接着在 redis-cli 终端输入 set hello world(hello 是 key,world 是 value)这个简单的指令,此时使用 bt 命令查看 redis-cli 的调用堆栈。操作过程如下:

```
127.0.0.1:6379> set hello world

Breakpoint 5, redisBufferWrite (c=0x44e050, done=0x7fffffffe09c) at hiredis.c:906
906             if (c->err)
(gdb) bt
#0  redisBufferWrite (c=0x44e050, done=0x7fffffffe09c) at hiredis.c:906
#1  0x0000000000426942 in redisGetReply (c=0x44e050, reply=0x7fffffffe0c8) at hiredis.c:948
#2  0x000000000040d565 in cliReadReply (output_raw_strings=0) at redis-cli.c:1192
#3  0x000000000040ddbd in cliSendCommand (argc=3, argv=0x4a4800, repeat=0) at redis-cli.c:1361
#4  0x000000000040f94a in issueCommandRepeat (argc=3, argv=0x4a4800, repeat=1) at redis-cli.c:1823
#5  0x000000000041010a in repl () at redis-cli.c:2018
#6  0x00000000004213fd in main (argc=0, argv=0x7fffffffe320) at redis-cli.c:8099
(gdb)
```

当然,待发送的数据需要被存储在一个全局静态变量 context 中,这是一个结构体类型,redisContext 结构被定义在 hiredis.h 文件中:

```
//hiredis.c 文件第 206 行
/* 用于一个连接的 context 对象 */
typedef struct redisContext {
    int err; /* 错误标志,0 表示无错误 */
    char errstr[128]; /* 错误信息 */
    int fd;
    int flags;
    char *obuf; /* 发送缓冲区 */
    redisReader *reader; /* 协议解析器 */

    enum redisConnectionType connection_type;
    struct timeval *timeout;
```

```
    struct {
        char *host;
        char *source_addr;
        int port;
    } tcp;

    struct {
        char *path;
    } unix_sock;

} redisContext;
```

redisContext->obuf 字段指向一个 sds 类型的对象, 这个对象用来存储当前需要发送的命令。这也解决了命令一次发不完时需要暂时缓存的问题。

在 redisGetReply 函数中发送完数据后立即调用 redisBufferRead 收取服务器的应答。redisGetReply 函数的实现如下:

```
//redis-cli.c 文件第 1186 行
int redisGetReply(redisContext *c, void **reply) {
    int wdone = 0;
    void *aux = NULL;

    if (redisGetReplyFromReader(c,&aux) == REDIS_ERR)
        return REDIS_ERR;

    if (aux == NULL && c->flags & REDIS_BLOCK) {

        do {
            if (redisBufferWrite(c,&wdone) == REDIS_ERR)
                return REDIS_ERR;
        } while (!wdone);

        /* Read until there is a reply */
        do {
            if (redisBufferRead(c) == REDIS_ERR)
                return REDIS_ERR;
            if (redisGetReplyFromReader(c,&aux) == REDIS_ERR)
                return REDIS_ERR;
        } while (aux == NULL);
    }

    if (reply != NULL) *reply = aux;
    return REDIS_OK;
}
```

redis-cli 拿到 redis-server 的应答后就可以进行解析并显示在终端了。

总结起来, redis-cli 使用了同步网络通信模式, 只不过通信的 socket 被设置为非阻塞模式, 这样做有如下三个好处:

◎ 使用 connect 连接服务器时, connect 函数不会阻塞, 可以立即返回, 之后可以调用 poll 函数检测 socket 是否可写来判断是否连接成功;

◎ 发送数据时，如果因为对端 TCP 窗口太小导致数据发送不出去，则 write 函数也会立即返回，不会阻塞，此时可以将未发送的数据暂存起来，用于下次继续发送；
◎ 接收数据时，如果当前没有数据可读，则 read 函数也不会阻塞，程序可以立即返回，继续响应用户的输入。

redis-cli 的代码不多，但是包含了网络编程中很多常用的经典写法，如果读者想提高自己的网络编程能力，则可以深入研究 redis-cli 的代码。

8.4 Redis 的通信协议格式

redis-server 的通信协议格式是典型的以特定内容为分隔符的代表，这里的分隔符是 \r\n。

8.4.1 请求命令格式

Redis 命令一般由命令关键字+命令参数组成（命令参数对于某些命令关键字不是必需的），其组装后的协议格式如下：

```
*<参数数量>\r\n
$<参数 1 的字节数>\r\n
<参数 1 的数据>\r\n
$<参数 2 的字节数>\r\n
<参数 2 的数据>\r\n
...
$<参数 n 的字节数>\r\n
<参数 n 的数据>\r\n
```

如上所示，命令关键字本身也是作为协议内容中的一个参数来传递的。举个例子，假设 redis-cli 向 redis-server 发送了一条 set hello world 命令，发送示例如下：

```
127.0.0.1:6379> set hello world
```

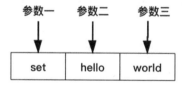

此时 redis-server 收到的数据格式如下：

```
*3\r\n
$3\r\n
set\r\n
$5\r\n
hello\r\n
$5\r\n
world\r\n
```

◎ 第 1 行的*3\r\n 以星号开始，数字 3 是接下来的参数个数，即 set、hello、world 共 3 个参数；

◎ 第 2 行的 $3\r\n 中的数字 3 是接下来第 1 个参数 set 关键字的字节数；
◎ 第 3 行 set\r\n 是第 1 个参数 set 命令的内容；
◎ 第 4 行 $5\r\n 是第 2 个参数 hello 的字节长度；
◎ 第 5 行 hello\r\n 是第 2 个参数 hello 的内容；
◎ 第 6 行 $5\r\n 数字 5 是第 3 个参数 world 的字节数；
◎ 第 7 行 world\r\n 是第 3 个参数的内容。

每一行都以\r\n 结束，以星号（*）开头表示参数数量，以$开头表示实际内容的字节数，实际的内容无特殊符号标记。

8.4.2 应答命令格式

Redis 的应答命令有多种类型，汇总如下。

（1）状态回复（status reply）的第 1 字节是 "+"。

（2）错误回复（error reply）的第 1 字节是 "-"。

（3）整数回复（integer reply）的第 1 字节是 ":"。

（4）批量回复（bulk reply）的第 1 字节是 "$"。

（5）多条批量回复（multi bulk reply）的第 1 字节是 "*"。

不同应答类型第 1 字节开始的标识符不一样，下面逐一介绍。

1. 状态回复

一个状态回复是以 "+" 开始且 "\r\n" 结尾的单行字符串，格式如下：

```
+状态信息\r\n
```

例如：

```
+OK
```

显示这个结果的客户端应该显示 "+" 号之后的所有内容，对于上面这个例子，客户端应该显示字符串 "OK"。

状态回复用于那些不需要返回内容数据的命令。

2. 错误回复

状态回复的第 1 字节是 "+"，错误回复的第 1 字节是 "-"，格式如下：

```
-错误信息\r\n
```

错误回复只在某些地方出现问题时产生。例如用户执行一个不存在的命令或者对不正确的数据类型执行命令等，一个客户端库应该在收到错误回复时产生一个异常，例如在 8.2.9 节介绍 redis-server 达到最大连接数时返回的错误信息就属于错误回复：

```
-ERR max number of clients reached\r\n
```

在 "-" 之后，直到遇到第 1 个 "\r\n" 为止，这中间的内容表示所返回错误的类型。

ERR 是一个通用错误，还有另一种叫作 WRONGTYPE 的错误类型。WRONGTYPE 错误类型表示一个特定的错误。一个自实现客户端可以根据错误类型自定义自己的处理逻辑。Redis 定义了非常多的 WRONGTYPE，例如：

```
-WRONGTYPE Operation against a key holding the wrong kind of value\r\n
```

以上错误信息被定义于 redismodule.h 文件第 139 行。

3. 整数回复

整数回复就是一个以 ":" 开头且 "\r\n" 结尾的字符串表示的整数。格式如下：

```
:整数值\r\n
```

例如，"\n" 和 ":1000\r\n" 都是整数回复。

又如 INCR key 和 LASTSAVE 命令均返回整数，前者返回键自增后的整数值，后者返回一个 UNIX 时间戳，返回值的唯一限制是这些数必须能够用 64 位有符号整数表示。

还有一种情况，整数回复用于表示布尔类型，例如 EXISTS key 和 SISMEMBER key member（前者判断一个 key 是否存在，后者判断 member 元素是否属于 key 集合的成员）。用返回值 1 表示真，用返回值 0 表示假。

其他一些命令例如 SADD、SREM、SETNX 只在操作真正被执行的时候才返回 1，否则返回 0。

Redis 返回整数回复的命令有 SETNX、DEL、EXISTS、INCR、INCRBY、DECR、DECRBY、DBSIZE、LASTSAVE、RENAMENX、MOVE、LLEN、SADD、SREM、SISMEMBER 和 SCARD 等。

4. 批量回复（Bulk Reply）

服务器使用批量回复来返回二进制安全的字符串，字符串的最大字节数为 512MB：

```
client: GET someKey\r\n
server: someValue\r\n
```

在 Redis 客户端发送的内容中：第 1 字节为 "$"；接下来是表示实际回复长度的数字值；后面跟着一个 "\r\n"；再后面跟着实际回复数据；末尾是另一个 "\r\n"。

对于 GET someKey 命令，服务器实际回复的内容如下：

```
$9\r\nsomeValue\r\n
```

如果被请求的 someKey 不存在，那么批量回复会使用 -1 这一特殊值作为长度值：

```
client: GET nonExistedKey\r\n
server: $-1\r\n
```

这种回复被称为空批量回复（NULL Bulk Reply）。

当请求对象不存在时，客户端应该返回空对象，而不是空字符串，对应 C/C++ 中的 NULL、Java 中的 null 及 Golang 中的 nil。

5. 多条批量回复（Multi Bulk Reply）

对于 LRANGE key start stop 这样的命令，需要返回多个值，这一目标可以通过多条批量回复来完成。多条批量回复是由多个回复组成的数组，数组中的每个元素都可以是任意类型的回复，包括多条批量回复本身。

多条批量回复的第 1 字节内容为 "*"，后面跟着一个整数值来表示多条批量回复的数量，接着是各个回复的长度和内容，长度前面以 "$" 为开头，格式如下：

```
client: LRANGE numberList 0 3\r\n
server: *4\r\n
server: $5\r\n
server: First\r\n
server: $6\r\n
server: Second\r\n
server: $5\r\n
server: Third\r\n
server: $6\r\n
server: Fourth\r\n
```

多条批量回复所使用的格式和客户端发送命令时使用的格式一模一样。服务器应答命令时所发送的多条批量回复不必是统一的类型，例如以下示例展示了一个多条批量回复，在回复中包含 3 个整型值和 1 个字符串：

```
*4\r\n
:1\r\n
:2\r\n
:3\r\n
$10\r\n
someString\r\n
```

在以上格式中，在回复的第 1 行，服务器发送 "*4\r\n"，表示这个多条批量回复包含 4 条回复，后面跟着的则是 4 条回复的内容。对于最后一条字符串回复类型，"$10" 表示字符串 someString 的长度。

当然，多条批量回复也可以是空白的（empty），例如：

```
client: LRANGE nokey 0 1
server: *0\r\n
```

无内容的多条批量回复（null multi bulk reply）也是存在的，例如命令 BLPOP key [key ...] timeout 在阻塞超时后，会返回一个无内容的多条批量回复，这个回复的计数值为 -1：

```
客户端: BLPOP key 1
服务器: *-1\r\n
```

客户端库应该区别对待空白的多条批量回复和无内容的多条批量回复：当 redis-server 返回一个无内容多条回复时，客户端库应该返回一个 null 对象，而不是一个空数组。

多条批量回复中的空元素

注意：多条批量回复中的元素可以将自身的长度设置为 -1，从而表示该元素不存在，并且不是一个空白字符串（empty string）。

例如，当 SORT key [BY pattern] [LIMIT offset count] [GET pattern [GET pattern ...]] [ASC | DESC] [ALPHA] [STORE destination]命令使用 GET pattern 选项对一个不存在的键进行操作时，就会出现多条批量回复中带有空白元素的情况。

以下示例展示了一个包含空元素的多条批量回复：

```
服务器： *3
服务器： $7
服务器： element
服务器： $-1
服务器： $4
服务器： item
```

上述回复中的第 2 个元素为空。

对于这个回复，客户端库应该返回类似于这样的回复：

```
//C/C++
["element", NULL, "item"]
或
//Java
["element", null, "item"]
```

或

```
//Golang
["element", nil, "item"]
```

8.4.3　多命令和流水线

客户端可以通过流水线在一次发送操作中发送多个命令，客户端可能得到：①在发送新命令之前不必读取前一个命令的回复；②多个命令的回复，会在最后一并返回。

8.4.4　特殊的 redis-cli 与内联命令

在某些情况下，当需要与 redis-server 通信但又找不到 redis-cli，而手上只有 telnet、nc 等命令可用时，我们可以通过 Redis 为这种情形而设的内联命令来发送命令。

以下是一个客户端和服务器使用内联命令进行交互的例子：

```
client: PING
server: +PONG
```

以下是另一个返回整数值的内联命令的例子：

```
client: EXISTS someKey
server: :0
```

因为没有了统一请求协议中的 "*" 项来声明参数的数量，所以在 telnet 会话中输入命令时，必须使用空格分割各个参数，服务器在接收到数据之后，会按空格对用户的输入进行解析并获取其中的命令参数。

以下是使用 nc 命令测试的结果：

```
[root@myaliyun src]# nc -v 127.0.0.1 6379
Ncat: Version 7.50 ( https://nmap.org/ncat )
Ncat: Connected to 127.0.0.1:6379.
PING
+PONG
EXISTS someKey
:0
GET HELLO WORLD
-ERR wrong number of arguments for 'get' command
GET HELLO
$-1
SET HELLO WORLD
+OK
GET HELLO
$5
WORLD
```

在以上代码中连接成功后，nc 客户端和 redis-server 的应答如下：

```
client: PING
server: PONG
client: EXISTS someKey
server: :0
client: GET HELLO WORLD
server: -ERR wrong number of arguments for 'get' command
client: GET HELLO
server: $-1
client: SET HELLO WORLD
server: +OK
client: GET HELLO
server: $5
server: WORLD
```

这里留给读者一个思考题：在以上操作中，client 和 server 每一行结尾的换行符是什么内容？是 "\r\n" 还是 "\n"？

8.4.5 Redis 对协议数据的解析逻辑

redis-server 对协议数据的解析逻辑位于 readQueryFromClient 函数中，在该函数中调用 connRead 函数收取数据并存入接收缓冲区 c->querybuf 中，然后调用 processInputBuffer 对数据进行解包：

```
//networking.c 文件第 1890 行
void readQueryFromClient(connection *conn) {
    //省略部分代码

    //收取数据并将其存入 c->querybuf 中
    nread = connRead(c->conn, c->querybuf+qblen, readlen);

    //省略部分代码

    //对收到的数据进行解包
    processInputBuffer(c);
}
```

processInputBuffer 函数的定义如下：

```c
//networking.c 文件第 1807 行
void processInputBuffer(client *c) {
    while(c->qb_pos < sdslen(c->querybuf)) {
        //省略部分代码

        /* 解析请求 type */
        if (!c->reqtype) {
            //判断收到的数据的第 1 字节是否是星号（*）
            if (c->querybuf[c->qb_pos] == '*') {
                //多条批量回复类型
                c->reqtype = PROTO_REQ_MULTIBULK;
            } else {
                //内联命令类型
                c->reqtype = PROTO_REQ_INLINE;
            }
        }

        if (c->reqtype == PROTO_REQ_INLINE) {
            if (processInlineBuffer(c) != C_OK) break;

            if (server.gopher_enabled &&
                ((c->argc == 1 && ((char*)(c->argv[0]->ptr))[0] == '/') || c->argc == 0))
            {
                processGopherRequest(c);
                resetClient(c);
                c->flags |= CLIENT_CLOSE_AFTER_REPLY;
                break;
            }
        } else if (c->reqtype == PROTO_REQ_MULTIBULK) {
            if (processMultibulkBuffer(c) != C_OK) break;
        } else {
            serverPanic("Unknown request type");
        }

        if (c->argc == 0) {
            resetClient(c);
        } else {

            if (c->flags & CLIENT_PENDING_READ) {
                c->flags |= CLIENT_PENDING_COMMAND;
                break;
            }

            /* 执行命令 */
            if (processCommandAndResetClient(c) == C_ERR) {

                return;
            }
        }
    }

    if (c->qb_pos) {
```

```
        sdsrange(c->querybuf,c->qb_pos,-1);
        c->qb_pos = 0;
    }
}
```

在以上代码中，在 processInputBuffer 函数中判断收到的数据的第 1 字节是否是星号（*），如果不是，则将其当作内联命令类型（PROTO_REQ_INLINE），反之当作多条批量回复（Multi Bulk Reply）类型（PROTO_REQ_MULTIBULK）。如果是内联命令类型，则调用 processInlineBuffer 函数根据内联命令的协议格式尝试解析数据；如果是多条批量回复类型，则调用 processMultibulkBuffer 函数尝试按多条批量回复类型解析数据，对协议按上文介绍的协议格式进行解析即可，这里不再具体分析了。

掌握了 Redis 的通信协议，我们就可以使用不同的编程语言来设计 Redis 客户端了。

第 9 章
服务器开发中的常用模块设计

本章讲解服务器开发中一些常用模块的设计思路和技巧。

9.1 断线自动重连的应用场景和逻辑设计

在有连接依赖关系的服务与服务之间或客户端与服务器之间，自动重连功能都是非常重要的功能。自动重连功能一般适用于以下 4 种场景。

（1）在一组服务之间，如果其中一些服务（主动连接方，以下简称 A）需要与另一些服务（被动连接方，以下简称 B）建立 TCP 长连接，而 A 没有自动连接 B 的功能，那么在部署或者测试这些服务时，必须先启动 B 再启动 A，因为一旦先启动 A，A 此时尝试连接 B（由于 B 还没有启动）就会失败，之后 A 再也不会连接 B 了（即使 B 随后被启动），从而导致整个系统不能正常工作。

（2）在部署或测试时，先启动 B 再启动 A，A 与 B 之间的连接在运行期间可能由于网络波动等原因断开，整个系统不能再正常工作。

（3）如果我们想升级 B，则更新完程序后重启 B 时也必须重启 A。如果这种依赖链较长（例如 A 连接 B，B 连接 C，C 连接 D，D 连接 E，等等），那么更新某个程序的效率会很低，更新成本非常高。

（4）对于客户端软件来说，如果因为用户的网络短暂故障导致客户端与服务器失去连接，那么等网络恢复后，较好的用户体验是客户端检测到用户网络变化时自动与服务器重连，以便用户及时收到最新的消息。

以上场景说明了断线自动重连机制的重要性，那么如何设计好的断线重连机制呢？

重连本身的功能开发很简单，其实就是调用套接字的 connect 函数不断重试，重试的技巧非常有讲究，如下所述。

（1）对于服务端程序，例如 A 连接 B，如果连接不上，整个系统则将无法工作。所以开发 A 服务时，重连的逻辑可以很简单：A 一旦发现与 B 断开了连接，就立即尝试与 B 重新连接，如果连接不上，则隔一段时间再重试（一般设置为 3 秒或 5 秒即可），一直到连接成功为止。当然，在此期间可以发送报警邮件或者输出错误日志，让开发人员或者运维人员尽快干预，尽快解决问题。

（2）对于客户端软件，以上做法也是可以的，但不是最优的做法。客户端所处的网络环境比服务器程序所处的网络环境一般要恶劣得多，以相等的时间间隔定时重连，一般作用不大（例如用户拔掉了网线）。因此对于客户端软件，出现断线时会尝试去重连，如果连接不上，则会隔一个比前一次时间间隔更长的时间间隔去重连，例如这个时间间隔可以是 2 秒、4 秒、8 秒、16 秒等。但是这样做也存在一个问题：随着重连次数的增加，重连的时间间隔会越来越大（也可以设置一个最大的重连时间间隔，之后恢复到之前较小的时间间隔），如果在某个时刻网络连接已经恢复正常（例如用户重新插上网线），程序就需要等待一个很长的时间间隔（如 16 秒）才能恢复连接，用户体验同样不好。解决办法：如果网络发生波动，程序就应该检测网络状态，如果网络状态恢复正常，就应该立即进行一次重连，而不是一成不变地按照设置的时间间隔重连。

操作系统提供了检测网络状态变化的 API 函数，例如对于 Windows，可以使用 IsNetworkAlive 函数去检测；对于 Android，在网络变化时会发送消息类型是 WifiManager.NETWORK_STATE_CHANGED_ACTION 的广播通知。

还需要注意的是，如果客户端网络断开，那么应该在界面的某个地方显式地告诉用户当前的连接状态，并告诉用户当前正在进行断线重连，应该有一个可以让用户放弃断线重连或者立即进行一次断线重连的功能。

综上所述，服务器程序之间的重连可以被设计成等时间间隔的定时重连。对于客户端程序，要结合依次放大重连时间间隔、网络状态变化时立即重连或用户主动发起重连三个因素来设计。

1. 不需要重连的情形

不需要重连的情形一般有：①用户使用客户端主动放弃重连；②因为一些业务上的要求，禁止客户端重连。

举个例子，如果某业务系统的账户在同一时刻不允许多个设备同时在线，某个账户在机器 A 上登录，接着又在机器 B 上登录，那么在机器 A 上登录的账户将被服务器踢下线，此时机器 A 上客户端的逻辑就应该被设计成禁止自动重连。

2. 技术上的断线重连和业务上的断线重连

技术上的断线重连指的是调用 connect 函数连接。在实际开发中，大多数系统只实现技术上的重连成功（即 connect 连接成功）是没有任何意义的，在网络连接成功后，还需要再次向服务器发送账号验证信息等（如登录数据包），在这些信息验签成功后，才叫作真正的重连成功。这里说的发送账号验证信息并验签成功就是指业务上的重连成功。在复杂的系统中可能需要连续进行几个验签流程。因此，我们若想设计完备的断线重连机制，则不仅要考虑技术上的重连，还要考虑业务上的重连。

9.2 保活机制与心跳包

在实际开发中，我们经常需要处理在下面两种情形中遇到的问题。

◎ 情形一：在通常情况下，服务器与某个客户端一般不在同一个网络中，它们之间可能经过数个路由器和交换机，如果其中某个必经路由器或者交换器出现了故障，并且在一段时间内没有恢复，则会导致这之间的链路不再畅通，而此时服务器与客户端之间也没有数据进行交换。由于 TCP 连接是状态机，所以对于这种情形，无论是客户端还是服务器，都无法感知与对方的连接是否正常，我们一般称这类连接为"死链"。

◎ 情形二：一个客户端在连接服务器以后，如果长时间没有和服务器有数据来往，则可能会被防火墙程序关闭连接，有时我们并不想被关闭连接。例如，对于一个即时通信软件，服务器没有消息时，我们确实不会和服务器有任何数据交换，但是如果连接被关闭，有新的消息到来，我们就再也无法收到消息了，这就违背了"即时通信"的设计要求。

对于情形一中的死链，只要我们此时在任意一端向对端发送一个数据包，即可检测链路是否正常，我们称这类数据包为"心跳包"，称这种操作为"心跳检测"。若一个连接长时间没有正常数据来往，也没有心跳包来往，就可以认为这个连接已不存在。为了节约服务器连接资源，可以通过关闭 socket 来回收连接资源。

情形二中的应用场景要求必须保持客户端与服务器之间的连接正常，这就是我们通常所说的"保活"。如上文所述，当服务器与客户端在一定时间内没有有效的业务数据来往时，我们只需向对端发送心跳包即可实现保活。

根据上面的分析，这里再强调一下，心跳检测一般有两个作用：保活和检测死链。

9.2.1 TCP keepalive 选项

操作系统的 TCP/IP 协议栈提供了 keepalive 选项用于 socket 的保活。在 Linux 上，我们可以通过代码启用一个 socket 的心跳检测（即每隔一定的时间间隔就发送一个心跳检测包给对端）：

```
//on 为 1 时表示打开 keepalive 选项，为 0 时表示关闭, 0 是默认值
int on = 1;
setsockopt(fd, SOL_SOCKET, SO_KEEPALIVE, &on, sizeof(on));
```

但是，keepalive 选项默认发送心跳检测数据包的时间间隔是 7200 秒（两小时），时间间隔实在太长，即使开启它，也不具有实用性。

当然，我们可以通过继续设置 keepalive 相关的三个选项来改变这个时间间隔，它们分别是 TCP_KEEPIDLE、TCP_KEEPINTVL 和 TCP_KEEPCNT，示例代码如下：

```
//发送 keepalive 报文的时间间隔
int val = 7200;
setsockopt(fd, IPPROTO_TCP, TCP_KEEPIDLE, &val, sizeof(val));
```

```
//两次重试报文的时间间隔
int interval = 75;
setsockopt(fd, IPPROTO_TCP, TCP_KEEPINTVL, &interval, sizeof(interval));

int cnt = 9;
setsockopt(fd, IPPROTO_TCP, TCP_KEEPCNT, &cnt, sizeof(cnt));
```

TCP_KEEPIDLE 选项设置了发送 keepalive 报文的时间间隔，发送时如果对端回复 ACK，则本端 TCP 协议栈认为该连接依然存活，继续等 7200 秒后再发送 keepalive 报文；如果对端回复 RESET，则说明对端进程已经重启，本端的应用程序应该关闭该连接。

如果对端没有任何回复，则本端进行重试，如果重试 9 次（TCP_KEEPCNT 值）仍然不可达，前后重试的间隔为 75 秒（TCP_KEEPINTVL 值），则向应用程序返回 ETIMEOUT（无任何应答）或 EHOST 错误信息。

我们可以使用以下命令查看 Linux 上这 3 个值的设置情况：

```
[root@mycentos ~]# sysctl -a | grep keepalive
net.ipv4.tcp_keepalive_intvl = 75
net.ipv4.tcp_keepalive_probes = 9
net.ipv4.tcp_keepalive_time = 7200
```

在 Windows 上设置 keepalive 及设置对应选项的代码略有不同，示例代码如下：

```
//开启 keepalive 选项
char on = 1;
setsockopt(socket, SOL_SOCKET, SO_KEEPALIVE, (char*)&on, sizeof(on));

//设置超时详细信息
DWORD cbBytesReturned;
tcp_keepalive klive;
//启用保活
klive.onoff = 1;
klive.keepalivetime = 7200;
//重试间隔为 10 秒
klive.keepaliveinterval = 1000 * 10;
WSAIoctl(socket, SIO_KEEPALIVE_VALS, &klive, sizeof(tcp_keepalive), NULL, 0,
&cbBytesReturned, NULL, NULL);
```

9.2.2 应用层的心跳包机制设计

我们在使用 keepalive 选项时，需要对每个连接中的 socket 进行设置，而这不一定是必需的，可能产生大量无意义的带宽浪费，而且 keepalive 选项不能与应用层很好地交互，因此在实际的服务开发中，还是建议在应用层设计自己的心跳包机制，如何设计呢？

从技术上来讲，心跳包其实就是一个预先规定好格式的数据包，在程序中启动一个定时器定时发送即可，这是最简单的实现思路。但是，若通信的两端有频繁的数据来往，什么时候到了下一个发心跳包的时间点，就什么时候发送一个心跳包，这其实是对流量的浪费。既然通信双方不断地有正常的业务数据包来往，那么这些数据包本身可以起到保活作用，为什么还要浪费流量发送这些心跳包呢？所以，对于用于保活的心跳包，最佳做法是记录最近一次收发数据包的时间,在每次收数据和发数据时都更新这个时间。而心跳检测计时器在每次检测时都将这个时间与当前系统时间做比较,如果时间间隔大于允许的最大时间间隔(在实际开发中根据需求,将其设置成 15~45 秒不等),则发送一次心跳包。总而言之,就是在与对端之间没有数据来往达到一定时间间隔时,才发送一次心跳包。

发送心跳包的伪代码如下:

```
bool CIUSocket::Send()
{
    int nSentBytes = 0;
    int nRet = 0;
    while (true)
    {
        nRet = ::send(m_hSocket, m_strSendBuf.c_str(), m_strSendBuf.length(), 0);
        if (nRet == SOCKET_ERROR)
        {
            if (::WSAGetLastError() == WSAEWOULDBLOCK)
                break;
            else
            {
                LOG_ERROR("Send data error, disconnect server:%s, port:%d.",
m_strServer.c_str(), m_nPort);
                Close();
                return false;
            }
        }
        else if (nRet < 1)
        {
            //一旦出现错误,就立刻关闭 socket
            LOG_ERROR("Send data error, disconnect server:%s, port:%d.",
m_strServer.c_str(), m_nPort);
            Close();
            return false;
        }

        m_strSendBuf.erase(0, nRet);
        if (m_strSendBuf.empty())
            break;

        ::Sleep(1);
    }

    {
        //记录最近一次发送数据包的时间
        std::lock_guard<std::mutex> guard(m_mutexLastDataTime);
        m_nLastDataTime = (long)time(NULL);
    }

    return true;
}


bool CIUSocket::Recv()
{
```

```cpp
    int nRet = 0;
    char buff[10 * 1024];
    while (true)
    {
        nRet = ::recv(m_hSocket, buff, 10 * 1024, 0);
        if (nRet == SOCKET_ERROR)
        {
            if (::WSAGetLastError() == WSAEWOULDBLOCK)
                break;
            else
            {
                //一旦出现错误，就立刻关闭socket
                LOG_ERROR("Recv data error, errorNO=%d.", ::WSAGetLastError());
                Close();
                return false;
            }
        }
        else if (nRet < 1)
        {
            LOG_ERROR("Recv data error, errorNO=%d.", ::WSAGetLastError());
            Close();
            return false;
        }

        m_strRecvBuf.append(buff, nRet);

        ::Sleep(1);
    }

    {
        std::lock_guard<std::mutex> guard(m_mutexLastDataTime);
        //记录最近一次收取数据包的时间
        m_nLastDataTime = (long)time(NULL);
    }

    return true;
}

void CIUSocket::RecvThreadProc()
{
    LOG_INFO("Recv data thread start...");

    int nRet;
    //上网方式
    DWORD   dwFlags;
    BOOL    bAlive;
    while (!m_bStop)
    {
        //若检测到数据，则收数据
        nRet = CheckReceivedData();
        //出错
        if (nRet == -1)
        {
            m_pRecvMsgThread->NotifyNetError();
```

```
        }
        //无数据
        else if (nRet == 0)
        {
            long nLastDataTime = 0;
            {
                std::lock_guard<std::mutex> guard(m_mutexLastDataTime);
                nLastDataTime = m_nLastDataTime;
            }

            if (m_nHeartbeatInterval > 0)
            {
                //若当前系统时间与上一次收发数据包的时间间隔超过了m_nHeartbeatInterval
                //则发送一次心跳包
                if ((long)time(NULL) - nLastDataTime >= m_nHeartbeatInterval)
                    SendHeartbeatPackage();
            }
        }
        //有数据
        else if (nRet == 1)
        {
            if (!Recv())
            {
                m_pRecvMsgThread->NotifyNetError();
                continue;
            }

            DecodePackages();
        }//end if
    }//end while-loop

    LOG_INFO("Recv data thread finish...");
}
```

同理，检测心跳包的一端，应该是在与对端没有数据来往达到一定时间间隔时才做一次心跳检测。

做心跳检测的一端的伪代码如下：

```
void BusinessSession::send(const char* pData, int dataLength)
{
    bool sent = TcpSession::send(pData, dataLength);

    //发送完数据后更新发送数据包的时间
    updateHeartbeatTime();
}

void BusinessSession::handlePackge(char* pMsg, int msgLength, bool& closeSession,
std::vector<std::string>& vectorResponse)
{
    //对数据的合法性进行校验
    if (pMsg == NULL || pMsg[0] == 0 || msgLength <= 0 || msgLength > MAX_DATA_LENGTH)
    {
        //非法刺探请求，不做任何应答，直接关闭连接
        closeSession = true;
```

```cpp
        return;
    }

    //更新收取数据包的时间
    updateHeartbeatTime();

    //省略数据包解析代码
}

void BusinessSession::updateHeartbeatTime()
{
    std::lock_guard<std::mutex> scoped_guard(m_mutexForlastPackageTime);
    m_lastPackageTime = (int64_t)time(nullptr);
}

bool BusinessSession::doHeartbeatCheck()
{
    const Config& cfg = Singleton<Config>::Instance();
    int64_t now = (int64_t)time(nullptr);

    std::lock_guard<std::mutex> lock_guard(m_mutexForlastPackageTime);
    if (now - m_lastPackageTime >= cfg.m_nMaxClientDataInterval)
    {
        //心跳包检测，超时，关闭连接
        LOGE("heartbeat expired, close session");
        shutdown();
        return true;
    }

    return false;
}

void TcpServer::checkSessionHeartbeat()
{
    int64_t now = (int64_t)time(nullptr);
    if (now - m_nLastCheckHeartbeatTime >= m_nHeartbeatCheckInterval)
    {
        m_spSessionManager->checkSessionHeartbeat();
        m_nLastCheckHeartbeatTime = (int64_t)time(nullptr);
    }
}

void SessionManager::checkSessionHeartbeat()
{
    std::lock_guard<std::mutex> scoped_lock(m_mutexForSession);
    for (const auto& iter : m_mapSessions)
    {
        //这里调用 BusinessSession::doHeartbeatCheck()
        iter.second->doHeartbeatCheck();
    }
}
```

一般是客户端主动向服务端发送心跳包，服务端做心跳检测来决定是否断开连接。从客户端的角度来说，客户端为了让自己得到服务端的正常服务，有必要主动和服务端保持正常的连接状态,而服务端不会局限于某个特定的客户端。如果客户端不能主动和其保持连接,就会主动回收与该客户端的连接。当然,服务端在收到客户端的心跳包时,应该给客户端一个心跳应答。

9.2.3 有代理的心跳包机制设计

前面设计了通用的心跳包机制,但是该机制在一种情况下不适用,请看下图。

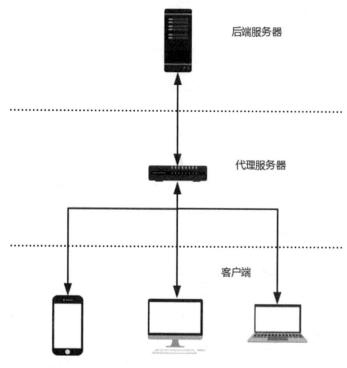

在我们的服务程序与客户端之间存在代理服务时,后端服务与代理服务之间是长连接,代理服务与客户端之间也是长连接,且后端服务器的业务类型是订阅类型,也就是说客户端一旦订阅某个类型的主题,就很少或者不再向服务器发送消息了,而后端服务会不断地将客户端订阅的特定主题数据下发给客户端(如股票交易中的行情服务)。

如果后端服务更新心跳包时间戳时使用了最后一次的上行数据或者下行数据的发包时间,则可能无法检测到客户端是否已经断开。因为在某一时刻,客户端和代理服务之间的网络连接可能已经断开,但代理服务与后端服务之间的连接是正常的,这很常见。对于大多数企业应用,后端服务和代理服务一般位于同一个内网环境下,二者之间的网络状态通常很好。既然后端服务与代理服务之间的连接正常,那么后端服务的下行数据会一直畅通,但此时客户端与代理服务可能已经断开好一会儿了,由于后端服务通过下行数据更新了最后一次心跳时间,所以会导致心跳检测机制误判后端服务与客户端的连接状态。在这种情形下,如果分析后端服务日志,就会发现后端一直有正常发包记录,而从客户端日志(或者从客户端的表现)来看,客户端的这路连接早已断开,在断开的这段时间内已经收不到服务器的数据包了，示意图如下。

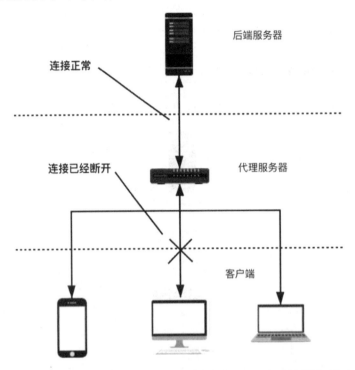

因此，我们应该将更新数据包的时间戳机制改为只通过后端服务的上行数据来统计，这样超过某段时间后若仍然没有上行数据，则说明客户端已经断开（在这段时间内，服务端既没有收到客户端的业务数据包，也没有收到客户端的心跳数据包）。

9.2.4 带业务数据的心跳包

上面介绍的心跳包是从纯技术角度来说的，在实际应用中，我们有时需要定时或者不定时地从服务端更新一些数据，我们可以把这类数据放在心跳包中，定时或者不定时更新。这类带业务数据的心跳包就不仅起到技术上的作用（保活和检测死链）了，也起到传递有效业务数据的作用，实现起来也很容易：在心跳包数据结构中加上需要的业务字段信息，然后在定时器中定时发送即可。

9.2.5 心跳包与流量

在通常情况下，与服务端保持连接的多个客户端中，同一时间段的活跃用户（这里指的是与服务器有频繁数据往来的客户端）一般不会太多。当连接数较多时，进出服务器程序的数据包通常是心跳包（为了保活），所以为了减轻网络带宽的压力和节省流量，尤其是针对一些 3G 或 4G 手机应用，我们在设计心跳包数据格式时应该尽量减小心跳包的数据大小。

9.2.6 心跳包与调试

如前文所述，对于心跳包，服务端的逻辑一般是在一定的时间间隔内没有收到客户端心跳包时主动断开连接。在开发和调试程序的过程中，我们可能需要将程序通过断点中断下来，这个过程可能是几秒到几十秒不等。等程序恢复执行时，连接可能因为心跳检测逻辑已经断开。在调试过程中，我们更多地关注业务数据处理逻辑是否正确，不想被一堆无意义的心跳包数据干扰。

鉴于以上原因，我们一般在调试模式下通过配置开关变量或者条件编译选项关闭心跳包检测逻辑。代码示例如下。

（1）通过开关变量控制心跳包检测逻辑开启：

```
if (config.heartbeatCheckEnabled)
{
   EnableHearbeatCheck();
}
```

（2）通过条件编译控制心跳包检测逻辑开启：

```
//这里设置了在非调试模式下才开启心跳包检测功能
#ifndef _DEBUG
   EnableHearbeatCheck();
#endif
```

9.2.7 心跳包与日志

在实际生产环境下，我们一般会将程序收到的和发出去的数据包写入日志中，但无业务信息的心跳包信息是个例外，一般会刻意地不将其写入日志中。这是因为心跳包数据一般较多，如果写入日志，则会导致日志文件变得很大，且充斥大量无意义的心跳包日志，所以一般在写日志时屏蔽心跳包信息的写入。

这里的建议是，可以将心跳包信息是否写入日志做成一个配置开关，一般处于关闭状态，在需要时再开启。例如，对于一个 WebSocket 服务，ping 和 pong 是心跳包数据，通过下面的示例代码可以按需输出心跳日志信息：

```
void BusinessSession::send(std::string_view strResponse)
{
   bool success = WebSocketSession::send(strResponse);
   if (success)
   {
      bool enablePingPongLog =
Singleton<Config>::Instance().m_bPingPongLogEnabled;

      //对其他消息正常打印，对心跳消息则根据 enablePingPongLog 变量的值按需打印
      if (strResponse != "pong" || enablePingPongLog)
      {
         LOGI("msg sent to client [%s], sessionId: %s, clientId: %s, accountId: %s, frontId: %s, msg: %s",
            getClientInfo(), m_strSessionId.c_str(), m_strClientID.c_str(),
m_strAccountID.c_str(), BusinessSession::m_strFrontId.c_str(),
```

```
strResponse.data());
        }
    }
}
```

需要说明的是，以上示例代码虽然使用 C++编写而成，但是本节介绍的心跳包机制设计思路和注意事项是有普适性的，同样适用于其他编程语言。

9.3 日志模块的设计

日志模块是应用程序的一个重要组件，本节将详细介绍设计日志模块时需要考虑的方方面面。

9.3.1 为什么需要日志

对于生产环境下的服务器或者产品，一般不允许开发人员直接调用调试器排查问题，这时，我们可以通过打印日志，将当时的程序行为上下文现场记录下来，然后从日志系统中找到某次不正常的行为的上下文信息。

本节将从技术和业务两个方面来介绍日志系统相关的设计与开发，即：如何从程序开发角度设计一款功能强大、性能优越、使用方便的日志系统；使用日志系统时应该记录哪些行为和数据，以做到既简洁、不啰嗦，又能方便、快捷地定位问题。

9.3.2 日志系统的技术实现

日志的最初原型是将程序运行的状态打印出来，对于 C/C++来说，就是利用 printf、std::cout 等控制台输出函数，将日志信息输出到控制台。

对于商业项目，为了方便排查问题，我们一般不将日志输出到控制台，而是输出到文件或者数据库系统中。不管输出到哪里，其基本思路都是相同的，这里以写文件为例进行详细介绍。

1. 同步写日志

同步写日志指在输出日志的地方将日志即时写入文件中，被广泛应用于相当多的客户端软件中。之所以使用这种方式，主要是因为它设计简单而且不影响使用体验。有的读者可能会有顾虑：客户端软件一般存在于界面上，而界面部分所属的逻辑就是程序的主线程，如果采用这种同步写日志方式，若写日志时的写文件是磁盘 I/O 操作，其他部分是 CPU 操作，则前者要慢很多，势必造成 CPU 等待且主线程卡在写文件处，使界面卡顿，使用体验不好。这种顾虑确实是存在的，但是我们一般不用担心，主要有两个原因。

◎ 对于客户端程序，即使在主线程（UI 线程）中同步写文件，其单次或者几次磁盘操作累加时间，与人（用户）的可感知时间相比，也是非常小的，也就是说用户根本感觉不到这种同步写文件造成的延迟。当然，如果在 UI 线程里面写日志，尤其是在一些高频操作中（如 Windows 的界面绘制消息 WM_PAINT 处理逻辑中），一定要控制写日志的长度和次数，否则会因为频繁写文件或一次写入数据过多造成界面卡顿。

◎ 客户端程序除了 UI 线程，还有与界面无关的其他工作线程，在这些线程中直接写文件一般不会对使用体验产生影响。

下面给一个具体的例子。

日志类的.h 文件：

```cpp
/**
 * IULog.h
 * zhangyl 2014.12.25
 */
#ifndef __LOG_H__
#define __LOG_H__

enum LOG_LEVEL
{
    LOG_LEVEL_INFO,
    LOG_LEVEL_WARNING,
    LOG_LEVEL_ERROR
};

//注意：如果在打印的日志信息中有中文，则要对格式化字符串用_T()宏包裹起来
#define LOG_INFO(...)     CIULog::Log(LOG_LEVEL_INFO, __FUNCSIG__, __LINE__, __VA_ARGS__)
#define LOG_WARNING(...)  CIULog::Log(LOG_LEVEL_WARNING, __FUNCSIG__, __LINE__, __VA_ARGS__)
#define LOG_ERROR(...)    CIULog::Log(LOG_LEVEL_ERROR, __FUNCSIG__, __LINE__, __VA_ARGS__)

class CIULog
{
public:
    static bool Init(bool bToFile, bool bTruncateLongLog, PCTSTR pszLogFileName);
    static void Uninit();

    static void SetLevel(LOG_LEVEL nLevel);

    //不输出线程 ID 号及所在函数的签名和行号
    static bool Log(long nLevel, PCTSTR pszFmt, ...);
    //输出线程 ID 号及所在函数的签名和行号，支持 ASCII 字符串
    static bool Log(long nLevel, PCSTR pszFunctionSig, int nLineNo, PCTSTR pszFmt, ...);
    //输出线程 ID 号及所在函数的签名和行号，支持 Unicode 字符串
    static bool Log(long nLevel, PCSTR pszFunctionSig, int nLineNo, PCTSTR pszFmt, ...);
private:
    CIULog() = delete;
    ~CIULog() = delete;

    CIULog(const CIULog& rhs) = delete;
    CIULog& operator=(const CIULog& rhs) = delete;
```

```cpp
    static void GetTime(char* pszTime, int nTimeStrLength);

private:
    //将日志写入文件还是控制台
    static bool             m_bToFile;
    static HANDLE           m_hLogFile;
    //是否截断长日志
    static bool             m_bTruncateLongLog;
    //日志级别
    static LOG_LEVEL        m_nLogLevel;
};

#endif //!__LOG_H__
```

日志的 cpp 文件如下：

```cpp
/**
 * IULog.cpp
 * zhangyl 2014.12.25
 */
#include "stdafx.h"
#include "IULog.h"
#include "EncodingUtil.h"
#include <tchar.h>

#ifndef LOG_OUTPUT
#define LOG_OUTPUT
#endif

#define MAX_LINE_LENGTH 256

bool CIULog::m_bToFile = false;
bool CIULog::m_bTruncateLongLog = false;
HANDLE CIULog::m_hLogFile = INVALID_HANDLE_VALUE;
LOG_LEVEL CIULog::m_nLogLevel = LOG_LEVEL_INFO;

bool CIULog::Init(bool bToFile, bool bTruncateLongLog, PCTSTR pszLogFileName)
{
#ifdef LOG_OUTPUT
    m_bToFile = bToFile;
    m_bTruncateLongLog = bTruncateLongLog;

    if (pszLogFileName == NULL || pszLogFileName[0] == NULL)
        return FALSE;

    TCHAR szHomePath[MAX_PATH] = {0};
    ::GetModuleFileName(NULL, szHomePath, MAX_PATH);
    for (int i = _tcslen(szHomePath); i >= 0; --i)
    {
        if (szHomePath[i] == _T('\\'))
        {
            szHomePath[i] = _T('\0');
            break;
        }
    }
```

```cpp
    TCHAR szLogDirectory[MAX_PATH] = { 0 };
    _stprintf_s(szLogDirectory, _T("%s\\Logs\\"), szHomePath);

    DWORD dwAttr = ::GetFileAttributes(szLogDirectory);
    if (!((dwAttr != 0xFFFFFFFF) && (dwAttr & FILE_ATTRIBUTE_DIRECTORY)))
    {
        TCHAR cPath[MAX_PATH] = { 0 };
        TCHAR cTmpPath[MAX_PATH] = { 0 };
        TCHAR* lpPos = NULL;
        TCHAR cTmp = _T('\0');

        _tcsncpy_s(cPath, szLogDirectory, MAX_PATH);

        for (int i = 0; i < (int)_tcslen(cPath); i++)
        {
            if (_T('\\') == cPath[i])
                cPath[i] = _T('/');
        }

        lpPos = _tcschr(cPath, _T('/'));
        while (lpPos != NULL)
        {
            if (lpPos == cPath)
            {
                lpPos++;
            }
            else
            {
                cTmp = *lpPos;
                *lpPos = _T('\0');
                _tcsncpy_s(cTmpPath, cPath, MAX_PATH);
                ::CreateDirectory(cTmpPath, NULL);
                *lpPos = cTmp;
                lpPos++;
            }
            lpPos = _tcschr(lpPos, _T('/'));
        }
    }

    m_hLogFile = ::CreateFile(pszLogFileName, GENERIC_READ | GENERIC_WRITE,
FILE_SHARE_READ, NULL, CREATE_NEW, FILE_ATTRIBUTE_NORMAL, NULL);
    if (m_hLogFile == INVALID_HANDLE_VALUE)
        return false;

#endif //end LOG_OUTPUT

    return true;
}

void CIULog::Uninit()
{
#ifdef LOG_OUTPUT
    if(m_hLogFile != INVALID_HANDLE_VALUE)
    {
```

```cpp
            ::CloseHandle(m_hLogFile);
            m_hLogFile = INVALID_HANDLE_VALUE;
        }
#endif //end LOG_OUTPUT
}

void CIULog::SetLevel(LOG_LEVEL nLevel)
{
    m_nLogLevel = nLevel;
}

bool CIULog::Log(long nLevel, PCTSTR pszFmt, ...)
{
#ifdef LOG_OUTPUT
    if (nLevel < m_nLogLevel)
        return false;

    char szTime[64] = { 0 };
    GetTime(szTime,ARRAYSIZE(szTime));
    std::string strDebugInfo(szTime);

    std::string strLevel("[INFO]");
    if (nLevel == LOG_LEVEL_WARNING)
        strLevel = "[Warning]";
    else if (nLevel == LOG_LEVEL_ERROR)
        strLevel = "[Error]";

    strDebugInfo += strLevel;

    //当前线程信息
    char szThreadID[32] = { 0 };
    DWORD dwThreadID = ::GetCurrentThreadId();
    sprintf_s(szThreadID, ARRAYSIZE(szThreadID), "[ThreadID: %u]", dwThreadID);
    strDebugInfo += szThreadID;

    //log 正文
    std::wstring strLogMsg;
    va_list ap;
    va_start(ap, pszFmt);
    int nLogMsgLength = _vsctprintf(pszFmt, ap);
    //容量必须算上最后一个\0
    if ((int)strLogMsg.capacity() < nLogMsgLength + 1)
    {
        strLogMsg.resize(nLogMsgLength + 1);
    }
    _vstprintf_s((TCHAR*)strLogMsg.data(), strLogMsg.capacity(), pszFmt, ap);
    va_end(ap);

    //string 内容正确但 length 不对，恢复其 length
    std::wstring strMsgFormal;
    strMsgFormal.append(strLogMsg.c_str(), nLogMsgLength);

    //如果日志开启截断，则长日志只取前 MAX_LINE_LENGTH 个字符
    if (m_bTruncateLongLog)
        strMsgFormal = strMsgFormal.substr(0, MAX_LINE_LENGTH);
```

```cpp
    std::string strLogMsgAscii;
    strLogMsgAscii = EncodeUtil::UnicodeToAnsi(strMsgFormal);

    strDebugInfo += strLogMsgAscii;
    strDebugInfo += "\r\n";

    if(m_bToFile)
    {
        if(m_hLogFile == INVALID_HANDLE_VALUE)
            return false;

        ::SetFilePointer(m_hLogFile, 0, NULL, FILE_END);
        DWORD dwBytesWritten = 0;
        ::WriteFile(m_hLogFile, strDebugInfo.c_str(), strDebugInfo.length(),
&dwBytesWritten, NULL);
        ::FlushFileBuffers(m_hLogFile);
        return true;
    }

    ::OutputDebugStringA(strDebugInfo.c_str());

#endif //end LOG_OUTPUT

    return true;
}
bool CIULog::Log(long nLevel, PCSTR pszFunctionSig, int nLineNo, PCTSTR pszFmt, ...)
{
#ifdef LOG_OUTPUT
    if (nLevel < m_nLogLevel)
        return false;

    //时间
    char szTime[64] = { 0 };
    GetTime(szTime, ARRAYSIZE(szTime));
    std::string strDebugInfo(szTime);

    //错误级别
    std::string strLevel("[INFO]");
    if (nLevel == LOG_LEVEL_WARNING)
        strLevel = "[Warning]";
    else if (nLevel == LOG_LEVEL_ERROR)
        strLevel = "[Error]";

    strDebugInfo += strLevel;

    //当前线程信息
    char szThreadID[32] = {0};
    DWORD dwThreadID = ::GetCurrentThreadId();
    sprintf_s(szThreadID, ARRAYSIZE(szThreadID), "[ThreadID: %u]", dwThreadID);
    strDebugInfo += szThreadID;

    //函数签名
    char szFuncSig[512] = { 0 };
```

```cpp
    sprintf_s(szFuncSig, "[%s:%d]", pszFunctionSig, nLineNo);
    strDebugInfo += szFuncSig;

    //log 正文
    std::wstring strLogMsg;
    va_list ap;
    va_start(ap, pszFmt);
    int nLogMsgLength = _vsctprintf(pszFmt, ap);
    //容量必须算上最后一个\0
    if ((int)strLogMsg.capacity() < nLogMsgLength + 1)
    {
        strLogMsg.resize(nLogMsgLength + 1);
    }
    _vstprintf_s((TCHAR*)strLogMsg.data(), strLogMsg.capacity(), pszFmt, ap);
    va_end(ap);

    //string 内容正确但 length 不对，恢复其 length
    std::wstring strMsgFormal;
    strMsgFormal.append(strLogMsg.c_str(), nLogMsgLength);

    //如果日志开启截断，则长日志只取前 MAX_LINE_LENGTH 个字符
    if (m_bTruncateLongLog)
        strMsgFormal = strMsgFormal.substr(0, MAX_LINE_LENGTH);

    std::string strLogMsgAscii;
    strLogMsgAscii = EncodeUtil::UnicodeToAnsi(strMsgFormal);

    strDebugInfo += strLogMsgAscii;
    strDebugInfo += "\r\n";

    if(m_bToFile)
    {
        if(m_hLogFile == INVALID_HANDLE_VALUE)
            return false;

        ::SetFilePointer(m_hLogFile, 0, NULL, FILE_END);
        DWORD dwBytesWritten = 0;
        ::WriteFile(m_hLogFile, strDebugInfo.c_str(), strDebugInfo.length(), &dwBytesWritten, NULL);
        ::FlushFileBuffers(m_hLogFile);
        return true;
    }

    ::OutputDebugStringA(strDebugInfo.c_str());

#endif //end LOG_OUTPUT

    return true;
}

bool CIULog::Log(long nLevel, PCSTR pszFunctionSig, int nLineNo, PCSTR pszFmt, ...)
{
#ifdef LOG_OUTPUT
    if (nLevel < m_nLogLevel)
        return false;
```

```cpp
//时间
char szTime[64] = { 0 };
GetTime(szTime, ARRAYSIZE(szTime));
std::string strDebugInfo(szTime);

//错误级别
std::string strLevel("[INFO]");
if (nLevel == LOG_LEVEL_WARNING)
    strLevel = "[Warning]";
else if (nLevel == LOG_LEVEL_ERROR)
    strLevel = "[Error]";

strDebugInfo += strLevel;

//当前线程信息
char szThreadID[32] = {0};
DWORD dwThreadID = ::GetCurrentThreadId();
sprintf_s(szThreadID, ARRAYSIZE(szThreadID), "[ThreadID: %u]", dwThreadID);
strDebugInfo += szThreadID;

//函数签名
char szFuncSig[512] = { 0 };
sprintf_s(szFuncSig, "[%s:%d]", pszFunctionSig, nLineNo);
strDebugInfo += szFuncSig;

//日志正文
std::string strLogMsg;
va_list ap;
va_start(ap, pszFmt);
int nLogMsgLength = _vscprintf(pszFmt, ap);
//容量必须算上最后一个\0
if ((int)strLogMsg.capacity() < nLogMsgLength + 1)
{
    strLogMsg.resize(nLogMsgLength + 1);
}
vsprintf_s((char*)strLogMsg.data(), strLogMsg.capacity(), pszFmt, ap);
va_end(ap);

//string 内容正确但 length 不对，恢复其 length
std::string strMsgFormal;
strMsgFormal.append(strLogMsg.c_str(), nLogMsgLength);

//如果日志开启截断，则长日志只取前 MAX_LINE_LENGTH 个字符
if (m_bTruncateLongLog)
    strMsgFormal = strMsgFormal.substr(0, MAX_LINE_LENGTH);

strDebugInfo += strMsgFormal;
strDebugInfo += "\r\n";

if(m_bToFile)
{
    if(m_hLogFile == INVALID_HANDLE_VALUE)
        return false;
```

```
            ::SetFilePointer(m_hLogFile, 0, NULL, FILE_END);
            DWORD dwBytesWritten = 0;
            ::WriteFile(m_hLogFile, strDebugInfo.c_str(), strDebugInfo.length(),
&dwBytesWritten, NULL);
            ::FlushFileBuffers(m_hLogFile);
            return true;
    }

    ::OutputDebugStringA(strDebugInfo.c_str());

#endif //end LOG_OUTPUT

    return true;
}

void CIULog::GetTime(char* pszTime, int nTimeStrLength)
{
    SYSTEMTIME st = {0};
    ::GetLocalTime(&st);
    sprintf_s(pszTime, nTimeStrLength, "[%04d-%02d-%02d %02d:%02d:%02d:%04d]",
st.wYear, st.wMonth, st.wDay, st.wHour, st.wMinute, st.wSecond, st.wMilliseconds);
}
```

在以上代码中根据日志级别定义了3个宏：LOG_INFO、LOG_WARNING、LOG_ERROR，要使用该日志模块，则只需在程序启动处调用CIULog::Init函数初始化日志：

```
SYSTEMTIME st = {0};
::GetLocalTime(&st);
TCHAR szLogFileName[MAX_PATH] = {0};
_stprintf_s(szLogFileName, MAX_PATH, _T("%s\\Logs\\%04d%02d%02d%02d%02d%02d.log"),
g_szHomePath, st.wYear, st.wMonth, st.wDay, st.wHour, st.wMinute, st.wSecond);
CIULog::Init(true, false, szLogFileName);
```

当然，最佳做法是在程序退出处调用CIULog::Uninit回收日志模块相关的资源：

```
CIULog::Uninit();
```

做好这些准备工作以后，如果想在程序的某个地方写一条日志，则只需这样写：

```
//打印一条INFO级别的日志
LOG_INFO("Request logon: Account=%s, Password=*****, Status=%d, LoginType=%d.",
pLoginRequest->m_szAccountName, pLoginRequest->m_szPassword,
pLoginRequest->m_nStatus, (long)pLoginRequest->m_nLoginType);
//打印一条WARNING级别的日志
LOG_WARN("Some warning...");
//打印一条ERROR级别的日志
LOG_ERROR("Recv data error, errorNO=%d.", ::WSAGetLastError());
```

CIULog这个日志模块类实际运行的输出效果如下：

```
[2018-11-09 23:52:54:0826][INFO][ThreadID: 7252][bool __thiscall
CIUSocket::Login(const char *,const char *,int,int,int,class
std::basic_string<char,struct std::char_traits<char>,class std::allocator<char> >
&):1107]Request logon: Account=zhangy, Password=*****, Status=76283204,
LoginType=1.
[2018-11-09 23:52:56:0352][INFO][ThreadID: 5828][void __thiscall
CIUSocket::SendThreadProc(void):794]Recv data thread start...
[2018-11-09 23:52:56:0385][INFO][ThreadID: 6032][void __thiscall
```

```
CSendMsgThread::HandleUserBasicInfo(const class CUserBasicInfoRequest
*):298]Request to get userinfo.
[2018-11-09 23:52:56:0355][INFO][ThreadID: 7140][void __thiscall
CIUSocket::RecvThreadProc(void):842]Recv data thread start...
[2018-11-09 23:52:57:0254][INFO][ThreadID: 7220][int __thiscall
CRecvMsgThread::HandleFriendListInfo(const class std::basic_string<char,struct
std::char_traits<char>,class std::allocator<char> > &):593]Recv user basic info,
info count=1.
```

1）出现的问题一

从上面的日志输出来看，这种同步的日志输出方式也存在时间顺序不正确的问题（时间戳大的日志比时间戳小的日志靠前）。这是由于多线程同时写日志到同一个文件时，产生日志的时间和实际写入磁盘的时间不是一个原子操作。下图展示了该问题出现的根源。

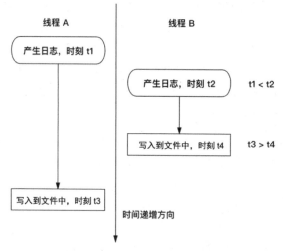

好在这种时间顺序不正确的问题只出现在不同的线程之间，对于同一个线程的不同时间的日志记录顺序肯定是正确的，所以并不影响我们使用日志。

2）出现的问题二

多个线程同时将日志写入同一个日志文件中还存在一个问题，就是假设线程 A 在某个时刻向日志文件中追加的内容为"AAAAA"，线程 B 在同一时刻向日志文件中追加的内容为"BBBBB"，线程 C 在同一时刻向日志文件中追加的内容为"CCCCC"，那么最终的日志文件中的内容会不会出现"AABBCCABCAACCBB"这种格式呢？

在类 UNIX 系统上（包括 Linux），同一个进程内针对同一个 FILE* 的操作是线程安全的，也就是说，在这类操作系统上得到的日志结果 A、B、C 的各个字母组一定是连续的，最终得到的日志内容可能是"AAAAACCCCCBBBBB"或"AAAAABBBBBCCCCC"等连续格式，绝不会出现 A、B、C 字母交错的现象。

而在 Windows 上，对 FILE* 的操作并不是线程安全的。笔者做了大量实验，在 Windows 上也没有出现 A、B、C 字母相间的现象。

这种同步日志的实现方式，一般用于低频写日志的软件系统中（如客户端软件），所以可以认为这种多线程同时写日志到一个文件中是可行的。

2. 异步写日志

当然，对于性能要求不高的应用（如大多数客户端程序、某些并发数量不多的服务），这种同步写日志的实现方式是可以满足要求的。但是，对于 QPS 要求很高或者对性能有一定要求的服务器程序，同步写日志会对服务的关键性逻辑的快速执行和及时响应带来一定的性能损失，因为写日志时等待磁盘 I/O 完成工作也需要一定时间。为了减小这种损失，我们可以采用异步写日志。

与同步写日志相反，异步写日志即在产生日志的地方不会将日志实时写入文件中，而是通过一些线程同步技术将日志暂存下来，然后通过一个或多个专门的日志写入线程将其写入磁盘中，这样的话，原来输出日志的线程就不存在等待写日志到磁盘的效率损耗了。异步写日志机制在本质上就是一个生产者/消费者模型，产生日志的线程是生产者，将日志写入文件的线程是消费者。当然，日志消费者线程数量在这里可以是一个或多个。在实际开发中，如果有多个日志消费线程，则我们又要考虑存在多个线程可能会造成写日志的时间顺序错位（将时间较晚的日志写在时间较早的日志前面）。所以，一般如无特殊要求，建议将日志消费者线程数量设置为 1。

总结起来，我们可以认为在异步写日志的逻辑中一般存在一组专门写日志的线程（一个或多个），程序的其他线程为这些日志线程生产日志。

至于其他线程如何将产生的日志交给日志线程，就是多线程之间资源同步的问题了。我们可以使用一个队列来存储其他线程产生的日志，日志线程从该队列中取出日志，然后将日志内容写入文件。最简单的方式是在日志生产线程将每次产生的日志信息都放入一个队列或日志写入线程从队列中取出日志时，都使用一个互斥体（mutex）将其保护起来。代码示例如下（C++ 11 代码）：

```cpp
/**
 * AsyncLogger.cpp
 * zhangyl 2018.11.10
 */
#include "stdafx.h"
#include <thread>
#include <mutex>
#include <list>
#include <string>
#include <sstream>
#include <iostream>

//保护队列的互斥体
std::mutex log_mutex;
std::list<std::string> cached_logs;
FILE* log_file = NULL;

bool init_log_file()
{
    //以追加模式写入文件，如果文件不存在，则创建
```

```cpp
    log_file = fopen("my.log", "a+");
    return log_file != NULL;
}

void uninit_log_file()
{
    if (log_file != NULL)
        fclose(log_file);
}

bool write_log_tofile(const std::string& line)
{
    if (log_file == NULL)
        return false;

    if (fwrite((void*)line.c_str(), 1, line.length(), log_file) != line.length())
        return false;

    //将日志立即冲刷到文件中
    fflush(log_file);

    return true;
}

void log_producer()
{
    int index = 0;
    while (true)
    {
        ++ index;
        std::ostringstream os;
        os << "This is log, index: " << index << ", producer threadID: " << std::this_thread::get_id() << "\n";
        //使用花括号括起来是为了减小锁的粒度
        {
            std::lock_guard<std::mutex> lock(log_mutex);
            cached_logs.emplace_back(os.str());
        }

        std::chrono::milliseconds duration(100);
        std::this_thread::sleep_for(duration);
    }
}

void log_consumer()
{
    std::string line;
    while (true)
    {
        //使用花括号括起来是为了减小锁的粒度
        {
            std::lock_guard<std::mutex> lock(log_mutex);
            if (!cached_logs.empty())
            {
                line = cached_logs.front();
```

```cpp
            cached_logs.pop_front();
        }
    }

    if (line.empty())
    {
        std::chrono::milliseconds duration(1000);
        std::this_thread::sleep_for(duration);

        continue;
    }

    write_log_tofile(line);

    line.clear();
    }
}

int main(int argc, char* argv[])
{
    if (!init_log_file())
    {
        std::cout << "init log file error." << std::endl;
        return -1;
    }

    std::thread log_producer1(log_producer);
    std::thread log_producer2(log_producer);
    std::thread log_producer3(log_producer);

    std::thread log_consumer1(log_consumer);
    std::thread log_consumer2(log_consumer);
    std::thread log_consumer3(log_consumer);

    log_producer1.join();
    log_producer2.join();
    log_producer3.join();

    log_consumer1.join();
    log_consumer2.join();
    log_consumer3.join();

    uninit_log_file();

    return 0;
}
```

以上代码分别模拟了 3 个生产日志的线程（log_producer1～3）和 3 个消费日志的线程（log_consumer1～3）。当然，可以继续优化以上代码，如果在当前缓存队列中没有日志记录，那么消费日志线程会做无用功。

1）优化方法一

可以使用条件变量，如果在当前队列中没有日志记录，就将日志消费线程挂起；在生产日志的线程产生新的日志后,置信(signal)条件变量,这样日志消费线程会被唤醒,以将日志从队列中取出来并写入文件中。代码如下:

```cpp
/**
 *AsyncLoggerLinux.cpp
 *zhangyl 2018.11.10
 */
#include "stdafx.h"
#include <thread>
#include <mutex>
#include <condition_variable>
#include <list>
#include <string>
#include <sstream>
#include <iostream>

std::mutex log_mutex;
std::condition_variable log_cv;
std::list<std::string> cached_logs;
FILE* log_file = NULL;

//其他逻辑与上面的相同,为了节省篇幅,省略之

void log_producer()
{
    int index = 0;
    while (true)
    {
        ++ index;
        std::ostringstream os;
        os << "This is log, index: " << index << ", producer threadID: " << std::this_thread::get_id() << "\n";
        //使用花括号括起来是为了减小锁的粒度
        {
            std::lock_guard<std::mutex> lock(log_mutex);
            cached_logs.emplace_back(os.str());
            log_cv.notify_one();
        }

        std::chrono::milliseconds duration(100);
        std::this_thread::sleep_for(duration);
    }
}

void log_consumer()
{
    std::string line;
    while (true)
    {
        //使用花括号括起来是为了减小锁的粒度
        {
            std::unique_lock<std::mutex> lock(log_mutex);
            while (cached_logs.empty())
            {
```

```cpp
            //无限等待
            log_cv.wait(lock);
        }

        line = cached_logs.front();
        cached_logs.pop_front();
    }

    if (line.empty())
    {
        std::chrono::milliseconds duration(1000);
        std::this_thread::sleep_for(duration);

        continue;
    }

    write_log_tofile(line);

    line.clear();
    }
}

int main(int argc, char* argv[])
{
    //其他逻辑与上面的相同，为了节省篇幅，省略之
}
```

2）优化方法二

除了条件变量，我们还可以使用信号量来设计异步日志系统。信号量是带有资源计数的线程同步对象，每产生一条日志，就将信号量资源计数自增1，日志消费线程默认等待这个信号量是否受信，如果受信，则每唤醒一个日志消费线程，信号量字数计数就将自动减1。通俗地说，就是生产者每生产1个资源，就将资源计数加1，而消费者每消费1个资源数量，就将资源计数减1；如果当前资源计数已经为0，则消费者将自动挂起。

由于 C++ 11 没有提供对不同平台的信号量对象的封装，所以这里分别给出 Windows 和 Linux 两个操作系统的实现代码，读者可以根据需要学习其中一个或两个。注意，为了保持代码风格一致，对于线程和读写文件相关的函数，在不同的操作系统上，我们都使用该操作系统相关的 API 接口，不再使用 C++ 11 相关的函数和类库。

Windows 平台的代码如下：

```cpp
/**
 * AsyncLogger.cpp, Windows 版本
 * zhangyl @date: 2018.11.10
 */
#include "stdafx.h"
#include <windows.h>
#include <list>
#include <string>
#include <iostream>
#include <sstream>
```

```cpp
std::list<std::string> cached_logs;
CRITICAL_SECTION g_cs;
HANDLE g_hSemaphore = NULL;
HANDLE g_hLogFile = INVALID_HANDLE_VALUE;

bool Init()
{
    InitializeCriticalSection(&g_cs);

    //假设资源数量上限是 0xFFFFFFFF
    g_hSemaphore = CreateSemaphore(NULL, 0, 0xFFFFFFFF, NULL);

    //如果文件不存在，则创建
    g_hLogFile = CreateFile(_T("my.log"), GENERIC_WRITE, FILE_SHARE_READ, NULL,
CREATE_ALWAYS, FILE_ATTRIBUTE_NORMAL, NULL);
    if (g_hLogFile == INVALID_HANDLE_VALUE)
        return false;

    return true;
}

void Uninit()
{
    DeleteCriticalSection(&g_cs);

    if (g_hSemaphore != NULL)
        CloseHandle(g_hSemaphore);

    if (g_hLogFile != INVALID_HANDLE_VALUE)
        CloseHandle(g_hLogFile);
}

bool WriteLogToFile(const std::string& line)
{
    if (g_hLogFile == INVALID_HANDLE_VALUE)
        return false;

    DWORD dwBytesWritten;
    //对于比较长的日志，应该分段写入
    //因为单次写入可能只能写入部分内容，所以这里为了演示方便，逻辑从简
    if (!WriteFile(g_hLogFile, line.c_str(), line.length(), &dwBytesWritten, NULL)
|| dwBytesWritten != line.length())
        return false;

    //将日志立即冲刷到文件中
    FlushFileBuffers(g_hLogFile);

    return true;
}

DWORD CALLBACK LogProduceThreadProc(LPVOID lpThreadParameter)
{
    int index = 0;
    while (true)
    {
```

```cpp
        ++index;
        std::ostringstream os;
        os << "This is log, index: " << index << ", producer threadID: " << GetCurrentThreadId() << "\n";

        EnterCriticalSection(&g_cs);
        cached_logs.emplace_back(os.str());
        LeaveCriticalSection(&g_cs);

        ReleaseSemaphore(g_hSemaphore, 1, NULL);

        Sleep(100);
    }

    return 0;
}
DWORD CALLBACK LogConsumeThreadProc(LPVOID lpThreadParameter)
{
    std::string line;
    while (true)
    {
        //无限等待
        WaitForSingleObject(g_hSemaphore, INFINITE);

        EnterCriticalSection(&g_cs);
        if (!cached_logs.empty())
        {
            line = cached_logs.front();
            cached_logs.pop_front();
        }
        LeaveCriticalSection(&g_cs);

        if (line.empty())
        {
            Sleep(1000);

            continue;
        }

        WriteLogToFile(line);

        line.clear();
    }
}

int main(int argc, char* argv[])
{
    if (!Init())
    {
        std::cout << "init log file error." << std::endl;
        return -1;
    }

    HANDLE hProducers[3];
```

```
    for (int i = 0; i < sizeof(hProducers) / sizeof(hProducers[0]); ++i)
    {
        hProducers[i] = CreateThread(NULL, 0, LogProduceThreadProc, NULL, 0, NULL);
    }

    HANDLE hConsumers[3];
    for (int i = 0; i < sizeof(hConsumers) / sizeof(hConsumers[0]); ++i)
    {
        hConsumers[i] = CreateThread(NULL, 0, LogConsumeThreadProc, NULL, 0, NULL);
    }

    //等待消费者线程退出
    for (int i = 0; i < sizeof(hProducers) / sizeof(hProducers[0]); ++i)
    {
        WaitForSingleObject(hProducers[i], INFINITE);
    }

    //等待生产者线程退出
    for (int i = 0; i < sizeof(hConsumers) / sizeof(hConsumers[0]); ++i)
    {
        WaitForSingleObject(hConsumers[i], INFINITE);
    }

    Uninit();

    return 0;
}
```

在以上代码中，多线程向队列中增加日志记录和从队列中取出日志记录时使用了 Windows 上的临界区（CRITICAL_SECTION，关键段）对象对队列进行保护。

Linux 平台的代码如下：

```
/**
 *AsyncLogger.cpp，Linux 版本
 *zhangyl 2018.11.10
 */
#include <unistd.h>
#include <list>
#include <stdio.h>
#include <pthread.h>
#include <semaphore.h>
#include <string>
#include <iostream>
#include <sstream>


std::list<std::string> cached_logs;
pthread_mutex_t log_mutex = PTHREAD_MUTEX_INITIALIZER;
sem_t           log_semphore;
FILE* plogfile = NULL;

bool init()
{
    pthread_mutex_init(&log_mutex, NULL);
```

```cpp
    //初始信号量资源数量是0
    sem_init(&log_semphore, 0, 0);

    //如果文件不存在，则创建
    plogfile = fopen("my.log", "a++");

    return plogfile != NULL;
}

void uninit()
{
    pthread_mutex_destroy(&log_mutex);
    sem_destroy(&log_semphore);

    if (plogfile != NULL)
        fclose(plogfile);
}

bool write_log_to_file(const std::string& line)
{
    if (plogfile == NULL)
        return false;

    //对于比较长的日志，应该分段写入，因为单次写入可能只能写入部分内容
    //这里为了演示方便，逻辑从简
    if (fwrite((void*)line.c_str(), 1, line.length(), plogfile) != line.length())
        return false;

    //将日志立即冲刷到文件中
    fflush(plogfile);

    return true;
}

void* producer_thread_proc(void* arg)
{
    int index = 0;
    while (true)
    {
        ++ index;
        std::ostringstream os;
        os << "This is log, index: " << index << ", producer threadID: " << pthread_self() << "\n";

        pthread_mutex_lock(&log_mutex);
        cached_logs.push_back(os.str());
        pthread_mutex_unlock(&log_mutex);

        sem_post(&log_semphore);

        usleep(100000);
    }
}

void* consumer_thread_proc(void* arg)
```

```cpp
{
    std::string line;
    while (true)
    {
        //无限等待
        sem_wait(&log_semphore);

        pthread_mutex_lock(&log_mutex);
        if (!cached_logs.empty())
        {
            line = cached_logs.front();
            cached_logs.pop_front();
        }
        pthread_mutex_unlock(&log_mutex);

        if (line.empty())
        {
            sleep(1);

            continue;
        }

        write_log_to_file(line);

        line.clear();
    }
}

int main(int argc, char* argv[])
{
    if (!init())
    {
        std::cout << "init log file error." << std::endl;
        return -1;
    }

    pthread_t producer_thread_id[3];
    for (int i = 0; i < sizeof(producer_thread_id) / sizeof(producer_thread_id[0]); ++i)
    {
        pthread_create(&producer_thread_id[i], NULL, producer_thread_proc, NULL);
    }

    pthread_t consumer_thread_id[3];
    for (int i = 0; i < sizeof(consumer_thread_id) / sizeof(consumer_thread_id[0]); ++i)
    {
        pthread_create(&consumer_thread_id[i], NULL, consumer_thread_proc, NULL);
    }

    //等待消费者线程退出
    for (int i = 0; i < sizeof(producer_thread_id) / sizeof(producer_thread_id[0]); ++i)
    {
        pthread_join(producer_thread_id[i], NULL);
```

```
    }

    //等待生产者线程退出
    for (int i = 0; i < sizeof(consumer_thread_id) / sizeof(consumer_thread_id[0]);
++i)
    {
        pthread_join(consumer_thread_id[i], NULL);
    }

    uninit();

    return 0;
}
```

我们使用 g++ 编译器编译以上代码，使用以下命令生成可移植性文件 AsyncLoggerLinux：

g++ -g -o AsyncLoggerLinux AsyncLoggerLinux.cpp -lpthread

接着执行生成的 AsyncLoggerLinux 文件，生成的日志效果如下：

```
This is log, index: 1, producer threadID: 140512358795008
This is log, index: 1, producer threadID: 140512367187712
This is log, index: 1, producer threadID: 140512375580416
This is log, index: 2, producer threadID: 140512358795008
This is log, index: 2, producer threadID: 140512367187712
This is log, index: 2, producer threadID: 140512375580416
This is log, index: 3, producer threadID: 140512358795008
This is log, index: 3, producer threadID: 140512367187712
This is log, index: 3, producer threadID: 140512375580416
This is log, index: 4, producer threadID: 140512358795008
This is log, index: 4, producer threadID: 140512367187712
This is log, index: 4, producer threadID: 140512375580416
This is log, index: 5, producer threadID: 140512358795008
This is log, index: 5, producer threadID: 140512367187712
This is log, index: 5, producer threadID: 140512375580416
//省略更多的输出
```

　　如果需要编写同时运行在 Windows 和 Linux 平台的代码，则可以使用信号量。由于 C++ 11 本身并没有提供现成的信号量库，所以我们可以自己利用 std::mutex、std::condition_variable 模拟信号量的功能。若有兴趣，可以自行尝试。

　　以上就是异步写日志的基本原理，在这个原理的基础上可以增加很多特性。下面继续介绍在实际生产环境下设计日志系统的一些常用技巧。

3. 日志的级别

　　常见环境下日志系统的日志级别从高到低一般分为 ERROR、WARNING、INFO，有些系统会有更细粒度的级别，例如在 INFO 下还有 TRACE、VERBOSE，在 ERROR 上还有 FATAL 等。这里将日志按级别从高到低排序：FATAL、ERROR、WARNING、INFO、TRACE、VERBOSE。对各个级别说明如下。

　　（1）FATAL 一般用于标记那些无法让程序继续运行的错误，对于这种级别的日志处理，程序不仅记录一条日志，还会主动"崩溃"或者退出。对于 C++ 程序，这很容易实现，只要向内存地址为 0 处随便写入一个数据即可，操作系统会检测到这是非法操作并终止程序，代码示例如下：

```
int* p = 0;
*p = 0;
```

哪些是 FATAL 级别的呢？可以记住这样一条规则：若程序中的一些关键性逻辑必须是正常且正确执行的，但其没有执行，则出现的错误就是"致命错误"。举个例子，某服务程序提供 8888 端口对外服务，但是程序启动时由于端口占用等原因，无法在这个端口上启动监听，此时程序就没有必要继续跑下去了。在监听端口失败时，可以打印一条 Fatal Error，让程序退出。注意，这类日志级别一般用于程序开发或者灰度发布的早期阶段，便于我们尽早排查程序中存在的逻辑漏洞和错误。在程序稳定运行一段时间后，尤其是发布到生产环境后，就尽量不要在程序中有这样的日志级别了。因为在生产环境下，产品的崩溃尤其是服务器程序的宕机、退出，对有一定用户量的公司造成的损失和影响是非常恶劣的。

（2）ERROR 用得较多，一般用于记录在程序中产生的错误，例如某次请求超时、数据无应答、下载头像失败、写入数据库失败等。

（3）WARNING 用于记录一些程序产生的但不太影响使用的错误。我们可以通过该类型的日志记录进一步优化我们的程序，精益求精。

（4）INFO 级别的日志也很常用，一般用于记录程序运行过程中的各种状态信息、重要事件等。当然，由于这个级别的日志可能写入得非常频繁，所以一般也是写满磁盘的常见原因之一。

（5）TRACE 与 VERBOSE、INFO 类似，一般在调试和测试阶段记录程序非常细粒度的代码执行情况，如果存在这类日志，则写入也会比较频繁。这类级别的日志不是日志系统必需的。

在实际项目中，我们通过设置日志的级别来控制输出哪些级别的日志。举个例子，在开发阶段为了快速定位问题，需要开启 TRACE 和 VERBOSE 级别的日志，在项目发布和生产以后，若有大量的 TRACE 和 VERBOSE 级别的日志，则很容易写满磁盘，同时不利于定位有效信息。我们一般将日志级别设置得更高一些，例如将级别设置成 INFO 甚至更高，这在逻辑上非常容易实现，伪代码如下：

```
enum LOG_LEVEL
{
    VERBOSE  = 0,
    TRACE    = 1,
    INFO     = 2,
    WARNING  = 3,
    ERROR    = 4,
    FATAL    = 5
};

bool writeToLogFile(int level, const std::string& line)
{
```

```
    //当前设置的日志级别大于需要输出的日志级别
    if (m_nCurrentLogLevel > level)
        return false;

    //将一行日志写入文件中

    return true;
}
```

日志的级别一般被写在每行日志内容的前面，方便我们使用过滤工具或命令（如 Linux 的 grep 命令）过滤。例如，某个 my.log 日志文件的内容如下：

```
20181110 12:38:20 INFO   get notify msg cache, userid: 449, m_mapNotifyMsgCache.size():
16, cached size: 0 - MsgCacheManager.cpp:49
20181110 12:38:20 INFO   get chat msg cache, userid: 449, m_listChatMsgCache.size():
3968, cached size: 0 - MsgCacheManager.cpp:81
20181110 12:38:20 INFO  Request from client: userid=449, cmd=1003, seq=0, data=,
datalength=0, buflength=15 - ChatSession.cpp:166
20181110 12:38:20 INFO  Response to client: userid=449, cmd=msg_type_getofriendlist,
data={"code": 0, "msg": "ok", "userinfo":[{"members":[],"teamname":"My Friends"}]
} - ChatSession.cpp:571
20181110 12:38:33 ERROR TcpConnection::handleError
[FLAMINGO-SERVER:0.0.0.0:8270#831] - SO_ERROR = 104 Connection reset by peer -
TcpConnection.cpp:403
20181110 12:38:33 INFO   client disconnected: 27.38.22.252:34988 -
ChatServer.cpp:106
20181110 12:38:33 INFO   current online user count: 1 - ChatServer.cpp:111
20181110 12:38:33 INFO   TcpServer::removeConnectionInLoop [FLAMINGO-SERVER] -
connection FLAMINGO-SERVER:0.0.0.0:8270#831 - TcpServer.cpp:104
20181110 12:38:33 INFO   Remove channel, channel = 0xFEC5D0, fd = 6 - EventLoop.cpp:243
20181110 14:45:19 INFO   TcpServer::newConnection [FLAMINGO-SERVER] - new connection
[FLAMINGO-SERVER:0.0.0.0:8270#832] from 122.54.229.114:38811 - TcpServer.cpp:75
```

我们可以使用 cat my.log | grep ERROR 过滤所有 ERROR 级别的日志。

4. 每行日志都应该包含哪些基本信息

这里说的基本信息，指的是日志技术中与业务不相关的基本信息。除了前面提到的日志级别，为了快速排查和定位问题，一般在每行日志的头部都应该包含该行写入的时间、打印日志所在的线程 ID、打印日志所在文件的名称和行号，示例如下：

```
[2018-11-09 23:53:01:0271][INFO][ThreadID: 6032][void __thiscall
CSendMsgThread::HandleGroupBasicInfo(const class CGroupBasicInfoRequest
*):336]Request to get group members, groupid=268435552.
```

如何实现打印这些信息，在同步写日志部分的示例代码中已经有详细的实现，这里不再赘述。

5. 日志文件的命名规则

对日志文件的命名比较自由，常用的方式是以日志创建时的系统时间戳为名称，如 20201111164800567.log，表示创建日志的时间是 2020 年 11 月 11 日 16 时 48 分 00 秒 567 毫秒。对于一组服务来说，这种日志命名方式可能让我们难以区分哪个日志文件对应哪个服务，因此，我们可以在日志文件名中加上服务的名称或者程序的 AppID，例如

chatserver-20201111164800567.log 或 11840-20201111164800567.log。

6. 日志文件的大小控制

由于磁盘空间有限，如果一个日志文件的体积太大，就难以打开，即使打开了，也可能由于包含的信息巨大，非常不利于查找我们需要的信息。所以在实际开发中对单个日志文件的体积也会做一些限制，在超过某个大小时（如 10MB），就重新创建一个新的日志文件来继续写。这个大小上限叫作 rollSize，在技术上也很好实现，伪代码如下：

```cpp
bool writeToLogFile(int level, const std::string& line)
{
    if (m_WrittenSize + line.length() > m_rollSize)
    {
        //关闭当前日志文件，创建一个新日志
        //将已经写的字节数大小清零
        m_WrittenSize = 0;
    }

    //将一行日志写入文件中

    //累加已经写入的大小
    m_WrittenSize += line.length();

    return true;
}
```

当然，即使是这样，对累加的日志文件如果没有一种清理机制，终究也会写满磁盘。在很多企业内部都有专门的运维人员定时清理过期的日志，或者对磁盘空间大小做监控，一旦磁盘空间过小，就会给相关人员告警。如果所在的公司不存在这样的策略，则要养成定期检查磁盘空间的习惯，及时清理无效的日志，为新的日志腾出足够的空间。

一些客户端软件会将日志写到本地，如果不对其及时清理，则也会占据用户大量的磁盘空间，所以这类软件会有在程序启动时或者在固定日期清理日志的功能。flamingo 的客户端软件在每次程序启动时，都会将上一次产生的日志清理掉。

```cpp
//节选自 Startup.cpp
int WINAPI _tWinMain(HINSTANCE hInstance, HINSTANCE /*hPrevInstance*/, LPTSTR lpstrCmdLine, int nCmdShow)
{
    //无关代码省略
    CIniFile iniFile;
    CString strIniFilePath(g_szHomePath);
    strIniFilePath += _T("config\\flamingo.ini");
    bool bNeedClear = true;
    if (iniFile.ReadInt(_T("app"), _T("clearexpirelog"), 0, strIniFilePath) == 0)
        bNeedClear = false;

    if (bNeedClear)
    {
        //清理过期的日志文件
        ClearExpiredLog(_T("log"));
    }
```

```
    //无关代码省略
}
```

ClearExpiredLog 函数的实现如下：

```
void ClearExpiredLog(PCTSTR pszFileSuffixName)
{
    if (pszFileSuffixName == NULL)
        return;

    WIN32_FIND_DATA win32FindData = { 0 };
    TCHAR szLogFilePath[MAX_PATH] = { 0 };
    _stprintf_s(szLogFilePath, MAX_PATH, _T("%sLogs\\*.%s"), g_szHomePath, pszFileSuffixName);
    HANDLE hFindFile = ::FindFirstFile(szLogFilePath, &win32FindData);
    if (hFindFile == INVALID_HANDLE_VALUE)
        return;

    do
    {
        if (_tcsicmp(win32FindData.cFileName, _T(".")) != 0 ||
            _tcsicmp(win32FindData.cFileName, _T("..")) != 0)
        {
            memset(szLogFilePath, 0, sizeof(szLogFilePath));
            _stprintf_s(szLogFilePath, MAX_PATH, _T("%sLogs\\%s"), g_szHomePath, win32FindData.cFileName);
            //这里不用检测是否删除成功，因为最新的一个log是我们需要的，它正被此进程占用，所以删不掉
            ::DeleteFile(szLogFilePath);
        }

        if (!::FindNextFile(hFindFile, &win32FindData))
            break;

    } while (true);

    ::FindClose(hFindFile);
}
```

ClearExpiredLog 函数的逻辑也很容易理解，遍历指定目录下指定格式的文件（这里是*.log），然后将其删除。当然，最后的日志文件是程序正在使用的日志文件，无法删除这个文件是我们预期的效果。

9.3.3 在 C/C++中输出网络数据包日志

众所周知，对于 C/C++，一般假定一个字符串最后有一个 ASCII 码值为 0 的字符（表示成 "\0" 或者 "NULL"），表示该字符串的结束标识。大多数 API 函数处理这个字符串时，遇到这个结束标识就停止处理。例如使用 strlen 函数计算宽字符串（wchar_t 类型）"hello"，不会得到预想的结果 5，只会得到 1：

```
wchar_t str = L"hello";
//length 值是 1 而不是 5
```

```
int length = strlen(str);
```

这是因为对于宽字符，strlen 函数仍按单字符处理，其序列实际上如下图所示。

| h | \0 | e | \0 | l | \0 | l | \0 | o | \0 |

strlen 函数在处理到第 2 个字符时，发现它是一个"\0"，就会中止处理，因此得到的结果是 1（strlen 函数在计算字符串长度时不计算字符串末尾的"\0"）。

我们在设计日志系统时，免不了要使用像 strlen 这样的字符串处理函数，但是如果输出网络数据包，则在网络数据包的字段与字段的间隙中一般都会存在这种 ASCII 码为 0 的字节。因此，我们如果在处理网络数据包时不注意这种现象，就会造成输出的日志不完整（被截断）。举个例子：

```
struct msg
{
    char        compressflag;       //压缩标志，如果为1，则启用压缩，反之不启用压缩
    int32_t     originsize;         //包体压缩前的大小
    int32_t     compresssize;       //包体压缩后的大小
    char        reserved[16];       //保留字段
};

msg m = { 0 };
m.compressflag = 1;
m.originsize = 100;
m.compresssize = 30;

printf("%s", (const char*)&m);
```

这里使用 printf 函数模拟输出（这里是输出到控制台而不是输出到文件），将结构体 m 当作字符串输出时，遇到 m 的内容是"\0"时，将会截断。

解决这类问题的方式通常有两种。第 1 种方式是通过使用一些安全函数指定完整的字符串长度，而不是让函数通过"\0"来判断字符串是否结束，例如：

```
std::string str;
//错误，不要让其自动判断结束
//str.append((const char*)&m);
//正确的做法
str.append((const char*)&m, sizeof(m));
```

另一种方式就是逐字节地去处理，尤其适用于处理网络通信双方收到的数据包。笔者在 flamingoserver 代码中实现了一个打印网络数据包的函数 LOG_DEBUG_BIN，使用方式如下：

```
LOG_DEBUG_BIN((unsigned char*)inbuf, buflength);
```

打印效果如下：

```
000000    01fc000000d300000002000000000000  0000000000000000f3789c2d4e3d0bc2
000001    30102df8472473871617db7f9336310d  6d9292b44315c10f0427271717051174
000002    707495fe18ad157f838339f586e3debb  77ef9de3386fc756e7057d9a8d50ac08
000003    4561d773bb48186627a452644169a8e6  c4e29efb9d25165687860996ac0281e4
000004    71fa27dbcdb6395c9af9b1592cdbf50e  b6031cd3a2cac1ba6f615c9a420920ad
```

```
000005   1cf68c4a42350a2137e2ba4808ae50e8 0781e7f99e6f49c399c445a9e1e071de
000006   3febd5fd5ab7ebd320c3824ba66e9319 f86042343506fe0698274a52598a08bc
000007   7f94c03c83d0f1076b63553d
```

其基本原理就是逐字节地输出，实现代码如下：

```
const char* ullto4Str(int n)
{
    static char buf[64 + 1];
    memset(buf, 0, sizeof(buf));
    sprintf(buf, "%06u", n);
    return buf;
}

char g_szchar[17] = "0123456789abcdef";

char* FormLog(int &index, char* szbuf, size_t size_buf, unsigned char* buffer, size_t size)
{
    size_t len = 0;
    size_t lsize = 0;
    int headlen = 0;
    char szhead[64 + 1];
    memset(szhead, 0, sizeof(szhead));
    while (size > lsize && len + 10 < size_buf)
    {
        if (lsize % 32 == 0)
        {
            if (0 != headlen)
            {
                szbuf[len++] = '\n';
            }

            memset(szhead, 0, sizeof(szhead));
            strncpy(szhead, ullto4Str(index++), sizeof(szhead) - 1);
            headlen = strlen(szhead);
            szhead[headlen++] = ' ';

            strcat(szbuf, szhead);
            len += headlen;
        }
        if (lsize % 16 == 0 && 0 != headlen)
            szbuf[len++] = ' ';
        szbuf[len++] = g_szchar[(buffer[lsize] >> 4) & 0xf];
        szbuf[len++] = g_szchar[(buffer[lsize]) & 0xf];
        lsize++;
    }
    szbuf[len++] = '\n';
    szbuf[len++] = '\0';
    return szbuf;
}
```

通过输出网络数据包可以非常方便地定位网络通信问题，另一款开源的日志工具 yaolog 也提供了类似的功能，我们可以下载它进行学习和研究。

9.3.4 调试时的日志

读者可能对这个小标题感到困惑,那么什么是"调试时的日志"呢?

一个设计良好的日志模块,对其输出媒介应该是可以灵活适应的。例如,对于正式运行的程序,程序可以将日志写入文件中;而对于被 attach 到调试器中的程序,程序可以将日志输出到控制台(例如使用 GDB 调试 Linux 程序)或调试器的输出面板(例如使用 Visual Studio 的 Output 面板)。调试时在文件中查找需要的信息是件麻烦的事情,不如在调试器中查看方便。这也是我们在设计日志系统时需要考虑的功能。在调试程序时,输出到调试器控制台或者输出面板的日志就是"调试时的日志"。

实现在调试时将日志输出到调试器控制台或输出面板这个功能并不难,我们只需设置一个控制变量用于标识是否将日志输出到控制台,或者干脆在写文件的地方直接使用 printf、std::cout、OutputDebugString 这样的控制台函数输出即可。在程序发布时,将日志输出到控制台一般是没有意义的,但是会对程序的性能产生影响,所以建议采用变量(可以是宏变量)来控制是否将日志输出到控制台。

以下是一段将日志同时输出到指定文件和控制台的代码:

```
void asyncOutput(const char* msg, int len)
{
    //将日志输出到某个地方,例如文件中
    g_asyncLog.append(msg, len);
    if (logToConsole)
        std::cout << msg << std::endl;
}
```

读到这里,读者应该能明白以上代码中加粗行的作用了吧。

9.3.5 统计程序性能日志

在实际开发中,有时为了优化程序的执行速度,我们需要计算某段关键性的或者存在性能瓶颈的代码执行时间,将其记录到日志中,我们称这类日志为"性能日志"。这个功能实现的原理也很简单,即在某段代码的开始和结束处分别记下系统时间,然后计算二者的时间差。需要注意,操作系统提供了有不同精确度的时间计算函数,而对一段代码的单次执行时间往往很短(一般是纳秒、微秒或者毫秒级),所以选择计算时间的系统函数时要尽量选一些精度高一点的。下面是笔者实现的一个计算 Windows 程序某段代码执行耗时的工具代码:

```
/**
* 性能计数器,统计一段代码的执行耗时并打印出来,PerformanceCounter.h
* zhangyl 2017.07.27
*/
#ifndef __PERMANCE_COUNTER_H__
#define __PERMANCE_COUNTER_H__

//#ifdef PERMANCECOUNTER_EXPORTS
//#define PERMANCECOUNTER_API __declspec(dllexport)
```

```cpp
//#else
//#define PERMANCECOUNTER_API  __declspec(dllimport)
//#endif

#define PERMANCECOUNTER_API

#define BEGIN_PERFORMANCECOUNTER        CPerformanceCounter::Begin(__FUNCSIG__, __LINE__);
#define END_PERFORMANCECOUNTER          CPerformanceCounter::End(__FUNCSIG__, __LINE__);

class PERMANCECOUNTER_API CPerformanceCounter
{
public:
    static bool Init(bool bToFile, PCSTR pszLogFileName);
    static void Uninit();

    //输出线程ID号和所在函数的签名、行号
    static bool Begin(PCSTR pszFunctionSig, int nLineNo);
    static bool End(PCSTR pszFunctionSig, int nLineNo);

private:
    CPerformanceCounter() = delete;
    ~CPerformanceCounter() = delete;

    CPerformanceCounter(const CPerformanceCounter& rhs) = delete;
    CPerformanceCounter& operator=(const CPerformanceCounter& rhs) = delete;

    static void GetTime(char* pszTime, int nTimeStrLength);

private:
    //将日志写入文件还是控制台
    static bool           m_bToFile;
    static HANDLE         m_hPerformanceFile;
    static LARGE_INTEGER m_liFreq;
    static LARGE_INTEGER m_liBegin;
    static LARGE_INTEGER m_liEnd;
};

#endif //!__PERMANCE_COUNTER_H__

/**
 * 性能计数器，统计一段代码的执行耗时并打印出来，PerformanceCounter.cpp
 * zhangyl 2017.07.27
 */
#include "PerformanceCounter.h"
#include "EncodingUtil.h"
#include <tchar.h>

#ifndef PERFORMANCE_OUTPUT
#define PERFORMANCE_OUTPUT
#endif

bool CPerformanceCounter::m_bToFile = false;
HANDLE CPerformanceCounter::m_hPerformanceFile = INVALID_HANDLE_VALUE;
```

```cpp
LARGE_INTEGER CPerformanceCounter::m_liFreq;
LARGE_INTEGER CPerformanceCounter::m_liBegin;
LARGE_INTEGER CPerformanceCounter::m_liEnd;

bool CPerformanceCounter::Init(bool bToFile, PCTSTR pszLogFileName)
{
#ifdef PERFORMANCE_OUTPUT
    m_bToFile = bToFile;

    if (0 >= _tcslen(pszLogFileName))
        return FALSE;

    TCHAR szHomePath[MAX_PATH] = { 0 };
    ::GetModuleFileName(NULL, szHomePath, MAX_PATH);
    for (int i = _tcslen(szHomePath); i >= 0; --i)
    {
        if (szHomePath[i] == _T('\\'))
        {
            szHomePath[i] = _T('\0');
            break;
        }
    }

    TCHAR szLogDirectory[MAX_PATH] = { 0 };
    _stprintf_s(szLogDirectory, _T("%s\\logs\\"), szHomePath);

    DWORD dwAttr = ::GetFileAttributes(szLogDirectory);
    if (!((dwAttr != 0xFFFFFFFF) && (dwAttr & FILE_ATTRIBUTE_DIRECTORY)))
    {
        TCHAR cPath[MAX_PATH] = { 0 };
        TCHAR cTmpPath[MAX_PATH] = { 0 };
        TCHAR* lpPos = NULL;
        TCHAR cTmp = _T('\0');

        _tcsncpy_s(cPath, szLogDirectory, MAX_PATH);

        for (int i = 0; i < (int)_tcslen(cPath); i++)
        {
            if (_T('\\') == cPath[i])
                cPath[i] = _T('/');
        }

        lpPos = _tcschr(cPath, _T('/'));
        while (lpPos != NULL)
        {
            if (lpPos == cPath)
            {
                lpPos++;
            }
            else
            {
                cTmp = *lpPos;
                *lpPos = _T('\0');
                _tcsncpy_s(cTmpPath, cPath, MAX_PATH);
                ::CreateDirectory(cTmpPath, NULL);
```

```cpp
            *lpPos = cTmp;
            lpPos++;
        }
        lpPos = _tcschr(lpPos, _T('/'));
    }
}

    m_hPerformanceFile = ::CreateFile(pszLogFileName, GENERIC_READ | GENERIC_WRITE,
FILE_SHARE_READ, NULL, CREATE_NEW, FILE_ATTRIBUTE_NORMAL, NULL);
    if (m_hPerformanceFile == INVALID_HANDLE_VALUE)
    {
        if (GetLastError() == ERROR_ALREADY_EXISTS)
        {
            return true;
        }
        return false;
    }

    ::QueryPerformanceFrequency(&m_liFreq);

#endif //end PERFORMANCE_OUTPUT

    return true;
}

void CPerformanceCounter::Uninit()
{
#ifdef PERFORMANCE_OUTPUT
    if (m_hPerformanceFile != INVALID_HANDLE_VALUE)
    {
        ::CloseHandle(m_hPerformanceFile);
        m_hPerformanceFile = INVALID_HANDLE_VALUE;
    }
#endif //end PERFORMANCE_OUTPUT
}

bool CPerformanceCounter::Begin(PCSTR pszFunctionSig, int nLineNo)
{
#ifdef PERFORMANCE_OUTPUT
    //时间
    char szTime[64] = { 0 };
    SYSTEMTIME st = { 0 };
    ::GetLocalTime(&st);
    sprintf_s(szTime, ARRAYSIZE(szTime), "[%04d-%02d-%02d %02d:%02d:%02d:%04d]",
st.wYear, st.wMonth, st.wDay, st.wHour, st.wMinute, st.wSecond, st.wMilliseconds);
    std::string strDebugInfo(szTime);

    //当前线程信息
    char szThreadID[32] = { 0 };
    DWORD dwThreadID = ::GetCurrentThreadId();
    sprintf_s(szThreadID, ARRAYSIZE(szThreadID), "[ThreadID: %u]", dwThreadID);
    strDebugInfo += szThreadID;

    //函数签名
    char szFuncSig[512] = { 0 };
```

```cpp
    sprintf_s(szFuncSig, "[%s:%d]", pszFunctionSig, nLineNo);
    strDebugInfo += szFuncSig;

    //正文
    strDebugInfo += "Performance counter begin:\r\n";

    if (m_bToFile)
    {
        if (m_hPerformanceFile == INVALID_HANDLE_VALUE)
            return false;

        ::SetFilePointer(m_hPerformanceFile, 0, NULL, FILE_END);
        DWORD dwBytesWritten = 0;
        ::WriteFile(m_hPerformanceFile, strDebugInfo.c_str(),
strDebugInfo.length(), &dwBytesWritten, NULL);
        ::FlushFileBuffers(m_hPerformanceFile);
    }
    else
        ::OutputDebugStringA(strDebugInfo.c_str());

    ::QueryPerformanceCounter(&m_liBegin);

#endif //end PERFORMANCE_OUTPUT

    return true;
}

bool CPerformanceCounter::End(PCSTR pszFunctionSig, int nLineNo)
{
#ifdef PERFORMANCE_OUTPUT

    ::QueryPerformanceCounter(&m_liEnd);

    //时间
    char szTime[64] = { 0 };
    SYSTEMTIME st = { 0 };
    ::GetLocalTime(&st);
    sprintf_s(szTime, ARRAYSIZE(szTime), "[%04d-%02d-%02d %02d:%02d:%02d:%04d]",
st.wYear, st.wMonth, st.wDay, st.wHour, st.wMinute, st.wSecond, st.wMilliseconds);
    std::string strDebugInfo(szTime);

    //当前线程信息
    char szThreadID[32] = { 0 };
    DWORD dwThreadID = ::GetCurrentThreadId();
    sprintf_s(szThreadID, ARRAYSIZE(szThreadID), "[ThreadID: %u]", dwThreadID);
    strDebugInfo += szThreadID;

    //函数签名
    char szFuncSig[512] = { 0 };
    sprintf_s(szFuncSig, "[%s:%d]", pszFunctionSig, nLineNo);
    strDebugInfo += szFuncSig;

    //正文
    double dbInterval = (double)(LONGLONG)((LONGLONG)m_liEnd.QuadPart -
(LONGLONG)m_liBegin.QuadPart) / (double)m_liFreq.QuadPart;
```

```cpp
    char szInterval[64] = { 0 };
    //精确到纳秒
    sprintf_s(szInterval, ARRAYSIZE(szInterval), "Performance counter
end: %2.12fms", (dbInterval * 1000));

    strDebugInfo += szInterval;

    strDebugInfo += "\r\n";

    if (m_bToFile)
    {
        if (m_hPerformanceFile == INVALID_HANDLE_VALUE)
            return false;

        ::SetFilePointer(m_hPerformanceFile, 0, NULL, FILE_END);
        DWORD dwBytesWritten = 0;
        ::WriteFile(m_hPerformanceFile, strDebugInfo.c_str(),
strDebugInfo.length(), &dwBytesWritten, NULL);
        ::FlushFileBuffers(m_hPerformanceFile);
        return true;
    }

    ::OutputDebugStringA(strDebugInfo.c_str());

#endif //end PERFORMANCE_OUTPUT

    return true;
}
```

在系统启动时，先初始化以上性能日志工具：

```cpp
//初始化性能计数器
TCHAR szPerformanceFileName[MAX_PATH] = { 0 };
_stprintf_s(szPerformanceFileName, MAX_PATH,
        _T("%s\\Logs\\%04d%02d%02d%02d%02d%02d.perf"), g_szHomePath, st.wYear,
st.wMonth, st.wDay, st.wHour, st.wMinute, st.wSecond);
CPerformanceCounter::Init(true, szPerformanceFileName);
```

然后在需要统计时间的代码片段前后分别加上 BEGIN_PERFORMANCECOUNTER 和 END_PERFORMANCECOUNTER 即可：

```cpp
BEGIN_PERFORMANCECOUNTER
//加载皮肤列表配置文件
CSkinManager::GetInstance()->LoadConfigXml();
END_PERFORMANCECOUNTER
```

这样在程序的 logs 目录下就会生成一个性能统计日志文件 20181111183849.perf，其内容如下：

```
[2018-11-11 18:38:49:0386][ThreadID: 5264][int __stdcall wWinMain(struct
HINSTANCE__ *,struct HINSTANCE__ *,wchar_t *,int):206]Performance counter begin:
[2018-11-11 18:38:49:0566][ThreadID: 5264][int __stdcall wWinMain(struct
HINSTANCE__ *,struct HINSTANCE__ *,wchar_t *,int):208]Performance counter end:
35.529896907216ms
```

通过上面的输出结果，可以统计出加载皮肤列表配置文件的逻辑大约耗时 35 毫秒。

很多商业软件在正式对外发布给用户使用之前，都会使用这种方式去评测一些高频执行代码的执行速度，如果不满足预期的要求，就要对代码做进一步优化。

9.3.6 根据类型将日志写入不同的文件中

我们知道，通过日志文件排查系统问题时，即使日志可以根据级别来过滤，但是当日志数量较多时，过滤和查找一些特定的信息也不是一件轻松的事情。笔者曾做过一个交易所的金融项目，当时是将日志根据业务分成了 error、info 和 runlog 三大类，将每一类都写到不同的文件中。例如在某个时候，某个服务可能会生成三个文件：xxxx.error.log、xxxx.info.log 和 xxxx.runlog。

在 xxxx.error.log 中记录了这个程序运行过程中的所有出错信息；在 xxxx.info.log 中记录了程序中的各种关键运行与状态信息；在 xxxx.runlog 中记录了程序运行时的各种细粒度的信息。这样分类的好处就是，我们的开发人员和运维人员每天只需检测日志目录是否存在 error.log 这样的文件，如果存在，则说明程序或者业务一定有错误，对其进行重点排查就可以了。而 runlog 一般是不开启的，由于其输出非常详细，所以一旦开启，就会产生大量的日志记录，所以在一般情况下用不上，只有当线上出现问题且需要定位和排查疑难杂症时才会将其打开，一旦解决问题，就再次把它关闭。

可能有读者会问，一般一个日志功能的开关是被配置在配置文件中的，在程序启动时加载，一旦加载就固定住了，如何在程序运行时动态改变这个值呢？这就是热加载技术。其实现方法也有很多，例如可以在程序中专门开启一个线程去检测日志文件的变更，如有变更，则使用最新的变更信息。还有一种方式就是使用监控端口，这种方式一般用于服务器程序中，在服务器程序中单独再开启一个监控端口，通过类似 nc、telnet 这样的命令连接到这个监控端口上，然后输入预先设置好的命令开关相应的日志功能。例如，flamingoserver 在程序启动后会开启一个 8888 监控端口，我们使用 nc 命令连接上去后，会显示如下图所示的界面。

```
[zhangyl@iZ238vnojlyZ logs]$ nc -v 120.55.94.78 8888
Ncat: Version 6.40 ( http://nmap.org/ncat )
Ncat: Connected to 120.55.94.78:8888.
1. help-show help info
2. ul-show online user list
3. su-show userinfo specified by userid: su [userid]
4. elpb-enable log package binary data
5. dlpb-disable log package binary data
```

连接成功以后，输入 elpb 命令就会将该服务器上来往的网络数据包内容以十六进制格式输出到日志文件中，输入 dlpb 时就会停止该输出。

9.3.7 集中式日志服务与分布式日志服务

将日志输出到一个或多个文件中，是存储日志的一种策略，但是对于许多大型企业来说，由于存在多个部门及成千上万的服务程序，这种日志存储方式会造成日志文件散落于各处，多而乱，非常不宜于管理，尤其是将这些日志文件交给专门的运维部门时，需要花费大量的人力和物力去维护。而且，一个功能复杂的软件可能被根据功能设计成多个模块，如下图所示。

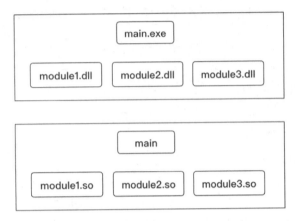

每个模块都可能是不同的部门或者人员开发的，模块与模块之间可能只提供接口契约，所以每个模块都可能有自己的日志输出，这样一个程序就可能产生多个日志文件。定位一个问题时往往要穿梭于各个模块之间，非常麻烦。

所以对于一组服务，我们有时会专门开发一个 Log 服务，其他服务会将日志信息通过网络发送给 Log 服务，查询日志时可以使用 Tag 或 AppID 区分不同服务的日志。这种日志系统的实现原理如下图所示。

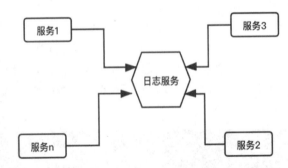

当然，也有另外一些系统，每个服务在这个 Log 系统上都申请了一个日志采集功能，该 Log 系统会去宿主服务的指定目录下采集日志。如下所示是一个集中式的日志管理系统示意图，左侧是服务名和一些可用于过滤日志的字段筛选条件，右侧中间是过滤后的日志。

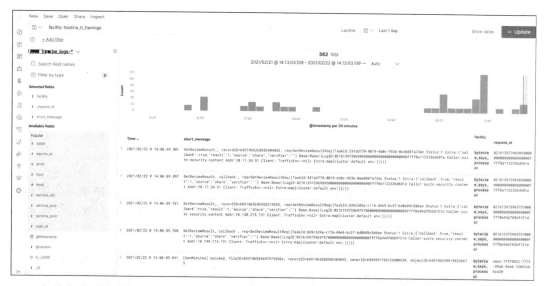

这种集中式日志系统的实现原理如下图所示。

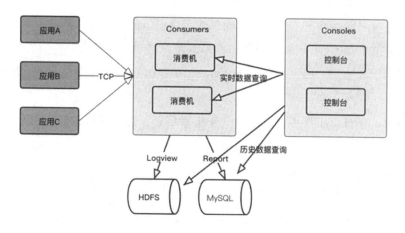

9.3.8 从业务层面看在一条日志中应该包含什么内容

不少开发者在输出程序日志时可能存在这样的问题：虽然每条日志的内容不少，但多数是一些无意义的信息；尽管某条日志也能反映程序故障并暴露问题，但对解决问题毫无作用，尤其是在服务器程序被发布到生产环境下，日志将问题暴露出来后，由于日志能反映当时故障现场的信息太少，导致从日志信息中只能看到出现了问题，却没法获知产生问题的原因，更不用说解决问题了。那么在一条日志中应该包含哪些业务信息才能帮助我们快速定位问题呢？

来看几个源于真实项目的反面案例。

第1个案例：

```
//示例一
CIULog::Log(LOG_WARNING, __FUNCSIG__, _T("Be cautious! Unhandled net data! req_ans_command=%d."), header.cmd);
```

这条日志记录只打印了一条警告信息和命令号（cmd），对产生这个警告的具体输入参数和当时的环境也没有任何记录，即使产生问题，事后也无法追踪。

第 2 个案例：

```
if (!HttpRequest(osURL.str().c_str(), strResponse, true, strCookie.c_str(), NULL,
false, 10))
{
    LogError("QueryTickets failed");
    return false;
}
```

这条日志对 HTTP 请求出错输出了一条日志 "QueryTickets failed"，但没有交待产生错误的参数和原因，如果在生产环境下出现这样的错误日志，那么我们大概率无法排查出错的原因。我们猜测出错的原因可能有以下三种。

（1）传递给 HttpRequest 函数的参数存在无效的值。这些无效的值可能是内部产生的，也可能是上游调用方带过来的（这种情形属于外部原因，需要上游调用方纠正），我们可以通过校验传给 HttpRequest 函数的参数来避免出错。

（2）调用时的网络状况不佳。这不能算作程序的 bug，一般不需要解决，如果因为网络原因导致 HttpRequest 调用失败频繁，就需要排查网络连接状况。

（3）程序设计上的 bug。例如 HttpRequest 函数的实现逻辑存在 bug，需要程序开发者解决。

无论如何，在日志中都不能只输出一条出错信息，这会导致我们无法得知为何出错，也无法定位出错的原因。

来看一条比较好的日志输出信息：

```
Log().Errorf("rpc.GetAndRenewXXXSession failed, userID: %d, userAgent: %s,
TenantID: %d, Platform: %s, version: %s, AppID: %s, sessionKey: %s, error: %s",
                userID, userAgent, tenantID, platform, version, appID,
sessionKey, err)
```

这条日志在某个调用失败后打印出当时的各个参数信息（userID、userAgent、TenantID、Platform、version、AppID、sessionKey）和当时的出错原因（err），这样可以利用这些信息去确认或者手动重试以定位问题。

总之，日志记录应该尽量精炼、详细，反映当时出错的现场参数、产生的环境等信息。例如一个注册失败的请求至少要描述出当时注册的用户名、密码、用户状态（比如是否已注册）、请求的注册地址，等等。日志报错不一定是程序的 bug，可能是业务上的错误等。不用担心服务器的磁盘空间，对于快速定位问题来说，用这些磁盘空间是值得的。

另外，在一个大的系统中存在很多基础模块，在调用这些基础模块的过程中也可能产生这些错误，由于基础模块仅仅是技术上的被调用者，一般不太明确出错的上下文，所以此时我们可以向上抛出错误，即被调用者将错误抛给调用者，由调用者输出具体的错误日志和上下文环境信息。以下是一段说明该问题的伪代码：

```cpp
//业务模块
class SomeService
{
public:
    bool getAllUserInfo(std::vector<UserInfo>& retrievedUserInfo, int userType)
    {
        if (!m_SomeDao.getAllUserInfo(retrievedUserInfo, userType))
        {
            //在这里可以记录完整的错误信息及产生错误的上下文信息
        }
    }

private:
    SomeDao    m_SomeDao;

};

//数据操作模块
class SomeDao {
public:
    bool getAllUserInfo(std::vector<UserInfo>& retrievedUserInfo, int userType)
    {
        //利用 userType 操作数据库
        //这里如果出错，则可以只记录技术上的错误或者直接将错误抛给调用者 SomeService
        //因为调用者（业务层）更清楚到底发生了什么
    }
};
```

9.3.9　在日志中不要出现敏感信息

前面强调了输出日志时应该交代清楚发生事件的上下文信息，但需要注意的是，在商业系统研发中，在日志输出信息里不应该出现敏感的信息，例如关键服务的 IP 地址、端口号、用户的用户名和密码、sessionKey 等。

java.net.Socket 在连接不上服务器时抛出 IOException 就是一个典型的例子，抛出该异常是为了避免隐私信息泄露，不会在异常信息中打印目标主机的 IP 地址和端口号。

另外，如果我们的日志库要被提供给他人使用或者为公司开发公共日志组件，则一般会有一个敏感信息过滤和报警机制。如果使用我们日志组件的人在日志中输出了敏感信息（这些敏感信息可能源于公司的规定或相关法律法规），我们就可以通过一定的机制（如邮件通知）报告给当事人以建议其修改，并对日志中的敏感信息进行加密或者替换。

看一个例子：SomeLogger 是某公司的一个日志库，按照该公司的规定，sessionKey 为敏感信息，但某个开发者仍然在日志中输出了该信息，该开发者会因此收到相应的报警信息，而且在输出后的日志中，sessionKey 被替换成了 "[REDACTED_SESSION_KEY]" 字样。代码如下：

```
SomeLogger.Infof("Get settings success, userID: %d, userAgent: %s, Platform: %s,
version: %s, AppID: %s, sessionKey: %s, settings: %s",
                        userID, userAgent, platform, version, appID, sessionKey,
```

string(settingsJsonBytes))

效果如下：

```
Get settings success, userID: 3032193, userAgent: Mozilla/5.0 (Macintosh; Intel Mac
OS X 10_15_3) , Platform: Mac, version: 1.1.0, AppID: 4888, sessionKey:
[REDACTED_SESSION_KEY], settings:
{\"data\":{\"heartbeat_expired_time\":73,"disable_log\":true}}
```

9.3.10 开发过程中的日志递进缩减策略

一个程序（App 或者服务）从最初的编码开发到上线，为了节约磁盘空间，也为了减少非必要的日志信息对我们寻找重点信息的干扰，一般在开发至上线期间，随着程序 bug 的减少，逻辑不断完善。我们除了要把日志的级别不断调高，原来需要输出到日志中的一些细粒度信息的日志代码也应该不断减少。这就是开发过程中的日志递进缩减策略。

这是一种非常有用的策略，在开发早期为了尽快定位问题、发现 bug 或者走通联调流程，在我们的程序中会增加输出各种上下文信息的大量日志逻辑，等功能一块块完善和流程一步步走通，这些日志逻辑不仅会产生大量的冗余信息，也会降低程序效率，我们可以将这些日志逻辑从程序代码中逐步去掉。

设计日志系统看起来简单，但设计出一套能快速定位和排查问题的日志系统就没有预想中那么容易了，希望读者好好体会本章所要表达的思想。

9.4 错误码系统的设计

对于服务器系统来说，设计一套好的错误码是非常有必要的，可以在用户请求出问题时迅速定位并解决问题。比如，在我们所用的宽带网络无法连接时，宽带服务器会给出诸如 651、678 等错误码，宽带维修人员可以利用这些错误码快速定位宽带故障的原因所在；又如，我们日常浏览网页时，若页面不存在，则一般返回 404 等错误码。

9.4.1 错误码的作用

错误码的作用一般如下。

1）迅速定位是用户输入的问题还是服务器自身的问题

用户输入的问题，指用户的不当操作。这些不当操作可能是因为客户端软件本身的逻辑错误或漏洞导致的，也可能是使用客户端的人的非法操作导致的，而客户端软件在设计上因为考虑不周所以缺乏有效性校验，这时都可能产生非法数据，并且直接发送给服务器。一个好的服务端系统不能假设客户端的请求数据一定是合法的，必须对传过来的数据做有效性校验。服务器没有义务一定对非法请求做出应答，因此请求的最终结果是服务器不应答或给了客户端其不想要的应答。

还是使用上面的例子，宽带用户输入了无效的用户名或密码造成服务器拒绝访问；用户在浏览器中输入了一个无效的网址等。此类错误都是需要用户（指人，或者代指使用某个服务器的所有下游服务和客户端）自己解决的。如果错误码可以反映出错的原因，那么在服务器的实际运维过程中，当用户反馈这一类故障时，我们通过错误码就可以准确、快速地定位问题所在。

2）快速定位是哪个步骤或服务出了问题

假设某单个服务在收到某个客户端（这里的客户端可以是真正意义上的客户端程序，也可以是发起请求的上游服务）请求后，需要经历多个处理步骤才能完成，而这中间任何一个处理步骤都可能出现问题，如果在不同的处理步骤出错时会返回不同的错误码，那么一旦处理出错，就可以通过错误码知道是哪个步骤出了问题。其次，稍微复杂一点的业务往往是由一组服务构成的。如果将错误码分段，对不同的服务使用不同的错误码段，那么我们通过错误码就能准确得知是哪个服务出了问题。

9.4.2 错误码系统设计实践

来看一个具体的例子。假设有如下一个智能邮件系统，结构如下图所示。

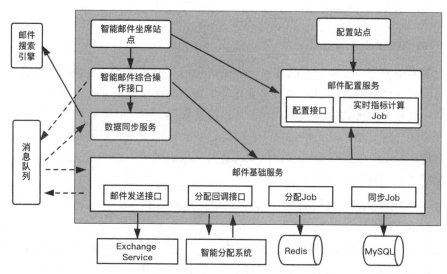

上图中的智能邮件坐席站点和配置站点是客户端，智能邮件综合操作接口和邮件配置服务是直接对客户端提供服务的两个前置服务，这两个前置服务依赖后面的数个服务。由于这里要说明的是技术问题，而不是业务问题，所以不再介绍每个服务的用途。在这个系统中，客户端在得到前置服务的某个不正确应答时，会同时得到一个错误码。现在假设按以下规则来设计错误码。

服务名称	正值错误码范围	负值错误码范围
智能邮件综合操作接口	100～199	-100～-199
数据同步服务	200～299	-200～-299
邮件配置服务	300～399	-300～-399
邮件基础服务	400～499	-400～-499

我们在设计这套系统时，做如下规定。

（1）所有正值错误码都表示所在服务的上游服务发来的请求不满足业务要求。举个例子，假设某次智能邮件坐席站点客户端得到了一个错误码101，我们就可以先确定错误产生的服务器是智能邮件综合操作接口服务；其次，产生该错误的原因是智能邮件坐席站点客户端发送给智能邮件综合操作接口服务的请求不满足要求，我们通过这个错误码甚至可以进一步确定发送的请求哪里不符合要求。例如，可以这样定义，如下所述。

- ◎ 100：用户名不存在。
- ◎ 101：密码无效。
- ◎ 102：发送的邮件收件人非法。
- ◎ 103：邮件正文含有非法字符。

（2）所有负值错误码都表示程序内部错误。例如，可以这样定义，如下所述。

- ◎ -100：数据库操作错误。
- ◎ -101：网络错误。
- ◎ -102：内存分配失败。
- ◎ -103：无法连接数据同步服务。

1. 对负值错误码的特殊处理

通过前面的介绍，读者应该能看出正值错误码与负值错误码的区别，即正值错误码一般是由请求服务的客户产生的，如果出现这样的错误，则应该由客户自己解决问题；而负值错误码一般是在服务内部产生的，如果是正值错误码，则错误码和错误信息一般可以被直接返回给客户端；如果是负值错误码，则我们一般只将错误码返回给客户端，不带具体的错误信息。对于负值错误码的错误信息，我们可以将其统一显示成"网络错误"或者其他比较友好的错误提示。这样做的原因如下。

- ◎ 客户端即使拿到这样的错误信息，也不能对排查和解决问题提供任何帮助，因为这些错误是程序内部的错误或者bug。
- ◎ 这类错误有可能是业务系统的设计缺陷导致的。若将此类缺陷直接暴露给客户，则除了没有任何正面作用，还可能被不法分子利用。

之所以带上错误码，是为了方便内部排查和定位问题。当然，在企业服务内部一般有大量的监控系统，可能不会再暴露这样的错误码了。

2. 扩展

前面介绍了利用错误码分段来定位问题的技术思想，其实，我们在开发一组服务时，也可以对业务类型通过编号来分段，这样通过业务号就能知道其归属于哪个服务了。

9.5 监控端口

在实际项目中，出于定位问题和统计数据的要求，往往会给一些正在运行的服务开一些对外的"口子"，技术人员和运维人员可以给正在运行的服务发送一些指令以得到一些想要的结果。

发送的这些指令既可以是短连接形式的（如发送一个 HTTP 请求），也可以是长连接形式的，为了方便使用，我们通常不会专门开发一些工具去做这些事，而是利用一些现成的命令如 nc、telnet 去做。

这里在即时通信软件 Flamingo 的 chatserver 中实现了这样一个监控功能，使用的是长连接形式，其中，chatserver 地址是 127.0.0.1，端口号是 8888，可以用 nc -v 127.0.0.1 8888 连上服务器，效果如下：

```
[root@myserver ~]# nc -v 127.0.0.1 8888
Ncat: Version 7.70 ( https://nmap.org/ncat )
Ncat: Connected to 127.0.0.1:8888.
1. help-show help info
2. ul-show online user list
3. su-show userinfo specified by userid: su [userid]
4. elpb-enable log package binary data
5. dlpb-disable log package binary data
```

连接成功后，服务器会输出 5 条命令，每条命令都有相应的解释，输入对应的命令就可以执行相应的操作。

◎ help 命令用于查看整个监控端口支持哪些命令。
◎ ul 命令用于显示当前内存中在线的用户信息。
◎ su 命令用于输出指定用户的信息，需要使用 su userid 格式来指定某个用户的 userid。
◎ elpb 命令用于启用将网络包的二进制格式写入日志文件的功能，用于排查网络包问题。

dlpb 命令用于禁用将网络包的二进制格式写入日志文件的功能，用于排查网络包问题。当然，elpb 和 dlpb 命令可以合二为一，通过不同的参数来指定。

执行上述命令后，效果如下：

```
[root@myserver ~]# nc -v 127.0.0.1 8888
Ncat: Version 7.70 ( https://nmap.org/ncat )
Ncat: Connected to 127.0.0.1:8888.
1. help-show help info
2. ul-show online user list
3. su-show userinfo specified by userid: su [userid]
4. elpb-enable log package binary data
5. dlpb-disable log package binary data
ul
No user online.
help
1. help-show help info
```

```
2. ul-show online user list
3. su-show userinfo specified by userid: su [userid]
4. elpb-enable log package binary data
5. dlpb-disable log package binary data
elpb
OK.
dlpb
OK.
```

监控命令的实际用途就是在服务运行过程中查询和修改服务内存中的一些信息,其实现原理也很简单,即在某个端口(在这里的例子中是 8888)上开启一个监听,然后接收连接,处理连接上来的端口发送的监控命令。由于这里使用的是 nc 命令作为监控端口的客户端工具,因此其每条指令都以\n 结束,我们把\n 作为每个数据包的结束标志。

实现逻辑如下:

```
void MonitorSession::onRead(const std::shared_ptr<TcpConnection>& conn, Buffer*
pBuffer, Timestamp receivTime)
{
    std::string buf;
    std::string substr;
    size_t pos;
    size_t totalsize = 0;
    while (true)
    {
        //xxx\nyyy\nuuuu\njjjjj
        buf.clear();
        buf = pBuffer->toStringPiece();
        while (true)
        {
            pos = buf.find("\n");
            if (pos != std::string::npos)
            {
                if (pos == 0)
                    substr = "\n";
                else
                    substr = buf.substr(0, pos);
                totalsize += substr.length();
                buf = buf.substr(pos + 1);
                LOGI("recv cmd: %s", substr.c_str());
                //LOGI << "buf: " << substr;
                process(conn, substr);
            }
            else
            {
                if (totalsize > 0)
                    pBuffer->retrieve(totalsize);
                return;
            }
        }//end inner while-loop
    }//end outer while-loop
}
```

在 MonitorSession::onRead 方法中每收到一段含有\n 标志的数据,就会将其当作客户端的一个监控命令，然后调用 process 方法对这条命令进行解析（代码第 25 行）。process 方法的实现如下：

```cpp
bool MonitorSession::process(const std::shared_ptr<TcpConnection>& conn, const std::string& inbuf)
{
    if (inbuf == "\n")
        return false;

    std::vector<std::string> v;
    StringUtil::split(inbuf, v, " ");

    if (v.empty())
        return false;
    else
    {
        if (v[0] == g_helpInfo[0].cmd)
            showHelp();
        else if (v[0] == g_helpInfo[1].cmd)
        {
            if (v.size() >= 2)
                showOnlineUserList(v[1]);
            else
                showOnlineUserList("");
        }
        else if (v[0] == g_helpInfo[2].cmd)
        {
            if (v.size() < 2)
            {
                char tip[32] = { "please specify userid.\n" };
                send(tip, strlen(tip));
            }
            else
            {
                showSpecifiedUserInfoByID(atoi(v[1].c_str()));
            }

        }
        else if (v[0] == g_helpInfo[3].cmd)
        {
            //开启日志数据包打印二进制字节
            Singleton<ChatServer>::Instance().enableLogPackageBinary(true);

            char tip[32] = { "OK.\n" };
            send(tip, strlen(tip));
        }
        else if (v[0] == g_helpInfo[4].cmd)
        {
            //开启日志数据包打印二进制字节
            Singleton<ChatServer>::Instance().enableLogPackageBinary(false);

            char tip[32] = { "OK.\n" };
            send(tip, strlen(tip));
        }
```

```
            else
            {
                //客户端发送不支持的命令
                char tip[32] = { "cmd not support\n" };
                send(tip, strlen(tip));
            }
        }

        return true;
    }
```

process 方法其实用于简单的字符串匹配，匹配哪条指令就执行哪条指令，执行完后将执行结果或状态返回给监控客户端（调用 send 方法）。我们在 g_helpInfo 这个全局数组中定义了所有指令名和指令的说明信息。g_helpInfo 数组的定义如下：

```
struct HelpInfo
{
    std::string cmd;
    std::string tip;
};

const HelpInfo g_helpInfo[] = {
    { "help", "show help info" },
    { "ul",   "show online user list" },
    { "su", "show userinfo specified by userid: su [userid]" },
    { "elpb", "enable log package binary data" },
    { "dlpb", "disable log package binary data" }
};
```

如果客户端发送了一个不支持的命令，服务器就会返回"cmd not support"信息。

关于监控端口，还有以下注意事项。

◎ 为了服务器的安全性，通常应该将监控端口的 IP 地址和端口号配置成只允许内网访问，不建议将其暴露在公网上。

◎ 监控端口输出的信息不应该暴露敏感信息，例如用户的密码，如果必须要访问用户的敏感信息，就必须通过进一步的指令权限认证才可以访问。

◎ 监控端口在获取内存中的某些数据时，如果这些数据会被多个线程访问，则应尽快将这些数据复制一份副本出来，以减小锁的粒度，尽量做到不影响程序中正常运行的线程处理这些数据的效率。